Pei-Chu Hu
Chung-Chun Yang

Value Distribution Theory Related to Number Theory

Birkhäuser Verlag
Basel · Boston · Berlin

Authors:

Pei-Chu Hu
Department of Mathematics
Shandong University
Jinan 250100, Shandong
P.R. China
e-mail: pchu@sdu.edu.cn

Chung-Chun Yang
Department of Mathematics
The Hong Kong University of Science and Technology
Clear Water Bay Road
Kowloon, Hong Kong
P.R. China
e-mail: mayang@ust.hk

2000 Mathematics Subject Classification 11Dxx, 11E95, 11F03, 11G05, 11Mxx, 11P05; 30D35, 30G06; 32A22

A CIP catalogue record for this book is available from the
Library of Congress, Washington D.C., USA

Bibliographic information published by Die Deutsche Bibliothek
Die Deutsche Bibliothek lists this publication in the Deutsche Nationalbibliografie; detailed bibliographic data is available in
the Internet at <http://dnb.ddb.de>.

ISBN 3-7643-7568-X Birkhäuser Verlag, Basel – Boston – Berlin

© 2006 Birkhäuser Verlag, P.O. Box 133, CH-4010 Basel, Switzerland
Part of Springer Science+Business Media
Printed on acid-free paper produced from chlorine-free pulp. TCF ∞
Cover design: Micha Lotrovsky, CH-4106 Therwil, Switzerland
Printed in Germany
ISBN-10: 3-7643-7568-X   e-ISBN: 3-7643-7569-8
ISBN-13: 978-3-7643-7568-3

9 8 7 6 5 4 3 2 1   www.birkhauser.ch

Contents

Preface

In 1879, Picard established the well-known and beautiful result that a transcendental entire function assumes all values infinitely often with one exception. Since then Hadamard (1893), Borel (1897) and Blumenthal (1910) had tried to give Picard's result a quantitative description and extend it to meromorphic functions. It was R. Nevanlinna, who achieved such an attempt in (1925) by establishing the so-called value distribution theory of meromorphic functions which was praised by H. Weyl (1943) as "One of the few great mathematical events of our century". Moreover, part of the significance of Nevanlinna's approach is that the concept of exceptional values can be given a geometric interpretation in terms of geometric objects like curves and mappings of subspaces of holomorphic curves from a complex plane $\mathbb{C}$ to a projective space $\mathbb{P}^n$. In the years since these results were achieved, mathematicians of comparable stature have made efforts to derive an analogous theory for meromorphic mappings and p-adic meromorphic functions. Besides the value distribution, the theory has had many applications to the analyticity, growth, existence, and unicity properties of meromorphic solutions to differential or functional equations. More recently, it has been found that there is a profound relation between Nevanlinna theory and number theory. C.F. Osgood [310], [311] first noticed a similarity between the 2 in Nevanlinna's defect relation and the 2 in Roth's theorem. S. Lang [230] pointed to the existence of a structure to the error term in Nevanlinna's second main theorem, conjectured what could be essentially the best possible form of this error term in general (based on his conjecture on the error term in Roth's theorem), and gave a quite detailed discussion in [235]. P.M. Wong [433] used a method of Ahlfors to prove Lang's conjecture in the one-dimensional case. As for higher dimensions, this problem was studied by S. Lang and W. Cherry [235], A. Hinkkanen [159], and was finally completed by Z. Ye [443]. The best possible form of error terms has been used in our present work to produce some sharp results.

In 1987, P. Vojta [415] gave a much deeper analysis of the situation, and compared the theory of heights in number theory with the characteristic functions of Nevanlinna theory. In his dictionary, the second main theorem, due to H. Cartan, corresponds to Schmidt's subspace theorem. Further, he proposed the general conjecture in number theory by comparing the second main theorem in Carlson-Griffiths-King's theory, or particularly influenced by Griffiths' conjecture, which also can be translated into a problem of non-Archimedean holomorphic curves posed by Hu and Yang [176]. Along this route, Shiffman's conjecture on hypersurface targets in value distribution theory corresponds to a subspace theorem for homogeneous polynomial forms in Diophantine approximation. Vojta's $(1, 1)$-form conjecture is an analogue of an inequality of characteristic functions of holomorphic curves for line bundles. Being influenced by Mason's theorem, Oesterlé and Masser formulated the abc-conjecture. The generalized abc-conjectures for integers are counterparts of Nevanlinna's third main theorem and its variations in value distribution theory, and so on.

Roughly speaking, a significant analogy between Nevanlinna theory and Diophantine approximation seems to be that the sets $X(\kappa)$ of κ-rational points of a projective variety X defined over number fields κ are finite if and only if there are no non-constant holomorphic curves into X. Mordell's conjecture (Faltings' theorem) and Picard's theorem are classic examples in this direction. In higher-dimensional spaces, this corresponds to a conjecture due to S. Lang, that is, Kobayashi hyperbolic manifolds (which do not contain non-constant holomorphic curves) are Mordellic. Bloch-Green-Griffiths' conjecture on degeneracy of holomorphic curves into pseudo-canonical projective varieties is an analogue of the Bombieri-Lang conjecture on pseudo canonical varieties. We have introduced these problems and the related developments in this book. Generally, topics or problems in number theory are briefly introduced and translated as analogues of topics in value distribution theory. We have omitted the proofs of theorems in number theory. However, we have discussed the problems of value distribution in detail. In this book, we will not discuss value distribution theory of moving targets, say, K. Yamanoi's work [437], and their counterparts in number theory.

When a holomorphic curve f into X is not constant, we have to distinguish whether it is degenerate in Nevanlinna theory, that is, whether its image is contained or not in a proper subvariety. If it is degenerate, usually it is difficult to deal with it in value distribution theory. If f is non-degenerate, we can study its value distributions and measure its growth well by a characteristic function $T_f(r)$. Similarly, we should distinguish whether or not certain rational points are degenerate. Related to the degeneracy, it seems that for each number field κ, $X(\kappa)$ is contained in a proper Zariski closed subset if and only if there are no algebraically non-degenerate holomorphic curves into X. To compare with Nevanlinna theory, therefore, we need to rule out degenerate κ-rational points that are contained in a subspace of lower dimension, and give a proper measure for non-degenerate κ-rational points. By integrating heights over non-degenerate κ-rational points, we can obtain quantitative measurements $T_\kappa(r)$.

They have the following basic properties:

(i) f is constant if and only if $T_f(r)$ is bounded; there are no non-degenerate κ-rational points if and only if $T_\kappa(r)$ is bounded .

(ii) f is rational if and only if $T_f(r) = O(\log r)$; there are only finitely many non-degenerate κ-rational points if and only if $T_\kappa(r) = O(\log r)$.

It has been observed that there exist non-constant holomorphic curves into elliptic curves such that they must be surjective. Thus it is possible that there are infinitely many rational points on some elliptic curves. However, since non-constant holomorphic curves into elliptic curves have normal properties, say, they are surjective, then distribution of rational points on elliptic curves should be "normal". Really, elliptic curves are modular according to the Shimura-Taniyama-Weil conjecture, which was proved by Breuil, Conrad, Diamond, and Taylor [37] by extending work of Wiles [431], Taylor and Wiles [390]. Moreover, as a result of studies of the analogous results between Nevanlinna's value distribution theory and Diophantine approximation, some novel ideas and generalizations have been developed or derived in the two topics, with many open problems posed for further investigations.

The book consists of seven chapters: In Chapter 1, we introduce some basic notation and terminology on fields and algebraic geometry which are mainly used to explain clearly the topics in Chapter 3 related to number theory. Chapter 2 is a foundation of value distribution theory which is used in Chapter 4, Chapter 6 and Chapter 7 to introduce the analogues related to number theory in Nevanlinna theory, say, *abc*-problems, meromorphic solutions of Fermat's equations and the Waring problem, Green-Griffiths' conjecture, Griffiths' and Lang's conjectures, Riemann's ζ-function, and so on. Chapter 5 contains value distribution theory over non-Archimedean fields and some applications related to topics in number theory. Moreover, a few equidistribution formulae illustrating the differences with the classical Nevanlinna theory have been exhibited. Each chapter of this book is self-contained and this book is appended with a comprehensive and up-dated list of references. The book will provide not just some new research results and directions but challenging open problems in studying Diophantine approximation and Nevanlinna theory. One of the aims of this book is to make timely surveys on these new results and their related developments; some of which are newly obtained by the authors and have not been published yet. It is hoped that the publication of this book will stimulate, among our peers, further researches on Nevanlinna's value distribution theory, Diophantine approximation and their applications.

We gratefully acknowledge support for the related research and for writing of the present book from the Natural Science Fund of China (NSFC) and the Research Grant Council of Hong Kong during recent years. Also the authors would like to thank the staff of Birkhäuser, in particular, the Head of Editorial Department STM, Dr. Thomas Hempfling, and last but not least, we want to express our thanks to Dr. Michiel Van Frankenhuijsen for his thorough reviewing, valuable criticism and concrete suggestions.

Pei-Chu Hu
Chung-Chun Yang

Chapter 1

Heights

In this chapter, we will introduce some basic notation, terminology and propositions on fields and algebraic geometry, which will be used in this book.

1.1   Field extensions

We will denote the fields of complex, real, and rational numbers by $\mathbb{C}$, $\mathbb{R}$, and $\mathbb{Q}$, respectively, and let $\mathbb{Z}$ be the ring of integers. If κ is a set, we will write

$$\kappa^n = \{(x_1, \dots, x_n) \mid x_i \in \kappa\} = \kappa \times \cdots \times \kappa \ (n\text{-times}).$$

If κ is partially ordered, write

$$\kappa(s, r) = \{x \in \kappa \mid s < x < r\}, \quad \kappa(s, r] = \{x \in k \mid s < x \le r\},$$
$$\kappa[s, r) = \{x \in \kappa \mid s \le x < r\}, \quad \kappa[s, r] = \{x \in \kappa \mid s \le x \le r\},$$

$$\kappa_+ = \kappa[0, \infty), \quad \kappa^+ = \kappa(0, \infty).$$

For example, $\mathbb{Z}[s, r]$ means the set of integers i satisfying $s \le i \le r$, $\mathbb{R}^+$ is the set of positive real numbers, and so on.

For later discussions, we will need some notions of rings. When we speak of a *ring*, we shall always mean a commutative ring with a multiplicative identity. A vector subspace I in a ring A is called an *ideal* if $AI \subset I$, that is, if $xy \in I$ for all $x \in A$ and $y \in I$. If $I \ne A$, we say that the ideal is *proper*. A proper ideal I in A is said to be *prime* if $xy \in I$ for $x, y \in A$ means $x \in I$ or $y \in I$. The set of multiples of a particular element $a \in A$, or equivalently, the set of elements divisible by a, forms an ideal called the *principal ideal* generated by a. Elements x, y of a ring are said to be *zero divisors* if $x \ne 0$, $y \ne 0$, and $xy = 0$. We define a ring to be *entire* (or a *domain*, or an *integral domain*) if $1 \ne 0$, and if there are no zero divisors in the ring. A *field* is a domain in which every non-zero element is a *unit*, i.e., has a multiplicative inverse. Any domain A has a *quotient field* or *field of fractions* κ, which is a field containing A as a subring, and any element in κ may be written (not necessarily uniquely) as a ratio of two elements of A. A ring is called *Noetherian* if every ideal in the ring is finitely generated. Fields are Noetherian rings. A basic fact is the following *Hilbert basis theorem*:

Theorem 1.1. *If A is a Noetherian ring, then the ring $A[X_1, \ldots, X_n]$ of polynomials in n variables over A is a Noetherian ring.*

Proof. See Fulton [109], Atiyah-Macdonald [7], Theorem 7.5 or Lang [227], Section 6.2. □

Proposition 1.2 (cf. [109]). *The following conditions on a ring A are equivalent:*

(i) *The set of non-units in A forms an ideal.*

(ii) *A has a unique maximal ideal which contains every proper ideal of A.*

A ring satisfying the conditions of Proposition 1.2 is called a *local ring*. A sequence $x_1, \ldots, x_r$ of elements of a ring A is called a *regular sequence* for A if x_1 is not a zero divisor in A, and for all $i = 2, \ldots, r$, x_i is not a zero divisor in the ring $A/(x_1, \ldots, x_{i-1})$, where

$$(x_1, \ldots, x_{i-1}) = \{a_1 x_1 + \cdots + a_{i-1} x_{i-1} \mid a_j \in A\}$$

is the ideal generated by $x_1, \ldots, x_{i-1}$. If A is a local ring with maximal ideal $\mathbf{m}$, then the *depth* of A is the maximum length of a regular sequence $x_1, \ldots, x_r$ for A with all $x_i \in \mathbf{m}$. We say that a local Noetherian ring A is *Cohen-Macaulay* if depth$A = \dim A$. Now we list some properties of Cohen-Macaulay rings.

Proposition 1.3. *Let A be a local Noetherian ring with maximal ideal $\mathbf{m}$.*

(a) *If A is regular, that is, $\dim_\kappa \mathbf{m}/\mathbf{m}^2 = \dim A$, where $\kappa = A/\mathbf{m}$ is the residue class field, then it is Cohen-Macaulay.*

(b) *If A is Cohen-Macaulay, then any localization $A_{\mathfrak{p}}$ of A at a prime ideal $\mathfrak{p}$ is also Cohen-Macaulay, where $A_{\mathfrak{p}} = \{a/b \mid a, b \in A, b \notin \mathfrak{p}\}$.*

(c) *If A is Cohen-Macaulay, then a set of elements $x_1, \ldots, x_r \in \mathbf{m}$ forms a regular sequence for A if and only if $\dim A/(x_1, \ldots, x_r) = \dim A - r$.*

(d) *If A is Cohen-Macaulay, and $x_1, \ldots, x_r \in \mathbf{m}$ is a regular sequence for A, then $A/(x_1, \ldots, x_r)$ is also Cohen-Macaulay.*

(e) *If A is Cohen-Macaulay, and $x_1, \ldots, x_r \in \mathbf{m}$ is a regular sequence, let I be the ideal $(x_1, \ldots, x_r)$. Then the natural mapping $(A/I)[t_1, \ldots, t_r] \to \oplus_{n \geq 0} I^n/I^{n+1}$, defined by sending $t_i \mapsto x_i$, is an isomorphism. In other words, I/I^2 is a free A/I-module of rank r, and for each $n \geq 1$, the natural mapping $S^n(I/I^2) \to I^n/I^{n+1}$ is an isomorphism, where S^n denotes the nth symmetric power.*

Proof. Matsumura [260]: (a) p. 121; (b) p. 104; (c) p. 105; (d) p. 104; (e) p. 110 or Hartshorne[148], Theorem 8.21A. □

An element x in a ring A is *irreducible* if for any factorization $x = ab$, $a, b \in A$, either a or b is a unit. A domain A is a *unique factorization domain* if every non-zero element in A can be factored uniquely, up to units and the ordering of the factors, into a product of irreducible elements. A subring A of the field κ is called a *valuation ring* if it has the property that for any $x \in \kappa$ we have $x \in A$ or $x^{-1} \in A$.

Proposition 1.4 (cf. [109]). *Let A be a domain which is not a field. Then the following conditions are equivalent:*

(I) *A is Noetherian and local, and the maximal ideal is principal.*

(II) *There is an irreducible element $t \in A$ such that every non-zero $z \in A$ may be written uniquely in the form $z = ut^n$, u a unit in A, n a non-negative integer.*

A ring A satisfying the conditions of Proposition 1.4 is called a *discrete valuation ring*. An element t as in Proposition 1.4 is called a *uniformizing parameter* for A; any other uniformizing parameter is of the form ut, u a unit in A. Let κ be the quotient field of A. Then any non-zero element $z \in \kappa$ has a unique expression $z = ut^n$, u a unit in A, $n \in \mathbb{Z}$, The exponent n is called the *order* of z, and written $n = \mathrm{ord}(z)$; we define $\mathrm{ord}(0) = +\infty$. Then

$$A = \{z \in \kappa \mid \mathrm{ord}(z) \geq 0\},$$

and

$$\mathbf{m} = \{z \in \kappa \mid \mathrm{ord}(z) > 0\}$$

is the maximal ideal in A.

Let κ be a field and A a subring of κ. An element α of κ is said to be *integral* over A if it satisfies a monic equation over A:

$$\alpha^n + a_1\alpha^{n-1} + \cdots + a_n = 0,$$

where $a_i \in A$ for $i = 1, \ldots, n$. If every element of κ integral over A lies in A, then A is said to be *integrally closed* in κ. An integral domain is called *integrally closed* if it is integrally closed in its field of fractions. The set of all elements of κ integral over A is called the *integral closure* of A in κ; it always includes A itself because any α in A satisfies the equation $x - \alpha = 0$ and so is integral over A. Moreover, the integral closure of A is a subring of κ.

Let κ be a field and denote the multiplicative group $\kappa - \{0\}$ by κ_*. If κ is a subfield of a field K, then we also say that K is an *extension field* of κ, which will be denoted by K/κ. The field K can always be regarded as a vector space over κ. The dimension $\dim_\kappa K$ of K as a κ-vector space is called the *degree* of K over κ. It will be denoted by

$$[K : \kappa] = \dim_\kappa K.$$

If $[K : \kappa] < \infty$, K is called a *finite extension* of κ, otherwise, an *infinite extension* of κ. The following proposition is a basic fact of extension fields:

Proposition 1.5. *Let κ be a field and $F \subset K$ extension fields of κ. Then*

$$[K : \kappa] = [K : F][F : \kappa]. \tag{1.1.1}$$

If $\{x_i\}_{i \in I}$ is a basis for F over κ and $\{y_j\}_{j \in J}$ is a basis for K over F, then $\{x_i y_j\}_{(i,j) \in I \times J}$ is a basis for K over κ.

Let κ be a subfield of a field K. Take an element α in K. The field extension of κ, which is generated by α, will be denoted by $\kappa(\alpha)$, that is, $\kappa(\alpha)$ is the smallest field containing κ and α. We denote the ring generated by α over κ by $\kappa[\alpha]$. It consists of all elements of K that can be written as polynomials in α with coefficients in κ:

$$a_n \alpha^n + \cdots + a_1 \alpha + a_0, \quad a_i \in \kappa. \tag{1.1.2}$$

The field $\kappa(\alpha)$ is isomorphic to the field of fractions of $\kappa[\alpha]$. Its elements are ratios of elements of the form (1.1.2). The element α is said to be *algebraic over κ* if it is the root of some non-zero polynomial with coefficients in κ, otherwise, *transcendental over κ*. The lowest degree irreducible monic polynomial with coefficients in κ such that $f(\alpha) = 0$ is called the *minimal polynomial of α over κ*. The degree of the polynomial is also called the *degree of α over κ*, which is determined as follows:

Proposition 1.6. *Let α be algebraic over κ. Then $\kappa(\alpha) = \kappa[\alpha]$, and $\kappa(\alpha)$ is finite over κ. The degree $[\kappa(\alpha) : \kappa]$ is equal to the degree of the minimal polynomial for α over κ.*

A field extension K of κ is called an *algebraic extension*, or K is said to be *algebraic over κ*, if all its elements are algebraic over κ. One important case of a tower of field extensions is that K is a given extension of κ and α is an element of K. The field $\kappa(\alpha)$ is an intermediate field:

$$\kappa \subset \kappa(\alpha) \subset K.$$

Thus, one has

$$[K : \kappa] = [K : \kappa(\alpha)][\kappa(\alpha) : \kappa].$$

Note that $[\kappa(\alpha) : \kappa]$ is the degree of α over κ if α is algebraic, otherwise $[\kappa(\alpha) : \kappa] = \infty$. Hence one shows the property:

Proposition 1.7. *If K is a finite extension of κ, then K is algebraic over κ.*

Let κ be a subfield of K and let $\alpha_1, \ldots, \alpha_n$ be elements of K. We denote by $\kappa(\alpha_1, \ldots, \alpha_n)$ the smallest subfield of K containing κ and $\alpha_1, \ldots, \alpha_n$. Its elements consist of all quotients

$$\frac{P(\alpha_1, \ldots, \alpha_n)}{Q(\alpha_1, \ldots, \alpha_n)}$$

where P, Q are polynomials in n variables with coefficients in κ, and $Q(\alpha_1, \ldots, \alpha_n) \neq 0$. We say that K is *finitely generated* over κ if there is a finite family of elements $\alpha_1, \ldots, \alpha_n$ of K such that $K = \kappa(\alpha_1, \ldots, \alpha_n)$. We exhibit an example of such fields as follows:

Proposition 1.8. *If K is a finite extension of κ, then K is finitely generated over κ.*

A field extension K of κ is said to be an *algebraic closure* of κ if K is algebraic over κ, and K is *algebraically closed*, that is, every polynomial $f(x) \in K[x]$ of positive degree has a root in K.

Proposition 1.9. *Every field κ has an algebraic closure.*

Let K be an extension of a field κ and let

$$\sigma : \kappa \longrightarrow L$$

be an embedding (i.e., an injective homomorphism) of κ into a field L. Then σ induces an isomorphism of κ with its image $\sigma(\kappa)$. An embedding τ of K in L will be said to be *over σ* if the restriction of τ to κ is equal to σ. We also say that τ *extends* σ. If σ is the identity, then we say that τ is an embedding of K *over κ*.

Assume that L is algebraically closed. We analyze the extensions of σ to algebraic extensions K of κ. First of all, we consider the special case $K = \kappa(\alpha)$, where α is algebraic over κ. Let P_α be the minimal polynomial of α over κ. Let β be a root of $\sigma(P_\alpha)$ in L. Given an element of $\kappa(\alpha) = \kappa[\alpha]$, we can write it in the form $f(\alpha)$ with some polynomial f over κ. We define an extension of σ by mapping

$$f(\alpha) \mapsto \sigma(f)(\beta).$$

This is in fact well defined, i.e., independent of the choice of polynomial f used to express our element in $\kappa[\alpha]$. Indeed, if $g(X)$ is in $\kappa[X]$ such that $g(\alpha) = f(\alpha)$, then $(g-f)(\alpha) = 0$, whence $P_\alpha(X)$ divides $g(X) - f(X)$. Hence $\sigma(P_\alpha)(X)$ divides $\sigma(g)(X) - \sigma(f)(X)$, and thus $\sigma(g)(\beta) = \sigma(f)(\beta)$. It is clear that the mapping is a homomorphism inducing σ on κ, and that it is an extension of σ to $\kappa(\alpha)$. Hence we get:

Proposition 1.10. *The number of possible extensions of σ to $\kappa(\alpha)$ is $\leq$ the number of roots of P_α, and is equal to the number of distinct roots of P_α.*

We are interested in extensions of σ to arbitrary algebraic extensions of κ. By using Zorn's lemma, one can prove the following result:

Proposition 1.11. *Let κ be a field, K an algebraic extension of κ, and $\sigma : \kappa \longrightarrow L$ an embedding of κ into an algebraically closed field L. Then there exists an extension of σ to an embedding of K in L. If K is algebraically closed and L is algebraic over $\sigma(\kappa)$, then any such extension of σ is an isomorphism of K onto L.*

As a corollary, we have a certain uniqueness for an algebraic closure of a field κ.

Corollary 1.12. *If L, L' are two algebraic closures of a field κ, there is an isomorphism $\lambda : L \longrightarrow L'$, which is the identity mapping on κ.*

Let K be an algebraic extension of a field κ and let

$$\sigma : \kappa \longrightarrow L$$

be an embedding of κ into an algebraically closed field L. Let S_σ be the set of extensions of σ to an embedding of K in L. Assume that L is algebraic over $\sigma(\kappa)$, hence is equal to an algebraic closure of $\sigma(\kappa)$. Let L' be another algebraically closed field, and let $\tau : \kappa \longrightarrow L'$ be an embedding. Let S_τ be the set of embeddings of K in L' extending τ. We also assume that L' is an algebraic closure of $\tau(\kappa)$. By Proposition 1.11, there exists an isomorphism $\lambda : L \longrightarrow L'$ extending the mapping $\tau \circ \sigma^{-1}$ applied to the field $\sigma(\kappa)$. If $\sigma^* \in S_\sigma$ is an extension of σ to an embedding of K in L, then $\lambda \circ \sigma^*$ is an extension of τ to an embedding of K in L', because for the restriction to κ we have

$$\lambda \circ \sigma^* = \tau \circ \sigma^{-1} \circ \sigma = \tau.$$

Thus λ induces a mapping from S_σ into S_τ. It is clear that the inverse mapping is induced by λ^{-1}, and hence that S_σ, S_τ are in bijection under the mapping $\sigma^* \mapsto \lambda \circ \sigma^*$. In particular, the cardinality of S_σ, S_τ are the same. Thus this cardinality depends only on the extension K/κ, and will be denoted by $[K : \kappa]_s$. We shall call it the *separable degree* of K over κ. A basic fact is listed as follows:

Proposition 1.13. *Let κ be a field and $F \subset K$ be algebraic extensions of κ. Then*

$$[K : \kappa]_s = [K : F]_s [F : \kappa]_s.$$

Furthermore, if K is finite over κ, then $[K : \kappa]_s$ is finite and divides $[K : \kappa]$.

Let κ be a field and let f be a polynomial over κ of degree ≥ 1. By a *splitting field* K of f we shall mean an extension K of κ such that f splits into linear factors in K, i.e.,

$$f(x) = c(x - \alpha_1) \cdots (x - \alpha_n)$$

with $\alpha_i \in K$, $i = 1, \ldots, n$, and such that

$$K = \kappa(\alpha_1, \ldots, \alpha_n)$$

is generated by all roots of f. A splitting field of a polynomial f over κ is unique in the sense of isomorphism:

Proposition 1.14. *Let K be a splitting field of a polynomial f over κ. If F is another splitting field of f, there is an isomorphism $\sigma : F \longrightarrow K$, which is the identity mapping on κ. If $\kappa \subset K \subset \overline{\kappa}$, where $\overline{\kappa}$ is an algebraic closure of κ, then any embedding of F in $\overline{\kappa}$ inducing the identity on κ must be an isomorphism of F onto K.*

Let I be a set of indices and let $\{f_i\}_{i \in I}$ be a family of polynomials over κ of degrees ≥ 1. By a *splitting field* for this family we shall mean an extension K of

κ such that every f_i splits into linear factors in K, and K is generated by total roots of the polynomials f_i, $i \in I$. Let $\overline{\kappa}$ be an algebraic closure of κ, and let K_i be a splitting field of f_i in $\overline{\kappa}$. Then the smallest subfield of $\overline{\kappa}$ containing all fields K_i, $i \in I$ is a splitting field for the family $\{f_i\}_{i \in I}$.

Corollary 1.15. *Let K be a splitting field for the family $\{f_i\}_{i \in I}$. If F is another splitting field for $\{f_i\}_{i \in I}$, any embedding of F into $\overline{\kappa}$ inducing the identity on κ gives an isomorphism of F onto K.*

The following result gives characteristic conditions that an algebraic extension of κ is a splitting field of a family of polynomials over κ.

Theorem 1.16 (cf. [227]). *Let K be an algebraic extension of κ, contained in an algebraic closure $\overline{\kappa}$ of κ. Then the following conditions are equivalent:*

1) *K is the splitting field of a family of polynomials over κ.*

2) *Every embedding σ of K in $\overline{\kappa}$ over κ is an automorphism of K.*

3) *Every irreducible polynomial over κ which has a root in K splits into linear factors in K.*

An extension K of κ satisfying one of the hypotheses 1)-3) in Theorem 1.16 will be said to be *normal*. If K is an algebraic extension of κ, then there exists a smallest normal extension E of κ containing K. The field E can be given by taking the intersection of all normal extensions of κ containing K.

Let κ be a field, and $0 \neq f(x) \in \kappa[x]$. If $f(x)$ has no multiple root in an algebraic closure $\overline{\kappa}$ of κ, then f is called a *separable polynomial*. Let K be an extension of κ, and $\alpha \in K$. If α is algebraic over κ and the minimal polynomial of it over κ is separable, then α is called a *separable algebraic element* over κ. We see that this condition is equivalent to saying that $[\kappa(\alpha) : \kappa]_s = [\kappa(\alpha) : \kappa]$ according to the following criterion:

Proposition 1.17. *Let K be an algebraic closure of κ. Take $\alpha \in K$ and let P_α be the minimal polynomial of α over κ. If char $\kappa = 0$, then all roots of P_α have multiplicity 1 (P_α is separable). If char $\kappa = p > 0$, then there exists an integer $\mu \geq 0$ such that every root of P_α has multiplicity p^μ. We have*

$$[\kappa(\alpha) : \kappa] = p^\mu [\kappa(\alpha) : \kappa]_s,$$

and α^{p^μ} is separable over κ.

If all elements in K are separable algebraic over κ, then K is called a *separable algebraic extension* of κ. If K is separable over κ, we can choose a smallest normal extension E of κ containing K such that E is separable over κ. One has the following condition determining a separable algebraic extension:

Proposition 1.18. *Let K be a finite extension of a field κ. Then K is separable over κ if and only if $[K : \kappa]_s = [K : \kappa]$.*

Let K be an extension of a field κ. If $\alpha \in K$ is a separable algebraic element over κ, then $\kappa[\alpha]$ is a separable algebraic extension of κ. Further, if K is generated by a family of separable algebraic elements $\{\alpha_i\}_{i \in I}$ over κ, then K is separable over κ. Then one has the *theorem of the primitive element*:

Theorem 1.19. *Let K be a finite extension of a field κ. There exists an element $\alpha \in K$ such that $K = \kappa(\alpha)$ if and only if there exists only a finite number of fields F such that $\kappa \subset F \subset K$. If K is separable over κ, then there exists such an element α.*

Proof. Zariski and Samuel [448], Ch. II, Theorem 19, p. 84.   □

A field κ of characteristic $p > 0$ is called *perfect* if $\{x^p \mid x \in \kappa\} = \kappa$. Every field of characteristic zero is also called perfect. It is well known that if κ is perfect, every algebraic extension of κ is separable and perfect (cf. [227]). If K is an extension field of κ which is not algebraic, the *transcendence degree* of K/κ is the maximum number of elements of K that are algebraically independent over κ. A subset S of K which is algebraically independent over κ and is maximal with respect to the inclusion ordering will be called a *transcendence base* of K over κ. If K is a finitely generated extension of κ, $K = \kappa(x)$, it is said to be *separably generated* if we can find a transcendence base $t = (t_1, \ldots, t_r)$ of K/κ such that K is separably algebraic over $\kappa(t)$. Such a transcendence base is said to be a *separating transcendence base* for K over κ.

Proposition 1.20 (cf. [63]). *Let κ be a perfect field and K an extension of κ of transcendence degree 1. Then there exists $x \in K$ such that $K/\kappa(x)$ is a separable extension. The element x is called a separating element of the extension.*

Let κ be a field and let G be a group of automorphisms of κ. Let $F(G)$ be the set of invariants of G, namely,

$$F(G) = \{x \mid x \in \kappa, \ \sigma(x) = x \text{ for all } \sigma \in G\}.$$

Then the set $F(G)$ is a subfield of κ, which is called the *invariant field* of G, or the *fixed field* of G. Let K be an algebraic extension of κ and let $G_{K/\kappa}$ be the group of automorphisms of K over κ. The field K is called a *Galois extension* of κ if K is a normal separable extension of κ. If K is a Galois extension of κ, then $G_{K/\kappa}$ is called the *Galois group* of K over κ. For the convenience of the reader, we shall now state the main theorem of Galois theory for finite Galois extensions.

Theorem 1.21 (cf. [272]). *Let K be a finite Galois extension of κ. Then we have*

(1) *Let E be an intermediate field between K and κ, namely $\kappa \subset E \subset K$. Then*

    (α) *K is a Galois extension of E. $G_{K/E}$ is a subgroup of $G_{K/\kappa}$.*

    (β) *The order of the group $G_{K/E}$ is $[K : E]$.*

    (γ) *The invariant field $F(G_{K/E})$ of $G_{K/E}$ is E.*

(2) *Let H be a subgroup of $G_{K/\kappa}$. Then*

    (δ) *The invariant field $F(H)$ is an intermediate field between K and κ.*

    (ε) *The order of H is $[K : F(H)]$.*

    (ζ) *The Galois group $G_{K/F(H)} = H$.*

Therefore, there is a bijective mapping between the set of subfields E of K containing κ, and the set of subgroups H of $G_{K/\kappa}$, given by $E = F(H)$ (resp., $H = G_{K/E}$). The E is Galois over κ if and only if H is normal in $G_{K/\kappa}$, and if that is the case, then the mapping $\sigma \mapsto \sigma|E$ induces an isomorphism of $G_{K/\kappa}/H$ onto the Galois group $G_{E/\kappa}$ of E over κ.

Let K be a finite field extension of κ. Take $\alpha \in K$. Then α induces a κ-linear mapping

$$\mathbf{A}_\alpha : K \longrightarrow K$$

defined by $\mathbf{A}_\alpha(x) = \alpha x$. Let $\{w_1, \ldots, w_n\}$ be a base of K over κ. Write

$$\mathbf{A}_\alpha(w_i) = \alpha w_i = \sum_{j=1}^{n} a_{ij} w_j.$$

The characteristic polynomial

$$\chi_\alpha(x) = \det(xI - A_\alpha)$$

of the matrix form $A_\alpha = (a_{ij})$ of $\mathbf{A}_\alpha$ is called the *field polynomial* of α. The field polynomial χ_α is independent of the base $\{w_1, \ldots, w_n\}$ selected for K over κ. Obviously, α is a root of its field polynomial.

Lemma 1.22. *Let K be a finite field extension of a field F which is a finite field extension of κ, and $\alpha \in F$. Let the field polynomial of α as an element of K be $\chi_\alpha(x)$, the field polynomial of α as an element of F be $G(x)$, and the minimal polynomial of α over κ be $P_\alpha(x)$. Then we have*

(A) $K = \kappa(\alpha)$ *if and only if* $\chi_\alpha(x) = P_\alpha(x)$.

(B) $\chi_\alpha(x) = G(x)^{[K:F]}$.

Proof. (A) Since α is a root of χ_α, therefore we have $P_\alpha(x) \mid \chi_\alpha(x)$. Since their degrees are equal, and both are monic polynomials, then they must be equal.

    (B) Let $[F : \kappa] = s$. Let $\{u_1, \ldots, u_s\}$ be a basis for F over κ, and $\{v_1, \ldots, v_m\}$ be a basis for K over F. Let K_i be the subspace

$$K_i = \oplus_{j=1}^{s} u_j v_i \kappa.$$

Then we have

$$K = \oplus_{i=1}^{m} K_i.$$

Each K_i is an invariant subspace of $\mathbf{A}_\alpha$ with the characteristic polynomial of the restriction of $\mathbf{A}_\alpha$ to K_i equaling $G(x)$. Therefore, we have $\chi_\alpha(x) = G(x)^{[K:F]}$. $\quad\square$

Let $P_\alpha \in \kappa[x]$ be the minimal polynomial of α over κ. Let χ_α be the field polynomial of α. Then one has

$$\chi_\alpha(0) = (-1)^{[K:\kappa]} \det(A_\alpha) = P_\alpha(0)^{[K:\kappa(\alpha)]},$$

and hence

$$\det(A_\alpha) = \left\{(-1)^d P_\alpha(0)\right\}^{[K:\kappa(\alpha)]} = \left\{(-1)^d P_\alpha(0)\right\}^{\frac{[K:\kappa]}{d}},$$

where $d = \deg(P_\alpha) = [\kappa(\alpha) : \kappa]$. We will denote the element of κ by $\mathbf{N}_{K/\kappa}(\alpha)$, called the *norm of α over κ*. Define the *trace* $\mathbf{Tr}_{K/\kappa}(\alpha)$ of α over κ as $\text{trace}(A_\alpha) = \sum a_{ii}$. In other words, in the following field polynomial χ_α of α, we have

$$\chi_\alpha(x) = x^n - \mathbf{Tr}_{K/\kappa}(\alpha)x^{n-1} + \cdots + (-1)^n \mathbf{N}_{K/\kappa}(\alpha),$$

where $n = [K : \kappa]$. The norm of K over κ

$$\mathbf{N}_{K/\kappa} : K \longrightarrow \kappa$$

is a multiplicative homomorphism of K_* into κ_*, namely

$$\mathbf{N}_{K/\kappa}(\alpha\beta) = \mathbf{N}_{K/\kappa}(\alpha)\mathbf{N}_{K/\kappa}(\beta) \in \kappa_*, \quad \alpha, \beta \in K_*.$$

The trace of K over κ,

$$\mathbf{Tr}_{K/\kappa} : K \longrightarrow \kappa,$$

determines a κ-linear mapping of K to κ. When $\alpha \in \kappa$, the formula

$$\mathbf{Tr}_{K/\kappa}(\alpha) = [K : \kappa]\alpha, \quad \mathbf{N}_{K/\kappa}(\alpha) = \alpha^{[K:\kappa]} \tag{1.1.3}$$

is trivial. By Lemma 1.22, if K is a finite field extension of a field F which is a finite field extension of κ, and $\alpha \in F$, then we have

$$\mathbf{Tr}_{K/\kappa}(\alpha) = [K : F]\mathbf{Tr}_{F/\kappa}(\alpha), \quad \mathbf{N}_{K/\kappa}(\alpha) = \mathbf{N}_{F/\kappa}(\alpha)^{[K:F]}. \tag{1.1.4}$$

Let K be a finite extension of a field κ. Let $[K : \kappa]_s = r$, and let

$$p^\mu = \frac{[K : \kappa]}{[K : \kappa]_s}$$

if the characteristic of κ is $p > 0$, and 1 otherwise. Let $\bar{\kappa}$ be an algebraic closure of κ and let $\sigma_1, \ldots, \sigma_r$ be the distinct embeddings of K in $\bar{\kappa}$. Then for $\alpha \in K$, one has

$$\mathbf{N}_{K/\kappa}(\alpha) = \prod_{i=1}^r \sigma_i\left(\alpha^{p^\mu}\right), \quad \mathbf{Tr}_{K/\kappa}(\alpha) = p^\mu \sum_{i=1}^r \sigma_i(\alpha). \tag{1.1.5}$$

When $K = \kappa(\alpha)$, it is easy to show that (1.1.5) holds by using Proposition 1.10. Generally, the mappings of K into κ defined by (1.1.5) are transitive, in other words, if we have three fields $\kappa \subset F \subset K$, then (cf. [227])

$$\mathbf{Tr}_{F/\kappa} \circ \mathbf{Tr}_{K/F} = \mathbf{Tr}_{K/\kappa}, \quad \mathbf{N}_{F/\kappa} \circ \mathbf{N}_{K/F} = \mathbf{N}_{K/\kappa}. \tag{1.1.6}$$

Thus (1.1.5) follows from (1.1.4) and (1.1.6) applied to $F = \kappa(\alpha)$.

Theorem 1.23. *Let K be a finite separable extension of a field κ. Then $\mathbf{Tr}_{K/\kappa} : K \longrightarrow \kappa$ is a non-zero functional. The mapping $(x, y) \mapsto \mathbf{Tr}_{K/\kappa}(xy)$ of $K \times K \longrightarrow \kappa$ is bilinear, and identifies K with its dual space.*

Proof. Trivially, $\mathbf{Tr}_{K/\kappa} : K \longrightarrow \kappa$ is a non-zero functional. For each $x \in K$, the mapping

$$\mathbf{Tr}_{K/\kappa,x} : K \longrightarrow \kappa$$

such that $\mathbf{Tr}_{K/\kappa,x}(y) = \mathbf{Tr}_{K/\kappa}(xy)$ is obviously a κ-linear mapping, and the mapping

$$x \mapsto \mathbf{Tr}_{K/\kappa,x}$$

is a κ-homomorphism of K into its dual space K^*. If $\mathbf{Tr}_{K/\kappa,x}$ is the zero mapping, then $\mathbf{Tr}_{K/\kappa}(xK) = 0$. If $x \neq 0$ then $xK = K$. Hence the kernel of $x \mapsto \mathbf{Tr}_{K/\kappa,x}$ is 0. Hence we get an injective homomorphism of K into K^*. Since these spaces have the same finite dimension, it follows that we get an isomorphism. $\qquad\square$

Let $w_1, \ldots, w_n$ be a basis of K over κ. Then $\mathbf{Tr}_{K/\kappa,w_1}, \ldots, \mathbf{Tr}_{K/\kappa,w_n}$ is a basis of K^*. Thus we can find $v_1, \ldots, v_n \in K$ to satisfy

$$\mathbf{Tr}_{K/\kappa}(w_i v_j) = \delta_{ij}.$$

Obviously, $v_1, \ldots, v_n$ forms a basis of K over κ.

1.2   Fields with valuations

We will introduce basic properties of absolute values over a field and their extensions to extension fields. In particular, classification of absolute values will be exhibited, which includes the first and second theorems of Ostrowski.

1.2.1   Absolute values

Definition 1.24. *An absolute value on a field κ is a function*

$$|\cdot| : \kappa \longrightarrow \mathbb{R}_+$$

that satisfies the following conditions:

1) $|x| = 0$ *if and only if* $x = 0$;

2) $|xy| = |x||y|$ *for all* $x, y \in \kappa$;

3) $|x + y| \leq |x| + |y|$ *for all* $x, y \in \kappa$.

If instead of 3) *the absolute value satisfies the stronger condition*

4) $|x + y| \leq \max\{|x|, |y|\}$ *for all* $x, y \in \kappa$,

then the absolute value is called ultrametric or non-Archimedean. Otherwise, it is called Archimedean.

Let κ be a field with an absolute value $|\cdot|$. If $|\cdot|$ is non-Archimedean, in fact we have

$$|x + y| = \max\{|x|, |y|\}, \quad |x| \neq |y|.$$

The absolute value $|\cdot|$ is said to be *trivial* if

$$|x| = \begin{cases} 1 & : \quad x \in \kappa_*, \\ 0 & : \quad x = 0, \end{cases}$$

and *dense* if the set

$$|\kappa| = \{|x| \mid x \in \kappa\}$$

is dense in $\mathbb{R}_+$.

Generally, an absolute value $|\cdot|$ on κ induces a *distance function d* defined by

$$d(x, y) = |x - y|,$$

for any two elements $x, y \in \kappa$, and hence induces a topology on κ. For a positive real number r and a point $x \in \kappa$, denote the open and closed balls of radius r centered at x, respectively, by

$$\kappa(x; r) = \{y \in \kappa \mid d(x, y) < r\}, \quad \kappa[x; r] = \{y \in \kappa \mid d(x, y) \leq r\},$$

and denote the circle by

$$\kappa\langle x; r\rangle = \{y \in \kappa \mid d(x, y) = r\} = \kappa[x; r] - \kappa(x; r).$$

By using the distance, then $|\cdot|$ is non-Archimedean if and only if the induced metric satisfies

$$d(x, y) \leq \max\{d(x, z), d(z, y)\}, \quad x, y, z \in \kappa.$$

For this case, we know that

$$y \in \kappa[x; r] \Longrightarrow \kappa[y; r] = \kappa[x; r];$$

$$y \in \kappa\langle x; r\rangle \Longrightarrow \kappa(y; r) \subset \kappa\langle x; r\rangle.$$

Thus $\kappa[x; r]$, $\kappa(x; r)$ and $\kappa\langle x; r\rangle$ all are open and closed. Such sets are usually called *clopen*.

Example 1.25. *Let $p \in \mathbb{Z}^+$ be a prime number. For $x = a/b \in \mathbb{Q}_*$, there exist integers $v_p(x), a', b'$ such that*

$$x = p^{v_p(x)}\frac{a'}{b'}, \quad p \nmid a'b'.$$

Define

$$|x|_p = \begin{cases} p^{-v_p(x)} & : \quad x \neq 0, \\ 0 & : \quad x = 0. \end{cases}$$

The function $|\cdot|_p$ is a non-Archimedean absolute value on $\mathbb{Q}$, called the p-adic absolute value and was first introduced by Hensel in 1994.

According to the standard theory in p-adic analysis, the completion of $\mathbb{Q}$ relative to the topology induced by $|\cdot|_p$ is just the field $\mathbb{Q}_p$ of *p-adic numbers*, and the absolute value $|\cdot|_p$ on $\mathbb{Q}$ extends to a non-Archimedean absolute value on $\mathbb{Q}_p$, which is also denoted by $|\cdot|_p$. The set of values of $\mathbb{Q}$ and $\mathbb{Q}_p$ under $|\cdot|_p$ is the same, which is equal to the set

$$\{p^n \mid n \in \mathbb{Z}\} \cup \{0\}.$$

Let $|\cdot|_\infty$ denote the ordinary Archimedean absolute value on $\mathbb{Q}$ and set

$$M_{\mathbb{Q}} = \{\infty\} \cup \{\text{primes}\}.$$

For any $x \in \mathbb{Q}_*$, we have the *product formula*

$$\prod_{v \in M_{\mathbb{Q}}} |x|_v = 1. \tag{1.2.1}$$

Theorem 1.26 (Ostrowski [314]). *Let $|\cdot|$ be a non-trivial absolute value on $\mathbb{Q}$. If $|\cdot|$ is Archimedean, then there exists α with $0 < \alpha \leq 1$ such that*

$$|x| = |x|_\infty^\alpha, \quad x \in \mathbb{Q}.$$

If $|\cdot|$ is non-Archimedean, then there exist a prime p and real $\beta > 0$ such that

$$|x| = |x|_p^\beta, \quad x \in \mathbb{Q}.$$

A proof can be found in [338]. Usually, Theorem 1.26 is called *Ostrowski's first theorem*.

There is another characterization for non-Archimedean absolute values. We begin by noting that for any field κ there is a mapping $\iota : \mathbb{Z} \longrightarrow \kappa$ defined by

$$\iota(n) = \begin{cases} \underbrace{1 + 1 + \cdots + 1}_{n} & : \quad n > 0, \\ 0 & : \quad n = 0, \\ -\underbrace{(1 + 1 + \cdots + 1)}_{-n} & : \quad n < 0, \end{cases}$$

where 1 is the unit of κ, and its kernel $\iota^{-1}(0)$ is an ideal generated by an integer $p \geq 0$. The number p is called the *characteristic* of the field κ, which is either 0 or a prime number. In the first case, κ contains as a subfield an isomorphic image of the rational numbers, and in the second case, it contains an isomorphic image of $\mathbb{F}_p = \mathbb{Z}/p\mathbb{Z}$. We will identify $\mathbb{Z}$ with its image under the mapping in κ. Then an absolute value $|\cdot|$ on κ is Archimedean if and only if

$$\sup_{n\in\mathbb{Z}} |n| = +\infty,$$

and is non-Archimedean if and only if

$$\sup_{n\in\mathbb{Z}} |n| = 1.$$

Two absolute values on κ are said to be *equivalent* (or *dependent*) if they induce the same topology on κ. An equivalence class of non-trivial absolute values on a field κ will be called a *prime* or *place* of κ, and sometimes be denoted by a German letter $\mathfrak{p}$. A place $\mathfrak{p}$ is called *non-Archimedean* or *finite* (resp., *Archimedean* or *infinite*) if its absolute value is non-Archimedean (resp., Archimedean). We have the following more accessible criterion (cf. [117]):

Lemma 1.27. *Let $|\cdot|_1$ and $|\cdot|_2$ be absolute values on a field κ. The following statements are equivalent:*

(i) *$|\cdot|_1$ and $|\cdot|_2$ are equivalent absolute values;*

(ii) *$|x|_1 < 1$ if and only if $|x|_2 < 1$ for any $x \in \kappa$;*

(iii) *there exists a positive real number α such that for each $x \in \kappa$, one has $|x|_1 = |x|_2^\alpha$.*

Lemma 1.27 and Theorem 1.26 shows that $M_\mathbb{Q}$ is just the set of all places on $\mathbb{Q}$.

Definition 1.28. *A valuation on a field κ is a function v from κ into $\mathbb{R} \cup \{+\infty\}$ satisfying:*

(a) *$v(x) = +\infty$ if and only if $x = 0$.*

(b) *$v(xy) = v(x) + v(y)$ for all $x, y \in \kappa$.*

(c) *$v(x + y) \geq \min\{v(x), v(y)\}$ for all $x, y \in \kappa$.*

A valuation v is said to be *trivial* if

$$v(x) = \left\{ \begin{array}{lll} 0 & : & x \in \kappa_*, \\ +\infty & : & x = 0. \end{array} \right.$$

Two valuations v_1 and v_2 are *equivalent* if and only if there is a real positive constant λ such that $v_1(x) = \lambda v_2(x)$ for all $x \in \kappa$.

Let v be a valuation on a field κ. For a real constant $c > 1$, a non-Archimedean absolute value

$$|x|_v = c^{-v(x)}, \quad x \in \kappa$$

is well defined. Conversely, if $|\cdot|$ is a non-Archimedean absolute value on κ, a valuation $v_c : \kappa \longrightarrow \mathbb{R} \cup \{+\infty\}$ is defined by

$$v_c(x) = \begin{cases} -\log_c |x| & : \quad x \in \kappa_*, \\ +\infty & : \quad x = 0, \end{cases}$$

and is named the *(additive) valuation associated to the absolute value*, where $\log_c$ is the real logarithm function of base c. Thus a non-Archimedean absolute value can be characterized by a valuation.

Example 1.29. *Let $p \in \mathbb{Z}^+$ be a prime number. If $x \in \mathbb{Q}$, set*

$$v_p(x) = \begin{cases} -\log_p |x|_p & : \quad x \neq 0, \\ +\infty & : \quad x = 0. \end{cases}$$

The function v_p on $\mathbb{Q}$ is called the p-adic valuation on $\mathbb{Q}$.

The image of κ_* by a valuation v is a subgroup of the additive group $(\mathbb{R}, +)$ called the *valuation group of κ*. The valuation of κ is said to be *discrete* (resp., *dense*) if its valuation group is a discrete (resp., dense) subgroup of $\mathbb{R}$. For the trivial valuation, the valuation group consists of 0 alone. If v is a non-trivial valuation, its valuation group Γ either has a least positive element λ or Γ has no least positive element. For the former, $\Gamma = \lambda\mathbb{Z}$. For the latter, Γ is clearly dense in $\mathbb{R}$. For a discrete valuation we can always find an equivalent one with precise valuation group $\mathbb{Z}$; such a valuation is said to be *normalized* or an *order function* on κ. The function v_p on $\mathbb{Q}$ is an order function. Krull [217] has observed that a valuation uses only the addition and the ordering of the real numbers. This means that we can take the values to lie in any ordered additive group. For discussion of the general extensions, see Cohn [63].

Lemma 1.30. *If κ is algebraically closed, then the valuation v associated with a non-trivial absolute value $|\cdot|$ is dense and hence the absolute value $|\cdot|$ also is dense.*

Proof. Since $|\cdot|$ is non-trivial, then there is an element $a \in \kappa_*$ with $|a| \neq 0, 1$. We may assume $0 < |a| < 1$. Since κ is algebraically closed, then the equation

$$z^n - a^q \ (n \in \mathbb{Z}^+, \ q \in \mathbb{Z})$$

always has solutions. Thus, we have

$$v(z) = \frac{q}{n} v(a) \in v(\kappa_*).$$

Obviously, the set

$$\left\{ \frac{q}{n} v(a) \mid n \in \mathbb{Z}^+, \ q \in \mathbb{Z} \right\}$$

is dense in $\mathbb{R}$. Thus, the valuation v is dense. $\qquad\square$

If $|\cdot|_v$ is a non-Archimedean absolute value on the field κ given by a valuation v, the subset

$$\mathfrak{O}_\kappa = \kappa[0;1] = \{x \in \kappa \mid |x|_v \leq 1\}$$

is a subring of κ that is called the *valuation ring* of v; its elements are the *valuation integers* or *v-integers*. Its subset $\kappa(0;1)$ is an ideal of $\mathfrak{O}_\kappa$, which is called the *valuation ideal* of v. Furthermore, $\kappa(0;1)$ is a maximal ideal in $\mathfrak{O}_\kappa$, and every element of the complement $\kappa\langle 0;1 \rangle$ is invertible in $\mathfrak{O}_\kappa$. Thus, $\mathfrak{O}_\kappa$ is a local ring of κ. The field

$$\mathbb{F}(\kappa) = \mathbb{F}_v(\kappa) = \mathfrak{O}_\kappa/\kappa(0;1)$$

is called the *residue class field* of κ. The characteristic of $\mathbb{F}(\kappa)$ is named the *residue characteristic* of κ. In particular, the valuation ring

$$\mathbb{Z}_p = \mathfrak{O}_{\mathbb{Q}_p} = \mathbb{Q}_p[0;1] = \mathbb{Q}_p(0;p)$$

is both open and closed, which is called the ring of *p-adic integers*.

1.2.2   Extensions of absolute values

Definition 1.31. *Let V be a vector space over a field κ and let $|\cdot|$ be a non-trivial absolute value on κ. A norm on V (compatible with the absolute value of κ) is a function*

$$|\cdot| : V \longrightarrow \mathbb{R}_+$$

that satisfies the following conditions:

(α) $|X| = 0$ *if and only if $X = 0$;*

(β) $|X + Y| \leq |X| + |Y|$ *for all $X,Y \in V$;*

(γ) $|aX| = |a| \cdot |X|$ *for all $a \in \kappa$ and all $X \in V$.*

A vector space V with a norm is called a normed vector space over κ.

Let V be a normed vector space. Then any norm $|\cdot|$ induces a metric d,

$$d(X,Y) = |X - Y|,$$

which makes V a topological space. We say two norms $|\cdot|_1$ and $|\cdot|_2$ on V are *equivalent* if they define the same topology on V. Equivalently, there are positive real numbers c_1 and c_2 such that

$$c_1|\cdot|_1 \leq |\cdot|_2 \leq c_2|\cdot|_1.$$

If V is finite-dimensional and if κ is complete under a non-trivial absolute value, then any two norms on V are equivalent, and V is complete with respect to the metric induced by any norm. Further, if κ is locally compact, then V is also locally compact.

Let K be an extension of a field κ. An absolute value w of K is said to be an *extension* of an absolute value v of κ if

$$|x|_w = |x|_v, \quad x \in \kappa.$$

We will denote the relation between v and w by $w|v$. Obviously, if $w|v$, then $|\cdot|_w$ is also a norm on K as a κ-vector space, and, further, if v is non-Archimedean, then w has to be non-Archimedean because this depends only on the absolute values of the elements of $\mathbb{Z}$, which are in κ.

Theorem 1.32. *Let K be a field extension of κ. Then an absolute value on κ has an extension to K. If K is an algebraic extension of κ, and if κ is complete, then there is a unique absolute value on K extending the absolute value of κ. If K is finite over κ, then K is complete.*

Proof. We prove only the uniqueness. For the rest, see [227]. Suppose $|\cdot|_1$ and $|\cdot|_2$ are two absolute values on K that extend the absolute value $|\cdot|$ of κ. For any $x \in K$, it is easy to show that $x^n \to 0$ with respect to the topology induced by $|\cdot|_1$ (resp., $|\cdot|_2$) if and only if $|x|_1 < 1$ (resp., $|x|_2 < 1$). Note that $|\cdot|_1$ and $|\cdot|_2$ are equivalent as norms on the κ-vector space K, i.e., they define the same topology. Therefore, we have convergence with respect to $|\cdot|_1$ if and only if we have convergence with respect to $|\cdot|_2$, or $|x|_1 < 1$ if and only if $|x|_2 < 1$. Hence $|\cdot|_1$ and $|\cdot|_2$ also are two equivalent absolute values. Thus, there exists a positive real number α such that for each $x \in K$, one has $|x|_1 = |x|_2^\alpha$, and hence $\alpha = 1$ since $|x|_1 = |x|_2 = |x|$ when $x \in \kappa$, i.e., the two absolute values are the same. $\square$

Let K be a field extension of κ. We shall now examine ways of extending a valuation on κ to K. By induction it will be enough to look at simple extensions, $K = \kappa(\alpha)$, and we shall treat the cases where α is algebraic or transcendental separately. Thus our first task is to extend a non-Archimedean valuation v on κ to the rational function field $\kappa(t)$, where t is transcendental over κ. We shall give a simple construction of a valuation w on $\kappa(t)$ which extends v. Given a polynomial

$$f(t) = a_n t^n + \cdots + a_1 t + a_0 \in \kappa[t],$$

define

$$w(f) = \min_{0 \le i \le n} v(a_i). \tag{1.2.2}$$

It is clear that this reduces to v on κ, and moreover satisfies

$$w(f + g) \ge \min\{w(f), w(g)\};$$

further, the argument used to prove Gauss' lemma shows that

$$w(fg) = w(f) + w(g),$$

cf. Lemma 1.97. Hence we obtain a valuation w on $\kappa[t]$; it extends in a unique way to a valuation on $\kappa(t)$, still denoted w, by the rule

$$w\left(\frac{f}{g}\right) = w(f) - w(g), \quad f, g \in \kappa[t]. \tag{1.2.3}$$

The valuation thus defined is called the *Gaussian extension* of v to $\kappa(t)$. Clearly v and w have the same value groups.

Consider next a finite extension K/κ, and so an algebraic extension. We will assume that κ is complete under an absolute value $|\cdot|_v$. There can be at most one valuation extension $|\cdot|_w$ to K, by Theorem 1.32. Observe first that if K is a normal extension of κ, and σ is an automorphism of K over κ, then σ induces an absolute value $|\cdot|_{w,\sigma}$ on K defined by

$$|x|_{w,\sigma} = |\sigma(x)|_w, \quad x \in K.$$

Since the restriction of σ on κ is the identity, then $|\cdot|_{w,\sigma}$ also is an extension of the absolute value of κ, and hence

$$|x|_{w,\sigma} = |x|_w, \quad x \in K.$$

Since K is algebraic over κ, if σ is an embedding of K in $\bar{\kappa}$ over κ, where $\bar{\kappa}$ is an algebraic closure of κ, then the same conclusion remains valid, as one sees immediately by embedding K in a normal extension of κ. In particular, if α is algebraic over κ, and if $\sigma_1, \ldots, \sigma_r$ are the distinct embeddings of $\kappa(\alpha)$ in $\bar{\kappa}$, then one has

$$\left|\mathbf{N}_{\kappa(\alpha)/\kappa}(\alpha)\right|_v = \left|\prod_{i=1}^{r} \sigma_i(\alpha)\right|_v^{[\kappa(\alpha):\kappa]/r} = |\alpha|_w^{[\kappa(\alpha):\kappa]},$$

and taking the root, we get (cf. [63], [95], [117]):

Theorem 1.33. *Let κ be a field which is complete under a non-trivial absolute value $|\cdot|_v$ and let K/κ be a finite field extension. Then v has a unique extension w to K, given by*

$$|\alpha|_w = \left|\mathbf{N}_{K/\kappa}(\alpha)\right|_v^{1/[K:\kappa]}. \tag{1.2.4}$$

We recall from standard field theory that a general field extension is obtained by taking a purely transcendental extension, followed by an algebraic extension. Thus if we are given a complete field κ with respect to a non-Archimedean valuation, we can form the rational function field $\kappa(t)$ with the Gaussian extension. Repeating this process and applying Theorem 1.33 to the final algebraic extension, we have found an extension of a complete valuation to any finitely generated extension. It is not difficult to extend the argument to extensions that are not finitely generated.

Example 1.34. *Let $K/\mathbb{Q}_p$ be a finite extension. The unique absolute value on K that extends the p-adic absolute value on $\mathbb{Q}_p$ is called the p-adic absolute value on K, which also is denoted by $|\cdot|_p$. It makes K a locally compact topological field, and makes K complete. From the formula for the absolute value on K, we immediately see that for any $\alpha \in K_*$,*

$$|\alpha|_p = p^{-w(\alpha)}, \quad w(\alpha) = \frac{1}{[K:\mathbb{Q}_p]} v_p(\mathbf{N}_{K/\mathbb{Q}_p}(\alpha)) \in \frac{1}{[K:\mathbb{Q}_p]}\mathbb{Z}.$$

The unique rational number $w(\alpha)$ is also denoted by $v_p(\alpha)$. Setting $v_p(0) = +\infty$, one obtains a valuation v_p on K, which also is called the p-adic valuation on K.

Now we show that the p-adic absolute value on $\mathbb{Q}_p$ can be extended to an algebraic closure $\overline{\mathbb{Q}}_p$ of $\mathbb{Q}_p$. In fact, given any $x \in \overline{\mathbb{Q}}_p$, then x is in the finite extension $\mathbb{Q}_p(x)$, and hence one can define $|x|_p$ by using the unique extension of the p-adic absolute value to $\mathbb{Q}_p(x)$. Therefore, one obtains a function

$$|\cdot| : \overline{\mathbb{Q}}_p \longrightarrow \mathbb{R}_+$$

that extends the p-adic absolute value on $\mathbb{Q}_p$, and it is easy to prove that this function is an absolute value. The unique absolute value on $\overline{\mathbb{Q}}_p$ is also called the p-adic absolute value. However, $\overline{\mathbb{Q}}_p$ is not complete with respect to the p-adic absolute value. The completion of $\overline{\mathbb{Q}}_p$ relative to the topology induced by $|\cdot|_p$ is a field that is denoted by $\mathbb{C}_p$, and the absolute value $|\cdot|_p$ on $\overline{\mathbb{Q}}_p$ extends to a non-Archimedean absolute value on $\mathbb{C}_p$, which is also denoted by $|\cdot|_p$, such that $\overline{\mathbb{Q}}_p$ is dense in $\mathbb{C}_p$, and $\mathbb{C}_p$ is complete, algebraically closed. The image of $\mathbb{C}_{p*}$ under v_p is $\mathbb{Q}$.

Let κ be a field with a non-trivial absolute value v. The completion of κ relative to the topology induced by v is a field which is denoted by κ_v. If v is non-Archimedean, then κ_v is a finite extension of $\mathbb{Q}_p$ for some prime p. If v is Archimedean, then $\kappa_v = \mathbb{R}$ or $\mathbb{C}$, which is just contents of *Ostrowski's second theorem*:

Theorem 1.35. *Let K be a field with an Archimedean absolute value for which it is complete. Then K is isomorphic either to $\mathbb{R}$ or to $\mathbb{C}$.*

We know that a non-trivial absolute value v of κ can be extended to κ_v, and then uniquely to its algebraic closure $\overline{\kappa}_v$. Let K be an extension of κ. Given two embeddings $\sigma, \tau : K \longrightarrow \overline{\kappa}_v$, we shall say that they are *conjugate over κ_v* if there exists an automorphism λ of $\overline{\kappa}_v$ over κ_v such that $\sigma = \lambda \circ \tau$.

Proposition 1.36 (cf. [225]). *Let K be an algebraic extension of κ. Two embeddings $\sigma, \tau : K \longrightarrow \overline{\kappa}_v$ give rise to the same absolute value on K if and only if they are conjugate over κ_v.*

If v is non-Archimedean, we have the canonical isomorphism

$$\mathbb{F}(\kappa) \cong \mathbb{F}(\kappa_v).$$

Let K be a finite extension of κ, and let w be an absolute value of K extending v. Then the residue class field $\mathbb{F}(K)$ is an extension of $\mathbb{F}(\kappa)$, and its degree is called the *residue class degree* of the extension K/κ. The valuation group $v(\kappa_*)$ of κ is a subgroup of the valuation group $w(K_*)$ of K, whose index is called the *ramification index* of the extension K/κ. By the definition, the index is the number of cosets of the subgroup $v(\kappa_*)$ in $w(K_*)$. The extension is said to be *ramified* if its ramification index > 1, *unramified* otherwise.

Proposition 1.37 (cf. [63], [225]). *Let K/κ be an extension of valuated fields with ramification index e and residue class degree d, where the valuation w of K extends that of κ, v say. If K/κ is finite, then*

$$ed \le [K : \kappa], \tag{1.2.5}$$

and if v is trivial or discrete, then so is w. If v is discrete and κ is complete, then equality holds in (1.2.5).

The following result describes the extensions of an incomplete field:

Theorem 1.38 (cf. [63]). *Let κ be a field with a discrete valuation v and let K/κ be a separable extension of degree n. Then there are at most n extensions of v to K, say $w_1, \ldots, w_r$, where $r \le n$. If e_i denotes the ramification index and d_i the residue class degree of w_i, then*

$$\sum_{i=1}^{r} e_i d_i = n.$$

Further, if the completion of κ under v is κ_v and that of K under w_i is K_{w_i}, then

$$K \otimes \kappa_v \cong K_{w_1} \times \cdots \times K_{w_r}.$$

Finally, we consider the absolute value defined by an irreducible polynomial $p(x)$ over a field κ with the associated valuation v given by $v(\varphi) = \nu$ if

$$\varphi = p^{\nu} \frac{f}{g}, \quad p \nmid fg.$$

This valuation is trivial on κ. We also say that v is a valuation of $\kappa(x)$ over κ. We can now determine all valuations of $\kappa(x)$ over κ:

Proposition 1.39 (cf. [63]). *Let κ be any field. Any general valuation on the rational function field $\kappa(x)$ over κ which is non-trivial is either associated to an irreducible polynomial over κ, or to x^{-1}.*

Consider the case $\kappa = \mathbb{C}$; here the irreducible polynomials are $x - a$ ($a \in \mathbb{C}$), so every complex number defines a valuation on $\mathbb{C}(x)$; in addition we have a valuation corresponding to x^{-1}. These valuations just correspond to the points on the Riemann sphere, and the representation

$$\varphi(x) = \begin{cases} (x-a)^\nu \frac{f(x)}{g(x)}, & p(x) = x - a; \\ x^{-\nu} \frac{f(x)}{g(x)}, & p(x) = x^{-1} \end{cases}$$

may be regarded as indicating the leading term of φ at the point $x = a$ or $x = \infty$.

1.3   Discriminant of field extensions

In this section, discriminants of field extensions will be defined. Mainly, we will show an important formula of discriminants on a tower of field extensions.

1.3.1   Discriminant

Let κ be a field, and K a finite field extension of κ with a basis $\{w_1, \ldots, w_n\}$. Then the *discriminant* of the basis $\{w_1, \ldots, w_n\}$, $D_{K/\kappa}(w_1, \ldots, w_n)$, is defined as

$$D_{K/\kappa}(w_1, \ldots, w_n) = \det(\mathbf{Tr}_{K/\kappa}(w_i w_j)).$$

Let $\{w_1', \ldots, w_n'\}$ be another basis of K over κ. Set

$$w_i' = \sum_{k=1}^n b_{ik} w_k.$$

Then we have

$$D_{K/\kappa}(w_1', \ldots, w_n') = \det\left(\mathbf{Tr}_{K/\kappa}(w_i' w_j')\right)$$

$$= \det\left(\sum_{k,l} b_{ik} b_{jl} \mathbf{Tr}_{K/\kappa}(w_k w_l)\right)$$

$$= \{\det(b_{ik})\}^2 \det(\mathbf{Tr}_{K/\kappa}(w_k w_l))$$

$$= \{\det(b_{ik})\}^2 D_{K/\kappa}(w_1, \ldots, w_n).$$

Theorem 1.40. *Let K be a finite extension of κ, and $\{w_1, \ldots, w_n\}$ a basis of K over κ. Then $D_{K/\kappa}(w_1, \ldots, w_n) \neq 0$ if and only if K is a separable algebraic extension of κ.*

Proof. ($\Rightarrow$) Let κ_K^S be a separable closure of κ in K. If $\kappa_K^S = K$, then K is separable algebraic over κ, and we are done. Otherwise, $\kappa_K^S \neq K$, K is purely inseparable over κ_K^S, and

$$[K : \kappa_K^S] = p^r, \quad r \geq 1.$$

Let α be any element in K. We claim that $\mathbf{Tr}_{K/\kappa}(\alpha) = 0$, and so $D_{K/\kappa}(w_1, \dots, w_n) = 0$. We distinguish two cases, (i) $\alpha \in \kappa_K^S$, (ii) $\alpha \notin \kappa_K^S$.

Case (i). Let the field polynomial of α as an element in κ_K^S (resp. K) be $G(x)$ (resp. $\chi_\alpha(x)$), and write

$$G(x) = x^m + a_1 x^{m-1} + \cdots + a_m.$$

It follows from Lemma 1.22, that

$$\chi_\alpha(x) = G(x)^{p^r} = x^{mp^r} + 0 x^{mp^r - 1} + \cdots .$$

Therefore, $\mathbf{Tr}_{K/\kappa}(\alpha) = 0$.

Case (ii). There is an integer $l \geq 1$ such that the minimal polynomial $P_\alpha(x)$ of α over κ is in $\kappa[x^{p^l}]$. It follows from Lemma 1.22 that the field polynomial $\chi_\alpha(x)$ of α as an element in K is of the form

$$\chi_\alpha(x) = P_\alpha(x)^s \in \kappa[x^{p^l}].$$

Therefore, $\mathbf{Tr}_{K/\kappa}(\alpha) = 0$.

($\Leftarrow$) Since K is a finite separable algebraic extension of κ, then there exists an element $\alpha \in K$ such that $K = \kappa[\alpha]$. Let us take $\{1, \alpha, \alpha^2, \dots, \alpha^{n-1}\}$ as a basis of K over κ. Let $\overline{K}$ be an algebraic closure of K, and $P_\alpha(x)$ be the minimal polynomial of α over κ. Note that the field polynomial of α is $P_\alpha(x)$. Let $P_\alpha(x)$ be split completely in $\overline{K}$:

$$P_\alpha(x) = \prod_{i=1}^{n} (x - \alpha_i), \ \alpha_1 = \alpha, \ \alpha_i \neq \alpha_j \ (i \neq j).$$

Then we have

$$\mathbf{Tr}_{K/\kappa}(\alpha) = \sum_{i=1}^{n} \alpha_i.$$

We claim

$$\mathbf{Tr}_{K/\kappa}(\alpha^j) = \sum_{i=1}^{n} \alpha_i^j.$$

Let the splitting field of $P_\alpha(x)$ in $\overline{K}$ be E which is a Galois extension of κ with Galois group $G_{E/\kappa}$. In the collection $\{\alpha_1^j, \alpha_2^j, \dots, \alpha_n^j\}$, some elements may be identical. It is easy to see that each element appears with the same multiplicities. Picking all distinct elements from it to form a set $\{\beta_1, \dots, \beta_m\}$. Then we have that $m \mid n$, and the polynomial $Q(x)$ defined as

$$Q(x) = \prod_{i=1}^{m} (x - \beta_i)$$

is the minimal polynomial of α^j over κ. Then we have

$$\mathbf{Tr}_{\kappa[\alpha^j]/\kappa}(\alpha^j) = \sum_{i=1}^{m} \beta_i,$$

$$\mathbf{Tr}_{K/\kappa}(\alpha^j) = \frac{n}{m}\sum_{i=1}^{m} \beta_i = \sum_{i=1}^{n} \alpha_i^j.$$

Therefore, we have the following computation:

$$D_{K/\kappa}(1, \alpha, \ldots, \alpha^{n-1}) = \begin{vmatrix} n & \sum_i \alpha_i & \cdots & \sum_i \alpha_i^{n-1} \\ \sum_i \alpha_i & \sum_i \alpha_i^2 & \cdots & \sum_i \alpha_i^n \\ \cdots\cdots\cdots\cdots\cdots\cdots\cdots\cdots \\ \sum_i \alpha_i^{n-1} & \sum_i \alpha_i^n & \cdots & \sum_i \alpha_i^{2n-2} \end{vmatrix}$$

$$= \begin{vmatrix} 1 & 1 & \cdots & 1 \\ \alpha_1 & \alpha_2 & \cdots & \alpha_n \\ \cdots\cdots\cdots\cdots\cdots \\ \alpha_1^{n-1} & \alpha_2^{n-1} & \cdots & \alpha_n^{n-1} \end{vmatrix} \cdot \begin{vmatrix} 1 & \alpha_1 & \cdots & \alpha_1^{n-1} \\ 1 & \alpha_2 & \cdots & \alpha_2^{n-1} \\ \cdots\cdots\cdots\cdots\cdots \\ 1 & \alpha_n & \cdots & \alpha_n^{n-1} \end{vmatrix}$$

$$= \left\{ \prod_{i>j}(\alpha_i - \alpha_j) \right\}^2 \neq 0. \qquad \square$$

Let K be a finite separable algebraic extension of κ, of degree n. Let $\sigma_1, \ldots, \sigma_n$ be the distinct embeddings of K in $\overline{\kappa}$ over κ, where $\overline{\kappa}$ is an algebraic closure of κ. Then the discriminant of a basis $\{w_1, \ldots, w_n\}$ of K over κ satisfies

$$D_{K/\kappa}(w_1, \ldots, w_n) = \{\det(\sigma_i(w_j))\}^2.$$

In fact, by (1.1.5), we have

$$\mathbf{Tr}_{K/\kappa}(w_i w_j) = \sum_{m=1}^{n} \sigma_m(w_i w_j) = \sum_{m=1}^{n} \sigma_m(w_i)\sigma_m(w_j),$$

which means

$$(\mathbf{Tr}_{K/\kappa}(w_i w_j)) = {}^t((\sigma_i(w_j))((\sigma_i(w_j)),$$

and hence the claim follows.

Let K be a finite separable algebraic extension of κ, and let $\kappa \subset F \subset K$. Every element of K is separable over F, and every element of F is an element of K, so separable over κ. Hence each step in the tower is separable. Then there exist elements $\alpha \in K$, $\beta \in F$ such that $K = F[\alpha]$, $F = \kappa[\beta]$. Set

$$p = [K : F], \quad q = [F : \kappa], \quad n = pq = [K : \kappa].$$

Let us take $\{1, \alpha, \alpha^2, \ldots, \alpha^{p-1}\}$ (resp. $\{1, \beta, \ldots, \beta^{q-1}\}$) as a basis of K (resp. F) over F (resp. κ). Let $\overline{K}$ be an algebraic closure of K, and $P(x)$ (resp. $Q(x)$) be the minimal polynomial of α (resp. β) over F (resp. κ). Note that the field polynomial of α (resp. β) is $P(x)$ (resp. $Q(x)$). Let $P(x)$ and $Q(x)$ be split completely in $\overline{K}$ as follows

$$P(x) = \prod_{i=1}^{p} (x - \alpha_i), \ \alpha_1 = \alpha, \ \alpha_i \neq \alpha_j \ (i \neq j),$$

$$Q(x) = \prod_{i=1}^{q} (x - \beta_i), \ \beta_1 = \beta, \ \beta_i \neq \beta_j \ (i \neq j).$$

Then we have

$$\mathbf{Tr}_{K/F}(\alpha) = \sum_{i=1}^{p} \alpha_i, \quad \mathbf{Tr}_{K/F}(\alpha^l) = \sum_{i=1}^{p} \alpha_i^l,$$

$$\mathbf{Tr}_{F/\kappa}(\beta) = \sum_{i=1}^{q} \beta_i, \quad \mathbf{Tr}_{F/\kappa}(\beta^m) = \sum_{i=1}^{q} \beta_i^m.$$

Therefore

$$\mathbf{Tr}_{K/\kappa}(\beta^m) = \mathbf{Tr}_{F/\kappa} \circ \mathbf{Tr}_{K/F}(\beta^m) = \mathbf{Tr}_{F/\kappa}(p\beta^m) = p \sum_{i=1}^{q} \beta_i^m.$$

In the collection $\{\alpha_i^l \beta_j^m \mid 1 \leq i \leq p, 1 \leq j \leq q\}$, some elements may be identical. It is easy to see that each element appears with the same multiplicities. Pick all distinct elements from it to form a set $\{\gamma_1, \ldots, \gamma_s\}$. Then we have that $s \mid n$, and the polynomial $R(x)$ defined as

$$R(x) = \prod_{i=1}^{s} (x - \gamma_i)$$

is the minimal polynomial of $\alpha^l \beta^m$ over κ. Then we have

$$\mathbf{Tr}_{k[\alpha^l \beta^m]/\kappa}(\alpha^l \beta^m) = \sum_{i=1}^{s} \gamma_i,$$

$$\mathbf{Tr}_{K/\kappa}(\alpha^l \beta^m) = \frac{n}{s} \sum_{i=1}^{s} \gamma_i = \sum_{i=1}^{p} \sum_{j=1}^{q} \alpha_i^l \beta_j^m.$$

Abbreviate

$$\Phi = (1, \alpha, \ldots, \alpha^{p-1}), \quad \Psi = (1, \beta, \ldots, \beta^{q-1}).$$

$$\Phi\Psi = (\Phi, \beta\Phi, \ldots, \beta^{q-1}\Phi).$$

Then $\Phi\Psi$ is a basis of K over κ. Consider the Vandermonde matrices

$$A = \begin{pmatrix} 1 & \alpha_1 & \cdots & \alpha_1^{p-1} \\ 1 & \alpha_2 & \cdots & \alpha_2^{p-1} \\ \cdots\cdots\cdots\cdots\cdots \\ 1 & \alpha_p & \cdots & \alpha_p^{p-1} \end{pmatrix}, \quad B = \begin{pmatrix} 1 & \beta_1 & \cdots & \beta_1^{q-1} \\ 1 & \beta_2 & \cdots & \beta_2^{q-1} \\ \cdots\cdots\cdots\cdots\cdots \\ 1 & \beta_q & \cdots & \beta_q^{q-1} \end{pmatrix}.$$

We have the following computation:

$$D_{K/\kappa}(\Phi\Psi) = \begin{vmatrix} n & q\sum\alpha_i & \cdots & \sum\alpha_i^{p-1}\beta_j^{q-1} \\ q\sum\alpha_i & q\sum\alpha_i^2 & \cdots & \sum\alpha_i^p\beta_j^{q-1} \\ \cdots\cdots\cdots\cdots\cdots\cdots\cdots\cdots \\ \sum\alpha_i^{p-1}\beta_j^{q-1} & \sum\alpha_i^p\beta_j^{q-1} & \cdots & \sum\alpha_i^{2p-2}\beta_j^{2q-2} \end{vmatrix}$$

$$= \begin{vmatrix} A & \beta_1 A & \cdots & \beta_1^{q-1}A \\ A & \beta_2 A & \cdots & \beta_2^{q-1}A \\ \cdots\cdots\cdots\cdots\cdots\cdots \\ A & \beta_q A & \cdots & \beta_q^{q-1}A \end{vmatrix}^2$$

$$= \det(A)^{2q}\det(B)^{2p} = D_{K/F}(\Phi)^q D_{F/\kappa}(\Psi)^p.$$

Therefore we obtain a relation

$$D_{K/\kappa}(\Phi\Psi) = D_{F/\kappa}(\Psi)^{[K:F]} D_{K/F}(\Phi)^{[F:\kappa]}. \tag{1.3.1}$$

Further if $D_{K/F}(\Phi) \in \kappa$, then

$$D_{K/F}(\Phi)^{[F:\kappa]} = \mathbf{N}_{F/\kappa}(D_{K/F}(\Phi)).$$

Therefore we also have

$$D_{K/\kappa}(\Phi\Psi) = D_{F/\kappa}(\Psi)^{[K:F]} \mathbf{N}_{F/\kappa}(D_{K/F}(\Phi)). \tag{1.3.2}$$

1.3.2   Dedekind domain

For any subring $\mathfrak{o}$ of a field κ we define a *fractional ideal* of $\mathfrak{o}$ as an $\mathfrak{o}$-module $\mathfrak{A}$ in κ such that

$$u\mathfrak{o} \subseteq \mathfrak{A} \subseteq v\mathfrak{o}$$

for some $u, v \in \kappa_*$. An ordinary ideal $\mathfrak{A}$ of $\mathfrak{o}$ is a fractional ideal precisely if it is non-zero; this is also called an *integral* ideal. Clearly a fractional ideal is integral if and only if it is contained in $\mathfrak{o}$.

The usual ideal multiplication can be carried out for fractional ideals by defining

$$\mathfrak{A}_1\mathfrak{A}_2 = \left\{ \sum_\nu x_\nu y_\nu \mid x_\nu \in \mathfrak{A}_1,\ y_\nu \in \mathfrak{A}_2 \right\}.$$

Then $u_i \mathfrak{o} \subseteq \mathfrak{A}_i \subseteq v_i \mathfrak{o}$ imply that $u_1 u_2 \mathfrak{o} \subseteq \mathfrak{A}_1 \mathfrak{A}_2 \subseteq v_1 v_2 \mathfrak{o}$. Hence the product is again a fractional ideal. This multiplication is clearly associative, with $\mathfrak{o}$ as unit element, so the set F of all fractional ideals is a monoid. Moreover, there is a generalized inverse:

$$(\mathfrak{o} : \mathfrak{A}) = \{x \in \mathfrak{o} \mid x\mathfrak{A} \subseteq \mathfrak{o}\}.$$

If $u\mathfrak{o} \subseteq \mathfrak{A} \subseteq v\mathfrak{o}$, then $v^{-1}\mathfrak{o} \subseteq (\mathfrak{o} : \mathfrak{A}) \subseteq u^{-1}\mathfrak{o}$, and for any $c \in \mathfrak{o}$ we see that $x\mathfrak{A} \subseteq \mathfrak{o}$ implies

$$cx\mathfrak{A} \subseteq c\mathfrak{o} \subseteq \mathfrak{o}.$$

Therefore $(\mathfrak{o} : \mathfrak{A})$ is again a fractional ideal. Further, we have

$$(\mathfrak{o} : \mathfrak{A})\mathfrak{A} \subseteq \mathfrak{o}, \tag{1.3.3}$$

as is easily verified. If the equality in (1.3.3) holds, we also write $\mathfrak{A}^{-1}$ in place of $(\mathfrak{o} : \mathfrak{A})$ and call $\mathfrak{A}$ *invertible*. For example, any non-zero principal ideal $a\mathfrak{o}$ is invertible: $(a\mathfrak{o})^{-1} = a^{-1}\mathfrak{o}$.

Let S be a family of non-trivial and pairwise inequivalent absolute values of a field κ. Then the members of S are equivalence classes of absolute values, also called *prime divisors* or simply *places*. Here we shall denote the members of S by small Gothic (or Fraktur) letters: $\mathfrak{p}, \mathfrak{q}, \ldots$ and write $v_{\mathfrak{p}}, v_{\mathfrak{q}}, \ldots$ for the corresponding valuation. With each $\mathfrak{p} \in S$ we associate its valuation ring

$$\mathfrak{o}_{\mathfrak{p}} = \{x \in \kappa \mid v_{\mathfrak{p}}(x) \geq 0\}.$$

Suppose that $\mathfrak{o}$ is the ring of integers for S, i.e.,

$$\mathfrak{o} = \bigcap_{\mathfrak{p} \in S} \mathfrak{o}_{\mathfrak{p}}.$$

We define the *divisor group* D of κ with respect of S as the free Abelian group on S as generating set. The typical element is written

$$\mathfrak{a} = \prod \mathfrak{p}^{\alpha_{\mathfrak{p}}},$$

where the $\alpha_{\mathfrak{p}}$ are integers, almost all zero, and $\mathfrak{a}$ is called a *divisor*. Our aim will be to explore the relations between D and the monoid F of all fractional ideals. We define a mapping $\phi : F \longrightarrow D$ by the rule: for any $\mathfrak{A} \in F$ we put

$$v_{\mathfrak{p}}(\mathfrak{A}) = \min\{v_{\mathfrak{p}}(x) \mid x \in \mathfrak{A}\}. \tag{1.3.4}$$

If $u\mathfrak{o} \subseteq \mathfrak{A} \subseteq v\mathfrak{o}$, then clearly $v_{\mathfrak{p}}(u) \geq v_{\mathfrak{p}}(\mathfrak{A}) \geq v_{\mathfrak{p}}(v)$ for all $\mathfrak{p}$; this shows that $v_{\mathfrak{p}}(\mathfrak{A}) = 0$ for almost all $\mathfrak{p}$. We can therefore define

$$\phi(\mathfrak{A}) = \prod \mathfrak{p}^{v_{\mathfrak{p}}(\mathfrak{A})}. \tag{1.3.5}$$

It is easy to show that ϕ is a homomorphism. For if $c \in \mathfrak{A}\mathfrak{B}$, say,

$$c = \sum a_i b_i, \ a_i \in \mathfrak{A}, \ b_i \in \mathfrak{B},$$

then

$$v_{\mathfrak{p}}(c) \geq \min_i \{v_{\mathfrak{p}}(a_i) + v_{\mathfrak{p}}(b_i)\} \geq v_{\mathfrak{p}}(\mathfrak{A}) + v_{\mathfrak{p}}(\mathfrak{B}),$$

therefore

$$v_{\mathfrak{p}}(\mathfrak{A}\mathfrak{B}) \geq v_{\mathfrak{p}}(\mathfrak{A}) + v_{\mathfrak{p}}(\mathfrak{B}),$$

and here equality holds, as we see by taking $c = ab$, where a, b are chosen in $\mathfrak{A}, \mathfrak{B}$ so as to attain the minimum in (1.3.4). Hence

$$v_{\mathfrak{p}}(\mathfrak{A}\mathfrak{B}) = v_{\mathfrak{p}}(\mathfrak{A}) + v_{\mathfrak{p}}(\mathfrak{B}),$$

and it follows that (1.3.5) is a homomorphism.

An integral domain whose fractional ideals form a group under ideal multiplication is called a *Dedekind domain*. The following fact comes from Cohn [63], Theorem 4.5.

Theorem 1.41. *If $\mathfrak{o}$ is a Dedekind domain with field of fractions κ, then $\mathfrak{o}$ can be defined as the intersection of principal valuation rings for a family of absolute values.*

If $\mathfrak{o}$ is a Dedekind domain with field of fractions κ, then the mapping ϕ from the fractional ideals to the divisors is an isomorphism. E. Noether gives the following equivalence conditions of Dedekind domain with the more usual form (see Cohn [63], Theorem 4.6):

Theorem 1.42. *An integral domain $\mathfrak{o}$ with field of fractions κ is a Dedekind domain if and only if it satisfies the following three conditions:*

(i) *$\mathfrak{o}$ is Noetherian;*

(ii) *$\mathfrak{o}$ is integrally closed in κ;*

(iii) *every non-zero prime ideal in $\mathfrak{o}$ is maximal.*

The following result comes from Cohn [63], Theorem 5.1.

Theorem 1.43. *Let A be a Dedekind domain with field of fractions κ, let K be a finite separable algebraic extension of κ and denote by B the integral closure of A in K. Then B is again a Dedekind domain.*

Given a Dedekind domain A with field of fractions κ, we have a homomorphism θ_κ of κ_* into the divisor group $D = D_\kappa$ of κ given by

$$\theta_\kappa(x) = \prod \mathfrak{p}^{v_{\mathfrak{p}}(x)}. \tag{1.3.6}$$

The kernel of this mapping is the group of units in A, because the mapping ϕ from the fractional ideals to the divisors is an isomorphism.

If K is a finite separable extension of κ and θ_K is the corresponding mapping defined by (1.3.6) for K, then for $\mathfrak{P}|\mathfrak{p}$ we have

$$v_{\mathfrak{P}}(x) = e_{\mathfrak{P}} v_{\mathfrak{p}}(x), \ x \in K,$$

where $e_{\mathfrak{P}}$ is the ramification index for $\mathfrak{P}$. Hence for any $x \in K$ we have

$$\theta_K(x) = \prod_{\mathfrak{P}} \mathfrak{P}^{v_{\mathfrak{P}}(x)} = \prod_{\mathfrak{p}} \left(\prod_{\mathfrak{P}|\mathfrak{p}} \mathfrak{P}^{v_{\mathfrak{P}}(x)} \right) = \prod_{\mathfrak{p}} \left(\prod_{\mathfrak{P}|\mathfrak{p}} \mathfrak{P}^{e_{\mathfrak{P}}} \right)^{v_{\mathfrak{p}}(x)}.$$

Thus we obtain a commutative diagram $\gamma \circ \theta_{\kappa} = \theta_K \circ \iota$ as shown below by defining a mapping $\gamma : D_{\kappa} \longrightarrow D_K$,

$$\gamma(\mathfrak{p}) = \prod_{\mathfrak{P}|\mathfrak{p}} \mathfrak{P}^{e_{\mathfrak{P}}}, \tag{1.3.7}$$

where $\iota : \kappa \longrightarrow K$ is the inclusion. The mapping (1.3.7) is an embedding of D_{κ} in D_K, sometimes called the *conorm mapping*.

In the other direction we have the norm mapping. Given a place $\mathfrak{p}$ of κ, let the divisors of $\mathfrak{p}$ in K be $\mathfrak{P}_1$, ..., $\mathfrak{P}_r$ and let K_1, ..., K_r be the corresponding extensions of $\kappa_{\mathfrak{p}}$, the completion of κ (cf. Theorem 1.38). As we saw, we have

$$[K_i : \kappa_{\mathfrak{p}}] = n_i = e_i d_i,$$

hence

$$v_{\mathfrak{P}_i}(x) = \frac{1}{n_i} v_{\mathfrak{P}_i}\left(\mathbf{N}_{K_i/\kappa_{\mathfrak{p}}}(x) \right) = \frac{e_i}{n_i} v_{\mathfrak{p}}\left(\mathbf{N}_{K_i/\kappa_{\mathfrak{p}}}(x) \right),$$

and so we obtain

$$v_{\mathfrak{p}}\left(\mathbf{N}_{K/\kappa}(x) \right) = \sum v_{\mathfrak{p}}\left(\mathbf{N}_{K_i/\kappa_{\mathfrak{p}}}(x) \right) = \sum d_i v_{\mathfrak{P}_i}(x).$$

In terms of divisors this may be written as

$$\mathbf{N}_{K/\kappa}(\mathfrak{P}_i) = \mathfrak{p}^{d_i}. \tag{1.3.8}$$

Hence we find that

$$\mathbf{N}_{K/\kappa}(\gamma(\mathfrak{p})) = \mathbf{N}_{K/\kappa}(\mathfrak{P}_1^{e_1} \cdots \mathfrak{P}_r^{e_r}) = \mathfrak{p}^{\sum e_i d_i} = \mathfrak{p}^n,$$

where we have used Theorem 1.38.

1.3.3 Different

Let K be a finite separable algebraic extension of a field κ. Take A to be a Dedekind domain with κ as its field of fractions and B its integral closure in K. Then B is again a Dedekind domain. Given $X \subseteq K$, we define its *complementary set* or simply *complement* as

$$X' = \{y \in K \mid \mathbf{Tr}_{K/\kappa}(yX) \subseteq A\}.$$

Clearly X' is a subgroup of the additive group of K. Moreover, we have

$$\mathbf{Tr}_{K/\kappa}(yz \cdot X) = \mathbf{Tr}_{K/\kappa}(y \cdot zX),$$

hence

$$(zX)' = z^{-1}X'.$$

We also have

$$X \subset Y \Rightarrow X' \supset Y', \tag{1.3.9}$$

$$zX \subset X \Rightarrow zX' \subset X'. \tag{1.3.10}$$

Let $u_1, \ldots, u_n$ be a basis of K/κ. Since $\mathbf{Tr}_{K/\kappa}$ is non-singular, that is, $\mathbf{Tr}_{K/\kappa}(ax) = 0$ for all $x \in K$ implies $a = 0$, the mapping $K \to \kappa^n$ defined by

$$x \mapsto (\mathbf{Tr}_{K/\kappa}(u_1 x), \ldots, \mathbf{Tr}_{K/\kappa}(u_n x))$$

is injective, and hence an isomorphism. Thus we can find $v_1, \ldots, v_n$ in K to satisfy

$$\mathbf{Tr}_{K/\kappa}(u_i v_j) = \delta_{ij} = \begin{cases} 1 & : \ i = j, \\ 0 & : \ \text{otherwise.} \end{cases}$$

The v's form the *dual basis* for the u's.

Lemma 1.44 (cf. [63]). *Let κ, K, A, B be as above and $[K : \kappa] = n$. If U is a free A-submodule of K of rank n, then its complement U' is also free of rank n, and we obtain a basis of U' by taking the dual of a basis of U.*

Theorem 1.45 (cf. [63]). *Let A be a principal ideal domain with a field of fractions κ, K a finite separable algebraic extension of κ, and B the integral closure of A in K. Then B and generally every fractional ideal of B is a free A-module of rank $[K : \kappa]$.*

Proof. Let $u_1, \ldots, u_n$ be a basis of K over κ, $n = [K : \kappa]$. On multiplying the u's by suitable elements we may assume that $u_i \in B$ $(i = 1, \ldots, n)$. Then the A-module U spanned by the u's satisfies $U \subset B$, hence $U' \supset B'$. Since $\mathbf{Tr}_{K/\kappa}(B \cdot B) \subset A$, we have $B \subset B'$ and so

$$U \subset B \subset B' \subset U'.$$

By Lemma 1.44, U' is free of rank n, hence B as submodule of U' is also free of rank at most n. In fact the rank must be n because B contains U of rank n. Now the result follows for any fractional ideal $\mathfrak{A}$ because we have $cB \subset \mathfrak{A} \subset dB$, hence

$$d^{-1}B' \subset \mathfrak{A}' \subset c^{-1}B',$$

and so the theorem is proved. $\square$

If $\mathfrak{A}$ is an invertible ideal, we have the following explicit formula for the complement:

$$\mathfrak{A}' = B'\mathfrak{A}^{-1}. \tag{1.3.11}$$

In fact, we have

$$\mathbf{Tr}_{K/\kappa}(B'\mathfrak{A}^{-1}\mathfrak{A}) = \mathbf{Tr}_{K/\kappa}(B'B) \subset A,$$

hence $B'\mathfrak{A}^{-1} \subset \mathfrak{A}'$, and conversely,

$$\mathbf{Tr}_{K/\kappa}(\mathfrak{A}'\mathfrak{A}B) \subset \mathbf{Tr}_{K/\kappa}(\mathfrak{A}'\mathfrak{A}) \subset A,$$

hence $\mathfrak{A}'\mathfrak{A} \subset B'$ and so $\mathfrak{A}' \subset B'\mathfrak{A}^{-1}$, which proves equality in (1.3.11).

Theorem 1.46 (cf. [63]). *Let A be a Dedekind ring with a field of fractions κ, K a finite separable algebraic extension of κ, and B the integral closure of A in K. Then its complement B' is an invertible ideal whose complement is integral:*

$$\mathfrak{d} = (B')^{-1}, \tag{1.3.12}$$

and for any fractional ideal $\mathfrak{A}$ of B, its inverse is related to its complement by the formula:

$$\mathfrak{A}^{-1} = \mathfrak{d}\mathfrak{A}'. \tag{1.3.13}$$

Proof. We have seen that $B' \supset B$ and it is clear that $BB' \subset B'$, by (1.3.10). Moreover, B' is finitely generated as an A-module, by $c_1, \ldots, c_r$ say. Writing the c_i as fractions with a common denominator in B, say $c_i = a_i b^{-1}$, where $a_i, b \in B$, we have $(B')^{-1} \subset B$ and so $\mathfrak{d}$ defined by (1.3.12) is an integral ideal. Now (1.3.13) follows on multiplying both sides of (1.3.11) by $\mathfrak{d}$. $\square$

The ideal $\mathfrak{d} = \mathfrak{d}_{K/\kappa}$ defined by (1.3.12) is called the *different* of the extension B over A (or also of K over κ). In the particular case where B is generated by a single element over A, there is an explicit formula for $\mathfrak{d}$ which also explains the name.

Theorem 1.47 (cf. [63]). *Let A be a Dedekind domain, $B = A[\alpha]$ a separable extension generated by a single element α integral over A, and let f be the minimal polynomial for α over A. Then $B' = B/(f'(\alpha))$, where f' denotes the derivative, and so*

$$\mathfrak{d} = (f'(\alpha)). \tag{1.3.14}$$

Let K/κ again be a finite separable extension with rings of integers B, A, and consider any A-module U of K which is free of rank $n = [K : \kappa]$ as an A-module. Let $\sigma_1, \ldots, \sigma_n$ be the distinct embeddings of K in $\bar{\kappa}$ over κ, where $\bar{\kappa}$ is an algebraic closure of κ. We define the discriminant of U relative to any A-basis $\{u_1, \ldots, u_n\}$ of U as

$$D_{K/\kappa}(U) := D_{K/\kappa}(u_1, \ldots, u_n) = \det(\mathbf{Tr}_{K/\kappa}(u_i u_j)) = \{\det P\}^2,$$

where $P = (\sigma_i(u_j))$. If we replace $u_1, \ldots, u_n$ by $\alpha u_1, \ldots, \alpha u_n$, where $\alpha \in K$, then P becomes PC where $C = \mathrm{diag}(\sigma_1(\alpha), \ldots, \sigma_n(\alpha))$. It is clear that $\det C$ is the product of the conjugates of α, i.e., $\mathbf{N}_{K/\kappa}(\alpha)$. Hence we have

$$D_{K/\kappa}(U\alpha) = \{\det PC\}^2 = \mathbf{N}_{K/\kappa}(\alpha)^2 D_{K/\kappa}(U). \tag{1.3.15}$$

Let $\mathfrak{A} = \prod \mathfrak{P}_i^{\alpha_i}$ be any fractional ideal in B. For a given place $\mathfrak{P}_i$ we defined its *norm* in (1.3.8) as

$$\mathbf{N}_{K/\kappa}(\mathfrak{P}_i) = \mathfrak{p}^{d_i}, \tag{1.3.16}$$

where $\mathfrak{p}$ is the place of κ which $\mathfrak{P}_i$ divides and d_i is the corresponding residue degree. Thus we may define the *norm* of $\mathfrak{A}$ by

$$\mathbf{N}_{K/\kappa}(\mathfrak{A}) = \prod \mathbf{N}_{K/\kappa}(\mathfrak{P}_i)^{\alpha_i}. \tag{1.3.17}$$

We apply this result to (1.3.15), taking $U = B$, $U\alpha = B'$. Locally, i.e., at a place $\mathfrak{P}_i$, we can do this because in the corresponding valuation ring every ideal is principal; we thus obtain

$$D_{K/\kappa}(B') = \mathbf{N}_{K/\kappa}(B')^2 D_{K/\kappa}(B). \tag{1.3.18}$$

Suppose now that $\{u_i\}, \{v_i\}$ are dual bases for B, B' respectively and put $Q = (\sigma_i(v_j))$. Then

$$\sum_l \sigma_l(u_i)\sigma_l(v_j) = \mathbf{Tr}_{K/\kappa}(u_i v_j) = \delta_{ij},$$

hence $PQ = I$, and since

$$D_{K/\kappa}(B) = \{\det P\}^2, \quad D_{K/\kappa}(B') = \{\det Q\}^2,$$

we have

$$D_{K/\kappa}(B)D_{K/\kappa}(B') = 1.$$

Combining this with (1.3.18), we find

$$D_{K/\kappa}(B')^2 = \mathbf{N}_{K/\kappa}(B')^2.$$

Now the *discriminant* $D_{K/\kappa}$ of K/κ is $D_{K/\kappa}(B)$, hence we have

$$D_{K/\kappa} = D_{K/\kappa}(B')^{-1} = \mathbf{N}_{K/\kappa}(B')^{-1}.$$

Thus in terms of fractional ideals we have the equation

$$D_{K/\kappa} = \mathbf{N}_{K/\kappa}(\mathfrak{d}), \tag{1.3.19}$$

which identifies the discriminant as the norm of the different.

Theorem 1.48 (cf. [63]). *Let A be a Dedekind domain with field of fractions κ, let $\kappa \subset F \subset K$ be finite separable algebraic extensions and C, B the integral closures of A in F, K respectively. Then*

$$\mathfrak{d}_{K/\kappa} = \mathfrak{d}_{K/F}\mathfrak{d}_{F/\kappa}, \tag{1.3.20}$$

$$D_{K/\kappa} = \mathbf{N}_{F/\kappa}(D_{K/F})D_{F/\kappa}^{[K:F]}. \tag{1.3.21}$$

Proof. In order to establish equation (1.3.20) we must show that

$$B'_{K/\kappa} = B'_{K/F}C'_{F/\kappa}, \tag{1.3.22}$$

where the prime means the complement in the extension indicated by the suffix. We have

$$\mathbf{Tr}_{K/\kappa} = \mathbf{Tr}_{F/\kappa} \circ \mathbf{Tr}_{K/F},$$

hence

$$\mathbf{Tr}_{K/\kappa}\left(B'_{K/F}C'_{F/\kappa}B\right) = \mathbf{Tr}_{F/\kappa}\left(C'_{F/\kappa}\mathbf{Tr}_{K/F}\left(B'_{K/F}B\right)\right)$$
$$\subset \mathbf{Tr}_{F/\kappa}\left(C'_{F/\kappa}C\right) \subset A.$$

This shows that the left-hand side of (1.3.22) includes the right-hand side. To obtain the reverse inclusion, take $\gamma \in B'_{K/\kappa}$; then

$$\mathbf{Tr}_{F/\kappa}\left(C \cdot \mathbf{Tr}_{K/F}(\gamma B)\right) \subset \mathbf{Tr}_{K/\kappa}(\gamma B) \subset A,$$

hence $\mathbf{Tr}_{K/F}(\gamma B) \subset C'_{F/\kappa}$ and so

$$\mathbf{Tr}_{K/F}\left(C'^{-1}_{F/\kappa} \cdot \gamma B\right) = C'^{-1}_{F/\kappa}\mathbf{Tr}_{K/F}(\gamma B) \subset C.$$

Therefore $C'^{-1}_{F/\kappa}\gamma \subset B'_{K/F}$ and so $\gamma \in B'_{K/F}C'_{F/\kappa}$, which shows that equality holds in (1.3.22). This proves (1.3.20), and now (1.3.21) follows by taking norms. $\square$

The following *Dedekind discriminant theorem* provides more precise information about the discriminant divisors.

Theorem 1.49. *Let K/κ be a finite separable algebraic extension, A a Dedekind domain with field of fractions κ and B the integral closure of A in K, and express the discriminant as a product of prime divisors in A:*

$$D_{K/\kappa} = \prod \mathfrak{p}^{\delta_\mathfrak{p}}.$$

For any prime divisor $\mathfrak{p}$ of A let $\mathfrak{P}_1, \ldots, \mathfrak{P}_r$ be its extensions to B and write e_i, d_i for the ramification index and residue class degree of the $\mathfrak{P}_i$-adic valuation. Then

$$\delta_{\mathfrak{p}} \geq \sum d_i(e_i - 1).$$

For the proof, we refer to Cohn [63], Theorem 6.6.

1.4 Product formula

We will show that the product formula (1.2.1) over $\mathbb{Q}$ can be extended to finite extension fields of $\mathbb{Q}$ with a proper modification.

Let κ be a field with a non-trivial absolute value v. Let κ_v be the completion of κ for v. Let K be an extension of κ. If w is an absolute value on K extending an absolute value v on κ, we write $w|v$. If $w|v$ and if $[K : \kappa]$ is finite, then we shall call $[K_w : \kappa_v]$ the *local degree*, which satisfy

$$\sum_{w|v} [K_w : \kappa_v] \leq [K : \kappa].$$

Proposition 1.50 (cf. [225]). *If K is a finite separable extension of κ, then*

$$[K : \kappa] = \sum_{w|v} [K_w : \kappa_v].$$

Whenever v is a non-trivial absolute value on κ such that for any finite extension K of κ we have

$$[K : \kappa] = \sum_{w|v} [K_w : \kappa_v],$$

we shall say that v is *well behaved*. Suppose we have a tower of finite extensions, $\kappa \subset K \subset E$. Let w range over the absolute values of K extending v, and u over those of E extending v. If $u|w$, then E_u contains K_w. Since

$$\sum_{u|v} [E_u : \kappa_v] = \sum_{w|v} \sum_{u|w} [E_u : K_w][K_w : \kappa_v]$$

$$= \sum_{w|v} [K_w : \kappa_v] \sum_{u|w} [E_u : K_w],$$

thus we have

$$\sum_{u|v} [E_u : \kappa_v] \leq \sum_{w|v} [K_w : \kappa_v][E : K]$$

$$\leq [K : \kappa][E : K] = [E : \kappa].$$

From this we immediately see that if v is well behaved, K finite over κ, and w extends v on K, then w is well behaved.

Proposition 1.51 (cf. [225]). *Let K be a finite extension of κ, and assume that v is well behaved. Take $\alpha \in K$. Then*

$$\mathbf{N}_{K/\kappa}(\alpha) = \prod_{w|v} \mathbf{N}_{K_w/\kappa_v}(\alpha),$$

$$\mathbf{Tr}_{K/\kappa}(\alpha) = \sum_{w|v} \mathbf{Tr}_{K_w/\kappa_v}(\alpha).$$

Let κ be a field. An absolute value v on κ is said to be *proper* if it is non-trivial, well behaved, and if, κ having characteristic 0, its restriction to $\mathbb{Q}$ is either the ordinary absolute value or a p-adic absolute value. A set M_κ of absolute values on κ is said to be *proper* if every absolute value in it is proper, if any two distinct absolute values are not equivalent, and if, given $x \in \kappa_*$, there exists only a finite number of $v \in M_\kappa$ such that $|x|_v \neq 1$. In particular, if M_κ is proper, there can be only a finite number of Archimedean absolute values in M_κ. If K is an algebraic extension of κ, we shall denote by M_K the set of absolute values on K extending some absolute value in M_κ. If K is finite over κ, then M_K is proper if M_κ is proper.

Let M_κ be a proper set of absolute values on κ. For each $v \in M_\kappa$, let n_v be a real number > 0. We shall say that M_κ satisfies the *product formula with multiplicities n_v* if for each $x \in \kappa_*$, we have

$$\prod_{v \in M_\kappa} |x|_v^{n_v} = 1.$$

We shall say that M_κ satisfies the *product formula* if all $n_v = 1$. When we deal with a fixed set of multiplicities n_v, then we write for convenience

$$\|x\|_v = |x|_v^{n_v} \tag{1.4.1}$$

so that the product formula reads

$$\prod_{v \in M_\kappa} \|x\|_v = 1.$$

Suppose now that we have a field $\mathbf{F}$ with a proper set $M_\mathbf{F}$ of absolute values satisfying the product formula with multiplicities 1. Let κ be a finite extension of $\mathbf{F}$, and let M_κ be the set of absolute values on κ which extend the absolute values of $M_\mathbf{F}$. Then M_κ is also a proper set of absolute values on κ. If $t \in M_\mathbf{F}$ and $v \in M_\kappa$ with $v|t$, set $n_v = [\kappa_v : \mathbf{F}_t]$. Then for any $\alpha \in \kappa_*$, we get by Proposition 1.51:

$$1 = \prod_{t \in M_\mathbf{F}} \left| \mathbf{N}_{\kappa/\mathbf{F}}(\alpha) \right|_t = \prod_{t \in M_\mathbf{F}} \prod_{v|t} |\alpha|_v^{n_v} = \prod_{v \in M_\kappa} |\alpha|_v^{n_v}.$$

This shows that M_κ satisfies the product formula with multiplicities n_v.

The classical example is that of the rational numbers $\mathbb{Q}$. For each prime number p we have the p-adic absolute value $|\cdot|_p$. The ordinary Archimedean absolute value will be said to be *at infinity*, denoted by $|\cdot|_\infty$. Thus the set

$$M_\mathbb{Q} = \{\infty\} \cup \{\text{primes}\} = \{\infty, 2, 3, 5, \dots\}$$

is proper, and satisfies the product formula.

We assume that κ is a number field, i.e., a finite extension of $\mathbb{Q}$. By the arguments as above, the set M_κ of absolute values extending those of $M_\mathbb{Q}$ is a proper set of absolute values on κ satisfying the product formula with multiplicities n_v, where $n_v = [\kappa_v : \mathbb{Q}_p]$ if $v|p$ for some $p \in M_\mathbb{Q}$ such that

$$\sum_{v|p} n_v = [\kappa : \mathbb{Q}].$$

For convenience, we also write

$$|\|x\||_v = \|x\|_v^{1/[\kappa:\mathbb{Q}]}, \quad x \in \kappa. \tag{1.4.2}$$

If we have a tower of finite extensions, $\mathbb{Q} \subset \kappa \subset K$, and if $w|v$ for $w \in M_K, v \in M_\kappa$, then K_w contains κ_v with

$$n_w = [K_w : \kappa_v]n_v,$$

and hence

$$\sum_{w|v} \frac{n_w}{n_v} = \sum_{w|v} [K_w : \kappa_v] = [K : \kappa]. \tag{1.4.3}$$

The set M_κ will be called the *canonical set*. Let S_κ^∞ be the subset of Archimedean absolute values in M_κ. It is called the set of *absolute values at infinity*. If S is a finite subset of M_κ containing the set S_κ^∞, denote by $\mathcal{O}_{\kappa,S}$ the ring of *S-integers* of κ, i.e.,

$$\mathcal{O}_{\kappa,S} = \{z \in \kappa \mid \|z\|_\rho \le 1, \ \rho \notin S\}. \tag{1.4.4}$$

If $v \in M_\kappa$ is one of the absolute values extending the ordinary absolute value on $\mathbb{Q}$, then κ_v is either $\mathbb{R}$ or $\mathbb{C}$. We also say that v is *real* or *complex*, accordingly. The multiplicity n_v is given by

$$n_v = [\kappa_v : \mathbb{Q}_\infty] = \begin{cases} 1 & : \text{if } \kappa_v = \mathbb{R}, \\ 2 & : \text{if } \kappa_v = \mathbb{C}. \end{cases}$$

In other words, if κ has r_1 real embeddings mapping $x \in \kappa$ respectively into $x^{(1)}, \dots, x^{(r_1)}$ and r_2 pairs of complex conjugate embeddings mapping x into

$$x^{(r_1+1)}, \overline{x^{(r_1+1)}}, \dots, x^{(r_1+r_2)}, \overline{x^{(r_1+r_2)}}$$

where $r_1 + 2r_2 = [\kappa : \mathbb{Q}]$, then the absolute values dividing ∞ are

$$\left|x^{(1)}\right|, \dots, \left|x^{(r_1)}\right|, \left|x^{(r_1+1)}\right|, \dots, \left|x^{(r_1+r_2)}\right|.$$

The first r_1 of these have $n_v = 1$ and the last r_2 have $n_v = 2$.

If $v \in M_\kappa$ is one of the absolute values extending the p-adic absolute value on $\mathbb{Q}$, its multiplicity is

$$n_v = [\kappa_v : \mathbb{Q}_p] = e_{\mathfrak{p}} d_{\mathfrak{p}},$$

where $\mathfrak{p}$ is the prime containing v, and $e_{\mathfrak{p}}$ and $d_{\mathfrak{p}}$ are the ramification index and residue class degree, respectively. We see that the number of elements in $\mathbb{F}(\kappa)$, which we denote by $\mathbf{N}\mathfrak{p}$, is

$$\mathbf{N}\mathfrak{p} = p^{d_{\mathfrak{p}}},$$

since $\mathbb{F}(\kappa)$ is of degree $d_{\mathfrak{p}}$ over $\mathbb{F}_p$.

Proposition 1.52. *Let $\sigma : \kappa \longrightarrow K$ be an isomorphism. If $v \in M_K$, for $x \in \kappa$ put $|x|_w = |\sigma(x)|_v$. Then $w \in M_\kappa$, and this gives a one-to-one mapping $M_K \longrightarrow M_\kappa$, and in this correspondence $n_v = n_w$.*

1.5 Hermitian geometry

We will introduce some technical lemmas, basic operators and their gauges on a projective space associated to a vector space. A good reference is Stoll [385] for the complex case.

1.5.1 Gauges of elementary operators

Let V be a vector space of finite dimension $n + 1 > 0$ over a field κ. Write the *projective space* $\mathbb{P}(V) = V/\kappa_*$ and let $\mathbb{P} : V_* \longrightarrow \mathbb{P}(V)$ be the standard *projection*, where $V_* = V - \{0\}$. If $A \subset V$, abbreviate

$$\mathbb{P}(A) = \mathbb{P}(A \cap V_*).$$

The dual vector space V^* of V consists of all κ-linear functions $\alpha : V \longrightarrow \kappa$, and we shall call

$$\langle \xi, \alpha \rangle = \alpha(\xi)$$

the *inner product* of $\xi \in V$ and $\alpha \in V^*$. If $\alpha \neq 0$, the n-dimensional linear subspace

$$E[a] = E[\alpha] = \mathrm{Ker}(\alpha) = \alpha^{-1}(0)$$

depends on $a = \mathbb{P}(\alpha) \in \mathbb{P}(V^*)$ only, and $\ddot{E}[a] = \mathbb{P}(E[a])$ is a *hyperplane* in $\mathbb{P}(V)$. Thus $\mathbb{P}(V^*)$ bijectively parameterizes the hyperplanes in $\mathbb{P}(V)$.

Identify $V^{**} = V$ by $\langle \xi, \alpha \rangle = \langle \alpha, \xi \rangle$ and $\left(\bigwedge_{k+1} V \right)^* = \bigwedge_{k+1} V^*$ by

$$\langle \xi_0 \wedge \cdots \wedge \xi_k, \alpha_0 \wedge \cdots \wedge \alpha_k \rangle = \det(\langle \xi_i, \alpha_j \rangle),$$

where $\bigwedge_{k+1} V$ is the *exterior product of V of order $k+1$*, and where $\xi_i \in V$, $\alpha_i \in V^*$ for $i = 0, \ldots, k$. Take $k, l \in \mathbb{Z}[0, n]$ and take $\xi \in \bigwedge_{k+1} V$ and $\alpha \in \bigwedge_{l+1} V^*$, where

$$\mathbb{Z}[m, n] = \{ i \in \mathbb{Z} \mid m \le i \le n \}.$$

If $k \ge l$, the *interior product* $\xi \angle \alpha \in \bigwedge_{k-l} V$ is uniquely defined by

$$\langle \xi \angle \alpha, \beta \rangle = \langle \xi, \alpha \wedge \beta \rangle$$

for all $\beta \in \bigwedge_{k-l} V^*$. If $k = l$, then

$$\xi \angle \alpha = \langle \xi, \alpha \rangle \in \kappa = \bigwedge_0 V$$

by definition. If $k < l$, we define the *interior product* $\xi \angle \alpha \in \bigwedge_{l-k} V^*$ such that if $\eta \in \bigwedge_{l-k} V$,

$$\langle \eta, \xi \angle \alpha \rangle = \langle \xi \wedge \eta, \alpha \rangle.$$

Take a base $e = (e_0, \ldots, e_n)$ of V and take a valuation v on κ. For $\xi = \xi_0 e_0 + \cdots + \xi_n e_n \in V$, define the norm

$$|\xi|_{v,e} = \begin{cases} \left(|\xi_0|_v^2 + \cdots + |\xi_n|_v^2 \right)^{\frac{1}{2}} & : \quad \text{if } v \text{ is Archimedean,} \\ \max_{0 \le i \le n} \{ |\xi_i|_v \} & : \quad \text{if } v \text{ is non-Archimedean.} \end{cases}$$

Obviously, the norm depends on the base e, and will be called a *norm over the base e*. If $| \cdot |_{v,e'}$ is another norm over a base $e' = (e_0', \ldots, e_n')$, it is easy to prove that there exist positive constants c_v and c_v' such that

$$c_v |\xi|_{v,e} \le |\xi|_{v,e'} \le c_v' |\xi|_{v,e}$$

hold for all $\xi \in V$, i.e., norms over bases are equivalent, where $c_v = 1$, $c_v' = 1$ for all but finitely many $v \in M_\kappa$. We will abbreviate

$$|\xi|_v = |\xi|_{v,e}.$$

Further if κ is a number field with a proper set M_κ satisfying the product formula of multiplicities n_v, we will use notations

$$\|\xi\|_v = |\xi|_v^{n_v}, \ \|\|\xi\|\|_v = \|\xi\|_v^{1/[\kappa:\mathbb{Q}]}. \tag{1.5.1}$$

Let $\epsilon = (\epsilon_0, \ldots, \epsilon_n)$ be the dual base of $e = (e_0, \ldots, e_n)$. Then the norm on V induces a norm on V^* defined by

$$|\alpha|_v = \begin{cases} \left(|\alpha_0|_v^2 + \cdots + |\alpha_n|_v^2\right)^{\frac{1}{2}} & : \quad \text{if } v \text{ is Archimedean,} \\ \max_{0 \le i \le n}\{|\alpha_i|_v\} & : \quad \text{if } v \text{ is non-Archimedean,} \end{cases}$$

where $\alpha = \alpha_0\epsilon_0 + \cdots + \alpha_n\epsilon_n$. *Schwarz inequality*

$$|\langle \xi, \alpha \rangle|_v \le |\xi|_v \cdot |\alpha|_v$$

holds for $\xi \in V$, $\alpha \in V^*$. The *distance* from $x = \mathbb{P}(\xi)$ to $\ddot{E}[a]$ with $a = \mathbb{P}(\alpha) \in \mathbb{P}(V^*)$ is defined by

$$0 \le |x, a|_v = \frac{|\langle \xi, \alpha \rangle|_v}{|\xi|_v \cdot |\alpha|_v} \le 1. \tag{1.5.2}$$

Further if κ is a number field with a proper set M_κ satisfying the product formula of multiplicities n_v, we will use the normalization

$$\|\alpha\|_v = |\alpha|_v^{n_v}, \quad \|x, a\|_v = |x, a|_v^{n_v}, \tag{1.5.3}$$

and the notations

$$\||\alpha\||_v = \|\alpha\|_v^{1/[\kappa:\mathbb{Q}]}, \quad \||x, a\||_v = \|x, a\|_v^{1/[\kappa:\mathbb{Q}]}. \tag{1.5.4}$$

In particular, if $V = \kappa^{n+1}$, we may take the standard base $e_0, e_1, \ldots, e_n$, where $e_j = (0, \ldots, 0, 1, 0, \ldots, 0) \in \mathbb{Z}_+^{n+1}$ in which 1 is the $(j+1)$th component of e_j. Take $\xi \in \kappa^{n+1} - \{0\}$ and write

$$\xi = \xi_0 e_0 + \xi_1 e_1 + \cdots + \xi_n e_n = (\xi_0, \xi_1, \ldots, \xi_n).$$

We usually denote $\mathbb{P}(\xi)$ by $[\xi_0, \xi_1, \ldots, \xi_n]$ which are called the *homogeneous coordinates* of $\mathbb{P}(\kappa^{n+1})$, and abbreviate

$$\mathbb{P}^n = \mathbb{P}^n(\kappa) = \mathbb{P}(\kappa^{n+1}).$$

We can embed κ^n into $\mathbb{P}^n$ by using the mapping $(\xi_1, \ldots, \xi_n) \mapsto [1, \xi_1, \ldots, \xi_n]$, and obtain the disjoint union $\mathbb{P}^n = \kappa^n \cup \mathbb{P}^{n-1}$. Particularly, $\mathbb{P}^0 = \mathbb{P}(\kappa)$ consists of one point denoted by ∞, and so $\mathbb{P}^1 = \kappa \cup \{\infty\}$. If v is non-Archimedean, set

$$\chi_v(x, a) = \begin{cases} \frac{|x-a|_v}{|x|_v^\vee |a|_v^\vee} & : \quad x, a \in \kappa, \\ \frac{1}{|x|_v^\vee} & : \quad a = \infty, \end{cases} \tag{1.5.5}$$

where, by definition, $r^\vee = \max\{1, r\}$ ($r \in \mathbb{R}$). If v is Archimedean, set

$$\chi_v(x, a) = \begin{cases} \frac{|x-a|_v}{(1+|x|_v^2)^{1/2}(1+|a|_v^2)^{1/2}} & : \quad x, a \in \kappa, \\ \frac{1}{\sqrt{1+|x|_v^2}} & : \quad a = \infty. \end{cases} \tag{1.5.6}$$

If we set $\frac{1}{0} = \infty$, $\frac{1}{\infty} = 0$, then we have

$$\chi_v \left(\frac{1}{x}, \frac{1}{a} \right) = \chi_v(x, a), \quad x, a \in \kappa \cup \{\infty\}.$$

Identify $\kappa^{n+1} = (\kappa^{n+1})^*$ such that

$$\langle \xi, \alpha \rangle = \xi_0 \alpha_0 + \cdots + \xi_n \alpha_n$$

for $\xi = (\xi_0, \ldots, \xi_n), \alpha = (\alpha_0, \ldots, \alpha_n) \in \kappa^{n+1}$. It is easy to show that

$$\chi_v(x, a) = \begin{cases} |[1, x], [-a, 1]|_v & : \quad x, a \in \kappa, \\ |[1, x], [1, 0]|_v & : \quad a = \infty. \end{cases} \tag{1.5.7}$$

Take non-negative integers a and b with $a \leq b$. Let J_a^b be the set of all increasing injective mappings $\lambda : \mathbb{Z}[0, a] \longrightarrow \mathbb{Z}[0, b]$. Then $J_b^b = \{\iota\}$, where ι is the inclusion mapping. If $a < b$, there exists one and only one $\lambda^\perp \in J_{b-a-1}^b$ for each $\lambda \in J_a^b$ such that $\mathrm{Im}\lambda \cap \mathrm{Im}\lambda^\perp = \emptyset$. The mapping $\perp : J_a^b \longrightarrow J_{b-a-1}^b$ is bijective. A permutation $(\lambda, \lambda^\perp)$ of $\mathbb{Z}[0, b]$ is defined by

$$(\lambda, \lambda^\perp)(i) = \begin{cases} \lambda(i) & : \quad i \in \mathbb{Z}[0, a], \\ \lambda^\perp(i - a - 1) & : \quad i \in \mathbb{Z}[a + 1, b] . \end{cases}$$

The signature of the permutation is denoted by $\mathrm{sign}(\lambda, \lambda^\perp)$.

The norm on V also induces norms on $\bigwedge_{k+1} V$ and $\bigwedge_{k+1} V^*$. Take $\xi \in \bigwedge_{k+1} V$, $\alpha \in \bigwedge_{k+1} V^*$ and write

$$\xi = \sum_{\lambda \in J_k^n} \xi_\lambda e_\lambda, \quad \alpha = \sum_{\lambda \in J_k^n} \alpha_\lambda \epsilon_\lambda,$$

where

$$e_\lambda = e_{\lambda(0)} \wedge \cdots \wedge e_{\lambda(k)}.$$

Then we can define the norms

$$|\xi|_v = |\xi|_{v,e} = \begin{cases} \left(\sum_{\lambda \in J_k^n} |\xi_\lambda|_v^2 \right)^{\frac{1}{2}} & : \quad \text{if } v \text{ is Archimedean}, \\ \max_{\lambda \in J_k^n} \{|\xi_\lambda|_v\} & : \quad \text{if } v \text{ is non-Archimedean}, \end{cases}$$

and

$$|\alpha|_v = |\alpha|_{v,e} = \begin{cases} \left(\sum_{\lambda \in J_k^n} |\alpha_\lambda|_v^2 \right)^{\frac{1}{2}} & : \quad \text{if } v \text{ is Archimedean}, \\ \max_{\lambda \in J_k^n} \{|\alpha_\lambda|_v\} & : \quad \text{if } v \text{ is non-Archimedean}. \end{cases}$$

Generally, let $V_1, \ldots, V_m$ and W be normed vector spaces over κ. Let

$$\odot : V_1 \times \cdots \times V_m \longrightarrow W$$

be an m-linear mapping over κ. If $\xi = (\xi_1, \ldots, \xi_m) \in V_1 \times \cdots \times V_m$, we write

$$\odot(\xi) = \xi_1 \odot \cdots \odot \xi_m,$$

and say that ξ is *free for the operation* $\odot$ if $\odot(\xi) \neq 0$. Take $x_j \in \mathbb{P}(V_j)$ $(j = 1, \ldots, m)$. We will say that $x_1, \ldots, x_m$ are *free for* $\odot$ if there exist $\xi_j \in V_j$ such that $x_j = \mathbb{P}(\xi_j)$ and $\xi = (\xi_1, \ldots, \xi_m)$ is free for the operation $\odot$. For free $x_1, \ldots, x_m$, we can define

$$x_1 \odot \cdots \odot x_m = \mathbb{P}(\xi_1 \odot \cdots \odot \xi_m).$$

Also, the *gauge of $x_1, \ldots, x_m$ for* $\odot$ is defined to be

$$|x_1 \odot \cdots \odot x_m|_v = \frac{|\xi_1 \odot \cdots \odot \xi_m|_v}{|\xi_1|_v \cdots |\xi_m|_v}$$

which is well defined. If $x_1, \ldots, x_m$ are not free for $\odot$, we define $|x_1 \odot \cdots \odot x_m|_v = 0$.

In the following, we will prove some elementary but useful inequalities about multi-vectors and give several gauges. Take a positive number r. For each $v \in M_\kappa$, we define

$$\varsigma_{v,r} = \begin{cases} \sqrt{r} & \text{if } v \text{ is Archimedean,} \\ 1 & \text{if } v \text{ is non-Archimedean.} \end{cases} \qquad (1.5.8)$$

First and easiest to prove is of course the following *generalized Schwarz's inequality*:

Lemma 1.53. *Take $k, l \in \mathbb{Z}[0, n]$ and take $\xi \in \bigwedge_{k+1} V$ and $\alpha \in \bigwedge_{l+1} V^*$. Then*

$$|\xi \angle \alpha|_v \leq \varsigma_{v, \binom{n-p}{q-p}} |\xi|_v \cdot |\alpha|_v,$$

where $p = \min\{k, l\}, q = \max\{k, l\}$.

Proof. Without loss of generality, we may assume $k \geq l$, and write

$$\xi = \sum_{\lambda \in J_k^n} \xi_\lambda e_\lambda, \qquad \alpha = \sum_{\lambda \in J_l^n} \alpha_\lambda \epsilon_\lambda.$$

First of all, we consider the non-Archimedean case. If $l = k$, noting that

$$\xi \angle \alpha = \langle \xi, \alpha \rangle = \sum_{\lambda \in J_k^n} \xi_\lambda \alpha_\lambda,$$

we have

$$|\xi \angle \alpha|_v \leq \max_{\lambda \in J_k^n} |\xi_\lambda \alpha_\lambda|_v \leq \left(\max_{\lambda \in J_k^n} |\xi_\lambda|_v \right) \cdot \left(\max_{\lambda \in J_k^n} |\alpha_\lambda|_v \right) = |\xi|_v \cdot |\alpha|_v$$

and so the inequality follows. If $l < k$, by Laplace's theorem of determinant expansion

$$
\begin{aligned}
\langle(e_0 \wedge \cdots \wedge e_k)\angle\alpha, \beta\rangle &= \langle e_0 \wedge \cdots \wedge e_k, \alpha \wedge \beta\rangle \\
&= \sum_{\nu \in J_l^k} \operatorname{sign}(\nu, \nu^\perp) \langle e_\nu, \alpha \rangle \langle e_{\nu^\perp}, \beta \rangle \\
&= \left\langle \sum_{\nu \in J_l^k} \operatorname{sign}(\nu, \nu^\perp) \langle e_\nu, \alpha \rangle e_{\nu^\perp}, \beta \right\rangle
\end{aligned}
$$

holds for any $\beta \in \bigwedge_{k-l} V^*$, that is,

$$
(e_0 \wedge \cdots \wedge e_k)\angle\alpha = \sum_{\nu \in J_l^k} \operatorname{sign}(\nu, \nu^\perp) \langle e_\nu, \alpha \rangle e_{\nu^\perp} = \sum_{\nu \in J_l^k} \operatorname{sign}(\nu, \nu^\perp) \alpha_\nu e_{\nu^\perp}. \quad (1.5.9)
$$

Then

$$
|(e_0 \wedge \cdots \wedge e_k)\angle\alpha|_v = \max_{\nu \in J_l^k}\{|\langle e_\nu, \alpha \rangle|_v\} \le |\alpha|_v.
$$

Thus, we have

$$
|\xi\angle\alpha|_v = \left| \sum_{\lambda \in J_k^n} \xi_\lambda e_\lambda \angle\alpha \right|_v \le \max_{\lambda \in J_k^n}\{|\xi_\lambda|_v|e_\lambda\angle\alpha|_v\} \le |\xi|_v \cdot |\alpha|_v.
$$

Finally, assume that v is Archimedean. We have

$$
|\xi\angle\alpha|_v = \left| \sum_{\lambda \in J_k^n} \xi_\lambda e_\lambda \angle\alpha \right|_v \le \sum_{\lambda \in J_k^n} |\xi_\lambda|_v|e_\lambda\angle\alpha|_v \le |\xi|_v \left(\sum_{\lambda \in J_k^n} |e_\lambda\angle\alpha|_v^2 \right)^{1/2}.
$$

For $\lambda \in J_k^n$, set

$$
J_l^\lambda = \{\nu \in J_l^n \mid \nu \subset \lambda\},
$$

where $\nu \subset \lambda$ means $\{\nu(0), \ldots, \nu(l)\} \subset \{\lambda(0), \ldots, \lambda(k)\}$. By (1.5.9), we obtain

$$
|e_\lambda\angle\alpha|_v^2 = \sum_{\nu \in J_l^\lambda} |\alpha_\nu|^2. \quad (1.5.10)
$$

Since, by applying (1.5.10),

$$
\begin{aligned}
\sum_{\lambda \in J_k^n} |e_\lambda\angle\alpha|_v^2 &= \sum_{\lambda \in J_k^n} \sum_{\nu \in J_l^\lambda} |\alpha_\nu|^2 = \sum_{\nu \in J_l^n} \sum_{\nu \subset \lambda \in J_k^n} |\alpha_\nu|^2 \\
&= \binom{n-l}{k-l} \sum_{\nu \in J_l^n} |\alpha_\nu|^2 = \binom{n-l}{k-l} |\alpha|_v^2,
\end{aligned}
$$

the inequality in Lemma 1.53 follows. $\qquad \square$

Now assume $\xi \neq 0$ and $\alpha \neq 0$ and set $x = \mathbb{P}(\xi) \in \mathbb{P}\left(\bigwedge_{k+1} V\right)$ and $a = \mathbb{P}(\alpha) \in \mathbb{P}\left(\bigwedge_{l+1} V^*\right)$. We can define the *gauge of x and a for $\angle$*,

$$|x\angle a|_v = \frac{|\xi\angle\alpha|_v}{|\xi|_v \cdot |\alpha|_v}. \tag{1.5.11}$$

In particular, if $k = l = 0$, then $|x\angle a|_v = |x, a|_v$. The projective space $\mathbb{P}\left(\bigwedge_{n+1} V^*\right)$ consists of one and only one point denoted by ∞.

Lemma 1.54. *For all $x \in \mathbb{P}(V)$, $|x\angle\infty|_v = 1$.*

Proof. Take $\xi \in V - \{0\}$ with $x = \mathbb{P}(\xi)$. Put

$$\xi = \xi_0 e_0 + \xi_1 e_1 + \cdots + \xi_n e_n.$$

Then

$$\xi_j = \langle\xi, \epsilon_j\rangle, \quad j = 0, 1, \ldots, n.$$

For $j \in \mathbb{Z}[0, n]$, setting

$$\hat{\epsilon}_j = (-1)^j \epsilon_0 \wedge \cdots \wedge \epsilon_{j-1} \wedge \epsilon_{j+1} \wedge \cdots \wedge \epsilon_n,$$

we have

$$|\xi\angle(\epsilon_0 \wedge \cdots \wedge \epsilon_n)|_v = \left|\sum_{j=0}^{n}\langle\xi, \epsilon_j\rangle\hat{\epsilon}_j\right|_v = \left|\sum_{j=0}^{n}\xi_j\hat{\epsilon}_j\right|_v = |\xi|_v.$$

Since $\infty = \mathbb{P}(\epsilon_0 \wedge \cdots \wedge \epsilon_n)$, then

$$|x\angle\infty|_v = \frac{|\xi\angle(\epsilon_0 \wedge \cdots \wedge \epsilon_n)|_v}{|\xi|_v \cdot |\epsilon_0 \wedge \cdots \wedge \epsilon_n|_v} = 1. \qquad \square$$

Next we show a more subtle inequality (cf. Wu [436]):

Lemma 1.55. *Take $p, q \in \mathbb{Z}[1, n]$ with $p + q \leq n + 1$. If $\xi \in \bigwedge_p V$ and $\eta \in \bigwedge_q V$, then*

$$|\xi \wedge \eta|_v \leq \varsigma_{v,\binom{p+q}{p}}|\xi|_v|\eta|_v.$$

Proof. First of all, note that a norm on the p-fold tensor product $\otimes_p V$ of V can be defined as follows: Taking a base $\{e_0, \ldots, e_n\}$ of V and writing an element $\xi \in \otimes_p V$ by

$$\xi = \sum \xi_{i_1 \cdots i_p} e_{i_1} \otimes \cdots \otimes e_{i_p}, \tag{1.5.12}$$

then

$$|\xi|_{v,\otimes} = \begin{cases} \left(\sum |\xi_{i_1 \cdots i_n}|_v^2\right)^{\frac{1}{2}} & : \quad \text{if } v \text{ is Archimedean,} \\ \max\{|\xi_{i_1 \cdots i_n}|_v\} & : \quad \text{if } v \text{ is non-Archimedean.} \end{cases}$$

Let $\mathcal{J}_p$ be the permutation group on $\mathbb{Z}[1, p]$. For each $\lambda \in \mathcal{J}_p$, a linear isomorphism $\lambda : \otimes_p V \longrightarrow \otimes_p V$ is uniquely defined by

$$\lambda(\xi_1 \otimes \cdots \otimes \xi_p) = \xi_{\lambda^{-1}(1)} \otimes \cdots \otimes \xi_{\lambda^{-1}(p)}, \quad \xi_j \in V \ (j = 1, \ldots, p).$$

The linear mapping

$$A_p = \frac{1}{p!} \sum_{\lambda \in \mathcal{J}_p} \text{sgn}(\lambda)\lambda : \otimes_p V \longrightarrow \otimes_p V$$

is called the *anti-symmetrizer of* $\otimes_p V$ with $\text{Im} A_p = \bigwedge_p V$, where $\text{sgn}(\lambda)$ is the sign of the permutation λ, that is,

$$\text{sgn}(\lambda) = \begin{cases} 1 & \text{if } \lambda \text{ is even permutation,} \\ -1 & \text{if } \lambda \text{ is odd permutation.} \end{cases}$$

For the tensor (1.5.12), we have

$$A_p(\xi) = \sum \xi_{i_1 \cdots i_p} e_{i_1} \wedge \cdots \wedge e_{i_p} \in \bigwedge_p V,$$

and hence it is easy to show that $|A_p(\xi)|_v \leq \varsigma_{v,p!}|\xi|_{v,\otimes}$, where the elementary inequality

$$(a_1 + \cdots + a_n)^2 \leq n(a_1^2 + \cdots + a_n^2) \ (a_i \in \mathbb{R}_+)$$

is used for the proof of the Archimedean case. In particular, if $\xi \in \bigwedge_p V$, then $A_p(\xi) = \xi$. We can obtain the equality $|\xi|_v = c'_p |\xi|_{v,\otimes}$, where

$$c'_p = \begin{cases} \sqrt{p!} & \text{if } v \text{ is Archimedean,} \\ |p!|_v & \text{if } v \text{ is non-Archimedean.} \end{cases}$$

Further, if $\eta \in \bigwedge_q V$, noting that

$$\xi \wedge \eta = A_{p+q}(\xi \otimes \eta), \quad |\xi \otimes \eta|_{v,\otimes} = |\xi|_{v,\otimes}|\eta|_{v,\otimes},$$

then we have

$$|\xi \wedge \eta|_v \leq \sqrt{(p+q)!}|\xi \otimes \eta|_{v,\otimes} = \binom{p+q}{p}^{\frac{1}{2}} |\xi|_v |\eta|_v$$

if v is Archimedean. If v is non-Archimedean, writing

$$\eta = \sum \eta_{j_1 \cdots j_q} e_{j_1} \wedge \cdots \wedge e_{j_q},$$

then

$$\xi \wedge \eta = \sum \xi_{i_1 \cdots i_p} \eta_{j_1 \cdots j_q} e_{i_1} \wedge \cdots \wedge e_{i_p} \wedge e_{j_1} \wedge \cdots \wedge e_{j_q}$$

$$= p! q! \sum_{i_1 < \cdots < i_p} \sum_{j_1 < \cdots < j_q} \xi_{i_1 \cdots i_p} \eta_{j_1 \cdots j_q} e_{i_1} \wedge \cdots \wedge e_{i_p} \wedge e_{j_1} \wedge \cdots \wedge e_{j_q}$$

and hence

$$|\xi \wedge \eta|_v \leq |p!q!|_v \max |\xi_{i_1 \cdots i_p} \eta_{j_1 \cdots j_q}|_v \leq |p!q!|_v |\xi|_{v,\otimes} |\eta|_{v,\otimes} = |\xi|_v |\eta|_v.$$

Therefore Lemma 1.55 is proved.   □

Take $x_j \in \mathbb{P}(V)$ $(j = 0, \ldots, k \leq n)$ and take $\xi_j \in V$ such that $x_j = \mathbb{P}(\xi_j)$. The *gauge of* $x_0, \ldots, x_k$ *for* $\wedge$ is well defined to be

$$|x_0 \wedge \cdots \wedge x_k|_v = \frac{|\xi_0 \wedge \cdots \wedge \xi_k|_v}{|\xi_0|_v \cdots |\xi_k|_v} \tag{1.5.13}$$

which satisfies

$$0 \leq |x_0 \wedge \cdots \wedge x_k|_v \leq \varsigma_{v,2}\varsigma_{v,3} \cdots \varsigma_{v,k+1} = \varsigma_{v,(k+1)!}.$$

When $k = n$ this is a form of Hadamard's determinant inequality (see [142], [143]).

Lemma 1.56. *For* $x \in \mathbb{P}(V)$, $a_j \in \mathbb{P}(V^*), j = 0, 1, \ldots, n$, *then*

$$|a_0 \wedge \cdots \wedge a_n|_v \leq \varsigma_{v,(n+1)!} \max_{0 \leq j \leq n} |x, a_j|_v.$$

Proof. If $|a_0 \wedge \cdots \wedge a_n|_v = 0$, the inequality is trivial. If $|a_0 \wedge \cdots \wedge a_n|_v > 0$, then $a_0 \wedge \cdots \wedge a_n = \infty$. Thus Lemma 1.54 implies $|x\angle(a_0 \wedge \cdots \wedge a_n)|_v = 1$. For each $j \in \mathbb{Z}[0, n]$, take $\alpha_j \in V^* - \{0\}$ with $\mathbb{P}(\alpha_j) = a_j$. Also take $\xi \in V - \{0\}$ with $\mathbb{P}(\xi) = x$. We have

$$\begin{aligned}
|a_0 \wedge \cdots \wedge a_n|_v &= |a_0 \wedge \cdots \wedge a_n|_v \cdot |x\angle(a_0 \wedge \cdots \wedge a_n)|_v \\
&= \frac{|a_0 \wedge \cdots \wedge a_n|_v}{|\alpha_0|_v \cdots |\alpha_n|_v} \cdot \frac{|\xi\angle(a_0 \wedge \cdots \wedge a_n)|_v}{|\xi|_v |\alpha_0 \wedge \cdots \wedge \alpha_n|_v} \\
&= \frac{|\sum_{j=0}^n \langle \xi, \alpha_j \rangle \hat{\alpha}_j|_v}{|\xi|_v |\alpha_0|_v \cdots |\alpha_n|_v},
\end{aligned}$$

and hence

$$\begin{aligned}
|a_0 \wedge \cdots \wedge a_n|_v &\leq \varsigma_{v,n+1} \max_{0 \leq j \leq n} \frac{|\langle \xi, \alpha_j \rangle|_v |\hat{\alpha}_j|_v}{|\xi|_v |\alpha_0|_v \cdots |\alpha_n|_v} \\
&= \varsigma_{v,n+1} \max_{0 \leq j \leq n} |x, a_j|_v |a_0 \wedge \cdots \wedge a_{j-1} \wedge a_{j+1} \wedge \cdots \wedge a_n|_v \\
&\leq \varsigma_{v,(n+1)!} \max_{0 \leq j \leq n} |x, a_j|_v.
\end{aligned}$$

This finishes the proof.   □

1.5.2 Hypersurfaces

Let V be a normed vector space of dimension $n+1 > 0$ over a field κ. Take a positive integer d. Let $\mathcal{J}_d$ be the permutation group on $\mathbb{Z}[1,d]$ and let $\otimes_d V$ be the d-fold tensor product of V. For each $\lambda \in \mathcal{J}_d$, a linear isomorphism $\lambda : \otimes_d V \longrightarrow \otimes_d V$ is uniquely defined by

$$\lambda(\xi_1 \otimes \cdots \otimes \xi_d) = \xi_{\lambda^{-1}(1)} \otimes \cdots \otimes \xi_{\lambda^{-1}(d)}, \quad \xi_j \in V \ (j = 1, \ldots, d).$$

A vector $\xi \in \otimes_d V$ is said to be *symmetric* if $\lambda(\xi) = \xi$ for all $\lambda \in \mathcal{J}_d$. The set of all symmetric vectors in $\otimes_d V$ is a linear subspace of $\otimes_d V$, denoted by $\mathrm{II}_d V$, called the *d-fold symmetric tensor product of V*. Then

$$\dim \mathrm{II}_d V = \binom{n+d}{d}.$$

The linear mapping

$$S_d = \frac{1}{d!} \sum_{\lambda \in \mathcal{J}_d} \lambda : \otimes_d V \longrightarrow \otimes_d V$$

is called the *symmetrizer of* $\otimes_d V$ with $\mathrm{Im} S_d = \mathrm{II}_d V$. If $\xi \in \mathrm{II}_d V$ and $\eta \in \mathrm{II}_l V$, the *symmetric tensor product*

$$\xi \amalg \eta = S_{d+l}(\xi \otimes \eta)$$

is defined with $\xi \amalg \eta = \eta \amalg \xi$. Similarly, for $\xi_j \in V \ (j = 1, \ldots, d)$, we can define the *symmetric tensor product*

$$\xi_1 \amalg \cdots \amalg \xi_d = S_d(\xi_1 \otimes \cdots \otimes \xi_d).$$

Let $\xi^{\amalg d}$ be the *d*th *symmetric tensor power* of $\xi \in V$, and define

$$x^{\amalg d} = \mathbb{P}(\xi^{\amalg d})$$

for $x = \mathbb{P}(\xi)$. Thus a mapping $\varphi_d : \mathbb{P}(V) \longrightarrow \mathbb{P}(\mathrm{II}_d V)$ is well defined by setting $\varphi_d(x) = x^{\amalg d}$, which is called the *Veronese mapping*. We can identify $\mathrm{II}_d V^* = (\mathrm{II}_d V)^*$ by

$$\langle \xi_1 \amalg \cdots \amalg \xi_d, \alpha_1 \amalg \cdots \amalg \alpha_d \rangle = \frac{1}{d!} \sum_{\lambda \in \mathcal{J}_d} \langle \xi_1, \alpha_{\lambda(1)} \rangle \cdots \langle \xi_d, \alpha_{\lambda(d)} \rangle$$

for all $x_j \in V, \alpha_j \in V^*, j = 1, \ldots, d$.

Let $J_{n,d}$ be the set of all mappings $\lambda : \mathbb{Z}[0,n] \longrightarrow \mathbb{Z}[0,d]$ such that

$$|\lambda| = \lambda(0) + \cdots + \lambda(n) = d.$$

For $\lambda \in J_{n,d}, e = (e_0, \ldots, e_n) \in V^{n+1}$, define

$$\lambda! = \lambda(0)! \cdots \lambda(n)!, \quad e^{\amalg \lambda} = e_0^{\amalg \lambda(0)} \amalg \cdots \amalg e_n^{\amalg \lambda(n)} \in \amalg_d V.$$

If $e = (e_0, \ldots, e_n)$ is a base of V, then $\{e^{\amalg \lambda}\}_{\lambda \in J_{n,d}}$ is a base of $\amalg_d V$, and $\{\frac{d!}{\lambda!}\epsilon^{\amalg \lambda}\}_{\lambda \in J_{n,d}}$ is the dual base of $\amalg_d V^*$, where $\epsilon = (\epsilon_0, \ldots, \epsilon_n)$ is the dual of e. The norm $|\cdot|$ on V induces norms on $\amalg_d V$ and $\amalg_d V^*$ as follows: For $\eta \in \amalg_d V, \beta \in \amalg_d V^*$ with

$$\eta = \sum_{\lambda \in J_{n,d}} \frac{d!}{\lambda!}\eta_\lambda e^{\amalg \lambda}, \quad \beta = \sum_{\lambda \in J_{n,d}} \frac{d!}{\lambda!}\beta_\lambda \epsilon^{\amalg \lambda},$$

define

$$|\eta| = |\eta|_e = \begin{cases} \left(\sum_{\lambda \in J_{n,d}} \frac{d!}{\lambda!}|\eta_\lambda|^2\right)^{\frac{1}{2}} & : \quad \text{if } |\cdot| \text{ is Archimedean,} \\ \max_{\lambda \in J_{n,d}}\{|\eta_\lambda|\} & : \quad \text{if } |\cdot| \text{ is non-Archimedean,} \end{cases}$$

and

$$|\beta| = |\beta|_e = \begin{cases} \left(\sum_{\lambda \in J_{n,d}} \frac{d!}{\lambda!}|\beta_\lambda|^2\right)^{\frac{1}{2}} & : \quad \text{if } |\cdot| \text{ is Archimedean,} \\ \max_{\lambda \in J_{n,d}}\{|\beta_\lambda|\} & : \quad \text{if } |\cdot| \text{ is non-Archimedean,} \end{cases}$$

where e is orthonormal if $|\cdot|$ is Archimedean. Note that

$$\xi^{\amalg d} = \sum_{\lambda \in J_{n,d}} \frac{d!}{\lambda!}\xi_0^{\lambda(0)} \cdots \xi_n^{\lambda(n)} e^{\amalg \lambda}, \quad \alpha^{\amalg d} = \sum_{\lambda \in J_{n,d}} \frac{d!}{\lambda!}\alpha_0^{\lambda(0)} \cdots \alpha_n^{\lambda(n)} \epsilon^{\amalg \lambda},$$

where

$$\xi = \xi_0 e_0 + \cdots + \xi_n e_n \in V, \quad \alpha = \alpha_0 \epsilon_0 + \cdots + \alpha_n \epsilon_n.$$

Then we obtain a formula

$$|\xi^{\amalg d}| = |\xi|^d, \quad |\alpha^{\amalg d}| = |\alpha|^d. \tag{1.5.14}$$

Let $V_{[d]}$ be the vector space of all homogeneous polynomials of degree d on V. We obtain a linear isomorphism

$$\sim: \amalg_d V^* \longrightarrow V_{[d]}$$

defined by

$$\tilde{\alpha}(\xi) = \langle \xi^{\amalg d}, \alpha \rangle, \quad \xi \in V, \ \alpha \in \amalg_d V^*.$$

Thus if $\xi \neq 0$ and $\alpha \neq 0$, the distance $|x^{\amalg d}, a|$ is well defined for $x^{\amalg d} = \mathbb{P}(\xi^{\amalg d})$ and $a = \mathbb{P}(\alpha)$. If $\alpha \neq 0$, the n-dimensional subspace

$$E^d[a] = \operatorname{Ker}(\tilde{\alpha}) = \tilde{\alpha}^{-1}(0)$$

in V depends on a only, and $\ddot{E}^d[a] = \mathbb{P}(E^d[a])$ is a *hypersurface of degree d* in $\mathbb{P}(V)$. Thus $\mathbb{P}(\mathrm{II}_d V^*)$ bijectively parameterizes the hypersurfaces in $\mathbb{P}(V)$. Take a sequence $\{d_0, d_1, \ldots, d_q\}$ of positive integers. Let $\mathscr{A} = \{a_0, a_1, \ldots, a_q\}$ be a family of points $a_j \in \mathbb{P}\left(\mathrm{II}_{d_j} V^*\right)$. Take $\alpha_j \in \mathrm{II}_{d_j} V^* - \{0\}$ with $\mathbb{P}(\alpha_j) = a_j$, and define

$$\tilde{\alpha}_j(\xi) = \left\langle \xi^{\mathrm{II} d_j}, \alpha_j \right\rangle, \quad \xi \in V, \ j = 0, 1, \ldots, q.$$

According to Eremenko and Sodin [94], we will use the following notation:

Definition 1.57. *The family $\mathscr{A} = \{a_0, a_1, \ldots, a_q\}$ ($q \geq n$) is said to be admissible (or in general position) if, for every $\lambda \in J_n^q$, the system*

$$\tilde{\alpha}_{\lambda(i)}(\xi) = 0, \quad i = 0, 1, \ldots, n \tag{1.5.15}$$

has only the trivial solution $\xi = 0$ in V.

1.5.3 Nochka weights

Let V be a vector space of finite dimension $n + 1 > 0$ over a field κ. Let $\mathscr{A} = \{a_0, a_1, \ldots, a_q\}$ be a family of points $a_j \in \mathbb{P}(V^*)$. Take $\alpha_j \in V^* - \{0\}$ with $\mathbb{P}(\alpha_j) = a_j$. For $\lambda \in J_l^q$, set $\mathscr{A}_\lambda = \{a_{\lambda(0)}, \ldots, a_{\lambda(l)}\}$, and let $E(\mathscr{A}_\lambda)$ be the linear subspace generated by $\{\alpha_{\lambda(0)}, \ldots, \alpha_{\lambda(l)}\}$ in V^*. Define

$$J_l(\mathscr{A}) = \{\lambda \in J_l^q \mid \alpha_{\lambda(0)} \wedge \cdots \wedge \alpha_{\lambda(l)} \neq 0\}.$$

Then $\mathscr{A}$ is said to be in *general position* if $\dim E(\mathscr{A}_\lambda) = l + 1$ for any $\lambda \in J_l^q$ with $l \leq \min\{n, q\}$. If so the hyperplanes $\ddot{E}[a_0], \ldots, \ddot{E}[a_q]$ also are called in *general position*. Following Chen [56], we also use the concept of subgeneral position as follows:

Definition 1.58. *Let $\mathscr{A} = \{a_0, a_1, \ldots, a_q\}$ be a family of points $a_j \in \mathbb{P}(V^*)$. For $1 \leq n \leq u \leq q$, then $\mathscr{A}$ is said to be in u-subgeneral position if $E(\mathscr{A}_\lambda) = V^*$ for any $\lambda \in J_u^q$.*

In particular, if $u = n$ this concept agrees with the usual concept of hyperplanes in general position. The notion of subgeneral position will play a key role in the proof of Cartan's conjecture due to Nochka in Section 2.8. To prove Cartan's conjecture, Nochka used the following technical lemma:

Lemma 1.59. *Let $\mathscr{A} = \{a_0, a_1, \ldots, a_q\}$ be a family of points $a_j \in \mathbb{P}(V^*)$ in u-subgeneral position with $1 \leq n \leq u < q$. Then there exists a function $\omega : \mathscr{A} \longrightarrow \mathbb{R}(0, 1]$ and a real number $\theta \geq 1$ satisfying the properties:*

1) $0 < \omega(a_j)\theta \leq 1, \quad j = 0, 1, \ldots, q$;
2) $q - 2u + n = \theta(\sum_{j=0}^q \omega(a_j) - n - 1)$;
3) $1 \leq \frac{u+1}{n+1} \leq \theta \leq \frac{2u-n+1}{n+1}$;

4) $\sum_{j=0}^{k} \omega(a_{\sigma(j)}) \leq \dim E(\mathscr{A}_\sigma)$ *if* $\sigma \in J_k^q$ *with* $0 \leq k \leq u;$

5) *Let* $r_0, \ldots, r_q$ *be a sequence of real numbers with* $r_j \geq 1$ *for all* j. *Then for any* $\sigma \in J_k^q$ *with* $0 \leq k \leq u$, *setting* $\dim E(\mathscr{A}_\sigma) = l + 1$, *then there exists* $\lambda \in J_l(\mathscr{A})$ *such that*

$$\mathrm{Im}\lambda = \{\lambda(0), \ldots, \lambda(l)\} \subset \{\sigma(0), \ldots, \sigma(k)\}, \quad E(\mathscr{A}_\lambda) = E(\mathscr{A}_\sigma),$$

and

$$\prod_{j=0}^{k} r_{\sigma(j)}^{\omega(a_{\sigma(j)})} \leq \prod_{j=0}^{l} r_{\lambda(j)}.$$

The function ω and the real number θ are respectively called a *Nochka weight* and a *Nochka constant* of the family $\mathscr{A}$ in u-subgeneral position. If $u = n$, then $\theta = 1$ and $\omega(a_j) = 1$ for each $j = 0, 1, \ldots, q$. From Lemma 1.59, it follows that values of the function ω become small if u is large. Hence Nochka weight is a gauge of a subgeneral position leaving a general position. Nochka's original paper (see [299],[300],[301]) on the weights of Nochka was quite sketchy; a complete proof can be found in Chen's thesis [56] (or see Fujimoto [107], Hu and Yang [176]). Here we omit the proof since it is very long.

Let $\mathscr{A} = \{a_0, a_1, \ldots, a_q\}$ $(n \leq u \leq q)$ be in u-subgeneral position. Define the *gauge* $\Gamma(\mathscr{A})$ of $\mathscr{A}$ on a valuation v of κ by

$$\Gamma(\mathscr{A}) = \frac{1}{\varsigma_{v,(n+1)!}} \inf_{\lambda \in J_n(\mathscr{A})} \{|a_{\lambda(0)} \wedge \cdots \wedge a_{\lambda(n)}|_v\}$$

with $0 < \Gamma(\mathscr{A}) \leq 1$. Let $\#P$ be the *cardinality* of a set P.

Lemma 1.60. *For* $x \in \mathbb{P}(V)$, $0 < r \in \mathbb{R}$, *define*

$$\mathscr{A}(x, r) = \{j \in \mathbb{Z}[0, q] \mid |x, a_j|_v < r\}.$$

If $\Gamma(\mathscr{A}) \geq r$, *then* $\#\mathscr{A}(x, r) \leq u$.

Proof. Assume, to the contrary, that $\#\mathscr{A}(x, r) \geq u + 1$. Then $\lambda \in J_n(\mathscr{A})$ exists such that $\mathrm{Im}\lambda \subseteq \mathscr{A}(x, r)$. Hence

$$|x, a_{\lambda(j)}|_v < r, \quad j = 0, \ldots, n.$$

Then Lemma 1.56 implies

$$\begin{aligned} 0 < \Gamma(\mathscr{A}) &\leq |a_{\lambda(0)} \wedge \cdots \wedge a_{\lambda(n)}|_v / \varsigma_{v,(n+1)!} \\ &\leq \max_{0 \leq j \leq n} |x, a_{\lambda(j)}|_v < r \leq \Gamma(\mathscr{A}), \end{aligned}$$

which is impossible. Hence we have $\#\mathscr{A}(x, r) \leq u$. $\qquad\square$

Lemma 1.61. *Take $x \in \mathbb{P}(V)$ such that $|x, a_j|_v > 0$ for $j = 0, \ldots, q$. Then*

$$\prod_{j=0}^{q} \left(\frac{1}{|x, a_j|_v} \right)^{\omega(a_j)} \leq \left(\frac{1}{\Gamma(\mathscr{A})} \right)^{q-u} \max_{\lambda \in J_n(\mathscr{A})} \prod_{j=0}^{n} \frac{1}{|x, a_{\lambda(j)}|_v}, \qquad (1.5.16)$$

where $\omega : \mathscr{A} \longrightarrow \mathbb{R}(0,1]$ is the Nochka weight. In particular, if $u = n$ we also have

$$\prod_{j=0}^{q} \frac{1}{|x, a_j|_v} \leq \left(\frac{1}{\Gamma(\mathscr{A})} \right)^{q+1-n} \max_{\lambda \in J_{n-1}^q} \prod_{j=0}^{n-1} \frac{1}{|x, a_{\lambda(j)}|_v}. \qquad (1.5.17)$$

Proof. Take $r = \Gamma(\mathscr{A})$. Lemma 1.60 implies $\#\mathscr{A}(x, r) \leq u$. Thus $\sigma \in J_u^q$ exists such that $\mathscr{A}(x, r) \subset \operatorname{Im}\sigma$. Note that $E(\mathscr{A}_\sigma) = V^*$. By Lemma 1.59, there exists $\lambda \in J_n(\mathscr{A})$ with $\operatorname{Im}\lambda \subset \operatorname{Im}\sigma$ such that $E(\mathscr{A}_\lambda) = E(\mathscr{A}_\sigma)$, and such that

$$\prod_{j=0}^{u} \left(\frac{1}{|x, a_{\sigma(j)}|_v} \right)^{\omega(a_{\sigma(j)})} \leq \prod_{j=0}^{n} \frac{1}{|x, a_{\lambda(j)}|_v}. \qquad (1.5.18)$$

Set $C = \mathbb{Z}[0, q] - \operatorname{Im}\sigma$. Thus $|x, a_j|_v \geq r$ for $j \in C$. Hence

$$\prod_{j \in C} \left(\frac{1}{|x, a_j|_v} \right)^{\omega(a_j)} \leq \prod_{j \in C} \frac{1}{|x, a_j|_v} \leq \left(\frac{1}{r} \right)^{\#C} = \left(\frac{1}{\Gamma(\mathscr{A})} \right)^{q-u}.$$

Thus the inequality (1.5.16) follows.

If $u = n$, then $\sigma = \lambda$ and $\operatorname{Im}\lambda - \mathscr{A}(x, r) \neq \emptyset$, that is, there is some $j_0 \in \mathbb{Z}[0, n]$ such that $|x, a_{\lambda(j_0)}|_v \geq r$. Now (1.5.18) becomes

$$\prod_{j=0}^{n} \frac{1}{|x, a_{\sigma(j)}|_v} \leq \frac{1}{r} \prod_{j \neq j_0} \frac{1}{|x, a_{\lambda(j)}|_v},$$

and so (1.5.17) follows. $\qquad\square$

Finally, let κ be the field $\mathbb{C}$ of complex numbers. A positive definite Hermitian form

$$(,) : V \times V \longrightarrow \mathbb{C}$$

is called a *Hermitian product* or a *Hermitian metric* on V. It defines a *norm*

$$|\xi| = (\xi, \xi)^{\frac{1}{2}}, \quad \xi \in V.$$

A complex vector space together with a Hermitian product is called a *Hermitian vector space*. For each $\xi \in V$, one and only one dual vector $\xi^* \in V^*$ is defined by $(\eta, \xi) = \langle \eta, \xi^* \rangle$ for all $\eta \in V$. The mapping $\xi \mapsto \xi^*$ is an anti-linear isomorphism of V onto V^*. Here V^* becomes a Hermitian vector space by setting

$$(\xi^*, \eta^*) = (\eta, \xi), \quad \xi, \eta \in V.$$

Then $\xi^{**} = \xi$ and $V^{**} = V$, as Hermitian vector space. A Hermitian product is uniquely defined on $\bigwedge_{p+1} V$ by the requirement

$$(\xi_0 \wedge \cdots \wedge \xi_p, \eta_0 \wedge \cdots \wedge \eta_p) = \det((\xi_j, \eta_k)), \quad \xi_j, \eta_j \in V.$$

1.6 Basic geometric notions

Let κ be a field and let $\bar{\kappa}$ be an algebraic closure of κ. The space $\bar{\kappa}^n$ is called the *affine n-space* (over κ), which is usually denoted by $\mathbb{A}^n$ or $\mathbb{A}^n(\bar{\kappa})$. The set of *$\kappa$-rational points* of $\mathbb{A}^n$ is the set

$$\mathbb{A}^n(\kappa) = \{(z_1, \ldots, z_n) \in \mathbb{A}^n \mid z_i \in \kappa\}.$$

We will introduce some geometric notation in spaces $\mathbb{A}^n$ and $\mathbb{P}^n = \mathbb{P}(\mathbb{A}^{n+1})$.

1.6.1 Varieties

Let I_κ be an ideal in the polynomial ring in n variables $\kappa[z_1, \ldots, z_n]$. The Hilbert basis theorem says that I_κ is generated by a finite number of polynomials $P_1, \ldots, P_r$. Assume that $P_1, \ldots, P_r$ generate a prime ideal in the ring $\bar{\kappa}[z_1, \ldots, z_n]$. The set of zeros of I_κ,

$$Z = \{z \in \mathbb{A}^n \mid P(z) = 0 \text{ for all } P \in I_\kappa\},$$

is called an *affine algebraic variety* or *affine variety defined over* κ, which in fact is the set of common zeros of the finite collection of polynomials $P_1, \ldots, P_r$. Especially, if $r = 1$, the affine variety Z is usually called an *affine hypersurface*.

Remark. The condition that the polynomials generate a prime ideal is to insure what is called the irreducibility of the variety. Under our condition, it is not possible to express a variety as the finite union of proper subvarieties. In the definition of the affine algebraic variety Z, if the ideal I_κ is not assumed prime, then we obtain an *affine algebraic set* Z. The set Z is a finite union of its irreducible components (affine algebraic varieties).

Let Z be an affine variety, defined by an ideal I_κ in $\kappa[z_1, \ldots, z_n]$. The ring

$$\kappa[Z] = \kappa[z_1, \ldots, z_n]/I_\kappa$$

is called the *affine coordinate ring* of Z, or simply the *affine ring* of Z. This ring has no zero divisors, and its quotient field is called the *function field* of Z over κ, denoted by $\kappa(Z)$. An element of $\kappa(Z)$ is called a *rational function* on Z, which is the quotient of two polynomial functions on Z such that the denominator does not vanish identically on Z.

If K is a field containing κ, the set of zeros of I_κ with coordinate $(z_1, \ldots, z_n) \in K^n$ is called the set of *K-rational points* of Z, and is denoted by $Z(K)$. It is equal to the set of solutions of the finite number of equations

$$P_j(z_1, \ldots, z_n) = 0, \quad j = 1, \ldots, r, \quad (z_1, \ldots, z_n) \in K^n.$$

By a *projective variety* X over a field κ we mean the set of solutions in a projective space $\mathbb{P}^n = \mathbb{P}(\mathbb{A}^{n+1})$ of a finite number of equations

$$f_j(\xi_0, \ldots, \xi_n) = 0, \quad j = 1, \ldots, r$$

such that each f_j is a homogeneous polynomial in $n+1$ variables with coefficients in κ, and $f_1, \ldots, f_r$ generate a prime ideal in the polynomial ring $\bar{\kappa}[\xi_0, \ldots, \xi_n]$. If $r = 1$, the variety X is called a *projective hypersurface*, and the degree of f_1 is called the *degree* of the hypersurface. Further, if $r = 1$ and if f_1 is linear, X is called a *hyperplane*.

If K is a field containing κ, by $X(K)$ we mean the set of such zeros having some projective coordinates $[\xi_0, \ldots, \xi_n]$ with $\xi_i \in K$ for all $i = 0, \ldots, n$, called the set of K-*rational points* of X. The set of points in $X(\bar{\kappa})$ is called the set of *algebraic points* over κ. For a point $x = [\xi_0, \ldots, \xi_n] \in \mathbb{P}^n$, we denote by $\kappa(x)$ the field

$$\kappa(x) = \kappa(\xi_0, \ldots, \xi_n)$$

such that at least one of the projective coordinates is equal to 1, which is called the *field of definition* of the point x or the *residue class field of the point*. It does not matter which such coordinate is selected. If for instance $\xi_0 \neq 0$, then

$$\kappa(x) = \kappa\left(\frac{\xi_1}{\xi_0}, \ldots, \frac{\xi_n}{\xi_0}\right).$$

We can define the *Zariski topology* on $\mathbb{P}^n$ (resp., $\mathbb{A}^n$) by prescribing that a *closed set* is a finite union of varieties. A *Zariski open set* is defined to be the complement of a closed set. The *Zariski topology* on a variety X is the topology induced by the inclusion $X \subset \mathbb{P}^n$ (resp., $X \subset \mathbb{A}^n$). By a *quasi-projective variety*, we mean the open subset of a projective variety obtained by omitting a closed subset.

A projective variety X is covered by a finite number of affine varieties as follows. Set

$$z_{l,i+1} = \begin{cases} \frac{\xi_i}{\xi_l}, & 0 \leq i \leq l-1; \\ \frac{\xi_{i+1}}{\xi_l}, & l \leq i < n, \end{cases}$$

and let

$$P_{l,j}(z_{l,1}, \ldots, z_{l,n}) = f_j(z_{l,1}, \ldots, z_{l,l}, 1, z_{l,l+1}, \ldots, z_{l,n}).$$

Then the polynomials $P_{l,1}, \ldots, P_{l,r}$ generate a prime ideal in $\bar{\kappa}[z_{l,1}, \ldots, z_{l,n}]$, and the set of solutions of the equations

$$P_{l,j}(z_{l,1}, \ldots, z_{l,n}) = 0, \ j = 1, \ldots, r$$

is an affine variety, which is an open subset of X, denoted by U_l. It consists of those points $[\xi_0, \ldots, \xi_n] \in X$ such that $\xi_l \neq 0$. The projective variety X is covered by the open sets $U_0, \ldots, U_n$. The function fields $\kappa(U_0), \ldots, \kappa(U_n)$ are all equal, and are generated by the restrictions to X of the quotients ξ_i/ξ_l (for all i, l such that ξ_l is not identically 0 on X). The *function field* of X over κ is defined to be $\kappa(U_l)$ (for any l), and is denoted by $\kappa(X)$. A rational function can also be expressed as a quotient of two homogeneous polynomial functions $P(\xi_0, \ldots, \xi_n)/Q(\xi_0, \ldots, \xi_n)$ where P, Q have the same degree.

A variety X is *complete* or *proper* if for any variety Y, the projection $X \times Y \longrightarrow Y$ is closed, i.e., the image of every closed subset is closed. Projective varieties are complete.

1.6.2   Function fields

We here give a more intrinsic definition of the function field of a variety X. Let x be a point on X. A function $f : X \longrightarrow \bar{\kappa}$ is *regular at x* if there is an open affine neighborhood $U \subset X$ of x, say $U \subset \mathbb{A}^n$, and two polynomials $P, Q \in \bar{\kappa}[z_1, \ldots, z_n]$ such that $Q(x) \neq 0$ and $f = P/Q$ on U. We say that f is *regular on X* if it is regular at every point of X. We denote by $\mathcal{O}(X)$ the ring of all regular functions on X. A regular function on a projective variety is constant (see [148], I.3.4(a)).

Note that the property of being regular is open, that is, if f is regular at x, then it is regular at every point in some neighborhood of x. This suggests looking at the collection of functions that are regular at a given point. If x is a point on a variety X, we define the *local ring of X at x* to be the ring of germs of functions that are regular at x. This ring is denoted by $\mathcal{O}_X(x)$, or simply by $\mathcal{O}(x)$ if no confusion is likely to arise. In other words, an element of $\mathcal{O}(x)$ is a pair (U, f) where U is an open subset of X containing x, and f is a regular function on U, and where we identify two such pairs (U, f) and (W, g) if there is a neighborhood $V \subset U \cap W$ of x such that $f = g$ on V. Note that $\mathcal{O}(x)$ is indeed a local ring: its maximal ideal $\mathbf{m}(x)$ is the set of germs of regular functions which vanish at x. For if $f(x) \neq 0$, then $1/f$ is regular in some neighborhood of x. A variety is called *normal* if the local ring of every point is integrally closed. A non-singular variety is normal.

By a *subvariety* of a variety X we shall always mean a closed subvariety unless otherwise specified. Let $Y \subset X$ be a subvariety of a variety X. The *local ring of X along Y*, denoted by $\mathcal{O}_X(Y)$, is the set of pairs (U, f), where U is an open subset of X with $U \cap Y \neq \emptyset$ and $f \in \mathcal{O}(U)$ is a regular function on U, where we identify two pairs $(U, f) = (W, g)$ if $f = g$ on $U \cap W$. The ring $\mathcal{O}_X(Y)$ is a local ring, its unique maximal ideal being given by

$$\mathbf{m}_X(Y) = \{ f \in \mathcal{O}_X(Y) \mid f(x) = 0 \text{ for all } x \in Y \}.$$

The *function field* of X, denoted by $\bar{\kappa}(X)$, is defined to be $\mathcal{O}_X(X)$, the local ring of X along X. The elements of $\bar{\kappa}(X)$ are called *rational functions* on X over $\bar{\kappa}$. In other words, $\bar{\kappa}(X)$ is the set of pairs (U, f), where U is a non-empty open subset of X and f is a regular function on U, subject to the identification $(U, f) = (W, g)$ if $f = g$ on $U \cap W$. It is easy to check that $\bar{\kappa}(X)$ is a field that contains every local ring $\mathcal{O}_X(Y)$ of X for any subvariety $Y \subset X$, and that we have

$$\mathcal{O}_X(Y)/\mathbf{m}_X(Y) \cong \bar{\kappa}(Y).$$

The function fields of $\mathbb{A}^n$ and $\mathbb{P}^n$ are both equal to $\bar{\kappa}(z_1, \ldots, z_n)$, the field of rational functions in n variables.

A mapping $\varphi : X \longrightarrow Y$ between varieties is a *morphism* if it is continuous, and if for every open set $V \subset Y$ and every regular function g on V, the function $g \circ \varphi$ is regular on $\varphi^{-1}(V)$. Note that the image of a projective variety by a morphism is a projective variety (see [158], Theorem A.1.2.3).

A mapping $\varphi : X \longrightarrow Y$ between varieties is *regular* at a point $x \in X$ if it is a morphism on some open neighborhood of x. One can show that φ is regular at x if there is an affine neighborhood $U \subset \mathbb{A}^m$ of x in X and an affine neighborhood $V \subset \mathbb{A}^n$ of $\varphi(x)$ in Y such that φ sends U into V and such that φ can be defined on U by n polynomials in m variables. That these definitions are equivalent comes from the fact that a morphism of affine varieties is defined globally by polynomials, as can be deduced readily from Theorem 1.62 below. If $\varphi : X \longrightarrow Y$ is regular at each point of X, then φ is said to be a *regular mapping*.

A regular mapping $\varphi : X \longrightarrow Y$ is an *isomorphism* if it has an inverse, that is, if there exists a regular mapping $\psi : Y \longrightarrow X$ such that both $\varphi \circ \psi : Y \longrightarrow Y$ and $\psi \circ \varphi : X \longrightarrow X$ are the identity mappings. In this case we say that X and Y are *isomorphic*. An isomorphism from X to itself is also called an *automorphism* on X. The group $\mathrm{Aut}(X)$ of automorphisms of X is an extremely interesting object. For example, some examples of $\mathrm{Aut}(\mathbb{A}^2)$ are simple to construct: the affine linear mappings, and *elementary mappings* of the form

$$y_1 = \alpha x_1 + f(x_2), \quad y_2 = \beta x_2 + \gamma, \qquad (1.6.1)$$

where α, β, γ are constants with $\alpha\beta \neq 0$, and f a polynomial. It is known that the whole group $\mathrm{Aut}(\mathbb{A}^2)$ is generated by these automorphisms in the sense that every element of $\mathrm{Aut}(\mathbb{A}^2)$ is a finite composition of the affine linear mappings and the elementary mappings (cf. [198]). A famous unsolved problem related to automorphisms of $\mathbb{A}^n$ is the *Jacobian conjecture*. This asserts that, if the ground field κ has characteristic 0, a mapping given by

$$y_i = f_i(x_1, \ldots, x_n), \quad i = 1, \ldots, n$$

with $f_i \in \bar{\kappa}[x_1, \ldots, x_n]$ is an automorphism of $\mathbb{A}^n$ if and only if the Jacobian determinant $\det\left(\frac{\partial f_i}{\partial y_j}\right)$ is a non-zero constant (cf. [16]). The necessity is easy. For the case $n = 2$, this conjecture is proved when the degrees of f_1 and f_2 are not too large (the order of 100).

A *rational mapping* from a variety X to a variety Y is a mapping that is a morphism on some non-empty open subset of X. Let $\varphi : X \longrightarrow Y$ be a rational mapping. Then there is a largest open subset Ω on which φ is a morphism. This open subset is called the *domain of definition* of φ, denoted dom(φ). The rational mapping φ is said to be *dominant* if $\varphi(U)$ is dense in Y for some (and consequently every) non-empty open set $U \subset X$ on which it is a morphism. A *birational mapping* is a rational mapping that has a rational inverse. Two varieties are said to be *birationally equivalent* if there is a birational mapping between them.

Theorem 1.62. *Let Z and Z' be affine varieties. Then*

(i) $\mathcal{O}(Z) \cong \bar{\kappa}[Z]$.

(ii) *A morphism $\varphi : Z \longrightarrow Z'$ induces a ring homomorphism $\varphi^* : \bar{\kappa}[Z'] \longrightarrow \bar{\kappa}[Z]$ defined by $g \mapsto g \circ \varphi$. The natural mapping*

$$\mathrm{Mor}(Z, Z') \longrightarrow \mathrm{Hom}_{\bar{\kappa}}(\bar{\kappa}[Z'], \bar{\kappa}[Z])$$

defined by $\varphi \mapsto \varphi^*$ is a bijection.

Proof. Hartshorne [148], I.3.2. □

If $\varphi : Z \longrightarrow Z'$ is a morphism between affine varieties, we may view $\bar{\kappa}[Z]$ as a $\bar{\kappa}[Z']$-module by means of φ^*. The morphism φ is called *finite* if $\bar{\kappa}[Z]$ is a finitely generated $\bar{\kappa}[Z']$-module. A morphism $\varphi : X \longrightarrow Y$ between varieties is *finite* if for every affine open subset $V \subset Y$, the set $\varphi^{-1}(V)$ is affine and the mapping $\varphi : \varphi^{-1}(V) \longrightarrow V$ is finite.

A mapping φ between affine varieties is dominant if and only if φ^* is injective, so we say that φ is *finite surjective* if it is finite and φ^* is injective. If $\varphi : X \longrightarrow Y$ is a finite mapping, then it is a closed mapping and all fibers $\varphi^{-1}(y)$ consist of a finite number of points. Further, there is an integer d and a non-empty open $V \subset \varphi(X)$ such that

$$\#\varphi^{-1}(y) = d, \ y \in V.$$

The degree d can be described algebraically as the degree of the associated field extension, and we define this quantity to be the *degree of the finite mapping φ*,

$$\deg(\varphi) = [\bar{\kappa}(X) : \varphi^*\bar{\kappa}(Y)].$$

An *algebraic group* defined over κ is a variety G defined over κ, a point $e \in G(\kappa)$, and morphisms $m : G \times G \longrightarrow G$ and $i : G \longrightarrow G$ satisfying the axioms of a group law:

(α) $m(e, x) = m(x, e) = x$.

(β) $m(i(x), x) = m(x, i(x)) = e$.

(γ) $m(m(x, y), z) = m(x, m(y, z))$.

An *Abelian variety* is a projective variety that is also an algebraic group.

1.6.3 Dimensions

The *dimension* of a variety X is defined to be the transcendence degree of its function field $\bar{\kappa}(X)$ over $\bar{\kappa}$ (cf. [109], [158]), denoted by $\dim X$. There is another definition of dimension. Consider a maximal chain of subvarieties

$$Y_0 \subset Y_1 \subset \cdots \subset Y_m = X,$$

where Y_0 is a point and $Y_i \neq Y_{i+1}$ for all i. Then all such chains have the same number of elements m, and m is the *dimension* of X (cf. [158], [232]). In particular, we have the following useful corollary.

Proposition 1.63. *Let X be a variety, and let Y be a subvariety of X. If $Y \neq X$, then $\dim Y < \dim X$.*

Proof. Hindry-Silverman [158], Corollary A.1.3.3 or Shafarevich [342], I.6, Theorem 1.   □

If $Y \subset X$ is a closed subvariety of X, then the number $\dim X - \dim Y$ is called the *codimension* of Y in X, and written $\mathrm{codim}(Y)$ or $\mathrm{codim}_X(Y)$. Not surprisingly, both $\mathbb{A}^n$ and $\mathbb{P}^n$ have dimension n. Similarly, the dimension of a hypersurface in $\mathbb{A}^n$ or $\mathbb{P}^n$ is $n - 1$. In fact, a kind of converse is true.

Theorem 1.64. *A variety of dimension $n - 1$ is birational equivalent to a hypersurface in $\mathbb{A}^n$ or $\mathbb{P}^n$.*

Proof. See Hindry and Silverman [158], or Hartshorne [148], Ch. I, Proposition 4.9. The main idea is that the function field $\bar{\kappa}(X)$ of the variety X of dimension $n - 1$ over $\bar{\kappa}$ is a finitely generated extension of $\bar{\kappa}$ so that $\bar{\kappa}(X)$ is separably generated (see Zariski and Samuel [448], Ch. II, Theorem 31, p. 105, or Matsumura [260], Ch. 10, Corollary, p. 194). Hence we can find a transcendence base $\{x_1, \ldots, x_{n-1}\} \subset \bar{\kappa}(X)$ such that $\bar{\kappa}(X)$ is a finite separable extension of $\bar{\kappa}(x_1, \ldots, x_{n-1})$. Then by Theorem 1.19, we can find one further element $x_n \in \bar{\kappa}(X)$ such that $\bar{\kappa}(X) = \bar{\kappa}(x_1, \ldots, x_{n-1}, x_n)$. Now x_n is algebraic over $\bar{\kappa}(x_1, \ldots, x_{n-1})$, so it satisfies a polynomial equation with coefficients which are rational functions in $x_1, \ldots, x_{n-1}$. Clearing denominators, we obtain an irreducible polynomial $f(x_1, \ldots, x_n) = 0$. This defines a hypersurface in $\mathbb{A}^n$ with function field $\bar{\kappa}(X)$, which is birational to X. Its projective closure is a hypersurface in $\mathbb{P}^n$.   □

The *dimension* of an algebraic subset V is the maximum of the dimensions of its irreducible components. If all the irreducible components of V have the same (finite) dimension d, then V is said to be of *pure dimension* d. If V is an algebraic subset of $\mathbb{A}^n$ (or $\mathbb{P}^n$) of dimension $n - r$, defined by r equations

$$f_j = 0, \ j = 1, \ldots, r,$$

then we say that V is a *complete intersection*.

Theorem 1.65. *Any affine variety of dimension d can be realized as an irreducible component of some affine complete intersection of pure dimension d.*

Proof. C. Musili [284], Theorem 25.7.   □

To conform with the usual terminology, a variety of dimension 1 is called a *curve*, and a variety of dimension 2 is called a *surface*. If κ is a subfield of $\mathbb{C}$, then $X(\mathbb{C})$ is a complex analytic space of complex analytic dimension 1. Now a curve is also sometimes called a *Riemann surface*.

In order to compute the dimension of a variety, we need to know how the dimension behaves for intersections of algebraic sets, which is answered by the *affine* (or *projective*) *dimension theorem*:

Theorem 1.66. *Let X and Y be varieties in $\mathbb{A}^n$ (or $\mathbb{P}^n$) of dimensions l and m, respectively. Then every component of $X \cap Y$ has dimension at least $l + m - n$.*

Proof. Hindry-Silverman [158] or Shafarevich [342] or Hartshorne [148], Ch. I, Proposition 7.1 and Theorem 7.2.   □

Theorem 1.67. *Let $\varphi : X \longrightarrow Y$ be a surjective morphism of varieties. Then*

(I) $\dim \varphi^{-1}(y) \geq \dim X - \dim Y$ *for all $y \in Y$.*

(II) *There is a non-empty open subset $V \subset Y$ such that for all $y \in V$,*

$$\dim \varphi^{-1}(y) = \dim X - \dim Y.$$

Proof. Shafarevich [342], I.6, Theorem 7.   □

Let Z be an affine variety in affine space $\mathbb{A}^n$, with coordinates $(z_1, \ldots, z_n)$, and defined over a field κ. Let $a = (a_1, \ldots, a_n)$ be a point of Z. Suppose κ algebraically closed and $a_i \in \kappa$ for all i. Let

$$P_j(z_1, \ldots, z_n) = 0, \ j = 1, \ldots, r$$

be a set of defining equations for Z. We say that the point a is *regular* (or *non-singular* or *smooth*) if

$$\operatorname{rank}\left(\frac{\partial P_j}{\partial z_i}(a) \right) = n - m,$$

where m is the dimension of Z, otherwise, is *singular*. We say that Z is *non-singular* or *smooth* if every point on Z is regular. A projective variety is called *non-singular* if all the affine open sets $U_0, \ldots, U_n$ above are non-singular.

Theorem 1.68. *Let φ be a rational mapping from a smooth variety X to a projective variety. Then*

$$\operatorname{codim}(X - \operatorname{dom}(\varphi)) \geq 2.$$

Proof. See Shafarevich [342], II.3, Theorem 3.   □

Let C be a curve, so its function field $\bar{\kappa}(C)$ is of transcendence degree 1. It follows that $\bar{\kappa}(C)$ is algebraic over any subfield $\bar{\kappa}(x)$ generalized by a non-constant function $x \in \bar{\kappa}(C)$. Hence we may write $\bar{\kappa}(C) = \bar{\kappa}(x, y)$, where x and y are non-constant functions on C satisfying an algebraic relation

$$P(x, y) = 0.$$

Let $C_0 \subset \mathbb{A}^2$ denote the affine plane curve defined by P, and let $C_1 \subset \mathbb{P}^2$ be the projective plane curve defined by the homogenized equation

$$Z^{\deg(P)} P\left(\frac{X}{Z}, \frac{Y}{Z}\right) = 0.$$

Clearly, C is birational to both C_0 and C_1. Any curve birational to C is called a *model of* C, so we can say that every curve has a plane affine model and a plane projective model. Theorem 1.68 yields immediately the following result:

Theorem 1.69. *A rational mapping from a smooth curve to a projective variety extends to a morphism defined on the whole curve.*

Theorem 1.70. *Any algebraic curve is birational to a unique (up to isomorphism) smooth projective curve, which is called a normalization of the algebraic curve.*

Proof. See Fulton [109], VII.5, Theorem 3, Hartshorne [148], I, Corollary 6.11, or Hindry and Silverman [158], Theorem A.4.1.4.   □

1.6.4 Differential forms

Let x be a point on a variety X. The *tangent space to X at x* is the $\bar{\kappa}$-vector space

$$T_x(X) = \operatorname{Hom}_{\bar{\kappa}}(\mathbf{m}(x)/\mathbf{m}(x)^2, \bar{\kappa}).$$

In other words, the tangent space is defined to be the dual of the vector space $\mathbf{m}(x)/\mathbf{m}(x)^2$. We naturally call $\mathbf{m}(x)/\mathbf{m}(x)^2$ the *cotangent space to X at x*, denoted by $T_x^*(X)$. It is not difficult to check that $T_x(X)$ and $T_x^*(X)$ are vector spaces over $\bar{\kappa}$ since

$$\mathcal{O}(x)/\mathbf{m}(x) \cong \bar{\kappa}.$$

Theorem 1.71. *Let X be a variety. Then $\dim T_x(X) \geq \dim X$ for all $x \in X$. Furthermore, there is a non-empty open set $U \subset X$ such that $\dim T_x(X) = \dim X$ for $x \in U$.*

Proof. See Hartshorne [148], I.5, Proposition 2A and Theorem 3 or Shafarevich [342], II.1, Theorem 3.   □

According to Jacobian criterion (see [148], I.5), a point x in an affine variety Z is regular if and only if $\dim T_x(Z) = \dim Z$. An *ordinary singularity* in a curve is a singularity whose tangent cone is composed of distinct lines. The *multiplicity* of an ordinary singularity is the number of lines in its tangent cone.

Consider a rational mapping $\varphi : X \longrightarrow Y$ between two varieties that is regular at x, and let $y = \varphi(x)$. According to Hartshorne [148], I.4, Theorem 4, the mapping

$$\varphi^* : \mathcal{O}_Y(y) \longrightarrow \mathcal{O}_X(x), \quad g \mapsto g \circ \varphi$$

is a homomorphism of local rings, in particular,

$$\varphi^*(\mathbf{m}(y)) \subset \mathbf{m}(x), \quad \varphi^*(\mathbf{m}(y)^2) \subset \mathbf{m}(x)^2,$$

and hence it induces a $\bar{\kappa}$-linear mapping

$$\varphi^* : T_y^*(Y) \longrightarrow T_x^*(X).$$

The *tangent mapping* or *differential* of φ at x

$$d\varphi(x) : T_x(X) \longrightarrow T_y(Y)$$

is defined to be the transpose of the mapping φ^*.

Let X be a variety. Take a function $f \in \bar{\kappa}(X)_*$ and fix a point x in the domain of f. We obtain a tangent mapping

$$df(x) : T_x(X) \longrightarrow T_{f(x)}(\mathbb{A}^1) = \bar{\kappa},$$

so $df(x)$ is a linear form on $T_x(X)$, that is, $df(x) \in T_x^*(X)$. Obviously, the classical rules

$$d(f + g) = df + dg, \ d(fg) = f\,dg + g\,df \tag{1.6.2}$$

are valid. Thus we may view df as a mapping that associates to each point $x \in \text{dom}(f)$ a cotangent vector in $T_x^*(X)$. According to Hindry and Silverman [158], such a mapping is called an *abstract differential* 1-*form*. A *regular differential* 1-*form* on X is an abstract differential 1-form ω such that for all $x \in X$ there is a neighborhood U of x and regular functions $f_i, g_i \in \mathcal{O}(U)$ such that

$$\omega(x) = \sum g_i(x)df_i(x), \ x \in U.$$

We denote the set of regular differential 1-forms on X by $\Omega^1[X]$. It is clearly a $\bar{\kappa}$-vector space, and in fact, it is an $\mathcal{O}(X)$-module.

Let x be a non-singular point on a variety X of dimension n. Functions $t_1, \ldots, t_n \in \mathcal{O}(x)$ are called *local parameters* at x if each $t_i \in \mathbf{m}(x)$, and if the images of $t_1, \ldots, t_n$ form a basis of $T_x^*(X)$. The functions $t_1, \ldots, t_n$ give *local coordinates* on X if $u_i := t_i - t_i(x)$ give local parameters at all x in X. It is easy to see that $t_1, \ldots, t_n$ are local parameters if and only if n linear forms $dt_1(x), \ldots, dt_n(x)$

on $T_x(X)$ are linearly independent. Since $\dim T_x(X) = n$, this in turn is equivalent to saying that in $T_x(X)$,

$$\bigcap_i \ker(dt_i(x)) = \{0\}.$$

According to Shafarevich [342], III.5, Theorem 1, any non-singular point x of a variety X has local parameters $t_1, \ldots, t_n$ defined on a neighborhood U of x such that

$$\Omega^1[U] = \bigoplus_{i=1}^n \mathcal{O}(U)dt_i. \tag{1.6.3}$$

The abstract differential 1-forms considered were mappings sending each point $x \in X$ to an element of $T_x^*(X)$. We now consider more general *abstract differential r-forms* that send $x \in X$ to a skew symmetric r-linear form on $T_x(X)$, that is, to an element of the rth exterior product $\bigwedge_r T_x^*(X)$ of $T_x^*(X)$, or equivalently, to a linear mapping $\bigwedge_r T_x(X) \longrightarrow \bar{\kappa}$. A *regular differential r-form* ω on X is an abstract differential r-form such that for all $x \in X$ there is a neighborhood U containing x and functions $f_i, g_{i_1,\ldots,i_r} \in \mathcal{O}(U)$ such that

$$\omega = \sum g_{i_1,\ldots,i_r} df_{i_1} \wedge \cdots \wedge df_{i_r}.$$

We denote the set of regular differential r-forms on X by $\Omega^r[X]$. It is clearly an $\mathcal{O}(X)$-module. The analogue of (1.6.3) is true. If $t_1, \ldots, t_n$ are local coordinates on U, then

$$\Omega^r[U] = \bigoplus_{i_1 < \cdots < i_r} \mathcal{O}(U)dt_{i_1} \wedge \cdots \wedge dt_{i_r}. \tag{1.6.4}$$

We now introduce a new object, consisting of an open set $U \subset X$ and a differential r-form $\omega \in \Omega^r[U]$. On pairs (U, ω) we introduce the equivalence relation $(\omega, U) \sim (\omega', U')$ if $\omega = \omega'$ on $U \cap U'$. Note that the set of points at which a regular differential form is 0 is closed (see Shafarevich [342], III.5.4, Lemma). It is enough to require that $\omega = \omega'$ on some open subset of $U \cap U'$. The transitivity of the equivalence relation follows from this. An equivalence class under this relation is called a *rational differential r-form* on X. We denote the set of rational differential r-forms on X by $\Omega^r(X)$, which is a vector space of dimension $\binom{n}{r}$ over $\bar{\kappa}(X)$ (see Shafarevich [342], III.5.4, Theorem 3). Obviously,

$$\Omega^0(X) = \bar{\kappa}(X).$$

If $t_1, \ldots, t_n$ is a separable transcendence basis of $\bar{\kappa}(X)$, then the forms

$$\{dt_{i_1} \wedge \cdots \wedge dt_{i_r} \mid 1 \leq i_1 < \cdots < i_r \leq n\}$$

form a basis of $\Omega^r(X)$ over $\bar{\kappa}(X)$ (see Shafarevich [342], III.5.4, Theorem 4). Each element $\omega \in \Omega^r(X)$ has a largest open $U \subset X$ such that ω defines a regular r-form on U, called the *domain of regularity* of ω.

Let $\varphi : X \longrightarrow Y$ be a morphism of smooth varieties. Then there is a mapping

$$\varphi^* : \Omega^r[Y] \longrightarrow \Omega^r[X]$$

defined by the formula

$$\varphi^* \left(\sum g_{i_1,\dots,i_r} df_{i_1} \wedge \cdots \wedge df_{i_r} \right) = \sum (g_{i_1,\dots,i_r} \circ \varphi) d(f_{i_1} \circ \varphi) \wedge \cdots \wedge d(f_{i_r} \circ \varphi).$$

1.6.5 Divisors

Next we describe divisors on an algebraic variety X. There are two kinds. The *group of Weil divisors on X* is the free Abelian group generated by the subvarieties of codimension 1 on X. It is denoted by $\mathrm{Div}(X)$. In other words, a Weil divisor can be written as a linear combination

$$D = \sum_i n_i Y_i,$$

where Y_i is a subvariety of codimension 1, and $n_i \in \mathbb{Z}$. The *support of the divisor D* is the union of all those Y_i's for which the multiplicity n_i is non-zero. It is denoted by $\mathrm{supp}(D)$. If all $n_i \geq 0$ then D is called *effective* or *positive*. We write $D \geq 0$ for D effective.

Let Y be a subvariety of codimension 1 of X. For any regular point $x \in X$, Y can be given in a neighborhood $U \subset X$ of x as the zeros of a regular function $g \in \mathcal{O}(U)$. Moreover, any function $f \in \mathcal{O}(x)$ vanishing on $U \cap Y$ is divisible by g. The function g is called a *local defining function*, $g = 0$ is called the *local equation* of Y at x, and is unique, up to multiplication by a function non-zero at x (see Shafarevich [342], II.3, Theorem 1). We recall that $\mathcal{O}_X(Y)$ is the local ring of functions regular in a neighborhood of some point of Y. In particular, if X is non-singular along Y, then $\mathcal{O}_X(Y)$ is a discrete valuation ring. Take $f \in \mathcal{O}_X(Y) - \{0\}$. Let $x \in X$ be a regular point, and g a local defining function for Y near x. Since $f \in \mathcal{O}_X(x)$ and $\mathcal{O}_X(x)$ is a discrete valuation ring, there exist a unit u in $\mathcal{O}_X(x)$ and an non-negative integer d such that

$$f = ug^d.$$

Note that the integer d is independent of the choice of regular points in $X \cap Y$ and will be called the *order of f along Y*, denoted by $\mathrm{ord}_Y(f)$. We can extend ord_Y to $\bar{\kappa}(X)_*$ in the usual way. Its main properties are summarized as follows:

Proposition 1.72. *Fix $f \in \bar{\kappa}(X)_*$. The order function $\mathrm{ord}_Y : \bar{\kappa}(X)_* \longrightarrow \mathbb{Z}$ has the following properties:*

(a) $\operatorname{ord}_Y(fg) = \operatorname{ord}_Y(f) + \operatorname{ord}_Y(g)$ *for all* $g \in \bar{\kappa}(X)_*$.

(b) $\operatorname{ord}_Y(f+g) \geq \min\{\operatorname{ord}_Y(f), \operatorname{ord}_Y(g)\}$ *for all* $g \in \bar{\kappa}(X)_*$ *with* $f + g \neq 0$.

(c) *There are only finitely many Y's with* $\operatorname{ord}_Y(f) \neq 0$.

(d) $\operatorname{ord}_Y(f) \geq 0$ *if and only if* $f \in \mathcal{O}_X(Y)$. *Similarly,* $\operatorname{ord}_Y(f) = 0$ *if and only if f is a unit in* $\mathcal{O}_X(Y)$.

(e) *Assume further that X is projective. Then the following are equivalent:*

  (e1) $\operatorname{ord}_Y(f) \geq 0$ *for all Y.*

  (e2) $\operatorname{ord}_Y(f) = 0$ *for all Y.*

  (e3) $f \in \bar{\kappa}_*$.

Proof. Hindry-Silverman [158], Lemma A.2.1.2 or Shafarevich [342], III.1.1, (2). $\qquad \square$

Let $f \in \bar{\kappa}(X)_*$ be a rational function on X. The *divisor* of f is the divisor

$$(f) = \sum_Y \operatorname{ord}_Y(f) Y.$$

Usually we say that f has a *zero of order d along Y* if $\operatorname{ord}_Y(f) = d > 0$, and that f has a *pole of order d along Y* if $\operatorname{ord}_Y(f) = -d < 0$. A divisor is said to be *principal* if it is the divisor of a function. Two divisors D and D' are said to be *linearly equivalent*, denoted by $D \sim D'$, if their difference is a principal divisor. The *divisor class group of X* is the group of divisor classes modulo linear equivalence. It is denoted by $\operatorname{Cl}(X)$. A divisor class is called *effective* if it contains an effective divisor.

A *Cartier divisor* on a variety X is an (equivalence class of) collections of pairs $\{(U_i, f_i)\}_{i \in I}$ satisfying the following conditions:

(A) The U_i's are Zariski open sets that cover X.

(B) The f_i's are non-zero rational functions $f_i \in \bar{\kappa}(U_i)_* = \bar{\kappa}(X)_*$.

(C) $f_i f_j^{-1} \in \mathcal{O}(U_i \cap U_j)^*$ (i.e., $f_i f_j^{-1}$ has no poles or zeros on $U_i \cap U_j$).

Two collections $\{(U_i, f_i)\}_{i \in I}$ and $\{(V_j, g_j)\}_{j \in J}$ are considered to be equivalent (define the same divisor) if $f_i g_j^{-1} \in \mathcal{O}(U_i \cap V_j)^*$ for all $i \in I$ and $j \in J$. The *support* of a Cartier divisor $\{(U_i, f_i)\}_{i \in I}$ is the set of zeros and poles of the f_i's. A pair (U_i, f_i) is said to *represent the divisor locally*, or on the open set U_i. The Cartier divisor is said to be *effective* if for all representing pairs (U_i, f_i) the rational function f_i is regular at all points of U_i, that is, f_i has no poles on U_i. We then view the Cartier divisor as a hypersurface on X, defined locally on U_i by the equation $f_i = 0$. The Cartier divisors form a group, denoted by $\operatorname{CaDiv}(X)$. Indeed, if Cartier divisors are respectively $\{(U_i, f_i)\}_{i \in I}$ and $\{(V_j, g_j)\}_{j \in J}$, then their sum is

$$\{(U_i, f_i)\}_{i \in I} + \{(V_j, g_j)\}_{j \in J} = \{(U_i \cap V_j, f_i g_j)\}_{(i,j) \in I \times J}.$$

Associated to a function $f \in \bar{\kappa}(X)_*$ is its *principal Cartier divisor*, denoted by

$$\mathrm{div}(f) = \{(X, f)\}.$$

Two divisors are said to be *linearly equivalent* if their difference is a principal Cartier divisor. The group of Cartier divisor classes of X is the group of divisor classes modulo linear equivalence. It is called the *Picard group* of X and is denoted by $\mathrm{Pic}(X)$.

We now compare the two types of divisors. Let Y be an irreducible subvariety of codimension 1 in X, and let D be a Cartier divisor defined by $\{(U_i, f_i)\}_{i \in I}$. We define the order of D along Y, denoted by $\mathrm{ord}_Y(D)$, as follows. Choose one of the open sets U_i such that $U_i \cap Y \neq \emptyset$ and set

$$\mathrm{ord}_Y(D) = \mathrm{ord}_Y(f_i).$$

It is easily seen that $\mathrm{ord}_Y(D)$ is independent of the choice of (U_i, f_i), so that we obtain a map from Cartier divisors to Weil divisors by sending D to $\sum \mathrm{ord}_Y(D)Y$. In general, this mapping is neither surjective nor injective. For example, see Fulton [110], Examples 2.1.2 and 2.1.3 or Hartshorne [148], II.6.11.3.

Theorem 1.73. *If X is a smooth variety, then the natural mappings*

$$\mathrm{CaDiv}(X) \longrightarrow \mathrm{Div}(X), \quad \mathrm{Pic}(X) \longrightarrow \mathrm{Cl}(X)$$

are isomorphisms.

Proof. Hartshorne [148], II.6.11. □

In the sequel we will consider only Cartier divisors when the variety in question might be singular, and we will freely identify Weil and Cartier divisors when we work with smooth varieties.

Let X be a smooth variety of dimension n, and let ω be a non-zero rational differential n-form on X. We cover X by affine open subsets U_i of X with local coordinates $t_1^{(i)}, \ldots, t_n^{(i)}$. In U_i, we can write

$$\omega = g^{(i)} dt_1^{(i)} \wedge \cdots \wedge dt_n^{(i)}.$$

In particular, we have the expression

$$dt_\alpha^{(i)} = \sum_{\beta=1}^{n} h_{\alpha\beta} dt_\beta^{(j)}, \ \alpha = 1, \ldots, n. \tag{1.6.5}$$

Since $dt_1^{(i)}(x), \ldots, dt_n^{(i)}(x)$ form a basis of $T_x^*(X)$ for each $x \in U_i$, it follows from (1.6.5) that the *Jacobian determinant* of the functions $t_1^{(i)}, \ldots, t_n^{(i)}$ with respect to

$t_1^{(j)}, \ldots, t_n^{(j)}$ satisfies

$$\frac{D\left(t_1^{(i)}, \ldots, t_n^{(i)}\right)}{D\left(t_1^{(j)}, \ldots, t_n^{(j)}\right)} := \det(h_{\alpha\beta}) \neq 0.$$

Substituting (1.6.5) in the expression for ω and simple calculations in the exterior algebra shows that on the intersection $U_i \cap U_j$, we get

$$g^{(j)} = g^{(i)} \frac{D\left(t_1^{(i)}, \ldots, t_n^{(i)}\right)}{D\left(t_1^{(j)}, \ldots, t_n^{(j)}\right)}.$$

Since the Jacobian determinant is regular and nowhere zero in $U_i \cap U_j$, the collection of pairs $(U_i, g^{(i)})$ defines a divisor on X. This divisor is called the *divisor* of ω, and is denoted by $\mathrm{div}(\omega)$.

Any other non-zero rational differential n-form ω' on X has the form $\omega' = f\omega$ for some rational function $f \in \bar{\kappa}(X)_*$. It follows that

$$\mathrm{div}(\omega') = \mathrm{div}(\omega) + \mathrm{div}(f),$$

so that the divisor class associated to an n-form is independent of the chosen form. This divisor class is called the *canonical class* of X. By abuse of language, any divisor in the canonical class is called a *canonical divisor* and is denoted by K, as well as its class, or by K_X if we wish to emphasize the dependence on X.

Let $\varphi : X \longrightarrow Y$ be a morphism of varieties, let $D \in \mathrm{CaDiv}(Y)$ be a Cartier divisor defined by $\{(V_j, g_j)\}_{j \in J}$, and assume that $\varphi(X)$ is not contained in the support of D. Then the Cartier divisor $\varphi^*(D) \in \mathrm{CaDiv}(X)$ is the divisor defined by

$$\varphi^*(D) = \{(\varphi^{-1}(V_j), g_j \circ \varphi)\}_{j \in J}.$$

Let $\varphi : X \longrightarrow Y$ be a finite mapping of smooth projective varieties, let Z be an irreducible divisor on X, and let $Z' = \varphi(Z)$ be the image of Z under φ. Note that the dimension theorem (Theorem 1.67) tells us that Z' is an irreducible divisor on Y. Let s_Z be a generator of the maximal ideal of $\mathcal{O}_X(Z)$, and similarly let $s_{Z'}$ be a generator of the maximal ideal of $\mathcal{O}_Y(Z')$, that is, s_Z and $s_{Z'}$ are local equations for Z and Z'. The *ramification index of f along Z* is defined to be the integer

$$e_Z = e_Z(\varphi) = \mathrm{ord}_Z(s_{Z'} \circ \varphi),$$

where we recall that $\mathrm{ord}_Z : \mathcal{O}_X(Z) \longrightarrow \mathbb{Z}$ is the valuation on $\mathcal{O}_X(Z)$. Equivalently,

$$s_{Z'} \circ \varphi = u s_Z^{e_Z}, \quad u \in \mathcal{O}_X(Z)^*.$$

The mapping φ is said to be *ramified along Z* if $e_Z(\varphi) \geq 2$. We have the following *Hurwitz formula*:

Theorem 1.74. *Let $\varphi : X \longrightarrow Y$ be a finite mapping between smooth projective varieties.*

(1) *The mapping φ is ramified only along a finite number of irreducible divisors.*

(2) *If we assume further either that the characteristic of κ is 0 or that the characteristic of κ does not divide any of the ramification indices, then we have the formula*

$$K_X \sim \varphi^*(K_Y) + \sum_Z (e_Z(\varphi) - 1)Z.$$

Proof. Hindry-Silverman [158], Proposition A.2.2.8.   □

1.6.6   Linear systems

Let D be a divisor on a variety X. The *associated vector space* or *Riemann-Roch space* of D is defined to be the subset of rational functions

$$\mathcal{L}(D) = \mathcal{L}(X, D) = \{f \in \bar{\kappa}(X)_* \mid D + (f) \geq 0\} \cup \{0\}. \tag{1.6.6}$$

This set is a vector space over $\bar{\kappa}$ under the usual algebraic operations on functions. Indeed, if $D = \sum n_i Y_i$ then $f \in \bar{\kappa}(X)_*$ belongs to $\mathcal{L}(D)$ if and only if

$$\mathrm{ord}_Y(f) \geq \begin{cases} -n_i, & Y = Y_i; \\ 0, & Y \neq Y_i \text{ for all } i, \end{cases}$$

and because of this, our assertion follows at once from (b) in Proposition 1.72. The dimension of $\mathcal{L}(D)$ is denoted by $\ell(D)$ (which is called the *dimension* of D by some authors).

Theorem 1.75. *Let D be a divisor on a projective variety. Then $\ell(D)$ is finite.*

Proof. See, for example, Hartshorne [148], Theorem III.2, Hindry and Silverman [158], Corollary A.3.2.7, or Shafarevich [342], III.2.3, Theorem 5.   □

We know $\ell(D) = \ell(D')$ if $D \sim D'$ (see [342], III.1.5, Theorem 3). Thus we see that it makes sense to speak of the dimension $\ell(c)$ of a divisor class c, that is, the common dimension of all the divisors of this class. This number has the following meaning. If $D \in c$ and $f \in \mathcal{L}(D)$, then the divisor

$$D_f = D + (f) \in c$$

is effective. Conversely, any effective divisor $D' \in c$ is of the form D_f for some $f \in \mathcal{L}(D)$. Obviously, if X is projective, f is uniquely determined by D_f up to a constant factor. Thus we can set up a one-to-one correspondence between effective divisors in the class c and points of $\mathbb{P}(\mathcal{L}(D)) \cong \mathbb{P}^{\ell(D)-1}$.

The following definition slightly generalizes this construction. A *linear system* L on a variety X is a subset of effective divisors all linearly equivalent to a fixed divisor D and parametrized by a linear subvariety of $\mathbb{P}(\mathcal{L}(D))$. The *dimension* of the linear system L is the dimension of the linear subvariety. The set of *base points* of L is the intersection of the supports of all divisors in L. We will say that L is *base-point free* if this intersection is empty. The set of effective divisors linearly equivalent to D is a linear system, called the *complete linear system* of D. It is denoted by $|D|$. If $|D|$ is base-point free, the divisor D is also said to be *base-point free*.

Let L be a linear system of dimension n parametrized by a projective space $\mathbb{P}(V) \subset \mathbb{P}(\mathcal{L}(D))$, where V is a subspace of $\mathcal{L}(D)$ of dimension $n+1$ over $\bar{\kappa}$. Let B_L be the set of base points of L. When $x \in X - B_L$, the subspace of V

$$V_x = \{f \in V \mid f(x) = 0\}$$

has dimension n. Thus there exists unique element $\varphi_L(x) \in \mathbb{P}(V^*)$ such that

$$E[\varphi_L(x)] = V_x.$$

It is easy to show that if $L \neq \emptyset$, then

$$\varphi_L : X - B_L \longrightarrow \mathbb{P}(V^*)$$

is regular, which further extends a rational mapping

$$\varphi_L : X \longrightarrow \mathbb{P}(V^*) \tag{1.6.7}$$

called the *dual classification mapping*.

Next we explain it clearly. Select a basis $f_0, \dots, f_n$ of V and let $e_0, \dots, e_n$ be the dual basis in V^*. Choose $\tilde{\varphi}_L(x) \in V^* - \{0\}$ such that $\mathbb{P}(\tilde{\varphi}_L(x)) = \varphi_L(x)$. Thus we can write

$$\tilde{\varphi}_L(x) = \sum_{i=0}^{n} \tilde{\varphi}_i(x) e_i.$$

By the definition,

$$E[\varphi_L(x)] = \left\{ \xi = \sum_{i=0}^{n} \xi_i f_i \in V \;\middle|\; \langle \xi, \tilde{\varphi}_L(x) \rangle = \sum_{i=0}^{n} \xi_i \tilde{\varphi}_i(x) = 0 \right\}.$$

Since $E[\varphi_L(x)] = V_x$, then $\xi \in E[\varphi_L(x)]$ means that

$$\xi(x) = \sum_{i=0}^{n} \xi_i f_i(x) = 0,$$

that is, $[f_0(x), \dots, f_n(x)]$ can serve as the homogeneous coordinates of $\varphi_L(x)$. Therefore we can identify

$$\varphi_L = [f_0, \dots, f_n] : X \longrightarrow \mathbb{P}^n. \tag{1.6.8}$$

We will abbreviate

$$\varphi_D = \varphi_{|D|}.$$

A linear system L on a projective variety X is *very ample* if the associated mapping φ_L is an embedding, that is, φ_L is a morphism that maps X isomorphically onto its image $\varphi_L(X)$. A divisor D is said to be *very ample* if the complete linear system $|D|$ is very ample, and to be *ample* if some positive multiple of D is very ample.

Proposition 1.76. *Let X be a projective variety. Given a divisor D and an ample divisor E, there exists a positive integer n such that $D + nE$ is very ample. In particular, every divisor D is linearly equivalent*

$$D \sim E_1 - E_2,$$

where E_1 and E_2 are very ample.

Proof. Lang [232], Proposition 1.1, or Hindry-Silverman [158], Theorem A.3.2.3. $\square$

The *dimension* of a divisor D on a projective variety X is the quantity

$$\dim D = \max_{m \geq 1} \dim \varphi_{mD}(X);$$

that is, it is the maximal dimension of the image of X under the dual classification mapping φ_{mD}, which is also called D-*dimension* of X. If $\mathcal{L}(mD)$ is always empty, then let $\dim D = -1$ by convention (some authors instead prefer to set $\dim D = -\infty$ in this situation). If $\dim D = \dim X$, then we say that D is *pseudo ample*, which means that there exists some positive integer m such that φ_{mD} is an imbedding of some non-empty Zariski open subset of X into a locally closed subset of $\mathbb{P}(\mathcal{L}(mD))$. Usually, $\dim K_X$ is called the *Kodaira dimension* of X. It is a result of Kodaira that:

Theorem 1.77. *On a non-singular projective variety, a divisor D is pseudo ample if and only if there exists some positive integer m such that $mD \sim E + Z$, where E is ample and Z is effective.*

Proof. See [210], Appendix; [415], Proposition 1.2.7; [208], Lemma 7.3.6 and Lemma 7.3.7; or Lemma 2.30. $\square$

A non-singular projective variety X is defined to be *canonical* if the canonical class K_X is ample, *very canonical* if K_X is very ample, and *pseudo canonical* if K_X is pseudo ample. Instead of pseudo canonical, a variety has been called of *general type*. This new notion comes from Lang and Griffiths (cf. [232]). Generally, a projective variety (possibly singular) is called *pseudo canonical* if X is birationally equivalent to a projective non-singular pseudo canonical variety.

If κ has characteristic 0, *resolution of singularities* is known, and due to Hironaka. This means that given X a projective variety, there exists a birational morphism

$$\varphi : \tilde{X} \longrightarrow X$$

such that $\tilde{X}$ is projective and non-singular and φ is an isomorphism over the Zariski open subset of X consisting of the regular points. The non-singular projective variety $\tilde{X}$ is called a *normalization* of X.

An important characterization of a subvariety of an Abelian variety being pseudo canonical was given by Ueno [401] (or see Iitaka [191], [192]):

Theorem 1.78. *Let X be a subvariety of an Abelian variety over an algebraically closed field. Then X is pseudo canonical if and only if the group of translations which preserve X is finite.*

We have quite generally *Ueno's theorem* (see [401], Theorem 3.10):

Theorem 1.79. *Let X be a subvariety of an Abelian variety A, and let B be the connected component of the group of translations preserving X. Then the quotient $\varphi : X \longrightarrow X/B$ is a morphism, whose image is a pseudo canonical subvariety of the Abelian quotient A/B, and whose fibers are translations of B. In particular, if X does not contain any translations of Abelian subvarieties of dimension ≥ 1, then X is pseudo canonical.*

Proof. See Iitaka [193], Theorem 10.13, and Mori [281], Theorem 3.7.   □

The mapping φ is called the *Ueno fibration of X*. Lang [232] formulated the Kawamata theorem [200] into the following *Kawamata's structure theorem*:

Theorem 1.80. *Let X be a pseudo canonical subvariety of an Abelian variety A in characteristic 0. Then there exists a finite number of proper subvarieties Z_i with Ueno fibrations $\varphi_i : Z_i \longrightarrow Y_i$ whose fibers have dimension ≥ 1, such that every translate of an Abelian subvariety of A of dimension ≥ 1 contained in X is actually contained in the union of the subvarieties Z_i.*

The union of the subvarieties Z_i is called the *Ueno-Kawamata fibrations in X* when X is pseudo canonical. Note that the set of Z_i is empty if and only if X does not contain any translations of an Abelian subvariety of dimension ≥ 1.

Let C be a smooth projective curve. A divisor on C is simply a finite formal sum

$$D = \sum n_P P,$$

and we can define the *degree* of D to be

$$\deg(D) = \sum n_P.$$

The following *Riemann-Roch theorem* which allows us to compute the dimension $\ell(D)$ in most cases, is of inestimable value in the study of algebraic curves.

Theorem 1.81. *Let C be a smooth projective curve. There exists an integer $g \geq 0$ such that for all divisors $D \in \mathrm{Div}(C)$,*

$$\ell(D) - \ell(K_C - D) = \deg(D) - g + 1.$$

Proof. See Serre [340], II.9, Théorème 3, Hartshorne [148], IV, Theorem 1.3 or Fulton [109], VIII.6.   □

The Riemann-Roch theorem implies immediately

$$\ell(K_C) = g, \ \deg(K_C) = 2g - 2.$$

Further, if C is of degree n, we also have (cf. [158])

$$g = \frac{(n-1)(n-2)}{2}.$$

The integer g is called the *genus* of smooth projective curve C. When C is not necessarily smooth or projective, its *genus* is defined to be the genus of a normalization of C.

Theorem 1.82. *Let C be a projective plane curve of degree n with only ordinary singularities. Then its genus is given by the formula*

$$g = \frac{(n-1)(n-2)}{2} - \sum_{P \in S} \frac{\delta_P(\delta_P - 1)}{2},$$

where S is the set of singular points and δ_P the multiplicity of C at P.

Proof. See Fulton [109], VIII.3, Proposition 5.   □

We now describe a useful formula, called the *Riemann-Hurwitz formula*, that can frequently be used to compute the genus of a curve.

Theorem 1.83. *Let C be a curve of genus g, let C' be a curve of genus g', and let $\varphi : C \longrightarrow C'$ be a finite separable mapping of degree $d \geq 1$. For each point $P \in C$, write e_P for the ramification index of φ at P, and assume either that $\mathrm{char}(\bar{\kappa}) = 0$ or else that $\mathrm{char}(\bar{\kappa})$ does not divide any of the e_P's. Then*

$$2g - 2 = d(2g' - 2) + \sum_{P \in C} (e_P - 1). \tag{1.6.9}$$

Proof. See Hindry and Silverman [158], Theorem A.4.2.5.   □

Since the number $2 - 2g$ is just the *Euler characteristic* $\chi(C)$ of C, then (1.6.9) also assumes the following form:

$$\chi(C) = d\chi(C') + \sum_{P \in C} (1 - e_P). \tag{1.6.10}$$

The formula (1.6.10) for functions may be regarded as a logarithmic analogue of the formula (1.3.21) for numbers.

1.7   Weil functions

In this section, we discuss a class of functions on varieties, called Weil functions, which have logarithmic singularities on a given divisor, and are parameterized by a proper set of absolute values over a number field. Associated to Weil functions of divisors, proximity functions, valence functions and heights are well defined by the divisors, up to $O(1)$.

Let κ be a number field with a proper set of absolute values M_κ. Take $v \in M_\kappa$. Let κ_v be the completion of κ for v and extend $|\cdot|_v$ to an absolute value on the algebraic closure $\overline{\kappa}_v$. Let D be a Cartier divisor on a variety X, given by a collection $\{(U_i, f_i)\}_{i \in I}$. A *local Weil function* for D relative to v is a function

$$\lambda_{D,v} : X(\overline{\kappa}_v) - \mathrm{supp} D \longrightarrow \mathbb{R}$$

with the following form:

$$\lambda_{D,v}(x) = -\log\left|\|f_i(x)\|\right|_v + h_i(x),$$

where h_i is a continuous function on $U_i(\overline{\kappa}_v)$. We sometimes think of $\lambda_{D,v}$ as a function of $X(\kappa) - \mathrm{supp} D$ or $X(\overline{\kappa}) - \mathrm{supp} D$ by implicitly choosing an embedding $\overline{\kappa} \mapsto \overline{\kappa}_v$.

Let s denote the collection $\{f_i\}_{i \in I}$. Then v induces a *metric* of s if there exists a set of continuous positive functions

$$\rho_{i,v} : U_i(\overline{\kappa}_v) \longrightarrow \mathbb{R}^+$$

satisfying

$$\rho_{i,v} = \left|f_i f_j^{-1}\right|_v \rho_{j,v}$$

on $U_i(\overline{\kappa}_v) \cap U_j(\overline{\kappa}_v)$. Thus we can define

$$|s(x)|_v = \frac{|f_i(x)|_v}{\rho_{i,v}(x)}, \quad x \in U_i(\overline{\kappa}_v)$$

so that

$$\lambda_{D,v}(x) = -\log\left|\|s(x)\|\right|_v$$

is a local Weil function for D at v.

We define an M_κ-*constant* γ to be a real-valued function

$$\gamma : M_\kappa \longrightarrow \mathbb{R}$$

such that $\gamma(v) = 0$ for almost all $v \in M_\kappa$ (all but a finite number of v in M_κ). If w is an extension of an element v in M_κ to the algebraic closure $\overline{\kappa}$, then we define

$$\gamma(w) = \gamma(v).$$

Thus γ is extended to a function of $M_{\overline{\kappa}}$ into $\mathbb{R}$.

Let X be a variety defined over κ. A subset A of $X(\overline{\kappa}) \times M_{\overline{\kappa}}$ is said to be *affine bounded* if there exists a coordinated affine open subset U of $X(\overline{\kappa})$ with coordinates $(x_1, \ldots, x_m)$ and an $M_{\overline{\kappa}}$-constant γ such that for all $(x, v) \in E$ we have

$$\max_i |x_i|_v \leq e^{\gamma(v)}.$$

If there is only one absolute value and κ is algebraically closed, this notion coincides with the notion of a bounded set of points on an affine variety. The subset E is called *bounded* if it is contained in the finite union of affine bounded subsets. In particular, if X is a projective variety, then $X(\overline{\kappa}) \times M_{\overline{\kappa}}$ is bounded (see [225]).

A function

$$h : X(\overline{\kappa}) \times M_{\overline{\kappa}} \longrightarrow \mathbb{R}$$

is called *bounded from above* if there exists an M_κ-constant γ such that

$$h(x, v) \leq \gamma(v), \ (x, v) \in X(\overline{\kappa}) \times M_{\overline{\kappa}}.$$

We define similarly *bounded from below* and *bounded*. We say that h is *locally bounded* if it is bounded on every bounded subset of $X(\overline{\kappa}) \times M_{\overline{\kappa}}$; and define *locally bounded from above* or *below* similarly. The function h is called *continuous* if for each $v \in M_{\overline{\kappa}}$ the function

$$x \mapsto h(x, v) = h_v(x)$$

is continuous on $X(\overline{\kappa})$.

Let D be a divisor on X. According to Lang [225], by a (*global*) *Weil function* associated with D we mean a function

$$\lambda_D : (X(\overline{\kappa}) - \operatorname{supp}D) \times M_{\overline{\kappa}} \longrightarrow \mathbb{R}$$

having the following property. Let (U, f) be a pair representing D. Then there exists a locally bounded continuous function

$$h : U(\overline{\kappa}) \times M_{\overline{\kappa}} \longrightarrow \mathbb{R}$$

such that for any point in $U(\overline{\kappa}) - \operatorname{supp}D$ we have

$$\lambda_D(x, v) = -\log |||f(x)|||_v + h(x, v).$$

The function h is then uniquely determined by λ_D and the pair (U, f).

We sometimes think of λ_D as a function over κ, that is, λ_D is defined on $(X(\kappa) - \operatorname{supp}D) \times M_\kappa$. If E is a finite extension of the number field κ and λ_D is a Weil function for D over κ, then

$$\lambda_D(x, w) = \frac{[E_w : \kappa_v]}{[E : \kappa]}\lambda_D(x, v).$$

is a Weil function for D over E. Thus, if $x \in X(\kappa)$, then

$$\lambda_D(x, v) = \sum_{w|v} \lambda_D(x, w) + O(1),$$

where $O(1)$ means a *bounded function* of x.

If f is a rational function on X, a Weil function associated with the principal divisor (f) is given by

$$\lambda_f(x, v) = -\log |||f(x)|||_v.$$

Proposition 1.84. *Weil functions satisfy the following properties:*

(a) *If λ_D and $\lambda_{D'}$ are Weil functions for D and D', then $\lambda_D + \lambda_{D'}$ is a Weil function for $D + D'$ and $-\lambda_D$ is a Weil function for $-D$.*

(b) *Assume that $X(\overline{\kappa}) \times M_{\overline{\kappa}}$ is bounded. If D is an effective divisor, then its Weil functions are bounded from below.*

(c) *Assume that X is projective. If λ, λ' are Weil functions with the same divisor, then $\lambda - \lambda'$ is bounded.*

(d) *Let $\varphi : X' \longrightarrow X$ be a morphism of varieties and let D be a divisor on X not containing the image of φ. If λ_D is a Weil function for D on X, then $\lambda_D \circ \varphi$ is a Weil function for $\varphi^* D$ on X'.*

(e) *Let $D_1, \ldots, D_r$ and D be divisors on X such that $D_i - D$ are effective divisors with no points in common. Then*

$$\lambda_D = \inf_i \lambda_{D_i}$$

is a Weil function for D.

See [225] for the proof of Proposition 1.84. The existence of a Weil function associated with a given divisor on a projective variety is also due to S. Lang [225]:

Theorem 1.85. *Let X be a projective variety. Let D be a divisor on X. Then there exists a Weil function having this divisor.*

Let S be a finite set of places containing S_{κ}^{∞}. Let $\lambda = \lambda_D$ be a Weil function for a divisor D on X. For any point $x \in X(\overline{\kappa})$ not in the support of D, the *proximity function* for λ is defined by

$$m_\lambda(x) = m_{\lambda,S}(x) = \sum_{v \in S} \lambda(x, v).$$

Then

$$N_\lambda(x) = N_{\lambda,S}(x) = \sum_{v \in M_\kappa - S} \lambda(x, v)$$

serves as the *valence function* for λ. The associated *height* is defined by

$$h_\lambda(x) = \sum_{v \in M_\kappa} \lambda(x, v) = m_\lambda(x) + N_\lambda(x).$$

A Weil function λ' for another divisor D' on X is said to be *linearly equivalent* to λ if there exists a rational function f such that

$$\lambda' - \lambda = \lambda_f + \gamma,$$

where γ is an M_κ-constant. Thus if λ' is linearly equivalent to λ, by the product formula we have

$$h_{\lambda'}(x) = h_\lambda(x) + O(1), \quad x \notin \mathrm{supp}D \cup \mathrm{supp}D'.$$

Now we can extend the definition of h_λ to $\mathrm{supp}D$ as follows. For each point $x \in \mathrm{supp}D$, there exists a rational function f such that x does not lie in the support of $D - (f)$. Put $\lambda' = \lambda - \lambda_f$. We then define

$$h_\lambda(x) = h_{\lambda'}(x).$$

This value is independent of the choice of f.

Suppose that X is projective. If λ, λ' are Weil functions with the same divisor D, Proposition 1.84, (c) implies that there exists a positive M_κ-constant γ such that

$$|\lambda'(x, v) - \lambda(x, v)| \leq \gamma(v)$$

for all $x \in X(\overline{\kappa})$ and all $v \in M_\kappa$. Then

$$|m_{\lambda'} - m_\lambda| \leq \sum_{v \in S} \gamma(v),$$

$$|N_{\lambda'} - N_\lambda| \leq \sum_{v \in M_\kappa - S} \gamma(v),$$

and hence

$$|h_{\lambda'} - h_\lambda| \leq \sum_{v \in M_\kappa} \gamma(v).$$

Thus the functions $m_\lambda(x)$, $N_\lambda(x)$ and $h_\lambda(x)$ are well determined by the divisor D, up to $O(1)$, and will be denoted by $m(x, D)$, $N(x, D)$ and $h(x, D)$, respectively. Further, if D is an effective divisor, by Proposition 1.84 (b), we may assume

$$m(x, D) \geq 0, \quad N(x, D) \geq 0, \quad h(x, D) \geq 0$$

by using a proper Weil function of D.

Let D be an effective divisor on X and let $\mathcal{R}$ be a subset of $X(\bar{\kappa}) - \mathrm{supp}D$. Then $\mathcal{R}$ is a set of (S, D)-*integralizable points* if there exists a global Weil function λ_D and a M_κ-constant γ such that for all $x \in \mathcal{R}$, all $v \in M_\kappa - S$, and all embeddings of $\bar{\kappa}$ in $\bar{\kappa}_v$,

$$\lambda_D(x, v) \le \gamma(v).$$

As easy consequences of the properties of Weil functions, one finds (cf. [415], p.11): If K is a finite extension field of κ, and if T is the set of places of K lying over places in S, then $\mathcal{R} \subset X(\bar{\kappa})$ is a set of (S, D)-integralizable points if and only if it is a set of (T, D)-integralizable points.

1.8   Heights in number fields

Based on the product formula in Section 1.4, one defines new heights on projective spaces defined over number fields, and further defines heights on varieties associated to divisors, which will be compared with the heights defined by using Weil functions of the divisors. The corresponding first main theorems can also be exhibited.

We assume that κ is a number field. Let M_κ be a proper set of absolute values on κ with multiplicities n_v. Let $V_{\bar{\kappa}}$ be a vector space of finite dimension $n + 1 > 0$ over $\bar{\kappa}$. Set $V = V_\kappa$ and take $\xi \in V_*$. Then $|\xi|_v = 1$ for all but finitely many $v \in M_\kappa$. We can define the *height* of ξ by

$$H_\kappa(\xi) = \prod_{v \in M_\kappa} |\xi|_v^{n_v}.$$

Write $\xi = \xi_0 e_0 + \cdots + \xi_n e_n$ for a fixed basis $e = (e_0, \ldots, e_n)$ of V. If, e.g., $\xi_0 \neq 0$, then $|\xi|_v \ge |\xi_0|_v$ for each v, which implies $H_\kappa(\xi) \ge 1$. Also $|\lambda\xi|_v = |\lambda|_v |\xi|_v$, so

$$H_\kappa(\lambda\xi) = H_\kappa(\xi), \quad \lambda \in \kappa_*$$

by the product formula.

If we have a tower of finite extensions $\mathbb{Q} \subset \kappa \subset K$ and if $\xi \in V_*$ is defined over κ, then

$$H_K(\xi) = \prod_{w \in M_K} |\xi|_w^{n_w} = \prod_{v \in M_\kappa} \prod_{w \in M_K, w|v} |\xi|_v^{n_w}.$$

By using (1.4.3), we have

$$H_K(\xi) = \prod_{v \in M_\kappa} |\xi|_v^{n_v [K:\kappa]} = H_\kappa(\xi)^{[K:\kappa]},$$

and so

$$H_K(\xi)^{\frac{1}{[K:\mathbb{Q}]}} = H_\kappa(\xi)^{\frac{[K:\kappa]}{[K:\mathbb{Q}]}} = H_\kappa(\xi)^{\frac{1}{[\kappa:\mathbb{Q}]}}.$$

The *absolute height* is defined by

$$H(\xi) = H_\kappa(\xi)^{\frac{1}{[\kappa:\mathbb{Q}]}},$$

which does not depend on finite field extensions of $\mathbb{Q}$, that is, we obtain the function

$$H : V_{\bar{\mathbb{Q}}} \longrightarrow \mathbb{R}[1, +\infty).$$

We often use the *absolute (logarithmic) height* $h(\xi)$ which is defined by

$$h(\xi) = \log H(\xi) = \frac{1}{[\kappa : \mathbb{Q}]} \log H_\kappa(\xi).$$

Definition 1.86. *Two heights H_1 and H_2 (resp. logarithmic heights h_1 and h_2) are called equivalent if*

$$cH_1 < H_2 < c'H_1 \ (resp. \ h_2 = h_1 + O(1))$$

holds for some positive constants c and c'.

Hence if the base of V is changed, we obtain a equivalent height. Take $x \in \mathbb{P}(V)$. Then there exists $\xi \in V_*$ such that $x = \mathbb{P}(\xi)$. The *relative (multiplicative) height* of x is defined by

$$H_\kappa(x) = H_\kappa(\xi) = \prod_{v \in M_\kappa} \|\xi\|_v.$$

Similarly, the *absolute height*

$$H(x) = H(\xi) = H_\kappa(\xi)^{\frac{1}{[\kappa:\mathbb{Q}]}}$$

and the *absolute (logarithmic) height* of x

$$h(x) = h(\xi) = \frac{1}{[\kappa : \mathbb{Q}]} \sum_{v \in M_\kappa} \log \|\xi\|_v = \sum_{v \in M_\kappa} \log \||\xi\||_v,$$

are defined respectively. By the product formula, this does not depend on the choice of ξ.

Let σ be an isomorphism of κ over $\mathbb{Q}$ (i.e., leaving $\mathbb{Q}$ fixed). Let x be a point as above, with coordinates $(\xi_0, \ldots, \xi_n)$ rational over κ. Then we can define the point $\sigma(x)$, rational over $\sigma(\kappa)$, and having coordinates $(\sigma(\xi_0), \ldots, \sigma(\xi_n))$. By transport of structure, we get immediately

$$H_\kappa(x) = H_{\sigma(\kappa)}(\sigma(x))$$

whence in particular,

$$H(x) = H(\sigma(x)).$$

Proposition 1.87 ([225]). *Take $a \in \mathbb{P}(V_{\bar{\kappa}}^*)$, and let λ be a Weil function whose divisor is the hyperplane $\ddot{E}[a]$. Then $h_\lambda - h$ is bounded on $\mathbb{P}(V_{\bar{\kappa}})$.*

Proof. Let $H_0, \ldots, H_n$ be the hyperplane corresponding to the coordinate functions. There exist rational functions f_i such that

$$(f_i) = H_i - \ddot{E}[a].$$

For any point $x \notin \ddot{E}[a]$, it is easy to obtain

$$h(x) = \sum_{v \in M_\kappa} \sup_i \log \||f_i(x)\||_v + O(1),$$

or equivalently,

$$h(x) = -\sum_{v \in M_\kappa} \inf_i \lambda_{f_i}(x, v) + O(1),$$

where λ_{f_i} is a Weil function associated with the principal divisor (f_i). We conclude the proof by applying Proposition 1.84 (a), (c) and (e). $\qquad \square$

In particular, we define the *height* of an element $x \in \kappa$ to be the height of the point $[1, x]$ in $\mathbb{P}^1(\kappa) = \mathbb{P}(\kappa^2)$, so that we have

$$H_\kappa(x) = \prod_{v \in M_\kappa} |[1, x]|_v^{n_v} = \left(\prod_{v \in S_\kappa^\infty} \left(\sqrt{1 + |x|_v^2}\right)^{n_v}\right)\left(\prod_{v \in M_\kappa - S_\kappa^\infty} (\max\{1, |x|_v\})^{n_v}\right),$$

and we see that if $x \neq 0$, then

$$H_\kappa(x) = H_\kappa(x^{-1}).$$

To really see what the logarithmic height is telling you it is perhaps best to look at the simplest example: let $x = a/b \in \mathbb{Q}$ denote a rational number in lowest terms. It is then easy to see that

$$h(x) = \frac{1}{2}\log(a^2 + b^2).$$

It is then clear that there are only finitely many rational numbers with bounded height. More generally, any point $x \in \mathbb{P}(\mathbb{Q}^{n+1})$ has a set of coordinates $(\xi_0, \ldots, \xi_n)$ which are relatively prime integers, and we then see that

$$H_\mathbb{Q}(x) = \sqrt{\xi_0^2 + \cdots + \xi_n^2}.$$

In particular, the set of points in $\mathbb{P}(\mathbb{Q}^{n+1})$ of height $\leq$ a fixed number is finite. Such a fact is also true in a number field (see Theorem 1.102).

Let S be a finite set of places containing S_κ^∞. Take $a \in \mathbb{P}(V_{\bar{\kappa}}^*)$. Abbreviate the *proximity function* and the *valence function* for hyperplane $\ddot{E}[a]$,

$$m(x, a) = m_S(x, a) = m\left(x, \ddot{E}[a]\right),$$

$$N(x, a) = N_S(x, a) = N\left(x, \ddot{E}[a]\right).$$

By Proposition 1.87, one obtains the *first main theorem*:

$$m(x, a) + N(x, a) = h(x) + O(1), \quad x \notin \ddot{E}[a], \tag{1.8.1}$$

and therefore,

$$m(x, a) \le h(x) + O(1), \quad x \notin \ddot{E}[a].$$

Proposition 1.88. *For $x \in \mathbb{P}(V_{\bar{\kappa}}) - \ddot{E}[a]$, the proximity function $m(x, a)$ is given by*

$$m(x, a) = \sum_{v \in S} \log \frac{1}{|||x, a|||_v} + O(1).$$

Proof. By Theorem 1.85, there exists a Weil function λ having the divisor $\ddot{E}[a]$. Define

$$\lambda'(x, v) = \begin{cases} \log \frac{1}{|||x,a|||_v} & : v \in S, \\ \lambda(x, v) & : v \notin S. \end{cases}$$

It is easy to check that λ' is a Weil function for $\ddot{E}[a]$, and so we complete the proof according to the discussion in Section 1.7. $\square$

Note that

$$\sum_{v \in M_\kappa} \log \frac{1}{|||x, a|||_v} = h(x) + h(a), \quad x \notin \ddot{E}[a]. \tag{1.8.2}$$

Thus (1.8.1), (1.8.2) and Proposition 1.88 imply

$$N(x, a) = \sum_{v \in M_\kappa - S} \log \frac{1}{|||x, a|||_v} + O(1). \tag{1.8.3}$$

Let X be a projective variety defined over κ and let D be a very ample divisor on X. Let $\varphi_D : X \longrightarrow \mathbb{P}(V^*)$ be the associated dual classification mapping, where $V = \mathcal{L}(D)$. Then the *absolute (multiplicative) height* of $x \in X$ for D is defined by

$$H_D(x) = H(\varphi_D(x)),$$

and the *absolute (logarithmic) height* of x for D is defined as

$$h_D(x) = h(\varphi_D(x)) \ge 0.$$

Up to equivalence, the height depends only on D.

Theorem 1.89 ([225]). *Let X be a projective variety over κ and let λ be a Weil function of a very ample divisor D on X. Then $h_\lambda - h_D$ is bounded on $X(\overline{\kappa})$.*

Proof. Note that there exists $a \in \mathbb{P}(V)$ such that $D = \varphi_D^* \ddot{E}[a]$. Let λ' be a Weil function of $\ddot{E}[a]$. By Proposition 1.84, (d), $\lambda' \circ \varphi_D$ is a Weil function for D. Proposition 1.87 implies that

$$h_{\lambda' \circ \varphi_D}(x) - h_D(x) = h_{\lambda'}(\varphi_D(x)) - h(\varphi_D(x))$$

is bounded on $X(\overline{\kappa})$. Therefore Theorem 1.89 follows from Proposition 1.84, (c). □

Take $s \in V$ with $(s) = D$ and set $a = \mathbb{P}(s)$. For $x \notin D$, the *proximity function* $m(x, D)$ and the *valence function* $N(x, D)$ are given respectively by

$$m(x, D) = m(\varphi_D(x), a) = \sum_{v \in S} \log \frac{1}{|||\varphi_D(x), a|||_v},$$

and

$$N(x, D) = N(\varphi_D(x), a) = \sum_{v \in M_\kappa - S} \log \frac{1}{|||\varphi_D(x), a|||_v}.$$

By Theorem 1.89, one obtains the *first main theorem*:

$$m(x, D) + N(x, D) = h_D(x) + O(1). \tag{1.8.4}$$

Lemma 1.90. *If D and D' are two very ample divisors on X, then*

$$m(x, D + D') = m(x, D) + m(x, D'),$$

$$N(x, D + D') = N(x, D) + N(x, D').$$

Proof. Set $V' = \mathcal{L}(D')$. Take $s \in V, s' \in V'$ with $(s) = D, (s') = D'$ and set $a = \mathbb{P}(s), a' = \mathbb{P}(s')$. Then $(s \otimes s') = D + D'$, and hence

$$|\varphi_{D+D'}(x), b|_v = \frac{|(s \otimes s')(x)|_v}{|s \otimes s'|_v} = \frac{|s(x)|_v |s'(x)|_v}{|s|_v |s'|_v} = |\varphi_D(x), a|_v |\varphi_{D'}(x), a'|_v,$$

where $b = \mathbb{P}(s \otimes s') \in \mathbb{P}(V \otimes V')$. Hence the lemma follows. □

Furthermore, the formula (1.8.4) and Lemma 1.90 imply

Lemma 1.91 ([225]). *If D and D' are two very ample divisors on X, then*

$$h_{D+D'} = h_D + h_{D'} + O(1).$$

Now, given any divisor D, we can write $D = E - E'$ where E and E' are very ample, and define

$$h_D = h_E - h_{E'}.$$

This definition depends on the choices of E and E', but by Lemma 1.91, h_D is well defined up to equivalence. If λ is a Weil function for D, Proposition 1.84, (a) and Theorem 1.89 imply

$$h_\lambda = h_D + O(1).$$

Further we have

$$m(x, D) = m(x, E) - m(x, E'),$$

and

$$N(x, D) = N(x, E) - N(x, E').$$

Hence the *first main theorem*

$$m(x, D) + N(x, D) = h_D(x) + O(1) \tag{1.8.5}$$

holds.

Let D be a divisor on a non-singular projective variety X over κ. let λ be a Weil function of the divisor D. Theorem 1.89 and Proposition 1.84, (a) imply that $h_\lambda - h_D$ is bounded on $X(\overline{\kappa})$. If D is effective, by Proposition 1.84, (b), we have $h_D \geq -O(1)$, in other words we can choose h_D in its equivalence class such that $h_D \geq 0$. If D is ample, by the definition there exists $m \in \mathbb{Z}^+$ such that mD is very ample, and so

$$0 \leq h_{mD} = mh_D + O(1).$$

Without loss of generality we may assume $h_D \geq 0$. If D is pseudo ample, by Theorem 1.77 there exists some positive integer m such that $mD \sim E + Z$, where E is ample and Z is effective. Hence

$$h_{mD} = h_E + h_Z + O(1).$$

Thus we can choose h_D in its equivalence class such that $h_D \geq 0$.

1.9   Functorial properties of heights

By using heights, one can describe completely growth of morphisms between projective spaces (cf. Theorem 1.95). An interesting question due to Lehmer will be introduced.

Let κ be a number field with a proper set of absolute values M_κ satisfying the product formula with multiplicities n_v. Let X be a variety defined over κ. Let $\varphi : X \longrightarrow \mathbb{P}^n$ be a morphism of X into a projective space, defined over κ. Then for each point x of X, if $\varphi(x)$ is a point of $\mathbb{P}^n(\kappa)$, rational over κ, we can thus define its *relative height*

$$H_{\kappa,\varphi}(x) = H_\kappa(\varphi(x)).$$

If x is algebraic over $\mathbb{Q}$, then there exists a finite extension κ of $\mathbb{Q}$ over which it is rational, and hence the relative height $H_{\kappa,\varphi}(x)$ can be defined too.

We can then define its *absolute height*

$$H_\varphi(x) = H(\varphi(x)) = H_\kappa(\varphi(x))^{\frac{1}{[\kappa:\mathbb{Q}]}}$$

and the *absolute (logarithmic) height*

$$h_\varphi(x) = \log H_\varphi(x).$$

Thus $H_{\kappa,\varphi}$ is a function on $X(\kappa)$ while H_φ is a function on $X(\bar{\kappa}) = X(\bar{\mathbb{Q}})$.

Let X be a projective variety defined over κ. Suppose its points are represented by homogeneous coordinates $[x_0, \ldots, x_m]$. Let (a_{ij}) $(i = 0, \ldots, n,\ j = 0, \ldots, m)$ be a matrix with coefficients in κ, and put

$$y_i = a_{i0}x_0 + \cdots + a_{im}x_m.$$

Then the mapping

$$[x_0, \ldots, x_m] \longmapsto [y_0, \ldots, y_n]$$

defines a rational mapping $\varphi : X \longrightarrow \mathbb{P}^n$. If $x \in X$ is a point with homogeneous coordinates $[x_0, \ldots, x_m]$ such that not all y_i are equal to 0 in the above formula, then φ is defined at x. A mapping φ obtained in the manner just described is called a *linear projection*, defined over κ.

Lemma 1.92 ([225]). *Let X be a projective variety defined over κ. Let $\varphi : X \longrightarrow \mathbb{P}^n$ be a linear projection defined over κ. There exists a number $c > 0$ depending only on φ such that if x is a point of X, algebraic over κ, such that not all the coordinates y_i above are 0, then*

$$H_\varphi(x) \le cH(x), \tag{1.9.1}$$

where $H(x)$ is the height in the given projective embedding of X.

Proof. Let S_κ be the subset of M_κ containing all those absolute values v for which some $|a_{ij}|_v$ is not 1, and all Archimedean absolute values. Then S_κ is a finite set. If our coordinates are in a finite extension K of κ, then for any $w \in M_K$ extending some $v \in M_\kappa$, one has

$$\max_i |y_i|_w \le c_v \max_j |x_j|_w,$$

where

$$c_v = \begin{cases} (m+1)\max_{i,j}\{1, |a_{ij}|_v\}, & v \in S_\kappa; \\ 1, & v \notin S_\kappa. \end{cases}$$

Hence the inequality (1.9.1) in which

$$c = \left(\prod_{v \in M_\kappa} \prod_{w|v} (\varsigma_{v,n+1} c_v)^{n_w} \right)^{\frac{1}{[K:\mathbb{Q}]}}$$

follows easily. $\qquad\qquad\qquad\qquad\qquad\qquad\qquad\qquad\qquad\qquad\qquad\qquad\quad$ $\square$

Lemma 1.93 ([225]). *Let $f_0, \ldots, f_n$ be homogeneous polynomials of degree d with coefficients in κ, and in $m + 1$ variables $X_0, \ldots, X_m$. Let A be the set of points $x = [x_0, \ldots, x_m]$ in projective space $\mathbb{P}^m(\overline{\kappa})$ such that not all polynomials $f_i(x)$ vanish, $i = 0, \ldots, n$. Let $f : A \longrightarrow \mathbb{P}^n(\overline{\kappa})$ be the morphism defined by $x \mapsto [f_0(x), \ldots, f_n(x)]$. Then*

$$h_f(x) \le dh(x) + c$$

for some constant c independent of $x \in A$.

Proof. Trivial estimates using the triangle inequality show that for any point $x \in \mathbb{P}^m(\kappa)$, $x \in A$ we have

$$H_\kappa(f(x)) \le C^{[\kappa:\mathbb{Q}]} H_\kappa(x)^d.$$

Taking the $[\kappa : \mathbb{Q}]$ root and the logarithm yield the lemma. $\qquad\qquad\qquad$ $\square$

For the next discussion in this section, we will need *Hilbert's Nullstellensatz*:

Theorem 1.94. *Take polynomials $P_1, \ldots, P_r$ and P in $\overline{\kappa}[X_0, \ldots, X_m]$. If P vanishes at all the common zeros of $P_1, \ldots, P_r$, then there exist polynomials $Q_1, \ldots, Q_r$ in $\overline{\kappa}[X_0, \ldots, X_m]$ such that*

$$P^s = Q_1 P_1 + \cdots + Q_r P_r$$

holds for some natural number s.

Proof. Van der Waerden [409] or Lang [234] or Atiyah-Macdonald [7], p. 85 or Zariski-Samuel [448], vol. 2, p. 164. $\qquad\qquad\qquad\qquad\qquad\qquad\qquad\qquad\qquad$ $\square$

Theorem 1.95 ([225], [362]). *Let $f : \mathbb{P}^m \longrightarrow \mathbb{P}^n$ be a morphism of degree d, defined over κ. Then*

$$h_f = dh + O(1).$$

Proof. One inequality was proved in Lemma 1.93. Next we prove the inequality in the other direction. It is well known that the morphism $f : \mathbb{P}^m \longrightarrow \mathbb{P}^n$ is given in homogeneous coordinates by $[f_0, \ldots, f_n]$, where the polynomials $f_0, \ldots, f_n$ in the variables $X_0, \ldots, X_m$ have no common zero except the origin. By Theorem 1.94, there exist polynomials $b_{ij} \in \overline{\kappa}[X_0, \ldots, X_m]$ and a non-negative integer l such that

$$X_i^{d+l} = \sum_{j=0}^n b_{ij} f_j.$$

Disregarding the monomials in b_{ij} of degree $\neq l$, we can assume without loss of generality that b_{ij} is homogeneous of degree l. Extending κ if necessary, we may assume that the b_{ij} and f_j have coefficients in κ.

It is also convenient to clear denominators, so we pick an element a in κ, integral at all valuations of M_κ such that a is a denominator for all the coefficients of the polynomials b_{ij}. Multiplying by a, we may assume without loss of generality that we have the equation

$$aX_i^{d+l} = \sum_{j=0}^{n} b_{ij} f_j,$$

where the coefficients of b_{ij} are integral in κ. Take

$$x = [x_0, \dots, x_m] \in \mathbb{P}^m(\kappa)$$

with x_i integral in κ. Take $\alpha \in \{0, \dots, m\}$ such that

$$|x_\alpha|_v = \max_i |x_i|_v.$$

If $v \in M_\kappa$ is non-Archimedean, then

$$|a|_v |x_\alpha|_v^{d+l} \leq |x_\alpha|_v^l \max_j |f_j(x)|_v,$$

whence

$$|a|_v^{n_v} \max_i |x_i|_v^{(d+l)n_v} \leq \max_i |x_i|_v^{ln_v} \max_j |f_j(x)|_v^{n_v}.$$

If $v \in M_\kappa$ is Archimedean, then

$$|a|_v |x_\alpha|_v^{d+l} \leq (n+1)\binom{m+l}{l} C_\alpha |x_\alpha|_v^l \max_j |f_j(x)|_v,$$

where C_α is a constant giving the bound for the coefficients of the polynomials $b_{\alpha j}$ $(j = 0, \dots, n)$. We then obtain

$$|a|_v^{n_v} \max_i |x_i|_v^{(d+l)n_v} \leq C^{n_v} \max_i |x_i|_v^{ln_v} \max_j |f_j(x)|_v^{n_v}.$$

Taking the product yields

$$H_\kappa(x)^{d+l} \leq \left\{ (m+1)^{(d+l)/2} C \right\}^{[\kappa:\mathbb{Q}]} H_\kappa(x)^l H_\kappa(f(x)).$$

This proves the theorem. □

Corollary 1.96. *Let P and Q be two coprime polynomials in $\bar{\kappa}[X]$. Then we have*

$$h\left(\frac{P(x)}{Q(x)}\right) = \max\{\deg(P), \deg(Q)\}h(x) + O(1). \tag{1.9.2}$$

In many references, the *height* of an element $x \in \kappa$ is defined by

$$\bar{H}_\kappa(x) = \prod_{v \in M_\kappa} \max\{1, |x|_v^{n_v}\},$$

so that we see that if $x \neq 0$, then

$$\bar{H}_\kappa(x) = \bar{H}_\kappa(x^{-1}).$$

Furthermore, we have trivially

$$\bar{H}_\kappa(x_1 \cdots x_n) \leq \bar{H}_\kappa(x_1) \cdots \bar{H}_\kappa(x_n)$$

and for $x \in \kappa$,

$$\bar{H}_\kappa(x^n) = \bar{H}_\kappa(x)^n.$$

The *absolute logarithmic height* of an element $x \in \kappa$ is often defined by

$$\bar{h}(x) = \frac{1}{[\kappa : \mathbb{Q}]} \log \bar{H}_\kappa(x) = \frac{1}{[\kappa : \mathbb{Q}]} \sum_{v \in M_\kappa} \max\{0, \log |x|_v^{n_v}\}.$$

Obviously, we have

$$\bar{h} \leq h \leq \bar{h} + \frac{1}{2} \log 2.$$

For $x \neq 0$, $\bar{h}(x) = 0$ if and only if x is a root of unity (see [225]).

Suppose that α is algebraic of degree d over the rational numbers, and let

$$f(X) = a_d X^d + a_{d-1} X^{d-1} + \cdots + a_0 = 0, \ a_d > 0,$$

be its irreducible equation, with coefficients $a_i \in \mathbb{Z}$, and $\gcd(a_0, \ldots, a_d) = 1$. If $\kappa = \mathbb{Q}(\alpha)$, then one has the formula

$$\bar{H}_\kappa(\alpha) = a_d \prod_{i=1}^{d} \max\{1, |\alpha_i|\},$$

where $|\cdot|$ is the complex absolute value, and $\alpha_1, \ldots, \alpha_d$ are the distinct conjugates of α in $\mathbb{C}$. It is an old conjecture of Lehmer [240] that when α is of degree d over $\mathbb{Q}$, and is not 0 or a root of unity, then

$$\bar{h}(\alpha) \geq \frac{\log \alpha_0}{d} \tag{1.9.3}$$

where $\alpha_0 = 1.1762808 \cdots$ is the larger real root of the 10th-degree polynomial

$$x^{10} + x^9 - x^7 - x^6 - x^5 - x^4 - x^3 + x + 1.$$

Note that

$$\bar{h}\left(2^{1/d}\right) = \frac{\log 2}{d}.$$

The example shows that (1.9.3) would be best possible on the order of d. The best result in this direction is due to Dobrowolski [81] and says that if $d \geq 3$, then

$$\bar{h}(\alpha) > \frac{c}{d}\left(\frac{\log\log d}{\log d}\right)^3 \tag{1.9.4}$$

with an absolute constant $c > 0$.

In contrast, there is the following result of Zhang [449]: Suppose α is algebraic but not 0, 1, $(1 \pm \sqrt{-3})/2$. Then

$$\bar{h}(\alpha) + \bar{h}(1 - \alpha) \geq c > 0 \tag{1.9.5}$$

with an absolute constant $c > 0$. Zagier [445] gave a more natural proof and determined the best value of the constant

$$c = \frac{1}{2}\log\frac{1 + \sqrt{5}}{2}.$$

1.10 Gauss' lemma

By using absolute height h on a projective space $\mathbb{P}^n$ defined over a number field κ, we can define a closed ball $\mathbb{P}^n[O; r]$ of center O and radius $r > 0$. Further, by using Gauss' lemma, one can show that the number of κ-rational points in $\mathbb{P}^n[O; r]$ is finite.

We assume that κ is a number field. Let M_κ be a proper set of absolute values on κ with multiplicities n_v. An element f in the ring $\kappa[X_1, \ldots, X_n]$ can be written as a sum

$$f = \sum_{i \in I} a_i \varphi_i,$$

where I is a finite set of distinct elements in $\mathbb{Z}_+^n$, $a_i \in \kappa$, and

$$\varphi_i(X_1, \ldots, X_n) = X_1^{i_1} \cdots X_n^{i_n}, \quad i = (i_1, \ldots, i_n) \in \mathbb{Z}_+^n.$$

Then we define

$$|f|_v = \begin{cases} \left(\sum_{i \in I} |a_i|_v^2\right)^{\frac{1}{2}} & : \quad \text{if } v \text{ is Archimedean,} \\ \max_{i \in I}\{|a_i|_v\} & : \quad \text{if } v \text{ is non-Archimedean.} \end{cases}$$

The *absolute height* $H(f)$ of f is defined to be the height $H(P)$ of the point P having the a_i (in any order) as coordinates, and we define its *relative height* $H_\kappa(f)$ in a similar way. If σ is an isomorphism of κ over $\mathbb{Q}$, then we get the polynomial

$$\sigma(f) = \sum_{i \in I} \sigma(a_i)\varphi_i,$$

and thus, as for points, we have $H(f) = H(\sigma(f))$.

If v is non-Archimedean, Gauss' lemma for valuations then asserts that $|\cdot|_v$ is a valuation.

Lemma 1.97 (cf. [225]). *Take $f, g \in \kappa[X_1, \ldots, X_n]$. If v is a non-trivial non-Archimedean valuation, then $|fg|_v = |f|_v |g|_v$.*

Lemma 1.98. *Let $|\cdot|$ be an absolute value which coincides with the ordinary one on $\mathbb{Q}$. Let $f \in \kappa[X]$ be a polynomial of degree d, and let*

$$f(X) = \prod_{i=1}^{d}(X - \alpha_i)$$

be a factorization in $\bar{\kappa}$. We assume that our absolute value is extended to $\bar{\kappa}$. Then

$$5^{-\frac{d}{2}}\prod_{i=1}^{d}\sqrt{1 + |\alpha_i|^2} \leq |f| \leq 2^{\frac{d-1}{2}}\prod_{i=1}^{d}\sqrt{1 + |\alpha_i|^2}.$$

Proof. The right inequality is trivially proved by induction, estimating the coefficients in a product of a polynomial $g(X)$ by $(X - \alpha)$. We prove the other by induction on the number of indices i such that $|\alpha_i| > 2$. If $|\alpha_i| \leq 2$ for all i, our assertion is obvious. Suppose that

$$f(X) = g(X)(X - \alpha)$$

with $|\alpha| > 2$ and suppose that our assertion is true for

$$g(X) = X^d + b_{d-1}X^{d-1} + \cdots + b_0.$$

We have

$$f(X) = X^{d+1} + \sum_{i=0}^{d}(b_{i-1} - \alpha b_i)X^i,$$

where $b_d = 1$, $b_{-1} = 0$. Then

$$|f| = \left(1 + \sum_{i=0}^{d}|b_{i-1} - \alpha b_i|^2\right)^{\frac{1}{2}} \geq \left(1 + \sum_{i=0}^{d}(|\alpha b_i| - |b_{i-1}|)^2\right)^{\frac{1}{2}}$$

$$\geq \left(1 + \sum_{i=0}^{d}(|\alpha| - 1)(|\alpha||b_i|^2 - |b_{i-1}|^2)\right)^{\frac{1}{2}} \geq (|\alpha| - 1)|g| \geq \frac{\sqrt{1 + |\alpha|^2}}{\sqrt{5}}|g|$$

and our lemma is now obvious. $\qquad\square$

Lemma 1.99. *Let $|\cdot|$ be an absolute value which coincides with the ordinary one on $\mathbb{Q}$. Take $d \in \mathbb{Z}_+$. If f and g are two polynomials in $\kappa[X_1, \ldots, X_n]$ such that $\deg(f) + \deg(g) \le d$, then*

$$10^{-d^n/2}|fg| \le |f||g| \le 10^{d^n/2}|fg|.$$

Proof. Let us first assume that f, g are polynomials in one variable, so we can write

$$f(X) = a \prod_{i=1}^{p}(X - \alpha_i),$$

$$g(X) = b \prod_{j=1}^{q}(X - \beta_j).$$

Without loss of generality, we may assume $a = b = 1$, and that we have extended our absolute value to $\bar{\kappa}$. By Lemma 1.98, we get

$$|f||g| \le 2^{\frac{d}{2}-1}\left(\prod_{i=1}^{p}\sqrt{1+|\alpha_i|^2}\right)\prod_{j=1}^{q}\sqrt{1+|\beta_j|^2} \le 2^{\frac{d}{2}-1}5^{\frac{d}{2}}|fg| \le 10^{\frac{d}{2}}|fg|.$$

Similarly, we can obtain

$$|f||g| \ge 10^{-\frac{d}{2}}|fg|.$$

Now let f be a polynomial in n variables $X_1, \ldots, X_n$ of degree $\le d$. Then the polynomial in one variable

$$S_d(f)(Y) = f\left(Y, Y^d, \ldots, Y^{d^{n-1}}\right)$$

has the same set of non-zero coefficients as f. Thus, if f and g are two polynomials in n variables $X_1, \ldots, X_n$ such that the sum of their degrees is $\le d$, then

$$S_d(fg) = S_d(f)S_d(g)$$

has the same non-zero coefficients as fg. From this our reduction of the n-variable case to the 1-variable case is clear. $\qquad\square$

From Lemma 1.99 we can deduce analogous results for heights.

Lemma 1.100. *Take $d \in \mathbb{Z}_+$. If f and g are two polynomials in $\kappa[X_1, \ldots, X_n]$ such that $\deg(f) + \deg(g) \le d$, then*

$$10^{-\frac{d^n}{2}}H(fg) \le H(f)H(g) \le 10^{\frac{d^n}{2}}H(fg).$$

Proof. Set

$$\|f\|_v = |f|_v^{n_v}, \ v \in M_\kappa.$$

We have

$$H_\kappa(fg) = \prod_{v \in M_\kappa} \|fg\|_v \geq \prod_{v \in M_\kappa} c_v \|f\|_v \|g\|_v,$$

where $c_v = 1$ if v is non-Archimedean, and otherwise, $c_v = 10^{-\frac{d^n}{2} n_v}$. Since

$$\sum_{v \in S_\kappa^\infty} n_v = [\kappa : \mathbb{Q}],$$

whence

$$\prod_{v \in M_\kappa} c_v = 10^{-\frac{d^n}{2}[\kappa:\mathbb{Q}]}$$

and the inequality on the left follows immediately. The one on the right follows in a similar way.   □

Let α be algebraic over $\mathbb{Q}$, and let $f(X)$ be its irreducible polynomial over $\mathbb{Q}$. Then

$$f(X) = \prod_{j=1}^d (X - \alpha_j),$$

where d is the degree of α over $\mathbb{Q}$ and α_j are the conjugates of α. In view of the above results we get:

Proposition 1.101. *Take $d \in \mathbb{Z}^+$. There exist two numbers $c_1, c_2 > 0$ depending on d, such that if α is algebraic over $\mathbb{Q}$ of degree d, and $f(X)$ is its irreducible polynomial over $\mathbb{Q}$, then*

$$c_1 H(\alpha)^d \leq H(f) \leq c_2 H(\alpha)^d.$$

Theorem 1.102 ([304], [305]). *Let d_0, r_0 be two fixed positive numbers. Then the set of points x in $\mathbb{P}^n(\kappa)$ algebraic over $\mathbb{Q}$, and such that*

$$[\mathbb{Q}(x) : \mathbb{Q}] < d_0, \ H(x) < r_0$$

is finite.

Proof. Take a point $x = [\xi_0, \dots, \xi_n] \in \mathbb{P}^n(\kappa)$ of degree d and consider the polynomial

$$f(X_0, \dots, X_n) = \xi_0 X_0 + \dots + \xi_n X_n.$$

Then $H(f)$ is the height of the point x. Let

$$g = \prod_\sigma \sigma(f)$$

be the product being taken over all distinct isomorphisms σ of κ over $\mathbb{Q}$. Then g has coefficients in $\mathbb{Q}$, and $H(g)$, $H(f)^d$ have the same order of magnitude. We have already seen in Section 1.8 that the number of points with height less than a fixed number, in a projective space, and rational over $\mathbb{Q}$ is bounded. Consequently Theorem 1.102 follows.   □

Finally, we introduce a quantitative result related to Theorem 1.102. Let κ be a number field. A real function ν on $\mathbb{P}^n$ will be said to be a *weight function* for κ if $\nu(x) = 0$ for each $x \notin \mathbb{P}^n(\kappa)$. Denote the *center* of absolute height h on $\mathbb{P}^n$ by

$$O = \{x \in \mathbb{P}^n \mid h(x) = 0\}.$$

If $x = [\xi_0, \ldots, \xi_n] \in \mathbb{P}^n(\kappa)$ with $\xi_i \neq 0$ for some i, Kronecker's theorem (see [158], Corollary B.2.3.1) shows that $\max_{0 \leq j \leq n} |\xi_j/\xi_i|_v = 1$ hold for all $v \in M_\kappa$ if and only if the ratio ξ_j/ξ_i is a root of unity or zero for every $0 \leq j \leq n$. Thus $h(x) = 0$ if and only if $\xi_j = 0$ $(j \neq i)$.

Let ν be a weight function on $\mathbb{P}^n$ for κ. Take a subset $A \subset \mathbb{P}^n$. For $r \geq 0$, set

$$A[O; r] = \{x \in A \mid h(x) \leq r\}.$$

Based on Theorem 1.102, we can define the *spherical image* of κ for ν by

$$n_\nu(r) = \sum_{x \in \mathbb{P}^n[O;r]} \nu(x). \tag{1.10.1}$$

Fix $r_0 > 0$. For $r > r_0$, we define the *characteristic function* of κ for ν by

$$N_\nu(r) = N_\nu(r, r_0) = \int_{r_0}^{r} n_\nu(t) \frac{dt}{t}. \tag{1.10.2}$$

A basic weight function of κ is the characteristic function

$$\chi_\kappa(x) = \begin{cases} 1, & \text{if } x \in \mathbb{P}^n(\kappa); \\ 0, & \text{if } x \notin \mathbb{P}^n(\kappa). \end{cases} \tag{1.10.3}$$

Theorem 1.103 ([333]). *Let κ be a number field and set $[\kappa : \mathbb{Q}] = d$. Then there exists a constant c such that*

$$n_{\chi_\kappa}(r) = ce^{d(n+1)r} + \begin{cases} O\left(re^{dr}\right) & \text{if } d = 1, \ n = 1, \\ O\left(e^{(dn+d-1)r}\right) & \text{otherwise.} \end{cases}$$

Recall that we will use $O(1)$ to denote a *bounded function*. Generally, if $h(r)$ is a non-negative function, we will denote

$$O(h(r)) := O(1)h(r).$$

We also use the symbol $o(h(r))$ to denote a function such that $o(h(r))/h(r) \to 0$ as $r \to \infty$.

Chapter 2

Nevanlinna Theory

In this chapter, corresponding to height theory we will introduce Nevanlinna theory in higher-dimensional spaces, say, the first main theorems and second main theorems, which is a basic tool for our study, and discuss the Griffiths and Lang conjectures. We also introduce Kobayashi hyperbolic varieties and some related problems. Some connections between height theory and Nevanlinna theory will be exhibited in the sequent.

2.1   Notions in complex geometry

We will introduce basic notation, terminology and technical results of complex geometry used in value distribution theory.

2.1.1   Holomorphic functions

Take a positive integer m and let

$$\mathbb{R}^{2m} = \{(x_1, y_1, \ldots, x_m, y_m) \mid x_j, y_j \in \mathbb{R}\}$$

denote the $2m$-dimensional Euclidean space. Let Ω be an open set in $\mathbb{R}^{2m}$ and let $C^k(\Omega)$ be the space of k times continuously differentiable complex-valued functions in Ω, where $0 \leq k \leq \infty$. Let $i = \sqrt{-1}$ be the imaginary unit. We identify $\mathbb{C}^m$ with $\mathbb{R}^{2m}$ by setting

$$\mathbb{C}^m = \{(z_1, \ldots, z_m) \mid z_j = x_j + iy_j; \ x_j, y_j \in \mathbb{R}\}.$$

Thus we can regard Ω as an open set in $\mathbb{C}^m$, and so the differential df of a complex-valued function f in $C^1(\Omega)$ can be expressed as a linear combination of the differentials dz_j and $d\bar{z}_j$,

$$df = \sum_{j=1}^{m} \frac{\partial f}{\partial z_j} dz_j + \sum_{j=1}^{m} \frac{\partial f}{\partial \bar{z}_j} d\bar{z}_j, \tag{2.1.1}$$

where we have used the notation

$$dz_j = dx_j + i dy_j \quad , \quad d\bar{z}_j = dx_j - i dy_j,$$
$$\frac{\partial}{\partial z_j} = \frac{1}{2}\left(\frac{\partial}{\partial x_j} - i \frac{\partial}{\partial y_j}\right) \quad , \quad \frac{\partial}{\partial \bar{z}_j} = \frac{1}{2}\left(\frac{\partial}{\partial x_j} + i \frac{\partial}{\partial y_j}\right).$$

With the notation

$$\partial f = \sum_{j=1}^{m} \frac{\partial f}{\partial z_j} dz_j, \quad \bar{\partial} f = \sum_{j=1}^{m} \frac{\partial f}{\partial \bar{z}_j} d\bar{z}_j, \tag{2.1.2}$$

we may also write (2.1.1) in the form

$$df = \partial f + \bar{\partial} f. \tag{2.1.3}$$

Differential forms which are linear combinations of the differentials dz_j are said to be of *type* $(1,0)$, and those which are linear combinations of $d\bar{z}_j$ are said to be of *type* $(0,1)$. Thus ∂f (resp. $\bar{\partial} f$) is the component of df of type $(1,0)$ (resp. $(0,1)$).

A function $f \in C^1(\Omega)$ is called *holomorphic* or *analytic* in Ω if df is of type $(1,0)$, that is, if f satisfies the *Cauchy-Riemann equations*

$$\bar{\partial} f = 0.$$

The set of all holomorphic functions in Ω is denoted by $\mathcal{A}(\Omega)$. The differential operators ∂ and $\bar{\partial}$ are obviously linear and satisfy the product rule. Hence $\mathcal{A}(\Omega)$ is a ring.

Theorem 2.1 (cf.[162]). *Let f be a complex-valued function defined in the open set $\Omega \subset \mathbb{C}^m$. The following three conditions are equivalent:*

(1) *f is a holomorphic function in Ω;*

(2) *for every point $a = (a_1, \ldots, a_m) \in \Omega$ there exists a neighborhood $U \subset \Omega$ of a such that f can be expressed in U as a convergent series*

$$f(z) = \sum_{k_1, \ldots, k_m = 0}^{\infty} c_{k_1 \cdots k_m} (z_1 - a_1)^{k_1} \cdots (z_m - a_m)^{k_m};$$

(3) *f is holomorphic in each variable z_j when the other variables are given arbitrary fixed values.*

The following *Weierstrass preparation theorem* describes a local property of holomorphic functions in $\mathbb{C}^m$.

Theorem 2.2. *Let f be holomorphic in a neighborhood Ω of 0 in $\mathbb{C}^m$ and assume that $f(0, z_m)/z_m^\mu$ is holomorphic and $\neq 0$ at 0. Then one can find a polydisc $\Delta \subset \Omega$ such that every g which is holomorphic and bounded in Δ can be written in the form*

$$g = qf + r, \qquad (2.1.4)$$

where q and r are holomorphic in Δ, r is a polynomial in z_m of degree $< \mu$ (with coefficients depending on $z' = (z_1, \ldots, z_{m-1})$) and

$$\sup_{z \in \Delta} |q(z)| \leq c \sup_{z \in \Delta} |g(z)|, \qquad (2.1.5)$$

where c is independent of g. The expression (2.1.4) is unique. The coefficients of power series expansions of q and r are finite linear combinations of those in the expansion of g.

Proof. Hörmander [162], Theorem 6.1.1. $\square$

A set $D \subset \mathbb{C}^m$ is called a *polydisc* if there are discs $D_1, \ldots, D_m$ in $\mathbb{C}$ such that

$$D = \prod_{j=1}^{m} D_j = \{(z_1, \ldots, z_m) \in \mathbb{C}^m \mid z_j \in D_j, \ j = 1, \ldots, m\}.$$

In particular, when $g(z) = z_m^\mu$, Theorem 2.2 shows that one can write f in one and only one way in the form $f = hW$, where h and W are holomorphic in a neighborhood of 0, $h(0) \neq 0$, and W is a *Weierstrass polynomial*, that is,

$$W(z) = z_m^\mu + \sum_{j=0}^{\mu-1} a_j(z') z_m^j,$$

where a_j are holomorphic functions in a neighborhood of 0 vanishing when $z' = 0$. Note that, conversely, every f which can be represented as above must satisfy the hypotheses of Theorem 2.2.

Let $f : \Omega \longrightarrow \mathbb{C}^n$ be a *holomorphic mapping*, that is,

$$f = (f_1, \ldots, f_n),$$

where each component f_j is holomorphic in Ω. If $g \in C^1(W)$ for some open set W containing the image $f(\Omega)$ of f, the function $f^*g = g \circ f$ is in $C^1(\Omega)$ and we have

$$d(f^*g) = \sum_{j=1}^{n} \frac{\partial g}{\partial f_j} df_j + \sum_{j=1}^{n} \frac{\partial g}{\partial \overline{f}_j} d\overline{f}_j.$$

Since df_j is of type $(1,0)$ and $d\bar{f}_j$ of type $(0,1)$ in Ω, it follows that

$$\partial(f^*g) = \sum_{j=1}^{n} \frac{\partial g}{\partial f_j} df_j, \quad \bar{\partial}(f^*g) = \sum_{j=1}^{n} \frac{\partial g}{\partial \bar{f}_j} d\bar{f}_j.$$

Hence f^*g is holomorphic if g is holomorphic. More generally, the decomposition of d as $\partial + \bar{\partial}$ and the notion of holomorphic function are invariant under holomorphic mappings.

The definition of the ∂ and $\bar{\partial}$ operators can be extended to arbitrary differential forms. Take non-negative integers a and b with $a \leq b$. Let $J_{1,a}^{b}$ be the set of all increasing injective mappings

$$\lambda : \mathbb{Z}[1,a] \longrightarrow \mathbb{Z}[1,b].$$

A differential form ω on Ω is said to be of *type* (p,q) if it can be written in the form

$$\omega = \sum_{\alpha \in J_{1,p}^{m}} \sum_{\beta \in J_{1,q}^{m}} f_{\alpha,\beta} dz_\alpha \wedge d\bar{z}_\beta, \tag{2.1.6}$$

where the coefficients $f_{\alpha,\beta}$ are differentiable on Ω, and where we have used the notation

$$dz_\alpha = dz_{\alpha(1)} \wedge \cdots \wedge dz_{\alpha(p)}, \quad d\bar{z}_\beta = d\bar{z}_{\beta(1)} \wedge \cdots \wedge d\bar{z}_{\beta(q)}.$$

Every differential form can be written in one and only one way as a sum of forms of type (p,q); $0 \leq p, q \leq m$. If ω is of type (p,q), the *exterior differential*

$$d\omega = \sum_{\alpha \in J_{1,p}^{m}} \sum_{\beta \in J_{1,q}^{m}} df_{\alpha,\beta} \wedge dz_\alpha \wedge d\bar{z}_\beta$$

can be written

$$d\omega = \partial\omega + \bar{\partial}\omega,$$

where

$$\partial\omega = \sum_{\alpha \in J_{1,p}^{m}} \sum_{\beta \in J_{1,q}^{m}} \partial f_{\alpha,\beta} \wedge dz_\alpha \wedge d\bar{z}_\beta,$$

$$\bar{\partial}\omega = \sum_{\alpha \in J_{1,p}^{m}} \sum_{\beta \in J_{1,q}^{m}} \bar{\partial} f_{\alpha,\beta} \wedge dz_\alpha \wedge d\bar{z}_\beta$$

are of type $(p+1,q)$ and $(p,q+1)$, respectively. Since

$$0 = d^2\omega = \partial^2\omega + (\partial\bar{\partial} + \bar{\partial}\partial)\omega + \bar{\partial}^2\omega$$

and all terms are of different types, we obtain

$$\partial^2 = 0, \ \partial\bar{\partial} + \bar{\partial}\partial = 0, \ \bar{\partial}^2 = 0. \tag{2.1.7}$$

We shall use the notation $A^{p,q}(\Omega)$ for the space of differential forms of type (p,q) with coefficients belonging to $C^\infty(\Omega)$.

If $f : \Omega \longrightarrow \mathbb{C}^n$ is a holomorphic mapping, and if

$$\eta = \sum_{\alpha \in J^n_{1,p}} \sum_{\beta \in J^n_{1,q}} g_{\alpha,\beta} dw_\alpha \wedge d\bar{w}_\beta$$

is a form defined in an open neighborhood of the image of f, we can define a form $f^*\eta$ in Ω by

$$f^*\eta = \sum_{\alpha \in J^n_{1,p}} \sum_{\beta \in J^n_{1,q}} f^* g_{\alpha,\beta} df^* w_\alpha \wedge \overline{df^* w}_\beta,$$

where

$$df^* w_k = d(w_k \circ f), \quad \overline{df^* w}_k = d(\overline{w_k \circ f})$$

are differential forms in Ω of type $(1,0)$ and $(0,1)$ for $k = 1, \ldots, n$, respectively, since w_k is holomorphic. Hence $f^*\eta$ is of type (p,q) if η is of type (p,q), and since

$$d(f^*\eta) = f^*(d\eta),$$

it follows that

$$\partial(f^*\eta) = f^*(\partial\eta), \quad \bar{\partial}(f^*\eta) = f^*(\bar{\partial}\eta).$$

2.1.2 Complex manifolds

A Hausdorff topological space M with a countable basis is called a *manifold of dimension m* if every point in M has a neighborhood which is homeomorphic to an open set in $\mathbb{R}^m$. The concept of complex manifolds is defined by means of a family of such homeomorphisms:

Definition 2.3. *A manifold M of dimension $2m$ is called a complex (analytic) manifold of (complex) dimension m if there exists a family $\mathcal{F} = \{(U_\alpha, \varphi_\alpha)\}$, called a complex structure of M, which satisfies the following conditions:*

(i) *$\{U_\alpha\}$ is an open covering of M;*

(ii) *$\varphi_\alpha : U_\alpha \longrightarrow U'_\alpha$ is a homeomorphism onto an open subset U'_α of $\mathbb{C}^m$;*

(iii) *$\varphi_\beta \circ \varphi_\alpha^{-1} : \varphi_\alpha(U_\alpha \cap U_\beta) \longrightarrow \mathbb{C}^m$ is a holomorphic mapping if $U_\alpha \cap U_\beta \neq \emptyset$.*

The pair $(U_\alpha, \varphi_\alpha)$ is called a *holomorphic coordinate atlas* of M, and $\varphi_\alpha = (z_1, \ldots, z_m)$ is said to be a *local holomorphic coordinate system* on U_α or at every point of U_α. Take a point $p \in U_\alpha$. If $f_1, \ldots, f_m$ are holomorphic functions in a neighborhood of $\varphi_\alpha(p)$ in $\mathbb{C}^m$, then $(f_1(z), \ldots, f_m(z))$ is another system of coordinates at p if and only if

$$\det \left(\frac{\partial f_k}{\partial z_j}(\varphi_\alpha(p)) \right)^m_{k,j=1} \neq 0.$$

This follows from the implicit function theorem (see [162], Theorem 2.1.2). Usually, a complex manifold of dimension 1 is called a *curve* or *Riemann surface*, and a complex manifold of dimension 2 is called a *surface*.

Given another complex manifold N, a continuous mapping $f : M \longrightarrow N$ is called *holomorphic* or *analytic* if for every $p \in M$, there exist a holomorphic coordinate atlas (U, φ) around p and a holomorphic coordinate atlas (V, ψ) around $f(p)$ such that $f(U) \subset V$, and such that

$$\psi \circ f \circ \varphi^{-1} : \varphi(U) \longrightarrow \psi(V)$$

is holomorphic. Let $\mathrm{Hol}(M, N)$ be the set of holomorphic mappings from M into N. In particular, if $N = \mathbb{C}$ we have now defined the concept of *holomorphic functions* in a complex manifold M; the set of such functions with the topology of uniform convergence on compact subsets of M will be denoted by $\mathcal{A}(M)$. The elements in $\mathcal{A}(M)$ are also called *entire functions* on M. Let $\mathcal{A}^*(M)$ be the subset of $\mathcal{A}(M)$ such that $f \in \mathcal{A}^*(M)$ if and only if f vanishes nowhere.

If a homeomorphism $f : M \longrightarrow N$ between complex manifolds M and N is holomorphic such that $f^{-1} : N \longrightarrow M$ is also holomorphic, then f is said to be *biholomorphic* or an *analytic isomorphism*. Such complex manifolds are called *analytic isomorphic*. If f is a biholomorphic self-mapping on M, then f is said to be an *automorphism* on M. Let $\mathrm{Aut}(M)$ be the *group of automorphisms* on M. The group operation is composition.

It is clear that every open subset of a complex manifold M has a complex structure, so the concept of a holomorphic function (mapping) on an open subset is also well defined. Note that if f is holomorphic in $U'_\alpha \subset \mathbb{C}^m$, then $f \circ \varphi_\alpha$ is holomorphic in U_α. Hence by the definition of a complex manifold, holomorphic functions do exist locally.

Let M be a complex manifold of dimension m, $p \in M$ any point, and $z = (z_1, \dots, z_m)$ a holomorphic coordinate system around p. There are three different notions of a tangent space to M at p. First of all, if we consider M as a real manifold of dimension $2m$, we have the usual *real tangent space* $T_p(M)$ to M at p, which can be realized as the space of $\mathbb{R}$-linear derivations on the ring of real-valued C^∞ functions in a neighborhood of p. Setting $z_j = x_j + iy_j$, then $T_p(M)$ is spanned by $\left\{ \frac{\partial}{\partial x_j}, \frac{\partial}{\partial y_j} \right\}$ over $\mathbb{R}$. Secondly, we have the *complexified tangent space*

$$T_p(M)_{\mathbb{C}} = T_p(M) \otimes_{\mathbb{R}} \mathbb{C}.$$

Elements in $T_p(M)_{\mathbb{C}}$ are called *complex tangent vectors* at p. It can be realized as the space of $\mathbb{C}$-linear derivations on the ring of complex-valued C^∞ functions in a neighborhood of p, which can be spanned by $\left\{ \frac{\partial}{\partial z_j}, \frac{\partial}{\partial \bar{z}_j} \right\}$ over $\mathbb{C}$. Finally, there is the *holomorphic tangent space* $\mathbf{T}_p(M)$, which is the subspace of $T_p(M)_{\mathbb{C}}$ spanned by $\{\frac{\partial}{\partial z_j}\}$ over $\mathbb{C}$. The subspace $\overline{\mathbf{T}}_p(M) = \mathbb{C}\{\frac{\partial}{\partial \bar{z}_j}\}$ is called the *antiholomorphic tangent space* to M at p; clearly

$$T_p(M)_{\mathbb{C}} = \mathbf{T}_p(M) \oplus \overline{\mathbf{T}}_p(M).$$

A complex tangent vector (field) is of *type* $(1,0)$ (resp. $(0,1)$) if it belongs to $\mathbf{T}_p(M)$ (resp. $\overline{\mathbf{T}}_p(M)$).

Observe that for complex manifolds M, N, any smooth mapping $f : M \longrightarrow N$ induces the (real) differential of f at p,

$$df(p) : T_p(M) \longrightarrow T_{f(p)}(N),$$

and hence a mapping

$$T_p(M)_{\mathbb{C}} \longrightarrow T_{f(p)}(N)_{\mathbb{C}},$$

also denoted by $df(p)$, but do not in general induce a linear mapping from $\mathbf{T}_p(M)$ to $\mathbf{T}_{f(p)}(N)$. In fact, a smooth mapping $f : M \longrightarrow N$ is holomorphic if and only if

$$df(p)(\mathbf{T}_p(M)) \subset \mathbf{T}_{f(p)}(N)$$

for all $p \in M$. If so, we denote the induced mapping by

$$f'(p) : \mathbf{T}_p(M) \longrightarrow \mathbf{T}_{f(p)}(N),$$

which is called the *holomorphic differential* of f at p.

Let $z = (z_1, \ldots, z_m)$ be local holomorphic coordinates centered at $p \in M$, and $w = (w_1, \ldots, w_n)$ holomorphic coordinates centered at $f(p) \in N$. Locally, the holomorphic mapping $f : M \longrightarrow N$ can be expressed by holomorphic functions

$$w_k = f_k(z_1, \ldots, z_m), \ k = 1, \ldots, n.$$

Then the Jacobi's matrix $J(f'(p))$ corresponding to the linear mapping $f'(p)$ is given by

$$J(f'(p)) = \left(\frac{\partial f_k}{\partial z_j}(p)\right)_{1 \leq k \leq n, 1 \leq j \leq m}.$$

We may define the *rank* of f by

$$\text{rank}_p(f) = \text{rank} J(f'(p)), \ \text{rank}(f) = \sup_{p \in M} \text{rank}_p(f).$$

In particular, if $m = n$ it is not difficult to show that the determinant of the $2m \times 2m$ real Jacobi's matrix $J(df(p))$ corresponding to the linear mapping $df(p)$ satisfies

$$\det J(df(p)) = |\det J(f'(p))|^2.$$

Hence holomorphic mappings are *orientation preserving*, which further means that complex manifolds are *orientable*.

Let M be a complex manifold of dimension m. If we consider M as a real manifold of dimension $2m$, we have the space $A(M, \mathbb{R})$ of smooth differential forms on M which may be graded as follows:

$$A(M, \mathbb{R}) = \sum_{r=0}^{2m} A^r(M, \mathbb{R}),$$

where $A^r(M, \mathbb{R})$ is the space of smooth differential r-forms on M. Denote the subspace of *closed r-forms* by

$$Z^r(M, \mathbb{R}) = \{\omega \in A^r(M, \mathbb{R}) \mid d\omega = 0\}.$$

Since

$$d^2 = 0, \quad d\left(A^{r-1}(M, \mathbb{R})\right) \subset Z^r(M, \mathbb{R}),$$

the quotient groups

$$H_{\mathrm{DR}}^r(M, \mathbb{R}) = Z^r(M, \mathbb{R})/d\left(A^{r-1}(M, \mathbb{R})\right)$$

of closed forms modulo *exact forms* are called the *de Rham cohomology groups of M.*

An element of the complexification $A(M)$ of $A(M, \mathbb{R})$ is called a *complex differential form* on M. The space $A(M)$ may be graded as follows:

$$A(M) = \sum_{r=0}^{2m} A^r(M),$$

where $A^r(M)$ is just the complexification of $A^r(M, \mathbb{R})$. An element of $A^r(M)$ is called a *complex r-form* on M. Every complex r-form ω may be written uniquely as $\omega_1 + i\omega_2$, where ω_1 and ω_2 are *real r-forms*, that is, $\overline{\omega_j} = \omega_j$ for $j = 1, 2$. If we denote by $T_x^*(M)_{\mathbb{C}}$ the complexification of the dual space $T_x^*(M)$ (cotangent space to M at x) of $T_x(M)$ at $x \in M$, then a complex r-form ω on M gives an element $\omega(x)$ of $\bigwedge_r T_x^*(M)_{\mathbb{C}}$ at each point x of M; in other words, a skew-symmetric r-linear mapping

$$\omega(x) : T_x(M)_{\mathbb{C}} \times \cdots \times T_x(M)_{\mathbb{C}} \longrightarrow \mathbb{C}$$

at each point x of M. The differential operator d on $A(M, \mathbb{R})$ can be easily extended to $A(M)$. Let $Z^r(M)$ denote the space of closed complex r-forms on M, and let

$$H_{\mathrm{DR}}^r(M) = Z^r(M)/d\left(A^{r-1}(M)\right)$$

be the corresponding quotient; clearly

$$H_{\mathrm{DR}}^r(M) = H_{\mathrm{DR}}^r(M, \mathbb{R}) \otimes \mathbb{C}.$$

Each space $A^r(M)$ can be split into a direct sum

$$A^r(M) = \sum_{p+q=r} A^{p,q}(M).$$

An element of $A^{p,q}(M)$ is called a (complex) *form of type* (p,q). Obviously, $A^{p,q}(M)$ is a $C^\infty(M)$-module. The differential operator d on $A(M)$ has the basic property

$$dA^{p,q}(M) \subset A^{p+1,q}(M) + A^{p,q+1}(M).$$

As a consequence, one can split d into $\partial + \bar{\partial}$, where

$$\partial : A^{p,q}(M) \longrightarrow A^{p+1,q}(M), \ \bar{\partial} : A^{p,q}(M) \longrightarrow A^{p,q+1}(M),$$

which also satisfy the relations in (2.1.7). Let $Z_{\bar{\partial}}^{p,q}(M)$ denote the space of $\bar{\partial}$-closed forms of type (p,q). Since $\bar{\partial}^2 = 0$ on $A^{p,q}(M)$, and we have

$$\bar{\partial}(A^{p,q}(M)) \subset Z_{\bar{\partial}}^{p,q+1}(M),$$

accordingly, we define the *Dolbeault cohomology groups* to be

$$H_{\bar{\partial}}^{p,q}(M) = Z_{\bar{\partial}}^{p,q}(M)/\bar{\partial}(A^{p,q-1}(M)).$$

Associated to the differential operator $d = \partial + \bar{\partial}$ on $A(M)$, we also use the operator

$$d^c = \frac{i}{4\pi}(\bar{\partial} - \partial).$$

Then

$$dd^c = \frac{i}{2\pi}\partial\bar{\partial}.$$

A form $\omega \in A^{p,0}(M)$ with $\bar{\partial}\omega = 0$ is called a *holomorphic form* of degree p. Note that all coefficients of a holomorphic form are holomorphic. Usually, the space of holomorphic forms of degree p on M is denoted by $\Omega^p(M)$. Thus we obtain a sheaf Ω^p on M associated to the qth Čech cohomology group $H^q(M, \Omega^p)$ of Ω^p on M. Specially, $\mathcal{O} = \mathcal{A} = \Omega^0$ denotes the sheaf whose sections are given locally by holomorphic functions, that is, $\mathcal{O}(U) = \mathcal{A}(U)$ for an open set $U \subset M$. Dolbeault's theorem (cf. [127]) shows

$$H^q(M, \Omega^p) \cong H_{\bar{\partial}}^{p,q}(M).$$

Note that a form ψ of type (p,q) attaches to each $x \in M$ a bilinear mapping over $\mathbb{C}$,

$$\psi(x) : \bigwedge_p \mathbf{T}_x(M) \times \bigwedge_q \bar{\mathbf{T}}_x(M) \longrightarrow \mathbb{C}.$$

For the case $q = p$, the (p, p)-form ψ is said to be *positive* (resp. *non-negative*) as long as

$$\psi(x)(y_1 \wedge \cdots \wedge y_p, \overline{iy_1} \wedge \cdots \wedge \overline{iy_p}) > 0 \quad (\text{resp. } \geq 0)$$

for any set of linearly independent vectors $y_1, \ldots, y_p$ in $\mathbf{T}_x(M)$ at each point x of M. If ψ is positive (resp. non-negative), we will write $\psi > 0$ (resp., $\psi \geq 0$). We also write $\psi > \eta$ (resp. $\psi \geq \eta$) for another (p, p)-form η if $\psi - \eta > 0$ (resp. $\psi - \eta \geq 0$). Especially, if ψ is a $(1, 1)$-form given locally by

$$\psi(x) = i \sum_{k,l} a_{kl}(x) dz_k \wedge d\bar{z}_l,$$

then $\psi > 0$ (resp. $\psi \geq 0$) if and only if the matrix $(a_{kl}(x))$ is positive definite (resp. positive semidefinite) at each point x of M.

Let M be an m-dimensional complex manifold. If p is a point on M, we define the *local ring of holomorphic functions of M at p* to be the ring of germs of functions that are holomorphic in a neighborhood of p. This ring is denoted by $\mathcal{O}_{M,p}$, or simply by $\mathcal{O}_p$ if no confusion is likely to arise. In other words, an element of $\mathcal{O}_p$ is a pair (U, f) where U is an open subset of M containing p, and f is a holomorphic function on U, and where we identify two such pairs (U, f) and (W, g) if there is a neighborhood $V \subset U \cap W$ of p such that $f = g$ on V. Note that $\mathcal{O}_p$ is indeed a local ring: its maximal ideal $\mathbf{m}_p$ is the set of germs of holomorphic functions which vanish at p. For if $f(p) \neq 0$, then $1/f$ is holomorphic in some neighborhood of p. Further, $\mathcal{O}_p$ is a unique factorization domain (see [162], Theorem 6.2.2), and a Noetherian ring (see [162], Theorem 6.3.3).

A subset X in the complex manifold M is called an *analytic subset* if for every point $p \in M$ there is an open neighborhood $U \subset M$ and a family of functions $f_\lambda \in \mathcal{A}(U)$, $\lambda \in \Gamma$, such that

$$U \cap X = \{z \in U \mid f_\lambda(z) = 0, \ \lambda \in \Lambda\}.$$

The definition implies that X is closed. It turns out that Λ can always be taken finite (see [162], Theorem 6.5.2), that is, for every point $p \in M$ there is an open neighborhood $U \subset M$ and a finite number of functions $f_1, \ldots, f_r \in \mathcal{A}(U)$ such that

$$I_p(X) = \{(U, f) \in \mathcal{O}_p \mid f|_{X \cap U} = 0\} \tag{2.1.8}$$

is the ideal in $\mathcal{O}_p$ generated by $f_1, \ldots, f_r$. In particular,

$$U \cap X = \{z \in U \mid f_1(z) = \cdots = f_r(z) = 0\}.$$

Those $f_1, \ldots, f_r$ are called the *local defining functions*, and $f_1 = \cdots = f_r = 0$ is called the *local equation* of X at p. If we always have $r = 1$, then X is called an *analytic hypersurface*. Let X be an analytic subset of M. Then X is called

reducible if there are non-empty distinct analytic subsets X_1 and X_2 of M such that
$$X = X_1 \cup X_2, \ X \neq X_j \ (j = 1, 2),$$
otherwise, X is called *irreducible*. A point $x \in X$ is called *regular* or *non-singular* if there is a neighborhood U of x in M such that $X \cap U$ is a complex submanifold, otherwise, is *singular*. We say that X is *non-singular* or *smooth* if every point on X is regular.

Let $\{X_\lambda\}_{\lambda \in \Lambda}$ be a family of analytic hypersurfaces of M. Assume that it is *locally finite*, i.e., for any compact subset K of M, $\{\lambda \in \Lambda \mid \bar{X}_\lambda \cap K \neq \emptyset\}$ is finite. Then a formal sum
$$D = \sum_{\lambda \in \Lambda} n_\lambda X_\lambda$$
with coefficients $n_\lambda \in \mathbb{Z}$ is called a *divisor* on M. Without loss of generality, we may assume that X_λ's are irreducible and mutually distinct, and that every $n_\lambda \neq 0$. Then we define the *support* supp(D) of the divisor D by
$$\mathrm{supp}(D) = \bigcup_{\lambda \in \Lambda} X_\lambda.$$
If all $n_\lambda \geq 0$ then D is called *effective* or *positive*. We write $D \geq 0$ for D effective. If $U \subset M$ is an open subset, we define the intersection of D with U by
$$D \cap U = \sum_{\lambda \in \Lambda} n_\lambda (X_\lambda \cap U).$$

A divisor D on M is said to have *normal crossings* if D is locally given by an equation $z_1 \cdots z_k = 0$, where $(z_1, \ldots, z_m)$ are local holomorphic coordinates on M. If D has normal crossings, and if after expressing $D = \sum X_j$ as a sum of irreducible components, all X_j are non-singular, then we say that D has *simple normal crossings*. In case $M = \mathbb{P}(V)$ is a complex projective space associated to a complex vector space V of dimension $m + 1$, and if $D = \ddot{E}[a_1] + \cdots + \ddot{E}[a_q]$ is a linear combination of hyperplanes, then D has normal crossings if and only if the family $a_j (j = 1, \ldots, q)$ are in general position in $\mathbb{P}(V^*)$.

Definition 2.4. *A complex manifold M of dimension m is said to be a Stein manifold if*

(α) *M is countable at infinity, that is, there exists a countable number of compact subsets $K_1, K_2, \ldots$ such that every compact subset of M is contained in some K_j.*

(β) *M is holomorphic convex, that is,*
$$\hat{K} = \left\{ z \in M \ \middle| \ |f(z)| \leq \sup_{p \in K} |f(p)| \text{ for every } f \in \mathcal{A}(M) \right\}$$
is a compact subset of M for every compact subset K of M.

(γ) *If z_1 and z_2 are different points in M, then $f(z_1) \neq f(z_2)$ for some $f \in \mathcal{A}(M)$.*

(δ) *For every $z \in M$, one can find m functions $f_1, \ldots, f_m \in \mathcal{A}(M)$ which form a coordinate system at z.*

In fact, if a complex manifold M is countable at infinity, a topology in $\mathcal{A}(M)$ is then defined by the seminorms

$$|f| = \sup_{j \geq 1} \sup_{p \in K_j} |f(p)|, \quad f \in \mathcal{A}(M),$$

and the completeness is obvious.

Let M be a complex manifold of dimension m. Then M is said to have *holomorphic rank n* if there exist n analytically independent holomorphic functions on M, that is, there exist holomorphic functions $f_1, \ldots, f_n$ on M such that

$$df_1 \wedge \cdots \wedge df_n \not\equiv 0$$

on each connected component of M; equivalently, a holomorphic mapping

$$f = (f_1, \ldots, f_n) : M \longrightarrow \mathbb{C}^n$$

of strict rank n exists (see [5] or [380]). Such a mapping f is called *regular* if it has rank m at every point in M, that is, if for any point in M there is a coordinate system formed by m of the functions $f_1, \ldots, f_n$. If the inverse image of every compact subset of $\mathbb{C}^n$ is a compact subset of M, the mapping is called *proper*. It is clear that the image of a proper mapping f is closed, for every compact set is mapped on a compact set. If, in addition, the mapping is regular and one-to-one, the image is a complex submanifold of $\mathbb{C}^n$ which is isomorphic to M.

Theorem 2.5 (cf. [162]). *If M is a Stein manifold of dimension m, there exists an element $f \in \mathcal{A}(M)^{2m+1}$ which defines a one-to-one regular proper mapping of M into $\mathbb{C}^{2m+1}$. In particular, M has holomorphic rank m.*

Consider $\mathbb{C}^m$ as a vector space and let $\omega_1, \ldots, \omega_{2m}$ be any basis of $\mathbb{C}^m$ over the field $\mathbb{R}$. Let Λ be the subgroup of $\mathbb{C}^m$ generated by $\omega_1, \ldots, \omega_{2m}$:

$$\Lambda = \{ n_1 \omega_1 + \cdots + n_{2m} \omega_{2m} \mid n_j \in \mathbb{Z} \}.$$

The quotient group $\mathbb{C}^m / \Lambda$ is a connected compact complex manifold with the natural projection $\pi : \mathbb{C}^m \longrightarrow \mathbb{C}^m / \Lambda$ being holomorphic. We call $\mathbb{C}^m / \Lambda$ an m-dimensional *complex torus*. If $z_1, \ldots, z_m$ is the natural coordinate system in $\mathbb{C}^m$, then the holomorphic 1-forms $dz_1, \ldots, dz_m$ can be considered as forms on a complex torus $\mathbb{C}^m / \Lambda$. Every holomorphic 1-form on $\mathbb{C}^m / \Lambda$ is a linear combination of $dz_1, \ldots, dz_m$ with constant coefficients. In fact, every holomorphic 1-form on $\mathbb{C}^m / \Lambda$ is a linear combination of $dz_1, \ldots, dz_m$ with holomorphic functions as coefficients and since $\mathbb{C}^m / \Lambda$ is compact, these coefficients functions are constant functions.

An m-dimensional complex manifold M is said to be *complex parallelizable* if there exist m holomorphic vector fields $Z_1, \ldots, Z_m$ which are linearly independent at every point of M. Every complex torus is complex parallelizable. More generally, let G be a complex Lie group of complex dimension m. Taking m linearly independent complex vectors of type $(1,0)$ at the identity element of G and extending them by left translations, we obtain m left invariant holomorphic vector fields $Z_1, \ldots, Z_m$ on G which are linearly independent at every point of G. If Γ is a discrete subgroup of G, then $Z_1, \ldots, Z_m$ induce m holomorphic vector fields on the quotient complex manifold G/Γ which are linearly independent at every point of G/Γ, showing that G/Γ is complex parallelizable (see [209], Vol. II, Chapter IX, Example 2.3). H. C. Wang [425] proved the converse:

Theorem 2.6. *Every compact complex parallelizable manifold may be written as a quotient space G/Γ of a complex Lie group G by a discrete subgroup Γ.*

For the complex case, here we explain Theorem 1.70 on normalization of algebraic curves clearly. Let $f(x, y)$ be an irreducible polynomial of two variables x and y over $\mathbb{C}$ and consider the equation

$$f(x, y) = 0. \tag{2.1.9}$$

If the degree of f is d, then put

$$F(x, y, z) = z^d f\left(\frac{x}{z}, \frac{y}{z}\right),$$

so that F is homogeneous of degree d. Recall that the set C of solutions $[x, y, z] \in \mathbb{P}^2$ of the equation

$$F(x, y, z) = 0 \tag{2.1.10}$$

is called an *algebraic curve* on $\mathbb{P}^2$. The degree of F is said to be the *degree* of the curve. If the point (x_0, y_0) lies on the affine curve (2.1.9), then $[x_0, y_0, 1]$ lies on C. Conversely, if $[x_0, y_0, z_0]$ lies on C with $z_0 \neq 0$, then $(x_0/z_0, y_0/z_0)$ lies on the affine curve (2.1.9). Points on C with $z = 0$ are called the *points at infinity* of the affine curve.

In affine space, we say the *singular points* on C correspond to

$$f(x, y) = \frac{\partial f}{\partial x}(x, y) = \frac{\partial f}{\partial y}(x, y) = 0.$$

Without loss of generality, we may assume that $p = (0, 0)$ is a point on the affine curve (2.1.9). Write

$$f(x, y) = f_k(x, y) + f_{k+1}(x, y) + \cdots + f_d(x, y),$$

where $f_j(x, y)$ $(j = k, \ldots, d)$ is a homogeneous polynomial of degree j with $f_k(x, y) \not\equiv 0$, and $k \geq 1$ since $f(0, 0) = 0$. Note that p is a singular point if and only if $k \geq 2$. Recall that when $k \geq 2$, the singular point p is called a *k-fold point*, say, *double point* for the case $k = 2$, *triple point* when $k = 3$, and so on. Now there are k tangent lines (counting multiplicity) at p given by the equation

$$f_k(x, y) = 0.$$

Further, if k tangent lines at p are distinct, then p is called an *ordinary k-fold point* of the affine curve (2.1.9).

A point $p \in C$ is *singular* if

$$F(p) = \frac{\partial F}{\partial x}(p) = \frac{\partial F}{\partial y}(p) = \frac{\partial F}{\partial z}(p) = 0.$$

By Euler's formula of homogeneous functions, we know

$$\deg(F)F(x, y, z) = x\frac{\partial F}{\partial x}(x, y, z) + y\frac{\partial F}{\partial y}(x, y, z) + z\frac{\partial F}{\partial z}(x, y, z).$$

Hence a point $p \in C$ is singular if and only if

$$\frac{\partial F}{\partial x}(p) = \frac{\partial F}{\partial y}(p) = \frac{\partial F}{\partial z}(p) = 0.$$

It is a simple lemma to prove that if an affine point is non-singular, then the corresponding projective point is also non-singular, and conversely. If the algebraic curve C defined by (2.1.10) is irreducible, then C has at most finitely many singular points (cf. [126]). Thus we may state Theorem 1.70 in the following form:

Theorem 2.7 (cf. [126]). *Let C be an irreducible algebraic curve in $\mathbb{P}^2$ and let S be the set of singular points of C. There exist a compact Riemann surface M and a holomorphic mapping $\eta : M \longrightarrow \mathbb{P}^2$ such that $\eta(M) = C$, $\eta^{-1}(S)$ is finite, and $\eta : M - \eta^{-1}(S) \longrightarrow C - S$ is one-to-one.*

Recall that (M, η) is called the *normalization* (or *resolution of singularity*) of C. We also know that if (M', η') is another normalization of C, then there exists a biholomorphic mapping $\sigma : M \longrightarrow M'$ such that $\eta = \eta' \circ \sigma$. The genus of M is also called that of C. Conversely, we also have the following basic fact:

Theorem 2.8 (cf. [127]). *Any compact Riemann surface M can be realized as a normalization of some plane algebraic curve C whose only singularities are ordinary double points, that is, there exists a holomorphic mapping*

$$\eta : M \longrightarrow \mathbb{P}^2$$

such that $\eta(M)$ is an algebraic curve which has at most ordinary double points.

2.1.3   Meromorphic mappings

Let M and N be connected complex manifolds of dimensions m and n, respectively. Let S be a thin analytic subset of M, where *thin* means that $A = M - S$ is dense in M. Let $f_A : A \longrightarrow N$ be a holomorphic mapping. Recall that a continuous mapping is said to be *proper* if the inverse images of compact sets are compact. The mapping f_A is said to be *meromorphic* on M and denoted by $f : M \longrightarrow N$ if the closure $\overline{G(f_A)}$ of the graph

$$G(f_A) = \{(x, f_A(x)) \mid x \in M\}$$

of f_A in $M \times N$ is analytic in $M \times N$ and if the projection $\pi_M : \overline{G(f_A)} \longrightarrow M$ is proper. We set $G(f) = \overline{G(f_A)}$ which is called the *graph* of the meromorphic mapping f determined by f_A. A meromorphic mapping $f : M \longrightarrow N$ from an affine algebraic variety M into a projective algebraic variety N is said to be *rational* if it can be extended to a meromorphic mapping $\tilde{f} : \bar{M} \longrightarrow N$, where $\bar{M}$ is a projective closure of M. If N is embedded, then $\tilde{f}$ is given by rational functions.

Assume that $f : M \longrightarrow N$ is meromorphic and that $\pi_N : G(f) \longrightarrow N$ is the projection. For each $x \in M$, the set

$$\Sigma_f(x) = \pi_N(\pi_M^{-1}(x)) = \{y \in N \mid (x, y) \in G(f)\}$$

is analytic and not empty. The *indeterminacy*

$$I_f = \{x \in M \mid \#\Sigma_f(x) > 1\}$$

is analytic and contained in S. If $x \in I_f$ and $y \in \Sigma_f(x)$, then $\dim_y \Sigma_f(x) > 0$. The holomorphic mapping $f_A : A \longrightarrow N$ extends to a holomorphic mapping $f_{M-I_f} : M - I_f \longrightarrow N$. We also write $f(x) = \Sigma_f(x)$ for all $x \in M$. Here $\dim I_f \leq m - 2$. The *rank* of f is defined by

$$\mathrm{rank}_x f = \dim_x M - \dim_x f^{-1}(f(x)), \quad x \in M - I_f, \tag{2.1.11}$$

$$\mathrm{rank} f = \max_{x \in M - I_f} \mathrm{rank}_x f. \tag{2.1.12}$$

If $N = \mathbb{P}(V)$, where V is a Hermitian vector space of dimension $n + 1 > 1$, another equivalent definition of a meromorphic mapping exists. Assume M, S, A as above and let $f_A : A \longrightarrow \mathbb{P}(V)$ be a holomorphic mapping. Let $U \neq \emptyset$ be a connected open subset of M. A holomorphic vector function $\tilde{f} : U \longrightarrow V$ is said to be a *representation* of f_A on U if $\tilde{f} \not\equiv 0$ and if

$$\mathbb{P}(\tilde{f}(x)) = f_A(x), \quad x \in U - \tilde{f}^{-1}(0).$$

For each $x \in U$, we also say that $\tilde{f}$ is a representation of f_A at x. Further, the representation is said to be *reduced* if $\dim \tilde{f}^{-1}(0) \leq m - 2$. Then the mapping f_A

is meromorphic on M if and only if there is a representation of f_A at every point of M. If f_A is meromorphic, then there is even a reduced representation of f_A at every point of M. A (reduced) representation of f_A also is called a (reduced) representation of the meromorphic mapping $f : M \longrightarrow \mathbb{P}(V)$ determined by f_A. If $\tilde{f} : U \longrightarrow V$ is a reduced representation of f on an open subset U of M, then

$$U \cap I_f = \tilde{f}^{-1}(0).$$

If $M = \mathbb{C}^m$ and if $f : \mathbb{C}^m \longrightarrow \mathbb{P}(V)$ is meromorphic, there exists a reduced representation $\tilde{f} : \mathbb{C}^m \longrightarrow V$ of f.

A meromorphic mapping $f : M \longrightarrow \mathbb{P}^1$ into the Riemann sphere $\mathbb{P}^1$ with $f(M) \neq \infty$ is called a *meromorphic function* on M. All meromorphic functions on M naturally form a field, called the meromorphic function field of M, which is denoted by $\mathcal{M}(M)$. The transcendence degree of $\mathcal{M}(M)$, denoted $\dim_{\mathrm{alg}}(M)$, is called the *algebraic dimension* of M. Then $\dim_{\mathrm{alg}}(M) \leq m$ when M is compact. If M is compact with $\dim_{\mathrm{alg}}(M) = m$, then M is called a *Moishezon space*. A Moishezon space does not differ very much from a projective variety.

Take $f \in \mathcal{M}(M)$ and let $\tilde{f} = (h, g) : U \longrightarrow \mathbb{C}^2$ be a reduced representation of f on an open subset U of M. Then

$$f(z) = \mathbb{P}(\tilde{f}(z)) = \frac{g(z)}{h(z)}, \quad z \in U - h^{-1}(0).$$

Since $f(M) \neq \infty$, then $h^{-1}(0)$ is thin. Further, g, h and U can be taken such that

$$\dim_z h^{-1}(0) \cap g^{-1}(0) \leq m - 2, \quad z \in h^{-1}(0) \cap g^{-1}(0), \tag{2.1.13}$$

and so

$$I_f \cap U = h^{-1}(0) \cap g^{-1}(0).$$

Therefore we obtain analytic hypersurfaces Z_f and P_f of M by defining locally

$$Z_f \cap U = \{g = 0\}, \ P_f \cap U = \{h = 0\},$$

which are called the *zeros* and the *poles* of f, respectively. Note that Z_f and P_f have no common irreducible component. Put

$$X = Z_f \cup P_f = \bigcup_\lambda Y_\lambda,$$

where Y_λ are irreducible components of X. According to the arguments in Section 1.6.5, we also can define the *order* of $f \in \mathcal{A}(U)$ along an irreducible analytic hypersurface Y in M, also denoted by $\mathrm{ord}_Y(f)$. Thus we obtain the *zero divisor* $(f)_0$ and the *pole divisor* $(f)_\infty$ of f by defining locally

$$(f)_0 \cap U = \sum_{\mathrm{ord}_{Y_\lambda}(g) > 0} \mathrm{ord}_{Y_\lambda}(g)(Y_\lambda \cap U),$$

$$(f)_\infty \cap U = \sum_{\mathrm{ord}_{Y_\lambda}(h) > 0} \mathrm{ord}_{Y_\lambda}(h)(Y_\lambda \cap U).$$

The *divisor* of f is defined by

$$(f) = (f)_0 - (f)_\infty.$$

The question arises whether or not an arbitrarily given divisor D on a complex manifold M is always a principal divisor. That is, whether there is a meromorphic function f on M such that $(f) = D$. Cousin answered this question affirmatively if M satisfies certain assumptions, and we therefore call this problem *Cousin's problem*. However, it was later shown by Oka that Cousin's problem is not necessarily solvable under certain topological conditions on M.

Theorem 2.9. *Cousin's problem is solvable for complex space* $\mathbb{C}^m$.

Proof. Siegel [360], Chapter 5, Section 5, Theorem 2. $\qquad\square$

In the case of a single variable, Theorem 2.9 yields Weierstrass' theorem about the existence of entire functions with prescribed zeros. This is the reason for calling Theorem 2.9 the *theorem of Weierstrass and Cousin*. As consequence, for each $f \in \mathcal{M}(\mathbb{C}^m)$ there exist two relatively prime entire functions g and $h \neq 0$ on $\mathbb{C}^m$ satisfying $f = \frac{g}{h}$.

Let Γ be a discrete subgroup of $\mathrm{Aut}(M)$. A meromorphic function f on M is called a *(multiplicative) automorphic function* for Γ if each $\gamma \in \Gamma$ determines an element $j_\gamma \in \mathcal{A}^*(M)$ such that

$$f(\gamma(z)) = j_\gamma(z)f(z), \ z \in M.$$

In particular, f is called a *multiplicative function* if all j_γ are constants, an *automorphic function* if $j_\gamma = 1$ for each $\gamma \in \Gamma$, and called an *automorphic form of weight* k if

$$j_\gamma(z) = J_\gamma(z)^{-k}, \ \gamma \in \Gamma,$$

where J_γ is the Jacobian determinant of γ. Usually, automorphic forms of weight k are assumed to be holomorphic, and satisfy some additional conditions at "infinite points".

For $g \in \mathcal{A}(M)$, the *zero multiplicity* of g at a point $x \in M$ is defined to be the order of vanishing of g at x, denoted by $\mu_g^0(x)$. In terms of local coordinates $z = (z_1, \ldots, z_m)$, that is the greatest integer μ such that all partial derivatives

$$\partial^\nu g(x) = 0, \quad |\nu| \leq \mu - 1,$$

where we denote the *length of a multi-index* $\nu = (i_1, \ldots, i_m) \in \mathbb{Z}_+^m$ by $|\nu| = i_1 + \cdots + i_m$, and write

$$\partial^\nu g = \frac{\partial^{|\nu|} g}{\partial z_1^{i_1} \cdots \partial z_m^{i_m}}.$$

For a meromorphic function $f \in \mathcal{M}(M)$, define the *a-multiplicity* μ_f^a of f as follows: Write locally as $f = g/h$ with g, h holomorphic and

$$\dim g^{-1}(0) \cap h^{-1}(0) \leq m - 2$$

on an open subset U of M and define

$$\mu_f^a|_U = \begin{cases} \mu_{g-ah}^0 & \text{if} \quad a \in \mathbb{C}, \\ \mu_h^0 & \text{if} \quad a = \infty. \end{cases}$$

If $s : M \longrightarrow V$ is a holomorphic vector function, we also can assign a multiplicity μ_s to s. Take $x \in M$. Then there exist an open connected neighborhood U of x, a holomorphic vector function t and a holomorphic function h on U such that $s = ht$ on U and such that $\dim t^{-1}(0) \leq m - 2$. Then $\mu_s(x) = \mu_h^0(x)$ is well defined.

2.1.4 Holomorphic vector bundles

We begin with the general notion of holomorphic vector bundles.

Definition 2.10. *Let N be a complex manifold of dimension n. We call a triple (E, π, N) a holomorphic vector bundle of rank q if the following conditions are satisfied:*

(a) *E is an $(n + q)$-dimensional complex manifold;*

(b) *$\pi : E \longrightarrow N$ is a surjective holomorphic mapping, called the projection;*

(c) *For every $x \in N$, the fiber $E_x = \pi^{-1}(x)$ is a complex vector space of complex dimension q;*

(d) *For every $x \in N$, there exist an open neighborhood U of x and a biholomorphic mapping*

$$\varphi_U : U \times \mathbb{C}^q \longrightarrow \pi^{-1}(U) = E|_U,$$

called a local trivialization of E over U, such that

$$\pi \circ \varphi_U(p, y) = p, \ p \in U, \ y \in \mathbb{C}^q,$$

and $\varphi_{U,p} : \mathbb{C}^q \longrightarrow E_p$ is a linear isomorphism for each $p \in U$, where

$$\varphi_{U,p}(y) = \varphi_U(p, y), \ y \in \mathbb{C}^q.$$

If no confusion occurs, we usually write $\pi : E \longrightarrow N$ or E instead of (E, π, N). In particular, if $q = 1$, then E is called a *holomorphic line bundle* over N. Let Ω be an open subset of N. A holomorphic mapping $s : \Omega \longrightarrow E$ is called a *holomorphic (cross) section* over Ω if

$$\pi \circ s(p) = p, \ p \in \Omega.$$

The vector space of holomorphic sections of E over Ω is denoted by $\Gamma(\Omega, E)$.

Let $\{e_1, \ldots, e_q\}$ be the standard base in $\mathbb{C}^q$, that is,

$$e_j = (0, \ldots, 0, \underset{j\text{th}}{1}, 0, \ldots, 0), \ j = 1, \ldots, q.$$

We can obtain q holomorphic sections on U

$$s_{Uj}(p) = \varphi_U(p, e_j), \ p \in U, \ j = 1, \ldots, q, \tag{2.1.14}$$

such that $\{s_{U1}(p), \ldots, s_{Uq}(p)\}$ is a base of E_p at every $p \in U$. Such $(s_{U1}, \ldots, s_{Uq})$ is called a *holomorphic local frame* of E over U. Conversely, if there is a holomorphic local frame over an open subset $U \subset N$, there is a local trivialization of E over U.

Let $\mathcal{B} = \{U, W, Z, \ldots\}$ be an open covering of N such that for each element in $\mathcal{B}$, say U, there is a biholomorphic mapping φ_U of $U \times \mathbb{C}^q$ onto $E|_U$ with the properties listed above. Then

$$g_{UW} = \varphi_U^{-1} \circ \varphi_W : U \cap W \longrightarrow GL(q, \mathbb{C}) \tag{2.1.15}$$

can be regarded as a holomorphic mapping into the group $GL(q, \mathbb{C})$ of invertible $q \times q$ matrices with complex coefficients, and we have

(e) $g_{UW} g_{WU}$ is the identity in $U \cap W$,

(f) $g_{UW} g_{WZ} g_{ZU}$ is the identity in $U \cap W \cap Z$.

A system of such $q \times q$ matrices g_{UW} with coefficients holomorphic in $U \cap W$ is called a *system of transition matrices*. Conversely, if there exist an open covering $\mathcal{B} = \{U, W, Z, \ldots\}$ of N and a family $\{g_{UW}\}$ of holomorphic mappings (2.1.15) satisfying the conditions (e) and (f), then there is a holomorphic vector bundle (E, π, N) such that $\{g_{UW}\}$ is the system of transition matrices.

Let $\pi : E \longrightarrow N$ be a holomorphic vector bundle over N and Ω an open subset of N. Let $s : \Omega \longrightarrow E$ be a holomorphic section over Ω. If we have a covering $\mathcal{B}$ as above, this means that $\varphi_U^{-1} \circ s = s_U$ is a holomorphic mapping of $\Omega \cap U$ into $\mathbb{C}^q$ such that in $\Omega \cap U \cap W \neq \emptyset$,

$$s_U = g_{UW} s_W. \tag{2.1.16}$$

Conversely, any system of holomorphic mappings s_U of $\Omega \cap U$ into $\mathbb{C}^q$ with these properties corresponds to precisely one holomorphic section of E over Ω.

An E-valued (r, t)-form ω on Ω means to give for every $U \in \mathcal{B}$ an q-tuple ω_U of forms in $A^{r,t}(\Omega \cap U)$ such that

$$\omega_U = g_{UW} \omega_W$$

in $\Omega \cap U \cap W \neq \emptyset$. The vector space of E-valued (r,t)-forms ω on Ω is denoted by $A^{r,t}(\Omega, E)$. Since g_{UW} is holomorphic, it follows that

$$\bar{\partial}\omega_U = g_{UW}\bar{\partial}\omega_W,$$

hence the q-tuples $\bar{\partial}\omega_U$ of forms of type $(r, t+1)$ define an element $\bar{\partial}\omega$ in $A^{r,t+1}(\Omega, E)$. The operator

$$\bar{\partial} : A^{r,t}(\Omega, E) \longrightarrow A^{r,t+1}(\Omega, E)$$

satisfies $\bar{\partial}^2 = 0$. Set

$$A^k(\Omega, E) = \sum_{r+t=k} A^{r,t}(\Omega, E),$$

and so $A^0(\Omega, E)$ is just the C^∞-sections of E over Ω.

Let $Z_{\bar{\partial}}^{r,t}(N, E)$ denote the space of $\bar{\partial}$-closed E-valued differential forms of type (r, t) on N, and we define the *Dolbeault cohomology groups* $H_{\bar{\partial}}^{r,t}(E)$ of E to be

$$H_{\bar{\partial}}^{r,t}(E) = Z_{\bar{\partial}}^{r,t}(N, E)/\bar{\partial}A^{r,t-1}(N, E).$$

Note that $\Omega^r(N, E) = Z_{\bar{\partial}}^{r,0}(N, E)$ is just E-valued holomorphic forms of degree r on N. We obtain a sheaf $\Omega^r(E)$ on N whose sections are given locally by E-valued holomorphic forms of degree r, that is, $\Omega^r(E)(U) = \Omega^r(U, E)$ for an open set $U \subset N$. Specially, $\mathcal{O}(E) = \Omega^0(E)$ denotes the sheaf whose sections are given locally by holomorphic sections of E, that is, $\mathcal{O}(E)(U) = \Gamma(U, E)$ for an open set $U \subset M$. Thus the tth Čech cohomology group $H^t(N, \Omega^r(E))$ of $\Omega^r(E)$ on N is well defined. Similarly, Dolbeault's theorem (cf. [127]) holds:

$$H^t(N, \Omega^r(E)) \cong H_{\bar{\partial}}^{r,t}(E).$$

Let (E, π, N) and (E', π', N) be holomorphic vector bundles. A holomorphic mapping $f : E \longrightarrow E'$ is called a *bundle homomorphism* if $\pi' \circ f = \pi$ and the restriction

$$f|_{E_p} : E_p \longrightarrow E'_p$$

is a linear mapping for any $p \in N$. If f is moreover a biholomorphic mapping, then f is called a *bundle isomorphism*. In this case, E and E' are said to be *isomorphic*. A holomorphic vector bundle is said to be *trivial* if it is isomorphic to the holomorphic vector bundle $(N \times \mathbb{C}^q, \pi, N)$ with the natural projection $\pi : N \times \mathbb{C}^q \longrightarrow N$.

Let $\pi : E \longrightarrow N$ be the holomorphic vector bundle given in Definition 2.10. The *dual holomorphic vector bundle* $\pi^* : E^* \longrightarrow N$ is defined as follows: For any $p \in N$, let E_p^* be the dual vector space of E_p and set

$$E^* = \bigcup_{p \in N} E_p^*; \quad \pi^*(E_p^*) = p, \ p \in N.$$

Then E^* naturally becomes an $(n + q)$-dimensional complex manifold. In fact, using the holomorphic local frame defined by (2.1.14), for every $p \in U$, define $t_{Uk}(p) \in E_p^*$ uniquely by the dual relation

$$\langle s_{Uj}, t_{Uk} \rangle = \delta_{jk} = \begin{cases} 1, & j = k; \\ 0, & j \neq k. \end{cases} \tag{2.1.17}$$

Then $\{t_{U1}(p), \ldots, t_{Uq}(p)\}$ is a base of E_p^* at every $p \in U$. Define a bijective mapping

$$\psi_U : U \times \mathbb{C}^q \longrightarrow E^*|_U$$

by setting

$$\psi_U(p, \lambda_U) = \sum_{k=1}^{q} \lambda_{Uk} t_{Uk}(p).$$

Note that $\varphi_{U,p} : \mathbb{C}^q \longrightarrow E_p$ is a linear isomorphism for each $p \in U$, and so

$$\varphi_U(p, y_U) = \sum_{j=1}^{q} y_{Uj} s_{Uj}(p).$$

For any $p \in U \cap W \neq \emptyset$, we see that the relation

$$\varphi_U(p, y_U) = \varphi_W(p, y_W), \quad y_U, y_W \in \mathbb{C}^q$$

holds if and only if

$$y_U = g_{UW} y_W,$$

where we regard y's as column vectors. Thus when

$$\psi_U(p, \lambda_U) = \psi_W(p, \lambda_W), \quad \lambda_U, \lambda_W \in \mathbb{C}^q,$$

we have

$$\lambda_W y_W = \lambda_U y_U = \lambda_U g_{UW} y_W,$$

where we regard λ's as row vectors, and so

$$\lambda_U = {}^t g_{UW}^{-1} \lambda_W,$$

where we regard λ's as column vectors. Therefore the mapping $\psi_U^{-1} \circ \psi_W$ is given by

$$\psi_U^{-1} \circ \psi_W = {}^t g_{UW}^{-1} = {}^t g_{WU} : U \cap W \longrightarrow GL(q, \mathbb{C}).$$

By this observation, (E^*, π^*, N) becomes a holomorphic vector bundle in the natural manner so that ψ_U is a local trivialization of E over U.

Example 2.11. *Let N be a complex manifold. As usual, we have the holomorphic tangent bundle*

$$\mathbf{T}(N) = \bigcup_{p \in N} \mathbf{T}_p(N),$$

such that the system $\{J_{UW}\}$ of transition matrices consists of the Jacobi matrices J_{UW} of holomorphic coordinate transforms on $U \cap W$. The system of transition matrices of the holomorphic cotangent bundle

$$\mathbf{T}^*(N) = \bigcup_{p \in N} \mathbf{T}_p^*(N)$$

is just $\{\,{}^t J_{UW}^{-1}\}$. Hence $\mathbf{T}^*(N)$ is the dual holomorphic vector bundle of $\mathbf{T}(N)$.

Example 2.12. *Let E and E' be two holomorphic vector bundles over N with the systems $\{g_{UW}\}$ and $\{g'_{UW}\}$ of transition matrices, respectively. Then we have*

(g) *Define*

$$h_{UW} = \begin{pmatrix} g_{UW} & 0 \\ 0 & g'_{UW} \end{pmatrix}.$$

Then $\{h_{UW}\}$ satisfies the conditions (e) and (f), and so determines a holomorphic vector bundle over N called the direct sum of E and E', denoted $E \oplus E'$.

(h) *The tensor product $g_{UW} \otimes g'_{UW}$ of g_{UW} and g'_{UW} satisfies the conditions (e) and (f), and so determines a holomorphic vector bundle over N called the tensor product of E and E', denoted $E \otimes E'$.*

Furthermore, the exterior product bundle $\bigwedge_r E$ $(1 \le r \le q)$ is naturally defined as a holomorphic vector bundle over N. In particular, $\bigwedge_q E$ is called the *determinant bundle* of E and denoted by $\det(E)$. It is a holomorphic line bundle such that it is trivial over U and the system of transition functions is given by $\{\det(g_{UW})\}$.

Example 2.13. *A complex submanifold $F \subset E$ is called a holomorphic vector subbundle if F is itself a holomorphic vector bundle of rank r $(0 \le r \le q)$ over N whose fiber structure is compatible with that of E. Then there is the associated quotient bundle E/F which is a holomorphic vector bundle of rank $q - r$, and there is the exact sequence*

$$0 \longrightarrow F \longrightarrow E \longrightarrow E/F \longrightarrow 0.$$

Let $\pi : E \longrightarrow N$ be a holomorphic vector bundle over a connected complex manifold N. For $s \in \Gamma(N, E)$, let $s^{-1}(0)$ be the zero set. If $\dim_p s^{-1}(0) \le \dim N - 2$, then s is said to be *reduced at* $p \in N$. If $\dim s^{-1}(0) \le \dim N - 2$, then s is *reduced*. If $s \not\equiv 0$, the *zero divisor* (s) of s is defined by the following property: choosing a reduced section $t \in \Gamma(U, E)$ over an open connected subset U of N

and a holomorphic function f over U such that $s|_U = ft$, then $(s) \cap U = (f)$. If U is a Cousin II domain, such t and f exist. Let μ_s be the *multiplicity* of the divisor (s). Obviously, $\mu_s \geq 0$ and $\mathrm{supp}\mu_s \subset s^{-1}(0)$. For line bundles one has $\mathrm{supp}\mu_s = s^{-1}(0)$.

A *connection* of E means a mapping

$$\nabla : A^0(N, E) \longrightarrow A^1(N, E)$$

satisfying the following conditions:

(j) $\nabla(s + t) = \nabla s + \nabla t, \ s, t \in A^0(N, E)$;

(k) $\nabla(as) = da \otimes s + a\nabla s, \ a \in C^\infty(N), \ s \in A^0(N, E)$.

Assume that ∇ is a connection of E. Take $Z \in \Gamma(N, \mathbf{T}(N)), \ s \in A^0(N, E)$. By using the dual relation $\langle, \rangle$ between $\mathbf{T}(N)$ and $\mathbf{T}^*(N)$, the *covariant derivative* $\nabla_Z s \in A^0(N, E)$ *of s in the direction of Z* is defined by

$$\nabla_Z s = \langle Z, \nabla s \rangle.$$

Locally, by using the holomorphic local frame (2.1.14), we can write

$$\nabla s_{U\alpha} = \sum_{\beta=1}^{q} \omega_{U\alpha}^{\beta} \otimes s_{U\beta}, \ \omega_{U\alpha}^{\beta} \in A^1(U).$$

Let $s_U = {}^t(s_{U1}, \ldots, s_{Uq})$ denote the column vector and set

$$\omega_U = (\omega_{U\alpha}^{\beta}), \ 1 \leq \alpha, \beta \leq q.$$

Then we can rewrite the expression into the following matrix form:

$$\nabla s_U = \omega_U \otimes s_U.$$

The matrix ω_U is called the *connection matrix* of ∇ for s_U. If $U \cap W \neq \emptyset$, then

$$s_W = A_{WU} s_U, \qquad (2.1.18)$$

where A_{WU} is a $q \times q$ matrix consisting of holomorphic functions on $U \cap W$ such that $\det(A_{WU}) \in \mathcal{A}^*(U \cap W)$. Since

$$\begin{aligned}
\nabla s_W &= dA_{WU} \otimes s_U + A_{WU} \nabla s_U \\
&= (dA_{WU} + A_{WU}\omega_U) \otimes s_U \\
&= (dA_{WU} A_{WU}^{-1} + A_{WU}\omega_U A_{WU}^{-1}) \otimes s_W,
\end{aligned}$$

we obtain the transformation formula:

$$\omega_W = dA_{WU} A_{WU}^{-1} + A_{WU}\omega_U A_{WU}^{-1}. \qquad (2.1.19)$$

Differentiating (2.1.19), we have

$$d\omega_W A_{WU} - \omega_W \wedge dA_{WU} = dA_{WU} \wedge \omega_U + A_{WU} d\omega_U. \tag{2.1.20}$$

Substituting

$$dA_{WU} = \omega_W A_{WU} - A_{WU}\omega_U$$

into (2.1.20), we obtain

$$\Omega_W A_{WU} = A_{WU}\Omega_U, \tag{2.1.21}$$

where

$$\Omega_U = d\omega_U - \omega_U \wedge \omega_U \tag{2.1.22}$$

is the *curvature matrix* of ∇ on U.

Let E^* be the dual holomorphic vector bundle of E. The dual pairing

$$\langle \, , \, \rangle : E_x \times E_x^* \longrightarrow \mathbb{C}$$

induces a dual pairing

$$\langle \, , \, \rangle : A^0(N, E) \times A^0(N, E^*) \longrightarrow A^0(N).$$

Given a connection ∇ in E, we define a connection, also denoted by ∇, in E^* by the following formula:

$$d\langle s, \eta \rangle = \langle Ds, \eta \rangle + \langle s, D\eta \rangle, \;\; s \in A^0(N, E), \;\; \eta \in A^0(N, E^*). \tag{2.1.23}$$

By using the local holomorphic frame $(t_{U1}, \ldots, t_{Uq})$ of E^* defined by (2.1.17), we have

$$\nabla t_{U\alpha} = -\sum_{\beta=1}^{q} \omega_{U\beta}^{\alpha} \otimes t_{U\beta}. \tag{2.1.24}$$

If $\eta = \sum \eta_\alpha t_{U\alpha}$ is an arbitrary section of E^* over U, then (2.1.24) implies

$$\nabla \eta = \sum_{\alpha} \left(d\eta_\alpha - \sum_{\beta} \omega_{U\alpha}^{\beta} \eta_\beta \right) \otimes t_{U\alpha}. \tag{2.1.25}$$

Example 2.14. *Take $E = \mathbf{T}(N)$ and so $E^* = \mathbf{T}^*(N)$. Let $z_1, \ldots, z_n$ be local holomorphic coordinates on U. Hence*

$$s_{U\alpha} = \frac{\partial}{\partial z_\alpha} = \partial_\alpha, \; \alpha = 1, \ldots, n$$

define a holomorphic local frame on U with the dual holomorphic local frame

$$t_{U\alpha} = dz_\alpha, \; \alpha = 1, \ldots, n.$$

Write

$$\omega_{U\alpha}^{\beta} = \sum_{\gamma} \Gamma_{\alpha\gamma}^{\beta} dz_{\gamma}, \ 1 \le \alpha, \beta \le n. \qquad (2.1.26)$$

Take $f \in A^0(N)$. Then (2.1.25) implies

$$\nabla \partial f = \sum_{\alpha} \left(d\partial_{\alpha} f - \sum_{\beta,\gamma} \Gamma_{\alpha\gamma}^{\beta} \partial_{\beta} f dz_{\gamma} \right) \otimes dz_{\alpha}. \qquad (2.1.27)$$

Let X and Y be holomorphic vector fields on U and set

$$X = \sum_{\alpha} \xi_{\alpha} \partial_{\alpha}, \ Y = \sum_{\beta} \eta_{\beta} \partial_{\beta}.$$

Then we have

$$\nabla \partial f(X, Y) = \sum_{\alpha} \eta_{\alpha} \left(X \partial_{\alpha} f - \sum_{\beta,\gamma} \Gamma_{\alpha\gamma}^{\beta} \partial_{\beta} f \xi_{\gamma} \right) = XYf - \nabla_X Y f.$$

The tensor field $\nabla \partial f$ is called the Hessian of f.

Here we compare the connection ∇ with the operator

$$\bar{\partial} : A^0(N, E) \longrightarrow A^{0,1}(N, E).$$

Take $s \in A^0(N, E)$ and write

$$s|_U = \sum_{\alpha=1}^{q} a_{\alpha} s_{U\alpha}, \ a_{\alpha} \in C^{\infty}(U).$$

Then

$$\bar{\partial} s|_U = \sum_{\alpha=1}^{q} \bar{\partial} a_{\alpha} \otimes s_{U\alpha}.$$

On the other hand, the connection ∇ has a decomposition

$$\nabla = \nabla^{1,0} + \nabla^{0,1}$$

such that

$$\nabla^{1,0} s \in A^{1,0}(N, E), \quad \nabla^{0,1} s \in A^{0,1}(N, E).$$

By using the above expression of the section s, we have

$$\nabla s = \sum_{\alpha} da_{\alpha} \otimes s_{U\alpha} + \sum_{\alpha,\beta} a_{\alpha} \omega_{U\alpha}^{\beta} \otimes s_{U\beta}.$$

If ω_U is of type $(1,0)$, i.e., the matrix ω_U consists of $(1,0)$-forms, then

$$\nabla^{0,1}s = \sum_\alpha \bar{\partial}a_\alpha \otimes s_{U\alpha} = \bar{\partial}s,$$

that is, ∇ agrees with $\bar{\partial}$ in the $(0,1)$-direction. Since $\bar{\partial}A_{UW} = 0$, the formula (2.1.19) shows that ω_W is a matrix of $(1,0)$-forms if and only if ω_U does so, that is, the property that ω_U is of type $(1,0)$ do not depend on choice of the holomorphic local frame. If the connection matrices of ∇ for holomorphic local frames consist of $(1,0)$-forms, then ∇ is called a *connection of type* $(1,0)$.

A function h of class C^∞ on $E \oplus E$ is called an *Hermitian metric* along the fibers of E if for each $p \in N$ the restriction

$$h|_p : E_p \oplus E_p \longrightarrow \mathbb{C}$$

defines an Hermitian metric on the vector space E_p. Define

$$|w|_h = \sqrt{h(w,w)}, \quad w \in E_p.$$

Also E together with h is called an *Hermitian vector bundle*. A connection ∇ of type $(1,0)$ of E with an Hermitian metric h is called an *Hermitian connection* if ∇ is *compatible with the metric* h, that is,

$$dh(s,t) = h(\nabla s, t) + h(s, \nabla t) \tag{2.1.28}$$

holds for any $s, t \in A^0(N, E)$. Such a connection exists uniquely. Locally, by using the holomorphic local frame (2.1.14) and setting

$$h_{\alpha\beta} = h(s_{U\alpha}, s_{U\beta}),$$

then (2.1.28) is equivalent to

$$dh_{\alpha\beta} = \sum_\gamma \omega_{U\alpha}^\gamma h_{\gamma\beta} + \sum_\gamma \overline{\omega}_{U\beta}^\gamma h_{\alpha\gamma}, \tag{2.1.29}$$

or the matrix form

$$dH = \omega_U H + H \,{}^t\overline{\omega}_U, \tag{2.1.30}$$

where

$$H = (h_{\alpha\beta}), \ 1 \le \alpha, \beta \le q.$$

Since ∇ is of type $(1,0)$, we have

$$\partial H = \omega_U H. \tag{2.1.31}$$

Conversely, we can define uniquely the Hermitian connection by (2.1.31).

Differentiating (2.1.31) and using (2.1.30), we see that the curvature matrix of the Hermitian connection satisfies

$$\Omega_U = -\partial\bar{\partial}H H^{-1} + \partial H H^{-1} \wedge \bar{\partial}H H^{-1}, \tag{2.1.32}$$

which consists of $(1,1)$-forms. Differentiating (2.1.30), we have

$$\Omega_U H + H\, {}^t\overline{\Omega}_U = 0. \tag{2.1.33}$$

The formula (2.1.21) shows that the *jth Chern form* $c_j(E,h) \in A^{2j}(N)$ of E for h exists such that

$$\det\left(I + \frac{i}{2\pi}\Omega_U\right) = \sum_{j=0}^{q} c_j(E,h)|_U, \tag{2.1.34}$$

where I is the $q \times q$ unit matrix, and i is the imaginary unit. The $2j$-form $c_j(E,h)$ is closed (see [209]). The formula (2.1.33) further shows that $c_j(E,h)$ is real. In particular, we have

$$c_1(E,h)|_U = \frac{i}{2\pi}\mathrm{tr}(\Omega_U) = -dd^c \log \det(H). \tag{2.1.35}$$

The jth Chern form $c_j(E,h)$ determines an element $c_j(E)$ in the de Rham cohomology group $H^{2j}_{\mathrm{DR}}(N,\mathbb{R})$ of N, called the *jth Chern class* of E. The *jth Chern class* $c_j(N)$ of N is defined to be the *jth Chern class* $c_j(\mathbf{T}(N))$ of the holomorphic tangent bundle $\mathbf{T}(N)$.

If $f : M \longrightarrow N$ is a holomorphic mapping, the *pullback bundle* $\tilde{\pi} : f^*(E) \longrightarrow M$ is defined only up to an isomorphism such that $f \circ \tilde{\pi} = \pi \circ \tilde{f}$, where $\tilde{\pi}$ is the bundle projection and $\tilde{f} : f^*(E) \longrightarrow E$ is a bundle homomorphism over f. The *standard model* is defined by

$$f^*(E) = \{(p, w) \in M \times E \mid f(p) = \pi(w)\},$$

where $\tilde{\pi} : f^*(E) \longrightarrow M$ with $\tilde{\pi}(p, w) = p$ is a vector bundle and where $\tilde{f} : f^*(E) \longrightarrow E$ with $\tilde{f}(p, w) = w$ is a bundle homomorphism over f. Anyway, the restriction

$$\tilde{f}_p = \tilde{f} : f^*(E)_p \longrightarrow E_{f(p)}$$

is an isomorphism. If $s \in \Gamma(N, E)$, then a *lifted section* $s_f \in \Gamma(M, f^*(E))$ of s for f is uniquely defined by

$$s_f(p) = \tilde{f}_p^{-1}(s(f(p))), \quad p \in M.$$

Assume $s \not\equiv 0$ and $f(M) \not\subset s^{-1}(0)$. The lifted section s_f exists with $s_f \not\equiv 0$. Also the pullback divisor f^*D of $D = (s)$ exists such that $\mu_{s_f} \geq \mu_{f^*D}$, where the equality holds for line bundles.

If $(f^*(E), \tilde{\pi}, \tilde{f})$ is the pullback of E under the holomorphic mapping $f : M \longrightarrow N$, then a Hermitian metric h in E induces an Hermitian metric f^*h along the fibers of $f^*(E)$ defined by $f^*h = h \circ (\tilde{f} \oplus \tilde{f})$. The Hermitian metric h along the fibers of E induces the dual metric h^* along the fibers of the dual vector bundle E^* and an Hermitian metric $h^{\wedge r}$ along the fibers of the exterior product $\bigwedge_r E$. If (E_i, h_i) $(1 \le i \le r)$ are Hermitian vector bundles, then Hermitian metrics $h_1 \oplus \cdots \oplus h_r$ along the fibers of the direct sum $E_1 \oplus \cdots \oplus E_r$ and $h_1 \otimes \cdots \otimes h_r$ along the fibers of the tensor product $E_1 \otimes \cdots \otimes E_r$ are defined.

Now we end this subsection by the following facts:

Theorem 2.15. *Let E be a holomorphic vector bundle over a Stein manifold M. For every $z_0 \in M$ and every $s_0 \in E_{z_0}$, one can find an holomorphic section s of E over M such that $s(z_0) = s_0$.*

Proof. See [162], Corollary 5.6.3.   □

Theorem 2.16. *Every holomorphic vector bundle on a non-compact Riemann surface is trivial.*

Proof. See [100], Theorem 30.4.   □

2.1.5 Holomorphic line bundles

Let N be a connected complex manifold and let $\pi : L \longrightarrow N$ be a holomorphic line bundle over N. Let $\mathcal{B} = \{U, W, Z, \dots\}$ be an open covering of N with the trivialization of L,

$$\varphi_U : U \times \mathbb{C} \longrightarrow L|_U = \pi^{-1}(U).$$

Let ξ_U be the holomorphic local frame of L over U defined by

$$\xi_U(p) = \varphi_U(p, 1), \ p \in U.$$

Then the system $\{g_{UW}\}$ of transition functions satisfies

$$\xi_W = \xi_U g_{UW}. \tag{2.1.36}$$

Let κ be an Hermitian metric along fibers of L. Then $\kappa_U = \kappa(\xi_U, \xi_U)$ is a positive C^∞-function on U satisfying

$$\kappa_U = |g_{UW}|^{-2} \kappa_W. \tag{2.1.37}$$

Conversely, positive C^∞-functions κ_U defined on U satisfying (2.1.37) determine an Hermitian metric along fibers of L. By (2.1.35), the (1th) *Chern form* $c_1(L, \kappa)$ of type $(1, 1)$ of L for κ is given by

$$c_1(L, \kappa)|_U = -dd^c \log \kappa_U.$$

Then

$$dc_1(L, \kappa) = 0, \quad c_1(L^*, \kappa^*) = -c_1(L, \kappa), \quad c_1(f^*L, f^*\kappa) = f^*(c_1(L, \kappa)).$$

If (L_i, κ_i) $(1 \le i \le p)$ are Hermitian line bundles, then

$$c_1(L_1 \otimes \cdots \otimes L_p, \kappa_1 \otimes \cdots \otimes \kappa_p) = \sum_{i=1}^{p} c_1(L_i, \kappa_i).$$

The line bundle L is said to be *non-negative* (respectively *positive*) if there exists a Hermitian metric κ along the fibers of L such that $c_1(L, \kappa) \ge 0$ (respectively $c_1(L, \kappa) > 0$). Write $L \ge 0$ (respectively $L > 0$).

Given a holomorphic section $s \in \Gamma(N, L)$, that is, a collection $s = \{s_U\}$ of holomorphic functions $s_U \in \mathcal{A}(U)$ satisfying

$$s_U = g_{UW} s_W. \tag{2.1.38}$$

By using the relation (2.1.36), we have $s_U \xi_U = s_W \xi_W$ on $U \cap W \ne \emptyset$. Hence we often write $s|_U = s_U \xi_U$. We can define the norm $|s|_\kappa$ of s on U as follows:

$$|s|_\kappa^2 = \kappa(s_U \xi_U, s_U \xi_U) = \kappa_U |s_U|^2,$$

and obtain the relation

$$c_1(L, \kappa) = -dd^c \log |s|_\kappa^2.$$

The section s defines a divisor $D := (s)$ on N as follows:

$$D \cap U = (s_U). \tag{2.1.39}$$

More generally, a collection $s = \{s_U\}$ of meromorphic functions $s_U \in \mathcal{M}(U)$ satisfying (2.1.38) is called a *meromorphic section* of L, which also defines a divisor by using (2.1.39).

Let D be a divisor on N. For an arbitrary point $p \in N$, there are irreducible, mutually coprime holomorphic functions $f_1, \ldots, f_r$ in a neighborhood U of p such that $f_1 \cdots f_r = 0$ is the local equation of $\mathrm{supp}(D)$ in U. Taking U smaller if necessary, we may assume that the analytic hypersurfaces

$$X_\lambda = \{x \in U \mid f_\lambda(x) = 0\}, \ \lambda = 1, \ldots, r$$

are distinct and irreducible. Then we may write uniquely

$$D \cap U = \sum_{\lambda=1}^{r} n_\lambda X_\lambda, \ n_\lambda \in \mathbb{Z}.$$

Putting

$$s_U = f_1^{n_1} \cdots f_r^{n_r},$$

we obtain the relation (2.1.39). In this way, we can obtain an open covering $\mathcal{B} = \{U, W, Z, \dots\}$ of N and a family $\{s_U\}_{U \in \mathcal{B}}$ of meromorphic functions such that the relation (2.1.39) holds for $U \in \mathcal{B}$. Set

$$g_{UW} = \frac{s_U}{s_W} \tag{2.1.40}$$

on $U \cap W \neq \emptyset$. Then $g_{UW} \in \mathcal{A}^*(U \cap W)$ satisfy the cocycle conditions (e) and (f), and so determine a holomorphic line bundle over N, denoted by $[D]$, such that $[D]$ is trivial over U, $\{s_U\}_{U \in \mathcal{B}}$ forms a meromorphic section s of $[D]$ and $\{g_{UW}\}$ is the system of transition functions. The line bundle $[D]$ is uniquely determined by D, up to isomorphisms of line bundles. By the construction, we have

$$(s) = D, \quad [-D] = [D]^*, \tag{2.1.41}$$

$$[D + D'] \cong [D] \otimes [D'], \ D, D' \in \mathrm{Div}(N). \tag{2.1.42}$$

Let V be a complex vector space of dimensions $n + 1 \geq 1$. Then the trivial bundle $\mathbb{P}(V) \times V$ contains the *tautological bundle*

$$H^{-1} = \{(x, \eta) \in \mathbb{P}(V) \times V \mid \eta \in E(x)\}$$

as a holomorphic subbundle, where $E(x) = H_x^{-1}$ is the fiber over $x \in \mathbb{P}(V)$. The quotient bundle $Q(V) = (\mathbb{P}(V) \times V)/H^{-1}$ has fiber dimension n. An exact sequence

$$0 \to H^{-1} \to \mathbb{P}(V) \times V \underset{\phi}{\to} Q(V) \to 0 \tag{2.1.43}$$

is defined as the classifying sequence. Taking the dual one obtains the dual classifying sequence

$$0 \to Q(V)^* \to \mathbb{P}(V) \times V^* \underset{\varepsilon}{\to} H \to 0, \tag{2.1.44}$$

where $H = (H^{-1})^*$ is a holomorphic line bundle, called the *hyperplane bundle* over $\mathbb{P}(V)$. If $\alpha \in V^*$, a global holomorphic section s_α of H over $\mathbb{P}(V)$ is defined by

$$s_\alpha(x) = \varepsilon(x, \alpha) = \alpha|E(x), \quad x \in \mathbb{P}(V). \tag{2.1.45}$$

If $\alpha \neq 0$ and $a = \mathbb{P}(\alpha)$, the section s_α is a holomorphic frame of H over $\mathbb{P}(V) - \ddot{E}[a]$, whose dual s_α^* is defined by

$$s_\alpha^*(\mathbb{P}(\xi)) = \frac{\xi}{\alpha(\xi)}, \quad \xi \in V - E[a].$$

Let ℓ be an Hermitian metric on V. Then ℓ induces Hermitian metrics ℓ along the fibers of $\mathbb{P}(V) \times V$, H^{-1} and H, and a *Fubini-Study form* Ω on $\mathbb{P}(V)$ ([103], [388]). Then

$$c_1(H, \ell) = \Omega, \quad c_1(H^{-1}, \ell) = -\Omega. \tag{2.1.46}$$

In particular, if $n = 1$, then $\mathbb{P}(V) \cong \mathbb{C} \cup \{\infty\}$. Restricted to $\mathbb{C}$, we have

$$\Omega(z) = -dd^c \log \chi(z, \infty)^2 = \frac{i}{2\pi} \cdot \frac{dz \wedge d\bar{z}}{(1 + |z|^2)^2}, \tag{2.1.47}$$

where χ is defined by (1.5.6).

Lemma 2.17 (Weyl[430], Stoll[384]). *Take* $a \in \mathbb{P}(V^*)$, *then*

$$\int\limits_{x \in \mathbb{P}(V)} \log \frac{1}{|x, a|^2} \Omega^n(x) = \sum_{j=1}^{n} \frac{1}{j}.$$

Next we recall some notations in Section 1.6.6 in terms of line bundles. Assume that N is a connected compact complex manifold of dimension n. Let L be a holomorphic line bundle over N with a Hermitian metric κ along its fibers. Then the Hodge theorem implies that $\Gamma(N, L)$ is a vector space of finite dimension $k+1$. Assume $k \geq 0$ and let $|L| = \mathbb{P}(\Gamma(N, L))$ be the *complete linear system* of L. Set

$$B_L = \bigcap_{s \in \Gamma(N, L)} s^{-1}(0).$$

Then $B_L \neq N$, and B_L is a (possibly empty) analytic subset of N, called the set of *base points* of the system $|L|$. Consider the *evaluation mapping*

$$e_L : N \times \Gamma(N, L) \longrightarrow L \tag{2.1.48}$$

defined by

$$e_L(x, s) = s(x), \quad (x, s) \in N \times \Gamma(N, L). \tag{2.1.49}$$

Obviously, $e_L(\{x\} \times \Gamma(N, L)) = L_x$ if $x \in N - B_L$. Let E be the kernel of $e_L|_{N-B_L}$. An exact sequence

$$0 \to E \to \{N - B_L\} \times \Gamma(N, L) \to L|_{N-B_L} \to 0 \tag{2.1.50}$$

is defined. Here E has fiber dimension k. If $x \in N - B_L$, one element $\varphi_L(x) \in \mathbb{P}(\Gamma(N, L)^*)$ exists such that $E[\varphi_L(x)] = E_x$. Since B_L is thin, the mapping

$$\varphi_L : N - B_L \longrightarrow \mathbb{P}(\Gamma(N, L)^*)$$

is holomorphic and extends to a meromorphic mapping $\varphi_L : N \longrightarrow \mathbb{P}(\Gamma(N, L)^*)$, which is called a *dual classification mapping* (cf. [380]). If $B_L = \emptyset$, then

$$L = \varphi_L^* H, \tag{2.1.51}$$

where H is the hyperplane bundle on $\mathbb{P}(\Gamma(N, L)^*)$. An Hermitian metric ℓ on $\Gamma(N, L)$ induces Hermitian metrics ℓ along the fibers of H and L such that

$$c_1(L, \ell) = \varphi_L^*(c_1(H, \ell)) = \varphi_L^*(\Omega), \tag{2.1.52}$$

where Ω is the Fubini-Study form on $\mathbb{P}(\Gamma(N, L)^*)$.

The line bundle L is said to be *very ample* if $B_L = \emptyset$, and $\varphi_L : N \longrightarrow \mathbb{P}(\Gamma(N,L)^*)$ gives a projective imbedding. If L^j is very ample for some $j \geq 1$, then L is said to be *ample*. By definition, L is *pseudo ample* if the image $\varphi_{L^j}(N)$ is n-dimensional for some positive integer j. If L is ample with L^j very ample, according to (2.1.52), a metric ℓ^j on L^j induces a metric ℓ on L satisfying

$$jc_1(L, \ell) = c_1(L^j, \ell^j) = \varphi_{L^j}^*(\Omega) > 0,$$

where Ω is the Fubini-Study form on $\mathbb{P}(\Gamma(N, L^j)^*)$, that is, L is positive. Conversely, the Kodaira imbedding theorem (cf. [127]) shows that L is ample if L is positive.

Generally, the so-called L-*dimension* $\kappa(N, L)$ of N is defined by

$$\kappa(N, L) = \max_{j \geq 1} \dim \varphi_{L^j}(N), \qquad (2.1.53)$$

provided $\Gamma(N, L^j) \neq \{0\}$ for some $j \geq 1$. If $\Gamma(N, L^j) = \{0\}$ for all $j \geq 1$, then by convention we set $\kappa(N, L) = -1$. We have in general

$$\kappa(N, L) \leq \dim_{\mathrm{alg}}(N) \leq n. \qquad (2.1.54)$$

The L-dimension $\kappa(N, L)$ is equal to the complex dimension n of N if and only if L is pseudo ample. In this case, N is clearly Moishezon space. Set

$$\mathcal{Z}(N, L) = \{j \geq 1 \mid \Gamma(N, L^j) \neq \{0\} \}.$$

Then $\mathcal{Z}(N, L)$ is a semigroup under addition. Let d denote the greatest common divisor of $\mathcal{Z}(N, L)$. One of the fundamental theorems on L-dimension states

Theorem 2.18. *Let N be a connected compact complex manifold and let L be a holomorphic line bundle over N. Then there exist positive numbers a, b and a positive integer j_0 such that*

$$aj^{\kappa(N,L)} \leq \dim \Gamma(N, L^{jd}) \leq bj^{\kappa(N,L)}, \ \ j \geq j_0.$$

Proof. Ueno [402] or Iitaka [190]. Iitaka [190] used the inequality in the above theorem to define the L-dimension.   □

Particularly, the line bundle L is pseudo ample if and only if

$$\limsup_{j \to +\infty} \frac{1}{j^n} \dim \Gamma(N, L^j) > 0, \qquad (2.1.55)$$

where n is the dimension of N. In fact, when L is pseudo ample, we have the meromorphic imbedding

$$\varphi_{L^k} : N \longrightarrow \mathbb{P}(\Gamma(N, L^k)^*)$$

for some positive integer k. Let z_0 be a point where φ_{L^k} is holomorphic. We choose a basis $\psi_0, \psi_1, \ldots, \psi_n, \ldots$ of $\Gamma(N, L^k)$ in such a way that

$$\psi_0(z_0) \neq 0, \ \psi_1(z_0) = \cdots = \psi_n(z_0) = \cdots = 0,$$

and $\psi_1/\psi_0, \ldots, \psi_n/\psi_0$ form a local coordinate system around z_0. For any positive integer m, the set

$$\{\psi_{i_1}\psi_{i_2}\cdots\psi_{i_m} \mid 0 \leq i_1 \leq i_2 \leq \cdots \leq i_m \leq n\}$$

is a linearly independent subset of $\Gamma(N, L^{km})$. Hence

$$\dim \Gamma(N, L^{km}) \geq \binom{n+m}{m} \geq cm^n,$$

where c is a positive number, and hence (2.1.55) follows. Conversely, if (2.1.55) holds, then L is pseudo ample (see Lemma 2.30 and the remark after Lemma 2.30).

Let N be a complex manifold of dimension n. Let $\mathbf{T}(N)$ be the holomorphic tangent bundle on N and let $\bar{\mathbf{T}}(N)$ be its conjugate. Then $\mathbf{T}(N) \oplus \bar{\mathbf{T}}(N) = T(N)_{\mathbb{C}}$ is the complexified differential tangent bundle. Let $\mathbf{T}^*(N)$ and $\bar{\mathbf{T}}^*(N)$ be the dual of $\mathbf{T}(N)$ and $\bar{\mathbf{T}}(N)$, respectively. The *canonical bundle* of N is defined by

$$K_N = \det(\mathbf{T}^*(N)) = \bigwedge_n \mathbf{T}^*(N).$$

According to S. Lang [229], recall that N is said to be *canonical* if K_N is ample, *very canonical* if K_N is very ample, and *pseudo canonical* if K_N is pseudo ample. Usually, a pseudo canonical manifold is said to be of *general type*. Generally, if N is a singular variety, we say that N is of *general type* or *pseudo canonical* if some desingularization has this property. The following fact is due to Kobayashi [208], Proposition 7.4.3:

Proposition 2.19. *For a compact complex manifold N of dimension n, the following are equivalent:*

(I) K_N *is pseudo ample.*

(II) $\dim K_N = n$, *i.e., the Kodaira dimension of N is equal to n.*

(III) $\limsup_{j \to \infty} \frac{1}{j^n} \dim \Gamma(N, K_N^j) > 0.$

2.1.6 Hermitian manifolds

Let N be a complex manifold of dimension n. An Hermitian metric along the fibers of holomorphic tangent bundle $\mathbf{T}(N)$ on N also is called an *Hermitian metric* of N, and N is called an *Hermitian manifold* if an Hermitian metric is given on N. An Hermitian metric h on N is given by a positive definite Hermitian structure

$$h|_p : \mathbf{T}_p(N) \times \mathbf{T}_p(N) \longrightarrow \mathbb{C}$$

on the holomorphic tangent space at p for each $p \in N$, depending smoothly on p, that is, such that for a local coordinate system $(U; z_1, \ldots, z_n)$ of N, the functions $h_{kl} = h(Z_k, Z_l)$ are C^∞, where

$$Z_j = \frac{\partial}{\partial z_j}, \quad j = 1, \ldots, n.$$

In terms of the local coordinates $z = (z_1, \ldots, z_n)$, the Hermitian metric is given by

$$h = \sum_{k,l} h_{kl} dz_k \otimes d\bar{z}_l.$$

A *coframe* for the Hermitian metric is an n-tuple $(\theta_1, \ldots, \theta_n)$ of forms of type $(1, 0)$ such that

$$h = \sum_k \theta_k \otimes \bar{\theta}_k,$$

i.e., such that, in terms of the Hermitian structure induced on $\mathbf{T}_p^*(N)$ by $h|_p$ on $\mathbf{T}_p(N)$, $(\theta_1(p), \ldots, \theta_n(p))$ is an orthonormal basis for $\mathbf{T}_p^*(N)$. From this description it is clear that coframes always exit locally. The dual of a coframe is a frame.

The real and imaginary parts of an Hermitian inner product on a complex vector space give an ordinary inner product and an alternating quadratic form, respectively, on the underlying real vector space. We see that for an Hermitian metric h on N,

$$ds_N^2 = 2\mathrm{Re}(h) : T_p(N) \otimes T_p(N) \longrightarrow \mathbb{R}$$

is a Riemann metric on N, called the induced *Riemann metric* of the Hermitian metric. When we speak of distance, area, or volume on a complex manifold with an Hermitian metric, we always refer to the induced Riemann metric. It is then customary to write

$$ds_N^2 = 2 \sum_{k,l} h_{kl} dz_k d\bar{z}_l,$$

where

$$dz_k d\bar{z}_l = \frac{1}{2}(dz_k \otimes d\bar{z}_l + d\bar{z}_l \otimes dz_k).$$

We also see that since the quadratic form

$$\varphi = 2\mathrm{Im}(h) : T_p(N) \otimes T_p(N) \longrightarrow \mathbb{R}$$

is alternating, it represents a real differential form of degree 2. In terms of the local coordinates $z = (z_1, \ldots, z_n)$, the form is given by

$$\varphi = -2i \sum_{k,l} h_{kl} dz_k \wedge d\bar{z}_l.$$

Usually we use the form

$$\omega = -\frac{1}{4\pi}\varphi = \frac{i}{2\pi}\sum_{k,l} h_{kl} dz_k \wedge d\bar{z}_l,$$

which is called the *associated* $(1,1)$-*form* (or *Kähler form*) of the metric. If the Kähler form is closed, that is, $d\omega = 0$, then N is called a *Kähler manifold*, and the metric is called the *Kähler metric*.

Let ∇ be the *Hermitian connection* of N, that is, the Hermitian connection of holomorphic tangent bundle $\mathbf{T}(N)$ for an Hermitian metric h. Write the *covariant derivatives* $\nabla_{Z_k} Z_m$ as follows:

$$\nabla_{Z_k} Z_m = \sum_j \Gamma^j_{mk} Z_j,$$

where Γ^j_{mk} are the *connection components* (or *connection coefficients* or *Christoffel symbols*) of ∇. By (2.1.31), we have

$$\Gamma^j_{mk} = \sum_{\beta=1}^n h^{\beta j}\frac{\partial h_{m\beta}}{\partial z_k},$$

where (h^{kj}) is the inverse matrix of (h_{kj}).

For another local coordinate system $(W; w_1, \ldots, w_n)$ of N, we set

$$X_j = \frac{\partial}{\partial w_j}, \quad j = 1, \ldots, n.$$

When $U \cap W \neq \emptyset$, we have

$$X = J_{WU} Z,$$

where $Z = {}^t(Z_1, \ldots, Z_n)$, and

$$J_{WU} = \left(\frac{\partial z_j}{\partial w_k}\right), \ 1 \leq j, k \leq n$$

is the Jacobian matrix of coordinate transformation. Then (2.1.19) implies

$$\omega_W = dJ_{WU} J_{WU}^{-1} + J_{WU}\omega_U J_{WU}^{-1}, \tag{2.1.56}$$

that is,

$$\omega_{Wi}^j = \sum_p d\left(\frac{\partial z_p}{\partial w_i}\right)\frac{\partial w_j}{\partial z_p} + \sum_{p,q}\frac{\partial z_p}{\partial w_i}\frac{\partial w_j}{\partial z_q}\omega_{Up}^q.$$

If we write

$$\omega_{Wi}^j = \sum_k \Xi_{ik}^j dw^k,$$

then the coordinate transformation of connection coefficients is given by

$$
\Xi_{ik}^j = \sum_{p,r,q} \Gamma_{pr}^q \frac{\partial w_j}{\partial z_q} \frac{\partial z_p}{\partial w_i} \frac{\partial z_r}{\partial w_k} + \sum_p \frac{\partial^2 z_p}{\partial w_i \partial w_k} \frac{\partial w_j}{\partial z_p}. \tag{2.1.57}
$$

Express the curvature matrix Ω_U on U of the Hermitian connection ∇ by

$$
\Omega_U = \left(\Omega_m^j \right), \ 1 \le m, j \le n.
$$

We can write

$$
\Omega_m^j = \sum_{k,l} K_{mkl}^j dz_k \wedge d\bar{z}_l,
$$

where K_{mkl}^j are the *curvature components*. By the definition (2.1.22), we obtain

$$
K_{mkl}^j = -\frac{\partial \Gamma_{mk}^j}{\partial \bar{z}_l}.
$$

Define

$$
K_{mjkl} = \sum_{\beta=1}^n K_{mkl}^\beta h_{\beta j}.
$$

Then

$$
K_{mjkl} = -\frac{\partial^2 h_{mj}}{\partial z_k \partial \bar{z}_l} + \sum_{p,q} h^{pq} \frac{\partial h_{mp}}{\partial z_k} \frac{\partial h_{qj}}{\partial \bar{z}_l}. \tag{2.1.58}
$$

If we set

$$
H = (h_{\alpha\beta}), \ 1 \le \alpha, \beta \le n,
$$

and write

$$
K_{kl} = \sum_j K_{jkl}^j,
$$

then

$$
-c_1(\mathbf{T}(N), h) = dd^c \log \det(H) = -\frac{i}{2\pi} \sum_{k,l} K_{kl} dz_k \wedge d\bar{z}_l
$$

is just the *Ricci form*. Fix $p \in N$ and take two non-zero tangent vectors

$$
Z = \sum_k \lambda_k Z_k \in \mathbf{T}_p(N), \quad W = \sum_k \eta_k Z_k \in \mathbf{T}_p(N).
$$

Then

$$
K(Z, W) = \frac{h(R(W, \overline{W})Z, Z)}{h(Z, Z)h(W, W)} = \frac{\sum K_{mjkl} \lambda_m \bar{\lambda}_j \eta_k \bar{\eta}_l}{(\sum h_{mj} \lambda_m \bar{\lambda}_j)(\sum h_{kl} \eta_k \bar{\eta}_l)} \tag{2.1.59}
$$

is what is called the *holomorphic bisectional curvature* determined by Z and W in Goldberg and Kobayashi [115]. In particular, $K(Z, Z)$ is called the *holomorphic sectional curvature* determined by Z.

Definition 2.20. *A variety N is called negatively curved if there exists an Hermitian metric h on N all of whose holomorphic sectional curvatures are bounded from above by a negative constant.*

For each $p \in \mathbb{Z}^+$, define

$$i_p = \left(\frac{\sqrt{-1}}{2\pi}\right)^p (-1)^{\frac{p(p-1)}{2}} p!. \tag{2.1.60}$$

A positive form Ψ of bidegree (n, n) and of class C^∞ on N is called a *volume form* on N. The volume form Ψ on N induces a metric κ_Ψ on the canonical bundle K_N as follows. If $U \neq \emptyset$ is open in N and if ξ and η are forms of bidegree $(n, 0)$ and class C^∞ on U, then

$$i_n \xi \wedge \overline{\eta} = \kappa_\Psi(\xi, \eta)\Psi. \tag{2.1.61}$$

The *Ricci form* of Ψ is defined to be the Chern form of the metric κ_Ψ on K_N, so that

$$\mathrm{Ric}(\Psi) = c_1(K_N, \kappa_\Psi). \tag{2.1.62}$$

In terms of holomorphic coordinates $z_1, \ldots, z_n$, such a form is one which can be written

$$\Psi(z) = \rho(z) \prod_{j=1}^{n} \frac{i}{2\pi} dz_j \wedge d\bar{z}_j,$$

where ρ is a positive C^∞ function. In practice one often has

$$\rho(z) = h(z)|g(z)|^{2q},$$

where h is a positive C^∞ function, $q > 0$ is some fixed rational number, and g is holomorphic not identically zero. According to S. Lang [228], such a form is called a *pseudo volume form*. It is a continuous (n, n)-form and is C^∞ outside a proper analytic subset. The Ricci form of Ψ can be given by

$$\mathrm{Ric}(\Psi) = dd^c \log \rho = dd^c \log h = -\frac{i}{2\pi} \sum_{k,l} K_{kl} dz_k \wedge d\bar{z}_l,$$

where

$$K_{kl} = -\frac{\partial^2 \log h}{\partial z_k \partial \bar{z}_l}, \quad 1 \leq k, l \leq n$$

define the *Ricci tensor* of N with respect to Ψ. Associated with the form Ψ, the *Griffiths function*

$$G(\Psi) = \frac{1}{n!}\mathrm{Ric}(\Psi)^n / \Psi \tag{2.1.63}$$

is defined. In particular, if $n = 1$ and if z is a complex coordinate, then the Griffiths function is given by

$$G(\Psi) = \frac{1}{\rho} \frac{\partial^2 \log \rho}{\partial z \partial \bar{z}}.$$

The function $-G(\Psi)$ is called the *Gauss curvature*.

2.1.7 Parabolic manifolds

Let M be a connected complex manifold of dimension m. A non-negative function

$$\tau : M \longrightarrow \mathbb{R}_+$$

of class C^∞ is said to be an *exhaustion* of M if it is proper, that is, $\tau^{-1}(K)$ is compact whenever K is. Assume that τ is an exhaustion of M. Define

$$v = dd^c \tau, \quad \omega = dd^c \log \tau, \quad \sigma = d^c \log \tau \wedge \omega^{m-1}. \tag{2.1.64}$$

Then it is easy to show the relations

$$\tau^{p+1} \omega^p = \tau v^p - p d\tau \wedge d^c \tau \wedge v^{p-1}, \tag{2.1.65}$$

$$d^c \log \tau \wedge \omega^p = \tau^{-p-1} d^c \tau \wedge v^p. \tag{2.1.66}$$

Denote the *center* for τ by

$$O = O_\tau = \tau^{-1}(0).$$

For $A \subseteq M$ and $r \geq s \geq 0$, define

$$A(O; r) = \{x \in A | \tau(x) < r^2\},$$
$$A[O; r] = \{x \in A | \tau(x) \leq r^2\},$$
$$A\langle O; r \rangle = \{x \in A | \tau(x) = r^2\},$$
$$A[O; s, r] = A[O; r] - A(O; s).$$

Further, if φ is a (p, p)-form on M, write

$$M[O; r; \varphi] = r^{2p-2m} \int_{M[O;r]} \varphi \wedge v^{m-p}, \tag{2.1.67}$$

$$M\langle O; r; \varphi \rangle = \int_{M\langle O;r \rangle} \varphi \wedge d^c \log \tau \wedge \omega^{m-p-1}, \tag{2.1.68}$$

$$M[O; s, r; \varphi] = \int_{M[O;s,r]} \varphi \wedge \omega^{m-p}, \tag{2.1.69}$$

as long as the integrals exist, where $0 \leq p \leq m$. For the case $p = 1$, we will write

$$T(r, s; \varphi) = \int_s^r \frac{M[O; t; \varphi]}{t} dt \ (r > s > 0). \tag{2.1.70}$$

We know that every open manifold, real or complex, always admits an exhaustion function τ such that τ has only isolated critical points in $M - M(O; r(\tau))$ for some $r(\tau)$. Let

$$\mathcal{C}_\tau = \{r^2 \in \mathbb{R}_+ \ |d\tau(x) = 0 \text{ for some } x \in M\langle O; r \rangle\} \tag{2.1.71}$$

be the set of critical values of τ, and define a set

$$\hat{\mathcal{R}}_\tau = \{r | r^2 \in \mathbb{R}_+ - \mathcal{C}_\tau\}. \tag{2.1.72}$$

Then $\mathcal{C}_\tau \cap \mathbb{R}[r(\tau), +\infty)$ is discrete. By Sard's theorem, $\mathcal{C}_\tau$ has measure zero. If $r \in \hat{\mathcal{R}}_\tau$, the boundary $\partial M(O; r) = M\langle O; r \rangle$ is a real, compact, $(2m-1)$-dimensional C^∞-submanifold of M, oriented to the exterior of $M\langle O; r \rangle$.

An exhaustion function τ of M is called *concave* if $v \leq 0$ on $M - M(O; r(\tau))$ for some $r(\tau) \geq 0$, i.e., the eigenvalues of the Levi form $dd^c\tau$ are non-positive from $r(\tau)$ upward, *convex* if $v \geq 0$ on $M - M(O; r(\tau))$ for some $r(\tau)$, *logarithmic concave* or *logarithmic convex* if $\log \tau$ is concave or convex, and *eventually parabolic* if there exists a number $r(\tau) \geq 0$ such that on $M - M[O; r(\tau)]$,

$$\omega \geq 0, \quad \omega^m \equiv 0, \quad v^m \not\equiv 0, \tag{2.1.73}$$

where $\omega^m = 0$ is just the Monge-Ampère homogeneous differential equation on $u = \log \tau$ in complex analysis. This equation is closely related to the study of the curvature of the manifold M. The formulae (2.1.73) will play important role in value distribution theory.

Assume that τ is eventually parabolic, and restricts τ on $M - M[O; r(\tau)]$. Then τ also is convex since

$$v = dd^c\tau \geq \tau dd^c \log \tau = \tau\omega \geq 0.$$

Further, we have

$$\tau v^m = m d\tau \wedge d^c\tau \wedge v^{m-1} = m\tau^m d\tau \wedge \sigma, \tag{2.1.74}$$

$$d\sigma = \omega^m = 0, \tag{2.1.75}$$

so Stokes theorem implies that

$$M\langle O; r; 1 \rangle = M\langle O; s; 1 \rangle \tag{2.1.76}$$

for $s, r \in \hat{\mathcal{R}}_\tau$ with $r(\tau) < s < r$. Denote this common number by ς. Obviously,

$$M[O; r; 1] = \varsigma > 0 \quad (r > r(\tau)). \tag{2.1.77}$$

Particularly, if $r(\tau) = 0$, the exhaustion is simply called *parabolic*. A parabolic exhaustion τ is said to be *strict* if $v > 0$ on M. According to Stoll [380], a complex manifold M is said to be *parabolic* if there exists an unbounded parabolic exhaustion on M.

Let V be a Hermitian vector space of dimension m, say, $V = \mathbb{C}^m$. Define an unbounded exhaustion function τ of V by

$$\tau : V \longrightarrow \mathbb{R}_+, \quad \tau(z) = |z|^2. \tag{2.1.78}$$

Then

$$v = dd^c\tau > 0, \quad \omega = dd^c \log \tau \geq 0.$$

One closed positive form Ω of bidegree (1.1) on $\mathbb{P}(V)$ exists such that $\mathbb{P}^*(\Omega) = \omega$ on V_*. The form Ω is just the Fubini-Study form. It determines the Fubini-Study Käehler metric on $\mathbb{P}(V)$. Obviously

$$\omega^m = \mathbb{P}^*(\Omega^m) = 0, \tag{2.1.79}$$

that is, the exhaustion τ is strict parabolic, and so V is a parabolic manifold. Further, if M has holomorphic rank m, i.e., there exists a holomorphic mapping $\pi : M \longrightarrow \mathbb{C}^m$ of rank m, it follows that $\tau = |\pi|^2$ is an unbounded parabolic exhaustion of M (cf. [380]), and hence M is parabolic.

Strict parabolic exhaustion functions are completely determined by Stoll (cf. [381], [382]) as follows:

Theorem 2.21. *If a complex manifold M of dimension m admits a strict unbounded parabolic exhaustion τ, then there exists a biholomorphic mapping $h : \mathbb{C}^m \longrightarrow M$ such that*

$$\tau(h(z)) = |z|^2 = \sum_{j=1}^{m} |z_j|^2$$

for all $z = (z_1, \ldots, z_m) \in \mathbb{C}^m$.

Alternative proofs were given by Burns [44] and Wong [432]. Bott-Chern [31] and Wu [435] used concave or convex exhaustion. Griffiths and King [128] considered a special parabolic exhaustion which has only finitely many critical values and such that $\log \tau$ has only finitely many logarithmic singularities. Such exhaustion exists on smooth affine algebraic varieties. The properties of parabolic exhaustion were discussed by Stoll [380]. For example, Stoll proved that a non-compact Riemann surface has a parabolic exhaustion if and only if every subharmonic function which is bounded above is constant. Classically the Riemann surfaces satisfying the latter property are called parabolic. Hence Stoll's result shows the consistency of the nomenclature in the non-compact case. However a compact Riemann surface is also parabolic in the classical sense but does not admit a parabolic exhaustion. For $dd^c \tau \geq 0$ would imply that τ is constant which contradicts $dd^c \tau \not\equiv 0$.

2.1.8   Inequalities on symmetric polynomials

We begin with a result (cf. Hardy, Littlewood, and Pólya [146], pp. 104–105, Beckenbach and Bellman [17] , p.11 or [271], pp. 70–71), which is an immediate consequence of Rolle's theorem:

Lemma 2.22. *If the polynomial*

$$P(x, y) = c_0 x^n + c_1 x^{n-1} y + \cdots + c_n y^n$$

has all its zeros x/y real, then the same is true of all polynomials obtained from it by differentiating with respect to x or y.

Consider now the polynomial

$$P(x, y) = \prod_{i=1}^{n} (x + x_i y),$$

where x_i are real. We find

$$P(x, y) = \sum_{k=0}^{n} \rho_k x^{n-k} y^k,$$

where $\rho_0 = 1$, and for $k = 1, 2, \ldots, n$, $\rho_k = \rho_k(x_1, \ldots, x_n)$ is the kth elementary symmetric polynomial. Writing

$$p_k = \rho_k / \binom{n}{k},$$

and by repeated differentiation with respect to x and y, we obtain

$$\frac{\partial^{l+m} P(x, y)}{\partial x^l \partial y^m} = \sum_{k=m}^{n-l} \binom{n}{k} p_k \frac{k!}{(k-m)!} \frac{(n-k)!}{(n-k-l)!} x^{n-k-l} y^{k-m}.$$

If we take $m = r - 1$ and $l = n - r - 1$ with $1 \leq r \leq n - 1$, we have

$$\frac{\partial^{l+m} P(x, y)}{\partial x^l \partial y^m} = \frac{n!}{2} \left(p_{r-1} x^2 + 2 p_r xy + p_{r+1} y^2 \right).$$

By Lemma 2.22, this polynomial has both its zeros real, so that we have

$$p_r^2 - p_{r-1} p_{r+1} \geq 0, \ 1 \leq r \leq n - 1. \tag{2.1.80}$$

This inequality holds for all real x_i, positive, negative, or zero.

Using (2.1.80), when all the x_i are positive, we get

$$\prod_{r=1}^{k} (p_{r-1} p_{r+1})^r \leq \prod_{r=1}^{k} p_r^{2r},$$

or

$$p_{k+1}^{\frac{1}{k+1}} \leq p_k^{\frac{1}{k}}, \ 1 \leq k \leq n - 1, \tag{2.1.81}$$

which is a result of Maclaurin [250]. Thus by (2.1.81), we obtain the inequalities for the elementary symmetric polynomials

$$\rho_j^{\frac{1}{j}} \leq c_{ij} \rho_i^{\frac{1}{i}}, \ 1 \leq i \leq j \leq n, \tag{2.1.82}$$

where

$$c_{ij} = \binom{n}{j}^{\frac{1}{j}} \binom{n}{i}^{-\frac{1}{i}}.$$

From (2.1.81) we obtain

$$p_n^{\frac{1}{n}} \leq p_1,$$

the geometric-mean–arithmetic-mean inequality.

2.2   Kobayashi hyperbolicity

There exists a rich theory and many research results on Kobayashi hyperbolicity. Here we only introduce simple notations and some open problems related to topics in this book. For more details, see Kobayashi [207], [208], and Lang [229].

2.2.1   Hyperbolicity

Let $\mathbb{D}$ be the open unit disc $\{z \in \mathbb{C} \mid |z| < 1\}$. We have the classic *Schwarz-Pick lemma* (cf. Kobayashi [207]):

Theorem 2.23. *Assume* $f \in \mathrm{Hol}(\mathbb{D}, \mathbb{D})$. *Then*

$$\frac{|f'(z)|}{1 - |f(z)|^2} \leq \frac{1}{1 - |z|^2}, \quad z \in \mathbb{D},$$

and equality at a single point z implies that $f \in \mathrm{Aut}(\mathbb{D})$.

We consider the *Hermitian metric* h on $\mathbb{D}$ given by

$$h = \frac{2}{(1 - |z|^2)^2} dz \otimes d\bar{z},$$

which induces the *Riemann metric*

$$ds_{\mathbb{D}}^2 = \frac{4}{(1 - |z|^2)^2} dz d\bar{z}.$$

Then the inequality in Theorem 2.23 may be written as

$$f^* ds_{\mathbb{D}}^2 \leq ds_{\mathbb{D}}^2,$$

or

$$d_h(f(z), f(z')) \leq d_h(z, z')$$

for the associated distance function d_h. The metric h (or $ds_{\mathbb{D}}^2$) is called the *Poincaré metric* or the *Poincaré-Bergman metric* of $\mathbb{D}$. We note that the Gaussian curvature of the metric h is equal to -1 everywhere. By a simple calculation we have

$$d_h(z, w) = \log \frac{1 + |\alpha|}{1 - |\alpha|} \quad (z, w \in \mathbb{D}),$$

where

$$\alpha = \frac{w - z}{1 - \bar{z}w}.$$

Definition 2.24. *Let M be a complex space. Let $x, y \in M$ be arbitrary points. A holomorphic chain α from x to y is the collection of holomorphic mappings $f_i \in \mathrm{Hol}(\mathbb{D}, M)$ and $p_i, q_i \in \mathbb{D}$ for $i = 0, \ldots, l$ such that*

$$f_0(p_0) = x, \quad f_i(q_i) = f_{i+1}(p_{i+1}) \quad (0 \leq i \leq l - 1), \quad f_l(q_l) = y.$$

Then the Kobayashi pseudo distance d_M is given by

$$d_M(x,y) = \inf_\alpha \left\{ \sum_{i=0}^l d_h(p_i, q_i) \right\}, \tag{2.2.1}$$

where the infimum is taken for all holomorphic chains α from x to y.

For the existence of a holomorphic chain from x to y, the reader is referred to S. Lang [229]. It is easy to see that for $x, y, z \in M$,

$$d_M(x,x) = 0, \ d_M(x,y) = d_M(y,x), \ d_M(x,z) \leq d_M(x,y) + d_M(y,z). \tag{2.2.2}$$

In general, a mapping $d_M : M \times M \longrightarrow \mathbb{R}_+$ satisfying the relation above is called a *pseudo distance* which may identically vanish. If M and N are two complex spaces, then the Kobayashi pseudo distances satisfy

$$d_N(f(x), f(y)) \leq d_M(x,y), \ f \in \mathrm{Hol}(M, N), \ \{x,y\} \subset M. \tag{2.2.3}$$

Example 2.25. *Let $M = \mathbb{C}$ with the Euclidean metric. Then $d_\mathbb{C}(x,y) = 0$ for all $x, y \in \mathbb{C}$. In fact, given two points $x, y \in \mathbb{C}$ and an arbitrarily small positive number ε, there is a mapping $f \in \mathrm{Hol}(\mathbb{D}, \mathbb{C})$ such that $f(0) = x$ and $f(\varepsilon) = y$. Hence $d_\mathbb{C}(x,y) \leq \log \frac{1+\varepsilon}{1-\varepsilon}$.*

It will be useful to consider the following generalization. Let M be a subset of a complex Hermitian manifold $\bar{M}$. We can define d_M on M by taking the mappings f_i to lie in M, and to be holomorphic as mappings into $\bar{M}$. Then we obtain a pseudo distance on M. We say that M is *Kobayashi hyperbolic* in $\bar{M}$ if this pseudo distance is a distance, that is if $x \neq y$ implies $d_M(x,y) \neq 0$. For simplicity, we shall say *hyperbolic* instead of Kobayashi hyperbolic. If M is a complex space imbedded in a complex manifold $\bar{M}$, then a mapping into M is analytic if and only if it is analytic viewed as a mapping into $\bar{M}$. Therefore the definition of the Kobayashi pseudo distance on M is intrinsic, independent of the imbedding of M into a manifold.

If M is hyperbolic, then directly from (2.2.3) and Example 2.25, there cannot be a non-constant holomorphic mapping $f : \mathbb{C} \longrightarrow M$. The converse is due to Brody [38]:

Theorem 2.26 ([38]). *Let M be a relatively compact complex subspace of a complex Hermitian manifold $\bar{M}$, and suppose M is not hyperbolic. Then there exists a non-constant holomorphic mapping $f : \mathbb{C} \longrightarrow \bar{M}$ such that*

$$\|f'(0)\| = 1; \quad \|f'(z)\| \leq 1, \ z \in \mathbb{C}.$$

Recall that the induced linear mapping

$$f'(z) : \mathbf{T}_z(\mathbb{C}) = \mathbb{C} \longrightarrow \mathbf{T}_{f(z)}(\bar{M}),$$

is the *holomorphic differential* at each $z \in \mathbb{C}$. Each complex tangent space has
its norm: $\mathbf{T}_{f(z)}(\bar{M})$ has the Hermitian norm, and $\mathbf{T}_z(\mathbb{C}) = \mathbb{C}$ has the Euclidean
norm. The *norm* of the linear mapping $f'(z)$ is defined as usual:

$$\|f'(z)\| = \sup_v \frac{\|f'(z)v\|}{\|v\|}, \quad v \in \mathbf{T}_z(\mathbb{C}), \ v \neq 0.$$

Based on this theorem, it is useful to define a complex space M to be *Brody
hyperbolic* if every holomorphic mapping of $\mathbb{C}$ into M is constant.

   Let M be any variety. Lang [228], [232] introduces the *holomorphic special
set* $\mathrm{Sp}_{\mathrm{hol}}(M)$ of M to be the Zariski closure of the union of all images of non-
constant holomorphic mappings $f : \mathbb{C} \longrightarrow M$. Thus M is hyperbolic if and only if
this special set is empty. In general the special set may be the whole variety. Here
we consider a smooth toroidal compactification M of $\mathcal{D}/\Gamma$, where $\mathcal{D}$ is a bounded
symmetric domain of $\mathbb{C}^m$ and $\Gamma \subset \mathrm{Aut}(\mathcal{D})$ is an arithmetic subgroup. In general,
Γ may not act freely on $\mathcal{D}$, but a subgroup of finite index will act without fixed
points, and we lose no essential generality in assuming that this is true for $\mathcal{D}$. It
is well known that $\mathcal{D}/\Gamma$ is negatively curved since in fact the *Bergman metric* on
$\mathcal{D}$ has negative holomorphic sectional curvatures $\leq -c < 0$ and is Γ-invariant.
Further, $\mathcal{D}/\Gamma$ is hyperbolic (see [207], [286]), and so $\mathrm{Sp}_{\mathrm{hol}}(M) \not\subset \mathcal{D}/\Gamma$. It is a
basic theorem of Baily-Borel [11] that $\mathcal{D}/\Gamma$ is quasi-projective. It is natural to ask
whether $\mathrm{Sp}_{\mathrm{hol}}(M) \subset M - \mathcal{D}/\Gamma$?

   Kiernan and Kobayashi [204] discuss the notion of M being *hyperbolic modulo
a subset S*, meaning that the Kobayashi pseudo distance in M satisfies $d_M(x, y) \neq$
0 unless $x = y$ or $x, y \in S$. According to S. Lang [228], the variety M is said to be
pseudo Brody hyperbolic if the special set $\mathrm{Sp}_{\mathrm{hol}}(M)$ is a proper subset; and M is
pseudo Kobayashi hyperbolic if there exists a proper algebraic subset S such that
M is hyperbolic modulo S. S. Lang [228] conjectures that the two definitions are
equivalent with $S = \mathrm{Sp}_{\mathrm{hol}}(M)$.

2.2.2   Measure hyperbolicity

Let M be a complex manifold with volume or pseudo volume form Ψ. Let $C_0(M)$
be the set of continuous functions with compact support. Then Ψ defines a positive
functional on $C_0(M)$ by

$$\varphi \mapsto \int_M \varphi\Psi.$$

Hence there is a unique regular positive measure μ_Ψ such that

$$\int_M \varphi d\mu_\Psi = \int_M \varphi\Psi, \quad \varphi \in C_0(M).$$

   For example, on the ball of radius r in $\mathbb{C}^m$ with center at 0:

$$\mathbb{C}^m(0; r) = \{(z_1, \ldots, z_m) \in \mathbb{C}^m \mid |z|^2 = |z_1|^2 + \cdots + |z_m|^2 < r^2\},$$

there is the standard positive $(1, 1)$-form

$$\omega = 2i \left\{ \sum_k \frac{1}{r^2 - |z|^2} dz_k \wedge d\bar{z}_k + \frac{4|z|^2}{(r^2 - |z|^2)^2} \partial|z| \wedge \bar{\partial}|z| \right\} \qquad (2.2.4)$$

with

$$\Theta_r \;=\; \frac{1}{m!} \omega^m = \frac{2^m r^2}{(r^2 - |z|^2)^{m+1}} \prod_{k=1}^m i dz_k \wedge d\bar{z}_k, \qquad (2.2.5)$$

$$\mathrm{Ric}(\Theta_r) \;=\; -(m+1) dd^c \log(r^2 - |z|^2) = \frac{m+1}{4\pi} \omega. \qquad (2.2.6)$$

Thus the Einstein-Kähler metric on $\mathbb{C}^m(0; r)$ is given by

$$h_r = \sum_{k=1}^m \frac{2}{r^2 - |z|^2} dz_k \otimes d\bar{z}_k + \frac{2}{(r^2 - |z|^2)^2} \left(\sum_{k=1}^m \bar{z}_k dz_k \right) \otimes \left(\sum_{k=1}^m z_k d\bar{z}_k \right) \quad (2.2.7)$$

such that holomorphic sectional curvatures are -1 everywhere.

Lemma 2.27. *Let M be a complex manifold of dimension m and let Ψ be a pseudo volume form on M such that $\mathrm{Ric}(\Psi)$ is positive, and such that there exists a constant $c > 0$ satisfying $cG(\Psi) \geq 1$. Then for all holomorphic mappings $f : \mathbb{C}^m(0; r) \longrightarrow M$, we have*

$$f^* \Psi \leq c \left(\frac{m+1}{4\pi} \right)^m \Theta_r.$$

Proof. See [207], Theorem 4.4; [208], Corollary 2.4.15; or [174]. □

For $r = 1$, we write Θ for Θ_1. The unit ball $\mathbb{C}^m(0; 1)$ will be denoted $\mathbb{B}^m$.

Definition 2.28. *Let M be a complex manifold of dimension m. Let A be a Borel measurable subset of M. A holomorphic chain α for A is the collection of holomorphic mappings $f_i \in \mathrm{Hol}(\mathbb{B}^m, M)$ and open sets U_i in $\mathbb{B}^m$ for $i = 1, 2, \ldots$ such that*

$$A \subset \bigcup_i f_i(U_i).$$

The space M is said to be covered by holomorphic chains if there exists a holomorphic chain for M. Then the Kobayashi measure μ_M is defined by

$$\mu_M(A) = \inf_\alpha \sum_{i=1}^\infty \mu_\Theta(U_i), \qquad (2.2.8)$$

where the infimum is taken for all holomorphic chains α for A, where μ_Θ is the regular measure on $\mathbb{B}^m$ induced by Θ. If $\mu_M(W) > 0$ for all non-empty open sets W in M, then M is called measure hyperbolic.

Since the open sets generate the σ-algebra of Borel measurable sets, it follows that if B is measurable in $\mathbb{B}^m$ and f is holomorphic, then $f(B)$ is measurable. Furthermore, a regular measure satisfies the property that the measure of a set is the infimum of the measures of the open sets containing it. Hence in the definition of the Kobayashi measure, instead of taking open sets U_i we could take measurable sets B_i in $\mathbb{B}^m$. A basic fact is that if μ is a measure on a complex manifold M of dimension m such that every holomorphic mapping $f : \mathbb{B}^m \longrightarrow M$ satisfies

$$\mu(f(B)) \leq \mu_\Theta(B)$$

for every Borel measurable set B in $\mathbb{B}^m$, then $\mu \leq \mu_M$ (cf. [207], Proposition 1.5; [208], Theorem 7.2.6). Thus the complex manifold M satisfying the conditions in Lemma 2.27 is measure hyperbolic (cf. [208], Theorem 7.4.1).

We let

$$n \to_{div} \infty$$

denote the property that n tends to infinity, ordered by divisibility. In speaking of estimates, we use the standard notation of number theorists

$$A(n) \ll B(n), \quad n \to \infty$$

to mean that there is a constant c such that $A(n) \leq cB(n)$ for all sufficiently large n. Here, sufficiently large may mean with respect to the divisibility ordering. We recall two lemmas from basic algebraic geometry (cf. [208], [228]).

Lemma 2.29. *Let X be a variety of dimension n. Let L be a holomorphic line bundle on X. Then*

$$\dim H^0(X, \mathcal{O}(L^m)) \ll m^n, \quad m \to \infty.$$

Proof. Let H be an ample line bundle such that $E = L \otimes H$ is ample. If m is large enough so that H^m is very ample, then the exact sequence of sheaves (cf. [127], p. 139)

$$0 \longrightarrow \mathcal{O}(L^m) \longrightarrow \mathcal{O}(E^m) \longrightarrow \mathcal{O}(E^m|_D) \longrightarrow 0$$

where D is a smooth effective divisor of X obtained as the zero set of a general holomorphic section of H^m, and $E^m|_D$ denotes the restriction of E^m to D, induces an exact sequence

$$0 \longrightarrow H^0(X, \mathcal{O}(L^m)) \longrightarrow H^0(X, \mathcal{O}(E^m)) \longrightarrow H^0(D, \mathcal{O}(E^m|_D))$$

which further implies

$$\dim H^0(X, \mathcal{O}(L^m)) \leq \dim H^0(X, \mathcal{O}(E^m)).$$

Furthermore, since E^m is ample, then Kodaira's vanishing theorem implies

$$\dim H^0(X, \mathcal{O}(E^m)) = \chi(X, E^m).$$

On the other hand (Hirzebruch [161], p. 150), we have

$$\chi(X, E^m) = a_0 + a_1 m + \cdots + a_n m^n,$$

where $a_0, a_1, \ldots, a_n$ are rational numbers determined by characteristic classes of X and E, thus proving the lemma. □

Lemma 2.30. *Let X be a non-singular variety of dimension n. Let L be a holomorphic line bundle on X such that*

$$\dim H^0(X, \mathcal{O}(L^m)) \gg m^n, \quad m \to_{div} \infty.$$

Then for a very ample line bundle E on X,

$$\dim H^0(X, \mathcal{O}(L^m \otimes E^*)) \gg m^n, \quad m \to_{div} \infty,$$

in particular, $H^0(X, \mathcal{O}(L^m \otimes E^*)) \neq \{0\}$.

Proof. Let D be a non-singular effective divisor of X obtained as the zero set of a general holomorphic section of E. We have the exact sequence of sheaves

$$0 \to \mathcal{O}(L^m \otimes E^*) \to \mathcal{O}(L^m) \to \mathcal{O}(L^m|_D) \to 0,$$

whence the exact cohomology sequence

$$0 \to H^0(X, \mathcal{O}(L^m \otimes E^*)) \to H^0(X, \mathcal{O}(L^m)) \to H^0(D, \mathcal{O}(L^m|_D)).$$

Applying Lemma 2.29 to this invertible sheaf on D we conclude that the dimension of the term on the right is $\ll m^{n-1}$, so for m large

$$\dim H^0(X, \mathcal{O}(L^m \otimes E^*)) \gg m^n,$$

and in particular is positive for m large, whence the lemma follows. □

Conversely, if L is a holomorphic line bundle on X, and if E is a very ample line bundle on X such that

$$H^0(X, \mathcal{O}(L^m \otimes E^*)) \neq \{0\}$$

for some positive integer m, then L is pseudo ample (see [208], Lemma 7.3.7). In fact, let α be a non-trivial holomorphic section of $L^m \otimes E^*$ and set

$$\Gamma_X = \{\alpha s \mid s \in \Gamma(X, E)\} \subset \Gamma(X, L^m).$$

A holomorphic projective imbedding of X is well defined by using only the subspace Γ_X, i.e., the sections of L^m that are divisible by α. The imbedding thus obtained is none other than the imbedding φ_E obtained by using $\Gamma(X, E)$. If we use $\Gamma(X, L^m)$, then we obtain only a meromorphic imbedding φ_{L^m} of X into a projective space.

Theorem 2.31 (Kodaira [213], Kobayashi-Ochiai [210]). *Let X be a non-singular pseudo canonical variety. Then X admits a pseudo volume form Ψ with $\mathrm{Ric}(\Psi)$ positive, and X is measure hyperbolic.*

Proof. Set $n = \dim X$. Since X is pseudo canonical, then

$$\dim H^0(X, \mathcal{O}(K_X^m)) \gg m^n$$

for m large, so we can apply Lemma 2.30. Let L be a very ample line bundle on X. We shall obtain a projective imbedding of X by means of some of the sections in $H^0(X, \mathcal{O}(K_X^m))$. By Lemma 2.30, for m large there exists a non-trivial holomorphic section α of $K_X^m \otimes L^*$. Let $\{s_0, \ldots, s_N\}$ be a basis of $H^0(X, \mathcal{O}(L))$. Then

$$\alpha \otimes s_0, \ldots, \alpha \otimes s_N$$

are linearly independent sections of $H^0(X, \mathcal{O}(K_X^m))$. Since $[s_0, \ldots, s_N]$ gives a projective imbedding of X into $\mathbb{P}^N$ because L is assumed very ample, it follows that $\alpha \otimes s_0, \ldots, \alpha \otimes s_N$ vanish simultaneously only at the zeros of α, but nevertheless give the same projective imbedding, which is determined only by their ratios. Then

$$\alpha \overline{\alpha} \otimes \sum_{j=0}^{N} s_j \otimes \overline{s}_j$$

may be considered as a section of

$$(K_X^m L^*)L \otimes (\overline{K}_X^m \overline{L}^*)\overline{L} = K_X^m \otimes \overline{K}_X^m,$$

and can be locally expressed in terms of complex coordinates in the form

$$|g(z)|^2 \sum_{j=0}^{N} |g_j(z)|^2 \Phi(z)^{\otimes m},$$

where as usual $\Phi(z)$ is the standard Euclidean volume form on $\mathbb{C}^n$, while $g(z)$, $g_0(z), \ldots, g_N(z)$ are local holomorphic functions representing $\alpha, s_0, \ldots, s_N$ respectively. Set

$$h(z) = \left(\sum_{j=0}^{N} |g_j(z)|^2 \right)^{\frac{1}{m}}.$$

Then there is a unique pseudo volume form Ψ on X which has the local expression

$$\Psi(z) = |g(z)|^{\frac{2}{m}} h(z) \Phi(z).$$

Furthermore $\mathrm{Ric}(\Psi)$ is positive, because $\mathrm{Ric}(\Psi)$ is the pull-back of the Fubini-Study form on $\mathbb{P}^N$ by the projective imbedding. $\square$

Its converse is due to Burt Totaro [400] (or see Kobayashi [208]), that is, if a non-singular projective variety X admits a pseudo volume form Ψ with $\mathrm{Ric}(\Psi)$ positive, then X is pseudo canonical. In fact, take an open cover $\{U_j\}$ of X with holomorphic coordinates $z_1^j, \ldots, z_n^j$ on U_j, where $n = \dim X$. We obtain a non-vanishing holomorphic section of K_X on each U_j:

$$\xi_j = dz_1^j \wedge \cdots \wedge dz_n^j.$$

According to (2.1.61), Ψ induces a "pseudo" metric $\kappa = \kappa_\Psi$ on K_X such that

$$i_n \xi_j \wedge \overline{\xi_j} = |\xi_j|_\kappa^2 \Psi.$$

Then our assumption on Ψ means

$$\Psi = h_j |g_j|^{2q} i_n \xi_j \wedge \overline{\xi_j},$$

where h_j is a positive C^∞ function on U_j, $q > 0$ is some fixed rational number, and g_j is holomorphic not identically zero. Hence we have

$$|\xi_j|_\kappa^{-2} = h_j |g_j|^{2q}.$$

Write $q = p/m$ for coprime positive integers p and m. If $\xi_k = \lambda_{jk} \xi_j$ on $U_j \cap U_k$, then

$$h_j^m |g_j|^{2p} = |\lambda_{jk}|^{2m} h_k^m |g_k|^{2p}.$$

Define

$$\chi_{jk} = \lambda_{jk}^m \left(g_k g_j^{-1} \right)^p$$

so that

$$h_j^{-m} = |\chi_{jk}|^{-2} h_k^{-m}.$$

Since $g_k g_j^{-1}$ is a holomorphic function on $U_j \cap U_k$ without zeroes, we can define a line bundle H by the system of transition functions $\{\chi_{jk}\}$. Then $\{h_j^{-m}\}$ define a metric ρ on H such that

$$c_1(H, \rho)|_{U_j} = -dd^c \log h_j^{-m} = m\mathrm{Ric}(\Psi)|_{U_j} > 0,$$

that is, H is positive, and so H is ample. Hence $E = H^l$ is very ample for some positive integer l. Since the transition functions for $K_X^{ml} \otimes E^*$ are given by $\{\lambda_{jk}^{ml} \chi_{jk}^{-l}\}$ and since

$$g_j^{pl} = \lambda_{jk}^{ml} \chi_{jk}^{-l} g_k^{pl},$$

$\{g_j^{pl}\}$ represents a holomorphic section of $K_X^{ml} \otimes E^*$. This shows that $\Gamma(X, K_X^{ml} \otimes E^*) \neq \{0\}$. Therefore K_X is pseudo ample according to the remark after Lemma 2.30.

2.2.3   Open problems

Problem 2.32. *Let M be a projective algebraic variety. Determine which of the following conditions are equivalent:*

(1) *M is Kobayashi hyperbolic;*

(2) *All subvarieties of M (including M itself) are pseudo canonical;*

(3) *Every subvariety of M is measure hyperbolic;*

(4) *M is negatively curved;*

(5) *M is Brody hyperbolic.*

Now we know that (1) $\Longleftrightarrow$ (5). Kobayashi has shown that (4) implies (1); otherwise all equivalences above remain unproved. He stated (1) $\Longrightarrow$ (4) as a problem; other implications in the above list are conjectures of Lang.

Problem 2.33. *Let M be a projective algebraic variety. Determine which of the following conditions are equivalent:*

(i) *M is pseudo Kobayashi hyperbolic;*

(ii) *M is pseudo canonical;*

(iii) *M is measure hyperbolic;*

(iv) *There exists a pseudo volume form Ψ for which $\mathrm{Ric}(\Psi) > 0$;*

(v) *M is pseudo Brody hyperbolic.*

Currently what is known is that (ii) $\Longleftrightarrow$ (iv) (see Kodaira [213], Totaro [400]), (ii) $\Longrightarrow$ (iii) (see Kobayashi-Ochiai [210]), (i) $\Longrightarrow$ (iii) (see Kobayashi [207]), and (iii) $\Longrightarrow$ (ii) for surfaces (see Mori-Mukai [282]). Kobayashi [207] posed (iii) $\Longrightarrow$ (ii) as a problem; other implications are conjectures of Lang.

If M is a non-singular projective variety over $\mathbb{C}$, Kobayashi and Ochiai [211] conjectured that if M is hyperbolic then the canonical class K_M is pseudo ample, but Lang [232] made the stronger conjecture:

Conjecture 2.34. *If M is non-singular and hyperbolic, then K_M is ample.*

Here we consider a non-singular hypersurface M of degree d in $\mathbb{P}^n(\mathbb{C})$. When $d \leq n - 1$, then M contains a line through every point (cf. [228]). The adjunction formula immediately implies

$$K_M = \left(K_{\mathbb{P}^n(\mathbb{C})} \otimes [M] \right) |_M = (H|_M)^{d-n-1},$$

where H is the hyperplane line bundle on $\mathbb{P}^n(\mathbb{C})$. Then $d \geq n + 2$ is precisely the condition that makes the canonical bundle K_M ample. When $n \geq 3$, since the Fermat hypersurface

$$x_0^d + \cdots + x_n^d = 0$$

contains a line

$$z \in \mathbb{C} \longmapsto [z, c_1 z, \ldots, c_r z, c_{r+1}, \ldots, c_{n-1}, 1] \in \mathbb{P}^n(\mathbb{C}),$$

where $1 \leq r \leq n - 2$, and $c_1, \ldots, c_{n-1}$ are numbers such that

$$1 + c_1^d + \cdots + c_r^d = 0, \quad c_{r+1}^d + \cdots + c_{n-1}^d + 1 = 0,$$

we see that in general the condition that M has ample canonical bundle does not imply M hyperbolic. However, in 1970, S. Kobayashi ([207], p. 132) made the following conjecture:

Conjecture 2.35. *A generic hypersurface of degree $\geq 2n+1$ of $\mathbb{P}^n(\mathbb{C})$ is hyperbolic, and that its complement is complete hyperbolic.*

This conjecture is still open, but there has been some progress on the existence of hyperbolic hypersurfaces of $\mathbb{P}^n(\mathbb{C})$. Examples of hyperbolic hypersurfaces were constructed by R. Brody and M. Green [39], M. Zaidenberg [446], A.M. Nadel [285], H.-K. Hà [138], M. McQuillan [266], J.-P. Demailly and J. El Goul [78], B. Shiffman and M. Zaidenberg [348] in dimension 2, M. Shirosaki [353], C. Ciliberto and M. Zaidenberg [62] in dimension 3, and finally by K. Masuda and J. Noguchi [257], Y.T. Siu and S.K. Yeung [367], B. Shiffman and M. Zaidenberg [349], and H. Fujimoto [108] in any dimension. J. El Goul [90] gave a construction of a hyperbolic surface of degree 14 and J.-P. Demailly [77] later reduced the degree in El Goul's construction to 11. Y.T. Siu and S.K. Yeung [367] also obtained an elegant hyperbolic surface of degree 11 by using their generalized Borel lemma (Theorem 4.34). M. Shirosaki [354] constructed a hyperbolic surface of degree 10. H. Fujimoto [108] improved Shirosaki's construction to give examples of degree 8. J. Duval [86] gave hyperbolic surfaces of degree 6 in $\mathbb{P}^3(\mathbb{C})$. We will introduce a method constructing hyperbolic surfaces of lower degrees in Section 4.9.

2.3 Characteristic functions

In this section, we will define characteristic (or order) functions of meromorphic functions (or mappings) in Nevanlinna theory, which are the counterparts of heights in number theory, and derive the corresponding first main theorem from Jensen's formula similar to the case in Section 1.8, the latter is an analogue of the product formula in Section 1.4.

Recall that $\mathbb{C}^m$ is a parabolic manifold. By (2.1.78) and (2.1.79), a strict unbounded parabolic exhaustion function τ on $\mathbb{C}^m$ is defined by

$$\tau(z) = |z|^2 = \sum_{j=1}^{m} |z_j|^2, \ z = (z_1, \ldots, z_m) \in \mathbb{C}^m. \tag{2.3.1}$$

Obviously, the center $O = \tau^{-1}(0)$ of τ contains only one element 0. For the case $M = \mathbb{C}^m$, we will use the notations

$$\mathbb{C}^m[0;r] = M[O;r], \ \ \mathbb{C}^m[0;r;\varphi] = M[O;r;\varphi], \ \ \mathbb{C}^m\langle 0;r;\varphi\rangle = M\langle O;r;\varphi\rangle,$$

and so on. Further, putting

$$z_j = x_j + \sqrt{-1}y_j, \ \ j = 1,\ldots,m,$$

we have

$$\upsilon = \frac{\sqrt{-1}}{2\pi}\sum_{j=1}^{m}dz_j \wedge d\bar{z}_j = \frac{1}{\pi}\sum_{j=1}^{m}dx_j \wedge dy_j, \tag{2.3.2}$$

and hence

$$\upsilon^m = \frac{m!}{\pi^m}dx_1 \wedge dy_1 \wedge \cdots \wedge dx_m \wedge dy_m.$$

Since the volume of ball $\mathbb{C}^m[0;r]$ of radius $r > 0$ in $\mathbb{R}^{2m}$ is $\frac{\pi^m r^{2m}}{m!}$, we obtain

$$\varsigma = \frac{1}{r^{2m}}\int_{\mathbb{C}^m[0;r]}\upsilon^m = \frac{m!}{\pi^m r^{2m}}\int_{\mathbb{C}^m[0;r]}dx_1dy_1\cdots dx_mdy_m = 1. \tag{2.3.3}$$

For $r > s > 0$, applying Stokes theorem to $d\sigma = \omega^m = 0$ on $\mathbb{C}^m[0;s,r]$, we have

$$\int_{\mathbb{C}^m\langle 0;r\rangle}\sigma = \int_{\mathbb{C}^m\langle 0;s\rangle}\sigma = \varsigma = 1. \tag{2.3.4}$$

Thus the volume of sphere $\mathbb{C}^m\langle 0;r\rangle$ of radius r in $\mathbb{R}^{2m}$ is $\frac{2\pi^m r^{2m-1}}{m!}$.

Let ν be a multiplicity on $\mathbb{C}^m$. For $t > 0$, the *counting function* n_ν is defined by

$$n_\nu(t) = t^{2-2m}\int_{A[0;t]}\nu\upsilon^{m-1} \tag{2.3.5}$$

where $A = \operatorname{supp}\nu$. Here if $m = 1$, we define

$$n_\nu(t) = \sum_{z\in A[0;t]}\nu(z). \tag{2.3.6}$$

Then $n_\nu(t) \to n_\nu(0)$ as $t \to 0$ and (cf. [380])

$$n_\nu(t) = \int_{A(0;t)}\nu\omega^{m-1} + n_\nu(0). \tag{2.3.7}$$

If ν is non-negative, then n_ν increases. Fix $r_0 > 0$. The *valence function* of ν is defined by

$$N_\nu(r) = N_\nu(r,r_0) = \int_{r_0}^{r}n_\nu(t)\frac{dt}{t} \ \ (r \geq r_0). \tag{2.3.8}$$

Similarly, we can define $n_\nu(t)$ and $N_\nu(r)$ for a multiplicity ν on a parabolic manifold.

Take $f \in \mathcal{M}(\mathbb{C}^m)$ and $a \in \mathbb{P}^1$. We will denote the *counting function*

$$n_{\mu_f^a}(t) = \begin{cases} n\left(t, \frac{1}{f-a}\right) & \text{if } a \in \mathbb{C}, \\ n(t, f) & \text{if } a = \infty \end{cases} \tag{2.3.9}$$

and the *valence function*

$$N_{\mu_f^a}(r) = \begin{cases} N\left(r, \frac{1}{f-a}\right) & \text{if } a \in \mathbb{C}, \\ N(r, f) & \text{if } a = \infty. \end{cases} \tag{2.3.10}$$

If $f \not\equiv 0$, we have *Jensen's formula* (cf. [396], [151], [128], [380]):

$$N\left(r, \frac{1}{f}\right) - N(r, f) = \int_{\mathbb{C}^m\langle 0;r\rangle} \log|f|\sigma - \int_{\mathbb{C}^m\langle 0;r_0\rangle} \log|f|\sigma. \tag{2.3.11}$$

The following simple result can be derived directly by using Jensen's formula.

Lemma 2.36. *If $f_1 (\not\equiv 0)$ and $f_2 (\not\equiv 0)$ are meromorphic functions in $\mathbb{C}^m$, then for $r > 0$ we have*

$$N\left(r, \frac{1}{f_1 f_2}\right) - N(r, f_1 f_2) = N\left(r, \frac{1}{f_1}\right) + N\left(r, \frac{1}{f_2}\right) - N(r, f_1) - N(r, f_2).$$

For every real number $\alpha \geq 0$, write

$$\log^+ \alpha := \max\{0, \log\alpha\} = \begin{cases} \log\alpha, & \alpha \geq 1, \\ 0, & 0 \leq \alpha < 1. \end{cases}$$

If $f \in \mathcal{M}(\mathbb{C}^m) - \{0\}$, define the *proximity function* of f by

$$m(r, f) = \int_{\mathbb{C}^m\langle 0;r\rangle} \log^+ |f|\sigma \geq 0. \tag{2.3.12}$$

The (*Nevanlinna*) *characteristic function* of f is defined by

$$T(r, f) = m(r, f) + N(r, f). \tag{2.3.13}$$

Then we can rewrite the Jensen formula (2.3.11) as follows:

$$T\left(r, \frac{1}{f}\right) = T(r, f) - \int_{\mathbb{C}^m\langle 0;r_0\rangle} \log|f|\sigma. \tag{2.3.14}$$

Take $a \in \mathbb{C}$. By applying (2.3.14) to $f - a$, we have

$$T\left(r, \frac{1}{f-a}\right) = T(r, f - a) - \int_{\mathbb{C}^m\langle 0;r_0\rangle} \log|f - a|\sigma.$$

Note that
$$T(r, f - a) \leq T(r, f) + \log^+ |a| + \log 2,$$
and
$$\begin{aligned} T(r, f) &= T(r, f - a + a) \\ &\leq T(r, f - a) + \log^+ |a| + \log 2. \end{aligned}$$

One obtains the *first main theorem* (cf. [292], [151], [128], [380])

$$m\left(r, \frac{1}{f - a}\right) + N\left(r, \frac{1}{f - a}\right) = T(r, f) + O(1). \tag{2.3.15}$$

Related to Zhang's inequality (1.9.5), an interesting question is whether there exists a minimal positive constant r^* such that when $r_0 > r^*$, there is a positive constant $c(r_0)$ which depends only on r_0 such that for each non-constant meromorphic function f on $\mathbb{C}$,

$$T(r, f) + T(r, 1 - f) \geq c(r_0), \quad r \geq r_0. \tag{2.3.16}$$

Generally, we consider a meromorphic mapping $f : \mathbb{C}^m \longrightarrow \mathbb{P}(V)$, where V is a Hermitian vector space of dimension $n + 1 > 1$. Let Ω be the Fubini-Study form on $\mathbb{P}(V)$. Then for $t > 0$,

$$A_f(t) = t^{2-2m} \int_{\mathbb{C}^m[0;t]} f^*(\Omega) \wedge \upsilon^{m-1} \tag{2.3.17}$$

is just the *spherical image* of f. The *(Ahlfors-Shimizu) characteristic function* of f is defined by

$$T_f(r) = T_f(r, r_0) = \int_{r_0}^{r} A_f(t) \frac{dt}{t} \quad (r \geq r_0). \tag{2.3.18}$$

The *(growth) order* of the mapping f is defined by

$$\mathrm{Ord}(f) = \limsup_{r \to +\infty} \frac{\log^+ T_f(r)}{\log r}. \tag{2.3.19}$$

If $\tilde{f} : \mathbb{C}^m \longrightarrow V$ is a global representation of f, then $\tilde{f}$ induces a multiplicity $\mu_{\tilde{f}}$. One has (cf. Stoll [384], (6.64))

$$T_f(r) = \int_{\mathbb{C}^m\langle 0;r\rangle} \log |\tilde{f}|\sigma - \int_{\mathbb{C}^m\langle 0;r_0\rangle} \log |\tilde{f}|\sigma - N_{\mu_{\tilde{f}}}(r). \tag{2.3.20}$$

In particular, if $\tilde{f}$ is reduced, then

$$T_f(r) = \int_{\mathbb{C}^m\langle 0;r\rangle} \log |\tilde{f}|\sigma - \int_{\mathbb{C}^m\langle 0;r_0\rangle} \log |\tilde{f}|\sigma. \tag{2.3.21}$$

Further if $\tilde{f}(0) \neq 0$, then limit $r_0 \to 0$ implies

$$T_f(r,0) = \int_{\mathbb{C}^m \langle 0; r \rangle} \log |\tilde{f}| \sigma - \log |\tilde{f}(0)|. \qquad (2.3.22)$$

If f is not constant, then $A_f(t) > 0$ when $t > 0$ and $T_f(r) \to \infty$ as $r \to \infty$ (see Proposition 2.78).

Take a base $e = (e_0, \ldots, e_n)$ of V and let

$$\tilde{f} = \tilde{f}_0 e_0 + \cdots + \tilde{f}_n e_n : \mathbb{C}^m \longrightarrow V \qquad (2.3.23)$$

be a representation of the meromorphic mapping $f : \mathbb{C}^m \longrightarrow \mathbb{P}(V)$. The base e is said to be *allowable* for f if $\tilde{f}_0 \not\equiv 0$. Thus the *jth coordinate function* of f is defined by

$$f_j = \frac{\tilde{f}_j}{\tilde{f}_0}, \quad j = 0, \ldots, n. \qquad (2.3.24)$$

Lemma 2.37 ([386]). *Let $f : \mathbb{C}^m \longrightarrow \mathbb{P}(V)$ be a meromorphic mapping and take a base $e = (e_0, \ldots, e_n)$ of V which is allowable for f. Let f_j be the jth coordinate function of f. Then*

$$T_f(r) \leq \sum_{j=1}^{n} T(r, f_j) + O(1), \qquad (2.3.25)$$

$$T_f(r) \geq T(r, f_j) + N_{\mu_{(\tilde{f}_0, \tilde{f}_j)}}(r) + O(1), \quad j \in \mathbb{Z}[1, n], \qquad (2.3.26)$$

where $\mu_{(\tilde{f}_0, \tilde{f}_j)}$ is the multiplicity of the zero divisor $D_{(\tilde{f}_0, \tilde{f}_j)}$ of $(\tilde{f}_0, \tilde{f}_j)$.

Proof. Let $\tilde{f} = \tilde{f}_0 e_0 + \cdots + \tilde{f}_n e_n : \mathbb{C}^m \longrightarrow V$ be a reduced representation of f. Then the inequality

$$\mu_{\tilde{f}_0}^0 \leq \sum_{j=1}^{n} \mu_{\tilde{f}_j}^{\infty} \qquad (2.3.27)$$

holds on $\mathbb{C}^m - I_f$. In fact, the estimate (2.3.27) holds obviously at $z_0 \in \mathbb{C}^m - I_f$ if $\tilde{f}_0(z_0) \neq 0$. When $\tilde{f}_0(z_0) = 0$, an index $i \in \mathbb{Z}[1, n]$ exists such that $\tilde{f}_i(z_0) \neq 0$ since $I_f = \tilde{f}^{-1}(0)$. Hence

$$\mu_{\tilde{f}_0}^0(z_0) = \mu_{f_i}^{\infty}(z_0) \leq \sum_{j=1}^{n} \mu_{f_j}^{\infty}(z_0).$$

Now (2.3.27) implies

$$N\left(r, \frac{1}{\tilde{f}_0}\right) \leq \sum_{j=1}^{n} N(r, f_j). \qquad (2.3.28)$$

Then

$$|\tilde{f}|^2 = \left| \sum_{j=0}^{n} \tilde{f}_j e_j \right|^2 \leq \left(\sum_{j=0}^{n} |\tilde{f}_j| |e_j| \right)^2$$

$$\leq c \left(\sum_{j=0}^{n} |\tilde{f}_j|^2 \right) \leq c |\tilde{f}_0|^2 \left(1 + \sum_{j=1}^{n} |f_j|^2 \right),$$

where $c = \sum_{j=0}^{n} |e_j|^2$. By (2.3.21), we obtain

$$T_f(r) = \mathbb{C}^m \langle 0; r; \log |\tilde{f}| \rangle - \mathbb{C}^m \langle 0; r_0; \log |\tilde{f}| \rangle$$

$$\leq \sum_{j=1}^{n} \mathbb{C}^m \langle 0; r; \log^+ |f_j| \rangle + \mathbb{C}^m \langle 0; r; \log |\tilde{f}_0| \rangle + O(1). \qquad (2.3.29)$$

Further the Jensen formula (2.3.11) implies

$$N\left(r, \frac{1}{\tilde{f}_0}\right) = \mathbb{C}^m \langle 0; r; \log |\tilde{f}_0| \rangle - \mathbb{C}^m \langle 0; r_0; \log |\tilde{f}_0| \rangle.$$

Thus, by (2.3.28) and (2.3.29), we obtain

$$T_f(r) \leq \sum_{j=1}^{n} \{m(r, f_j) + N(r, f_j)\} + O(1) = \sum_{j=1}^{n} T(r, f_j) + O(1),$$

which proves (2.3.25).

There exists a constant $c' > 0$ such that for all $\xi = x_0 e_0 + \cdots + x_n e_n \in V$ we have

$$|x_0|^2 + |x_1|^2 + \cdots + |x_n|^2 \leq c' |\xi|^2.$$

Because $f_j = \tilde{f}_j / \tilde{f}_0$, a representation $(\tilde{f}_0, \tilde{f}_j)$ of $f_j : \mathbb{C}^m \longrightarrow \mathbb{P}^1$ is given. Hence (2.3.20) yields

$$T(r, f_j) + N_{\mu_{(\tilde{f}_0, \tilde{f}_j)}}(r) = \mathbb{C}^m \left\langle 0; r; \log \sqrt{|\tilde{f}_0|^2 + |\tilde{f}_j|^2} \right\rangle + O(1)$$

$$\leq \mathbb{C}^m \langle 0; r; \log |\tilde{f}| \rangle + O(1)$$

$$= T_f(r) + O(1),$$

and so (2.3.26) follows. $\qquad\qquad\qquad\qquad\qquad\qquad\qquad\qquad\qquad\qquad\square$

Let $\tilde{f}$ be a reduced representation of f. Take $a \in \mathbb{P}(V^*)$, $\alpha \in V^*$ such that $a = \mathbb{P}(\alpha)$. Suppose $f(\mathbb{C}^m) \not\subseteq \ddot{E}[a]$. Then $F = \langle \tilde{f}, \alpha \rangle \ (\neq 0)$ is holomorphic on $\mathbb{C}^m$. Write

$$\mu_f^a = \mu_F^0.$$

The *counting function* and the *valence function* of f for a are defined respectively by

$$n_f(r, a) = n_{\mu_f^a}(r) \tag{2.3.30}$$

and

$$N_f(r, a) = N_{\mu_f^a}(r). \tag{2.3.31}$$

For $r > 0$, the *proximity function* (or *compensation function*) is defined by

$$m_f(r, a) = \int_{\mathbb{C}^m \langle 0; r \rangle} \log \frac{1}{|f, a|} \sigma \geq 0. \tag{2.3.32}$$

Then the *first main theorem* (cf. [128], [385]) states

$$T_f(r) = N_f(r, a) + m_f(r, a) - m_f(r_0, a). \tag{2.3.33}$$

The meromorphic mapping $f : \mathbb{C}^m \longrightarrow \mathbb{P}(V)$ is said to be *linearly non-degenerate*, if $f(\mathbb{C}^m) \not\subset \ddot{E}[a]$ for all $a \in \mathbb{P}(V^*)$. If f is linearly non-degenerate, Lemma 2.17 implies

$$\int_{a \in \mathbb{P}(V^*)} m_f(r, a) \Omega^n(a) = \frac{1}{2} \sum_{j=1}^{n} \frac{1}{j} \quad (r > 0). \tag{2.3.34}$$

Then (2.3.33) and (2.3.34) give us the *integral average theorem* (cf. [295], [128], [380])

$$T_f(r) = \int_{a \in \mathbb{P}(V^*)} N_f(r, a) \Omega^n(a) \quad (r > r_0). \tag{2.3.35}$$

Since f is linearly non-degenerate, then f is not constant, and hence $T_f(r) \to \infty$ as $r \to \infty$. The *defect* of f for a is defined by

$$\delta_f(a) = \liminf_{r \to \infty} \frac{m_f(r, a)}{T_f(r)} \geq 0. \tag{2.3.36}$$

The first main theorem (2.3.33) implies

$$\delta_f(a) = 1 - \limsup_{r \to \infty} \frac{N_f(r, a)}{T_f(r)} \leq 1.$$

The integral average theorem (2.3.35) gives

$$\int_{a \in \mathbb{P}(V^*)} \delta_f(a) \Omega^n(a) = 0. \tag{2.3.37}$$

Hence $\delta_f(a) = 0$ for almost all $a \in \mathbb{P}(V^*)$. As a direct consequence, one obtains the *Casorati-Weierstrass theorem* (cf. Griffiths-King [128], Stoll [384], p. 141)

Theorem 2.38. *A linearly non-degenerate meromorphic mapping from $\mathbb{C}^m$ into a complex projective space intersects almost all hyperplanes.*

Next we compare the functions $T_f(r)$ and $T(r, f)$ for $f \in \mathcal{M}(\mathbb{C}^m)$. Obviously, we have

$$N_f(r, a) = \begin{cases} N\left(r, \frac{1}{f-a}\right) & \text{if } a \in \mathbb{C}, \\ N(r, f) & \text{if } a = \infty. \end{cases} \tag{2.3.38}$$

By applying the first main theorem (2.3.33) to $a = \infty$ and (1.5.6), we have

$$T_f(r) = N(r, f) + \int_{\mathbb{C}^m \langle 0; r \rangle} \log \sqrt{1 + |f|^2} \sigma - \int_{\mathbb{C}^m \langle 0; r_0 \rangle} \log \sqrt{1 + |f|^2} \sigma. \tag{2.3.39}$$

It is easy to show the inequality

$$m(r, f) \leq \int_{\mathbb{C}^m \langle 0; r \rangle} \log \sqrt{1 + |f|^2} \sigma \leq m(r, f) + \log 2.$$

Thus we obtain

$$T_f(r) = T(r, f) + O(1). \tag{2.3.40}$$

Therefore the *defect* of f for a can be calculated by

$$\delta_f(a) = 1 - \limsup_{r \to \infty} \frac{N\left(r, \frac{1}{f-a}\right)}{T(r, f)}. \tag{2.3.41}$$

An element $a \in \mathbb{P}^1$ is said to be a *Picard (exceptional) value* of f if $a \notin f(\mathbb{C}^m)$. Obviously, $\delta_f(a) = 1$ if a is a Picard value of f.

2.4   Growth of rational functions

Corresponding to heights of morphisms discussed in Section 1.9, characteristic functions of rational functions (or mappings) have similar properties. For example, compare the formula (1.9.2) with (2.4.6).

Take $\xi \in \mathbb{C}^m - \{0\}$. Define a holomorphic mapping $j_\xi : \mathbb{C} \longrightarrow \mathbb{C}^m$ by $j_\xi(z) = z\xi$ and write

$$F_\xi = F \circ j_\xi$$

for a function F in $\mathbb{C}^m$. The *Crofton formula* for divisor (cf. [375]) reads as follows:

Theorem 2.39. *Take $0 < R \leq +\infty$. Let D be a divisor on $\mathbb{C}^m(0; R)$ and let $F : \mathbb{C}^m(0; R) \longrightarrow \mathbb{C}$ be a function such that $\mu_D F \omega^{m-1}$ is integrable over $A = \mathrm{supp}D$. Then*

$$J(\xi) = \sum_{0 < |z| < R} \mu_{j_\xi^* D}(z) F_\xi(z)$$

converges for almost all $\xi \in \mathbb{C}^m \langle 0; 1 \rangle$ such that

$$J(e^{i\varphi}\xi) = J(\xi), \quad i = \sqrt{-1}, \; \varphi \in \mathbb{R}.$$

A function $\ddot{J}$ exists almost everywhere on $\mathbb{P}^{m-1}$ such that $\ddot{J} \circ \mathbb{P} = J$ almost everywhere on $\mathbb{C}^m \langle 0; 1 \rangle$. Moreover,

$$\int_A F\omega^{m-1} = \int_{\mathbb{C}^m \langle 0;1 \rangle} J\sigma = \int_{\mathbb{P}^{m-1}} \ddot{J}\Omega^{m-1},$$

where Ω is the Fubini-Study form on $\mathbb{P}^{m-1}$.

For $0 < r < R$, the Crofton formula and (2.3.7) imply

$$n_{\mu_D}(r) = \int_{\mathbb{C}^m \langle 0;1 \rangle} n_{\mu_{j_\xi^* D}}(r)\sigma(\xi),$$

and hence

$$N_{\mu_D}(r) = \int_{\mathbb{C}^m \langle 0;1 \rangle} N_{\mu_{j_\xi^* D}}(r)\sigma(\xi). \tag{2.4.1}$$

Let $f : \mathbb{C}^m \longrightarrow \mathbb{P}(V)$ be a non-degenerate meromorphic mapping. Take $a \in \mathbb{P}(V^*)$. Note that

$$f_\xi = f \circ j_\xi : \mathbb{C} \longrightarrow \mathbb{P}(V), \quad \xi \in \mathbb{C}^m - \{0\}$$

is a holomorphic mapping. By (2.4.1), one obtains (cf. [379])

$$N_f(r, a) = \int_{\mathbb{C}^m \langle 0;1 \rangle} N_{f_\xi}(r, a)\sigma(\xi).$$

The following result is due to Stoll [378]:

Lemma 2.40. *Take $r > 0$. Let $F : \mathbb{C}^m \langle 0; r \rangle \longrightarrow \mathbb{C}$ be a function. Assume that $F\sigma$ is integrable over $\mathbb{C}^m \langle 0; r \rangle$. Then*

$$\int_{\mathbb{C}^m \langle 0;r \rangle} F\sigma = \int_{\mathbb{C}^m \langle 0;1 \rangle} \left(\frac{1}{2\pi} \int_0^{2\pi} F_\xi \left(re^{i\varphi} \right) d\varphi \right) \sigma(\xi).$$

By Lemma 2.40, one obtains (cf. [379])

$$m_f(r, a) = \int_{\mathbb{C}^m \langle 0;1 \rangle} m_{f_\xi}(r, a)\sigma(\xi).$$

The first main theorem (2.3.33) implies

$$T_f(r, r_0) = \int_{\mathbb{C}^m \langle 0;1 \rangle} T_{f_\xi}(r, r_0)\sigma(\xi). \tag{2.4.2}$$

Assume that f is holomorphic at $0 \in \mathbb{C}^m$. Then there exists a reduced representation $\tilde{f} : \mathbb{C}^m \longrightarrow V$ with $\tilde{f}(0) \neq 0$. For $\xi \in \mathbb{C}^m - \{0\}$, $\tilde{f}_\xi = \tilde{f} \circ j_\xi$ is a representation of f_ξ. Hence (2.3.20) and the above arguments show

$$T_{f_\xi}(r,0) \leq \frac{1}{2\pi} \int_0^{2\pi} \log|\tilde{f}_\xi\left(re^{i\varphi}\right)| \, d\varphi - \log|\tilde{f}(0)|. \qquad (2.4.3)$$

A non-negative pluri-subharmonic function u_r on $\mathbb{C}^m$ is defined by

$$u_r(\xi) = \frac{1}{2\pi} \int_0^{2\pi} \log|\tilde{f}_\xi\left(re^{i\varphi}\right)| \, d\varphi - \log|\tilde{f}(0)|.$$

Then (2.3.22) and Lemma 2.40 yield

$$T_f(r,0) = \int_{\mathbb{C}^m\langle 0;1\rangle} u_r \sigma.$$

The following result is due to Kneser [205]:

Theorem 2.41. *Take $0 < r < R \leq +\infty$ and let θ be a number with $0 < \theta < 1$. Let u be a pluri-subharmonic function on $\mathbb{C}^m(0;R)$. Then*

$$u(\zeta) \leq r^{2m-2} \int_{\mathbb{C}^m\langle 0;r\rangle} u(\eta) \frac{r^2 - |\zeta|^2}{|\eta - \zeta|^{2m}} \sigma(\eta), \quad \zeta \in \mathbb{C}^m(0;r).$$

If u is pluri-harmonic, the equality holds. If $\zeta \in \mathbb{C}^m[0;\theta r]$, then

$$u(\zeta) \leq \frac{1+\theta}{(1-\theta)^{2m-1}} \int_{\mathbb{C}^m\langle 0;r\rangle} u^+ \sigma,$$

where $u^+ = \max\{0, u\}$.

Take $\theta \in \mathbb{R}$ with $0 < \theta < 1$ and $\xi \in \mathbb{C}^m[0;1] - \{0\}$. Then (2.4.3) and Theorem 2.41 imply the following inequality (cf. [379]):

$$T_{f_\xi}(r,0) \leq \frac{1+\theta}{(1-\theta)^{2m-1}} T_f\left(\frac{r}{\theta},0\right). \qquad (2.4.4)$$

Lemma 2.42 (cf. [302]). *Let P be a non-constant polynomial in $\mathbb{C}^m$. We have*

$$T(r,P) = \deg(P) \log r + O(1). \qquad (2.4.5)$$

Proof. Write $d = \deg(P)$. Then there exists a positive constant C such that

$$|P(\zeta)| \leq C|\zeta|^d$$

holds when $|\zeta| > 1$. Hence for $r > 1$, we obtain

$$T(r,P) = m(r,P) \leq d \log r + O(1).$$

On the other hand, there exists a homogeneous polynomial H of degree d such that $\deg(P - H) < d$. Take $\xi \in \mathbb{C}^m \langle 0; 1 \rangle$ with $H(\xi) \neq 0$. Then $P_\xi = P \circ j_\xi$ is a polynomial of degree d in $\mathbb{C}$. It is well known that (cf. [168])

$$T(r, P_\xi) = d \log r + O(1).$$

However, the equality (2.4.2) and (2.3.40) imply

$$T(r, P) = \int_{\mathbb{C}^m \langle 0;1 \rangle} T(r, P_\xi) \sigma(\xi) + O(1) = d \log r + O(1).$$

Hence Lemma 2.42 follows. $\qquad\qquad\qquad\qquad\qquad\qquad\qquad\qquad\qquad\qquad\qquad\quad$ $\square$

If Q is another non-constant polynomial in $\mathbb{C}^m$ such that P and Q are co-prime, by using (2.3.40), (2.3.21), (2.4.2) and the method in Lemma 2.42 we can prove that

$$T\left(r, \frac{P}{Q}\right) = \max\{\deg(P), \deg(Q)\} \log r + O(1) \qquad (2.4.6)$$

holds.

An effective divisor D on $\mathbb{C}^m$ is said to be *algebraic* if D is the zero divisor of a polynomial. For the multiplicity ν of an effective divisor on $\mathbb{C}^m$, define

$$n_\nu(\infty) = \lim_{r \to \infty} n_\nu(r) = \lim_{r \to \infty} \frac{N_\nu(r)}{\log r}.$$

The following fact is due to Rutishauser [332] and Stoll [377].

Proposition 2.43. *An effective divisor D on $\mathbb{C}^m$ is algebraic if and only if the counting function n_{μ_D} is bounded. Moreover, if $n_{\mu_D}(\infty) = n < \infty$, then D is the divisor of a polynomial of degree n.*

Lemma 2.44. *Let $f : \mathbb{C}^m \longrightarrow \mathbb{P}^n$ be a meromorphic mapping with a reduced representative $(f_0, f_1, \ldots, f_n) : \mathbb{C}^m \longrightarrow \mathbb{C}^{n+1}$ defined by polynomials $f_0, f_1, \ldots, f_n$ in $\mathbb{C}^m$. Then*

$$A_f(\infty) = \lim_{r \to \infty} A_f(r) = \lim_{r \to \infty} \frac{T_f(r)}{\log r} = \max_{0 \le j \le n}\{\deg(f_j)\}. \qquad (2.4.7)$$

Proof. By using (2.3.21), it is easy to show the following inequality:

$$T_f(r) \le \max_{0 \le j \le n}\{\deg(f_j)\} \log r + O(1).$$

The inequality (2.3.26) in Lemma 2.37 yields

$$T_f(r) \ge T\left(r, \frac{f_j}{f_0}\right) + N_{\mu_{(f_0, f_j)}}(r) + O(1), \quad j = 1, \ldots, n.$$

We can choose polynomials f_{j0}, f_{0j} and h_j such that f_{j0}, f_{0j} are coprime, and

$$f_j = h_j f_{j0}, \quad f_0 = h_j f_{0j}.$$

Thus (2.4.6) implies

$$T\left(r, \frac{f_j}{f_0}\right) = \max\{\deg(f_{j0}), \deg(f_{0j})\} \log r + O(1).$$

Note that $\mu_{(f_0, f_j)} = \mu_{h_j}^0$. By Proposition 2.43, we have

$$\lim_{r \to \infty} \frac{N_{\mu_{(f_0, f_j)}}(r)}{\log r} = n\left(\infty, \frac{1}{h_j}\right) = \deg(h_j). \qquad (2.4.8)$$

Therefore, for $j = 1, \ldots, n$, we obtain

$$\lim_{r \to \infty} \frac{T_f(r)}{\log r} \geq \max\{\deg(f_{j0}), \deg(f_{0j})\} + \deg(h_j) = \max\{\deg(f_j), \deg(f_0)\},$$

and so (2.4.7) follows.   □

The following more general theorem given in [170] is essentially due to Valiron [406].

Theorem 2.45. *Let $f(z)$ be a non-constant meromorphic function in $\mathbb{C}^m$. Take*

$$\{a_0, \ldots, a_p, b_0, \ldots, b_q\} \subset \mathcal{M}(\mathbb{C}^m)$$

with $a_p \neq 0$ and $b_q \neq 0$ such that the rational function in w,

$$R(z, w) = \frac{\sum_{j=0}^p a_j(z) w^j}{\sum_{j=0}^q b_j(z) w^j}, \qquad (2.4.9)$$

is irreducible. Then the function $R_f(z) = R(z, f(z))$ satisfies the estimation

$$T(r, R_f) = \max\{p, q\} T(r, f) + O\left(\sum_{j=0}^p T(r, a_j) + \sum_{j=0}^q T(r, b_j)\right). \qquad (2.4.10)$$

Proof. See A.Z. Mokhon'ko [273], F. Gackstatter and I. Laine [111], P.C. Hu, P. Li and C. C. Yang [168].   □

2.5   Lemma of the logarithmic derivative

From the Poisson-Jensen formula (see Theorem 7.17), A. Gol'dberg and A. Grinshtein [114] estimated sharply growth of logarithmic derivatives of meromorphic functions in $\mathbb{C}$.

Lemma 2.46. *Let f be a non-constant meromorphic function in $\mathbb{C}$, and let $0 < \alpha < 1$. Then, for $r, \rho \in \mathbb{R}^+$ with $r < \rho$, we have*

$$\frac{1}{2\pi} \int_0^{2\pi} \left| \frac{f'(re^{i\theta})}{f(re^{i\theta})} \right|^\alpha d\theta \leq \left\{ \frac{\rho}{r(\rho - r)} \left(m(\rho, f) + m\left(\rho, \frac{1}{f}\right) \right) \right\}^\alpha$$
$$+ \frac{2 + 2^{3-\alpha}}{\cos \frac{\alpha\pi}{2}} \left\{ \frac{n(\rho, f) + n\left(\rho, \frac{1}{f}\right)}{r} \right\}^\alpha.$$

For a meromorphic function f in $\mathbb{C}^m$, following Stoll [375] (or Biancofiore and Stoll [24]) one restricts it to a complex line of $\mathbb{C}^m$ passing through the origin, applies the result of Gol'dberg and Grinshtein, and then averages over the set of all complex lines. Z. Ye [443] proved the following fact (or cf. [168]):

Lemma 2.47. *Let $\nu = (\nu_1, \ldots, \nu_m) \in \mathbb{Z}_+^m$ be a multi-index with length $|\nu| = \nu_1 + \cdots + \nu_m$. Let f be a non-constant meromorphic function in $\mathbb{C}^m$. Then for any α with $0 < \alpha|\nu| < \frac{1}{2}$, there is a constant $C > 1$ such that for any $r_0 < r < \rho < R$, we have*

$$\int_{\mathbb{C}^m \langle 0; r \rangle} \left| \frac{\partial^\nu f}{f} \right|^\alpha \sigma \leq C \left\{ \left(\frac{\rho}{r}\right)^{2m-1} \frac{T(R, f)}{\rho - r} \right\}^{\alpha|\nu|}.$$

Proof of the following lemma can be found in Hinkkanen [159] (also see [295], [151]).

Lemma 2.48. *Let $\varphi(r)$ and $\psi(r)$ be positive non-decreasing functions defined for $r \geq r_1 > 0$ and $r \geq r_2 > 0$, respectively, such that*

$$\int_{r_1}^\infty \frac{dr}{\varphi(r)} = \infty, \qquad \int_{r_2}^\infty \frac{dr}{r\psi(r)} < \infty. \tag{2.5.1}$$

Let $T(r)$ be a positive non-decreasing function defined for $r \geq r_3 \geq r_1$ and $T(r) \geq r_2$. Then if C is real with $C > 1$, one has

$$T\left(r + \frac{\varphi(r)}{\psi(T(r))}\right) \leq CT(r) \tag{2.5.2}$$

whenever $r \geq r_3$ and $r \notin E$, where

$$\int_E \frac{dr}{\varphi(r)} \leq \frac{1}{\psi(w)} + \frac{C}{C-1} \int_w^\infty \frac{dr}{r\psi(r)} < \infty \tag{2.5.3}$$

and $w = \max\{r_2, T(r_3)\}$.

Let a and b be real-valued functions on $\mathbb{R}^+$. We will use the notation

$$\| \ a(r) \leq b(r)$$

to denote that the inequality $a(r) \leq b(r)$ holds as $r \to \infty$ outside of a possible exceptional set E with $\int_E dr/\varphi(r) < \infty$. Next we assume

$$\varphi(r) = O(r) \tag{2.5.4}$$

and put

$$R = r + \frac{\varphi(r)}{\psi(T(r))}, \quad \rho = \frac{R+r}{2} = r + \frac{\varphi(r)}{2\psi(T(r))}. \tag{2.5.5}$$

Lemma 2.48 implies

$$\| \quad T(R) = T\left(r + \frac{\varphi(r)}{\psi(T(r))}\right) \leq CT(r). \tag{2.5.6}$$

On the other hand, for all large r,

$$\frac{\rho}{r} = O(1), \quad \frac{1}{\rho - r} = \frac{2\psi(T(r))}{\varphi(r)}. \tag{2.5.7}$$

Hence we have

$$\| \quad \left(\frac{\rho}{r}\right)^{2m-1} \frac{T(R)}{\rho - r} = O\left(\frac{T(r)\psi(T(r))}{\varphi(r)}\right). \tag{2.5.8}$$

We usually choose the functions ψ and φ as follows

$$\psi(r) = (\log r)^{1+\varepsilon} \ (\varepsilon > 0), \quad \varphi(r) = r. \tag{2.5.9}$$

Thus the inequality (2.5.8) assumes the following form:

$$\| \quad \left(\frac{\rho}{r}\right)^{2m-1} \frac{T(R)}{\rho - r} = O\left(\frac{T(r)(\log T(r))^{1+\varepsilon}}{r}\right). \tag{2.5.10}$$

The lemma of logarithmic derivative due to Nevanlinna [292] and Vitter [414] can be improved as follows:

Lemma 2.49. *Let ψ and φ be defined as in (2.5.1) satisfying $\varphi(r) = O(r)$ and assume that f is a non-constant meromorphic function in $\mathbb{C}^m$. Let $\nu = (\nu_1, \ldots, \nu_m) \in \mathbb{Z}_+^m$ be a multi-index. Then*

$$\| \quad m\left(r, \frac{\partial^\nu f}{f}\right) \leq |\nu| \log^+ \frac{T(r, f)\psi(T(r, f))}{\varphi(r)} + O(1).$$

Proof. By the concavity of the logarithmic function and Lemma 2.48, we obtain

$$m\left(r, \frac{\partial^\nu f}{f}\right) \leq \frac{1}{\alpha} \log^+ \mathbb{C}^m \left\langle 0; r; \left|\frac{\partial^\nu f}{f}\right|^\alpha\right\rangle + O(1)$$

$$\leq |\nu| \log^+ \left(\left(\frac{\rho}{r}\right)^{2m-1} \frac{T(R, f)}{\rho - r}\right) + O(1) \tag{2.5.11}$$

for any $R > \rho > r > r_0$. Therefore the lemma follows by applying the estimate (2.5.8) to the characteristic function $T(R, f)$. $\square$

If one applies the estimate (2.5.10) to $T(R, f)$, one obtains the following form of the lemma of the logarithmic derivative (cf. [443]):

Lemma 2.50. *Assume that f is a non-constant meromorphic function in $\mathbb{C}^m$. Let $\nu = (\nu_1, \ldots, \nu_m) \in \mathbb{Z}_+^m$ be a multi-index. Then for any $\varepsilon > 0$,*

$$\| \quad m\left(r, \frac{\partial^\nu f}{f}\right) \le |\nu| \log^+ T(r, f) + |\nu|(1 + \varepsilon) \log^+ \log T(r, f) + O(1).$$

Take multi-indices $\nu_i \in \mathbb{Z}_+^m$ $(i = 0, 1, \ldots, n)$ with $\nu_0 = 0$ and $|\nu_i| > 0$ $(i = 1, \ldots, n)$. For meromorphic functions $f_0, \ldots, f_n$ in $\mathbb{C}^m$, denote the *generalized Wronskian determinant* with respect to multi-indices $\nu_i \in \mathbb{Z}_+^m$ $(i = 1, \ldots, n)$ by

$$\mathbf{W}(f_0, \ldots, f_n) = \mathbf{W}_{\nu_1 \cdots \nu_n}(f_0, \ldots, f_n) = \begin{vmatrix} f_0 & f_1 & \cdots & f_n \\ \partial^{\nu_1} f_0 & \partial^{\nu_1} f_1 & \cdots & \partial^{\nu_1} f_n \\ \multicolumn{4}{c}{\cdots\cdots\cdots\cdots\cdots\cdots\cdots\cdots} \\ \partial^{\nu_n} f_0 & \partial^{\nu_n} f_1 & \cdots & \partial^{\nu_n} f_n \end{vmatrix},$$

(2.5.12)

and define the *generalized logarithmic Wronskian*

$$\mathbf{S}(f_0, \ldots, f_n) = \mathbf{S}_{\nu_1 \cdots \nu_n}(f_0, \ldots, f_n) = \frac{\mathbf{W}(f_0, \ldots, f_n)}{f_0 \cdots f_n}. \qquad (2.5.13)$$

Following Vitter [414] and Fujimoto [106], first of all, we introduce some basic properties of generalized Wronskians.

Lemma 2.51 (cf.[414], [106]). *Let $f_0, f_1, \ldots, f_n$ be linearly independent meromorphic functions in $\mathbb{C}^m$. Write $f = (f_0, f_1, \ldots, f_n)$. Then there are multi-indices $\nu_i \in \mathbb{Z}_+^m$ $(i = 1, \ldots, n)$ such that $0 < |\nu_i| \le i$ and $f, \partial^{\nu_1} f, \ldots, \partial^{\nu_n} f$ are linearly independent over $\mathbb{C}^m$.*

Proof. Set $\mathcal{F}_0 = \{f\}$, and for any positive integer k, write

$$\mathcal{F}_k = \{\partial^\nu f \mid \nu \in \mathbb{Z}_+^m, \ |\nu| = k\}.$$

We first claim that there is at least one element in $\mathcal{F}_1$ which is independent of f. Assume, to the contrary, that $\partial_{z_j} f$ and f are linearly dependent for $j = 1, \ldots, m$, that is, there exist constants c_j such that

$$\partial_{z_j} f = c_j f, \quad j = 1, \ldots, m. \qquad (2.5.14)$$

By using (2.5.14) and induction, it follows that any $g \in \bigcup_{k=1}^\infty \mathcal{F}_k$ and f are linearly dependent. Take $z_0 = (z_{01}, \ldots, z_{0m}) \in \mathbb{C}^m$ such that $\mu_{f_i}^\infty(z_0) = 0$ $(i = 0, 1, \ldots, n)$. Then f has the Taylor expansion near z_0,

$$\begin{aligned} f(z) &= f(z_0) + \sum_{j=1}^m \partial_{z_j} f(z_0)(z_j - z_{0j}) + \cdots \\ &= f(z_0) + f(z_0) \sum_{j=1}^m c_j(z_j - z_{0j}) + \cdots, \end{aligned}$$

where $z = (z_1, \ldots, z_m)$, that is, $f(z) = f(z_0)h(z)$ for a holomorphic function h near z_0. Hence

$$f_i(z) = f_i(z_0)h(z), \quad i = 0, 1, \ldots, n,$$

which implies that $f_0, f_1, \ldots, f_n$ are linearly dependent in a neighborhood of z_0, and therefore in $\mathbb{C}^m$ by the identity theorem. This is a contradiction. Hence the claim is proved.

Next suppose that

$$f, \partial^{\nu_1} f, \ldots, \partial^{\nu_s} f \ (|\nu_i| \leq i, 1 \leq s < n)$$

are linearly independent over $\mathbb{C}^m$. We will prove that there exists $\nu_{s+1} \in \mathbb{Z}_+^m$ such that $|\nu_{s+1}| \leq s+1$ and $f, \partial^{\nu_1} f, \ldots, \partial^{\nu_s} f, \partial^{\nu_{s+1}} f$ are linearly independent. Assume, to the contrary, that the claim is false. Then $f, \partial^{\nu_1} f, \ldots, \partial^{\nu_s} f$ forms a maximal linearly independent set in $\bigcup_{k=0}^{s+1} \mathcal{F}_k$, and hence in $\bigcup_{k=0}^{\infty} \mathcal{F}_k$ by induction. Thus the Taylor expansion of f at the point $z_0 \in \mathbb{C}^m$ gives

$$f(z) = f(z_0)h_0(z) + \partial^{\nu_1} f(z_0)h_1(z) + \cdots + \partial^{\nu_s} f(z_0)h_s(z),$$

where $h_i \ (i = 0, \ldots, s)$ are holomorphic functions near z_0. In particular, we have

$$f_i(z) = f_i(z_0)h_0(z) + \partial^{\nu_1} f_i(z_0)h_1(z) + \cdots + \partial^{\nu_s} f_i(z_0)h_s(z), \quad i = 0, \ldots, n.$$

Hence $f_0, f_1, \ldots, f_n$ are linearly dependent in a neighborhood of z_0, and so in $\mathbb{C}^m$ by the identity theorem. This is a contradiction. Hence the lemma is proved. $\square$

According to the proof of Lemma 2.51, we can choose carefully the multi-indices $\nu_i \in \mathbb{Z}_+^m$ such that they satisfy the following properties:

Corollary 2.52 (cf. [106]). *Let $f_0, f_1, \ldots, f_n$ be linearly independent meromorphic functions in $\mathbb{C}^m$. Write $f = (f_0, f_1, \ldots, f_n)$. Then there are multi-indices $\nu_i \in \mathbb{Z}_+^m$ such that*

$$0 < |\nu_i| \leq i \ (i = 1, \ldots, n), \quad |\nu_1| \leq \cdots \leq |\nu_n|,$$

$f, \partial^{\nu_1} f, \ldots, \partial^{\nu_n} f$ are linearly independent over $\mathbb{C}^m$, and we have the partition

$$\{\nu_1, \ldots, \nu_n\} = I_1 \cup I_2 \cup \cdots \cup I_s \ (1 \leq s \leq n)$$

such that

$$\emptyset \neq I_k \subset \{\nu \in \mathbb{Z}_+^m \mid |\nu| = k\}, \quad k = 1, \ldots, s,$$

and when $1 \leq k < s$, each element in $\{\partial^{\nu} f \mid \nu \in \mathbb{Z}_+^m, \ |\nu| = k, \ \nu \notin I_k\}$ can be expressed as a linear combination of the family $\{f, \partial^{\nu} f \mid \nu \in I_1 \cup I_2 \cup \cdots \cup I_k\}$. For such multi-indices, the identity

$$\mathbf{W}(hf_0, \ldots, hf_n) = h^{n+1} \mathbf{W}(f_0, \ldots, f_n)$$

holds for any non-zero meromorphic function h on $\mathbb{C}^m$.

Berenstein, Chang and Li [19] proved the following result:

Lemma 2.53. *Let* $f_0, f_1, \ldots, f_n$ *be* $n + 1$ *meromorphic functions in* $\mathbb{C}^m$. *Assume that there are* $\lambda \in J_{n-1}^n$ *and multi-indices* $\nu_i \in \mathbb{Z}_+^m$ *with* $0 < |\nu_i| \le i$, $1 \le i \le n - 1$ *such that*

$$\mathbf{W}_{\nu_1 \cdots \nu_{n-1}} \left(f_{\lambda(0)}, f_{\lambda(1)}, \ldots, f_{\lambda(n-1)} \right) \not\equiv 0.$$

Assume that for all $0 \le i \le n - 1$, $1 \le j \le m$, *one has*

$$\mathbf{W}_{\nu_1 \cdots \nu_{n-1}, \nu_i + \iota_j} \left(f_0, f_1, \ldots, f_{n-1}, f_n \right) \equiv 0,$$

where $\partial^{\nu_0} = 1$, $\iota_j = (0, \ldots, 0, 1, 0, \ldots, 0) \in \mathbb{Z}_+^m$ *in which* 1 *is the* j*th component of* ι_j. *Then* $f_0, f_1, \ldots, f_n$ *are linearly dependent.*

By Lemma 2.53, we can choose the subset I_s in Corollary 2.52 such that

$$\mathbf{W}_{\nu_1 \cdots \nu_n} \left(f_0, \ldots, f_n \right) \not\equiv 0.$$

If $\# I_s = 1$ in Corollary 2.52, then $\nu_n \in I_s$, and Lemma 2.53 shows that we may choose ν_n such that $\nu_n = \nu + \iota_j$ for some $\nu \in I_{s-1}, j \in \mathbb{Z}[1, m]$.

From now on, we will assume that the multi-indices $\nu_i \in \mathbb{Z}_+^m$ in Corollary 2.52 satisfy the above property. Set

$$l = |\nu_1| + \cdots + |\nu_n|, \quad w = |\nu_n|. \tag{2.5.15}$$

The integers l and w will be called the (*Wronskian*) *index* and the *Wronskian degree* of the family $\{f_0, f_1, \ldots, f_n\}$, respectively. Obviously, the numbers w and l satisfy the following properties:

$$1 \le w \le n \le l \le \frac{n(n+1)}{2}. \tag{2.5.16}$$

In particular, if $m = 1$, we have

$$w = n, \quad l = \frac{n(n+1)}{2}. \tag{2.5.17}$$

Set

$$\xi_i = (\xi_{i1}, \ldots, \xi_{in}) = \left(\frac{f_0}{f_i}, \frac{f_1}{f_i}, \ldots, \frac{f_{i-1}}{f_i}, \frac{f_{i+1}}{f_i}, \ldots, \frac{f_n}{f_i} \right), \quad i = 0, 1, \ldots, n.$$

Denote the union of the sets of poles of the ξ_i's by

$$P_f = \bigcup_{i=0}^n \bigcup_{j=1}^n \xi_{ij}^{-1}(\infty),$$

and define

$$
J_i = \begin{pmatrix}
\partial_{z_1}\xi_{i1} & \partial_{z_1}\xi_{i2} & \cdots & \partial_{z_1}\xi_{in} \\
\partial_{z_2}\xi_{i1} & \partial_{z_2}\xi_{i2} & \cdots & \partial_{z_2}\xi_{in} \\
\cdots\cdots\cdots\cdots\cdots\cdots\cdots\cdots\cdots\cdots \\
\partial_{z_m}\xi_{i1} & \partial_{z_m}\xi_{i2} & \cdots & \partial_{z_m}\xi_{in}
\end{pmatrix},
$$

$$
\gamma = \max_{z\in\mathbb{C}^m - P_f}\ \max_{0\le i\le n}\ \mathrm{rank}\,(J_i(z)). \tag{2.5.18}
$$

Note that

$$
\mathbf{W}(f_0, f_1, \ldots, f_n) = (-1)^i f_i^{n+1}\mathbf{W}(1, \xi_{i1}, \ldots, \xi_{in}).
$$

Then there exist integers $j_1, \ldots, j_\gamma$ with $1 \le j_1 < \cdots < j_\gamma \le m$ such that $f, \partial_{z_{j_1}} f, \ldots, \partial_{z_{j_\gamma}} f$ are linearly independent. Thus the multi-indices $\nu_i \in \mathbb{Z}_+^m$ ($i = 1, \ldots, n$) satisfy

$$
|\nu_i| = 1 \ (1 \le i \le \gamma), \quad 2 \le |\nu_i| \le i - \gamma + 1 \ (\gamma + 1 \le i \le n), \tag{2.5.19}
$$

and hence

$$
w \le n - \gamma + 1, \quad l \le \gamma + \frac{(n - \gamma)(n - \gamma + 3)}{2}. \tag{2.5.20}
$$

Lemma 2.47 implies directly the following result:

Lemma 2.54 (cf. [168]). *Let $f_0, \ldots, f_n$ be non-constant meromorphic functions in $\mathbb{C}^m$. Assume that there exists a positive non-decreasing function $T(r)$ in $\mathbb{R}^+$ such that*

$$
T(r, f_j) = O(T(r)), \quad j = 0, \ldots, n.
$$

Then for any real number α with $0 < \alpha|\nu_i| < \frac{1}{2(n+1)}$, there is a constant $C > 1$ such that for any $r_0 < r < \rho < R$, we have

$$
\mathbb{C}^m\ \langle 0; r; |\mathbf{S}(f_0, \ldots, f_n)|^\alpha\rangle \le C\left\{\left(\frac{\rho}{r}\right)^{2m-1}\frac{T(R)}{\rho - r}\right\}^{l\alpha}.
$$

Further we can easily get the following lemma from Lemma 2.54.

Lemma 2.55 (cf. [168]). *Given a family $\mathcal{F} = \{f_0, \ldots, f_q\}$ of meromorphic functions in $\mathbb{C}^m$ such that $q \ge n$ and $\mathbf{W}(f_{\lambda(0)}, \ldots, f_{\lambda(n)}) \not\equiv 0$ for some $\lambda \in J_n^q$. Assume that there exists a positive non-decreasing function $T(r)$ in $\mathbb{R}^+$ such that*

$$
T(r, f_j) = O(T(r)), \quad j = 0, \ldots, q.
$$

Then for any $r_0 < r < \rho < R$, we have

$$
\mathbb{C}^m\ \left\langle 0; r; \log\sum_{\lambda\in J_n^q} |\mathbf{S}(f_{\lambda(0)}, \ldots, f_{\lambda(n)})|\right\rangle \le l\log\left\{\left(\frac{\rho}{r}\right)^{2m-1}\frac{T(R)}{\rho - r}\right\} + O(1).
$$

In particular, if $T(r) = O(r^\mu)$ for some $\mu > 0$, we have

$$
\mathbb{C}^m\ \left\langle 0; r; \log\sum_{\lambda\in J_n^q} |\mathbf{S}(f_{\lambda(0)}, \ldots, f_{\lambda(n)})|\right\rangle \le l(\mu - 1)\log r + O(1).
$$

2.6   Second main theorem

To state the second main theorem in Nevanlinna theory, we need some notation. Let V be a complex vector space of dimensions $n+1 \geq 1$. We consider a linearly non-degenerate meromorphic mapping $f : \mathbb{C}^m \longrightarrow \mathbb{P}(V)$. Take an orthonormal basis $e = (e_0, \ldots, e_n)$ of V and let

$$\tilde{f} = \tilde{f}_0 e_0 + \cdots + \tilde{f}_n e_n : \mathbb{C}^m \longrightarrow V$$

be a reduced representation of f. Since f is linearly non-degenerate, it is equivalent to the fact that $\tilde{f}_0, \ldots, \tilde{f}_n$ are linearly independent in $\mathbb{C}^m$. Set

$$f_j = \frac{\tilde{f}_j}{\tilde{f}_0}, \quad j = 0, \ldots, n. \tag{2.6.1}$$

By Corollary 2.52, there are multi-indices $\nu_i \in \mathbb{Z}_+^m$ $(i = 1, \ldots, n)$ such that $0 < |\nu_i| \leq i$ and

$$\mathbf{W}(f_0, \ldots, f_n) = \mathbf{W}_{\nu_1 \cdots \nu_n}(f_0, \ldots, f_n) \not\equiv 0.$$

The multi-indices $\nu_i \in \mathbb{Z}_+^m$ in Corollary 2.52 do not depend on the choice of a reduced representation $\tilde{f}$ of f. The index $l = |\nu_1| + \cdots + |\nu_n|$ of the family $\{f_0, \ldots, f_n\}$ will be called the (*Wronskian*) *index* of f. The number $|\nu_n|$ is said to be the *Wronskian degree* of f.

Let γ be the rank of f. Then $1 \leq \gamma \leq \min\{m, n\}$. According to Lemma 2.51 and Corollary 2.52, the multi-indices $\nu_i \in \mathbb{Z}_+^m$ $(i = 1, \ldots, n)$ satisfy

$$|\nu_i| = 1 \; (1 \leq i \leq \gamma), \quad 2 \leq |\nu_i| \leq i - \gamma + 1 \; (\gamma + 1 \leq i \leq n), \tag{2.6.2}$$

and hence

$$w \leq n - \gamma + 1, \quad l \leq \gamma + \frac{(n - \gamma)(n - \gamma + 3)}{2}. \tag{2.6.3}$$

The *ramification term*

$$N_{\mathrm{Ram}}(r, f) = N\left(r, \frac{1}{\mathbf{W}(\tilde{f}_0, \ldots, \tilde{f}_n)}\right) \tag{2.6.4}$$

is well defined with respect to the multi-indices $\nu_i \in \mathbb{Z}_+^m$ in Corollary 2.52.

In particular, if f is a non-constant meromorphic function in $\mathbb{C}$, one can find two entire functions g and h without common zeros such that $hf = g$. Hence

$$\tilde{f} = (h, g) : \mathbb{C} \longrightarrow \mathbb{C}^2 - \{0\}$$

is a reduced representative of f. It follows that

$$\mathbf{W}(h, g) = hg' - gh' = h^2 f'.$$

Therefore we have

$$N_{\mathrm{Ram}}(r, f) = 2N(r, f) - N(r, f') + N\left(r, \frac{1}{f'}\right). \qquad (2.6.5)$$

Now we state and prove the *second main theorem* (cf. [295], [151], [384], [414], [159], [443]).

Theorem 2.56. *Let $f : \mathbb{C}^m \longrightarrow \mathbb{P}(V)$ be a linearly non-degenerate meromorphic mapping and let $\mathscr{A} = \{a_0, a_1, \ldots, a_q\}$ be a family of points $a_j \in \mathbb{P}(V^*)$ in general position. Let l be the index of f. Then*

$$(q - n)T_f(r) \;\leq\; \sum_{j=0}^{q} N_f(r, a_j) - N_{\mathrm{Ram}}(r, f)$$

$$+ l \log\left\{\left(\frac{\rho}{r}\right)^{2m-1} \frac{T_f(R)}{\rho - r}\right\} + O(1)$$

holds for any $r_0 < r < \rho < R$, and hence for any $\varepsilon > 0$,

$$(q - n)T_f(r) \;\leq\; \sum_{j=0}^{q} N_f(r, a_j) - N_{\mathrm{Ram}}(r, f)$$

$$+ l\left\{\log T_f(r) + (1 + \varepsilon) \log\log T_f(r) - \log r\right\} + O(1)$$

holds for all large r outside a set E with $\int_E d\log r < \infty$. In particular, if f is of finite order λ, then for any $\varepsilon > 0$, one has

$$(q - n)T_f(r) \;\leq\; \sum_{j=0}^{q} N_f(r, a_j) - N_{\mathrm{Ram}}(r, f)$$

$$+ l(\lambda + \varepsilon - 1) \log r + O(1).$$

Proof. We prove Theorem 2.56 for the case $q \geq n$. For the case $0 \leq q < n$, see the next section. Take $\tilde{a}_i \in V^* - \{0\}$ with $|\tilde{a}_i| = 1$ and $\mathbb{P}(\tilde{a}_i) = a_i$. Write

$$\tilde{a}_i = \tilde{a}_{i0}\epsilon_0 + \cdots + \tilde{a}_{in}\epsilon_n, \quad i = 0, \ldots, q,$$

where $\epsilon = (\epsilon_0, \ldots, \epsilon_n)$ is the dual of e. For $i = 0, 1, \ldots, q$, set

$$F_i = \left\langle \tilde{f}, \tilde{a}_i \right\rangle = \tilde{a}_{i0}\tilde{f}_0 + \tilde{a}_{i1}\tilde{f}_1 + \cdots + \tilde{a}_{in}\tilde{f}_n, \quad G_i = \frac{F_i}{\tilde{f}_0}.$$

Since f is linearly non-degenerate, then $F_i \not\equiv 0$. Because $\mathscr{A}$ is in general position, we have $c_\lambda = \det\left(\tilde{a}_{\lambda(i)j}\right) \neq 0$ for any $\lambda \in J_n^q$. We abbreviate the Wronskian

$$\mathbf{W} = \mathbf{W}\left(\tilde{f}_0, \tilde{f}_1, \ldots, \tilde{f}_n\right), \quad \mathbf{W}_\lambda = \mathbf{W}\left(F_{\lambda(0)}, F_{\lambda(1)}, \ldots, F_{\lambda(n)}\right).$$

It follows that $\mathbf{W}_\lambda = c_\lambda \mathbf{W}$. Lemma 1.61 implies

$$\prod_{j=0}^{q} \frac{1}{|f, a_j|} \leq \left(\frac{1}{\Gamma(\mathscr{A})}\right)^{q-n} \max_{\lambda \in J_n^q} \prod_{i=0}^{n} \frac{1}{|f, a_{\lambda(i)}|}.$$

Since

$$\prod_{i=0}^{n} \frac{1}{|f, a_{\lambda(i)}|} = \prod_{i=0}^{n} \frac{|\tilde{f}|}{|F_{\lambda(i)}|} = |\tilde{f}|^{n+1} \frac{|\mathbf{S}\left(F_{\lambda(0)}, F_{\lambda(1)}, \ldots, F_{\lambda(n)}\right)|}{|c_\lambda \mathbf{W}|}, \qquad (2.6.6)$$

by using Corollary 2.52 we have

$$\prod_{j=0}^{q} \frac{1}{|f, a_j|} \leq \left(\frac{1}{\Gamma(\mathscr{A})}\right)^{q-n+1} \frac{|\tilde{f}|^{n+1}}{|\mathbf{W}|} \sum_{\lambda \in J_n^q} |\mathbf{S}\left(G_{\lambda(0)}, G_{\lambda(1)}, \ldots, G_{\lambda(n)}\right)|,$$

which yields, for $r \geq r_0$,

$$\sum_{i=0}^{q} m_f(r, a_i) \leq (n+1)\mathbb{C}^m\langle 0; r; \log|\tilde{f}|\rangle - \mathbb{C}^m\langle 0; r; \log|\mathbf{W}|\rangle \qquad (2.6.7)$$

$$+ \mathbb{C}^m\left\langle 0; r; \log \sum_{\lambda \in J_n^q} |\mathbf{S}\left(G_{\lambda(0)}, G_{\lambda(1)}, \ldots, G_{\lambda(n)}\right)|\right\rangle + O(1).$$

According to the definition (2.3.13) of characteristic functions, it is easy to get the inequality

$$T(r, G_i) \leq \mathbb{C}^m\langle 0; r; \log^+|G_i|\rangle + N\left(r, \frac{1}{\tilde{f}_0}\right).$$

The Schwarz inequality yields $|F_i| \leq |\tilde{f}|$, and so $|G_i| \leq |\tilde{f}|/|\tilde{f}_0|$. Thus we have

$$T(r, G_i) \leq N\left(r, \frac{1}{\tilde{f}_0}\right) - \mathbb{C}^m\left\langle 0; r; \log|\tilde{f}_0|\right\rangle + \mathbb{C}^m\langle 0; r; \log|\tilde{f}|\rangle.$$

By Jensen's formula (2.3.11) and (2.3.21), the inequality

$$T(r, G_i) \leq T_f(r) + O(1)$$

holds for $i = 0, 1, \ldots, q$.

Applying (2.3.21), Jensen's formula (2.3.11) and Lemma 2.55 in (2.6.7), Theorem 2.56 follows from (2.6.7) and the first main theorem (2.3.33). The second inequality in Theorem 2.56 can be easily deduced from the method in § 2.5. □

The second main theorem was established by Nevanlinna [292] for meromorphic functions in $\mathbb{C}$, and was extended to holomorphic curves into projective spaces by Cartan [51]. The case of several complex variables was first proved by W. Stoll [375], [376] by using Ahlfors' theory of associated mappings (cf. [436]). The refined estimates of error terms in the second main theorem was obtained by Z. Ye [443] (or see Hinkkanen [159], Lang and Cherry [235] and Wong [433]). For the case $m = n = 1$, Z. Ye shows that Theorem 2.56 is sharp in the following sense:

Theorem 2.57 (cf. [235]). *Given $\varepsilon > 0$, there exists an entire function f of finite order on $\mathbb{C}$ and a finite set $\mathscr{A} = \{a_0, a_1, \ldots, a_q\}$ such that for all large r,*

$$(q-1)T(r,f) - \sum_{j=0}^{q} N\left(r, \frac{1}{f-a_j}\right) + N_{\mathrm{Ram}}(r,f) > (1-\varepsilon)\log T(r,f).$$

From Theorem 2.56, we obtain directly the following result of A. Bloch [28] and H. Cartan [49].

Corollary 2.58. *Let $\mathscr{A} = \{a_0, a_1, \ldots, a_{n+1}\}$ be a family of points $a_j \in \mathbb{P}(V^*)$ in general position, $n \geq 2$. If $f : \mathbb{C}^m \longrightarrow \mathbb{P}(V) - \bigcup_j \ddot{E}[a_j]$ is a non-constant meromorphic mapping, then its image lies in a hyperplane.*

Take $a \in \mathbb{P}(V^*)$. For a positive integer k, we define the *truncated multiplicity function of order k* on $\mathbb{C}^m$ by

$$\mu_{f,k}^a(z) = \min\{\mu_f^a(z), k\}, \quad z \in \mathbb{C}^m, \tag{2.6.8}$$

the *truncated counting function of order k*,

$$n_{f,k}(r,a) := n_{\mu_{f,k}^a}(r), \tag{2.6.9}$$

and the *truncated valence function of order k*,

$$N_{f,k}(r,a) := N_{\mu_{f,k}^a}(r). \tag{2.6.10}$$

Take $\tilde{a} \in V^* - \{0\}$ with $\mathbb{P}(\tilde{a}) = a$. Write $\tilde{a} = \tilde{a}_0\epsilon_0 + \cdots + \tilde{a}_n\epsilon_n$, where $\epsilon = (\epsilon_0, \ldots, \epsilon_n)$ is the dual of e and set

$$F = \left\langle \tilde{f}, \tilde{a} \right\rangle = \tilde{a}_0\tilde{f}_0 + \tilde{a}_1\tilde{f}_1 + \cdots + \tilde{a}_n\tilde{f}_n.$$

Then

$$N_{f,k}(r,a) = N_k\left(r, \frac{1}{F}\right). \tag{2.6.11}$$

Define the *truncated defect of order k* of f for a by

$$\delta_{f,k}(a) = 1 - \limsup_{r\to\infty} \frac{N_{f,k}(r,a)}{T_f(r)}. \tag{2.6.12}$$

Note that when $\tilde{f}$ is a reduced representation of f, we also have

$$N_f(r,a) = N\left(r, \frac{1}{F}\right). \tag{2.6.13}$$

If $n = 1$, that is, $\mathbb{P}(V^*) = \mathbb{P}^1$, we usually denote the truncated counting function

$$n_\nu(t) = \begin{cases} n_k\left(t, \frac{1}{f-a}\right) & \text{if } \nu = \mu_{f,k}^a, \\ \bar{n}\left(t, \frac{1}{f-a}\right) & \text{if } \nu = \mu_{f,1}^a, \end{cases} \tag{2.6.14}$$

and the truncated valence function

$$
N_\nu(t) = \begin{cases} N_k\left(t, \frac{1}{f-a}\right) & \text{if } \nu = \mu^a_{f,k}, \\ \overline{N}\left(t, \frac{1}{f-a}\right) & \text{if } \nu = \mu^a_{f,1}. \end{cases} \tag{2.6.15}
$$

In particular, if $a = \infty$, we often write

$$
\overline{N}\left(t, \frac{1}{f-a}\right) = \overline{N}(t, f), \quad N_k\left(t, \frac{1}{f-a}\right) = N_k(t, f),
$$

and so on.

We continue to study Theorem 2.56 and give applications of the index l and Wronskian degree w of f. Using the symbols in the proof of Theorem 2.56, we can define a meromorphic function by

$$
\mathbf{H} = \frac{F_0 F_1 \cdots F_q}{\mathbf{W}}. \tag{2.6.16}
$$

By (2.6.13) and the Jensen formula (2.3.11), we obtain

$$
\sum_{j=0}^q N_f(r, a_j) - N_{\mathrm{Ram}}(r, f) = N\left(r, \frac{1}{\mathbf{H}}\right) - N(r, \mathbf{H}). \tag{2.6.17}
$$

There exist three entire functions u, h_1 and h_2 on $\mathbb{C}^m$ such that $\mathbf{H} = \frac{h_1}{h_2}$, and

$$
\dim h_1^{-1}(0) \cap h_2^{-1}(0) \le m - 2, \quad \mathbf{W} = u h_2.
$$

If $\mu^0_{h_1}(x) > 0$ for some $x \in \mathbb{C}^m - (I_f \cup h_2^{-1}(0))$, then

$$
\#\{j \mid F_j(x) = 0\} \le n + 1
$$

since $\mathscr{A}$ is in general position. Without loss of generality, we assume $q \ge n$, and

$$
\{j \mid F_j(x) = 0\} \subset \{\lambda(0), \ldots, \lambda(n)\}
$$

for some $\lambda \in J_n^q$. Then one has

$$
\mu^0_u(x) = \mu^0_{\mathbf{W}}(x) = \mu^0_{\mathbf{W}_\lambda}(x) \ge \sum_{j=0}^q \max\left\{\mu^0_{F_j}(x) - w, 0\right\},
$$

where w is the Wronskian degree of f, which means

$$
\mu^0_{\mathbf{H}}(x) = \sum_{j=0}^q \mu^0_{F_j}(x) - \mu^0_{\mathbf{W}}(x) \le \sum_{j=0}^q \mu^0_{F_j, w}(x).
$$

Thus we can obtain the following estimate:

$$N\left(r, \frac{1}{\mathbf{H}}\right) \leq \sum_{j=0}^{q} N_{f,w}(r, a_j). \tag{2.6.18}$$

Similarly, we can prove

$$\mu_{\mathbf{H}}^0(x) = \mu_{F_0 \cdots F_q}^0(x) - \mu_{\mathbf{W}}^0(x) \leq \mu_{F_0 \cdots F_q, l}^0(x)$$

by estimating the multiplicity of zero of each term in the Laplace expansion of $\mathbf{W}$ at x, and hence

$$N\left(r, \frac{1}{\mathbf{H}}\right) \leq N_l\left(r, \frac{1}{F_0 \cdots F_q}\right). \tag{2.6.19}$$

Therefore Theorem 2.56 yields immediately the following truncated form of the *second main theorem*:

Corollary 2.59. *Let* $f : \mathbb{C}^m \longrightarrow \mathbb{P}(V)$ *be a linearly non-degenerate meromorphic mapping and let* $\mathscr{A} = \{a_0, a_1, \ldots, a_q\}$ *be a family of points* $a_j \in \mathbb{P}(V^*)$ *in general position. Let* l, w *be the index and Wronskian degree of* f, *respectively. Then*

$$(q-n)T_f(r) \leq \sum_{j=0}^{q} N_{f,w}(r, a_j) - N(r, \mathbf{H}) + l \log\left\{\left(\frac{\rho}{r}\right)^{2m-1} \frac{T_f(R)}{\rho - r}\right\} + O(1)$$

holds for any $r_0 < r < \rho < R$, *and hence for any* $\varepsilon > 0$,

$$(q-n)T_f(r) \leq \sum_{j=0}^{q} N_{f,w}(r, a_j) - N(r, \mathbf{H}) + l \log T_f(r)$$
$$+ l(1+\varepsilon) \log \log T_f(r) - l \log r + O(1)$$

holds for all large r *outside a set* E *with* $\int_E d\log r < \infty$. *In particular, if* f *is of finite order* λ, *then for any* $\varepsilon > 0$, *one has*

$$(q-n)T_f(r) \leq \sum_{j=0}^{q} N_{f,w}(r, a_j) - N(r, \mathbf{H}) + l(\lambda + \varepsilon - 1) \log r + O(1).$$

The Wronskian degree w of f occurring in Corollary 2.59 was observed by Fujimoto [106]. In [229], Chapter VII, Section 6 (or see [135]), it was observed that Cartan's proof [51] can be easily adjusted so that the term $-N(r, \mathbf{H})$ appears in the inequalities in Corollary 2.59.

Corollary 2.60. *Let* $f : \mathbb{C}^m \longrightarrow \mathbb{P}(V)$ *be a linearly non-degenerate meromorphic mapping and let* $\mathscr{A} = \{a_0, a_1, \ldots, a_q\}$ *be a family of points* $a_j \in \mathbb{P}(V^*)$ *in general position. Then*

$$\sum_{j=0}^{q} \delta_f(a_j) \leq \sum_{j=0}^{q} \delta_{f,w}(a_j) \leq n + 1,$$

where w *is the Wronskian degree of* f.

The inequality in Corollary 2.60 is usually called the *defect relation* of f. In particular, if $q = n + 1$, it means that there exists $j \in \{0, 1, \ldots, n+1\}$ such that $f(\mathbb{C}^m) \cap \ddot{E}[a_j] \neq \emptyset$ (Borel [29]). Further, this implies that a non-constant meromorphic function on $\mathbb{C}^m$ omits at most two values of $\mathbb{P}^1$, which is just the classic Picard's little theorem.

2.7 Notes on the second main theorem

We continue the situation of Section 2.6 to consider a linearly non-degenerate meromorphic mapping $f : \mathbb{C}^m \longrightarrow \mathbb{P}(V)$, where V is a complex vector space of dimensions $n + 1 \geq 1$, and use the symbols in Section 2.6. First of all, we consider the case $q = n$, and restate Theorem 2.56 as follows:

Theorem 2.61. *Let* $f : \mathbb{C}^m \longrightarrow \mathbb{P}(V)$ *be a linearly non-degenerate meromorphic mapping and let* $\mathscr{A} = \{a_0, a_1, \ldots, a_n\}$ *be a family of points* $a_j \in \mathbb{P}(V^*)$ *in general position. Let* l *be the index of* f. *Then*

$$N_{\mathrm{Ram}}(r, f) \leq \sum_{j=0}^{n} N_f(r, a_j) + l \log \left\{ \left(\frac{\rho}{r} \right)^{2m-1} \frac{T_f(R)}{\rho - r} \right\} + O(1)$$

holds for any $r_0 < r < \rho < R$.

Proof. We prove it again. Take an orthonormal base $e = (e_0, \ldots, e_n)$ of V and let

$$\tilde{f} = \tilde{f}_0 e_0 + \cdots + \tilde{f}_n e_n : \mathbb{C}^m \longrightarrow V$$

be a reduced representation of f. Let $\epsilon = (\epsilon_0, \ldots, \epsilon_n)$ be the dual of e. Without loss of generality, we may assume that

$$a_j = \mathbb{P}(\epsilon_j), \quad j = 0, 1, \ldots, n.$$

The ramification term

$$N_{\mathrm{Ram}}(r, f) = N\left(r, \frac{1}{\mathbf{W}(\tilde{f}_0, \ldots, \tilde{f}_n)}\right)$$

is well defined with respect to the multi-indices $\nu_i \in \mathbb{Z}_+^m$ in Corollary 2.52. Then the Jensen formula (2.3.11) implies

$$
\begin{aligned}
N_{\mathrm{Ram}}(r, f) &= \mathbb{C}^m \left\langle 0; r; \log |\mathbf{W}(\tilde{f}_0, \ldots, \tilde{f}_n)| \right\rangle + O(1) \\
&= \sum_{j=0}^{n} \mathbb{C}^m \left\langle 0; r; \log |\tilde{f}_j| \right\rangle + \mathbb{C}^m \left\langle 0; r; \log |\mathbf{S}(\tilde{f}_0, \ldots, \tilde{f}_n)| \right\rangle + O(1) \\
&= \sum_{j=0}^{n} N_f(r, a_j) + \mathbb{C}^m \left\langle 0; r; \log |\mathbf{S}(1, f_1, \ldots, f_n)| \right\rangle + O(1).
\end{aligned}
$$

Therefore Theorem 2.61 follows from Lemma 2.55 and Lemma 2.37. $\qquad \square$

When $0 \leq q < n$, Theorem 2.61 and the first main theorem (2.3.33) imply

$$N_{\mathrm{Ram}}(r, f) \leq \sum_{j=0}^{q} N_f(r, a_j) + (n - q)T_f(r) + l\log\left\{\left(\frac{\rho}{r}\right)^{2m-1}\frac{T_f(R)}{\rho - r}\right\} + O(1),$$

(2.7.1)

that is, Theorem 2.56 holds for $q \geq 0$. Further, we have

$$N_{\mathrm{Ram}}(r, f) \leq (n + 1)T_f(r) + l\log\left\{\left(\frac{\rho}{r}\right)^{2m-1}\frac{T_f(R)}{\rho - r}\right\} + O(1),$$

(2.7.2)

which is Theorem 5.12 in Stoll [383] with a good error term. If f is a non-constant meromorphic function on $\mathbb{C}^m$, we have

$$N_{\mathrm{Ram}}(r, f) \leq N(r, f) + N\left(r, \frac{1}{f}\right) + \log\left\{\left(\frac{\rho}{r}\right)^{2m-1}\frac{T(R, f)}{\rho - r}\right\} + O(1).$$

(2.7.3)

In particular, if f omits $q+1$ hyperplanes in general position with $0 \leq q \leq n$, then

$$N_{\mathrm{Ram}}(r, f) \leq (n - q)T_f(r) + l\log\left\{\left(\frac{\rho}{r}\right)^{2m-1}\frac{T_f(R)}{\rho - r}\right\} + O(1),$$

(2.7.4)

and so

$$\mathrm{Ram}_f = \liminf_{r \to \infty} \frac{N_{\mathrm{Ram}}(r, f)}{T_f(r)} \leq n - q.$$

Conversely, if $\mathrm{Ram}_f > n$, then f can not omit any hyperplane.

Secondly, if the condition of general position on $\mathscr{A}$ is omitted, then Theorem 2.56 assumes the following form (cf. [418], [328]):

Theorem 2.62. *Let $f : \mathbb{C}^m \longrightarrow \mathbb{P}(V)$ be a linearly non-degenerate meromorphic mapping and let $\mathscr{A} = \{a_0, a_1, \ldots, a_q\}$ be a family of points $a_j \in \mathbb{P}(V^*)$. Let l be the index of f. Then*

$$\mathbb{C}^m\left\langle 0; r; \max_{\lambda \in J_k(\mathscr{A})} \log \prod_{i=0}^{k} \frac{1}{|f, a_{\lambda(i)}|}\right\rangle \leq (n + 1)T_f(r) - N_{\mathrm{Ram}}(r, f)$$

$$+ l\log\left\{\left(\frac{\rho}{r}\right)^{2m-1}\frac{T_f(R)}{\rho - r}\right\} + O(1)$$

holds for $r_0 < r < \rho < R$, where the maximum is taken over all $\lambda \in J_k(\mathscr{A})$ with $0 \leq k \leq n$.

Proof. In fact, for $\lambda \in J_k(\mathscr{A})$, we can take $b_{\lambda,1}, \ldots, b_{\lambda,n-k}$ in $\mathbb{P}(V^*)$ such that

$$\mathscr{B}_\lambda = \{a_{\lambda(0)}, \ldots, a_{\lambda(k)}, b_{\lambda,1}, \ldots, b_{\lambda,n-k}\}$$

is a family in general position. Take $\tilde{b}_{\lambda,i} \in V^* - \{0\}$ with $|\tilde{b}_{\lambda,i}| = 1$ and $\mathbb{P}(\tilde{b}_{\lambda,i}) = b_{\lambda,i}$, and set

$$F_{\lambda,i} = \left\langle \tilde{f}, \tilde{b}_{\lambda,i} \right\rangle, \quad G_{\lambda,i} = \frac{F_{\lambda,i}}{\tilde{f}_0}.$$

According to the proof of Theorem 2.56, then there exists a positive constant c depending on $\mathscr{A}$ and $\{b_{\lambda,j}\}$ such that

$$\max_{\lambda \in J_k(\mathscr{A})} \prod_{i=0}^{k} \frac{1}{|f, a_{\lambda(i)}|} \leq \max_{\lambda \in J_k(\mathscr{A})} \left(\prod_{i=0}^{k} \frac{|\tilde{f}|}{|F_{\lambda(i)}|} \right) \left(\prod_{j=1}^{n-k} \frac{|\tilde{f}|}{|F_{\lambda,j}|} \right)$$

$$\leq c \frac{|\tilde{f}|^{n+1}}{|\mathbf{W}|} \max_{\lambda \in J_k(\mathscr{A})} \left| \mathbf{S} \left(F_{\lambda(0)}, \ldots, F_{\lambda(k)}, F_{\lambda,1}, \ldots, F_{\lambda,n-k} \right) \right|$$

$$\leq c \frac{|\tilde{f}|^{n+1}}{|\mathbf{W}|} \sum_{\lambda \in J_k(\mathscr{A})} \left| \mathbf{S} \left(G_{\lambda(0)}, \ldots, G_{\lambda(k)}, G_{\lambda,1}, \ldots, G_{\lambda,n-k} \right) \right|,$$

and so Theorem 2.62 follows from the arguments of Theorem 2.56. $\qquad \square$

Thirdly, for a positive integer d, by using the notations in Section 1.5.2, a non-constant meromorphic mapping $f : \mathbb{C}^m \longrightarrow \mathbb{P}(V)$ induces a meromorphic mapping

$$f^{\mathrm{II}d} = \varphi_d \circ f : \mathbb{C}^m \longrightarrow \mathbb{P}\left(\mathrm{II}_d V\right)$$

such that the characteristic function of $f^{\mathrm{II}d}$ satisfies

$$T_{f^{\mathrm{II}d}}(r) = dT_f(r),$$

where $\varphi_d : \mathbb{P}(V) \longrightarrow \mathbb{P}(\mathrm{II}_d V)$ is the Veronese mapping.

Take $a \in \mathbb{P}(\mathrm{II}_d V^*)$ such that the pair $(f^{\mathrm{II}d}, a)$ is free for the interior product $\angle$. Applying (2.3.33) to $f^{\mathrm{II}d}$ and a, we obtain the *first main theorem* for a hypersurface $\ddot{E}^d[a]$,

$$dT_f(r) = N_{f^{\mathrm{II}d}}(r,a) + m_{f^{\mathrm{II}d}}(r,a) - m_{f^{\mathrm{II}d}}(r_0,a). \qquad (2.7.5)$$

Further, if the mapping $f : \mathbb{C}^m \longrightarrow \mathbb{P}(V)$ is *algebraically non-degenerate*, that is, the image of f is not contained in any proper algebraic subvariety of $\mathbb{P}(V)$, one has the *second main theorem*:

Theorem 2.63. *Let $\mathscr{A}$ be a finite admissible family of $\mathbb{P}(\mathrm{II}_d V^*)$. Let $f : \mathbb{C}^m \longrightarrow \mathbb{P}(V)$ be an algebraically non-degenerate meromorphic mapping. Then*

$$\| \quad \sum_{a \in \mathscr{A}} \{dT_f(r) - N_{f^{\mathrm{II}d}}(r,a)\} \leq \{n+1+o(1)\}dT_f(r). \qquad (2.7.6)$$

Originally, Theorem 2.63 was a conjecture due to Shiffman [346]. For some discussion related to Theorem 2.63, we refer the reader to [173], [176], [175] and [330]. A proof will be introduced in Section 3.8.3.

Finally, as a simple application of Corollary 2.59, we can derive simply the following *abc-theorem* for entire functions over $\mathbb{C}^m$.

Theorem 2.64. *Let a, b and c be non-zero entire functions in $\mathbb{C}^m$ with $a + b = c$. Assume that a, b, c are not all constants, and $\dim I \leq m - 2$, where*

$$I = \{z \in \mathbb{C}^m \mid a(z) = b(z) = c(z) = 0\}.$$

Then the inequality

$$T(r) < \overline{N}\left(r, \frac{1}{abc}\right) + \log\left\{\left(\frac{\rho}{r}\right)^{2m-1} \frac{T(R)}{\rho - r}\right\} + O(1) \qquad (2.7.7)$$

holds for $r_0 < r < \rho < R$, where

$$T(r) = \max\left\{T\left(r, \frac{a}{c}\right), T\left(r, \frac{b}{c}\right)\right\}.$$

Proof. Write

$$f = \frac{a}{c}, \quad g = \frac{b}{c}.$$

Then both f and g are not constant by our assumptions, and satisfy $f + g = 1$. By the second main theorem (cf. Corollary 2.59), for $r_0 < r < \rho < R$ we obtain

$$
\begin{aligned}
T(r, f) &\leq \overline{N}(r, f) + \overline{N}\left(r, \frac{1}{f}\right) + \overline{N}\left(r, \frac{1}{f - 1}\right) \\
&\quad + \log\left\{\left(\frac{\rho}{r}\right)^{2m-1} \frac{T(R, f)}{\rho - r}\right\} + O(1).
\end{aligned}
$$

Noting that

$$\overline{N}\left(r, \frac{1}{f - 1}\right) = \overline{N}\left(r, \frac{1}{g}\right) = \overline{N}\left(r, \frac{1}{b}\right),$$

we obtain

$$
\begin{aligned}
T(r, f) &\leq \overline{N}\left(r, \frac{1}{c}\right) + \overline{N}\left(r, \frac{1}{a}\right) + \overline{N}\left(r, \frac{1}{b}\right) \\
&\quad + \log\left\{\left(\frac{\rho}{r}\right)^{2m-1} \frac{T(R)}{\rho - r}\right\} + O(1) \\
&= \overline{N}\left(r, \frac{1}{abc}\right) + \log\left\{\left(\frac{\rho}{r}\right)^{2m-1} \frac{T(R)}{\rho - r}\right\} + O(1).
\end{aligned}
$$

Symmetrically, this inequality is true for g. $\square$

If $m = 1$, Theorem 2.64 was proved by M. van Frankenhuysen [410]. However, the error term in van Frankenhuysen's result is not better than that of Theorem 2.64.

Take a positive integer k. For a polynomial f in $\mathbb{C}^m$, we have

$$n_k\left(\infty, \frac{1}{f}\right) = \lim_{r \to \infty} n_k\left(r, \frac{1}{f}\right) = \lim_{r \to \infty} \frac{N_k\left(r, \frac{1}{f}\right)}{\log r}.$$

This number will be simply denoted by $r_k(f)$. We also abbreviate

$$r(f) = r_1(f).$$

If the functions a, b and c in Theorem 2.64 are polynomials, dividing the inequality (2.7.7) by $\log r$ and letting $r \to \infty$, then (2.7.7) and (2.4.6) yield immediately *Stothers-Mason's theorem* ([252],[253],[254], [387], or cf. [231], [415]):

Theorem 2.65. *Let $a(z)$, $b(z)$, $c(z)$ be relatively prime polynomials in $\mathbb{C}$ and not all constants such that $a + b = c$. Then*

$$\max\{\deg(a), \deg(b), \deg(c)\} \leq r(abc) - 1. \tag{2.7.8}$$

In particular, if

$$a(z) = (1+z)^2, \ b(z) = -(1-z)^2, \ c(z) = 4z,$$

then the inequality in Theorem 2.65 becomes the equality $2 = 2$ (cf. [135]). Thus the inequality is sharp. Elementary proofs of Theorem 2.65 without using Nevanlinna theory are due to [119] and [231]. For applications of Theorem 2.65, see [253]; [288], pp. 183–185.

2.8 The Cartan-Nochka theorem

In this section, we use the estimates of the error terms of the second main theorem from Ye [443] to restate the results on Cartan's conjecture due to Nochka [301] (see Theorem 2.66 and Theorem 2.70).

Let V be a Hermitian vector space of dimension $n + 1$ over $\mathbb{C}$. First of all, we show the *second main theorem* of meromorphic mappings into $\mathbb{P}(V)$ for a family of $\mathbb{P}(V^*)$ in subgeneral position.

Theorem 2.66. *Let $\mathscr{A} = \{a_j\}_{j=0}^q$ be a finite family of points $a_j \in \mathbb{P}(V^*)$ in u-subgeneral position with $u \leq 2u - n < q$. Let $f : \mathbb{C}^m \longrightarrow \mathbb{P}(V)$ be a linearly non-degenerate meromorphic mapping. Let l be the index of f. Then*

$$(q - 2u + n)\, T_f(r) \ \leq \ \sum_{j=0}^q \theta \omega(a_j) N_f(r, a_j) - \theta N_{\mathrm{Ram}}(r, f)$$

$$+ l\theta \log\left\{\left(\frac{\rho}{r}\right)^{2m-1} \frac{T_f(R)}{\rho - r}\right\} + O(1) \tag{2.8.1}$$

holds for any $r_0 < r < \rho < R$, where $\theta \geq 1$ is a Nochka constant, and $\omega : \mathscr{A} \longrightarrow \mathbb{R}(0,1]$ is a Nochka weight.

Proof. We will adopt the notations that were used in the proof of Theorem 2.56, and without loss of generality, assume $\|\tilde{a}_j\| = 1$ for $j = 0, \ldots, q$. Lemma 1.61 implies

$$\prod_{j=0}^{q} \left(\frac{1}{\|f, a_j\|} \right)^{\omega(a_j)} \leq \left(\frac{1}{\Gamma(\mathscr{A})} \right)^{q-u} \max_{\lambda \in J_n(\mathscr{A})} \prod_{j=0}^{n} \frac{1}{\|f, a_{\lambda(j)}\|}.$$

Then from (2.6.6) and Corollary 2.52, we obtain

$$\prod_{j=0}^{q} \left(\frac{1}{\|f, a_j\|} \right)^{\omega(a_j)} \leq c \frac{\|\tilde{f}\|^{n+1}}{|\mathbf{W}|} \sum_{\lambda \in J_n^q} |\mathbf{S}\left(G_{\lambda(0)}, G_{\lambda(1)}, \ldots, G_{\lambda(n)} \right)|,$$

where c is a positive constant. According to the proof of Theorem 2.56, we obtain

$$\sum_{j=0}^{q} \omega(a_j) m_f(r, a_j) \leq (n+1) T_f(r) - N_{\mathrm{Ram}}(r, f)$$

$$+ l \log \left\{ \left(\frac{\rho}{r} \right)^{2m-1} \frac{T_f(R)}{\rho - r} \right\} + O(1). \qquad (2.8.2)$$

By (2.8.2) and the first main theorem (2.3.33), we obtain

$$\left(\sum_{j=0}^{q} \omega(a_j) - n - 1 \right) T_f(r) \leq \sum_{j=0}^{q} \omega(a_j) N_f(r, a_j) - N_{\mathrm{Ram}}(r, f)$$

$$+ l \log \left\{ \left(\frac{\rho}{r} \right)^{2m-1} \frac{T_f(R)}{\rho - r} \right\} + O(1). \, (2.8.3)$$

Thus (2.8.1) follows from (2.8.3) and 2) in Lemma 1.59. $\square$

Lemma 2.67. *Let $\mathscr{A} = \{a_j\}_{j=0}^{q}$ be a finite family of points $a_j \in \mathbb{P}(V^*)$ in u-subgeneral position with $u \leq 2u - n < q$. Let $f : \mathbb{C}^m \longrightarrow \mathbb{P}(V)$ be a linearly non-degenerate meromorphic mapping. Then for $z \in \mathbb{C}^m$,*

$$\sum_{j=0}^{q} \omega(a_j) \mu_{F_j}^0(z) - \mu_{\mathbf{W}}^0(z) \leq \sum_{j=0}^{q} \omega(a_j) \min \left\{ \mu_{F_j}^0(z), w \right\},$$

where $\omega : \mathscr{A} \longrightarrow \mathbb{R}(0,1]$ is a Nochka weight, and w is the Wronskian degree of f.

Proof. We follow the methods of Lemma 3.2.13 in Fujimoto [107] to prove Lemma 2.67. It suffices to show that

$$\mu_{\mathbf{W}}^0(z) \geq \sum_{j=0}^{q} \omega(a_j) \max \left\{ \mu_{F_j}^0(z) - w, 0 \right\} := \nu(z). \qquad (2.8.4)$$

In fact, since

$$\min\left\{\mu^0_{F_j}(z), w\right\} + \max\left\{\mu^0_{F_j}(z) - w, 0\right\} = \mu^0_{F_j}(z),$$

then (2.8.4) implies

$$\sum_{j=0}^{q} \omega(a_j)\mu^0_{F_j}(z) - \mu^0_{\mathbf{W}}(z) \leq \sum_{j=0}^{q} \omega(a_j)\mu^0_{F_j}(z) - \sum_{j=0}^{q} \omega(a_j)\max\left\{\mu^0_{F_j}(z) - w, 0\right\}$$

$$= \sum_{j=0}^{q} \omega(a_j)\min\left\{\mu^0_{F_j}(z), w\right\}.$$

To prove (2.8.4), take an arbitrary $z_0 \in \mathbb{C}^m - I_f$ and set

$$S = \left\{j \mid \mu^0_{F_j}(z_0) \geq w + 1\right\} \subset \{0, 1, \ldots, q\}.$$

We may assume that $S \neq \emptyset$. Then $\#S \leq u$. Otherwise, by the assumption of u-subgeneral position, there is $\lambda \in J^q_n$ with $\mathrm{Im}(\lambda) \subset S$ such that $\dim E(\mathscr{A}_\lambda) = n+1$, and so $\tilde{f}_0, \ldots, \tilde{f}_n$ are represented as linear combinations $F_{\lambda(0)}, \ldots, F_{\lambda(n)}$. Then $\tilde{f}_0, \ldots, \tilde{f}_n$ vanish at z_0, which is a contradiction.

Now we consider the sets S_i $(0 \leq i \leq l)$ such that

$$S_0 := \emptyset \neq S_1 \subset S_2 \subset \cdots \subset S_l := S$$

and $\mu^0_{F_j}(z_0)$ equals some constant w_i for each $j \in S_i - S_{i-1}$, where

$$w_1 > w_2 > \cdots > w_l.$$

Then we have

$$E(\mathscr{A}_{S_1}) \subset E(\mathscr{A}_{S_2}) \subset \cdots \subset E(\mathscr{A}_{S_l}).$$

For each i take a subset T_i of S_i such that

$$T_{i-1} \subset T_i, \ \dim E(\mathscr{A}_{S_i}) = \dim E(\mathscr{A}_{T_i}) = \#T_i.$$

Then we have

$$\#(T_i - T_{i-1}) = \dim E(\mathscr{A}_{S_i}) - \dim E(\mathscr{A}_{S_{i-1}}).$$

Abbreviate

$$w^*_i = w_i - w.$$

We have

$$\nu(z_0) = \sum_{j \in S} \omega(a_j)\left(\mu^0_{F_j}(z_0) - w\right) = \sum_{i=0}^{l} \sum_{j \in S_i - S_{i-1}} \omega(a_j)w^*_i$$

$$= (w^*_1 - w^*_2)\sum_{j \in S_1} \omega(a_j) + (w^*_2 - w^*_3)\sum_{j \in S_2} \omega(a_j) + \cdots + w^*_l\sum_{j \in S_l} \omega(a_j).$$

Thus 4) in Lemma 1.59 yields

$$\nu(z_0) \le (w_1^* - w_2^*)\dim E(\mathscr{A}_{S_1}) + (w_2^* - w_3^*)\dim E(\mathscr{A}_{S_2}) + \cdots + w_l^* \dim E(\mathscr{A}_{S_l})$$
$$= \#T_1 w_1^* + \#(T_2 - T_1)w_2^* + \cdots + \#(T_l - T_{l-1})w_l^*.$$

Assume $\#T_l = k+1$ with $0 \le k \le n$. We can choose $\sigma \in J_k^q$ such that $T_l = \operatorname{Im}(\sigma)$. Since $\dim E(\mathscr{A}_\sigma) = k+1$, after a suitable non-singular linear transformation of homogeneous coordinates we may assume that

$$\tilde{f}_0 = F_{\sigma(0)}, \ \ldots, \ \tilde{f}_k = F_{\sigma(k)}.$$

Then, with the Laplace expansion theorem for the determinant, the Wronskian $\mathbf{W}$ is expanded as the sum of the products of some minors of degree $n-k$ and some minors of degree $k+1$ whose components consist of the $\le w$th partial derivatives of the functions $F_{\sigma(0)}, \ldots, F_{\sigma(k)}$. This implies

$$\mu_{\mathbf{W}}^0(z_0) \ge \sum_{j \in T_l} \left(\mu_{F_j}^0(z_0) - w \right).$$

Since $\mu_{F_j}^0(z_0) = w_i$ for every $j \in T_i - T_{i-1}$, this quantity coincides with the last term of the above inequalities. This completes the proof of Lemma 2.67. □

Note that

$$\sum_{j=0}^q \omega(a_j)\mu_{F_j}^0 - \mu_{\mathbf{W}}^0 = \mu^+ - \mu^-,$$

where

$$\mu^+(z) = \max\left\{ \sum_{j=0}^q \omega(a_j)\mu_{F_j}^0(z) - \mu_{\mathbf{W}}^0(z), 0 \right\},$$

$$\mu^-(z) = \max\left\{ \mu_{\mathbf{W}}^0(z) - \sum_{j=0}^q \omega(a_j)\mu_{F_j}^0(z), 0 \right\}.$$

Set

$$\mathbf{H} = \frac{|F_0|^{\omega(a_0)} \cdots |F_q|^{\omega(a_q)}}{|\mathbf{W}(\tilde{f}_0, \ldots, \tilde{f}_n)|},$$

and formally write

$$N\left(r, \frac{1}{\mathbf{H}}\right) = N_{\mu^+}(r), \quad N(r, \mathbf{H}) = N_{\mu^-}(r).$$

Then Lemma 2.67 implies

$$N\left(r, \frac{1}{\mathbf{H}}\right) \le \sum_{j=0}^q \omega(a_j)N_{f,w}(r, a_j).$$

Corollary 2.68. *Assumptions as in Theorem* 2.66. *Then*

$$(q - 2u + n)T_f(r) \quad \le \quad \sum_{j=0}^{q} \theta\omega(a_j)N_{f,w}(r, a_j) - \theta N(r, \mathbf{H})$$

$$+ l\theta \log \left\{ \left(\frac{\rho}{r}\right)^{2m-1} \frac{T_f(R)}{\rho - r} \right\} + O(1), \qquad (2.8.5)$$

where l, w *are the index and Wronskian degree of* f, *respectively.*

Corollary 2.69. *Assumptions as in Theorem* 2.66. *Then*

$$\sum_{j=0}^{q} \omega(a_j)\delta_f(a_j) \le \sum_{j=0}^{q} \omega(a_j)\delta_{f,w}(a_j) \le n + 1,$$

$$\sum_{j=0}^{q} \delta_f(a_j) \le \sum_{j=0}^{q} \delta_{f,w}(a_j) \le 2u - n + 1.$$

Now we eliminate the restriction of non-degeneracy on f. Take a reduced representation $\tilde{f} : \mathbb{C}^m \longrightarrow V$ of a non-constant meromorphic mapping $f : \mathbb{C}^m \longrightarrow \mathbb{P}(V)$ and define a linear subspace of V^* as follows:

$$E[f] = \{\alpha \in V^* \mid \langle \tilde{f}, \alpha \rangle \equiv 0\}, \qquad (2.8.6)$$

which will be called the *null space* of f, and write

$$\ell_f = \dim E[f], \quad k = n - \ell_f.$$

The number k is non-negative, i.e., $0 \le \ell_f \le n$. In fact, if $k < 0$, that is, $\ell_f = n+1$, there is $\{\alpha_0, \ldots, \alpha_n\} \subset E[f]$ such that

$$\alpha_0 \wedge \cdots \wedge \alpha_n \neq 0; \quad \langle \tilde{f}, \alpha_j \rangle \equiv 0 \ (0 \le j \le n).$$

By Cramer's rule, $\tilde{f} \equiv 0$, which is impossible. Then V^* is decomposed into a direct sum

$$V^* = W^* \oplus E[f],$$

where W^* is a $k + 1$ dimensional subspace of V^*. Then f is said to be k-*flat*. In order to simplify our notation, we define $\ell_f = 0$ if f is linearly non-degenerate, that is, $E[f] = \{0\}$, and say that f is n-*flat*.

From now on, we assume that $\mathscr{A} = \{a_j\}_{j=0}^{q}$ is in general position and assume that f is non-constant and k-flat with $0 \le k \le n < q$ such that each pair (f, a_j) is free for $j = 0, \ldots, q$. We take an orthonormal base $\epsilon = (\epsilon_0, \ldots, \epsilon_n)$ of V^* such that $\epsilon_0, \ldots, \epsilon_k$ and $\epsilon_{k+1}, \ldots, \epsilon_n$ is a base of W^* and $E[f]$, respectively. Let

$e = (e_0, \ldots, e_n)$ be the dual base of ϵ. Let W be the vector space spanned by $e_0, \ldots, e_k$ over $\mathbb{C}$. Thus the reduced representation $\tilde{f} : \mathbb{C}^m \longrightarrow V$ is given by

$$\tilde{f} = \sum_{j=0}^{k} \tilde{f}_j e_j = \sum_{j=0}^{k} \langle \tilde{f}, \epsilon_j \rangle e_j$$

such that $\langle \tilde{f}, \epsilon_0 \rangle, \ldots, \langle \tilde{f}, \epsilon_k \rangle$ are holomorphic and linearly independent over $\mathbb{C}$. Hence a linearly non-degenerate meromorphic mapping $\hat{f} : \mathbb{C}^m \longrightarrow \mathbb{P}(W)$ is defined with a reduced representation

$$\tilde{\hat{f}} = \tilde{f} = \sum_{j=0}^{k} \langle \tilde{f}, \epsilon_j \rangle e_j : \mathbb{C}^m \longrightarrow W.$$

The mapping $\hat{f}$ will be called a *simplified mapping* of f. Therefore by (2.3.21), we obtain

$$T_{\hat{f}}(r) = \mathbb{C}^m \langle 0; r; \log \| \tilde{f} \| \rangle + O(1) = T_f(r) + O(1). \tag{2.8.7}$$

If $k = 0$, then $T_{\hat{f}}(r)$ is constant. The relation (2.8.7) will be impossible since $T_f(r) \to \infty$ as $r \to \infty$. Thus, we must have $k \geq 1$. Set

$$N_{\mathrm{Ram}}(r, f) = N_{\mathrm{Ram}}(r, \hat{f}). \tag{2.8.8}$$

The index l and Wronskian degree w of $\hat{f}$ will be called the (*Wronskian*) *index* and *Wronskian degree* of f. By (2.6.3), we have

$$1 \leq w \leq k - \gamma + 1, \quad k \leq l \leq \gamma + \frac{(k - \gamma)(k - \gamma + 3)}{2}, \tag{2.8.9}$$

where $\gamma = \mathrm{rank}(f)$.

Theorem 2.70. *Let $\mathscr{A} = \{a_j\}_{j=0}^{q}$ be a finite family of points $a_j \in \mathbb{P}(V^*)$ in general position. Take an integer k with $1 \leq k \leq n \leq 2n - k < q$. Let $f : \mathbb{C}^m \longrightarrow \mathbb{P}(V)$ be a non-constant meromorphic mapping that is k-flat such that each pair (f, a_j) is free for $j = 0, \ldots, q$. Let l, w be the index and Wronskian degree of f, respectively. Then*

$$(q - 2n + k)T_f(r) \leq \sum_{j=0}^{q} N_{f,w}(r, a_j) + l\theta \log \left\{ \left(\frac{\rho}{r} \right)^{2m-1} \frac{T_f(R)}{\rho - r} \right\} + O(1)$$

holds for any $r_0 < r < \rho < R$, where θ is a Nochka constant with

$$\frac{n+1}{k+1} \leq \theta \leq \frac{2n - k + 1}{k+1}.$$

Proof. Take $\tilde{a}_j \in V^* - \{0\}$ with $\mathbb{P}(\tilde{a}_j) = a_j$ and write

$$\tilde{a}_j = \sum_{i=0}^{n} \langle e_i, \tilde{a}_j \rangle \epsilon_i, \quad j = 0, \ldots, q.$$

Define

$$\tilde{\hat{a}}_j = \sum_{i=0}^{k} \langle e_i, \tilde{a}_j \rangle \epsilon_i \in W^* - \{0\}, \quad \hat{a}_j = \mathbb{P}\left(\tilde{\hat{a}}_j\right) \in \mathbb{P}(W^*), \quad j = 0, \ldots, q.$$

Then the family $\hat{\mathscr{A}} = \{\hat{a}_j\}_{j=0}^{q}$ is in n-subgeneral position. In fact, take $\sigma \in J_n^q$. Then $\tilde{a}_\sigma = \tilde{a}_{\sigma(0)} \wedge \cdots \wedge \tilde{a}_{\sigma(n)} \neq 0$ since $\mathscr{A}$ is in general position, and hence

$$\det(\langle e_i, \tilde{a}_{\sigma(j)} \rangle) \neq 0 \quad (0 \leq i, j \leq n).$$

Therefore, there is a $\lambda \in J_k^q$ with $\mathrm{Im}\lambda \subseteq \mathrm{Im}\sigma$ such that

$$\det(\langle e_s, \tilde{a}_{\lambda(t)} \rangle) \neq 0 \quad (0 \leq s, t \leq k).$$

We have

$$\tilde{\hat{a}}_\lambda = \det(\langle e_s, \tilde{a}_{\lambda(t)} \rangle)\epsilon_0 \wedge \cdots \wedge \epsilon_k \neq 0.$$

Hence $\lambda \in J_k(\hat{\mathscr{A}})$. Thus $\hat{\mathscr{A}}$ is in n-subgeneral position.

Note that

$$F_j = \langle \tilde{f}, \tilde{a}_j \rangle = \sum_{i=0}^{n} \langle \tilde{f}, \epsilon_i \rangle \langle e_i, \tilde{a}_j \rangle = \sum_{i=0}^{k} \langle \tilde{\hat{f}}, \epsilon_i \rangle \langle e_i, \tilde{a}_j \rangle = \langle \tilde{\hat{f}}, \tilde{\hat{a}}_j \rangle.$$

We obtain

$$\mu_f^{a_j} = \mu_{\hat{f}}^{\hat{a}_j}, \; j = 0, \ldots, q.$$

By applying Theorem 2.66 to $\hat{f}$, we have

$$(q - 2n + k)T_{\hat{f}}(r) \leq \sum_{j=0}^{q} \theta\omega(\hat{a}_j)N_{\hat{f}}(r, \hat{a}_j) - \theta N_{\mathrm{Ram}}(r, \hat{f})$$

$$+ l\theta \log\left\{\left(\frac{\rho}{r}\right)^{2m-1} \frac{T_{\hat{f}}(R)}{\rho - r}\right\} + O(1).$$

Thus Lemma 2.67 and the facts above yield

$$(q - 2n + k)T_f(r) \quad \leq \quad \sum_{j=0}^{q} \theta\omega(\hat{a}_j)N_{f,w}(r, a_j) - \theta N(r, \mathbf{H})$$

$$+ l\theta \log\left\{\left(\frac{\rho}{r}\right)^{2m-1} \frac{T_f(R)}{\rho - r}\right\} + O(1), \quad (2.8.10)$$

where

$$\mathbf{H} = \frac{|F_0|^{\omega(\hat{a}_0)} \cdots |F_q|^{\omega(\hat{a}_q)}}{|\mathbf{W}(\tilde{f}_0, \ldots, \tilde{f}_k)|},$$

and so Theorem 2.70 follows from 1) in Lemma 1.59.   □

Corollary 2.71. *With the assumptions as in Theorem 2.70,*

$$\sum_{j=0}^{q} \delta_f(a_j) \leq \sum_{j=0}^{q} \delta_{f,w}(a_j) \leq 2n - k + 1.$$

The *defect relation* in Corollary 2.71 refers to the Cartan conjecture which has been proved by Nochka [301].

Corollary 2.72. *Take an integer* k *with* $1 \leq k \leq n$. *Let* $\mathscr{A} = \{a_j\}_{j=0}^{2n-k+1}$ *be a finite family of points* $a_j \in \mathbb{P}(V^*)$ *in general position. If a meromorphic mapping* $f : \mathbb{C}^m \longrightarrow \mathbb{P}(V)$ *has its image in the complement of* $2n - k + 2$ *hyperplanes* $\ddot{E}[a_0]$, $\ldots$, $\ddot{E}[a_{2n-k+1}]$, *then this image is contained in a linear subspace of dimension* $\leq k - 1$.

Corollary 2.72 which strengthens Corollary 2.58 is due to H. Dufresnoy ([84], Théorème XVI), see also Fujimoto [104], [105], Green [120], Kobayashi [208], Lang [229].

Corollary 2.73. *If a meromorphic mapping* $f : \mathbb{C}^m \longrightarrow \mathbb{P}(V)$ *misses* $2n + 1$ *or more hyperplanes in general position, then it is a constant mapping.*

Corollary 2.74 (cf. [208]). *The complement of* $2n+1$ *or more hyperplanes in general position in* $\mathbb{P}(V)$ *is complete Kobayashi hyperbolic.*

P.J. Kiernan [203] proved that $\mathbb{P}(V)$ minus $2n$ hyperplanes in general position is not hyperbolic, and conjectured that the complement of $2n$ hyperplanes in any position in $\mathbb{P}(V)$ is never hyperbolic; he verified the conjecture for $n \leq 5$. His conjecture was proved by Snurnitsyn (cf [208]). Originally, Corollary 2.74 was a conjecture proposed by S. Kobayashi [207] and H. Wu [436] in the end of their book.

Eremenko and Sodin [94] obtained the following *second main theorem* of holomorphic curves which was conjectured by Shiffman [346] by an argument completely different from that of Shiffman [346]:

Theorem 2.75. *Let* $f : \mathbb{C} \longrightarrow \mathbb{P}(V)$ *be a holomorphic mapping and let* $\mathscr{A}$ *be a finite admissible family of* $\mathbb{P}(\amalg_d V^*)$ *such that for each* $a \in \mathscr{A}$, $(f^{\amalg d}, a)$ *is free for the interior product* $\angle$. *Then*

$$\| \sum_{a \in \mathscr{A}} \left\{ dT_f(r) - N_{f^{\amalg d}}(r, a) \right\} \leq \{2n + o(1)\} dT_f(r). \tag{2.8.11}$$

Based on Theorem 2.70 and Theorem 2.75, we suggest the following question:

Conjecture 2.76. *Let $f : \mathbb{C}^m \longrightarrow \mathbb{P}(V)$ be a non-constant meromorphic mapping. Let k be the minimal dimension of algebraic subvarieties containing the image of f in $\mathbb{P}(V)$. If $\mathscr{A}$ is a finite admissible family of $\mathbb{P}(\mathrm{II}_d V^*)$ such that for each $a \in \mathscr{A}$, $(f^{\mathrm{II}d}, a)$ is free for the interior product $\angle$, then*

$$\| \quad \sum_{a \in \mathscr{A}} \{dT_f(r) - N_{f^{\mathrm{II}d}}(r, a)\} \leq \{2n - k + 1 + o(1)\}dT_f(r). \qquad (2.8.12)$$

2.9 First main theorem for line bundles

Corresponding to heights for divisors in number theory, in this section we define characteristic functions of meromorphic mappings for divisors (or line bundles) in Nevanlinna theory. The first main theorem similar to (1.8.5) will be exhibited. We also introduce a counterpart of Theorem 1.102.

Let M and N be connected complex manifolds of dimensions m and n, respectively. Let $\pi : L \longrightarrow N$ be a holomorphic line bundle over N with an Hermitian metric κ along the fibers of L. Let $f : M \longrightarrow N$ be a meromorphic mapping. Assume that τ is an unbounded parabolic exhaustion of the complex manifold M. The *spherical image* of f for (L, κ) is defined by

$$A_f(r, L, \kappa) = M[O; r; f^*(c_1(L, \kappa))], \quad r > 0. \qquad (2.9.1)$$

Define the *characteristic function* or *order function* of f for (L, κ) by

$$T_f(r, r_0, L, \kappa) = \int_{r_0}^{r} A_f(t, L, \kappa) \frac{dt}{t}, \quad r \geq r_0. \qquad (2.9.2)$$

The following fact shows that characteristic functions of f have a similar property with heights associated to Weil functions.

Proposition 2.77 (cf. [128], [380]). *If N is compact, and if κ, κ' are Hermitian metrics along the fibers of L, then*

$$T_f(r, r_0, L, \kappa') - T_f(r, r_0, L, \kappa) = O(1). \qquad (2.9.3)$$

Proof. There exists a function u of class C^∞ on N such that

$$c_1(L, \kappa') = c_1(L, \kappa) + dd^c u.$$

First of all, assume $r_0, r \in \hat{\mathcal{R}}_\tau$ with $r > r_0 > 0$. Let B be the difference of two characteristic functions in (2.9.3). Then

$$B = \int_{r_0}^{r} \int_{M[O;t]} f^*(dd^c u) \wedge v^{m-1} t^{1-2m} dt.$$

Stokes's formula implies

$$B = \int_{r_0}^{r} \int_{M\langle O;t\rangle} f^*(d^c u) \wedge v^{m-1} t^{1-2m} dt$$

$$= \frac{1}{2} \int_{M[O;r_0,r]} \tau^{-m} d\tau \wedge d^c(u \circ f) \wedge v^{m-1}.$$

Note that if φ, ψ, χ have bidegree $(p,p), (q,q), (l,l)$ respectively with $p + q + l = m - 1$, then

$$d\varphi \wedge d^c \psi \wedge \chi = d\psi \wedge d^c \varphi \wedge \chi. \qquad (2.9.4)$$

Hence we have

$$B = \frac{1}{2} \int_{M[O;r_0,r]} \tau^{-m} d(u \circ f) \wedge d^c \tau \wedge v^{m-1}.$$

Further, by using (2.1.66), we obtain

$$B = \frac{1}{2} \int_{M[O;r_0,r]} d(u \circ f) \wedge \sigma$$

$$= \frac{1}{2} \int_{M\langle O;r\rangle} (u \circ f)\sigma - \frac{1}{2} \int_{M\langle O;r_0\rangle} (u \circ f)\sigma,$$

which implies

$$|B| \le \varsigma \max_{x \in N} |u(x)| < \infty.$$

Continuity shows the estimate for all $r > r_0 > 0$.   □

Abbreviate

$$A_f(r, L) = A_f(r, L, \kappa), \quad T_f(r, L) = T_f(r, r_0, L, \kappa).$$

Then when N is compact and r_0 is fixed, the function $T_f(r, L)$ is well defined, up to $O(1)$. According to the basic properties of Chern forms in Section 2.1.5, we have

$$T_f(r, L^*) = -T_f(r, L),$$

and

$$T_f(r, L_1 \otimes \cdots \otimes L_p) = T_f(r, L_1) + \cdots + T_f(r, L_p),$$

where $L_1, \ldots, L_p$ are Hermitian line bundles. In particular, if $L = N \times \mathbb{C}$ is the trivial line bundle, then

$$T_f(r, L) = 0.$$

Proposition 2.78 (cf. [380]). *Let κ be an Hermitian metric along the fibers of a line bundle L on N such that $c_1(L, \kappa) > 0$. If the meromorphic mapping $f : M \longrightarrow N$ is non-constant, then $A_f(r, L) > 0$ for $r > 0$. In particular,*

$$\lim_{r \to +\infty} \frac{T_f(r, L)}{\log r} = \lim_{r \to +\infty} A_f(r, L) > 0.$$

Proof. An open set in M can be defined by

$$M^+ = \{x \in M \mid v^m(x) > 0\} = \{x \in M \mid v(x) > 0\}$$

since $v = dd^c \tau \geq \tau dd^c \log \tau \geq 0$. We know that $O = \tau^{-1}(0) \neq \emptyset$ and $\varsigma > 0$ because τ is parabolic (cf. [380]). Hence

$$\int_{M[O;r]} v^m = \int_{M(O;r)} v^m = \varsigma r^{2m}$$

implies that $M^+(O;r)$ is a non-empty open set for all $r > 0$.

Let S be the set of all $x \in M$ such that the Jacobian of f at x has rank 0. Then S is thin analytic in M. Fix $r > 0$ and take $p \in M^+(O;r) - \{S \cup I_f\}$. We can choose local holomorphic systems $(U; z_1, \ldots, z_m)$ and $(W; w_1, \ldots, w_n)$ of p and $f(p)$ respectively such that $f(U) \subset W$, and

$$v(p) = \frac{\sqrt{-1}}{2\pi} \sum_{j=1}^{m} dz_j \wedge d\bar{z}_j.$$

Write

$$c_1(L, \kappa) = \frac{\sqrt{-1}}{2\pi} \sum_{k,l} h_{kl} dw_k \wedge d\bar{w}_l,$$

and so

$$f^*(c_1(L, \kappa)) = \frac{\sqrt{-1}}{2\pi} \sum_{i,j} g_{ij} dz_i \wedge d\bar{z}_j,$$

where

$$g_{ij} = \sum_{k,l} h_{kl} \circ f \frac{\partial f_k}{\partial z_i} \frac{\overline{\partial f_l}}{\partial z_j}, \quad f_k = w_k \circ f.$$

Then we have

$$f^*(c_1(L, \kappa))(p) \wedge v(p)^{m-1} = \left((m-1)! \sum_{j=1}^{m} g_{jj}(p) \right) \prod_{j=1}^{m} \frac{\sqrt{-1}}{2\pi} dz_j \wedge d\bar{z}_j > 0 \quad (2.9.5)$$

since $p \notin S$. By continuity, a neighborhood U_0 of p exists in $M^+(O;r) - \{S \cup I_f\}$ such that

$$f^*(c_1(L, \kappa)) \wedge v^{m-1}|_{U_0} > 0,$$

and hence $A_f(r, L) > 0$. This completes the proof of Proposition 2.78. □

Consequently, Proposition 2.78 implies that f is constant if and only if $T_f(r, L)$ is bounded. Note that the fact corresponding to this property in number theory is that the number of rational points with bounded heights is finite. Hence Proposition 2.78 is an analogue of Theorem 1.102 in Nevanlinna theory.

Take $s \in \Gamma(N, L)$ and let D be the divisor (s) of s. Assume that $f(M) \not\subset$ supp(D). The *counting function* and *valence function* of f for the divisor D on N are respectively defined by

$$n_f(r, D) = n_{\mu_{f^*D}}(r) \tag{2.9.6}$$

and

$$N_f(r, D) = N_{\mu_{f^*D}}(r). \tag{2.9.7}$$

For $0 < r \in \hat{\mathcal{R}}_\tau$, the *proximity function* or *compensation function* is defined by

$$m_f(r, D) = M \left\langle O; r; \log \frac{1}{|s \circ f|_\kappa} \right\rangle. \tag{2.9.8}$$

Since $f : M - I_f \longrightarrow N$ is holomorphic, the lifted section s_f is well defined on $M - I_f$. The zero divisor of s_f on $M - I_f$ continues uniquely to a divisor (s_f) on M since $\dim I_f \leq m - 2$. Let $\mu_f^s = \mu_{s_f}$ denote the *multiplicity* of (s_f) which is just equal to μ_{f^*D}. Let $G \neq \emptyset$ be an open, relative compact subset of M such that the boundary $\partial G = \overline{G} - G$ is either empty or a real $(2m - 1)$-dimensional C^∞ submanifold of M oriented to the exterior of G. Let ξ be a form of class C^1 and degree $2m - 1$ on M such that $\log |s \circ f|_\kappa^2 \xi$ is integrable over ∂G. Then the *singular Stokes formula*

$$\int_G \log |s \circ f|_\kappa^2 d\xi + \int_G d \log |s \circ f|_\kappa^2 \wedge \xi = \int_{\partial G} \log |s \circ f|_\kappa^2 \xi \tag{2.9.9}$$

holds.

Let χ be a form of class C^∞ and bidegree $(m - 1, m - 1)$ on M such that $\chi \wedge d^c \log |s \circ f|_\kappa^2$ is integrable over ∂G. Assume that $A \cap \text{supp}\chi|_{\partial G}$ has measure zero on $A = \text{supp}\mu_f^s$. Then the *residue formula*

$$- \int_G dd^c \log |s \circ f|_\kappa^2 \wedge \chi = \int_{A \cap G} \mu_f^s \chi - \int_{\partial G} d^c \log |s \circ f|_\kappa^2 \wedge \chi$$
$$+ \int_G d\chi \wedge d^c \log |s \circ f|_\kappa^2 \tag{2.9.10}$$

holds. If we take $\xi = d^c \chi$, then

$$d\chi \wedge d^c \log |s \circ f|_\kappa^2 = d \log |s \circ f|_\kappa^2 \wedge d^c \chi,$$

and hence the *Green residue formula*

$$- \int_G dd^c \log |s \circ f|_\kappa^2 \wedge \chi = \int_{A \cap G} \mu_f^s \chi - \int_{\partial G} d^c \log |s \circ f|_\kappa^2 \wedge \chi$$
$$+ \int_{\partial G} \log |s \circ f|_\kappa^2 d^c \chi - \int_G \log |s \circ f|_\kappa^2 dd^c \chi \tag{2.9.11}$$

follows from (2.9.9) and (2.9.10). Proofs of these formulae can be found in Stoll [380].

For $x > 0$, $y > 0$, define a function

$$\psi(x, y) = \begin{cases} \frac{1}{2m-2}\left(\frac{1}{x^{2m-2}} - \frac{1}{y^{2m-2}}\right) & : \quad m > 1, \\ \log y - \log x & : \quad m = 1, \end{cases}$$

and write

$$\psi_r(x) = \min\{\psi(x, r), \psi(r_0, r)\}, \quad r > r_0 > 0.$$

Use the symbols in (2.1.64), then on $M[O; r] - M(O; r_0)$, we have

$$2d^c(\psi_r \circ \sqrt{\tau}) \wedge v^{m-1} = -\sigma, \quad 2dd^c(\psi_r \circ \sqrt{\tau}) \wedge v^{m-1} = -\omega^m = 0. \qquad (2.9.12)$$

Take $r_0, r \in \hat{\mathcal{R}}_\tau$ and assume

$$\chi = (\psi_r \circ \sqrt{\tau})v^{m-1}.$$

Applying the residue formula (2.9.10) and Green residue formula (2.9.11) to $M(O; r_0)$ and $G = M(O; r) - M[O; r_0]$, respectively, and using the formula (2.9.12), one obtains

$$-\int_{M(O;r_0)} dd^c \log|s \circ f|_\kappa^2 \wedge \chi = \int_{A(O;r_0)} \mu_f^s \chi - \int_{M\langle O;r_0\rangle} d^c \log|s \circ f|_\kappa^2 \wedge \chi, \qquad (2.9.13)$$

where $d\chi = 0$ on $M(O; r_0)$ is used, and

$$-\int_G dd^c \log|s \circ f|_\kappa^2 \wedge \chi = \int_{A\cap G} \mu_f^s \chi + \int_{M\langle O;r_0\rangle} d^c \log|s \circ f|_\kappa^2 \wedge \chi$$
$$-\int_{M\langle O;r\rangle} \log|s \circ f|_\kappa \sigma + \int_{M\langle O;r_0\rangle} \log|s \circ f|_\kappa \sigma. \qquad (2.9.14)$$

Adding (2.9.13) and (2.9.14), one obtains the *first main theorem*

$$-\int_{M(O;r)} dd^c \log|s \circ f|_\kappa^2 \wedge \chi = \int_{A(O;r)} \mu_f^s \chi - \int_{M\langle O;r\rangle} \log|s \circ f|_\kappa \sigma$$
$$+\int_{M\langle O;r_0\rangle} \log|s \circ f|_\kappa \sigma. \qquad (2.9.15)$$

Lemma 2.79 ([384]). *Let A be a pure k-dimensional analytic subset of M. Let η be a form of bidegree (k, k) on M which is locally integrable over A. Take $0 \le s < r < \infty$. Let $h : \mathbb{R}[s, r] \longrightarrow \mathbb{C}$ be a function of class C^1. Then*

$$\int_{A[O;r]-A[O;s]} h \circ \sqrt{\tau}\eta = h(r)\int_{A[O;r]} \eta - h(s)\int_{A[O;s]} \eta$$
$$-\int_s^r \left(\int_{A[O;t]} \eta\right) h'(t)dt, \qquad (2.9.16)$$

where the integrals are sums if $k = 0$.

Proof. Define an auxiliary function

$$\gamma(x,y) = \begin{cases} 0 & : \quad s \le x \le y \le r, \\ 1 & : \quad s \le y < x \le r , \end{cases}$$

and set

$$I = h(r) \int_{A[O;r]} \eta - h(r) \int_{A[O;s]} \eta - \int_{A[O;r]-A[O;s]} h \circ \sqrt{\tau}\,\eta,$$

which can be expressed by the integral

$$I = \int_{A[O;r]-A[O;s]} \left(\int_{\sqrt{\tau}}^{r} h'(t)dt \right) \eta.$$

By using the simple equation

$$\int_{\sqrt{\tau}}^{r} h'(t)dt = \int_{s}^{r} \gamma\left(t, \sqrt{\tau}\right) h'(t)dt,$$

and exchanging order of the integrals, we obtain

$$I = \int_{s}^{r} \left(\int_{A[O;r]-A[O;s]} \gamma\left(t, \sqrt{\tau}\right) \eta \right) h'(t)dt$$

$$= \int_{s}^{r} \left(\int_{A[O;t]-A[O;s]} \eta \right) h'(t)dt$$

$$= \int_{s}^{r} \left(\int_{A[O;t]} \eta \right) h'(t)dt - (h(r) - h(s)) \int_{A[O;s]} \eta,$$

and so Lemma 2.79 follows. $\qquad\qquad\qquad\qquad\qquad\qquad\qquad\qquad\qquad\square$

Further, put $A = \mathrm{supp} f^* D$. Lemma 2.79 implies

$$N_f(r, D) = \int_{A[O;r]} \mu_{f^* D}(\psi_r \circ \sqrt{\tau}) \upsilon^{m-1}, \qquad (2.9.17)$$

$$T_f(r, L) = \int_{M[O;r]} (\psi_r \circ \sqrt{\tau}) f^*(c_1(L, \kappa)) \wedge \upsilon^{m-1}. \qquad (2.9.18)$$

Then (2.9.15) immediately yields the *first main theorem* (cf. [128], [344], [380])

$$T_f(r, L) = N_f(r, D) + m_f(r, D) - m_f(r_0, D). \qquad (2.9.19)$$

The identity (2.9.19) can be used to show that the compensation function extends to a continuous function on all positive real numbers such that (2.9.19) holds for all $0 < r_0 < r$. Obviously, the formula (2.9.19) is an analogue of (1.8.5) in Nevanlinna theory.

Remark. If u is a function of class C^2 on $M[O; r]$, according to the proof of the formula (2.9.15) and by Lemma 2.79 we can prove

$$\mathcal{T}(r, r_0; dd^c u) = \frac{1}{2} M\langle O; r; u\rangle - \frac{1}{2} M\langle O; r_0; u\rangle. \qquad (2.9.20)$$

A special case is given in Chapter 6 (see (6.1.3)). For the case $M = \mathbb{C}^m$, Noguchi and Ochiai ([302], Lemma 3.3.39) show that (2.9.20) holds for a plurisubharmonic function u on $\mathbb{C}^m[0; r]$.

Lemma 2.80. *Let N be a non-singular projective variety of dimension n. Let H be a very ample line bundle on N and let L be a pseudo ample line bundle on N. Then*

$$\limsup_{j \to +\infty} \frac{1}{j^n} \dim \Gamma(N, L^j \otimes H^*) > 0.$$

Lemma 2.80 follows easily from Lemma 2.30 (or see [229]). A meromorphic mapping $f : M \longrightarrow N$ into an algebraic variety N is called *algebraically non-degenerate* if the image of f is not contained in any proper algebraic subvariety of N, otherwise f is said to be *algebraically degenerate*.

Proposition 2.81. *Assume that N is a non-singular projective variety. Suppose that L is a pseudo ample line bundle on N. Then there exists a proper algebraic subset Z of N such that for any meromorphic mapping $f : M \longrightarrow N$ with $f(M) \not\subset Z$, there exists a positive constant c satisfying*

$$c \log r \le T_f(r, L) + O(1).$$

Proof. Since N is projective algebraic, there exists a very ample line bundle H on N. By Lemma 2.80, for j large there exists a non-trivial holomorphic section s of $L^j \otimes H^*$. An Hermitian metric κ along the fibers of $L^j \otimes H^*$ exists such that $|s|_\kappa \le 1$ because N is compact. Take $Z = \mathrm{supp}((s))$. When $f(M) \not\subset Z$, by (2.9.19), we have

$$N_f(r, (s)) \le T_f(r, L^j \otimes H^*) + O(1)$$
$$= jT_f(r, L) - T_f(r, H) + O(1),$$

which implies

$$T_f(r, H) \le jT_f(r, L) + O(1). \qquad (2.9.21)$$

Now Proposition 2.81 follows from Proposition 2.78. $\qquad\qquad\qquad\qquad\square$

In particular, Proposition 2.81 holds if the meromorphic mapping $f : M \longrightarrow N$ is algebraically non-degenerate. We end this section by the following result (see [128], [380]):

Proposition 2.82. *Let M be a smooth affine variety. Assume that N is a non-singular projective variety with a positive holomorphic line bundle L. Then a holomorphic mapping $f : M \longrightarrow N$ is rational if and only if*

$$T_f(r, L) = O(\log r).$$

2.10   Jacobian sections

We will use the notation of Jacobian sections in the proof of the second main theorem for line bundles. A good reference is Stoll [380], but for completeness, here we give a brief introduction.

Let M be a complex manifold of dimension m. Let N be a complex manifold of dimension n. Let $f : M \longrightarrow N$ be a holomorphic mapping. Then

$$K(f) = K_M \otimes f^*(K_N^*)$$

is called the *Jacobian bundle*, where K_N^* is the dual of K_N. A holomorphic section F of $K(f)$ over M is said to be a *Jacobian section*. The section F is called *effective* if $F^{-1}(0)$ is thin. The zero divisor (F) of F is called the *ramification divisor* of f for F.

Let F be a Jacobian section. Let U be an open subset in N such that $\tilde{U} = f^{-1}(U) \neq \emptyset$. The inner product

$$\langle , \rangle : K_N^* \oplus K_N \longrightarrow \mathbb{C}$$

is well defined and pulls back to

$$\langle , \rangle : f^*(K_N^*) \oplus f^*(K_N) \longrightarrow \mathbb{C}$$

which further induces a K_M-valued inner product

$$\langle , \rangle : K(f) \oplus f^*(K_N) \longrightarrow K_M.$$

The section F defines a linear mapping

$$F : \Gamma(U, K_N) \longrightarrow \Gamma(\tilde{U}, K_M)$$

by $F[\Psi] = \langle F, \Psi_f \rangle$ for all $\Psi \in \Gamma(U, K_N)$. If $(V; z_1, \ldots, z_m)$ and $(W; w_1, \ldots, w_n)$ are holomorphic coordinate charts on M and N respectively with $W \subset U$ and $f(V) \subset W$, then $\Psi = \Psi_W dw$ on W and $F = F_{VW} dz \otimes d^* w_f$ on V, where

$$dw = dw_1 \wedge \cdots \wedge dw_n, \quad d^*w = \frac{\partial}{\partial w_1} \wedge \cdots \wedge \frac{\partial}{\partial w_n},$$

and where Ψ_W and F_{VW} are holomorphic functions. Then on V we have

$$F[\Psi] = F_{VW}(\Psi_W \circ f)dz.$$

Further, a linear mapping

$$F : A^{2n}(U) \longrightarrow A^{2m}(\tilde{U})$$

is uniquely defined as follows: If $\Omega \in A^{2n}(U)$ is expressed by

$$\Omega = i_n \rho_W \, dw \wedge \overline{dw}$$

in the term of the holomorphic coordinate chart $(W; w_1, \ldots, w_n)$ on N, then

$$F[\Omega]|_{\tilde{W}} = i_m (\rho_W \circ f) F[dw] \wedge \overline{F[dw]} = i_m (\rho_W \circ f) |F_{VW}|^2 dz \wedge \overline{dz}. \qquad (2.10.1)$$

If κ is a Hermitian metric along the fibers of $K(f)$, then $|F|_\kappa^2 = \kappa_V |F_{VW}|^2$ with $\kappa_V = |dz \otimes d^* w_f|_\kappa^2 > 0$, and so $\Theta \in A^{2m}(\tilde{U})$ exists such that

$$F[\Omega] = |F|_\kappa^2 \Theta, \quad \Theta|_V = i_m (\rho_W \circ f) \kappa_V^{-1} dz \wedge \overline{dz}.$$

Obviously, $\Theta > 0$ if and only if $\Omega > 0$.

Now we discuss the existence of Jacobian sections. First assume

$$p = m - n \geq 0.$$

Then there exists a holomorphic section Df of $\left(\bigwedge_n \mathbf{T}^*(M) \right) \otimes f^*(K_N^*)$ such that

$$Df|_{\tilde{U}} = f^*(\Psi) \otimes \Psi_f^*$$

whenever U is open in N with $\tilde{U} = f^{-1}(U) \neq \emptyset$ and $\Psi \in \Gamma(U, K_N)$ vanishes nowhere, where Ψ^* is the dual frame to Ψ. If $(V; z_1, \ldots, z_m)$ and $(W; w_1, \ldots, w_n)$ are holomorphic coordinate charts on M and N respectively with $f(V) \subset W$, then

$$f^*(dw) = \sum_{\lambda \in J_{1,n}^m} \Delta_\lambda dz_\lambda,$$

where

$$\Delta_\lambda = \det \left(\frac{\partial w_i \circ f}{\partial z_{\lambda(j)}} \right), \quad dz_\lambda = dz_{\lambda(1)} \wedge \cdots \wedge dz_{\lambda(n)},$$

and so

$$Df|_V = \sum_{\lambda \in J_{1,n}^m} \Delta_\lambda dz_\lambda \otimes d^* w_f.$$

Hence zeros of Df is the set of all $x \in M$ such that the rank of the Jacobian matrix at x is smaller than n, which is called the *branching set* of f. We distinguish two cases:

(a) If $m = n$, Df is a Jacobian section which is effective if and only if f has strict rank n. In this case the divisor (Df) of Df also is called the *branching divisor* of f.

(b) If $m > n$, and if there exists a holomorphic form φ of degree p on M, then φ
   induces a Jacobian section

$$F_\varphi = \varphi \wedge Df.$$

The form is said to be *effective* for f if F_φ is effective, where the wedge product
extends from $\bigwedge_p \mathbf{T}^*(M)$, $\bigwedge_n \mathbf{T}^*(M)$ to $\bigwedge_p \mathbf{T}^*(M)$, $\left(\bigwedge_n \mathbf{T}^*(M)\right) \otimes f^*(K_N^*)$ and
so becomes $K(f)$-valued. Equivalently F_φ can be described by its action

$$F_\varphi[\Psi] = \varphi \wedge f^*(\Psi)$$

whenever $\Psi \in \Gamma(U, K_N)$ and U open in N with $\tilde{U} = f^{-1}(U) \neq \emptyset$.

We will unify Jacobian sections in the cases (a) and (b) by writing $F_\varphi = \varphi \wedge Df$,
where we think $\varphi = 1$ for the case (a). If $\Psi \in A^{2n}(U)$, it is not difficult to show
that

$$F_\varphi[\Psi] = \binom{m}{m-n} i_{m-n}\varphi \wedge \bar{\varphi} \wedge f^*(\Psi). \qquad (2.10.2)$$

If M is connected, and has holomorphic rank m, then there exists a holomorphic mapping

$$\beta : M \longrightarrow \mathbb{C}^m$$

of strict rank m. For $x \in M$, the differential

$$\beta'(x) : \mathbf{T}_x(M) \longrightarrow \mathbf{T}_{\beta(x)}(\mathbb{C}^m)$$

is a linear mapping. The branching set $\mathscr{B}$ of all $x \in M$ such that $\beta'(x)$ is not an
isomorphism is an analytic subset of M. Then there exists a thin analytic set B
of $\mathbb{C}^m$ with $\beta(\mathscr{B}) \subseteq B$ such that

$$\beta : M - \beta^{-1}(B) \longrightarrow \mathbb{C}^m - B$$

is a proper, surjective, local biholomorphic mapping, hence a covering space of
finite sheet number ν. Here ν is called the *sheet number* of β. We will use the
parabolic exhaustion $\tau = \|\beta\|^2$ of M and use the notations in (2.1.64).

Lemma 2.83 (Stoll [380]). *Let M and N be connected complex manifolds of dimension m and n respectively, and let $f : M \longrightarrow N$ be a holomorphic mapping of
rank n. Assume that a holomorphic mapping*

$$\beta = (\beta_1, \ldots, \beta_m) : M \longrightarrow \mathbb{C}^m$$

of strict rank m exists. Then there exists $\{j_1, \ldots, j_{m-n}\} \subset \mathbb{Z}[1, m]$ such that

$$\varphi = d\beta_{j_1} \wedge \cdots \wedge d\beta_{j_{m-n}}$$

satisfies that $F_\varphi^{-1}(0)$ is thin, and

$$i_{m-n}\varphi \wedge \bar{\varphi} \leq \upsilon^{m-n}.$$

Proof. The case $m = n$ is trivial. We may assume $p = m - n > 0$. Set

$$S = \{x \in M \mid \mathrm{rank}_x f < n\}.$$

Then S is a thin analytic set. Define

$$M_0 = M - \{\beta^{-1}(B) \cup \beta^{-1}(\beta(S))\}.$$

Take $z_0 \in M_0$. There exist a local holomorphic coordinate system $(W; w_1, \ldots, w_n)$ of $f(z_0)$ and an open connected neighborhood U of z_0 in M_0 with $f(U) \subset W$ such that $\beta : U \longrightarrow \beta(U)$ is biholomorphic. Define $f_j = w_j \circ f$ and set

$$df_j = \sum_{k=1}^{m} f_{jk} d\beta_k, \quad j = 1, \ldots, n.$$

Then

$$0 \neq df_1 \wedge \cdots \wedge df_n = \sum_{\nu \in J_{1,n}^m} \Delta_\nu d\beta_\nu,$$

where for $\nu \in J_{1,n}^m$,

$$\Delta_\nu = \det(f_{j\nu(k)}), \quad d\beta_\nu = d\beta_{\nu(1)} \wedge \cdots \wedge d\beta_{\nu(n)}.$$

Therefore $\nu \in J_{1,n}^m$ exists such that $\Delta_\nu \not\equiv 0$ on U. Then $\nu^\perp \in J_{1,p}^m$ is uniquely defined such that $(\nu^\perp, \nu)$ is a permutation of $\mathbb{Z}[1, m]$. Then $\varphi = d\beta_{\nu^\perp}$ is a holomorphic form of degree p on M such that

$$
\begin{aligned}
F_\varphi[dw_1 \wedge \cdots \wedge dw_n] &= \varphi \wedge f^*(dw_1 \wedge \cdots \wedge dw_n) \\
&= d\beta_{\nu^\perp} \wedge df_1 \wedge \cdots \wedge df_n = \Delta_\nu d\beta_{\nu^\perp} \wedge d\beta_\nu \\
&= \Delta_\nu \mathrm{sign}(\nu^\perp, \nu) d\beta_1 \wedge \cdots \wedge d\beta_m \not\equiv 0
\end{aligned}
$$

hold on U. Since $F_\varphi^{-1}(0) \cap U = \Delta_\nu^{-1}(0)$ is thin in U and since M is connected, then $F_\varphi^{-1}(0)$ is thin. Finally, since $\upsilon = dd^c \|\beta\|^2$ we have

$$\upsilon^p = p! \left(\frac{\sqrt{-1}}{2\pi}\right)^p \sum_{\gamma \in J_{1,p}^m} d\beta_{\gamma(1)} \wedge d\overline{\beta}_{\gamma(1)} \wedge \cdots \wedge d\beta_{\gamma(p)} \wedge d\overline{\beta}_{\gamma(p)}$$

which yields

$$\upsilon^p \geq i_p d\beta_{\nu^\perp} \wedge d\overline{\beta}_{\nu^\perp} = i_p \varphi \wedge \overline{\varphi},$$

and so Lemma 2.83 follows. $\square$

Secondly we consider the case

$$q = n - m > 0.$$

Note that a unique homomorphism

$$\hat{f} : f^* \left(\bigwedge_m \mathbf{T}^*(N) \right) \longrightarrow \bigwedge_m \mathbf{T}^*(M)$$

exists such that $\hat{f}(\xi_f) = f^*(\xi)$ for all $\xi \in \Gamma \left(U, \bigwedge_m \mathbf{T}^*(N) \right)$, where U is open in N with $\tilde{U} = f^{-1}(U) \neq \emptyset$. The interior product

$$\angle : K_N \oplus \bigwedge_q \mathbf{T}(N) \longrightarrow \bigwedge_m \mathbf{T}^*(N)$$

also pulls back to an interior product

$$\angle : f^*(K_N) \oplus f^* \left(\bigwedge_q \mathbf{T}(N) \right) \longrightarrow f^* \left(\bigwedge_m \mathbf{T}^*(N) \right).$$

Take a holomorphic section $\varphi \in \Gamma \left(M, f^* \left(\bigwedge_q \mathbf{T}(N) \right) \right)$, which is called a *holomorphic field on f over M of degree q*. Then φ induces a Jacobian section F_φ defined by

$$F_\varphi|_{\tilde{U}} = \hat{f}(\Psi_f \angle \varphi) \otimes \Psi_f^*,$$

where $\Psi \in \Gamma(U, K_N)$ vanishes nowhere for the open subset of N with $\tilde{U} = f^{-1}(U) \neq \emptyset$ and Ψ^* is the dual frame. Equivalently F_φ can be described by its action

$$F_\varphi[\Psi] = \hat{f}(\Psi_f \angle \varphi)$$

for all $\Psi \in \Gamma(U, K_N)$, where U is open in N with $\tilde{U} = f^{-1}(U) \neq \emptyset$. The section φ is said to be *effective* for f if F_φ is effective.

Theorem 2.84 ([380]). *Assume M is Stein and $q = n - m > 0$. Then there exists a holomorphic field φ on f over M of degree q such that φ is effective for f if and only if f has strict rank m.*

Proof. We may assume that M is connected. Let φ be an effective holomorphic field on f over M of degree q and let S be the set of all $x \in M$ such that the rank of the Jacobian of f at x is smaller than m. Take $z_0 \in M$. There exist local holomorphic coordinate systems $(U; z_1, \ldots, z_m)$ of z_0 in M and $(W; w_1, \ldots, w_n)$ of $f(z_0)$ in N with $f(U) \subset W$. Set

$$d^* w_k = \frac{\partial}{\partial w_k}, \quad k = 1, \ldots, n, \tag{2.10.3}$$

and write

$$\varphi = \sum_{\lambda \in J_{1,q}^n} \varphi_\lambda d^* w_{\lambda f}, \tag{2.10.4}$$

where

$$d^* w_\lambda = d^* w_{\lambda(1)} \wedge \cdots \wedge d^* w_{\lambda(q)}. \tag{2.10.5}$$

We also write

$$f^*(dw_\nu) = A_\nu dz, \quad \nu \in J_{1,m}^n. \tag{2.10.6}$$

Then

$$S \cap U = \bigcap_{\nu \in J_{1,m}^n} A_\nu^{-1}(0).$$

If $\lambda \in J_{1,q}^n$, then $\lambda^\perp \in J_{1,m}^n$ is uniquely defined such that $(\lambda^\perp, \lambda)$ is a permutation of $\{1, \ldots, n\}$. Then $\perp : J_{1,q}^n \longrightarrow J_{1,m}^n$ is bijective. Since $dw = dw_1 \wedge \cdots \wedge dw_n$ and

$$dw_f \angle \varphi = \sum_{\lambda \in J_{1,q}^n} \mathrm{sign}(\lambda^\perp, \lambda) \varphi_\lambda dw_{\lambda^\perp f},$$

then we have

$$F_\varphi[dw] = \hat{f}(dw_f \angle \varphi) = \sum_{\lambda \in J_{1,q}^n} \mathrm{sign}(\lambda^\perp, \lambda) \varphi_\lambda A_{\lambda^\perp} dz. \tag{2.10.7}$$

Thus we obtain

$$(F_\varphi) \cap U = \left(\sum_{\lambda \in J_{1,q}^n} \mathrm{sign}(\lambda^\perp, \lambda) \varphi_\lambda A_{\lambda^\perp} \right) \supseteq S \cap U.$$

Therefore $S \cap U$ is thin since the support of the divisor (F_φ) is thin, and so S is thin since M is connected. Hence f has rank m.

Assume that f has rank m on M. Take S, z_0, z, w, A_ν as above. Note that we can choose $z_0 \in M - S$ since S is thin. Then $\iota \in J_{1,m}^n$ exists such that $A_\iota(z_0) \neq 0$. Since M is Stein, Theorem 2.15 implies that global holomorphic sections $s_j \in \Gamma(M, f^*(\mathbf{T}(N)))$ exist such that

$$s_j(z_0) = d^* w_{jf}(z_0), \quad j = 1, \ldots, n. \tag{2.10.8}$$

Define

$$\varphi = s_{\iota^\perp} \in \Gamma\left(M, \bigwedge_q f^*(\mathbf{T}(N))\right)$$

with

$$\varphi(z_0) = s_{\iota^\perp}(z_0) = d^* w_{\iota^\perp f}(z_0).$$

Hence (2.10.4) holds;

$$\varphi_\lambda(z_0) = \begin{cases} 1, & \text{if } \lambda = \iota^\perp, \\ 0, & \text{if } \iota^\perp \neq \lambda \in J_{1,q}^n. \end{cases}$$

Also (2.10.7) holds with

$$F_\varphi[dw](z_0) = \mathrm{sign}(\iota, \iota^\perp) A_\iota(z_0) dz \neq 0.$$

Note that
$$(F_\varphi[dw]) \cap U = (F_\varphi) \cap U.$$
Hence $z_0 \in M - \operatorname{supp}((F_\varphi))$, that is, the analytic set $\operatorname{supp}((F_\varphi))$ is thin. Hence φ is effective for f. $\qquad\square$

If M is a non-compact Riemann surface, Theorem 2.16 and (2.1.14) imply that the sections defined by (2.10.8) can be chosen such that $(s_1, \ldots, s_n)$ is a global holomorphic frame of $f^*(\mathbf{T}(N))$ over M.

Lemma 2.85 ([167]). *Assume that $q = n - m > 0$, and that f has strict rank m. If there exist holomorphic vector fields $Z_1, \ldots, Z_n$ on N such that*
$$Z := Z_1 \wedge \cdots \wedge Z_n \not\equiv 0,$$
and if $f(M) \not\subseteq \operatorname{supp}((Z))$, then there exists $\lambda \in J_{1,q}^n$ such that a holomorphic field
$$\varphi = Z_\lambda f = (Z_{\lambda(1)} \wedge \cdots \wedge Z_{\lambda(q)})_f$$
on f over M of degree q is effective for f.

Proof. We may assume that M is connected. Under the conditions of Lemma 2.85, a lifted section $Z_f \in \Gamma(M, f^*(K_N^*))$ of Z for f exists with $Z_f \not\equiv 0$. Let S be the set of all $x \in M$ such that the rank of the Jacobian of f at x is smaller than m. Take $z_0 \in M - S$ such that $f(z_0) \notin \operatorname{supp}((Z))$ since S and $\operatorname{supp}((Z_f))$ are thin. There exist local holomorphic coordinate systems $(U; z_1, \ldots, z_m)$ of z_0 in M and $(W; w_1, \ldots, w_n)$ of $f(z_0)$ in N with $f(U) \subset W$ such that $Z_1, \ldots, Z_n$ form a frame of $\mathbf{T}(N)$ on W. Let $\psi_1, \ldots, \psi_n$ be the dual frame in $\mathbf{T}^*(N)$ and write
$$\psi_\nu = \psi_{\nu(1)} \wedge \cdots \wedge \psi_{\nu(m)}, \quad f^*(\psi_\nu) = B_\nu dz$$
for $\nu \in J_{1,m}^n$. Then $\iota \in J_{1,m}^n$ exists such that $B_\iota(z_0) \neq 0$. Take $\lambda = \iota^\perp \in J_{1,q}^n$ and define
$$\varphi = Z_\lambda f \in \Gamma\left(M, \bigwedge_q f^*(\mathbf{T}(N))\right).$$
Set $\Psi = \psi_1 \wedge \cdots \wedge \psi_n$ and note that
$$\Psi_f \angle \varphi = \operatorname{sign}(\iota, \iota^\perp) \psi_{\iota f},$$
then we have
$$F_\varphi[\Psi] = \hat{f}(\Psi_f \angle \varphi) = \operatorname{sign}(\iota, \iota^\perp) B_\iota dz.$$
Hence $z_0 \in M - \operatorname{supp}((F_\varphi))$ since
$$(F_\varphi[\Psi]) \cap U = (F_\varphi) \cap U,$$
that is, the analytic set $\operatorname{supp}((F_\varphi))$ is thin. Hence φ is effective for f. $\qquad\square$

Lemma 2.86 ([167]). *Take $M = \mathbb{C}^m$ and define*

$$Z_i = f'\left(\frac{\partial}{\partial z_i}\right), \quad i = 1, \ldots, m.$$

Assume $q = n - m > 0$. Then a holomorphic field φ on f over $\mathbb{C}^m$ of degree q such that φ is effective for f if and only if

$$Z = Z_{1f} \wedge \cdots \wedge Z_{mf} \wedge \varphi \not\equiv 0.$$

Further, we have $(F_\varphi) = (Z)$.

Proof. We will use the symbols in the proof of Theorem 2.84. Relative to the local holomorphic coordinates $w_1, \ldots, w_n$ on an open subset W of N, set

$$f_k = w_k \circ f, \quad k = 1, \ldots, n,$$

express φ by (2.10.4), and write

$$Z_i = \sum_{k=1}^{n} \frac{\partial f_k}{\partial z_i} d^* w_k, \quad i = 1, \ldots, m.$$

Then we have

$$Z_{1f} \wedge \cdots \wedge Z_{mf} = \sum_{\nu \in J_{1,m}^n} A_\nu d^* w_{\nu f},$$

where

$$A_\nu = \det\left(\frac{\partial f_{\nu(j)}}{\partial z_i}\right).$$

Hence we obtain

$$Z = \sum_{\lambda \in J_{1,q}^n} \operatorname{sign}(\lambda^\perp, \lambda) \varphi_\lambda A_{\lambda^\perp} d^* w_f.$$

On the other hand, we also have (2.10.6) and (2.10.7). Therefore we obtain

$$(F_\varphi) \cap U = (Z) \cap U,$$

and hence Lemma 2.86 follows. $\qquad\square$

Lemma 2.87 ([167]). *Let M and N be complex manifolds of dimensions m and n, respectively, and let θ and ψ be the associated 2-forms of Hermitian metrics on M and N, respectively. Assume $q = n - m > 0$. Let $f : M \longrightarrow N$ be a holomorphic mapping of strict rank m. Let φ be a holomorphic field on f over M of degree q. Define a non-negative function g by*

$$F_\varphi[\psi^n] = g^2 f^*(\psi^m).$$

Then $g \leq |\varphi|$.

Proof. In terms of a coframe $(\psi_1, \ldots, \psi_n)$ on N and a coframe $(\theta_1, \ldots, \theta_m)$ on M, we have

$$\psi = \frac{i}{2\pi} \sum_{\alpha=1}^{n} \psi_\alpha \wedge \bar{\psi}_\alpha, \quad \theta = \frac{i}{2\pi} \sum_{k=1}^{m} \theta_k \wedge \bar{\theta}_k.$$

Further define a function u_0 by

$$f^*(\psi^m) = u_0 \theta^m$$

and set

$$f^*(\psi_\nu) = A_\nu \theta_1 \wedge \cdots \wedge \theta_m, \quad \nu \in J_{1,m}^n,$$

where

$$\psi_\nu = \psi_{\nu(1)} \wedge \psi_{\nu(2)} \wedge \cdots \wedge \psi_{\nu(m)}.$$

Trivially, we see

$$u_0 = \sum_{\nu \in J_{1,m}^n} |A_\nu|^2 \not\equiv 0.$$

Relative to the dual frame $\psi_1^*, \ldots, \psi_n^*$ of $\psi_1, \ldots, \psi_n$, write

$$\varphi = \sum_{\lambda \in J_{1,q}^n} \varphi_\lambda \psi_{\lambda f}^*,$$

where

$$\psi_\lambda^* = \psi_{\lambda(1)}^* \wedge \psi_{\lambda(2)}^* \wedge \cdots \wedge \psi_{\lambda(q)}^*.$$

Set $\Psi = \psi_1 \wedge \cdots \wedge \psi_n$. We have

$$\Psi_f \angle \varphi = \sum_{\lambda \in J_{1,q}^n} \mathrm{sign}(\lambda^\perp, \lambda) \varphi_\lambda \psi_{\lambda^\perp f}.$$

Hence

$$F_\varphi[\Psi] = \hat{f}(\Psi_f \angle \varphi) = \sum_{\lambda \in J_{1,q}^n} \mathrm{sign}(\lambda^\perp, \lambda) \varphi_\lambda A_{\lambda^\perp} \theta_1 \wedge \cdots \wedge \theta_m,$$

which means

$$g^2 u_0 = \left| \sum_{\lambda \in J_{1,q}^n} \mathrm{sign}(\lambda^\perp, \lambda) \varphi_\lambda A_{\lambda^\perp} \right|^2.$$

By Schwarz's inequality, we obtain

$$\left| \sum_{\lambda \in J_{1,q}^n} \mathrm{sign}(\lambda^\perp, \lambda) \varphi_\lambda A_{\lambda^\perp} \right|^2 \leq u_0 |\varphi|^2,$$

and so $g \leq |\varphi|$ follows. $\square$

Finally we consider a holomorphic mapping

$$f : \mathbb{C}^m \longrightarrow \mathbb{C}^n/\Lambda.$$

Note that the complex m-space $\mathbb{C}^m$ with the metric

$$ds^2 = \sum_{j=1}^{m} dz_j d\bar{z}_j$$

(where $(z_1, \ldots, z_m)$ is the natural coordinate system) is a complete Kähler manifold. Let $(w_1, \ldots, w_n)$ be the natural coordinate system of $\mathbb{C}^n$ and $p : \mathbb{C}^n \longrightarrow \mathbb{C}^n/\Lambda$ be the natural projection. We fix the Kähler form ψ on $\mathbb{C}^n/\Lambda$ given by

$$p^*\psi = \frac{i}{2\pi} \sum_{j=1}^{n} dw_j \wedge d\bar{w}_j. \qquad (2.10.9)$$

The complex tori are the only compact complex parallelizable manifolds which admit Kähler metrics (see Wang [425]).

Lemma 2.88 ([167]). *Assume $q = n - m > 0$. Let $f : \mathbb{C}^m \longrightarrow \mathbb{C}^n/\Lambda$ be a holomorphic mapping of rank m. Then there exists a holomorphic field φ on f over $\mathbb{C}^m$ of degree q and effective for f such that $|\varphi| \leq c$ for some constant $c > 0$.*

Proof. Since $\mathbb{C}^n/\Lambda$ is complex parallelizable, there exist n holomorphic vector fields $Z_1, \ldots, Z_n$ which are linearly independent at every point of $\mathbb{C}^n/\Lambda$. We may assume

$$Z_j = \frac{\partial}{\partial w_j}, \quad j = 1, \ldots, n.$$

By Lemma 2.85, there exists $\lambda \in J_{1,q}^n$ such that a holomorphic field

$$\varphi = Z_\lambda f = (Z_{\lambda(1)} \wedge \cdots \wedge Z_{\lambda(q)})_f$$

on f over $\mathbb{C}^m$ of degree q is effective for f. We have

$$|\varphi| = |Z_\lambda| = 1,$$

and so Lemma 2.88 follows. $\qquad \square$

2.11 Stoll's theorems

Stoll [380] defines a Ricci function associated to a parabolic exhaustion of a complex manifold, which describes characteristics of curvature of the manifold in a sense. Here we introduce two Stoll's theorems related to the Ricci function. Finally, we estimate growth of integrals on logarithmic functions.

Let τ be a parabolic exhaustion of a complex manifold M of dimension m. Take a positive form Θ of degree $2m$ and class C^2 on M and define a non-negative function u on M by

$$v^m = u^2\Theta.$$

The $2m$-form $(\log u)v^m$ is locally integrable over M. Define

$$\mathcal{R}^0_\tau = \left\{ r \in \hat{\mathcal{R}}_\tau \mid \int_{M\langle O;r\rangle} (\log u)\sigma \text{ exists} \right\}. \tag{2.11.1}$$

The set $\mathcal{R}^0_\tau$ does not depend on the choice of Θ and $\mathbb{R}_+ - \mathcal{R}^0_\tau$ has measure zero. For $r, s \in \mathcal{R}^0_\tau$ with $r > s > 0$, the *Ricci function* of τ is defined by

$$\mathrm{Ric}_\tau(r,s) = \mathcal{T}(r,s;\mathrm{Ric}(\Theta)) + \int_{M\langle O;r\rangle} (\log u)\sigma - \int_{M\langle O;s\rangle} (\log u)\sigma. \tag{2.11.2}$$

The Ricci function Ric_τ of τ does not depend on the choice of Θ (see Stoll [380]).

Theorem 2.89 ([380]). *Let $\beta : M \longrightarrow \mathbb{C}^m$ be a proper holomorphic mapping of strict rank m. Let μ be the multiplicity of the branching divisor of β. Then $\tau = \|\beta\|^2$ is a parabolic exhaustion of M and for $r > r_0$,*

$$\mathrm{Ric}_\tau(r,r_0) = N_\mu(r). \tag{2.11.3}$$

Proof. Let $w = (w_1, \ldots, w_m)$ be the coordinates of $\mathbb{C}^m$ and set $dw = dw_1 \wedge \cdots \wedge dw_m$. Then

$$D\beta = \beta^*(dw) \otimes d^*w_\beta$$

is a Jacobian section. Hence

$$s = \beta^*(dw) \in \Gamma(M, K_M), \quad (s) = (D\beta), \quad v^m = i_m s \wedge \bar{s}.$$

Take a positive form Θ of class C^∞ and degree $2m$ on M. Then

$$i_m s \wedge \bar{s} = |s|^2_{\kappa_\Theta}\Theta.$$

Applying the first main theorem (2.9.19) to the identity $M \longrightarrow M$, we have

$$\mathcal{T}(r,s;\mathrm{Ric}(\Theta)) = N_\mu(r) + M\langle O;r_0; \log|s|_{\kappa_\Theta}\rangle - M\langle O;r; \log|s|_{\kappa_\Theta}\rangle. \tag{2.11.4}$$

Since $v^m = |s|^2_{\kappa_\Theta}\Theta$, the definition of Ric_τ gives (2.11.3). $\qquad\square$

If M is an affine algebraic variety, then a proper surjective holomorphic mapping $\beta : M \longrightarrow \mathbb{C}^m$ is well defined. Further, if M is smooth, then the branching divisor of β is affine algebraic, and so (2.11.3), Proposition 2.43 imply

$$\lim_{r\to\infty} \frac{\mathrm{Ric}_\tau(r,r_0)}{\log r} = \lim_{r\to\infty} \frac{N_\mu(r)}{\log r} = \lim_{r\to\infty} n_\mu(r) < \infty. \tag{2.11.5}$$

See Stoll [380], Theorem 20.4 or Griffiths-King [128].

Let F be an effective Jacobian section of f. Let (F) be the zero divisor of F and let $\mu_{(F)}$ be the multiplicity function of the divisor (F). Define the *ramification term* $N_{\mathrm{Ram}}(r, f)$ of f for F by

$$N_{\mathrm{Ram}}(r, f) = N_{\mu_{(F)}}(r). \tag{2.11.6}$$

The following *Stoll's theorem* (cf. [380]) will play an important role in the proof of the second main theorem below (Theorem 2.95).

Theorem 2.90. *Let M and N be complex manifolds of dimensions m and n, respectively. Let τ be a parabolic exhaustion of M. Let $f : M \longrightarrow N$ be a holomorphic mapping. Let F be an effective Jacobian section of f. Let Ω be a positive volume form of class C^∞ and degree $2n$ on N. A function h of class C^∞ on $M - F^{-1}(0)$ is defined by*

$$F[\Omega] = h^2 v^m.$$

If $r, r_0 \in \mathcal{R}_\tau^0$ with $r > r_0 > 0$, then

$$T_f(r, K_N) + N_{\mathrm{Ram}}(r, f) = \mathrm{Ric}_\tau(r, r_0) + \int_{M\langle O; r\rangle} (\log h)\sigma - \int_{M\langle O; r_0\rangle} (\log h)\sigma. \tag{2.11.7}$$

Proof. Take a positive form Θ of class C^∞ and degree $2m$ on M. Then Θ and Ω induce metrics κ_Θ and κ_Ω on the canonical bundles K_M and K_N, respectively, and further induce a metric $\kappa = \kappa_\Theta \otimes f^*(\kappa_\Omega^*)$ on the Jacobian bundle $K(f)$ such that

$$F[\Omega] = |F|_\kappa^2 \Theta.$$

Hence

$$v^m = h^{-2} F[\Omega] = h^{-2} |F|_\kappa^2 \Theta.$$

For $r, r_0 \in \mathcal{R}_\tau^0$ with $r > r_0 > 0$, then (2.11.2) implies

$$\mathrm{Ric}_\tau(r, r_0) = T(r, r_0; \mathrm{Ric}(\Theta)) + \int_{M\langle O; r\rangle} \log \frac{|F|_\kappa}{h}\sigma - \int_{M\langle O; r_0\rangle} \log \frac{|F|_\kappa}{h}\sigma. \tag{2.11.8}$$

By (2.10.1), it is easy to show that

$$\mathrm{Ric}(F[\Omega]) = f^*(\mathrm{Ric}(\Omega)). \tag{2.11.9}$$

However,

$$\mathrm{Ric}(F[\Omega]) = dd^c \log |F|_\kappa^2 + \mathrm{Ric}(\Theta) = -c_1(K(f), \kappa) + \mathrm{Ric}(\Theta).$$

Then

$$T(r, r_0; \mathrm{Ric}(\Theta)) = T_f(r, K_N) + T(r, r_0; c_1(K(f), \kappa)). \tag{2.11.10}$$

Applying the first main theorem (2.9.19) to the identity $M \longrightarrow M$, we obtain

$$T(r, r_0; c_1(K(f), \kappa)) = N_{\mathrm{Ram}}(r, f) - M\langle O; r; \log |F|_\kappa\rangle + M\langle O; r_0; \log |F|_\kappa\rangle. \tag{2.11.11}$$

Now (2.11.7) follows from (2.11.8), (2.11.10) and (2.11.11). $\qquad\square$

When $m = 1$, that is, M is a Riemann surface, we compare Theorem 2.90 with a theorem due to L.V. Ahlfors [2], H. Weyl and J. Weyl [430]. Now a multiplicity function $\mu_{df} : M \longrightarrow \mathbb{Z}_+$ is well defined as follows: $\mu_{df}(z_0) =$ the stationary index of f at z_0 if z_0 is a critical point of f (see Wu [436]), otherwise, $\mu_{df}(z_0) = 0$. Since τ is a parabolic exhaustion of M, then $\omega = dd^c \log \tau = 0$, that is, $\log \tau$ is harmonic. Let θ be the conjugate harmonic function of $\frac{1}{2} \log \tau$. Then

$$\zeta = \frac{1}{2} \log \tau + \sqrt{-1}\theta$$

is a coordinate function locally whenever τ is free of critical points (cf. Wu [436], Lemma 2.4). Let ψ be the associated $(1,1)$-form of an Hermitian metric on N. A non-negative function ρ_0 is defined by

$$f^*(\psi) = \rho_0^2 \frac{\sqrt{-1}}{2\pi} d\zeta \wedge d\bar{\zeta},$$

which satisfies the Ahlfors-Weyl formula (cf. Wu [436], or Yang-Hu [439])

$$\begin{aligned}
\mathcal{T}(r, r_0; \mathrm{Ric}(f^*(\psi))) &= -E(r) - N_{\mu_{df}}(r) \\
&\quad + M\langle O; r; \log \rho_0\rangle - M\langle O; r_0; \log \rho_0\rangle, \quad (2.11.12)
\end{aligned}$$

in which

$$E(r) = \int_{r_0}^{r} \chi\left(M[O; t]\right) \frac{dt}{t},$$

where $\chi\left(M[O; t]\right)$ denotes the Euler characteristic of $M[O; t]$.

We consider the non-negative function ρ defined by

$$f^*(\psi) = \rho^2 \upsilon. \qquad (2.11.13)$$

Note that

$$\upsilon = dd^c \tau = \frac{1}{\tau} d\tau \wedge d^c \tau = \tau \frac{\sqrt{-1}}{2\pi} d\zeta \wedge d\bar{\zeta}. \qquad (2.11.14)$$

Then $\rho_0 = \rho\sqrt{\tau}$. We know that (cf. [439], Part 1)

$$\mathrm{Ric}_\tau(r, r_0) = \varsigma \log \frac{r}{r_0} - E(r). \qquad (2.11.15)$$

Therefore, by (2.11.12) and (2.11.15), we obtain

$$\begin{aligned}
\mathcal{T}(r, r_0; \mathrm{Ric}(f^*(\psi))) &= \mathrm{Ric}_\tau(r, r_0) - N_{\mu_{df}}(r) \\
&\quad + M\langle O; r; \log \rho\rangle - M\langle O; r_0; \log \rho\rangle, \quad (2.11.16)
\end{aligned}$$

which is quite similar to the formula (2.11.7).

Define a non-negative function g on M by

$$F[\psi^n] = g^2 f^*(\psi). \qquad (2.11.17)$$

Taking $\Omega = \psi^n$ in Theorem 2.90, we have $h = \rho g$. Hence (2.11.7) and (2.11.16) yield

$$
\begin{aligned}
T_f(r, K_N) + N_{\mathrm{Ram}}(r, f) &= \mathcal{T}(r, r_0; \mathrm{Ric}(f^*(\psi))) + N_{\mu_{df}}(r) \\
&\quad + M\langle O; r; \log g \rangle - M\langle O; r_0; \log g \rangle.
\end{aligned}
$$

By (2.11.17), we obtain

$$
f^*(\mathrm{Ric}(\psi^n)) = \mathrm{Ric}(F[\psi^n]) = dd^c \log g^2 + \mathrm{Ric}(f^*(\psi)).
$$

Therefore

$$
\begin{aligned}
\mathcal{T}(r, r_0; dd^c \log g^2) &= N_{\mu_{df}}(r) - N_{\mathrm{Ram}}(r, f) \\
&\quad + M\langle O; r; \log g \rangle - M\langle O; r_0; \log g \rangle, \quad (2.11.18)
\end{aligned}
$$

which may be referred to as a kind of Jensen's formula.

The following lemma is due to R. Nevanlinna [294].

Lemma 2.91. *Let $h \geq 0$, $g \geq 0$ and $\alpha > 0$ be increasing continuous functions on $\mathbb{R}^+$. Assume g is of class C^1 on $\mathbb{R}^+$ and*

$$
\int_s^\infty \frac{dx}{\alpha(x)} < \infty
$$

for all $s > 0$. Then a measurable subset E of $\mathbb{R}^+$ exists such that

$$
0 \leq \int_E g'(x)dx < \infty
$$

and such that

$$
0 \leq h'(x) \leq g'(x)\alpha(h(x)), \quad x \in \mathbb{R}^+ - E.
$$

Proof. If E is not empty, then for $x \in E$ and $dx > 0$,

$$
dh(x) = h'(x)dx > \alpha(h(x))dg(x);
$$

therefore, $dg < \frac{dh}{\alpha(h)}$, from which the assertion follows by means of integration over E. $\qquad\square$

Lemma 2.92 ([166]). *Let $\eta \geq 0$ be a form of bidegree $(1,1)$ on M such that $\mathcal{T}(r, r_0; \eta)$ exists for $r > r_0 > 0$. Let u be a non-negative function on M satisfying*

$$
uv^m \leq \eta \wedge v^{m-1}.
$$

Then for any $\varepsilon > 0$,

$$
\| \int_{M\langle O;r\rangle} \log u\sigma \leq \varsigma(1 + 2\varepsilon) \log \mathcal{T}(r, r_0; \eta) + 4\varsigma\varepsilon \log r.
$$

Proof. Define

$$B(r, u) = \frac{1}{\varsigma} \int_{M\langle O;r\rangle} \log u\sigma. \qquad (2.11.19)$$

Note that for almost all $t > r_0$,

$$0 \le t^{2m-2} M[O; t; uv] = \int_{M[O;t]} uv^m = m \int_{M[O;t]} u\tau^{m-1} d\tau \wedge \sigma$$

$$= 2m \int_0^t \left\{ \int_{M\langle O;r\rangle} u\sigma \right\} r^{2m-1} dr = 2m \int_0^t M\langle O; r; u\rangle r^{2m-1} dr$$

$$\le t^{2m-2} M[O; t; \eta],$$

which means that $M\langle O; r; u\rangle$ exists for almost all $r > 0$.

Set

$$H(x) = \int_{r_0}^x \left\{ \int_0^t \exp(B(r, u)) r^{2m-1} dr \right\} t^{1-2m} dt.$$

Since

$$B(r, u) \le \log \frac{M\langle O; r; u\rangle}{\varsigma},$$

we have

$$H(x) \le \frac{1}{\varsigma} \int_{r_0}^x \left\{ \int_0^t M\langle O; r; u\rangle r^{2m-1} dr \right\} t^{1-2m} dt$$

$$= \frac{1}{2m\varsigma} T(x, r_0; uv) \le \frac{1}{2m\varsigma} T(x, r_0; \eta).$$

Applying Lemma 2.91 to the functions

$$h(x) = H(x), \ g(x) = \frac{x^{1+\varepsilon}}{1+\varepsilon}, \ \alpha(x) = x^\lambda$$

with $\varepsilon > 0$ and $\lambda > 1$, we obtain

$$\| \quad H'(x) = x^{1-2m} \int_0^x r^{2m-1} \exp(B(r, u)) dr$$

$$\le x^\varepsilon (H(x))^\lambda \le x^\varepsilon \left(\frac{1}{2m\varsigma} T(x, r_0; \eta) \right)^\lambda.$$

Keeping the same g and α and taking $h(x) = x^{2m-1} H'(x)$ in Lemma 2.91, we have

$$\| \quad r^{2m-1} \exp(B(r, u)) = h'(r) \le r^\varepsilon (h(r))^\lambda$$

$$\le r^\varepsilon \left(r^{2m-1+\varepsilon} \left(\frac{1}{2m\varsigma} T(r, r_0; \eta) \right)^\lambda \right)^\lambda$$

which implies

$$\| \quad B(r,u) \leq \lambda^2 \log \mathcal{T}(r, r_0; \eta) + c(\lambda, \varepsilon) \log r - \lambda^2 \log(2m\varsigma),$$

where

$$c(\lambda, \varepsilon) = \lambda(2m - 1 + \varepsilon) + 1 - 2m + \varepsilon.$$

Take $\delta \in \mathbb{R}^+$ such that

$$0 < \delta < \min\{1, \varepsilon\}, \quad \varepsilon(4 + \delta) + \delta(2m - 1) < 6\varepsilon,$$

and take

$$\lambda = 1 + \frac{\delta}{2}.$$

Then $\lambda^2 < 1 + 2\varepsilon$ and

$$c(\lambda, \varepsilon) = \frac{1}{2}\{\varepsilon(4 + \delta) + \delta(2m - 1)\} < 3\varepsilon.$$

Hence Lemma 2.92 follows if r is large enough.   □

2.12 Carlson-Griffiths-King theory

J. Carlson and P. Griffiths [47], P. Griffiths and J. King [128] studied value distribution theory of holomorphic mappings from affine varieties into projective varieties for line bundles. At the same time, P. Griffiths [125] (or cf. [345]) proposed a conjecture of second main theorem type. Lately, W. Stoll [380] extended this theory to parabolic spaces by using Jacobian sections. Y.T. Siu [363], [364], [365] studied the Griffiths conjecture by applying meromorphic connections. We will follow these methods to discuss the Griffiths conjecture.

2.12.1 Second main theorem for line bundles

First of all, we make a few general assumptions:

(A1) Let N be a compact complex manifold of dimension n.

(A2) Let L be a positive holomorphic line bundle over N.

(A3) Let $0 \neq s_j \in \Gamma(N, L), j = 1, \ldots, q$, be given such that the divisor $D = D_1 + \cdots + D_q$ has normal crossings in N, where $D_j = (s_j)$.

(A4) Let M be a parabolic connected complex manifold of dimension m.

(A5) Let $f : M \longrightarrow N$ be a holomorphic mapping.

Theorem 2.93. *Assume that (A1), (A2) and (A3) hold and further assume that each D_j $(1 \le j \le q)$ is smooth, that is, D has simple normal crossings. There are a positive number λ, a volume form Ω on N and a metric κ of L such that*

$$0 \le |s_j|_\kappa^2 < \frac{1}{e}, \quad j = 1, \ldots, q, \tag{2.12.1}$$

$$\lambda q c_1(L, \kappa) + \mathrm{Ric}(\Omega) > 0, \tag{2.12.2}$$

and such that the Ricci form

$$\psi = \mathrm{Ric}(\Psi) \tag{2.12.3}$$

of the following Carlson-Griffiths form

$$\Psi = \frac{\Omega}{\prod_{j=1}^{q} |s_j|_\kappa^{2\lambda} (\log |s_j|_\kappa^2)^2} \tag{2.12.4}$$

satisfies $\psi > 0$, $\int_N \psi^n < \infty$, and

$$\psi^n \ge \frac{\Omega}{\prod_{j=1}^{q} |s_j|_\kappa^2 (\log |s_j|_\kappa^2)^2}. \tag{2.12.5}$$

Proof. Here we follow Carlson-Griffiths [47], Griffiths-King [128] and Stoll [380] to give a proof. There is a metric κ_0 of L satisfying $c_1(L, \kappa_0) > 0$ since L is positive. Take a volume form Ω_0 on N. Since N is compact, there is a positive number λ such that

$$\chi := \lambda q c_1(L, \kappa_0) + \mathrm{Ric}(\Omega_0) > 0.$$

Further, a constant $c_0 > 1$ exists such that

$$-c_0 \chi < c_1(L, \kappa_0) < c_0 \chi.$$

Take $\delta > 0$ so small that

$$1 < 4q c_0 < -\log \max_{j,x} \{ \delta |s_j(x)|_{\kappa_0}^2 \}.$$

Then

$$0 < \delta |s_j|_{\kappa_0}^2 e \le \max_{j,x} \{ \delta |s_j(x)|_{\kappa_0}^2 \} e^{4q c_0} < 1.$$

Multiplying the metric κ_0 by δ, we obtain a metric κ such that $|s_j|_\kappa^2 < e^{-1}$ for each j, and

$$c_1(L, \kappa) = c_1(L, \kappa_0) > 0.$$

Define

$$\Psi_0 = \frac{\Omega_0}{\prod_{j=1}^{q} |s_j|_\kappa^{2\lambda} (\log |s_j|_\kappa^2)^2}.$$

It is easy to show that

$$\mathrm{Ric}(\Psi_0) = \chi + 2\sum_{j=1}^{q}\frac{c_1(L,\kappa)}{\log|s_j|_\kappa^2} + 2\sum_{j=1}^{q}\frac{d\log|s_j|_\kappa^2 \wedge d^c\log|s_j|_\kappa^2}{(\log|s_j|_\kappa^2)^2}. \qquad (2.12.6)$$

Since

$$\chi + 2\sum_{j=1}^{q}\frac{c_1(L,\kappa)}{\log|s_j|_\kappa^2} \geq \left\{1 + 2c_0\sum_{j=1}^{q}\frac{1}{\log|s_j|_\kappa^2}\right\}\chi \geq \frac{\chi}{2},$$

then

$$\mathrm{Ric}(\Psi_0) \geq \frac{\chi}{2} + 2\sum_{j=1}^{q}\frac{d\log|s_j|_\kappa^2 \wedge d^c\log|s_j|_\kappa^2}{(\log|s_j|_\kappa^2)^2} > 0.$$

The latter form is ≥ 0 because

$$dh \wedge d^c h = \frac{\sqrt{-1}}{2\pi}\partial h \wedge \bar\partial h \geq 0$$

for any real h.

We claim that for any $x \in N$, there exist an open neighborhood $U(x)$ of x and a constant $c(x)$ such that on $U(x)$,

$$\mathrm{Ric}(\Psi_0)^n \geq \left(c(x)\prod_{j=1}^{q}|s_j|_\kappa^{2\lambda-2}\right)\Psi_0 > 0. \qquad (2.12.7)$$

If $x \in N - \mathrm{supp}(D)$, this is trivial. Take $x \in \mathrm{supp}(D)$. Then $p \in \mathbb{Z}[1,q]$ and $\mu \in J_{1,p}^q$ exist such that $x \in \mathrm{supp}(D_j)$ if and only if $j \in \mathrm{Im}(\mu)$. Since D has normal crossings, then $p \leq n$, and also coordinates $(w_1,\ldots,w_n)$ in a neighborhood U of x exists such that $w(x) = (0,\ldots,0)$ and

$$U \cap \mathrm{supp}(D) = U \cap \bigcup_{\nu=1}^{p}\mathrm{supp}(D_{\mu(\nu)}),$$

$$U \cap D_{\mu(\nu)} = (w_\nu), \quad \nu = 1,\ldots,p.$$

Also U is taken so small that $L|U$ is trivial. We can write $s_{\mu(\nu)} = w_\nu\chi_\nu$, where χ_ν is a non-zero holomorphic section of L over U. Then

$$|s_{\mu(\nu)}|_\kappa^2 = b_\nu|w_\nu|^2,$$

where $b_\nu = |\chi_\nu|_\kappa^2 > 0$. Note that

$$\eta_\nu := d\log|s_{\mu(\nu)}|_\kappa^2 \wedge d^c\log|s_{\mu(\nu)}|_\kappa^2 = \frac{\sqrt{-1}}{2\pi}\frac{dw_\nu \wedge d\bar w_\nu}{|w_\nu|^2} + \varrho_\nu. \qquad (2.12.8)$$

The form

$$\varrho_\nu = \frac{\sqrt{-1}}{2\pi} \left(\frac{\partial b_\nu \wedge \bar\partial b_\nu}{b_\nu^2} + \frac{\partial b_\nu \wedge d\bar w_\nu}{b_\nu \bar w_\nu} + \frac{dw_\nu \wedge \bar\partial b_\nu}{w_\nu b_\nu} \right)$$

has the property that $|w_\nu|^2 \varrho_\nu$ is a smooth form whose coefficients vanish on $D_{\mu(\nu)}$.

Therefore on U we obtain

$$\mathrm{Ric}(\Psi_0)^n \;\geq\; \left(\frac{\chi}{2} + 2 \sum_{\nu=1}^{p} \frac{\eta_\nu}{(\log |s_{\mu(\nu)}|_\kappa^2)^2} \right)^n$$

$$\geq \frac{n!}{(n-p)!} \left(\frac{\chi}{2} \right)^{n-p} \wedge 2^p \prod_{\nu=1}^{p} \frac{\eta_\nu}{(\log |s_{\mu(\nu)}|_\kappa^2)^2}.$$

A form ϱ of class C^∞ and bidegree (p,p) exists on U such that $\varrho(x) = 0$ and such that

$$\prod_{\nu=1}^{p} |w_\nu|^2 \eta_\nu = \left(\frac{\sqrt{-1}}{2\pi} \right)^p dw_1 \wedge d\bar w_1 \wedge \cdots \wedge dw_p \wedge d\bar w_p + \varrho.$$

An open neighborhood $U(x)$ of x and a constant $c_1 > 0$ exists such that $\overline{U(x)}$ is compact and contained in U and such that on $U(x)$,

$$|w_\nu| \;<\; r < 1, \; \nu = 1, \ldots, p,$$

$$\frac{\chi}{2} \;>\; c_1 \sum_{\nu=p+1}^{n} \frac{\sqrt{-1}}{2\pi} dw_\nu \wedge d\bar w_\nu,$$

$$\left| \varrho \wedge \left(\frac{\chi}{2} \right)^{n-p} \right| \;<\; \frac{(n-p)!}{2} c_1^{n-p} \prod_{\nu=1}^{n} \frac{\sqrt{-1}}{2\pi} dw_\nu \wedge d\bar w_\nu.$$

Since $p \geq 1$, this implies

$$\mathrm{Ric}(\Psi_0)^n \geq c_1^{n-p} \left(\prod_{\nu=1}^{p} (|w_\nu| \log |s_{\mu(\nu)}|_\kappa^2)^{-2} \right) \prod_{\nu=1}^{n} \frac{\sqrt{-1}}{2\pi} dw_\nu \wedge d\bar w_\nu$$

on $U(x)$. A constant c_2 exists such that on $U(x)$,

$$c_1^{n-p} \prod_{\nu=1}^{n} \frac{\sqrt{-1}}{2\pi} dw_\nu \wedge d\bar w_\nu > c_2 \Omega_0.$$

Constants c_3 and c_4 exist such that

$$c_2 \prod_{\nu=1}^{p} b_\nu > c_3; \quad |s_j|_\kappa^2 (\log |s_j|_\kappa^2)^2 > c_4 \; (j \notin \mathrm{Im}(\mu)).$$

Therefore

$$\mathrm{Ric}(\Psi_0)^n \geq c_3 c_4^{q-p} \frac{\Omega_0}{\prod_{j=1}^{q} |s_j|_\kappa^2 (\log |s_j|_\kappa^2)^2} = \left(c_3 c_4^{q-p} \prod_{j=1}^{q} |s_j|_\kappa^{2\lambda-2} \right) \Psi_0.$$

Hence the claim (2.12.7) is satisfied with $c(x) = c_3 c_4^{q-p}$.

By the Heine-Borel theorem, there exist finitely many points $x_1, \ldots, x_k$ on N such that

$$N = U(x_1) \cup \cdots \cup U(x_k).$$

Define $c = \min_{1 \le j \le k} c(x_j)$. Then $c > 0$ is constant and $\Psi = c\Psi_0$ with $\Omega = c\Omega_0$ satisfies $\mathrm{Ric}(\Psi) = \mathrm{Ric}(\Psi_0) > 0$, (2.12.5) and (2.12.2) since $\mathrm{Ric}(\Omega) = \mathrm{Ric}(\Omega_0)$.

To prove $\int_N \psi^n < \infty$, it suffices to show that ψ^n is integrable over each neighborhood $U(x)$ constructed. This is trivial if $x \in N - \mathrm{supp}(D)$. Take $x \in \mathrm{supp}(D)$. Set $c_5 = 1 + 2qc_0$. Then (2.12.1) and (2.12.6) imply

$$\mathrm{Ric}(\Psi) \le c_5 \chi + 2 \sum_{j=1}^{q} \frac{d \log |s_j|_\kappa^2 \wedge d^c \log |s_j|_\kappa^2}{(\log |s_j|_\kappa^2)^2},$$

which yields immediately

$$\mathrm{Ric}(\Psi)^n \le \sum_{k_j} \frac{n!}{(n-k)!} (c_5 \chi)^{n-k} \wedge \prod_{j=1}^{q} \left(\frac{2d \log |s_j|_\kappa^2 \wedge d^c \log |s_j|_\kappa^2}{(\log |s_j|_\kappa^2)^2} \right)^{k_j},$$

where the summation runs over all $k_j \in \{0,1\}$ for $j = 1, \ldots, q$ such that

$$0 \le k_1 + \cdots + k_q = k \le n.$$

A constant $c_6 > 1$ exists such that $|\log b_\nu| < c_6$ on $U(x)$. If $\log |w_\nu| \le -c_6$, then

$$\log \frac{1}{|s_{\mu(\nu)}|_\kappa^2} = \log \frac{1}{|w_\nu|^2} + \log \frac{1}{b_\nu} \ge \log \frac{1}{|w_\nu|^2} - c_6 \ge \log \frac{1}{|w_\nu|} \ge \frac{1}{c_6} \log \frac{1}{|w_\nu|}.$$

If $\log |w_\nu| > -c_6$, then

$$\log \frac{1}{|s_{\mu(\nu)}|_\kappa^2} \ge 1 \ge \frac{1}{c_6} \log \frac{1}{|w_\nu|}.$$

Therefore on $U(x)$,

$$0 \le \frac{(\log |w_\nu|^2)^2}{(\log |s_{\mu(\nu)}|_\kappa^2)^2} \le 4c_6^2.$$

If $j \notin \mathrm{Im}(\mu)$, then $s_j(z) \ne 0$ for all $z \in \overline{U(x)}$. We see by our previous calculations that a constant c_7 exists such that

$$\mathrm{Ric}(\Psi)^n \le \frac{c_7}{\prod_{\nu=1}^{p} |w_\nu|^2 (\log |w_\nu|^2)^2} \prod_{\nu=1}^{n} \frac{\sqrt{-1}}{2\pi} dw_\nu \wedge d\bar{w}_\nu$$

which implies

$$\int_{U(x)} \mathrm{Ric}(\Psi)^n \le c_7 r^{2n-2p} |\log r^2|^{-p}.$$

$\square$

Lemma 2.94 ([47],[128],[380]). *Under the assumptions* (A1)–(A5), *there exist constants* $c_1 \leq 0$, $c_2 \geq 0$, $c_3 \geq 0$ *such that for* $r > r_0$,

$$
\begin{aligned}
c_1 &\leq \lambda q T_f(r, L) + T_f(r, K_N) - \mathcal{T}(r, r_0; f^*(\psi)) \\
&\leq q\varsigma \log\{T_f(r, L) + c_2\} + c_3.
\end{aligned}
\tag{2.12.9}
$$

Proof. By (2.12.4), we have

$$
\psi = \mathrm{Ric}(\Omega) + \lambda q c_1(L, \kappa) - \sum_{j=1}^{q} dd^c \log(\log |s_j|_\kappa^2)^2,
\tag{2.12.10}
$$

which implies

$$
\begin{aligned}
\mathcal{T}(r, r_0; f^*(\psi)) \;=\; & T_f(r, K_N) + \lambda q T_f(r, L) \\
& - \sum_{j=1}^{q} \mathcal{T}(r, r_0; dd^c \log(\log |s_j \circ f|_\kappa^2)^2).
\end{aligned}
\tag{2.12.11}
$$

By (2.9.20) (or see Stoll [380], Theorem A13), we obtain

$$
\begin{aligned}
\mathcal{T}(r, r_0; dd^c \log(\log |s_j \circ f|_\kappa^2)^2) = & M\langle O; r; \log |\log |s_j \circ f|_\kappa^2| \rangle \\
& - M\langle O; r_0; \log |\log |s_j \circ f|_\kappa^2| \rangle.
\end{aligned}
\tag{2.12.12}
$$

The condition (2.12.1) and the first main theorem (2.9.19) show that

$$
\begin{aligned}
0 \;\leq\;\; & M\langle O; r; \log |\log |s_j \circ f|_\kappa^2| \rangle \\
\leq\;\; & \varsigma \log\left(\frac{1}{\varsigma} M\left\langle O; r; \log \frac{1}{|s_j \circ f|_\kappa^2}\right\rangle\right) \\
\leq\;\; & \varsigma \log\{T_f(r, L) + m_f(r_0, D_j)\} - \varsigma \log \varsigma,
\end{aligned}
\tag{2.12.13}
$$

and hence (2.12.9) follows.   □

Now we can prove the following *second main theorem* (cf. [47], [128], [167], [344], [380]):

Theorem 2.95. *Assume that* (A1)–(A5) *hold and further assume that* D *has simple normal crossings. Suppose that* $f(M) \not\subseteq \mathrm{supp}(D)$ *and assume that* F *is an effective Jacobian section of* f. *Set* $b = \min\{m, n\}$ *and define a non-negative function* g *by*

$$
F[\psi^n] = g^2 f^*(\psi^b) \wedge v^{m-b}.
\tag{2.12.14}
$$

Then for any $\varepsilon > 0$,

$$
\begin{aligned}
\| \quad q T_f(r, L) + T_f(r, K_N) \leq & \sum_{j=1}^{q} N_f(r, D_j) - N_{\mathrm{Ram}}(r, f) + \mathrm{Ric}_\tau(r, r_0) \\
& + M\langle O; r; \log g \rangle + O(\log^+ T_f(r, L)) + O(\varepsilon \log r).
\end{aligned}
\tag{2.12.15}
$$

Proof. The first main theorem (2.9.19) implies

$$\sum_{j=1}^{q} \{T_f(r, L) - N_f(r, D_j)\} = \sum_{j=1}^{q} M \left\langle O; r; \log \frac{1}{|s_j \circ f|_\kappa} \right\rangle + O(1)$$

$$= M \langle O; r; \log Q \circ f \rangle + S(r, f),$$

where

$$Q^2 = \frac{1}{\prod_{j=1}^{q} |s_j|_\kappa^2 \, (\log |s_j|_\kappa^2)^2},$$

$$S(r, f) = \sum_{j=1}^{q} M \left\langle O; r; \log \left| \log |s_j \circ f|_\kappa^2 \right| \right\rangle.$$

By (2.12.13), we know

$$S(r, f) \le c \log^+ T_f(r, L) + O(1), \qquad (2.12.16)$$

where c is a positive constant which is independent of r.

Define a positive function G on N by

$$\psi^n = G^2 \Omega.$$

Then (2.12.5) yields $G \ge Q$. A function h of class C^∞ on $M - F^{-1}(0)$ also is defined by

$$F[\Omega] = h^2 v^m.$$

By Theorem 2.90, we obtain

$$qT_f(r, L) - \sum_{j=1}^{q} N_f(r, D_j) \le -T_f(r, K_N) - N_{\mathrm{Ram}}(r, f) + \mathrm{Ric}_\tau(r, r_0)$$

$$+ M \langle O; r; \log\{(G \circ f)h\} \rangle + c \log^+ T_f(r, L) + O(1).$$

Set

$$M^+ = \{x \in M \mid v(x) > 0\}.$$

For an integer i with $1 \le i \le m$, define a function ρ_i on M^+ by

$$f^*(\psi^i) \wedge v^{m-i} = \rho_i v^m. \qquad (2.12.17)$$

Then a pointwise relation among the ρ_i's is provided by the inequality (2.1.82) (or see [48], p. 239)

$$\rho_j^{\frac{1}{j}} \le c_{ij} \rho_i^{\frac{1}{i}} \quad (j \ge i). \qquad (2.12.18)$$

Since

$$(G \circ f)^2 h^2 v^m = (G \circ f)^2 F[\Omega] = F[G^2 \Omega] = F[\psi^n] = g^2 f^*(\psi^b) \wedge v^{m-b},$$

we have

$$\left(\frac{(G \circ f)h}{g}\right)^{\frac{2}{b}} = \rho_b^{\frac{1}{b}} \le c_{1b}\rho_1,$$

and hence

$$\left(\frac{(G \circ f)h}{g}\right)^{\frac{2}{b}} v^m \le c_{1b} f^*(\psi) \wedge v^{m-1}.$$

By Lemma 2.92, for any $\varepsilon > 0$ we obtain

$$\| \quad \int_{M\langle O;r\rangle} \log\left(\frac{(G \circ f)h}{g}\right)^{\frac{2}{b}} \sigma \le \varsigma(1 + 2\varepsilon)\log \mathcal{T}(r, r_0; f^*(\psi)) + O(\varepsilon \log r).$$

By Lemma 2.94, one has

$$\mathcal{T}(r, r_0; f^*(\psi)) \le c T_f(r, L) + c'.$$

Therefore

$$\| \quad M\langle O; r; \log\{(G \circ f)h\}\rangle \le M\langle O; r; \log g\rangle + O(\log^+ T_f(r, L)) + O(\varepsilon \log r).$$

Hence Theorem 2.95 is proved.   □

If the divisor $D = D_1 + \cdots + D_q$ has normal crossings in N, but D_j $(1 \le j \le q)$ may have singularities, by Hironaka's resolution of singularities [160] there exists a proper modification $\rho : \tilde{N} \longrightarrow N$, where $\tilde{N}$ is an algebraic manifold, such that the set $\tilde{D} = \rho^{-1}(D)$ is the union of a collection of smooth hypersurfaces with normal crossings. Let $\tilde{f} : M \longrightarrow \tilde{N}$ be the lift of $f : M \longrightarrow N$ given by $\rho \circ \tilde{f} = f$. By applying (2.12.15) to $\tilde{f}$, we conclude that (2.12.15) is valid for D (cf. Shiffman [344]).

Next we consider the case $m \ge n$ and make two additional assumptions:

(A6) Let $f : M \longrightarrow N$ be a holomorphic mapping of rank n.

(A7) Assume that M has holomorphic rank m, i.e., there exists a holomorphic mapping $\beta : M \longrightarrow \mathbb{C}^m$ of strict rank m. Let μ be the multiplicity of the branching divisor of β.

Under the conditions (A6) and (A7), a holomorphic form φ of degree $m - n$ on M exists such that the induced Jacobian section F_φ is effective for f. Note that

$$F_\varphi[\psi^n] = \binom{m}{m-n} i_{m-n}\varphi \wedge \bar{\varphi} \wedge f^*(\psi^n). \tag{2.12.19}$$

By Lemma 2.83, the function g defined by (2.12.14) for $F = F_\varphi$ satisfies

$$g^2 \le \binom{m}{m-n}. \tag{2.12.20}$$

Hence when $f(M) \not\subseteq \operatorname{supp}(D)$, the formulae (2.11.3) and (2.12.15) imply

$$\| \quad qT_f(r, L) + T_f(r, K_N) \leq \sum_{j=1}^{q} N_f(r, D_j) - N_{\mathrm{Ram}}(r, f) + N_\mu(r)$$

$$+ O(\log^+ T_f(r, L)) + O(\varepsilon \log r). \qquad (2.12.21)$$

For a positive integer k and a divisor D on N, we define the *truncated multiplicity function of order k of D on M* by

$$\mu_{f^*D,k}(z) = \min\{\mu_{f^*D}(z), k\}, \quad z \in M, \qquad (2.12.22)$$

and the *truncated valence function of order k of D*,

$$N_{f,k}(r, D) = N_{\mu_{f^*D,k}}(r). \qquad (2.12.23)$$

Usually, we write

$$\overline{N}_f(r, D) = N_{f,1}(r, D). \qquad (2.12.24)$$

Theorem 2.96 ([83]). *Assume that* (A1)–(A7) *hold. Then*

$$\| \quad qT_f(r, L) + T_f(r, K_N) \leq \sum_{j=1}^{q} \overline{N}_f(r, D_j) + N_\mu(r)$$

$$+ O(\log^+ T_f(r, L)) + O(\varepsilon \log r). \qquad (2.12.25)$$

A proof of Theorem 2.96 can be found in [168].

2.12.2 Griffiths' and Lang's conjectures

One of the major unsolved problems in value distribution theory is whether the inequality (2.12.21) for the case $m \geq n = \operatorname{rank} f$ holds for more general meromorphic mappings. Note that under the conditions (A1) and (A2), N is projective algebraic. It is natural to study algebraically non-degenerate meromorphic mappings. Conjecture 2.97 below restates this question, which has been previously asked by P. Griffiths [125] and M. Green (or cf. B. Shiffman [345]).

Conjecture 2.97. *Assume that* (A1)–(A4) *and* (A7) *hold. If* $f : M \longrightarrow N$ *is a meromorphic mapping which is algebraically non-degenerate, then there exists a number* $w = w(n, D)$ *such that*

$$\| \quad qT_f(r, L) + T_f(r, K_N) \leq \sum_{j=1}^{q} N_{f,w}(r, D_j) + N_\mu(r)$$

$$+ O(\log^+ T_f(r, L)) + O(\varepsilon \log r). \qquad (2.12.26)$$

In consideration of the analogues in number theory, S. Lang [229] weakens the condition of algebraic non-degeneracy in Conjecture 2.97 as follows:

Conjecture 2.98. *Assume that* (A7) *holds. Let D be a divisor with normal crossings on an algebraic manifold N. There exists a proper algebraic subset Z_D of N having the following property. Let $f : M \longrightarrow N$ be a meromorphic mapping such that $f(M) \not\subset Z_D$. Let E be an ample divisor. Then there exists a number $w = w(n, D)$ such that*

$$\| \quad T_f(r, [D]) + T_f(r, K_N) \leq N_{f,w}(r, D) + N_\mu(r)$$
$$+ O(\log^+ T_f(r, [E])) + O(\log r). \qquad (2.12.27)$$

When $M = \mathbb{C}$, $D = 0$ and K_N is ample, the inequality (2.12.27) becomes

$$\| \quad T_f(r, K_N) = O(\log r),$$

which implies that f is rational, so degenerate when $\dim N > 1$. Thus the following theorem (Green-Griffiths' conjecture) is a special case of Conjecture 2.98.

Theorem 2.99. *Let N be a projective algebraic variety with K_N ample. Then there are no algebraically non-degenerate holomorphic curves in N.*

Theorem 2.99 for the case $\dim N = 1$ follows from Liouville's theorem and the fact that the compact complex curves of genus greater than 1 are uniformized by the disc (see Theorem 4.57). Generally, a proof of Theorem 2.99 will be given in Chapter 6 (see Theorem 6.17). As a special case of Conjecture 2.97, B. Shiffman [345] listed Theorem 2.99 as a conjecture.

Qualitatively, Conjecture 2.98 also has the following simple consequence.

Conjecture 2.100. *Let N be a non-singular complex projective variety. Let K be the canonical divisor of N, and D a normal crossings divisor on N. Suppose that $K + D$ is pseudo ample. Then $N - D$ is pseudo Brody hyperbolic.*

Related to Conjecture 2.100, A. Levin [244] gave a *main Picard-type conjecture* as follows:

Conjecture 2.101. *Let N be a complex projective variety. Let $D = D_1 + \cdots + D_q$ be a divisor on N with the D_i's effective Cartier divisors for all i. Suppose that at most k D_i's meet at a point, so that the intersection of any $k + 1$ distinct D_i's is empty. Suppose that $\dim D_i \geq n_0 > 0$ for all i. If $q > k + \frac{k}{n_0}$, then there does not exist a holomorphic mapping $f : \mathbb{C} \longrightarrow N - D$ with Zariski-dense image.*

Picard's theorem is the case $k = n_0 = \dim N = 1$ of Conjecture 2.101. When N is a surface, $k \leq 2$, and the D_i's have no irreducible components in common, A. Levin [244] proved Conjecture 2.101. At the extreme of n_0, there is the following special case.

Conjecture 2.102. *Let N be a complex projective variety. Let $D = D_1 + \cdots + D_q$ be a divisor on N with the D_i's effective Cartier divisors for all i. Suppose that at most k D_i's meet at a point. If D_i is pseudo ample for all i and $q > k + \frac{k}{\dim N}$, then $N - D$ is pseudo Brody hyperbolic.*

When $q > 2k \dim N$, A. Levin [244] proved Conjecture 2.102. Further, A. Levin [244] proved that $N - D$ is complete hyperbolic if D_i is ample for all i. To relate Conjecture 2.100 with Conjecture 2.102, we recall the following theorem, which is a consequence of Mori theory (cf. [280], Lemma 1.7).

Theorem 2.103. *Let N be a non-singular complex projective variety of dimension n with the canonical divisor K. If $D_1, \ldots, D_{n+2}$ are ample divisors on N, then $K + D_1 + \cdots + D_{n+2}$ is ample.*

So when N is non-singular, the D_i's are ample, and $D = D_1 + \cdots + D_q$ has normal crossings, we see that Conjecture 2.102 is a consequence of Conjecture 2.100. Particularly, if N is the complex projective space $\mathbb{P}^n$, and if D_i is a hyperplane for each i, Corollary 2.72 with $k = n$ is a special case of Conjecture 2.102.

Here we make a remark for the case $m < n$. To prove Conjecture 2.97, based on Theorem 2.95 it is sufficient to obtain the estimate

$$\| \quad M \langle O; r; \log g \rangle \le O(\log^+ T_f(r, L)) + O(\varepsilon \log r). \qquad (2.12.28)$$

By using Lemma 2.87, it is sufficient to find a holomorphic field φ on f over M of degree $n - m$ which is effective for f such that the estimate

$$\| \quad M \langle O; r; \log |\varphi| \rangle \le O(\log^+ T_f(r, L)) + O(\varepsilon \log r) \qquad (2.12.29)$$

holds. Generally, we have the following expression (cf. Proof of Theorem 2.84):

$$\varphi = \eta_1 \wedge \cdots \wedge \eta_{n-m}$$

for some $\eta_j \in \Gamma(M, f^*(\mathbf{T}(N)))$ $(j = 1, \ldots, n - m)$. By Lemma 1.55, there exists a constant $c > 0$ such that

$$|\varphi| = |\eta_1 \wedge \cdots \wedge \eta_{n-m}| \le c |\eta_1| \cdots |\eta_{n-m}|.$$

To estimate $M \langle O; r; \log |\varphi| \rangle$, it is sufficient to find a bound of $M \langle O; r; \log |\eta_k| \rangle$ for each k. The difficult point for estimating $M \langle O; r; \log |\eta_k| \rangle$ is that the norm $|\eta_k|$ is measured by a "singular" metric induced by ψ. We will introduce two ways to construct an effective holomorphic field φ on f. One way is to use ample holomorphic vector fields on N as shown in Lemma 2.85. Another way is to use meromorphic connections which will be discussed in detail in Chapter 6.

Let $\psi_1, \ldots, \psi_n$ be a coframe in an open set of N such that the Hermitian metric of N is given by

$$ds_N^2 = 2 \sum_{\alpha=1}^n \psi_\alpha \bar\psi_\alpha.$$

Write

$$\psi = \mathrm{Ric}(\Psi) = \frac{i}{2\pi} \sum_{\alpha,\beta} \mathfrak{R}_{\alpha\beta} \psi_\alpha \wedge \bar{\psi}_\beta$$

and define the associated *scalar curvature*

$$\mathfrak{R} = \mathrm{tr}(\psi) = \sum_\alpha \mathfrak{R}_{\alpha\alpha}. \tag{2.12.30}$$

Since $\mathfrak{R}$ and $\det(\mathfrak{R}_{\alpha\beta})$ are respectively the sum and product of eigenvalues of the matrix $(\mathfrak{R}_{\alpha\beta})$, which are positive, the geometric-mean-arithmetic-mean inequality implies

$$\mathfrak{R} \geq n\{\det(\mathfrak{R}_{\alpha\beta})\}^{\frac{1}{n}} > 0.$$

Define a positive C^∞-function on N by

$$R = \mathrm{tr}(\lambda q c_1(L,\kappa) + \mathrm{Ric}(\Omega))$$

and set

$$\lambda_D = \prod_{j=1}^{q} \frac{1}{(\log|s_j|_\kappa^2)^2}.$$

Then the formula (2.12.10) implies

$$\mathfrak{R} = R + \frac{1}{4}\Delta \log \lambda_D, \tag{2.12.31}$$

where

$$\Delta \log \lambda_D = 4\mathrm{tr}(dd^c \log \lambda_D)$$

is just the Laplacian of $\log \lambda_D$.

Lemma 2.104 ([184]). *Assume that (A1)–(A3) hold. Then for any continuous section X of $\mathbf{T}(N)$ over N, there exists a positive constant c such that*

$$|X|^2 \leq c\mathfrak{R}, \tag{2.12.32}$$

where we used the metric induced by ψ.

Proof. By using (2.12.10), we can write $\psi = \psi' + \psi''$, where

$$\psi' = \lambda q c_1(L,\kappa) + \mathrm{Ric}(\Omega) + 2\sum_{j=1}^{q} \frac{c_1(L,\kappa)}{\log|s_j|_\kappa^2}$$

is continuous on N, and

$$\psi'' = 2\sum_{j=1}^{q} \frac{d\log|s_j|_\kappa^2 \wedge d^c \log|s_j|_\kappa^2}{(\log|s_j|_\kappa^2)^2} \geq 0.$$

In fact, according to the construction of Ψ, we have $\psi' > 0$ (see [380]). Denoting the norm $|X|_\psi$ induced by ψ, then $|X|_{\psi'}$ is bounded on N, and so

$$|X|_{\psi'}^2 \leq c' \mathrm{tr}(\psi') \leq c' \mathrm{tr}(\psi) = c' \mathfrak{R} \tag{2.12.33}$$

holds for a constant $c' > 0$.

Next we estimate $|X|_{\psi''}^2$. Take local holomorphic coordinates $w_1, \ldots, w_n$ and set

$$dw_\alpha = \sum_k a_{\alpha k} \psi_k, \quad \alpha = 1, \ldots, n.$$

Write

$$b_{j\alpha} = \frac{\partial \log \log |s_j|_\kappa^2}{\partial w_\alpha}, \quad c_{jk} = \sum_\alpha b_{j\alpha} a_{\alpha k}.$$

Then

$$\psi'' = \frac{i}{\pi} \sum_{j=1}^q \sum_{\alpha,\beta} b_{j\alpha} \bar{b}_{j\beta} dw_\alpha \wedge d\bar{w}_\beta = \frac{i}{\pi} \sum_{j=1}^q \sum_{k,l} c_{jk} \bar{c}_{jl} \psi_k \wedge \bar{\psi}_l.$$

Let $e_1, \ldots, e_n$ be a frame field which is dual to the coframe $(\psi_1, \ldots, \psi_n)$. Write

$$X = \sum_{\alpha=1}^n \xi_\alpha e_\alpha.$$

We have

$$|X|_{\psi''}^2 = 2 \sum_{j=1}^q \sum_{k,l} c_{jk} \bar{c}_{jl} \xi_k \bar{\xi}_l = 2 \sum_{j=1}^q \left| \sum_{k=1}^n c_{jk} \xi_k \right|^2.$$

The Schwarz inequality implies

$$|X|_{\psi''}^2 \leq 2 \sum_{j=1}^q \left(\sum_{k=1}^n |c_{jk}|^2 \right) \left(\sum_{k=1}^n |\xi_k|^2 \right).$$

Since X is continuous on N, there is a positive constant c'' such that

$$\sum_{k=1}^n |\xi_k|^2 \leq c'',$$

and hence

$$|X|_{\psi''}^2 \leq 2c'' \sum_{j=1}^q \sum_{k=1}^n |c_{jk}|^2 = c'' \mathrm{tr}(\psi'') \leq c'' \mathrm{tr}(\psi) = c'' \mathfrak{R}. \tag{2.12.34}$$

Since

$$|X|_\psi^2 = |X|_{\psi'}^2 + |X|_{\psi''}^2,$$

therefore (2.12.32) follows from (2.12.33) and (2.12.34). $\qquad \square$

Under the conditions (A1)–(A5), we define a function associated to D and f:

$$R_f(r) = M \langle O; r; \log \mathfrak{R} \circ f \rangle . \tag{2.12.35}$$

Theorem 2.105 ([184]). *Assume that* (A1)–(A5) *hold with* $m < n$, *and further assume that* D *has simple normal crossings. Assume that there exist holomorphic vector fields* $Z_1, \ldots, Z_n$ *on* N *such that*

$$Z := Z_1 \wedge \cdots \wedge Z_n \not\equiv 0.$$

Suppose that f *has strict rank* m *such that*

$$f(M) \not\subseteq \operatorname{supp}(D), \quad f(M) \not\subseteq \operatorname{supp}((Z)),$$

where (Z) *denotes the zero divisor of* Z. *Then for any* $\varepsilon > 0$, *we have*

$$\| \quad qT_f(r, L) + T_f(r, K_N) \leq \sum_{j=1}^{q} N_f(r, D_j) - N_{\mathrm{Ram}}(r, f) + \mathrm{Ric}_\tau(r, r_0)$$

$$+ \frac{n - m}{2} R_f(r) + O(\log^+ T_f(r, L)) + O(\varepsilon \log r). \tag{2.12.36}$$

Proof. By Lemma 2.85, there exists $\lambda \in J^n_{1, n-m}$ such that a holomorphic field

$$\varphi = Z_{\lambda f} = (Z_{\lambda(1)} \wedge \cdots \wedge Z_{\lambda(n-m)})_f$$

on f over M of degree $n - m$ is effective for f. By Lemma 1.55, there exists a constant $c > 0$ such that

$$\left| Z_{\lambda(1)} \wedge \cdots \wedge Z_{\lambda(n-m)} \right| \leq c \left| Z_{\lambda(1)} \right| \cdots \left| Z_{\lambda(n-m)} \right|,$$

and hence

$$M \langle O; r; \log |\varphi| \rangle \leq \frac{1}{2} \sum_{k=1}^{n-m} M \left\langle O; r; \log \left| Z_{\lambda(k)} \circ f \right|^2 \right\rangle + O(1). \tag{2.12.37}$$

By Lemma 2.104, we have

$$M \left\langle O; r; \log \left| Z_{\lambda(k)} \circ f \right|^2 \right\rangle \leq M \langle O; r; \log \mathfrak{R} \circ f \rangle + O(1). \tag{2.12.38}$$

Thus Theorem 2.105 follows from Theorem 2.95 and Lemma 2.87.   □

If N is a complex torus, then there exist n holomorphic vector fields $Z_1, \ldots, Z_n$ which are linearly independent at every point of N. If N is the complex projective space $\mathbb{P}^n(\mathbb{C})$, we define n holomorphic vector fields $Z_1, \ldots, Z_n$ over $\mathbb{P}^n(\mathbb{C})$ as follows. We consider the natural projection

$$\mathbb{P} : \mathbb{C}^{n+1} - \{0\} \longrightarrow \mathbb{P}^n(\mathbb{C})$$

defined by $(\xi_0, \ldots, \xi_n) \mapsto [\xi_0, \ldots, \xi_n]$, which induces the holomorphic differential

$$\mathbb{P}' : \mathbf{T}(\mathbb{C}^{n+1} - \{0\}) \longrightarrow \mathbf{T}(\mathbb{P}^n(\mathbb{C})).$$

Define

$$Z_i = \mathbb{P}'\left(\frac{\partial}{\partial \xi_i}\right), \quad i = 0, 1, \ldots, n.$$

Then $Z_0, \ldots, Z_n$ span the holomorphic tangent space at every point of $\mathbb{P}^n(\mathbb{C})$. For example, denoting the local coordinates

$$z_j = \frac{\xi_j}{\xi_0}, \quad j = 1, \ldots, n$$

on the domain $U_0 = \{\xi_0 \neq 0\}$ of $\mathbb{P}^n(\mathbb{C})$, we have

$$Z_i = \begin{cases} \frac{1}{\xi_0} \frac{\partial}{\partial z_i}, & \text{if } 1 \leq i \leq n; \\ -\sum_{j=1}^{n} \frac{\xi_j}{\xi_0^2} \frac{\partial}{\partial z_j}, & \text{if } i = 0. \end{cases}$$

Hence $Z_1, \ldots, Z_n$ satisfy the condition in Theorem 2.105.

Chapter 3

Topics in Number Theory

In this chapter, we will introduce some results and problems in number theory that have an analogue in Nevanlinna theory, say, the *abc*-conjecture, Roth's theorem, Schmidt's subspace theorem, Vojta's conjecture, and so on.

3.1 Elliptic curves

We think that if a variety admits non-constant holomorphic curves, it would have a rich and complicated distribution theory of rational points. It is well known that rational and elliptic curves admit non-constant holomorphic curves (see Section 3.1.5 and 4.8). These also are total compact Riemann surfaces satisfying the above property. The difference between rational and elliptic curves is that mappings of non-constant holomorphic curves into elliptic curves must be surjective (cf. Theorem 4.50), but the case for rational curves is not so. The phenomenon reflects a difference of distribution of rational points on two classes of curves. Generally speaking, distribution of rational points on elliptic curves is "normal". In this section, we will give an elementary exhibit of the more beautiful theory.

3.1.1 The geometry of elliptic curves

An algebraic curve E defined over a field κ is called an *elliptic curve* if a normalization of the curve has genus 1, usually written E/κ if E is defined over κ as a curve. In this section, we will discuss smooth elliptic curves. However, there are non-smooth elliptic curves. For example, the algebraic curve defined over $\mathbb{C}$ by the following equation of degree 4,

$$y^2(ax^2 + bx + c) + \alpha x^2 + \beta x + \gamma = 0,$$

has generally two ordinary double points. The genus formula implies that the curve has genus 1. Hence it is elliptic, but not smooth.

Proposition 3.1 (cf. [362]). *Let E be an elliptic curve defined over κ.*

(a1) *There exist functions $x, y \in \bar{\kappa}(E)$ such that the mapping $[x, y, 1] : E \longrightarrow \mathbb{P}^2$ gives an isomorphism of E/κ onto a curve given by a generalized Weierstrass equation*

$$y^2 + a_1 xy + a_3 y = x^3 + a_2 x^2 + a_4 x + a_6 \qquad (3.1.1)$$

with $a_i \in \kappa$.

(a2) *Any two generalized Weierstrass equations for E as in (a1) are related by a linear change of variables of the form*

$$
\begin{aligned}
x &= u^2 x' + r, \\
y &= u^3 y' + su^2 x' + t,
\end{aligned}
$$

with $u, r, s, t \in \kappa$, $u \neq 0$.

(a3) *Conversely, every smooth cubic curve given by a generalized Weierstrass equation as in (a1) is an elliptic curve defined over κ.*

As usual, the curve defined by the generalized Weierstrass equation in (a1) is the locus of the homogeneous coordinate equation

$$Y^2 Z + a_1 XYZ + a_3 Y Z^2 = X^3 + a_2 X^2 Z + a_4 X Z^2 + a_6 Z^3$$

in $\mathbb{P}^2$ with only one point $[0, 1, 0]$ on the line at ∞. The functions $x, y \in \bar{\kappa}(E)$ in (a1) are called *Weierstrass coordinate functions* on E, which have the property

$$\bar{\kappa}(E) = \bar{\kappa}(x, y), \ [\bar{\kappa}(E) : \bar{\kappa}(x)] = 2.$$

The point in E corresponding to $[0, 1, 0]$ under the mapping in (a1) is called a *base point of E*, or the *origin of E*, denoted by O. Obviously, O must be a pole of order 2 of x, and a pole of order 3 of y.

If $\mathrm{char}(\bar{\kappa}) \neq 2$, then we can simplify the equation by completing the square. Thus replacing y by $\frac{1}{2}(y - a_1 x - a_3)$ gives an equation of the form

$$y^2 = 4x^3 + b_2 x^2 + 2b_4 x + b_6,$$

where

$$
\begin{aligned}
b_2 &= 4a_2 + a_1^2, \\
b_4 &= 2a_4 + a_1 a_3, \\
b_6 &= 4a_6 + a_3^2.
\end{aligned}
$$

The following quantities are usually used:

$$b_8 = a_1^2 a_6 + 4a_2 a_6 - a_1 a_3 a_4 + a_2 a_3^2 - a_4^2,$$
$$c_4 = -24b_4 + b_2^2,$$
$$c_6 = -216b_6 + 36b_4 b_2 - b_2^3,$$
$$\Delta = -b_2^2 b_8 - 8b_4^3 - 27b_6^2 + 9b_2 b_4 b_6,$$
$$j = \frac{c_4^3}{\Delta},$$
$$\omega = \frac{dx}{2y + a_1 x + a_3} = \frac{dy}{3x^2 + 2a_2 x + a_4 - a_1 y}.$$

One easily verifies that they satisfy the relations

$$4b_8 = b_2 b_6 - b_4^2, \quad 1728\Delta = c_4^3 - c_6^2.$$

The quantity Δ is called the *discriminant* of the generalized Weierstrass equation, j is called the *j-invariant* of the elliptic curve E, and ω is the *invariant differential* associated with the generalized Weierstrass equation. The following properties are due to Silverman [362], Propositions 1.4 and 1.5.

Proposition 3.2.

(b1) *The curve given by a generalized Weierstrass equation is non-singular if and only if $\Delta \neq 0$.*

(b2) *The invariant differential ω on an elliptic curve associated to a generalized Weierstrass equation is regular and non-vanishing, i.e., $\mathrm{div}(\omega) = 0$.*

If further $\mathrm{char}(\bar\kappa) \neq 2, 3$, then replacing (x, y) by $((x - 3b_2)/36, y/108)$ eliminates the x^2 term, yielding the simple *Weierstrass equation*

$$y^2 = x^3 + ax + b$$

with $a = -27c_4$, $b = -54c_6$. The only change of variables preserving this form of the equation is

$$x = u^2 x', \quad y = u^3 y' \tag{3.1.2}$$

for some $u \in \bar\kappa_*$, and then

$$u^4 a' = a, \quad u^6 b' = b, \quad u^{12}\Delta' = \Delta.$$

We summarize the above discussion as follows:

Theorem 3.3 (cf. [126], [362]). *For each smooth elliptic curve E defined over a field κ of characteristic $\neq 2, 3$, there exists a coordinate system such that the affine equation of E may be expressed by a Weierstrass equation*

$$y^2 = x^3 + ax + b \tag{3.1.3}$$

with $a, b \in \kappa$, and

$$\Delta = -16(4a^3 + 27b^2) \neq 0, \ j = \frac{1728(-4a)^3}{\Delta}. \tag{3.1.4}$$

The only change of variables preserving this form of the equation is (3.1.2).

The condition that the discriminant Δ is non-zero is equivalent to the curve being smooth. It is also equivalent to the cubic $x^3 + ax + b$ having three different roots since

$$16(x_1 - x_2)^2(x_2 - x_3)^2(x_3 - x_1)^2 = \Delta,$$

where x_1, x_2, x_3 are the three zeros of the polynomial $x^3 + ax + b$.

Let E be an elliptic curve given by a generalized Weierstrass equation (3.1.1). Remember that $E \subset \mathbb{P}^2$ consists of the points $P = (x, y)$ satisfying the equation together with the point $O = [0, 1, 0]$ at infinity. Let $L \subset \mathbb{P}^2$ be a line. Then since the equation has degree 3, L intersects E at exactly three points, say P, Q, R. Note that if L is tangent to E, then P, Q, R may not be distinct. The fact that $L \cap E$ (counting multiplicity) consists of three points is a special case of Bezout's theorem. One can use this fact to define an addition law on E. Namely, given $P, Q \in E$, draw the line L through P and Q (tangent line to E if $P = Q$). Let R be the third point of intersection of L with E. Let L' be the line connecting R and O. Define $P + Q$ to be the third point of intersection of E with L'. The composition law makes E into an Abelian group with identity element O. Further,

$$E(\kappa) = \{\text{solutions } (x, y) \in \kappa^2 \text{ of } (3.1.1)\} \cup \{O\}$$

is a subgroup of E (see [362], Proposition 2.2).

Example 3.4 (cf. [429]). *Let $E/\mathbb{Q}$ be an elliptic curve defined by the equation (3.1.3). If $a = -1$, $b = 0$, then*

$$E(\mathbb{Q}) = \{(0, 0), (1, 0), (-1, 0), O\}.$$

Let E be an elliptic curve defined over κ and $m \in \mathbb{Z} - \{0\}$. The *m-torsion subgroup* of E, denoted $E[m]$, is the set of points of order m in E,

$$E[m] = \{P \in E \mid [m]P = O\},$$

where

$$[m]P = P + P + \cdots + P \ (m \text{ terms})$$

if $m > 0$, $[m]P = [-m](-P)$ if $m < 0$, and $[0]P = O$. The *torsion subgroup* of E, denoted E_{tors}, is the set of points of finite order,

$$E_{\text{tors}} = \bigcup_{m=1}^{\infty} E[m].$$

Then $E_{\text{tors}}(\kappa)$ will denote the points of finite order in $E(\kappa)$.

Proposition 3.5 ([362]). *Let E/κ be an elliptic curve and $m \in \mathbb{Z} - \{0\}$. Let p denote the characteristic of $\bar{\kappa}$.*

(c1) *If $p = 0$ or if m is prime to p, then*

$$E[m] \cong (\mathbb{Z}/m\mathbb{Z}) \times (\mathbb{Z}/m\mathbb{Z}).$$

(c2) *If $p > 0$, then either*

$$E[p^e] \cong \{O\}, \ e = 1, 2, 3, \ldots;$$

or

$$E[p^e] \cong \mathbb{Z}/p^e\mathbb{Z}, \ e = 1, 2, 3, \ldots.$$

Let E/κ be an elliptic curve and $\ell \in \mathbb{Z}$ a prime. The (ℓ-adic) *Tate module of E* is the group

$$T_\ell(E) = \varprojlim_n E[\ell^n],$$

the inverse limit being taken with respect to the natural mappings

$$[\ell] : E[\ell^{n+1}] \longrightarrow E[\ell^n].$$

Since each $E[\ell^n]$ is a $\mathbb{Z}/\ell^n\mathbb{Z}$-module, we see that the Tate module has a natural structure as a $\mathbb{Z}_\ell$-module. Note that since the multiplication mappings $[\ell]$ are surjective, the inverse limit topology on $T_\ell(E)$ is equivalent to the ℓ-adic topology it gains as a $\mathbb{Z}_\ell$-module. Proposition 3.5 implies immediately

Proposition 3.6 ([362]). *The Tate module has the following structure:*

(d1) $T_\ell(E) \cong \mathbb{Z}_\ell \times \mathbb{Z}_\ell$ *if* $\ell \neq \mathrm{char}(\kappa)$.
(d2) $T_p(E) \cong \{0\}$ *or* $\mathbb{Z}_p$ *if* $p = \mathrm{char}(\kappa) > 0$.

Let $m \geq 2$ be an integer (prime to $\mathrm{char}(\kappa)$ if $\mathrm{char}(\kappa) > 0$). Note that each element σ of the Galois group $G_{\bar{\kappa}/\kappa}$ acts on $E[m]$ since, if $[m]P = O$, then

$$[m]\sigma(P) = \sigma([m]P) = O.$$

We thus obtain a representation

$$G_{\bar{\kappa}/\kappa} \longrightarrow \mathrm{Aut}(E[m]) \cong GL(2, \mathbb{Z}/m\mathbb{Z}),$$

where the latter isomorphism involves choosing a basis for $E[m]$. The action of $G_{\bar{\kappa}/\kappa}$ on each $E[\ell^n]$ commutes with the multiplication mappings $[\ell]$ used to form the inverse limit, so $G_{\bar{\kappa}/\kappa}$ also acts on $T_\ell(E)$. The ℓ-adic representation (of $G_{\bar{\kappa}/\kappa}$ on E), denoted $\rho_{E,\ell}$, is the mapping

$$\rho_{E,\ell} : G_{\bar{\kappa}/\kappa} \longrightarrow \mathrm{Aut}(T_\ell(E))$$

giving the action of $G_{\bar{\kappa}/\kappa}$ on $T_\ell(E)$ as described above. If $\ell \neq \operatorname{char}(\kappa)$, by choosing a $\mathbb{Z}_\ell$-basis for $T_\ell(E)$ we obtain a representation

$$G_{\bar{\kappa}/\kappa} \longrightarrow GL(2, \mathbb{Z}_\ell);$$

and then the natural inclusion $\mathbb{Z}_\ell \subset \mathbb{Q}_\ell$ gives

$$G_{\bar{\kappa}/\kappa} \longrightarrow GL(2, \mathbb{Q}_\ell).$$

Remark. If κ is a local field, complete with respect to a discrete valuation v, we can find a generalized Weierstrass equation (3.1.1) for E/κ with all coefficients $a_i \in \mathfrak{O}_\kappa = \kappa[0; 1]$ since replacing (x, y) by $(u^{-2}x, u^{-3}y)$ causes each a_i to become $a_i u^i$, if we choose u divisible by a large power of a uniformizing parameter t for the valuation ring $\mathfrak{O}_\kappa$. Since v is discrete, we can look for an equation with $v(\Delta)$ as small as possible, called a *minimal (Weierstrass) equation* for E at v. The natural reduction mapping $\mathfrak{O}_\kappa \longrightarrow \mathbb{F}(\kappa) = \mathfrak{O}_\kappa/\kappa(0; 1)$ is denoted $z \mapsto \tilde{z}$. Now having chosen a minimal Weierstrass equation (3.1.1) for E/κ, we can reduce its coefficients modulo t to obtain a (possibly singular) curve over $\mathbb{F}(\kappa)$, namely

$$\tilde{E}: \quad y^2 + \tilde{a}_1 xy + \tilde{a}_3 y = x^3 + \tilde{a}_2 x^2 + \tilde{a}_4 x + \tilde{a}_6.$$

The curve $\tilde{E}/\mathbb{F}(\kappa)$ is called the *reduction of E modulo t*. Further, E is said to have *good* (or *stable*) *reduction* over κ if $\tilde{E}$ is non-singular, otherwise, it has *bad reduction*. In the case of having bad reduction, E is also said to have *multiplicative* (or *semi-stable*) *reduction* over κ if $\tilde{E}$ has an ordinary double point, otherwise, it has *additive* (or *unstable*) *reduction* over κ. If E has multiplicative reduction, then the reduction is said to be *split* (respectively *non-split*) if the slopes of the tangent lines at the double point are in $\mathbb{F}(\kappa)$ (respectively not in $\mathbb{F}(\kappa)$). One knows that E has good reduction if and only if $v(\Delta) = 0$ (cf. [362]).

3.1.2   Modular functions

We usually use the following form of Theorem 3.3 over $\mathbb{C}$ (cf. [359], [126]):

Theorem 3.7. *Every algebraic curve of genus 1 over $\mathbb{C}$ can be transformed birationally into a cubic curve of the special form*

$$y^2 = 4x^3 - Ax - B \tag{3.1.5}$$

with constants A, B satisfying $\Delta = A^3 - 27B^2 \neq 0$. Two such cubic curves are birationally equivalent if and only if they agree on the invariant

$$j = \frac{1728A^3}{A^3 - 27B^2}. \tag{3.1.6}$$

If this is the case, then the two curves go over into each other under an affine transformation of the form $x \mapsto u^2 x,\ y \mapsto u^3 y$, with constant $u \neq 0$.

Take $\omega_1, \omega_2 \in \mathbb{C}$ such that they are linearly independent over $\mathbb{R}$, that is, $\omega_i \neq 0$, $\omega_2/\omega_1 \notin \mathbb{R}$. Let Λ be the discrete subgroup of $\mathbb{C}$ generated by ω_1 and ω_2:

$$\Lambda = [\omega_1, \omega_2] = \{m\omega_1 + n\omega_2 \mid m, n \in \mathbb{Z}\},$$

which is called a *lattice* over $\mathbb{Z}$. Here we simply introduce meromorphic functions on the quotient space $\mathbb{C}/\Lambda$; or equivalently, meromorphic functions on $\mathbb{C}$ which are periodic with respect to the lattice Λ. An *elliptic function (relative to the lattice Λ)* is a meromorphic function f on $\mathbb{C}$ which satisfies

$$f(z + \omega) = f(z), \ z \in \mathbb{C}, \ \omega \in \Lambda.$$

The set of all such functions is clearly the field $\mathcal{M}(\mathbb{C}/\Lambda)$.

The *Eisenstein series of weight $2k$ (for Λ)* is the series

$$G_{2k} = G_{2k}(\Lambda) = \sum_{\omega \in \Lambda - \{0\}} \omega^{-2k},$$

which is absolutely convergent for all $k > 1$. The *Weierstrass $\wp$ function (relative to Λ)* is defined by the series

$$\wp(z) = \wp(z, \Lambda) = \frac{1}{z^2} + \sum_{\omega \in \Lambda - \{0\}} \left(\frac{1}{(z - \omega)^2} - \frac{1}{\omega^2} \right),$$

which converges absolutely and uniformly on every compact subset of $\mathbb{C} - \Lambda$. It defines an even elliptic function on $\mathbb{C}$ having a double pole with residue 0 at each lattice point and no other poles.

Theorem 3.8. *Every elliptic function is a rational combination of $\wp$ and $\wp'$, i.e.,*

$$\mathcal{M}(\mathbb{C}/\Lambda) = \mathbb{C}(\wp, \wp').$$

Proof. Siegel [358], Chapter 1, Section 14, Theorem 6, or Silverman [362]. □

It is standard notation to set

$$g_2 = g_2(\Lambda) = 60G_4, \quad g_3 = g_3(\Lambda) = 140G_6. \tag{3.1.7}$$

A basic theorem (cf. [358]) in elliptic function theory shows that $g_2^3 - 27g_3^2 \neq 0$, and the inverse function of the *elliptic integral of the first kind* in the Weierstrass normal form

$$z = \int_\infty^w \frac{d\zeta}{\sqrt{4\zeta^3 - g_2\zeta - g_3}}$$

formed with these g_2, g_3 coincides with the Weierstrass $\wp$ function which also is a unique even meromorphic function in $\mathbb{C}$ satisfying the differential equation

$$(\wp')^2 = 4\wp^3 - g_2\wp - g_3. \tag{3.1.8}$$

Conversely, one has the following *uniformization theorem:*

Theorem 3.9. Let $A, B \in \mathbb{C}$ satisfy $A^3 - 27B^2 \neq 0$. Then there exists a unique lattice $\Lambda \subset \mathbb{C}$ such that $g_2(\Lambda) = A$ and $g_3(\Lambda) = B$.

Proof. See Apostol [6], Theorem 2.9; Robert [325], I.3.13; Shimura [351], Section 4.2; Serre [341], VII Proposition 5, or Siegel [358], Chapter 1, Sections 11–13. □

By possibly reversing the order of ω_1 and ω_2, we can assume that the imaginary part of the ratio $\tau = \omega_2/\omega_1$ is positive. By (3.1.7), the quantity

$$j = j(\Lambda) = \frac{1728 g_2^3}{g_2^3 - 27 g_3^2}$$

associated with the algebraic curve

$$y^2 = 4x^3 - g_2 x - g_3$$

depends solely on the period lattice and is homogeneous of degree 0 in ω_1, ω_2, that is, it is the same if we replace ω_1, ω_2 by $c\omega_1, c\omega_2$ for any complex number $c \neq 0$. Thus we have $j(c\Lambda) = j(\Lambda)$, and we may define $j(\tau) = j(\Lambda)$. But $\mathbb{C}/\Lambda$ is a complex torus of dimension 1, and the above arguments show that j is the single invariant for isomorphism classes of such toruses. It follows that $j = j(\tau)$, considered in the upper half-plane $\mathbb{H}$, is a holomorphic function of τ alone which has the invariance property

$$j\left(\frac{a\tau + b}{c\tau + d}\right) = j(\tau)$$

with integers a, b, c, d and $ad - bc = 1$. Note that the transformation

$$\tau \mapsto \tau' = \frac{a\tau + b}{c\tau + d}, \quad \{a, b, c, d\} \subset \mathbb{Z}, \quad ad - bc = 1 \qquad (3.1.9)$$

maps $\mathbb{H}$ into itself. Such transformations form a group, called the *modular group* $SL(2, \mathbb{Z})$ or, more precisely, the *elliptic modular group*.

We denote the relation (3.1.9) between τ and τ' by

$$\tau = \tau' \bmod SL(2, \mathbb{Z}), \qquad (3.1.10)$$

and call two points τ and τ' *equivalent*. As usual, we define the *fundamental region* $\mathfrak{B}$ of $SL(2, \mathbb{Z})$ as a subset of $\mathbb{H}$ satisfying conditions:

(e1) If $\tau \in \mathbb{H}$, there exists $\tau' \in \mathfrak{B}$ such that (3.1.10) holds;
(e2) If $\tau, \tau' \in \mathfrak{B}$, then $\tau \neq \tau' \bmod SL(2, \mathbb{Z})$.

For example,

$$\mathfrak{B} = \left\{ x + iy \in \mathbb{H} \;\middle|\; x^2 + y^2 \geq 1, \; -\frac{1}{2} \leq x < \frac{1}{2} \right\} - l,$$

where

$$l = \left\{ x + iy \in \mathbb{H} \ \bigg| \ x^2 + y^2 = 1, \ 0 < x < \frac{1}{2} \right\}.$$

Now $\mathfrak{B}$ contains exactly one representative of each equivalence class. In view of Theorem 3.7, the function $j(\tau)$ has the important property of separating every two points of $\mathbb{H}$ by its values if these points are not equivalent with respect to the modular group, that is,

$$j(\tau) \neq j(\tau'), \quad \tau \neq \tau' \bmod SL(2, \mathbb{Z}),$$

which gives a holomorphic isomorphism (cf. [341])

$$j : \mathbb{H}/SL(2, \mathbb{Z}) \longrightarrow \mathbb{C}.$$

The space $\mathbb{H}/SL(2, \mathbb{Z})$ is a non-compact Riemann surface. Its natural compactification is $\mathbb{P}^1(\mathbb{C})$, obtained by adding a single extra point at infinity. Define

$$\mathbb{H}^* = \mathbb{H} \cup \mathbb{P}^1(\mathbb{Q}).$$

Here one should think of the points $[x, 1] \in \mathbb{P}^1(\mathbb{Q})$ as forming the usual copy of $\mathbb{Q}$ in $\mathbb{C}$; and the point $[1, 0] \in \mathbb{P}^1(\mathbb{Q})$ as a point at infinity. Notice that $SL(2, \mathbb{Z})$ acts on $\mathbb{P}^1(\mathbb{Q})$ in the usual manner,

$$\begin{pmatrix} a & b \\ c & d \end{pmatrix} : [x, y] \longmapsto [ax + by, cx + dy].$$

The quotient space $\mathbb{H}^*/SL(2, \mathbb{Z})$ can be given the structure of a Riemann surface, and one can show that the j-function then defines a holomorphic isomorphism

$$j : \mathbb{H}^*/SL(2, \mathbb{Z}) \longrightarrow \mathbb{P}^1(\mathbb{C}).$$

See Shimura [351], Sections 1.3, 1.4 and 1.5 for details. Since $SL(2, \mathbb{Z})$ acts transitively on $\mathbb{P}^1(\mathbb{Q})$, the net effect has been to add a single point, called a *cusp*, to $\mathbb{H}/SL(2, \mathbb{Z})$.

More generally, we consider an automorphic function $f(\tau)$ of one complex variable τ, which is meromorphic in $\mathbb{H}$ including the point of $\mathfrak{B}$ at infinity and which is invariant under the modular group. More precisely, the condition on the behavior at infinity states that there exists a Laurent expansion

$$f(\tau) = \sum_{n=-m}^{\infty} c_n(f) z^n$$

which converges for sufficiently small values of $|z|$ and contains only finitely many negative powers of z. Here the variable $z = e^{2\pi i \tau}$. Every function satisfying all these conditions is called a *modular function* or, more precisely, an *elliptic modular function.* The function $j(\tau)$ is an elliptic modular function with (cf. [6], Theorem 1.18, 1.19, 1.20, or [341], VII Proposition 4, 5, 8)

$$j(\tau) = \frac{1}{z} + 744 + \sum_{n=1}^{\infty} c(n)z^n, \ c(n) \in \mathbb{Z}.$$

The elliptic modular functions obviously form a field which consists precisely of the rational functions of $j(\tau)$ (see [358], [359], [360], [362]).

3.1.3 Cusp forms

Let N be a positive integer. One defines subgroups of $SL(2,\mathbb{Z})$ as follows:

$$\Gamma_0(N) = \left\{ \begin{pmatrix} a & b \\ c & d \end{pmatrix} \in SL(2,\mathbb{Z}) \middle| \ c \equiv 0(\mathrm{mod}N) \right\}, \tag{3.1.11}$$

$$\Gamma_1(N) = \left\{ \gamma \in SL(2,\mathbb{Z}) \middle| \ \gamma \equiv \begin{pmatrix} 1 & b \\ 0 & 1 \end{pmatrix} (\mathrm{mod}N) \right\}, \tag{3.1.12}$$

and

$$\Gamma(N) = \left\{ \gamma \in SL(2,\mathbb{Z}) \middle| \ \gamma \equiv \begin{pmatrix} 1 & 0 \\ 0 & 1 \end{pmatrix} (\mathrm{mod}N) \right\}. \tag{3.1.13}$$

More generally, a *congruence subgroup* of $SL(2,\mathbb{Z})$ is defined to be a subgroup Γ of $SL(2,\mathbb{Z})$ which contains $\Gamma(N)$ for some integer $N \geq 1$. If Γ is a congruence subgroup of $SL(2,\mathbb{Z})$, then Γ acts on $\mathbb{H}^*$, and we can form the quotient space $\mathbb{H}^*/\Gamma$, which has a natural structure as a Riemann surface (see Shimura [351], Sections 1.3 and 1.5). The action of Γ on $\mathbb{P}^1(\mathbb{Q})$ gives finitely many orbits; the images of these orbits in $\mathbb{H}^*/\Gamma$ are called the *cusps of* Γ.

We may view $1/N$ as a point of order N on the torus $\mathbb{C}/[1,\tau]$. Let Z_N be the cyclic group generated by $1/N$. Then we may consider the pair $(\mathbb{C}/[1,\tau], Z_N)$ as consisting of a torus and a cyclic subgroup of order N. One has the following parametrizations:

(f1) The association $\tau \mapsto (\mathbb{C}/[1,\tau], 1/N)$ gives a bijection between $\mathbb{H}/\Gamma_1(N)$ and isomorphism classes of toruses together with a point of order N.

(f2) The association $\tau \mapsto (\mathbb{C}/[1,\tau], Z_N)$ gives a bijection between $\mathbb{H}/\Gamma_0(N)$ and isomorphism classes of toruses together with a cyclic subgroup of order N.

Furthermore, there exist affine curves $Y_1(N)$ and $Y_0(N)$, defined over $\mathbb{Q}$, such that

$$Y_1(N)(\mathbb{C}) \approx \mathbb{H}/\Gamma_1(N), \ Y_0(N)(\mathbb{C}) \approx \mathbb{H}/\Gamma_0(N)$$

and such that $Y_1(N)$ parametrizes isomorphism classes of pairs (E,P) algebraically, where E is an elliptic curve and P is a point of order N, in the following sense. If κ is a field containing $\mathbb{Q}$, then a point of $Y_1(N)(\kappa)$ corresponds to such a pair

(E, P) with E defined over κ and P rational over κ. Similarly, $Y_0(N)$ parametrizes pairs (E, Z), where E is defined over κ and Z is invariant under the Galois group $G_{\bar{\kappa}/\mathbb{Q}}$. The affine curve $Y_1(N)$ can be compactified by adjoining the points which lie above $j = \infty$. Its completion, denoted by $X_1(N)$, is a smooth projective curve which contains $Y_1(N)$ as a dense Zariski open subset. Similarly, we have the completion $X_0(N)$ of $Y_0(N)$. Thus one obtains the holomorphic isomorphisms

$$X_1(N)(\mathbb{C}) \approx \mathbb{H}^* / \Gamma_1(N), \quad X_0(N)(\mathbb{C}) \approx \mathbb{H}^* / \Gamma_0(N).$$

See Shimura [351], or Silverman [362].

An automorphic form $f \in \mathcal{M}(\mathbb{H})$ of weight $\frac{k}{2}$ for a congruence subgroup Γ of $SL(2, \mathbb{Z})$ is said to be a *modular function of weight k for* Γ if f also is meromorphic at each of the cusps of $\mathbb{H}^* / \Gamma$ (see Shimura [351], Section 2.1 for the precise definition). A modular function is called a *modular form* if it is holomorphic on $\mathbb{H}$ and at each of the cusps of $\mathbb{H}^* / \Gamma$; and it is a *cusp form* if it is a modular form which vanishes at every cusp. Take

$$\gamma = \begin{pmatrix} a & b \\ c & d \end{pmatrix} \in \Gamma : \tau \longmapsto \gamma(\tau) = \frac{a\tau + b}{c\tau + d}.$$

Note that the Jacobian determinant J_γ of γ is just

$$\frac{d\gamma}{d\tau} = \frac{1}{(c\tau + d)^2}.$$

By the definition, a modular function f of weight k for Γ satisfies

$$f\left(\frac{a\tau + b}{c\tau + d}\right) = (c\tau + d)^k f(\tau), \quad \tau \in \mathbb{H}. \tag{3.1.14}$$

The Eisenstein series $G_{2k}(\tau) = G_{2k}(\Lambda)$ of weight $2k$ for the lattice $\Lambda = [1, \tau]$ is a modular form of weight $2k$ for $SL(2, \mathbb{Z})$. Its Fourier series is given by

$$G_{2k}(\tau) = 2\zeta(2k) + 2\frac{(2\pi i)^{2k}}{(2k-1)!} \sum_{n=1}^{\infty} \sigma_{2k-1}(n) e^{2\pi i n \tau},$$

where $\sigma_\alpha(n)$ is the *divisor function*

$$\sigma_\alpha(n) = \sum_{d|n} d^\alpha.$$

See [6], Theorem 1.18, 1.19, 1.20, or [341], VII Proposition 4, 5, 8. Related to the definition of $G_{2k}(\tau)$, here one introduces the function

$$E(\tau, s) = \frac{1}{2\zeta(2s)} \sum_{(m,n)\in\mathbb{Z}^2 - \{0\}} \frac{y^s}{|m + n\tau|^{2s}}, \quad \tau = x + iy \in \mathbb{H}. \tag{3.1.15}$$

This series converges absolutely and uniformly in any compact subset of the region $\mathrm{Re}(s) > 1$. Selberg [339] proved that $E(\tau, s)$ has a meromorphic continuation to the whole complex s-plane and satisfies the functional equation

$$E(\tau, s) = \frac{\xi(2s - 1)}{\xi(2s)} E(\tau, 1 - s). \tag{3.1.16}$$

The discriminant function $\Delta(\tau)$ is a classic cusp form of weight 12 for $SL(2, \mathbb{Z})$. Its Taylor expansion in $z = e^{2\pi i \tau}$ assumes the form

$$\Delta(\tau) = (2\pi)^{12} \sum_{n=1}^{\infty} \tau(n) z^n = (2\pi)^{12} z \prod_{n=1}^{\infty} (1 - z^n)^{24} \tag{3.1.17}$$

with $\tau(1) = 1$ and $\tau(n) \in \mathbb{Z}$ (see [6], Theorem 1.18, 1.19, 1.20, or [341], VII Proposition 4, 5, 8). The integer-valued function $n \mapsto \tau(n)$ is called the *Ramanujan τ-function*. Ramanujan also conjectured that the Hecke L-series associated to Δ has an *Euler product*:

$$L(\Delta, s) = \sum_{n=1}^{\infty} \frac{\tau(n)}{n^s} = \prod_p \left(1 - \frac{\tau(p)}{p^s} + \frac{1}{p^{2s-11}} \right)^{-1}. \tag{3.1.18}$$

This was proved by Mordell [276]. Further $L(\Delta, s)$ satisfies the functional equation

$$\Lambda_\Delta(s) = \Lambda_\Delta(12 - s), \tag{3.1.19}$$

where

$$\Lambda_\Delta(s) = (2\pi)^{-s} \Gamma(s) L(\Delta, s). \tag{3.1.20}$$

The study of modular forms is facilitated by the existence of certain linear operators. For each integer $n \geq 1$, we define the *Hecke operator* T_n on modular forms of weight k for $SL(2, \mathbb{Z})$ by the formula

$$T_n(f)(\tau) = n^{k-1} \sum_{d|n} \frac{1}{d^k} \sum_{i=0}^{d-1} f\left(\frac{n\tau + id}{d^2} \right).$$

For a more intrinsic definition, see Apostol [6], Section 6.8; Serre [341], VII, Section 5.1; or Shimura [351], Ch. 3. The Hecke operator satisfies the following basic properties:

(g1) If f is a modular form (respectively cusp form) of weight k for $SL(2, \mathbb{Z})$, then $T_n(f)$ is also.

(g2) For all integers m and n, $T_m T_n = T_n T_m$.

(g3) If m and n are relatively prime, then $T_{mn} = T_m T_n$.

See Apostol [6], Theorem 6.11 and 6.13; Serre [341], VII, Sections 5.1 and 5.3.

The Hecke operators defined above also act on the space of modular forms relative to congruence subgroups.

Proposition 3.10. *Let Γ be a congruence subgroup of $SL(2,\mathbb{Z})$, say $\Gamma \supset \Gamma(N)$ and let f be a modular form of weight k for Γ. Then for each integer $n \geq 1$ relatively prime to N, the function $T_n(f)$ is again a modular form of weight k for Γ. Further, if f is a cusp form, then so is $T_n(f)$.*

Proof. See Shimura [351], Proposition 3.37. $\square$

For a positive integer N, a cusp form of weight $k \geq 1$ for $\Gamma_1(N)$ is also called a *cusp form of weight* k and *level* N, which is a holomorphic function f on $\mathbb{H}$ such that

(h1) $f\left(\frac{a\tau+b}{c\tau+d}\right) = (c\tau + d)^k f(\tau)$ for all $\tau \in \mathbb{H}$ and all $\begin{pmatrix} a & b \\ c & d \end{pmatrix} \in \Gamma_1(N)$;

(h2) $|f(\tau)|^2 (\mathrm{Im}(\tau))^k$ is bounded on $\mathbb{H}$.

The space $S_k(N)$ of cusp forms of weight k and level N is a finite-dimensional complex vector space. If $f \in S_k(N)$, then it has a Taylor expansion in $e^{2\pi i \tau}$:

$$f(\tau) = \sum_{n=1}^{\infty} c_n(f) e^{2\pi i n \tau}.$$

We define the *L*-series of f to be

$$L(f,s) = \sum_{n=1}^{\infty} \frac{c_n(f)}{n^s}.$$

For all primes $p \nmid N$, the Hecke operators T_p can be simultaneously diagonalised on $S_k(N)$ and a simultaneous eigenvector is called an *eigenform*. If f is an eigenform, then the corresponding eigenvalues, $a_p(f)$, are algebraic integers,

$$c_1(f) \neq 0, \quad c_p(f) = a_p(f)c_1(f),$$

and one has

$$L(f,s) = c_1(f) \prod_p \left(1 - \frac{a_p(f)}{p^s} + \frac{1}{p^{2s-k+1}}\right)^{-1}. \tag{3.1.21}$$

Conversely, if (3.1.21) holds, then for each prime p, one has (see [351])

$$T_p(f) = a_p(f)f.$$

In particular, for each prime p we have

$$T_p(\Delta) = \tau(p)\Delta.$$

For any prime $\mathfrak{p}$ over p, we let D_p and I_p denote respectively the decomposition and inertial groups of p. Thus

$$D_p = \{\sigma \mid \sigma(\mathfrak{p}) = \mathfrak{p}\},$$

and I_p is the kernel of the reduction mapping $D_p \longrightarrow G_{\mathbb{F}_{\mathfrak{p}}/\mathbb{F}_p}$. This reduction mapping is surjective, and we let Frob_p denote an element of D_p that maps to the *Frobenius* $\alpha \mapsto \alpha^p$. It is well defined up to an element of I_p (and up to conjugation).

Let λ be a place of the algebraic closure of $\mathbb{Q}$ in $\mathbb{C}$ above a rational prime ℓ and let $\overline{\mathbb{Q}}_\lambda$ denote the algebraic closure of $\mathbb{Q}_\ell$ thought of as a $\overline{\mathbb{Q}}$ algebra via λ. If $f \in S_k(N)$ is an eigenform, then there is a unique continuous irreducible representation

$$\rho_{f,\lambda} : G_{\overline{\mathbb{Q}}/\mathbb{Q}} \longrightarrow \mathrm{GL}(2, \overline{\mathbb{Q}}_\lambda)$$

such that for any prime $p \nmid N\ell$, $\rho_{f,\lambda}$ is unramified at p and

$$\mathrm{tr}\rho_{f,\lambda}(\mathrm{Frob}_p) = a_p(f).$$

The existence of $\rho_{f,\lambda}$ is due to Shimura [351] if $k = 2$, to Deligne [75] if $k > 2$ and to Deligne and Serre [76] if $k = 1$. Its irreducibility is due to Ribet [321] if $k > 1$ and to Deligne and Serre [76] if $k = 1$. Moreover $\rho_{f,\lambda}$ is potentially semi-stable at ℓ in the sense of Fontaine.

Let $\rho : G_{\overline{\mathbb{Q}}/\mathbb{Q}} \longrightarrow \mathrm{GL}(2, \overline{\mathbb{Q}}_\ell)$ be a continuous representation which is unramified outside finitely many primes and for which the restriction of ρ to a decomposition group at ℓ is potentially semi-stable in the sense of Fontaine. It is known by work of Carayol and others that the following two conditions are equivalent:

(i1) $\rho \sim \rho_{f,\lambda}$ for some eigenform f and some place $\lambda | \ell$;

(i2) $\rho \sim \rho_{f,\lambda}$ for some eigenform f of level $N(\rho)$ and weight $k(\rho)$ and some place $\lambda | \ell$.

In (h2), $N(\rho)$ and $k(\rho)$ are respectively the *conductor* and the *weight* of ρ. When these equivalent conditions are met we call ρ *modular*.

3.1.4   Problems in elliptic curves

Next we assume that E is an elliptic curve over $\mathbb{Q}$ defined by (3.1.3), where $a, b \in \mathbb{Z}$. By (3.1.4), since $A = -4a$ and $B = -4b$ satisfies

$$A^3 - 27B^2 = -16(4a^3 + 27b^2) = \Delta \neq 0,$$

the uniformization theorem shows that there exists a unique lattice $\Lambda \subset \mathbb{C}$ such that

$$g_2 = g_2(\Lambda) = -4a, \quad g_3 = g_3(\Lambda) = -4b.$$

Hence the equation (3.1.3) has non-constant meromorphic solutions $x = \wp$, $y = \frac{1}{2}\wp'$.

Conjecture 3.11 (Lang, Stark [226]). *If $(x, y) \in \mathbb{Z}^2$ is a point on the elliptic curve E, then for $\varepsilon > 0$, there exists a number $C(\varepsilon)$ such that*

$$|x| \leq C(\varepsilon) \max\{|a|^3, |b|^2\}^{\frac{5}{3}+\varepsilon}. \tag{3.1.22}$$

Lang originally posed the conjecture with an unknown exponent; then Stark suggested that the exponent should be $5/3$.

Question 3.12. *Given polynomials a, b satisfying (3.1.4). If polynomials x, y satisfy (3.1.3), does the following relation hold*

$$\frac{3}{5} \deg(x) \leq \max\{3 \deg(a), 2 \deg(b)\}? \tag{3.1.23}$$

Theorem 3.13. *If E is an elliptic curve over a number field κ, then the commutative group $E(\kappa)$ is finitely generated.*

The theorem as stated here is actually due to Mordell [278] whose original statement is for rational points, while Weil has generalized it to arbitrary number fields and to Abelian varieties of higher dimension (see [158]). According to the Mordell-Weil theorem, we can write

$$E(\kappa) = \mathbb{Z}^r \oplus E_{\text{tors}}(\kappa),$$

where the *torsion subgroup* $E_{\text{tors}}(\kappa)$ is finite and the *rank* r of $E(\kappa)$ is a non-negative integer. A deep theorem of Mazur [261], [262] states which finite groups can occur as torsion subgroups of elliptic curves:

Theorem 3.14. *If E is an elliptic curve, then $E_{\text{tors}}(\mathbb{Q})$ is one of the following 15 groups:*

(A1) $\mathbb{Z}/n\mathbb{Z}$, with $1 \leq n \leq 10$ or $n = 12$,

(A2) $\mathbb{Z}/2m\mathbb{Z} \times \mathbb{Z}/2\mathbb{Z}$, with $1 \leq m \leq 4$.

Each of the groups in Theorem 3.14 occurs infinitely often as the torsion subgroup of an elliptic curve over $\mathbb{Q}$. For an example of each possible group, see exer. 8.12 in Silverman [362]. For arbitrary number fields, there is the following result of Manin [251]:

Theorem 3.15. *Let κ be a number field and $p \in \mathbb{Z}$ a prime. There is a constant $N = N(\kappa, p)$ so that for all elliptic curves E/κ, the p-primary component of $E(\kappa)$ has order dividing p^N.*

For those torsion subgroups which are allowed in Mazur's Theorem 3.14, it is a classical result that the elliptic curves E/κ having the specified torsion subgroup all lie in a 1-parameter family. See Exercise 8.13a, b in Silverman [362]. A complete list is given in Kubert [218]. Taken together, Theorem 3.14 and Theorem 3.15 provide the best evidence to date for the following longstanding conjecture (cf. Silverman [362]):

Conjecture 3.16. *Let κ be a number field. There is a constant N depending on κ so that for all elliptic curves E/κ,*

$$|E_{\text{tors}}(\kappa)| \leq N.$$

The rank of $E(\mathbb{Q})$ is called the *rank* of E and is written rank(E). The rank of the Mordell-Weil group is much more mysterious and much more difficult to compute. There are infinitely many elliptic curves E over $\mathbb{Q}$ with rank$(E) = 0$ (see [362], Corollary 6.2.1), but there are many elliptic curves E such that rank$(E) \geq 1$ (see [331]). The following conjecture is due to Lang [231], Silverman [362], or Hindry and Silverman [158]:

Conjecture 3.17. *There exist elliptic curves E over $\mathbb{Q}$ whose Mordell-Weil rank is arbitrarily large.*

Fix an elliptic curve E defined by (3.1.3) over $\mathbb{Q}$. For every prime number p not dividing $\Delta = -16(4a^3 + 27b^2)$, we can reduce a and b modulo p and view E as an elliptic curve over the finite field $\mathbb{F}_p$. For every prime number p not dividing Δ let

$$N_p = \#E(\mathbb{F}_p) = 1 + \#\{0 \leq x, y \leq p - 1 \mid y^2 \equiv x^3 + ax + b (\text{mod } p)\},$$

and set

$$a_p = 1 + p - N_p.$$

H. Hasse ([149], [150]) proved that

$$-2\sqrt{p} < a_p < 2\sqrt{p}. \tag{3.1.24}$$

Define the *Hasse-Weil L-function* of E by

$$L(E, s) = \prod_{p \nmid \Delta}\left(1 - \frac{a_p}{p^s} + \frac{1}{p^{2s-1}}\right)^{-1} \prod_{p \mid \Delta} l_p(E, s)^{-1} = \sum_{n=1}^{\infty} \frac{a_n}{n^s}, \tag{3.1.25}$$

where $l_p(E, s)$ is of the form

$$l_p(E, s) = 1 - \frac{a_p}{p^s}$$

for some well-defined integer $a_p = 1, -1$, or 0 (cf. [232], p. 97; [362], p. 240; [389], p. 196), which is defined as follows:

$$a_p = \begin{cases} 1, & \text{if } E \text{ has split multiplicative reduction over } \mathbb{Q} \text{ at } p; \\ -1, & \text{if } E \text{ has non-split multiplicative reduction over } \mathbb{Q} \text{ at } p; \\ 0, & \text{if } E \text{ has additive reduction over } \mathbb{Q} \text{ at } p. \end{cases}$$

The coefficients a_n are constructed easily from a_p for prime p. It follows from (3.1.24) that $L(E, s)$ converges absolutely and uniformly on compact subsets of the complex half-plane $\{s \in \mathbb{C} \mid \text{Re}(s) > 3/2\}$.

Let $N(E)$ be the *conductor* of the elliptic curve E,

$$N(E) = \prod_{p|\Delta} p^{f(p)},$$

in which $f(p)$ is 0 if $p \nmid \Delta$ and ≥ 1 if $p|\Delta$ (see [232], p. 97; [362], p. 361; [389], p. 196), called the *exponent of the conductor of E at p*. In particular, $f(p) = 1$ if E has multiplicative reduction over $\mathbb{Q}$ at p.

Let $\rho_{E,\ell}$ denote the representation of $G_{\overline{\mathbb{Q}}/\mathbb{Q}}$ on the ℓ-adic Tate module of $E(\overline{\mathbb{Q}})$. The following conditions are equivalent (cf. [37]):

(B1) The L-function $L(E, s)$ of E equals the L-function $L(f, s)$ for some eigenform f.

(B2) The L-function $L(E, s)$ of E equals the L-function $L(f, s)$ for some eigenform f of weight 2 and level $N(E)$.

(B3) For some prime ℓ, the representation $\rho_{E,\ell}$ is modular.

(B4) For all primes ℓ, the representation $\rho_{E,\ell}$ is modular.

(B5) There is a non-constant holomorphic mapping $X_1(N)(\mathbb{C}) \longrightarrow E(\mathbb{C})$ for some positive integer N.

(B6) There is a non-constant morphism $X_1(N(E)) \longrightarrow E$ which is defined over $\mathbb{Q}$.

The implications (B2) $\Rightarrow$ (B1), (B4) $\Rightarrow$ (B3) and (B6) $\Rightarrow$ (B5) are tautological. The implication (B1) $\Rightarrow$ (B4) follows from the characterization of $L(E, s)$ in terms of $\rho_{E,\ell}$. The implications (B3) $\Rightarrow$ (B2) follows from a theorem of Carayol [45]. The implications (B2) $\Rightarrow$ (B6) follows from a construction of Shimura [351] and a theorem of Faltings [98]. The implications (B5) $\Rightarrow$ (B3) follows from Mazur [263]. When these equivalent conditions are satisfied we call E *modular*.

Theorem 3.18. *If E is an elliptic curve over $\mathbb{Q}$, then E is modular.*

It has become a standard conjecture that all elliptic curves over $\mathbb{Q}$ are modular. Taniyama made a suggestion along the line (B1) as one of a series of problems collected at the Tokyo-Nikko conference in September 1955. However his formulation did not make clear whether the function f defined by coefficients of $L(E, s)$ should be a cusp form or some more general automorphic form. He also suggested that constructions as in (B5) and (B6) might help attack this problem at least for some elliptic curves. In private conversations with a number of mathematicians (including Weil) in the early 1960s, Shimura suggested that assertions along the lines of (B5) and (B6) might be true (see Shimura [352] and Weil [428]). However, it only became widely known through its publication in a paper of Weil [427] in 1967, in which Weil gave conceptual evidence for the conjecture. That assertion (B1) is true for CM elliptic curves follows at once from work of Hecke and Deuring. Shimura [350] went on to check the assertion (B5) for these curves. The

Shimura-Taniyama-Weil conjecture (Theorem 3.18) was finally proved by Breuil, Conrad, Diamond, and Taylor [37] by extending work of Wiles [431], Taylor and Wiles [390].

In 1985, Frey [101] made the remarkable observation that the Shimura-Taniyama-Weil conjecture should imply Fermat's last theorem. The precise mechanism relating the two was formulated by Serre as the ε-conjecture and this was then proved by Ribet in the summer of 1986, which enabled Ribet to show that the conjecture only for semistable elliptic curves implies Fermat's last theorem (see [322], [232]). However, one still needed to know that the curve in question would have to be modular, and this is accomplished by Wiles [431], Taylor and Wiles [390] via studying associated Galois representations of elliptic curves.

Theorem 3.18 implies the following long-standing conjecture of Hasse and Weil (cf. Silverman [362]): $L(E, s)$ has an analytic continuation to all of $\mathbb{C}$ and satisfies a functional equation

$$\Lambda_E(s) = w_E \Lambda_E(2 - s), \tag{3.1.26}$$

where $w_E = \pm 1$, called the *sign of the functional equation*, and

$$\Lambda_E(s) = N(E)^{\frac{s}{2}} (2\pi)^{-s} \Gamma(s) L(E, s). \tag{3.1.27}$$

Note that the Euler product (3.1.25) for $L(E, s)$ may not in general converge at $s = 1$. Goldfeld [116] proved that if there exist constants $C \in \mathbb{R}^+$ and $r \in \mathbb{R}$ such that

$$\prod_{p \leq x, p \nmid \Delta} \frac{N_p}{p} \sim C(\log x)^r,$$

then $r = \operatorname{ord}_{s=1} L(E, s)$, the order of vanishing of $L(E, s)$ at $s = 1$, and

$$\lim_{s \to 1} \frac{L(E, s)}{(s - 1)^r} = \sqrt{2} e^{r\gamma} C^{-1} \prod_{p \mid \Delta} l_p(E, 1)^{-1},$$

where γ is Euler's constant. In particular, if $r = 0$ then

$$L(E, 1) = \sqrt{2} \left(\prod_{p \nmid \Delta} \frac{N_p}{p} \times \prod_{p \mid \Delta} l_p(E, 1) \right)^{-1}.$$

Conjecture 3.19 (Birch and Swinnerton-Dyer [26]). *For every elliptic curve E,*

$$\operatorname{rank}(E) = \operatorname{ord}_{s=1} L(E, s).$$

The following theorem, a combination of work of Kolyvagin [214], [215], Gross and Zagier [129], and others, is the best result to date in the direction of the Birch and Swinnerton-Dyer conjecture.

Theorem 3.20.

(C1) $\mathrm{ord}_{s=1}L(E,s) = 0 \Longrightarrow \mathrm{rank}(E) = 0$,
(C2) $\mathrm{ord}_{s=1}L(E,s) = 1 \Longrightarrow \mathrm{rank}(E) = 1$.

The sign w_E in the functional equation (3.1.26) for $L(E,s)$ determines the parity of the integer $\mathrm{ord}_{s=1}L(E,s)$, that is, $\mathrm{ord}_{s=1}L(E,s)$ is even when $w_E = 1$, and is odd when $w_E = -1$. Thus the following *parity conjecture* is a consequence of the Birch and Swinnerton-Dyer conjecture.

Conjecture 3.21. *The integer* $\mathrm{rank}(E)$ *is even when* $w_E = 1$, *and is odd when* $w_E = -1$.

There may be many parametrizations $\varphi : X_0(N(E)) \longrightarrow E$. An interesting question is to find one of the ones of smallest degree, or at least to determine its degree. The following *modular parametrization conjecture* is due to Hindry and Silverman [158]:

Conjecture 3.22. *There is an absolute constant d such that for all elliptic curves $E/\mathbb{Q}$, there is a finite covering $\varphi : X_0(N(E)) \longrightarrow E$ such that $\deg(\varphi) \leq N(E)^d$.*

3.1.5 Hyperelliptic and rational curves

A curve of genus $g \geq 2$ is called a *hyperelliptic* if it is a double covering of the projective line $\mathbb{P}^1$. Next we assume $\mathrm{char}(\bar{\kappa}) \neq 2$. The function field of a hyperelliptic curve C is a quadratic extension of $\bar{\kappa}(\mathbb{P}^1)$, hence has the shape $\bar{\kappa}(x,y)$, where

$$y^2 = P(x) \tag{3.1.28}$$

for some polynomial $P(x) \in \bar{\kappa}[x]$. This equation gives an affine model for C. If the polynomial P has a double root, say α, then we can replace y by $(x - \alpha)y$ and cancel $(x - \alpha)^2$. So we may assume that C is given by an affine equation (3.1.28) for some $P(x) \in \bar{\kappa}[x]$ with distinct roots. Then the affine curve C is smooth.

Proposition 3.23 (cf. [126], [158]). *Every curve of genus 2 is hyperelliptic.*

A compact Riemann surface M defined on $\mathbb{C}$ of genus ≥ 2 is hyperelliptic if and only if there exists a meromorphic function f on M with only two poles. Any hyperelliptic Riemann surface of genus g can be realized as a smooth normalization of a plane algebraic curve (3.1.28) in $\mathbb{C}^2$, for P a polynomial of degree $2g + 2$. Conversely, for distinct $2g+2$ complex numbers $a_1, a_2, \ldots, a_{2g+2}$, a normalization of the plane algebraic curve

$$y^2 = \prod_{i=1}^{2g+2} (x - a_i)$$

is a hyperelliptic Riemann surface of genus g (cf. [126], [127]). Generally, if P is a polynomial of degree $n \geq 3$ such that its discriminant is not identically zero, the *hyperelliptic equation* (3.1.28) determines a Riemann surface of genus $g = [(n-1)/2]$ which can be derived from the Riemann-Hurwitz formula (1.6.9) (cf. [338]), where $[x]$ denotes the maximal integer $\leq x$.

A natural question is how one obtains all the irreducible algebraic curves which have a rational parametric representation in an independent variable z. In other words, if

$$f(x, y) = 0 \qquad (3.1.29)$$

is an irreducible equation of the curve, then we require two rational functions

$$x = \varphi(z), \quad y = \psi(z)$$

of a variable z, not both constant, which satisfy the equation (3.1.29) identically in z. An algebraic curve of this type is said to be a *rational algebraic curve*.

Theorem 3.24 (cf. [359]). *An algebraic curve is rational if and only if it is of genus* 0.

Hence algebraic curves of genus 0 admit a lot of holomorphic curves.

3.2   The *abc*-conjecture

Many results for Diophantine equations in integers are analogous to results for Diophantine equations in polynomials. Lang ([234], p. 196) said: "One of the most fruitful analogies in mathematics is that between the integers $\mathbb{Z}$ and the ring of polynomials $F[t]$ over a field F." Given Mason's wonderfully simple inequality for polynomial solutions to $a + b = c$ (namely Theorem 2.65), one wonders whether there is a similar result for integers.

In Theorem 2.65 we note that (cf. [288], p. 182 or [135])

$$r(f) = \deg(\mathrm{rad}(f)),$$

where $\mathrm{rad}(f)$ is the *radical* of a polynomial f on $\mathbb{C}$, which is the product of distinct irreducible factors of f. On the other hand, the *radical* of a non-zero integer A is defined to be

$$r(A) = \prod_{p \mid A} p,$$

i.e., the product of distinct primes dividing A. It was conjectured by Erdös and Woods that there exists an absolute constant $k > 2$ such that for every positive integers x and y, if $r(x + i) = r(y + i)$ for $i = 1, 2, \ldots, k$, then $x = y$. No examples with different x and y are known. It seems that the five-value theorem of Nevanlinna in value distribution theory is an analogue of the Erdös-Woods Conjecture, and so maybe $k = 5$ is possible.

There is a classical analogy between polynomials and integers, that is, prime factors of an integer are often considered to be an appropriate analogy to irreducible factors of a polynomial. Under that analogy, $r(f)$ of a polynomial f corresponds to $\log r(A)$ of an integer A, and in addition, the value $\log |A|$ of an integer A is a measure of how "large" the integer is, while the degree of a polynomial is a measure of how "large" the polynomial is (cf. [231] or [119]). Thus for polynomials we had an inequality formulated additively, whereas for integers we formulate the corresponding inequality multiplicatively.

After being influenced by Stothers-Mason's theorem (see Theorem 2.65), and based on considerations of Szpiro and Frey, Oesterlé [308] and Masser [256] formulated the *abc-conjecture* for integers as follows:

Conjecture 3.25. *Given $\varepsilon > 0$, there exists a number $C(\varepsilon)$ having the following property. For any non-zero relatively prime integers a, b, c such that $a + b = c$, we have*

$$\max\{|a|, |b|, |c|\} \leq C(\varepsilon) r \, (abc)^{1+\varepsilon} \, . \tag{3.2.1}$$

An interesting discussion in [119] illustrates how one is naturally led from Theorem 2.65 to the formulation of the *abc*-conjecture. In this setting Stewart and Tijdeman [372] proved that the conjecture would be false without the ε. In other words, it is not true that

$$\max\{|a|, |b|, |c|\} \leq Cr \, (abc) \, .$$

To prove or disprove the *abc*-conjecture would be an important contribution to number theory. For instance, some results that would follow from the *abc*-conjecture are in [288], pp. 185–188, [92], [412] (or see [119], [231], [234], [415]). Langevin ([236], [237]) proved that the *abc*-conjecture implies the Erdös-Woods conjecture with $k = 3$ except perhaps a finite number of counter examples.

Although the *abc*-conjecture seems completely out of reach, there are some results towards the truth of this conjecture. In 1986, C.L. Stewart and R. Tijdeman [372] proved

$$\max\{|a|, |b|, |c|\} < \exp\left\{Cr(abc)^{15}\right\} \, .$$

In 1991, C.L. Stewart and Kunrui Yu [373] obtained

$$\max\{|a|, |b|, |c|\} < \exp\left\{C(\varepsilon)r(abc)^{2/3+\varepsilon}\right\} \, .$$

In 1996, C.L. Stewart and Kunrui Yu [374] further proved

$$\max\{|a|, |b|, |c|\} < \exp\left\{C(\varepsilon)r(abc)^{1/3+\varepsilon}\right\} \, .$$

Next we show that the *abc*-conjecture is equivalent to the following:

Conjecture 3.26. *Let A, B be fixed non-zero integers. Take positive integers m, n with*

$$\alpha = 1 - \frac{1}{m} - \frac{1}{n} > 0. \tag{3.2.2}$$

Let $x, y, z \in \mathbb{Z}$ be variables such that x, y are relatively prime and

$$Ax + By = z \neq 0.$$

Assume that for a prime p (resp. q), $p \mid x$ (resp. $q \mid y$) implies $p^m \mid x$ (resp. $q^n \mid y$). Then for any $\varepsilon > 0$ there exists a number $C = C(\varepsilon, m, n, A, B)$ such that

$$\max\{|x|^\alpha, |y|^\alpha, |z|^\alpha\} \leq Cr(z)^{1+\varepsilon}. \tag{3.2.3}$$

Remark. We introduce a notation related to Conjecture 3.26. A positive integer A is *powerful* if for every prime p dividing A, p^2 also divides A. Every powerful number can be written as $a^2 b^3$, where a and b are positive integers. The Erdös-Mollin-Walsh conjecture asserts that there do not exist three consecutive powerful integers. The *abc*-conjecture implies the weaker assertion that the set of triples of consecutive powerful integers is finite.

Here we first show that Conjecture 3.25 implies Conjecture 3.26. To simplify notation in dealing with the possible presence of constants C, if a, b are positive functions, we write

$$a \ll b$$

to mean that there exists a constant $C > 0$ such that $a \leq Cb$. Thus $a \ll b$ means $a = O(b)$ in the big oh notation. By the *abc*-conjecture, we get

$$\max\{|x|, |y|, |z|\} \ll \left\{|x|^{\frac{1}{m}} |y|^{\frac{1}{n}} r(z)\right\}^{1+\varepsilon}.$$

If, say, $|Ax| \leq |By|$ then $|x| \ll |y|$. We substitute this estimate for x to get an inequality entirely in terms of y, namely

$$|y| \ll \left\{|y|^{\frac{1}{m}+\frac{1}{n}} r(z)\right\}^{1+\varepsilon}.$$

We first bring all powers of y to the left-hand side, and take care of the extra ε, so we obtain

$$|y|^\alpha \ll r(z)^{1+\varepsilon},$$

and then also

$$|x|^\alpha \ll r(z)^{1+\varepsilon}$$

because the situation is symmetric in x and y. Again by the *abc*-conjecture, we have

$$|z| \ll \left\{|x|^{\frac{1}{m}} |y|^{\frac{1}{n}} r(z)\right\}^{1+\varepsilon}.$$

By using the estimate for $|xy|$ coming from the product of the inequalities above we find

$$|z|^\alpha \ll r(z)^{1+\varepsilon}.$$

Conversely, Conjecture 3.25 can be derived from Conjecture 3.26. To do this, we see that Conjecture 3.26 contains obviously the following *generalized Szpiro conjecture* (cf. [231], [415]):

Conjecture 3.27. *Take integers x and y with $D = 4x^3 - 27y^2 \neq 0$ such that the common factor of x, y is bounded by M. Then for any $\varepsilon > 0$, there exists a constant $C = C(\varepsilon, M)$ satisfying*

$$\max\{|x|^3, y^2, |D|\} \leq Cr(D)^{6+\varepsilon}. \tag{3.2.4}$$

This is trivial if x, y are relatively prime. Suppose that x, y have some common factor, say d, bounded by M. Write

$$x = ud, \quad y = vd$$

with u, v relatively prime. Then

$$D = 4d^3 u^3 - 27d^2 v^2.$$

Now we can apply the inequality (3.2.3) with $A = 4d^3, m = 3; B = -27d^2, n = 2$, and we find the same inequality (3.2.4), with the constant depending also on M.

Further, it is well known that the generalized Szpiro conjecture implies the *abc*-conjecture (cf. [232], [415]). Here we introduce the proof roughly. Let $a + b = c$, and consider the Frey elliptic curve ([101], [102]),

$$y^2 = x(x-a)(x+b).$$

The discriminant of the right-hand side is the product of the differences of the roots squared, and so

$$D = (abc)^2.$$

We make a translation

$$\xi = x + \frac{b-a}{3}$$

to get rid of the x^2-term, so that the equation can be rewritten

$$y^2 = \xi^3 - \gamma_2 \xi - \gamma_3,$$

where

$$\gamma_2 = \frac{1}{3}(a^2 + ab + b^2), \quad \gamma_3 = \frac{1}{27}(a-b)(2a+b)(a+2b).$$

The discriminant does not change because the roots of the polynomial in ξ are translations of the roots of the polynomial in x. Then

$$D = 4\gamma_2^3 - 27\gamma_3^2.$$

One may avoid the denominator 27 by using the curve

$$y^2 = x(x - 3a)(x + 3b),$$

so that γ_2, γ_3 then come out to be integers, and one can apply the generalized Szpiro conjecture to the discriminant

$$D = 3^6(abc)^2 = 4\gamma_2^3 - 27\gamma_3^2,$$

and obtain

$$\max\left\{ \sqrt[3]{|abc|}, \sqrt{|\gamma_2|}, \sqrt[3]{|\gamma_3|} \right\} \ll r(abc)^{1+\varepsilon}.$$

A simple algebraic manipulation shows that the estimates on γ_2, γ_3 imply the desired estimates on $|a|, |b|$.

The following conjecture by Hall, Szpiro, and Lang-Waldschmidt (cf. [415], [231]) becomes a special case of Conjecture 3.26:

Conjecture 3.28. *Let A, B be fixed non-zero integers and take positive integers m and n satisfying (3.2.2). Let $x, y, z \in \mathbb{Z}$ be variables such that x, y are relatively prime and*

$$Ax^m + By^n = z \neq 0.$$

Then for any $\varepsilon > 0$ there exists a number $C = C(\varepsilon, m, n, A, B)$ such that

$$\max\{|x|^{m\alpha}, |y|^{n\alpha}, |z|^{\alpha}\} \leq Cr(z)^{1+\varepsilon}. \tag{3.2.5}$$

In particular, take non-zero integers x, y with $z = x^3 - y^2 \neq 0$. If x, y are relatively prime, then Conjecture 3.28 implies that there exists a constant $C = C(\varepsilon)$ such that

$$\max\left\{ |x|^{\frac{1}{2}}, |y|^{\frac{1}{3}} \right\} \leq C(\varepsilon)r(x^3 - y^2)^{1+\varepsilon}, \tag{3.2.6}$$

which further yields

$$|x|^{\frac{1}{2}} \ll |x^3 - y^2|^{1+\varepsilon}. \tag{3.2.7}$$

This is just the content of *Hall's conjecture*:

Conjecture 3.29 ([144]). *There exists a constant $C = C(\varepsilon)$ such that*

$$|x^3 - y^2| > C(\varepsilon)|x|^{\frac{1}{2}-\varepsilon} \tag{3.2.8}$$

holds for integers x, y with $x^3 \neq y^2$.

The inequality (3.2.8) is equivalent to the form (3.2.7). Danilov [71] has proved that $1/2$ is the best possible exponent. Actually, the original conjecture made by M. Hall Jr. [144] states the following: There exists a constant C such that

$$|x^3 - y^2| > C|x|^{\frac{1}{2}} \tag{3.2.9}$$

holds for integers x, y with $x^3 \neq y^2$. The final setting of the proofs in the simple abc context which we gave above had to await Mason and the abc-conjecture a decade later.

Another special case of Conjecture 3.28 is the following *Hall-Lang-Waldschmidt conjecture* (cf. [415]):

Conjecture 3.30. *For all integers* m, n, x, y *with* $x^m \neq y^n$,

$$\max\{|x|^{m\alpha}, |y|^{n\alpha}\} < C(\varepsilon) |x^m - y^n|^{1+\varepsilon}, \tag{3.2.10}$$

where $C(\varepsilon)$ *is a constant depending on* ε.

3.3   Mordell's conjecture and generalizations

In 1909, A. Thue [391] proved the following result: Take $m \in \mathbb{Z}$ and let

$$f(x, y) = a_0 x^d + a_1 x^{d-1} y + \cdots + a_d y^d$$

with $a_i \in \mathbb{Z}$ be a form of degree $d \geq 3$ which is irreducible over $\mathbb{Q}$. Then the *Thue equation*

$$f(x, y) = m \tag{3.3.1}$$

has only finitely many integer solutions (x, y). A simple proof by using Roth's theorem is due to Schmidt [338].

Theorem 3.31. *Let* $P(x)$ *be a polynomial of degree* $n \geq 2$ *over a number field* κ *such that its discriminant is not identically zero. Let* S *be a finite subset of* M_κ, *containing all of the Archimedean absolute values. Then a superelliptic equation*

$$y^d = P(x) \tag{3.3.2}$$

with $d \geq 2$ *and* $(d, n) \neq (2, 2)$ *has only finitely many solutions* $x, y \in \mathcal{O}_{\kappa,S}$.

The special case of an *elliptic equation*, that is, $d = 2$, $n = 3$, was done by Mordell [277], [278]. The general case is due to Siegel [356]. A proof can be found in [338]. Further, Siegel [357] proved that the number of integer points (x, y) of any irreducible curve, say (2.1.9), of genus > 0 is finite. The same conclusion holds for the affine curve C defined by (2.1.9) if $\#(M - C) > 2$, where M is a projective closure of C.

Mordell [278] had originally conjectured that on a curve of genus greater than 1 there are only finitely many rational points. Faltings [98] proved Mordell's conjecture by showing the following more general result:

Theorem 3.32. *If M is an irreducible algebraic curve of genus greater than 1 and κ is a number field of finite degree over $\mathbb{Q}$, then the set $M(\kappa)$ of κ-rational points on M is a finite set.*

Elkies [92] (or see [158]) showed that using an explicit version of the abc-conjecture (that is, with a value assigned to $C(\varepsilon)$ in (3.2.1) for each ε), one can deduce an explicit version of the Mordell-Faltings theorem.

Let M be a variety defined over an algebraically closed field of characteristic 0. We shall say that M is *Mordellic* if $M(\kappa)$ is finite for every finitely generated field κ over $\mathbb{Q}$. In this context, it is natural to define a variety M to be *pseudo Mordellic* if there exists a proper Zariski closed subset Y of M such that $M - Y$ is Mordellic. Since the counterpart of algebraic curves of genus > 1 in higher-dimensional spaces are Kobayashi hyperbolic varieties, accordingly the analogue of the Mordell-Faltings theorem is just Lang's conjecture (cf. [224], [228], [232]):

Conjecture 3.33. *A projective variety is hyperbolic if and only if it is Mordellic.*

The following problem is due to Shiffman [345].

Conjecture 3.34. *Let M be a projective algebraic variety that contains no rational or elliptic curves. Then there are no holomorphic curves in M.*

Let M be a projective variety. According to Lang [228], [232], the *algebraic special set* $\mathrm{Sp}_{\mathrm{alg}}(M)$ is defined to be the Zariski closure of the union of all images of non-constant rational mappings $A \longrightarrow M$, where A is an Abelian variety or $\mathbb{P}^1$. Thus $\mathrm{Sp}_{\mathrm{alg}}(M) = \emptyset$ if and only if every rational mapping of an Abelian variety or $\mathbb{P}^1$ into M is constant. In the case of subvarieties of Abelian varieties, a clear description of this special set is well known, that is, the Ueno-Kawamata fibrations in a subvariety of an Abelian variety constitute the special set (see Lang [232]). A variety M is said to be *algebraically hyperbolic* if $\mathrm{Sp}_{\mathrm{alg}}(M) = \emptyset$. Then one says that M is *pseudo algebraically hyperbolic* if $\mathrm{Sp}_{\mathrm{alg}}(M)$ is a proper subset. Ballico [15] proved that a generic hypersurface of large degree in $\mathbb{P}^n(\mathbb{C})$ is algebraically hyperbolic.

Conjecture 3.35 ([228], [232]).

(i) $\mathrm{Sp}_{\mathrm{alg}}(M) = \mathrm{Sp}_{\mathrm{hol}}(M)$.

(ii) *The complements of the special sets are Mordellic.*

(iii) *A projective variety M is Mordellic if and only if the special sets are empty.*

Recall that the holomorphic special set $\mathrm{Sp}_{\mathrm{hol}}(M)$ is the Zariski closure of the union of all images of non-constant holomorphic mappings $f : \mathbb{C} \longrightarrow M$. Observe that the claim (i) would give an algebraic characterization of hyperbolicity. The fundamental Diophantine condition conjecturally satisfied by pseudo canonical varieties is the following problem:

Conjecture 3.36. *Let M be a pseudo canonical variety defined over a number field κ. Then $M(\kappa)$ is not Zariski dense in M.*

Bombieri posed Conjecture 3.36 for pseudo canonical surfaces, and Lang (independently) formulated the general conjecture in its refined form of the exceptional Zariski closed subset. Conjecture 3.35 allows us to state the final form of the Bombieri-Lang conjecture.

Conjecture 3.37 ([228], [232]). *The following conditions are equivalent for a projective variety M.*

(1) *M is pseudo canonical;*

(2) *$\mathrm{Sp}_{\mathrm{alg}}(M)$ is a proper subset;*

(3) *M is pseudo Mordellic. The Zariski closed subset Y can be taken to be $\mathrm{Sp}_{\mathrm{alg}}(M)$.*

In the case of Abelian varieties, there is the following Lang's conjecture over finitely generated fields (cf. Lang [221], [232]):

Conjecture 3.38. *Let M be a subvariety of an Abelian variety over a field κ finitely generated over $\mathbb{Q}$. Then M contains a finite number of translations of Abelian subvarieties which contain all but a finite number of points of $M(\kappa)$.*

The following especially important case from Lang [222] has now been proved by Faltings [99]:

Theorem 3.39. *Let M be a subvariety of an Abelian variety, and suppose that M does not contain any translation of an Abelian subvariety of dimension ≥ 1. Then M is Mordellic.*

Hilbert's tenth problem asks whether there is a general algorithm to determine, given any polynomial in several variables, whether there exists a zero with all coordinates in $\mathbb{Z}$. It was solved in the negative by Yu. Matiyasevich [258] in 1970; for a general reference, see [259]. J. Richard Büchi attempted to prove a similar statement in which there may be any (finite) number of polynomials, but they must be quadratic and of a certain form:

$$\sum_{j=1}^{n} a_{ij} x_j^2 = b_i, \quad i = 1, \ldots, m$$

with $\{a_{ij}, b_i\} \subset \mathbb{Z}$, $\{m, n\} \subset \mathbb{Z}^+$. Büchi was able to show that a negative resolution of this question would be implied by the following "*n-squares problem*":

Conjecture 3.40. *For large enough n, the only integer solutions of the system of equations*

$$x_k^2 + x_{k-2}^2 = 2x_{k-1}^2 + 2, \quad k = 2, \ldots, n-1 \tag{3.3.3}$$

satisfy $\pm x_k = \pm x_{k-1} + 1$.

Let M be the projective variety in $\mathbb{P}^n(\mathbb{C})$ defined by the equations

$$x_k^2 + x_{k-2}^2 = 2x_{k-1}^2 + 2x^2, \quad k = 2, \ldots, n-1$$

in the homogeneous coordinates $[x, x_0, \ldots, x_{n-1}]$. Vojta [421] observed that for $n \geq 6$ the variety M is a pseudo canonical surface, and then showed:

Theorem 3.41.

(i) *For $n \geq 8$, the only curves on M of geometric genus 0 and 1 are the "trivial" lines*

$$\pm x_k = \pm x_0 + kx, \ \ k = 0, \ldots, n - 1.$$

(ii) *Let $n \geq 8$ be an integer and let $f : \mathbb{C} \longrightarrow M$ be a non-constant holomorphic curve. Then the image of f lies in one of the "trivial" lines.*

The statement (i) of Theorem 3.41 has a consequence that if Conjecture 3.36 is true, then Büchi's problem has a positive answer. Statement (ii) shows that the analogue of Büchi's problem for holomorphic functions has a positive answer.

3.4   Fermat equations and Waring's problem

Fermat's conjecture, now a theorem proved by Wiles [431], Taylor and Wiles [390], states that there cannot be non-zero integers x, y, z and an integer d, where $d \geq 3$, such that

$$x^d + y^d = z^d. \tag{3.4.1}$$

Related to *Fermat's equation* (3.4.1) is *Catalan's equation*

$$x^k - y^l = 1. \tag{3.4.2}$$

In 1844, Eugène Catalan [54] conjectured that this equation had only the trivial solution

$$(x, y, k, l) = (3, 2, 2, 3).$$

About 100 years before Catalan (1814-1894) sent his letter to Crelle, Euler had proven that 8 and 9 are the only consecutive integers among squares and cubes, that is, the only solution of the Diophantine equations

$$x^2 - y^3 = \pm 1, \ x > 0, \ y > 0.$$

To prove *Catalan's conjecture*, it obviously suffices to consider the equation

$$x^p - y^q = 1, \ x > 0, \ y > 0, \tag{3.4.3}$$

where p and q are different primes. The case $q = 2$ was solved in 1850 by V.A. Lebesgue [238]. Chao Ko [206] proved the case $p = 2$. In 1976, E.Z. Chein [55] published a new, very elegant proof.

Next we may assume that p and q are different odd primes. One of the early observations was that the number of solutions (x, y) to (3.4.3), for fixed exponents

p and q, is at most finite. This is a consequence of a general theorem about integer points on a curve, published by C.L. Siegel in 1929. For other results about the number of solutions, see the introductory section in Tijdeman [392].

By way of multiplicatization of the equation, rewrite (3.4.3) as

$$(x-1)\frac{x^p-1}{x-1} = y^q.$$

By considering the identity $x^p = ((x-1)+1)^p$ one easily finds that there are two possibilities: the greatest common divisor of the two factors on the left-hand side is either 1 or p. When the greatest common divisor equals 1, we obtain the equations

$$x-1 = a^q, \quad \frac{x^p-1}{x-1} = b^q, \quad y = ab,$$

where a and b are coprime and not divisible by p. In 1960, J.W.S. Cassels [53] showed that these equations yield a contradiction.

When the greatest common divisor equals p, we obtain the equations

$$x-1 = p^{q-1}a^q, \quad \frac{x^p-1}{x-1} = pb^q, \quad y = pab,$$

where again a and b are coprime and p does not divide b. Preda Mihăilescu [269], [270] showed that these equations also yield a contradiction. A deep theorem about cyclotomic fields plays a crucial role in his proof. For a survey about the proof of Catalan's conjecture, see [268].

Generally, the following conjecture was made by Pillai [317].

Conjecture 3.42. *Given integers $A > 0$, $B > 0$, $C > 0$, the equation*

$$Ax^k - By^l = C$$

in integers $x > 1$, $y > 1$, $k > 1$, $l > 1$ and with $(k,l) \neq (2,2)$ has only a finite number of solutions.

If k, l were fixed, this would be a special case of an algebraic Diophantine equation, the superelliptic equation. *Pillai's conjecture* can be derived from the *abc*-conjecture (see [338]). A more general application is as follows. Tijdeman [393] proved that for given non-zero integers A, B, C the Diophantine equation

$$Ax^k + By^l = Cz^n \tag{3.4.4}$$

has only finitely many solutions in positive integers $x > 1$, $y > 1$, $z > 1$, k, l, n subject to $\gcd(Ax, By, Cz) = 1$ and

$$\frac{1}{k} + \frac{1}{l} + \frac{1}{n} < 1. \tag{3.4.5}$$

On the other hand, Hindry has shown that for each triple k, l, n with

$$\frac{1}{k} + \frac{1}{l} + \frac{1}{n} \geq 1,$$

there exist A, B, C such that the above equation has infinitely many solutions x, y, z with $\gcd(Ax, By, Cz) = 1$.

Waring's problem states that for a fixed positive integer d, there exists an integer $n = n(d)$ such that an arbitrary positive integer N can be expressed as a sum of dth powers of n non-negative integers x_j, that is,

$$x_1^d + \cdots + x_n^d = N. \tag{3.4.6}$$

In 1909, Hilbert confirmed this problem. Further, it is natural to find the smallest number $n = g^*(d)$ such that each positive integer N can be expressed as a sum (3.4.6) of dth powers of n non-negative integers x_j. Lagrange proved that $g^*(2) \leq 4$. On the other hand, it is easy to show that 7 can not be expressed as a sum of squares of three integers, and so $g^*(2) = 4$. We also know that $g^*(3) = 9$. Dickson, Pillai and Niven obtained

$$g^*(d) = 2^d + \left[\frac{3^d}{2^d}\right] - 2$$

if d satisfies the following condition

$$d \geq 6, \quad 3^d - 2^d + 2 \leq (2^d - 1)\left[\frac{3^d}{2^d}\right].$$

Now the number $g^*(d)$ is completely determined except for $d = 4$.

When $d \geq 3$, let $n = g(d)$ be the smallest number such that for any positive integer $N \geq N_0(d)$, there exist non-negative integers $x_1, \ldots, x_n$ satisfying (3.4.6). Then $g(2) = 4$,

$$4 \leq g(3) \leq 7, \quad g(4) = 16 \text{ (Davenport)}.$$

Generally, it is known that the integer $g(d)$ satisfies the inequalities

$$d \leq g(d) \leq cd \log d,$$

where c is a constant. Hardy and Littlewood conjectured

$$g(d) = 4d \ (d = 2^m, \ m > 1); \quad g(d) \leq 2d + 1 \ (d \neq 2^m, \ m > 1).$$

3.5 Thue-Siegel-Roth's theorem

In this section, we introduce simply Roth's theorem, and show that *abc*-conjecture implies Roth's theorem. Further, following Vojta [415], we compare the analogy between Roth's theorem and Nevanlinna's second main theorem.

In a relatively early version of determining the best approximations of algebraic numbers by rational numbers, one had the Thue-Siegel-Dyson-Gelfond result (see [391], [355], [87], [113]): Given $\varepsilon > 0$ and an algebraic number α of degree n over $\mathbb{Q}$, there are only finitely many rational numbers x/y $(x, y \in \mathbb{Z},\ y > 0)$ such that

$$\left| \alpha - \frac{x}{y} \right| < \frac{1}{y^{\sqrt{2n}+\varepsilon}}.$$

In 1958, K.F. Roth received a Fields prize for his result:

Theorem 3.43 ([327]). *If α is algebraic and $\varepsilon > 0$, there are only finitely many rational numbers $\frac{x}{y}$ with*

$$\left| \alpha - \frac{x}{y} \right| < \frac{1}{y^{2+\varepsilon}}.$$

In 1842, Dirichlet [80] proved that given $\alpha \in \mathbb{R}$ and $N > 1$, there exist integers x, y with $1 \leq y \leq N$ and $|\alpha y - x| < 1/N$, which means that when α is irrational, there are infinitely many reduced fractions x/y with

$$\left| \alpha - \frac{x}{y} \right| < \frac{1}{y^2}.$$

Hence Dirichlet's theorem shows that Roth's result is best possible. An unknown conjecture (cf. Bryuno [42]; Lang [223]; Richtmyer, Devaney and Metropolis [323]) is the following: If α is algebraic and $\varepsilon > 0$, there are only finitely many rational numbers $\frac{x}{y}$ with

$$\left| \alpha - \frac{x}{y} \right| < \frac{1}{y^2 (\log y)^{1+\varepsilon}}.$$

In other words, given ε and α algebraic, the inequality

$$-\log \left| \alpha - \frac{x}{y} \right| \leq 2 \log y + (1 + \varepsilon) \log \log y$$

holds for all but a finite number of fractions x/y in lowest form. A theorem due to A. Khintchine [202] shows that this is true for almost all α. Note that this conjecture is similar to the case $n = 1$ of Theorem 2.56 in Nevanlinna theory, which will be further formulated lately.

Suppose that $F(x, y) \in \mathbb{Z}[x, y]$ is a binary homogenous form without repeated factors. In other words, if F has degree d, then $F(t, 1)$ is a polynomial of degree

$\geq d-1$, without repeated roots, and

$$F(x,y) = y^d F\left(\frac{x}{y}, 1\right).$$

For any coprime integers x and y, Roth's theorem yields

$$|F(x,y)| \gg |y|^d \prod_{F(\alpha,1)=0} \left|\alpha - \frac{x}{y}\right| \gg |y|^{d-2-\varepsilon}, \tag{3.5.1}$$

except at most finitely many rational numbers $\frac{x}{y}$. This statement is actually equivalent to Roth's theorem.

The abc-conjecture implies something that is somewhat stronger than Roth's theorem: For any coprime integers x and y,

$$r(F(x,y)) \gg \max\{|x|,|y|\}^{d-2-\varepsilon}. \tag{3.5.2}$$

Note that by taking

$$F(x,y) = xy(x+y),$$

the original abc-conjecture is recovered. Thus the conjecture (3.5.2) is equivalent to the abc-conjecture, although it appears far stronger. One sketchy proof of (3.5.2) following from the abc-conjecture is due to [119] (see also [411]).

In an algebraic number field, Roth's theorem can be formulated as follows (cf. [415]). The set of algebraic numbers, that is, the algebraic closure of $\mathbb{Q}$, is denoted by $\bar{\mathbb{Q}}$.

Theorem 3.44. *Let κ be a number field and let S be a finite subset of M_κ. Let ε and c be positive constants. For each $v \in S$, fix a number $a_v \in \bar{\mathbb{Q}}$. Then there are only finitely many $x \in \kappa$ such that*

$$\prod_{v \in S} \min\{1, \|x - a_v\|_v\} < \frac{c}{\bar{H}_\kappa(x)^{2+\varepsilon}}. \tag{3.5.3}$$

S. Lang [225] noted that if a_v, a_v' are two distinct elements of $\bar{\mathbb{Q}}$ for some v, and if x approximates a_v, then x stays away from a_v'. As x approaches a_v, its distance from a_v' approaches the distance between a_v and a_v'. Hence it would add no greater generality to the statement if we took a product over several a_v for each v. Based on this observation, we have the following fact:

Theorem 3.45. *Let S be a finite subset of M_κ. For each $v \in S$, let $P_v(X)$ be a polynomial in $\kappa[X]$ (one variable) and assume that the multiplicity of their roots is at most r for some integer $r > 0$. Let $c > 0$ be a number, and $\varepsilon > 0$. Then there are only finitely many $x \in \kappa$ such that*

$$\prod_{v \in S} \min\{1, \|P_v(x)\|_v\} < \frac{c}{\bar{H}_\kappa(x)^{r(2+\varepsilon)}}. \tag{3.5.4}$$

Proof. We may assume that P_v has leading coefficient 1 for each $v \in S$, and say

$$P_v(X) = \prod_{j=0}^{q_v}(X - a_{vj})^{r_{vj}}$$

is a factorization in $\bar{\mathbb{Q}}$. The expression on the left-hand side of our inequality is greater than or equal to

$$\prod_{v \in S} \prod_{j=0}^{q_v} \min\{1, \|x - a_{vj}\|_v\}^{r_{vj}},$$

which is itself greater than or equal to the expression obtained by replacing r_{vj} by r for all v and j. Now we are in the situation of Theorem 3.44; taking into account the above remark following it, the solutions x of the inequality

$$\prod_{v \in S} \prod_{j=0}^{q_v} \min\{1, \|x - a_{vj}\|_v\}^r < \frac{c}{\bar{H}_\kappa(x)^{r(2+\varepsilon)}} \qquad (3.5.5)$$

are only finite in number, hence the same is true for the solutions of original inequality. $\qquad \square$

Particularly, take $r = 1$, $q_v = q$ for each $v \in S$. The inequality (3.5.5) implies that all but finitely many $x \in \kappa$ satisfy

$$\sum_{v \in S} \sum_{j=0}^{q} \log^+ \frac{1}{\|\|x - a_{vj}\|\|_v} \le (2 + \varepsilon)h(x) + O(1), \qquad (3.5.6)$$

which is similar to Theorem 2.56 with $n = 1$, that is, the inequality (3.5.6) is an analogue of Nevanlinna's second main theorem in number theory.

3.6 Schmidt's subspace theorem

Following Vojta [415], we simply introduce Schmidt's subspace theorem, and compare the analogy with Cartan's second main theorem.

Let κ be a number field and let S be a finite subset of M_κ containing the set S_κ^∞. Let $V = V_\kappa$ be a vector space of finite dimension $n + 1 > 0$ over κ. Take a basis $e = (e_0, \dots, e_n)$ of V. We will identify $V \cong \mathbb{A}^{n+1}(\kappa)$ by the correspondence relation

$$\xi_0 e_0 + \cdots + \xi_n e_n \in V \longmapsto (\xi_0, \dots, \xi_n) \in \mathbb{A}^{n+1}(\kappa), \ \xi_j \in \kappa.$$

A point $\xi = \xi_0 e_0 + \cdots + \xi_n e_n \in V$ is said to be an *S-integral point* if $\xi_i \in \mathcal{O}_{\kappa,S}$ for all $0 \le i \le n$. An algebraic point $\xi \in V_{\bar{\kappa}}$ should be *integral* if its coordinates lie in the integral closure of $\mathcal{O}_{\kappa,S}$ in $\bar{\kappa}$. Denote by $\mathcal{O}_{V,S}$ the set of S-*integral points* of V, that is,

$$\mathcal{O}_{V,S} = \{\xi \in V \mid \|\xi\|_\rho \leq 1, \ \rho \notin S\}. \tag{3.6.1}$$

According to the identity $V \cong \mathbb{A}^{n+1}(\kappa)$, we have

$$\mathcal{O}_{V,S} \cong \mathcal{O}_{\kappa,S}^{n+1}. \tag{3.6.2}$$

Similarly, an affine variety $Z \subset \mathbb{A}^{n+1}$ defined over κ inherits a notion of integral point from the definition for $\mathbb{A}^{n+1}$. We state the *Schmidt subspace theorem* as follows:

Theorem 3.46. *For $\rho \in S$, $i \in \{0, \ldots, n\}$, take $\alpha_{\rho,i} \in V^* - \{0\}$ such that for each $\rho \in S$, $\alpha_{\rho,0}, \ldots, \alpha_{\rho,n}$ are linearly independent. Fix $\varepsilon > 0$. Let Q be the set of all $\xi \in \mathcal{O}_{V,S}$ satisfying*

$$\prod_{\rho \in S} \prod_{i=0}^{n} \|\langle \xi, \alpha_{\rho,i}\rangle\|_\rho < \left\{\max_{\rho \in S} \|\xi\|_\rho\right\}^{-\varepsilon}.$$

Then Q is contained in a finite union of hyperplanes of V.

Theorem 3.46 is a generalization of Roth's theorem due to Schmidt [337] and Schlickewei ([334],[335],[336]) in the non-Archimedean case. The following subspace theorem will turn out to be equivalent to Theorem 3.46 (see [338], [415], [176] or Section 3.8.2):

Theorem 3.47. *Take $\varepsilon > 0$, $q \geq n$. Assume that for each $\rho \in S$, the family*

$$\mathscr{A}_\rho = \{a_{\rho,0}, \ldots, a_{\rho,q}\} \subset \mathbb{P}(V^*)$$

is in general position. Then there exists a finite set $\{b_1, \ldots, b_s\}$ of $\mathbb{P}(V_{\bar{\kappa}}^*)$ such that the inequality

$$\sum_{\rho \in S} \sum_{j=0}^{q} \log \frac{1}{\||x, a_{\rho,j}\||_\rho} < (n + 1 + \varepsilon)h(x)$$

holds for all $x \in \mathbb{P}(V) - \bigcup_i \ddot{E}[b_i]$.

Theorem 3.47 is an analogue of Theorem 2.56. A.J. van der Poorten [408] generalized an idea of Schlickewei [335] to obtain the following result:

Theorem 3.48. *Let κ be a number field and let $n \geq 1$ be an integer. Let Γ be a finitely generated subgroup of κ_*. Then all but finitely many solutions of the equation*

$$u_0 + u_1 + \cdots + u_n = 1, \ u_i \in \Gamma, \tag{3.6.3}$$

lie in one of the diagonal hyperplanes H_I defined by the equation $\sum_{i \in I} x_i = 0$, where I is a subset of $\{0, 1, \ldots, n\}$ with at least two, but no more than n, elements.

The proof is due to Vojta [415]. Further, Vojta noted that such infinite families are restricted to finite unions of linear subspaces of dimension $\leq [n/2]$.

According to Vojta [415], a set I of κ-rational points of a variety is called *degenerate* if it is not dense in the Zariski topology. Here the κ-rational points in the complement of I are called *non-degenerate* relative to I. We will regard $\mathbb{P}(V)$ as a "holomorphic curve" of $\mathbb{P}(V_{\bar{\kappa}})$, that is, the image of a mapping from κ into $\mathbb{P}(V_{\bar{\kappa}})$, and establish the theory of value distribution by integrating the characteristic functions over non-degenerate κ-rational points relative to I.

We fix a proper subset $I = I_\kappa$ of $\mathbb{P}(V_{\bar{\kappa}})$ which will be chosen later, and define a weight function of κ (relative to I) as follows:

$$\mu_\kappa(x) = \mu_{\kappa,I}(x) = \left\{ \begin{array}{ll} 1, & \text{if } x \in \mathbb{P}(V) - I; \\ 0, & \text{otherwise.} \end{array} \right. \tag{3.6.4}$$

We can define the *spherical height* of κ (relative to I) by

$$A_{\kappa,I}(r) = n_{\mu_\kappa h}(r) = \sum_{h(x) \leq r} \mu_\kappa(x) h(x), \tag{3.6.5}$$

and the *spherical characteristic function* of κ (relative to I) by

$$T_{\kappa,I}(r) = T_{\kappa,I}(r, r_0) = \int_{r_0}^{r} A_{\kappa,I}(t) \frac{dt}{t}. \tag{3.6.6}$$

If there is no confusion, we will abbreviate

$$A_\kappa(r) = A_{\kappa,I}(r), \quad T_\kappa(r) = T_{\kappa,I}(r).$$

Obviously, there are infinitely many non-degenerate κ-rational points on $\mathbb{P}(V_{\bar{\kappa}})$ relative to I if and only if

$$\lim_{r \to \infty} \frac{T_\kappa(r)}{\log r} = \lim_{r \to \infty} A_\kappa(r) = \infty. \tag{3.6.7}$$

Take $a \in \mathbb{P}(V^*)$ such that $\ddot{E}[a](\kappa) \subset I$. By definition, we have

$$m(x,a) = \sum_{\rho \in S} \log \frac{1}{|||x,a|||_\rho}, \quad x \notin \ddot{E}[a],$$

and

$$N(x,a) = \sum_{\rho \in M_\kappa - S} \log \frac{1}{|||x,a|||_\rho}, \quad x \notin \ddot{E}[a].$$

Hence the *integrated proximity function*

$$m_\kappa(r,a) = m_{\kappa,I}(r,a) = \int_{r_0}^{r} \left(\sum_{h(x) \leq t} \mu_\kappa(x) m(x,a) \right) \frac{dt}{t} \tag{3.6.8}$$

and the *integrated valence function*

$$N_\kappa(r,a) = N_{\kappa,I}(r,a) = \int_{r_0}^{r} \left(\sum_{h(x)\leq t} \mu_\kappa(x) N(x,a) \right) \frac{dt}{t} \qquad (3.6.9)$$

are well defined. Therefore the first main theorem (1.8.1) yields the *integrated first main theorem*

$$m_\kappa(r,a) + N_\kappa(r,a) = T_\kappa(r) + O\left(N_{\mu_\kappa}(r)\right). \qquad (3.6.10)$$

By the definitions, we obtain the relation

$$T_\kappa(r) = O(\log r) \Longleftrightarrow N_{\mu_\kappa}(r) = O(\log r).$$

If there are infinitely many non-degenerate κ-rational points on $\mathbb{P}(V_{\bar\kappa})$ relative to I, we have

$$\lim_{r\to\infty} \frac{N_{\mu_\kappa}(r)}{T_\kappa(r)} = \lim_{r\to\infty} \frac{n_{\mu_\kappa}(r)}{A_\kappa(r)} = 0, \qquad (3.6.11)$$

and hence (3.6.10) has the following form:

$$m_\kappa(r,a) + N_\kappa(r,a) = T_\kappa(r) + o\left(T_\kappa(r)\right). \qquad (3.6.12)$$

When $T_\kappa(r) \to \infty$ as $r \to \infty$, define the *defect* of κ for a (relative to I) by

$$\delta_\kappa(a) = \delta_{\kappa,I}(a) = \liminf_{r\to\infty} \frac{m_\kappa(r,a)}{T_\kappa(r)}. \qquad (3.6.13)$$

In particular, if $A_\kappa(r) \to \infty$ as $r \to \infty$, the integrated first main theorem (3.6.12) implies

$$\delta_\kappa(a) = 1 - \limsup_{r\to\infty} \frac{N_\kappa(r,a)}{T_\kappa(r)} \qquad (3.6.14)$$

with $0 \leq \delta_\kappa(a) \leq 1$.

Next we consider a family $\mathscr{A} = \{a_0,\ldots,a_q\} \subset \mathbb{P}(V^*)$ in general position for $q \geq n$. Theorem 3.47 gives that

$$\sum_{j=0}^{q} m(x,a_j) < (n+1+\varepsilon)h(x) \qquad (3.6.15)$$

for all $x \in \mathbb{P}(V) - \bigcup_i \ddot{E}[b_i]$. Further, if I contains all κ-rational points of the variety

$$\bigcup_{i,j}(\ddot{E}[b_i] \cup \ddot{E}[a_j]),$$

integrating (3.6.15), we obtain the *integrated second main theorem*

$$\sum_{j=0}^{q} m_\kappa(r, a_j) < (n + 1 + \varepsilon)T_\kappa(r). \qquad (3.6.16)$$

If $T_\kappa(r) \to \infty$ as $r \to \infty$, then (3.6.16) immediately yields the *defect relation*:

$$\sum_{j=0}^{q} \delta_\kappa(a_j) \leq n + 1. \qquad (3.6.17)$$

Problem 3.49. *When $q \geq 2n$, does $T_\kappa(r) = O(\log r)$ hold?*

When $\bar{\kappa} = \mathbb{C}$ and $q \geq 2n$, Corollary 2.74 implies that the complement of hyperplanes defined by elements in $\mathscr{A}$ is complete hyperbolic. Based on Conjecture 3.33, it seems that Problem 3.49 is sure. It is known that if $f \in \mathcal{M}(\mathbb{C})$ assumes the extreme defect relation

$$\sum \delta_f(a) = 2,$$

then f has order $\frac{p}{2}$, where p is a positive integer or $+\infty$, which was originally conjectured by F. Nevanlinna [289] and R. Nevanlinna [293]. Is there an analogue of this result for a number field κ?

3.7 Vojta's conjectures

Let κ be a number field and let S be a finite subset of M_κ. P. Vojta [415] observed that some conditions of the second main theorem (2.12.21) in Carlson-Griffiths-King's theory may be relaxed somewhat, and then translated (2.12.21) into the following *main conjecture* in number theory.

Conjecture 3.50. *Let X be a non-singular complete variety over κ. Let K be the canonical divisor of X, and let D be a normal crossings divisor on X. If $\varepsilon > 0$, and if E is a pseudo ample divisor, then there exists a proper Zariski closed subset $Z = Z(X, D, \kappa, E, \varepsilon, S)$ such that for all $x \in X(\kappa) - Z$,*

$$m(x, D) + h_K(x) \leq \varepsilon h_E(x) + O(1). \qquad (3.7.1)$$

The requirement in Conjecture 3.50 that D have only normal crossings is necessary, since it is easy to produce counterexamples if this condition is dropped.

Example 3.51. *Each hyperplane E of $\mathbb{P}^n(\bar{\kappa})$ is very ample with*

$$\dim \mathcal{L}(E) = n + 1,$$

and hence the dual classification mapping φ is the identity. Thus $h_E(x) = h(x)$, and hence

$$h_K(x) = -(n + 1)h(x)$$

since $K = -(n + 1)E$. *Let* $D = \sum_i \ddot{E}[a_i]$ *be a sum of hyperplanes in general position. Then the conjecture reduces to*

$$\sum_i m(x, a_i) < (n + 1 + \varepsilon)h(x)$$

which follows from Schmidt's subspace theorem.

P. Vojta [420] proved that Conjecture 3.25 of Masser and Oesterlé would be derived from a weakening of Conjecture 3.50.

Conjecture 3.50 implies the Bombieri-Lang Conjecture 3.36. Recall that a variety X is said to be pseudo canonical if its canonical bundle K_X is pseudo ample. Indeed, if X is such a variety, we may assume X non-singular since both the notion of pseudo canonicity and non-denseness in the Zariski topology are birational invariants. Then Conjecture 3.50 with $D = 0$ implies that

$$h_K \leq \varepsilon h_E + O(1) \tag{3.7.2}$$

on an open dense set. But taking $E = K$ and $\varepsilon < 1$ implies that h_K is bounded, which is a contradiction unless $X(\kappa)$ is contained in a Zariski closed subset of X, that is, $X(\kappa)$ is degenerate.

Conjecture 3.50 also implies the Mordell conjecture since curves are pseudo canonical if and only if their genus is at least 2, and degeneracy on curves reduces to finiteness.

Conjecture 3.50 is known for curves. Thus for curves of genus 0 it is equivalent to Roth's theorem, and for curves of genus $g \geq 2$ it is a consequence of Faltings' theorem. As for curves of genus 1, it is equivalent to Siegel's theorem.

Let A denote an Abelian variety and let D be an ample divisor on A. Lang [222] conjectured that A has only finitely many (S, D)-integralizable points. Vojta [415] showed that Lang's conjecture follows from the main conjecture. Qualitatively, Conjecture 3.50 also has the following simple consequence.

Conjecture 3.52. *Let X be a non-singular projective variety defined over a number field κ. Let K be the canonical divisor of X, and D a normal crossings divisor on X. Suppose that $K + D$ is pseudo ample. Then $X - D$ is pseudo Mordellic.*

Related to Conjecture 3.52 and based on the analogy with Conjecture 2.101, A. Levin [244] proposed a *main Siegel-type conjecture* as follows:

Conjecture 3.53. *Let X be a projective variety defined over a number field κ. Let $D = D_1 + \cdots + D_q$ be a divisor on X with the D_i's effective Cartier divisors for all i. Suppose that at most m D_i's meet at a point, so that the intersection of any $m+1$ distinct D_i's is empty. Suppose that $\dim D_i \geq n_0 > 0$ for all i. If $q > m + \frac{m}{n_0}$ then there does not exist a Zariski-dense set of κ-rational (S, D)-integralizable points on X.*

Siegel's theorem [357] is the case $m = n_0 = \dim X = 1$ of Conjecture 3.53, or see [67] for a new proof of Siegel's theorem. When X is a surface, $m \leq 2$, and the D_i's have no irreducible components in common, A. Levin [244] proved Conjecture 3.53. At the extreme of n_0, there is the following special case.

Conjecture 3.54. *Let X be a projective variety defined over a number field κ. Let $D = D_1 + \cdots + D_q$ be a divisor on X with the D_i's effective Cartier divisors for all i. Suppose that at most m D_i's meet at a point. If D_i is pseudo ample for all i and $q > m + \frac{m}{\dim X}$ then $X - D$ is pseudo Mordellic.*

When $q > 2m \dim X$, A. Levin [244] proved Conjecture 3.54 based on a formulation of Corvaja-Zannier theorems in [67] and [69]. Further, A. Levin [244] proved that $X - D$ is Mordellic if D_i is ample for all i. By using Theorem 2.103, when X is non-singular, the D_i's are ample, and $D = D_1 + \cdots + D_q$ has normal crossings, we see that Conjecture 3.54 is a consequence of Conjecture 3.52.

Following P. Vojta [415] one introduces the *absolute (logarithmic) discriminant*

$$d_\kappa = \frac{1}{[\kappa : \mathbb{Q}]} \log D_{\kappa/\mathbb{Q}}. \tag{3.7.3}$$

If $\mathbb{Q} \subset \kappa \subset K$ are finite separable algebraic extensions, then (1.3.21) implies

$$d_K - d_\kappa = \frac{1}{[K : \mathbb{Q}]} \log \mathbf{N}_{\kappa/\mathbb{Q}}(D_{K/\kappa}). \tag{3.7.4}$$

By abuse of notation, let $d(x) = d_{\kappa(x)}$ if x is a closed point of a variety. Vojta compares the discriminant term as follows:

Theorem 3.55. *Let $\pi : X \longrightarrow W$ be a generically finite separable surjective morphism of complete non-singular varieties over a number field κ, with ramification divisor R. Let S be a finite set of absolute values. Then for all $P \in X(\bar{\kappa}) - R$, we have*

$$d(P) - d(\pi(P)) \leq N(P, R) + O(1).$$

Vojta's Theorem 3.55 is a generalization to the ramified case of a classical theorem of Chevalley-Weil. Further, P. Vojta [415] proposed the following *general conjecture*:

Conjecture 3.56. *Let X be a complete non-singular variety over κ. Let K be the canonical divisor of X, and let D be a normal crossings divisor on X. Let $\varepsilon > 0$ and let E be a pseudo ample divisor. If $\pi : X \longrightarrow W$ is a finite surjective morphism to a non-singular complete variety W, then there exists a proper Zariski closed subset $Z = Z(\pi, X, D, \kappa, E, \varepsilon, S)$ such that for all $x \in X - Z$ with $\pi(x) \in W(\kappa)$,*

$$m(x, D) + h_K(x) \leq d(x) + \varepsilon h_E(x) + O(1). \tag{3.7.5}$$

P. Vojta in [415] showed that Conjecture 3.25 of Masser and Oesterlé is an easy consequence of Conjecture 3.56, and in [420] noted that Conjecture 3.50 is possibly weaker than Conjecture 3.56. Conversely, van Frankenhuysen [412] proved that the abc-conjecture implies Vojta's general conjecture for curves, i.e., when X is one-dimensional. Lang [232] conjectures that Vojta's general conjecture is best possible for any curve of non-zero genus over a number field.

A. Levin [244] gave the following conjectural upper bound on the logarithmic discriminant in terms of heights.

Conjecture 3.57. *Let X be a non-singular projective variety of dimension n defined over a number field κ with canonical divisor K. Let E be a pseudo ample divisor on X. Let r be a positive integer and let $\varepsilon > 0$. Then there exists a proper Zariski closed subset Z such that*

$$d(x) \leq h_K(x) + (2[\kappa(x) : \kappa] + n - 1 + \varepsilon)h_E(x) + O(1) \tag{3.7.6}$$

for all $x \in X(\bar{\kappa}) - Z$ with $[\kappa(x) : \kappa] \leq r$.

If E is ample, A. Levin [244] conjectured that the set Z in Conjecture 3.57 is empty. It is a result of Silverman [361] that Conjecture 3.57 is true for $X = \mathbb{P}^n$ with $\varepsilon = 0$ and $r = \infty$, i.e., the inequality holds for all $x \in X(\bar{\kappa})$. For a curve, Conjecture 3.57 is true by a result of Song and Tucker [370] (cf. Eq. 2.0.3). They proved the stronger statement.

Theorem 3.58. *Let X be a non-singular projective curve defined over a number field κ with canonical divisor K. Let E be an ample divisor on X. Let r be a positive integer and let $\varepsilon > 0$. Then*

$$d(x) \leq d_a(x) \leq h_K(x) + (2[\kappa(x) : \kappa] + \varepsilon)h_E(x) + O(1) \tag{3.7.7}$$

for all $x \in X(\bar{\kappa})$ with $[\kappa(x) : \kappa] \leq r$.

In the inequality (3.7.7), $d_a(x)$ is the arithmetic discriminant of x. For the definition and properties, see Vojta [416]. Related to the arithmetic discriminant, Vojta [417] proved the following generalization of Falting's theorem on rational points on curves.

Theorem 3.59. *Let X be a non-singular projective curve defined over a number field κ with canonical divisor K. Let D be an effective divisor on X with no multiple components and E ample divisor on X. Let r be a positive integer and let $\varepsilon > 0$. Then the inequality*

$$m(x, D) + h_K(x) \leq d_a(x) + \varepsilon h_E(x) + O(1) \tag{3.7.8}$$

holds for all $x \in X(\bar{\kappa}) - D$ with $[\kappa(x) : \kappa] \leq r$.

Generalizing the main Siegel-type conjecture, A. Levin [244] further proposed a *general Siegel-type conjecture* as follows:

Conjecture 3.60. *Let X be a projective variety defined over a number field κ. Let $D = D_1 + \cdots + D_q$ be a divisor on X with the D_i's effective Cartier divisors for all i. Suppose that at most m D_i's meet at a point. Suppose that $\dim D_i \geq n_0 > 0$ for all i. Let d be a positive integer. If $q > m + \frac{m(2d-1)}{n_0}$ then there does not exist a Zariski-dense set of (S, D)-integralizable points on X of degree d over κ.*

According to the definition, the *degree* of a set $R \subset X(\bar{\kappa})$ over κ is defined to be

$$\deg_\kappa R = \sup_{x \in R}[\kappa(x) : \kappa].$$

Based on Conjecture 3.57, A. Levin [244] shows that Vojta's general conjecture implies general Siegel-type conjecture if D_i is ample for all i and D has normal crossings. Theorem 3.58 and Theorem 3.59 imply that Levin's general Siegel-type conjecture is true for curves.

In [415], Vojta also proposed the following $(1,1)$-*form conjecture*:

Conjecture 3.61. *Let X be a complete non-singular variety over a number field κ contained in $\mathbb{C}$ and let D be a normal crossings divisor on X. Let ω be a positive $(1,1)$-form on $X - D$ whose holomorphic sectional curvatures are bounded from above by $-c < 0$, i.e., for any non-constant holomorphic mapping $f : U \longrightarrow X$ ($U \subseteq \mathbb{C}$ is an open subset), one has*

$$\mathrm{Ric}(f^*\omega) \geq cf^*\omega.$$

Also assume that $\omega \geq c_1(L, \rho)$ for some metric ρ on a line bundle L on X. Let E be a pseudo ample divisor on X. Let S be a finite set of absolute values. Let I be a set of (S, D)-integralizable points of bounded degree over κ. Let $\varepsilon > 0$. Then for all points $P \in I$ we have

$$h_L(P) \leq \frac{1}{c}d(P) + \varepsilon h_E(P) + O(1).$$

Vojta [415] applies the $(1,1)$-form conjecture to deduce several number theoretic applications in which he proves that Conjecture 3.61 implies a conjecture of Shafarevich on the finiteness of curves with good reduction, proved by Faltings [97], [98].

Assume that $X \subset \mathbb{P}^n$ is a non-singular projective variety over κ. Let h be the absolute logarithmic height on $\mathbb{P}^n$. Fix $r > r_0 > 0$. Let D be a divisor on X. We fix a proper subset I of $X(\kappa)$, and modify a little the weight function μ_κ of κ on $\mathbb{P}^n$ defined by (3.6.4), which we denote by $\mu_{\kappa,X}$ and define as follows:

$$\mu_{\kappa,X}(x) = \mu_{\kappa,I,X}(x) = \begin{cases} 1, & \text{if } x \in X(\kappa) - I; \\ 0, & \text{otherwise.} \end{cases} \qquad (3.7.9)$$

Note that $X(\kappa) \cap X[O; r]$ is finite, where $O = X \cap h^{-1}(0)$ is the *center* of h in X. We can define the *spherical height* of κ for D by

$$A_\kappa(r, D) = A_{\kappa, I}(r, D) = \sum_{x \in X[O; r]} \mu_{\kappa, X}(x) h_D(x), \tag{3.7.10}$$

and the *spherical characteristic function* of κ for D by

$$T_\kappa(r, D) = T_{\kappa, I}(r, D) = \int_{r_0}^r A_\kappa(t, D) \frac{dt}{t}. \tag{3.7.11}$$

To integrate the inequality in Conjecture 3.50, we have to assume $\mathrm{supp}(D)(\kappa) \subset I$. Thus the *integrated proximity function*

$$m_\kappa(r, D) = \int_{r_0}^r \left(\sum_{x \in X[O; t]} \mu_{\kappa, X}(x) m(x, D) \right) \frac{dt}{t} \tag{3.7.12}$$

and the *integrated valence function*

$$N_\kappa(r, D) = \int_{r_0}^r \left(\sum_{x \in X[O; t]} \mu_{\kappa, X}(x) N(x, D) \right) \frac{dt}{t} \tag{3.7.13}$$

are well defined. Therefore the first main theorem (1.8.5) yields the *integrated first main theorem*

$$m_\kappa(r, D) + N_\kappa(r, D) = T_\kappa(r, D) + O\left(N_{\mu_\kappa, X}(r) \right). \tag{3.7.14}$$

If X has only finitely many non-degenerate κ-rational points relative to I, then

$$T_\kappa(r, D) = O(\log r), \quad N_{\mu_\kappa, X}(r) = O(\log r).$$

If X has infinitely many non-degenerate κ-rational points relative to I, and if D is pseudo ample, then h_D is unbounded on these non-degenerate κ-rational points, and so

$$\lim_{r \to \infty} \frac{N_{\mu_\kappa, X}(r)}{T_\kappa(r, D)} = \lim_{r \to \infty} \frac{n_{\mu_\kappa, X}(r)}{A_\kappa(r, D)} = 0.$$

Conjecture 3.50 and (3.7.14) imply the following problem:

Conjecture 3.62. *Let X be a non-singular projective variety over κ. Let K be the canonical divisor of X, and let D be a normal crossings divisor on X. If $\varepsilon > 0$, and if E is a pseudo ample divisor, then there exists a proper Zariski closed subset $Z = Z(X, D, \kappa, E, \varepsilon, S)$ containing $\mathrm{supp}(D)$ such that*

$$m_\kappa(r, D) + T_\kappa(r, K) \le \varepsilon T_\kappa(r, E) + O(N_{\mu_\kappa, X}(r))$$

holds for $I = Z(\kappa)$.

For the covering case in Conjecture 3.56, we choose a proper subset I of $X(\kappa)$, and have to modify the weight function (3.7.9) as follows:

$$\mu_{\kappa,\pi}(x) = \mu_{\kappa,I,\pi}(x) = \begin{cases} 1, & \text{if } x \in X - I, \ \pi(x) \in W(\kappa); \\ 0, & \text{otherwise.} \end{cases} \qquad (3.7.15)$$

We will assume that $W \subset \mathbb{P}^n$ is a non-singular projective variety. Let h be the absolute logarithmic height on $\mathbb{P}^n$. Fix $r > r_0 > 0$. Thus the set

$$W[O; r] = \{w \in W \mid h(w) \leq r\}$$

is well defined, where $O = W \cap h^{-1}(0)$ is the *center* of h in W. The *spherical height* of κ for (D, π) is modified as follows:

$$A_{\kappa,\pi}(r, D) = \sum_{\pi(x) \in W[O;r]} \mu_{\kappa,\pi}(x) h_D(x). \qquad (3.7.16)$$

The *spherical characteristic function* of κ for (D, π) still keeps the form

$$T_{\kappa,\pi}(r, D) = \int_{r_0}^{r} A_{\kappa,\pi}(t, D) \frac{dt}{t}. \qquad (3.7.17)$$

If $\mathrm{supp}(D)(\kappa) \subset I$, the *integrated proximity function*

$$m_{\kappa,\pi}(r, D) = \int_{r_0}^{r} \left(\sum_{\pi(x) \in W[O;t]} \mu_{\kappa,\pi}(x) m(x, D) \right) \frac{dt}{t} \qquad (3.7.18)$$

and the *integrated valence function*

$$N_{\kappa,\pi}(r, D) = \int_{r_0}^{r} \left(\sum_{\pi(x) \in W[O;t]} \mu_{\kappa,\pi}(x) N(x, D) \right) \frac{dt}{t} \qquad (3.7.19)$$

are well defined. Therefore the first main theorem (1.8.5) yields the *integrated first main theorem*

$$m_{\kappa,\pi}(r, D) + N_{\kappa,\pi}(r, D) = T_{\kappa,\pi}(r, D) + O\left(N_{\mu_{\kappa,\pi}}(r)\right). \qquad (3.7.20)$$

Further, set

$$d_{\kappa,\pi}(r) = \int_{r_0}^{r} \left(\sum_{\pi(x) \in W[O;t]} \mu_{\kappa,\pi}(x) d(x) \right) \frac{dt}{t}.$$

Conjecture 3.56 and (3.7.20) imply the following problem:

Conjecture 3.63. *Let X be a projective non-singular variety over κ. Let K be the canonical divisor of X, and let D be a normal crossings divisor on X. Let $\varepsilon > 0$ and let E be a pseudo ample divisor. If $\pi : X \longrightarrow W$ is a finite surjective morphism to a non-singular projective variety W, then there exists a proper Zariski closed subset $Z = Z(\pi, X, D, \kappa, E, \varepsilon, S)$ containing $\mathrm{supp}(D)$ such that*

$$m_{\kappa,\pi}(r, D) + T_{\kappa,\pi}(r, K) \le d_{\kappa,\pi}(r) + \varepsilon T_{\kappa,\pi}(r, E) + O\left(N_{\mu_{\kappa,\pi}}(r)\right)$$

holds for $I = Z(\kappa)$.

3.8   Subspace theorems on hypersurfaces

In this section, we will give an elegant example of formulation between Diophantine approximation and Nevanlinna theory, which shows how to translate Theorem 2.63 in value distribution into a subspace theorem on hypersurfaces, that is, Theorem 3.65.

3.8.1   Main results and problems

Let κ be a number field and let $V = V_\kappa$ be a normed vector space of dimension $n + 1 > 0$ over κ. Let E be a hyperplane on $\mathbb{P}(V_{\bar\kappa})$. Take a positive integer d. The dual classification mapping

$$\varphi_{dE} : \mathbb{P}(V_{\bar\kappa}) \longrightarrow \mathbb{P}(\amalg_d V_{\bar\kappa})$$

is just the *Veronese mapping*, that is,

$$\varphi_{dE}(x) = x^{\amalg d}.$$

Then the absolute (multiplicative) height of $x \in \mathbb{P}(V_{\bar\kappa})$ for dE is given by

$$H_{dE}(x) = H(\varphi_{dE}(x)) = H\left(x^{\amalg d}\right) = H(x)^d,$$

and the absolute (logarithmic) height of x for dE is given as

$$h_{dE}(x) = h(\varphi_{dE}(x)) = dh(x).$$

Take $\alpha \in \amalg_d V^*$ with $(\alpha) = dE$ and set $a = \mathbb{P}(\alpha)$. For $x \notin dE$, the *proximity function* $m(x, dE)$ is given by

$$m(x, dE) = m\left(x^{\amalg d}, a\right).$$

Similarly, the *valence function* is given by

$$N(x, dE) = N\left(x^{\amalg d}, a\right).$$

Thus equation (1.8.4) yields the following *first main theorem*:

$$m\left(x^{\mathrm{II}d}, a\right) + N\left(x^{\mathrm{II}d}, a\right) = dh(x) + O(1). \tag{3.8.1}$$

Fix $r > r_0 > 0$. If $\ddot{E}^d[a](\kappa) \subset I$, the *integrated proximity function*

$$m_\kappa(r, a) = m_{\kappa, I}(r, a) = \int_{r_0}^r \left(\sum_{h(x) \le t} \mu_\kappa(x) m\left(x^{\mathrm{II}d}, a\right) \right) \frac{dt}{t} \tag{3.8.2}$$

and the *integrated valence function*

$$N_\kappa(r, a) = N_{\kappa, I}(r, a) = \int_{r_0}^r \left(\sum_{h(x) \le t} \mu_\kappa(x) N\left(x^{\mathrm{II}d}, a\right) \right) \frac{dt}{t} \tag{3.8.3}$$

are well defined, where μ_κ is defined by (3.6.4). Similar to the proofs of (3.6.10), we obtain the *integrated first main theorem*

$$m_\kappa(r, a) + N_\kappa(r, a) = dT_\kappa(r) + O\left(N_{\mu_\kappa}(r)\right). \tag{3.8.4}$$

By using Theorem 3.47, we can obtain directly the following result:

Theorem 3.64. *Take $\varepsilon > 0$, $q \ge N = \binom{n+d}{d}$. Assume that for each $\rho \in S$, a family*

$$\{a_{\rho,1}, \ldots, a_{\rho,q}\} \subset \mathbb{P}(\mathrm{II}_d V^*)$$

is in general position. Then there exists a finite set $\{b_1, \ldots, b_s\}$ of $\mathbb{P}(\mathrm{II}_d V_{\bar{\kappa}}^*)$ such that the inequality

$$\sum_{\rho \in S} \sum_{j=1}^q \log \frac{1}{\||x^{\mathrm{II}d}, a_{\rho,j}\||_\rho} < d(N + \varepsilon)h(x) + O(1)$$

holds for all $x \in \mathbb{P}(V) - \bigcup_i \ddot{E}^d[b_i]$.

If the families in general position in Theorem 3.64 are replaced by admissible families, then N in the bound of the main inequality of Theorem 3.64 can be replaced by $n+1$, that is, we have the following *subspace theorem on hypersurfaces*:

Theorem 3.65 ([187]). *Take $\varepsilon > 0$, $q \ge n$. Assume that for $\rho \in S$, a family*

$$\mathscr{A}_\rho = \{a_{\rho,0}, \ldots, a_{\rho,q}\} \subset \mathbb{P}(\mathrm{II}_d V^*)$$

is admissible. Then there exist points

$$b_i \in \mathbb{P}(\mathrm{II}_{d_i} V_{\bar{\kappa}}^*) \ (1 \le d_i \in \mathbb{Z}, \ i = 1, \ldots, s < \infty)$$

such that the inequality

$$\sum_{\rho \in S} \sum_{j=0}^{q} \log \frac{1}{|||x^{\mathrm{II}d}, a_{\rho,j}|||_\rho} < d(n+1+\varepsilon)h(x) + O(1)$$

holds for all $x \in \mathbb{P}(V) - \bigcup_i \ddot{E}^{d_i}[b_i].$

Originally, Theorem 3.65 was a conjecture proposed by Hu and Yang [176] (or see [181]) based on analogy with Theorem 2.63. It extends the Schmidt subspace theorem. We will introduce the proof in Section 3.8.3 by using methods of P. Corvaja and U. Zannier [68]. In particular, Theorem 3.65 implies the following integrated form:

Theorem 3.66. *Take* $\varepsilon > 0$, $q \geq n$. *Assume that a family*

$$\mathscr{A} = \{a_0, \ldots, a_q\} \subset \mathbb{P}(\mathrm{II}_d V^*)$$

is admissible. Then for any $\varepsilon > 0$, *there exists a proper Zariski closed subset* $Z \subset \mathbb{P}(V_{\bar{\kappa}})$ *containing* $\bigcup_j \ddot{E}^d[a_j]$ *such that*

$$\sum_{j=0}^{q} m_\kappa(r, a_j) < d(n+1+\varepsilon)T_\kappa(r)$$

holds for $I = Z(\kappa)$.

Theorem 3.67 immediately yields the *defect relation*:

$$\sum_{j=0}^{q} \delta_\kappa(a_j) \leq n+1, \tag{3.8.5}$$

where the *defect* of κ for a_j is defined by

$$\delta_\kappa(a_j) = \liminf_{r \to \infty} \frac{m_\kappa(r, a_j)}{dT_\kappa(r)}. \tag{3.8.6}$$

In particular, if $A_\kappa(r) \to \infty$ as $r \to \infty$, the integrated first main theorem (3.8.4) implies

$$\delta_\kappa(a_j) = 1 - \limsup_{r \to \infty} \frac{N_\kappa(r, a_j)}{dT_\kappa(r)} \tag{3.8.7}$$

with $0 \leq \delta_\kappa(a_j) \leq 1$.

Theorem 3.46 implies the following result:

Theorem 3.67. *For* $\rho \in S$, $i \in \{1, \ldots, N\}$, *where* $N = \binom{n+d}{d}$, *take* $\alpha_{\rho,i} \in \mathrm{II}_d V^* - \{0\}$ *such that for each* $\rho \in S$, $\alpha_{\rho,1}, \ldots, \alpha_{\rho,N}$ *are linearly independent. Then for any* $\varepsilon > 0$ *there exists a finite set* $\{b_1, \ldots, b_s\}$ *of* $\mathbb{P}(\mathrm{II}_d V_{\bar{\kappa}}^*)$ *such that the inequality*

$$\prod_{\rho \in S} \prod_{i=1}^{N} \frac{1}{||\langle \xi^{\mathrm{II}d}, \alpha_{\rho,i} \rangle ||_\rho} \leq \left\{ \max_{\rho \in S} ||\xi||_\rho \right\}^\varepsilon$$

holds for all S-*integral points* $\xi \in \mathcal{O}_{V,S} - \bigcup_i E^d[b_i].$

Related to Theorem 3.67, we have:

Theorem 3.68 ([187]). *For $\rho \in S$, $i = 0, \ldots, n$, take $\alpha_{\rho,i} \in \amalg_d V^* - \{0\}$ such that the system*

$$\langle \xi^{\amalg d}, \alpha_{\rho,i} \rangle = 0, \quad i = 0, \ldots, n$$

has only the trivial solution $\xi = 0$ in $V_{\bar{\kappa}}$. Then for any $\varepsilon > 0$ there exist points

$$b_i \in \mathbb{P}(\amalg_{d_i} V_{\bar{\kappa}}^*) \ (1 \le d_i \in \mathbb{Z}, \ i = 1, \ldots, s < \infty)$$

such that the inequality

$$\prod_{\rho \in S} \prod_{i=0}^{n} \frac{1}{\|\langle \xi^{\amalg d}, \alpha_{\rho,i} \rangle\|_\rho} \le \left\{ \max_{\rho \in S} \|\xi\|_\rho \right\}^\varepsilon$$

holds for all S-integral points $\xi \in \mathcal{O}_{V,S} - \bigcup_i E^{d_i}[b_i]$.

Hu and Yang suggested Theorem 3.68 in [176] (or see [181]). We will prove that Theorem 3.65 is equivalent to Theorem 3.68.

Conjecture 3.50 corresponds to the following conjecture:

Conjecture 3.69. *Take $a_i \in \mathbb{P}(\amalg_d V_{\bar{\kappa}}^*)$ such that $\sum_i \ddot{E}^d[a_i]$ has normal crossings. Then for $\varepsilon > 0$ there exists a proper Zariski closed subset Z such that for all $x \in \mathbb{P}(V) - Z$,*

$$\sum_i m\left(x^{\amalg d}, a_i\right) \le (n + 1 + \varepsilon) h(x).$$

Conjecture 3.69 or Conjecture 3.62 implies the following integrated form:

Conjecture 3.70. *Take $a_i \in \mathbb{P}(\amalg_d V_{\bar{\kappa}}^*)$ such that $\sum_i \ddot{E}^d[a_i]$ have normal crossings. Then for any $\varepsilon > 0$, there exists a proper Zariski closed subset $Z \subset \mathbb{P}(V_{\bar{\kappa}})$ containing $\bigcup_i \ddot{E}^d[a_i]$ such that*

$$\sum_i m_\kappa(r, a_i) \le (n + 1 + \varepsilon) T_\kappa(r)$$

holds for $I = Z(\kappa)$.

Finally, we translate Conjecture 2.76 into the following conjecture:

Conjecture 3.71. *Take a positive real number $\varepsilon > 0$ and integers $q \ge n \ge r \ge 1$. Assume that for $\rho \in S$, a family*

$$\mathscr{A}_\rho = \{a_{\rho,0}, \ldots, a_{\rho,q}\} \subset \mathbb{P}(\amalg_d V^*)$$

is admissible. Then the set of points of $\mathbb{P}(V) - \bigcup \ddot{E}^d[a_{\rho,j}]$ satisfying

$$\sum_{\rho \in S} \sum_{j=0}^{q} \log \frac{1}{\||x^{\amalg d}, a_{\rho,j}|\|_\rho} \ge d(2n - r + 1 + \varepsilon) h(x) + O(1)$$

is contained in a finite union of subvarieties of dimension $\le r - 1$ of $\mathbb{P}(V_{\bar{\kappa}})$.

In [176] (or see [181]), we proposed this conjecture for the case $r = 1$.

3.8.2   Equivalence of Theorems 3.65 and 3.68

We will need some basic facts. Take a positive integer $q \geq n$ and take an admissible family $\mathscr{A} = \{a_0, a_1, \ldots, a_q\}$ of points $a_j \in \mathbb{P}\left(\amalg_{d_j} V_{\bar{\kappa}}^*\right)$. Let $|\cdot|$ be a norm defined over a base $e = (e_0, \ldots, e_n)$ of V. Write $\xi = \xi_0 e_0 + \cdots + \xi_n e_n$. By Theorem 1.94, for each $k \in \{0, \ldots, n\}$, the identity

$$\xi_k^s = \sum_{i=0}^n b_{ik}^\lambda(\xi) \tilde{\alpha}_{\lambda(i)}(\xi) \quad (\lambda \in J_n^q) \tag{3.8.8}$$

is satisfied for some natural number s with

$$s \geq d = \max_{0 \leq j \leq q} d_j,$$

where $b_{ik}^\lambda \in \bar{\kappa}[\xi_0, \ldots, \xi_n]$ are homogeneous polynomials of degree $s - d_{\lambda(i)}$ whose coefficients are integral-valued polynomials at the coefficients of $\tilde{\alpha}_{\lambda(i)}$ $(i = 0, \ldots, n)$. Write

$$b_{ik}^\lambda(\xi) = \sum_{\sigma \in J_{n,s-d_{\lambda(i)}}} b_{\sigma ik}^\lambda \xi_0^{\sigma(0)} \cdots \xi_n^{\sigma(n)}, \quad b_{\sigma ik}^\lambda \in \bar{\kappa}. \tag{3.8.9}$$

First of all, assume that the norm $|\cdot|$ is non-Archimedean. From (3.8.8) and (3.8.9), we have

$$|\xi_k|^s \leq \left(\max_{i,\sigma} |b_{\sigma ik}^\lambda| \cdot |\alpha_{\lambda(i)}| \right) \max_{0 \leq i \leq n} \left\{ \frac{|\tilde{\alpha}_{\lambda(i)}(\xi)|}{|\xi|^{d_{\lambda(i)}} |\alpha_{\lambda(i)}|} \right\} |\xi|^s. \tag{3.8.10}$$

Note that

$$\max_{0 \leq k \leq n} |\xi_k|^s = |\xi|^s, \quad |\tilde{\alpha}_j(\xi)| \leq |\xi|^{d_j} |\alpha_j|. \tag{3.8.11}$$

By maximizing the inequalities (3.8.10) over k, $0 \leq k \leq n$, and using (3.8.11), we obtain

$$1 \leq \max_{k,i,\sigma} |b_{\sigma ik}^\lambda| \cdot |\alpha_{\lambda(i)}|. \tag{3.8.12}$$

Define the *gauge*

$$\Gamma(\mathscr{A}) = \min_{\lambda \in J_n^q} \min_{k,i,\sigma} \left\{ \frac{1}{|b_{\sigma ik}^\lambda| \cdot |\alpha_{\lambda(i)}|} \right\}, \tag{3.8.13}$$

with $0 < \Gamma(\mathscr{A}) \leq 1$. From (3.8.10), (3.8.11) and (3.8.13), we obtain

$$\Gamma(\mathscr{A}) \leq \max_{0 \leq i \leq n} \left\{ \frac{|\tilde{\alpha}_{\lambda(i)}(\xi)|}{|\xi|^{d_{\lambda(i)}} |\alpha_{\lambda(i)}|} \right\},$$

that is,

$$\Gamma(\mathscr{A}) \leq \max_{0 \leq i \leq n} |x^{\amalg d_{\lambda(i)}}, a_{\lambda(i)}|, \quad \lambda \in J_n^q, \quad x \in \mathbb{P}(V_{\bar{\kappa}}). \tag{3.8.14}$$

If the norm $|\cdot|$ is Archimedean, now (3.8.8) and (3.8.9) imply

$$|\xi_k|^s \leq \left(\sum_{i=0}^{n} \sum_{\sigma} |b_{\sigma ik}^{\lambda}| \cdot |\alpha_{\lambda(i)}| \right) \max_{0 \leq i \leq n} \left\{ \frac{|\tilde{\alpha}_{\lambda(i)}(\xi)|}{|\xi|_*^{d_{\lambda(i)}} |\alpha_{\lambda(i)}|} \right\} |\xi|_*^s, \qquad (3.8.15)$$

where $|\xi|_* = \max_k |\xi_k|$. Without loss of generality, we may assume

$$|\xi| = (|\xi_0|^2 + \cdots + |\xi_n|^2)^{\frac{1}{2}}.$$

Since $|\xi| \leq \sqrt{n+1}|\xi|_*$, inequality (3.8.15) yields

$$1 \leq (n+1)^{\frac{d}{2}} \max_k \sum_{i=0}^{n} \sum_{\sigma} |b_{\sigma ik}^{\lambda}| \cdot |\alpha_{\lambda(i)}|. \qquad (3.8.16)$$

Define the *gauge*

$$\Gamma(\mathscr{A}) = (n+1)^{-\frac{d}{2}} \min_{\lambda \in J_n^q} \min_k \left\{ \sum_{i=0}^{n} \sum_{\sigma} |b_{\sigma ik}^{\lambda}| \cdot |\alpha_{\lambda(i)}| \right\}^{-1}, \qquad (3.8.17)$$

with $0 < \Gamma(\mathscr{A}) \leq 1$. From (3.8.15) and (3.8.17), we also obtain the inequality (3.8.14).

Lemma 3.72. *For* $x \in \mathbb{P}(V_{\bar{\kappa}}), 0 < r \in \mathbb{R}$, *define*

$$\mathscr{A}(x,r) = \{j \mid |x^{\amalg d_j}, a_j| < r, \ 0 \leq j \leq q\}. \qquad (3.8.18)$$

If $0 < r \leq \Gamma(\mathscr{A})$, *then* $\#\mathscr{A}(x,r) \leq n$.

Proof. Assume that $\#\mathscr{A}(x,r) \geq n+1$. Then $\lambda \in J_n^q$ exists such that

$$\{\lambda(0), \ldots, \lambda(n)\} \subseteq \mathscr{A}(x,r).$$

Hence

$$|x^{\amalg d_{\lambda(i)}}, a_{\lambda(i)}| < r \leq \Gamma(\mathscr{A}), \quad i = 0, \ldots, n,$$

which is impossible according to (3.8.14). $\qquad \square$

Lemma 3.73. *Take* $x \in \mathbb{P}(V_{\bar{\kappa}})$ *such that* $|x^{\amalg d_j}, a_j| > 0$ *for* $j = 0, \ldots, q$. *Then*

$$\prod_{j=0}^{q} \frac{1}{|x^{\amalg d_j}, a_j|} \leq \left(\frac{1}{\Gamma(\mathscr{A})} \right)^{q-n} \max_{\lambda \in J_n^q} \left\{ \prod_{i=0}^{n} \frac{1}{|x^{\amalg d_{\lambda(i)}}, a_{\lambda(i)}|} \right\} \qquad (3.8.19)$$

$$\leq \left(\frac{1}{\Gamma(\mathscr{A})} \right)^{q+1-n} \max_{\lambda \in J_{n-1}^q} \left\{ \prod_{i=0}^{n-1} \frac{1}{|x^{\amalg d_{\lambda(i)}}, a_{\lambda(i)}|} \right\}. \qquad (3.8.20)$$

Proof. Take $r = \Gamma(\mathscr{A})$. Lemma 3.72 implies $\#\mathscr{A}(x,r) \leq n$. Thus $\sigma \in J_n^q$ exists such that $\mathscr{A}(x,r) \subseteq \{\sigma(0), \ldots, \sigma(n)\}$. Note that $\mathrm{Im}\lambda - \mathscr{A}(x,r) \neq \emptyset$ for any $\lambda \in J_n^q$.

Then we have

$$\prod_{j=0}^{q} \frac{1}{|x^{\amalg d_j}, a_j|} \leq r^{n-q} \prod_{i=0}^{n} \frac{1}{|x^{\amalg d_{\sigma(i)}}, a_{\sigma(i)}|}$$

$$\leq \left(\frac{1}{\Gamma(\mathscr{A})}\right)^{q-n} \max_{\lambda \in J_n^q} \left\{ \prod_{i=0}^{n} \frac{1}{|x^{\amalg d_{\lambda(i)}}, a_{\lambda(i)}|} \right\}$$

$$\leq \left(\frac{1}{\Gamma(\mathscr{A})}\right)^{q+1-n} \max_{\lambda \in J_{n-1}^q} \left\{ \prod_{i=0}^{n-1} \frac{1}{|x^{\amalg d_{\lambda(i)}}, a_{\lambda(i)}|} \right\}. \qquad \square$$

We will need the following fact (see [176]):

Lemma 3.74. *For $x \in \mathbb{P}(V)$, we can choose $\xi \in \mathcal{O}_{V,S}$ such that $x = \mathbb{P}(\xi)$, and the relative height of x satisfies*

$$\max\left\{ \max_{\rho \in S} \|\xi\|_\rho, \prod_{\rho \in S} \|\xi\|_\rho \right\} \leq c H_\kappa(x) \leq c \left\{ \max_{\rho \in S} \|\xi\|_\rho \right\}^{\#S},$$

where c is a constant depending only on S but independent of x.

Obviously, Theorem 3.65 yields immediately Theorem 3.68 by taking $q = n$ and using Lemma 3.74. Conversely, Theorem 3.68 implies also Theorem 3.65. In fact, by Lemma 3.73 and Theorem 3.68, there exist points

$$b_i \in \mathbb{P}(\amalg_{d_i} V_{\bar{\kappa}}^*) \ (1 \leq d_i \in \mathbb{Z}, \ i = 1, \ldots, s < \infty)$$

such that the inequality

$$\prod_{\rho \in S} \prod_{j=0}^{q} \frac{1}{\||x^{\amalg d}, a_{\rho,j}\||_\rho} \leq \prod_{\rho \in S} \left\{ \left(\frac{1}{\Gamma(\mathscr{A}_\rho)}\right)^{q-n} \prod_{j=0}^{n} \frac{1}{\||x^{\amalg d}, a_{\rho,\sigma_\rho(j)}\||_\rho} \right\}$$

$$\leq c_1 \left(\prod_{\rho \in S} \||\xi\||_\rho^d\right)^{n+1} \left(\prod_{\rho \in S} \prod_{j=0}^{n} \frac{1}{\||\langle \xi^{\amalg d}, \alpha_{\rho,\sigma_\rho(j)}\rangle\||_\rho}\right)$$

$$\leq c_1 \left(\prod_{\rho \in S} \||\xi\||_\rho\right)^{d(n+1)} \left(\max_{\rho \in S} \||\xi\||_\rho\right)^{\varepsilon},$$

holds for all points $x = \mathbb{P}(\xi) \in \mathbb{P}(V) - \bigcup_i \ddot{E}^{d_i}[b_i]$, where c_1 is constant, and $\alpha_{\rho,j} \in V^* - \{0\}$ with $a_{\rho,j} = \mathbb{P}(\alpha_{\rho,j})$. By Lemma 3.74, there exists a constant c_2 such that

$$\prod_{\rho \in S} \prod_{j=0}^{q} \frac{1}{\||x^{\amalg d}, a_{\rho,j}\||_\rho} \leq c_2 H(x)^{d(n+1)+\varepsilon},$$

and hence Theorem 3.65 follows.

3.8.3   Proof of Theorem 3.65

In this section, we prove Theorem 3.65 based on methods of P. Corvaja and U. Zannier [68]. First of all, we state several lemmas from [68] (or see [330]). We shall use the *lexicographic ordering* on the p-tuples $\nu = (\nu(1), \dots, \nu(p)) \in \mathbb{Z}_+^p$, namely, $\mu > \nu$ if and only if for some $l \in \{1, \dots, p\}$ we have $\mu(k) = \nu(k)$ for $k < l$ and $\mu(l) > \nu(l)$.

Lemma 3.75. *Let A be a commutative ring and let $\{g_1, \dots, g_p\}$ be a regular sequence in A generating the ideal $I_p \subset A$. Suppose that for some $y, x_1, \dots, x_h \in A$ we have an equation*

$$g_1^{\nu(1)} \cdots g_p^{\nu(p)} y = \sum_{k=1}^{h} g_1^{\mu_k(1)} \cdots g_p^{\mu_k(p)} x_k,$$

where $\mu_k > \nu$ for $k = 1, \dots, h$. Then $y \in I_p$.

Proof. We prove Lemma 3.75 by induction on p. Since g_1 is not a zero divisor in A, the assertion is trivial for $p = 1$. Assume that $p > 1$ and that Lemma 3.75 is true up to $p - 1$. Since $\mu_k > \nu$ for $k = 1, \dots, h$, renumbering the indices $1, \dots, h$ we may assume that

$$\mu_k(1) \begin{cases} > \nu(1), & k = 1, \dots, s; \\ = \nu(1), & k = s+1, \dots, h \end{cases}$$

for some $0 \le s \le h$. Since g_1 is not a zero divisor in A we may write

$$g_2^{\nu(2)} \cdots g_p^{\nu(p)} y = g_1 b + \sum_{k=s+1}^{h} g_2^{\mu_k(2)} \cdots g_p^{\mu_k(p)} x_k, \quad b \in A.$$

Reducing modulo g_1, denoting the reduction with a bar, and working in the ring A/I_1, we have

$$\bar{g}_2^{\nu(2)} \cdots \bar{g}_p^{\nu(p)} \bar{y} = \sum_{k=s+1}^{h} \bar{g}_2^{\mu_k(2)} \cdots \bar{g}_p^{\mu_k(p)} \bar{x}_k.$$

Note that $(\mu_k(2), \dots, \mu_k(p)) > (\nu(2), \dots, \nu(p))$ for $k = s+1, \dots, h$ and that $\{\bar{g}_2, \dots, \bar{g}_p\}$ is a regular sequence in A/I_1. We may apply the inductive assumption with $p - 1$ in place of p and A/I_1 in place of A. Then $\bar{y}$ lies in the ideal of A/I_1 generated by $\bar{g}_2, \dots, \bar{g}_p$, i.e., $y \in I_p$, as required.   □

Let κ be a number field and let $\bar{\kappa}$ be an algebraic closure of κ. Let $V = V_{\bar{\kappa}}$ be a normed vector space of dimension $n + 1 > 0$ over $\bar{\kappa}$.

Lemma 3.76. *Let $\tilde{\beta}_1, \dots, \tilde{\beta}_p$ be homogeneous polynomials in $\bar{\kappa}[\xi_0, \dots, \xi_n]$. Assume that they define a subvariety of $\mathbb{P}(V)$ of dimension $n - p$. Then $\{\tilde{\beta}_1, \dots, \tilde{\beta}_p\}$ is a regular sequence.*

Proof. This follows from Hilbert's basis theorem and Proposition 1.3.   □

Lemma 3.77. *Let $\tilde{\beta}_1, \ldots, \tilde{\beta}_n$ be homogeneous polynomials in $\bar{\kappa}[\xi_0, \ldots, \xi_n]$. Assume that they define a subvariety of $\mathbb{P}(V)$ of dimension 0. Then, for all large N,*

$$\dim V_{[N]} / \{(\tilde{\beta}_1, \ldots, \tilde{\beta}_n) \cap V_{[N]}\} = \deg(\tilde{\beta}_1) \cdots \deg(\tilde{\beta}_n).$$

Proof. It is a classical fact from the theory of Hilbert polynomials that the dimension of the quotient in Lemma 3.77 is constant for large N, equal to the degree of the variety defined by $\tilde{\beta}_1, \ldots, \tilde{\beta}_n$ (see [148], Ch. I.7) which is just the product of the degrees of $\tilde{\beta}_i$ (see [342], Ch. IV).   □

Take $\rho \in S$ and take a positive integer d. Let $\mathscr{A}_\rho = \{a_{\rho,j}\}_{j=0}^q$ be a finite admissible family of points $a_{\rho,j} \in \mathbb{P}(\mathrm{II}_d V_\kappa^*)$ with $q \geq n$. Take $\alpha_{\rho,j} \in \mathrm{II}_d V_\kappa^* - \{0\}$ with $\mathbb{P}(\alpha_{\rho,j}) = a_{\rho,j}$, and define

$$\tilde{\alpha}_{\rho,j}(\xi) = \langle \xi^{\mathrm{II}d}, \alpha_{\rho,j} \rangle, \quad \xi \in V, \ j = 0, 1, \ldots, q.$$

Without loss of generality, assume $|\alpha_{\rho,j}|_\rho = 1$ for $j = 0, \ldots, q$. Lemma 3.73 implies

$$\prod_{j=0}^q \frac{1}{\|x^{\mathrm{II}d}, a_{\rho,j}\|_\rho} \leq \left(\frac{1}{\Gamma(\mathscr{A}_\rho)}\right)^{q+1-n} \max_{\lambda \in J_{n-1}^q} \prod_{i=0}^{n-1} \frac{1}{\|x^{\mathrm{II}d}, a_{\rho,\lambda(i)}\|_\rho} \qquad (3.8.21)$$

for $x \in \mathbb{P}(V_\kappa) - \cup_{j=0}^q \ddot{E}^d[a_{\rho,j}]$. P. Corvaja and U. Zannier [68] estimated the terms on the right-hand side of (3.8.21) as follows.

Now pick $\lambda \in J_{n-1}^q$. Since $\mathscr{A}_\rho$ is admissible, $\tilde{\alpha}_{\rho,\lambda(0)}, \ldots, \tilde{\alpha}_{\rho,\lambda(n-1)}$ define a subvariety of $\mathbb{P}(V)$ of dimension 0. Take a multi-index $\nu = (\nu(1), \ldots, \nu(n)) \in \mathbb{Z}_+^n$ with the length

$$|\nu| = \nu(1) + \cdots + \nu(n) \leq \frac{N}{d}.$$

For any $\gamma = (\gamma(1), \ldots, \gamma(n)) \in \mathbb{Z}_+^n$, abbreviate

$$\tilde{\alpha}_{\rho,\lambda}^\gamma = \tilde{\alpha}_{\rho,\lambda(0)}^{\gamma(1)} \cdots \tilde{\alpha}_{\rho,\lambda(n-1)}^{\gamma(n)}$$

and define the spaces

$$\mathbf{V}_{N,\nu} = \sum_{\gamma \geq \nu} \tilde{\alpha}_{\rho,\lambda}^\gamma V_{[N-d|\gamma|]}$$

with $\mathbf{V}_{N,0} = V_{[N]}$ and $\mathbf{V}_{N,\mu} \subset \mathbf{V}_{N,\nu}$ if $\mu > \nu$. Thus the $\mathbf{V}_{N,\nu}$ define a filtration of $V_{[N]}$.

Next we consider quotients between consecutive spaces in the filtration. Suppose that $\mathbf{V}_{N,\mu}$ follows $\mathbf{V}_{N,\nu}$ in the filtration:

$$V_{[N]} \supset \cdots \supset \mathbf{V}_{N,\nu} \supset \mathbf{V}_{N,\mu} \supset \cdots \supset \{0\}. \qquad (3.8.22)$$

Lemma 3.78. *There is an isomorphism*

$$\mathbf{V}_{N,\nu}/\mathbf{V}_{N,\mu} \cong V_{[N-d|\nu|]} / \{(\tilde{\alpha}_{\rho,\lambda(0)}, \ldots, \tilde{\alpha}_{\rho,\lambda(n-1)}) \cap V_{[N-d|\nu|]}\}.$$

Proof. A vector space homomorphism $\varphi : V_{[N-d|\nu|]} \longrightarrow \mathbf{V}_{N,\nu}/\mathbf{V}_{N,\mu}$ is defined as follows: For $\tilde{\beta} \in V_{[N-d|\nu|]}$, we define $\varphi(\tilde{\beta})$ as the class of $\tilde{\alpha}^{\nu}_{\rho,\lambda}\tilde{\beta}$ modulo $\mathbf{V}_{N,\mu}$. Obviously, φ is surjective.

Suppose $\tilde{\beta} \in \ker(\varphi) = \varphi^{-1}(0)$, which means that

$$\tilde{\alpha}^{\nu}_{\rho,\lambda}\tilde{\beta} \in \sum_{\gamma > \nu} \tilde{\alpha}^{\gamma}_{\rho,\lambda} V_{[N-d|\gamma|]}.$$

Thus we may write

$$\tilde{\alpha}^{\nu}_{\rho,\lambda}\tilde{\beta} = \sum_{\gamma > \nu} \tilde{\alpha}^{\gamma}_{\rho,\lambda}\tilde{\beta}_{\gamma}$$

for elements $\tilde{\beta}_{\gamma} \in V_{[N-d|\gamma|]}$. Lemma 3.75 implies that $\tilde{\beta}$ lies in the ideal generated by $\tilde{\alpha}_{\rho,\lambda(0)}, \ldots, \tilde{\alpha}_{\rho,\lambda(n-1)}$. Therefore

$$\tilde{\beta} = \sum_{i=0}^{n-1} \tilde{\eta}_i \tilde{\alpha}_{\rho,\lambda(i)},$$

where $\tilde{\eta}_i$ $(0 \leq i \leq n-1)$ are homogeneous with $\deg(\tilde{\eta}_i) = \deg(\tilde{\beta}) - d$, that is,

$$\tilde{\eta}_i \in V_{[N-d(|\nu|+1)]}, \quad i = 0, \ldots, n-1.$$

Hence $\tilde{\alpha}^{\nu}_{\rho,\lambda}\tilde{\beta}$ is a sum of terms in $\mathbf{V}_{N,\mu}$, which concludes the proof of the lemma. □

By Lemma 3.77 and Lemma 3.78, there exists an integer N_0 depending only on $\tilde{\alpha}_{\rho,\lambda(0)}, \ldots, \tilde{\alpha}_{\rho,\lambda(n-1)}$ such that

$$\Delta_{\nu} := \dim \mathbf{V}_{N,\nu}/\mathbf{V}_{N,\mu} \begin{cases} = d^n, & \text{if } d|\nu| < N - N_0; \\ \leq \dim V_{[N_0]}, & \text{otherwise.} \end{cases} \tag{3.8.23}$$

Now we choose inductively a suitable basis of $V_{[N]}$ in the following way. We start with the last non-zero $\mathbf{V}_{N,\mu}$ in the filtration (3.8.22) and pick any basis of it. Suppose $\mu > \nu$ are consecutive n-tuples such that $d|\nu|, d|\mu| \leq N$. It follows directly from the definition that we may pick representatives $\tilde{\alpha}^{\nu}_{\rho,\lambda}\tilde{\beta} \in \mathbf{V}_{N,\nu}$ of elements in the quotient space $\mathbf{V}_{N,\nu}/\mathbf{V}_{N,\mu}$, where $\tilde{\beta} \in V_{[N-d|\nu|]}$. We extend the previously constructed basis in $\mathbf{V}_{N,\mu}$ by adding these representatives. In particular, we have obtained a basis for $\mathbf{V}_{N,\nu}$ and our inductive procedure may go on unless $\mathbf{V}_{N,\nu} = V_{[N]}$, in which case we stop. In this way, we obtain a basis $\{\tilde{\psi}_1, \ldots, \tilde{\psi}_M\}$ of $V_{[N]}$, where $M = \dim V_{[N]}$.

For a fixed $k \in \{1, \ldots, M\}$, assume that $\tilde{\psi}_k$ is constructed with respect to $\mathbf{V}_{N,\nu}/\mathbf{V}_{N,\mu}$. We may write

$$\tilde{\psi}_k = \tilde{\alpha}^{\nu}_{\rho,\lambda}\tilde{\beta}, \quad \tilde{\beta} \in V_{[N-d|\nu|]}.$$

Then we have a bound

$$
\begin{aligned}
\|\tilde{\psi}_k(\xi)\|_\rho &= \|\tilde{\alpha}^\nu_{\rho,\lambda}(\xi)\|_\rho \|\tilde{\beta}(\xi)\|_\rho \\
&\leq c' \|\tilde{\alpha}^\nu_{\rho,\lambda}(\xi)\|_\rho \|\xi\|_\rho^{N-d|\nu|},
\end{aligned}
$$

where c' is a positive constant depending only on $\tilde{\psi}_k$, not on ξ. Observe that there are precisely Δ_ν such functions $\tilde{\psi}_k$ in our basis. Hence, taking the product over all functions in the basis, and then taking logarithms, we get

$$
\log \prod_{k=1}^M \|\tilde{\psi}_k(\xi)\|_\rho \leq \sum_{i=0}^{n-1} \sum_\nu \Delta_\nu \nu(i+1) \log \|\tilde{\alpha}^\nu_{\rho,\lambda(i)}(\xi)\|_\rho
$$

$$
+ \left(\sum_\nu \Delta_\nu (N - d|\nu|) \right) \log \|\xi\|_\rho + c, \qquad (3.8.24)
$$

where c is a positive constant depending only on $\tilde{\psi}_k$, not on ξ. Here ν in the summations is taken over the n-tuples in the filtration (3.8.22) with $|\nu| \leq N/d$.

Note that

$$
M = \dim V_{[N]} = \binom{n+N}{N} = \frac{N^n}{n!} + O\left(N^{n-1}\right), \qquad (3.8.25)
$$

$$
\sum_{t=0}^T \#J_{n-1,t} = \#J_{n,T} = \binom{n+T}{T}, \quad T \in \mathbb{Z}^+,
$$

and that, since the sum below is independent of j, we have that, for any positive integer T and for every $0 \leq j \leq n$,

$$
\begin{aligned}
\sum_{\nu \in J_{n,T}} \nu(j) &= \frac{1}{n+1} \sum_{\nu \in J_{n,T}} \sum_{j=0}^n \nu(j) = \frac{1}{n+1} \sum_{\lambda \in J_{n,T}} T \\
&= \frac{T}{n+1} \binom{n+T}{T} = \binom{n+T}{T-1} \\
&= \frac{T^{n+1}}{(n+1)!} + O\left(T^n\right). \qquad (3.8.26)
\end{aligned}
$$

Then, for N divisible by d and for every $0 \leq i \leq n-1$, (3.8.23) and (3.8.26) with $T = N/d$ yield

$$
\sum_\nu \Delta_\nu \nu(i+1) = d^n \sum_\nu \nu(i+1) + K_1 = d^n \binom{n+T}{T-1} + K_1, \qquad (3.8.27)
$$

where

$$
K_1 = \sum_{T-N_0/d \leq |\nu| \leq T} (\Delta_\nu - d^n) \nu(i+1) = O\left(T^n\right).
$$

Therefore, we obtain

$$\sum_\nu \Delta_\nu \nu(i+1) = \frac{N^{n+1}}{d(n+1)!} + O\left(N^n\right).$$

Further we have

$$\sum_\nu \Delta_\nu |\nu| = \sum_{i=0}^{n-1} \sum_\nu \Delta_\nu \nu(i+1) = nd^n \binom{n+T}{T-1} + nK_1, \tag{3.8.28}$$

and hence

$$\sum_\nu \Delta_\nu |\nu| = \frac{nN^{n+1}}{d(n+1)!} + O\left(N^n\right).$$

On the other hand, we have

$$\sum_\nu \Delta_\nu T = \sum_\nu d^n T + K_2 = d^n T \binom{n+T}{T} + K_2, \tag{3.8.29}$$

where

$$K_2 = \sum_{T-N_0/d \leq |\nu| \leq T} (\Delta_\nu - d^n) T = O\left(T^n\right).$$

Hence

$$\sum_\nu \Delta_\nu (T - |\nu|) = d^n \binom{n+T}{T-1} + K_2 - nK_1. \tag{3.8.30}$$

Therefore, by (3.8.24), (3.8.27) and (3.8.30), we have

$$
\begin{aligned}
\log \prod_{k=1}^M \|\tilde{\psi}_k(\xi)\|_\rho &\leq \left\{ d^n \binom{n+T}{T-1} + K_1 \right\} \log \prod_{i=0}^{n-1} \|\tilde{\alpha}_{\rho,\lambda(i)}(\xi)\|_\rho \\
&\quad + \left\{ d^n \binom{n+T}{T-1} + K_2 - nK_1 \right\} d \log \|\xi\|_\rho + c \\
&\leq K \left\{ \log \prod_{i=0}^{n-1} \|\tilde{\alpha}_{\rho,\lambda(i)}(\xi)\|_\rho + d \log \|\xi\|_\rho \right\} + c, \tag{3.8.31}
\end{aligned}
$$

where $K = K(d, n, N)$ is a positive constant such that

$$K = \frac{N^{n+1}}{d(n+1)!} \left(1 + O\left(\frac{1}{N}\right)\right). \tag{3.8.32}$$

Let $\tilde{\phi}_1, \dots, \tilde{\phi}_M$ be a fixed basis of $V_{[N]}$ such that when $\xi \in V - \{0\}$,

$$\Xi = (\tilde{\phi}_1(\xi), \dots, \tilde{\phi}_M(\xi)) \in \bar{\kappa}^M - \{0\}.$$

Then $\tilde{\psi}_k$ can be expressed as a linear form L_k in $\tilde{\phi}_1, \ldots, \tilde{\phi}_M$ so that $\tilde{\psi}_k(\xi) = L_k(\Xi)$. The linear forms $L_1, \ldots, L_M$ are linearly independent. By (3.8.31), we obtain

$$\log \prod_{k=1}^{M} \|L_k(\Xi)\|_\rho \ \leq \ K \left\{ \log \prod_{i=0}^{n-1} \|\tilde{\alpha}_{\rho,\lambda(i)}(\xi)\|_\rho + d \log \|\xi\|_\rho \right\} + c$$

$$= \ K \left\{ \log \prod_{i=0}^{n-1} \frac{\|\tilde{\alpha}_{\rho,\lambda(i)}(\xi)\|_\rho}{\|\xi\|_\rho^d} + (n+1)d \log \|\xi\|_\rho \right\} + c,$$

which implies

$$\log \prod_{i=0}^{n-1} \frac{\|\xi\|_\rho^d}{\|\tilde{\alpha}_{\rho,\lambda(i)}(\xi)\|_\rho} \ \leq \ \frac{1}{K} \left\{ \log \prod_{k=1}^{M} \frac{1}{\|L_k(\Xi)\|_\rho} + c \right\}$$

$$+ (n+1)d \log \|\xi\|_\rho, \qquad (3.8.33)$$

or, equivalently

$$\prod_{i=0}^{n-1} \frac{\|\xi\|_\rho^d}{\|\tilde{\alpha}_{\rho,\lambda(i)}(\xi)\|_\rho} \leq \left\{ e^c \prod_{k=1}^{M} \frac{1}{\|L_k(\Xi)\|_\rho} \right\}^{\frac{1}{K}} \|\xi\|_\rho^{(n+1)d}. \qquad (3.8.34)$$

Fix $\varepsilon > 0$. By Theorem 3.46, for all $\lambda \in J_{n-1}^q$, the set Q of all $\Xi \in \mathcal{O}_{V_{[N]},S}$ satisfying

$$\prod_{\rho \in S} \prod_{k=1}^{N} \|L_k(\Xi)\|_\rho < \left\{ \max_{\rho \in S} \|\Xi\|_\rho \right\}^{-\varepsilon}$$

is contained in a finite union of hyperplanes of $V_{[N]}$. Note that Q is just a finite union of hypersurfaces of degree N in V, say,

$$Q = \bigcup_{j=1}^{r} E^N[b_j], \quad b_j \in \mathbb{P}\left(\amalg_N V^*\right),$$

and that there is a positive constant $\tilde{c}$ such that

$$\|\Xi\|_\rho \leq \tilde{c} \|\xi\|_\rho^N, \quad \rho \in S.$$

Then we have

$$\prod_{\rho \in S} \prod_{i=0}^{n-1} \frac{\|\xi\|_\rho^d}{\|\tilde{\alpha}_{\rho,\lambda(i)}(\xi)\|_\rho} \leq \left\{ e^c \left(\max_{\rho \in S} \tilde{c} \|\xi\|_\rho^N \right)^\varepsilon \right\}^{\frac{1}{K}} \left(\prod_{\rho \in S} \|\xi\|_\rho \right)^{(n+1)d},$$

where

$$\xi \notin \bigcup_{\rho \in S} \bigcup_{j=0}^{q} E^d[a_{\rho,j}] \cup Q.$$

If we choose N large enough such that

$$d \mid N, \quad N_0 + 2d \leq N \leq K,$$

then Lemma 3.74 implies that there is a constant c depending only on S but independent of ξ such that

$$\prod_{\rho \in S} \prod_{i=0}^{n-1} \frac{\|\|\xi\|\|_\rho^d}{\|\|\tilde{\alpha}_{\rho,\lambda(i)}(\xi)\|\|_\rho} \leq cH(\xi)^{(n+1+\varepsilon)d}. \tag{3.8.35}$$

Therefore Theorem 3.65 follows from (3.8.21) and (3.8.35).

Remark on (3.8.35). If we take $\lambda \in J_n^q$, Lemma 3.72 means that there exists an index $i_0 \in \{0, 1, \ldots, n\}$ such that

$$\|x^{\mathrm{II}d}, a_{\rho,\lambda(i_0)}\|_\rho \geq \Gamma(\mathscr{A}_\rho), \quad x = \mathbb{P}(\xi).$$

Without loss of generality, we may assume $i_0 = n$. Thus from (3.8.34), according to the arguments leading up to (3.8.35) we can obtain

$$\prod_{\rho \in S} \prod_{i=0}^{n} \frac{1}{\|\tilde{\alpha}_{\rho,\lambda(i)}(\xi)\|_\rho} \leq c \left(\max_{\rho \in S} \|\xi\|_\rho \right)^\varepsilon. \tag{3.8.36}$$

Hence the above method yields also a proof of Theorem 3.68.

Now we exhibit the original subspace theorem of P. Corvaja and U. Zannier [68] as follows:

Theorem 3.79. *For $\rho \in S$, let $f_{i\rho}$, $i = 1, \ldots, n-1$, be polynomials in $k[X_1, \ldots, X_n]$ of degrees $\delta_{i\rho} > 0$. Put*

$$\delta_\rho = \max_i \delta_{i\rho}, \quad \mu = \min_{\rho \in S} \sum_{i=1}^{n-1} \frac{\delta_{i\rho}}{\delta_\rho}.$$

Fix $\epsilon > 0$ and consider the Zariski closure $\mathcal{H}$ in $\mathbb{P}^n$ of the set of solutions $x \in \mathcal{O}_{\kappa,S}^n$ of

$$\prod_{\rho \in S} \prod_{i=1}^{n-1} \|\|f_{i\rho}(x)\|\|_\rho^{\frac{1}{\delta_\rho}} \leq H([1, x])^{\mu - n - \epsilon}. \tag{3.8.37}$$

Suppose that, for $\rho \in S$, X_0 and the $\overline{f}_{i\rho}$, $i = 1, \ldots, n-1$, define a variety of dimension 0. Then $\dim \mathcal{H} \leq n - 1$.

Here for a polynomial $h \in k[X_1, \ldots, X_n]$, we denote by $\overline{h}$ the homogenized polynomial in $k[X_0, \ldots, X_n]$; namely, $\overline{h}$ is the unique homogeneous polynomial in $k[X_0, \ldots, X_n]$, of the same degree as h and such that $\overline{h}(1, X_1, \ldots, X_n) = h(X_1, \ldots, X_n)$. For the special cases

$$\delta_{i\rho} = d, \ 1 \leq i < n, \ \rho \in S,$$

by setting

$$\overline{f}_{0\rho}(X_0, X_1, \ldots, X_n) = X_0^d, \quad \rho \in S,$$

and applying the above argument to $\overline{f}_{i\rho}$ $(0 \leq i \leq n - 1)$ replacing $\tilde{\alpha}_{\rho,\lambda(0)}, \ldots,$ $\tilde{\alpha}_{\rho,\lambda(n-1)}$, the proof of Theorem 3.79 follows.

3.8.4 Proof of Theorem 2.63

The methods in Section 3.8.3 serve also as a proof of Theorem 2.63. Next we assume $\bar{\kappa} = \mathbb{C}$ and so V is a complex vector space of dimensions $n + 1 \geq 1$. We consider an algebraically non-degenerate meromorphic mapping $f : \mathbb{C}^m \longrightarrow \mathbb{P}(V)$. Fix a positive integer d. Then f induces a meromorphic mapping

$$f^{\mathrm{II}d} : \mathbb{C}^m \longrightarrow \mathbb{P}\left(\mathrm{II}_d V\right)$$

such that the characteristic function of $f^{\mathrm{II}d}$ satisfies

$$T_{f^{\mathrm{II}d}}(r) = d T_f(r).$$

Let $\mathscr{A} = \{a_j\}_{j=0}^q$ be a finite admissible family of points $a_j \in \mathbb{P}\left(\mathrm{II}_d V^*\right)$ with $q \geq n$. Take $\alpha_j \in \mathrm{II}_d V^* - \{0\}$ with $\mathbb{P}(\alpha_j) = a_j$, and define

$$\tilde{\alpha}_j(\xi) = \left\langle \xi^{\mathrm{II}d}, \alpha_j \right\rangle, \quad \xi \in V, \ j = 0, 1, \ldots, q.$$

Without loss of generality, assume $|\alpha_j| = 1$ for $j = 0, \ldots, q$. Write

$$F_j = \tilde{\alpha}_j \circ \tilde{f} = \left\langle \tilde{f}^{\mathrm{II}d}, \alpha_j \right\rangle, \quad j = 0, 1, \ldots, q,$$

where $\tilde{f} : \mathbb{C}^m \longrightarrow V$ is a reduced representation of f. Lemma 3.73 implies

$$\prod_{j=0}^q \frac{1}{|f^{\mathrm{II}d}, a_j|} \leq \left(\frac{1}{\Gamma(\mathscr{A})}\right)^{q+1-n} \max_{\lambda \in J_{n-1}^q} \prod_{i=0}^{n-1} \frac{1}{|f^{\mathrm{II}d}, a_{\lambda(i)}|},$$

which yields

$$\sum_{j=0}^q m_{f^{\mathrm{II}d}}(r, a_j) \leq I - (q + 1 - n) \log \Gamma(\mathscr{A}), \tag{3.8.38}$$

where

$$I = \mathbb{C}^m \left\langle 0; r; \ \max_{\lambda \in J_{n-1}^q} \log \prod_{i=0}^{n-1} \frac{1}{|f^{\mathrm{II}d}, a_{\lambda(i)}|} \right\rangle. \tag{3.8.39}$$

Hu and Yang observed the inequality (3.8.38) and estimated the corresponding term I over non-Archimedean fields (see [175], [176], or Theorem 5.22). For the complex case, M. Ru [330] estimated the integral I in (3.8.39) as follows.

According to the arguments in the proof of Theorem 3.65, replacing Ξ we now have a meromorphic mapping $g = g_N : \mathbb{C}^m \longrightarrow \mathbb{P}^{M-1}$ induced by

$$\tilde{g} = (\tilde{\phi}_1 \circ \tilde{f}, \ldots, \tilde{\phi}_M \circ \tilde{f}) : \mathbb{C}^m \longrightarrow \mathbb{C}^M,$$

which is linearly non-degenerate from the assumption of algebraic non-degeneracy of f, so that $\tilde{\psi}_k \circ \tilde{f} = L_k \circ \tilde{g}$. Now the inequality (3.8.33) becomes

$$\log \prod_{i=0}^{n-1} \frac{|\tilde{f}|^d}{|F_{\lambda(i)}|} \leq \frac{1}{K}\left\{\log \prod_{k=1}^{M} \frac{|\tilde{g}|}{|L_k \circ \tilde{g}|} - M\log|\tilde{g}| + c\right\}$$
$$+ (n+1)d\log|\tilde{f}|. \tag{3.8.40}$$

Since there are only finitely many choices $\lambda \in J_{n-1}^q$, we have a finite collection of linear forms $L_1, \ldots, L_u$. Hence (3.8.40) implies that

$$I = \mathbb{C}^m \left\langle 0; r; \max_{\lambda \in J_{n-1}^q} \log \prod_{i=0}^{n-1} \frac{|\tilde{f}|^d}{|F_{\lambda(i)}|} \right\rangle$$
$$\leq \frac{1}{K}\left\{\mathbb{C}^m \left\langle 0; r; \max_J \log \prod_{k\in J} \frac{|\tilde{g}|}{|L_k \circ \tilde{g}|} \right\rangle - M\mathbb{C}^m \langle 0; r; \log|\tilde{g}|\rangle + c\right\}$$
$$+ (n+1)d\,\mathbb{C}^m \left\langle 0; r; \log|\tilde{f}| \right\rangle, \tag{3.8.41}$$

where $\max_J$ is taken over all subsets J of $\{1, \ldots, u\}$ such that the linear forms L_k ($k \in J$) are linearly independent. Applying Theorem 2.62, we have

$$\mathbb{C}^m \left\langle 0; r; \max_J \log \prod_{k\in J} \frac{|\tilde{g}|}{|L_k \circ \tilde{g}|} \right\rangle \leq MT_g(r) - N_{\mathrm{Ram}}(r, g)$$
$$+ \frac{M(M-1)}{2}\log\left\{\left(\frac{\rho}{r}\right)^{2m-1} \frac{T_g(R)}{\rho - r}\right\} + O(1)$$

for any $r_0 < r < \rho < R$. The formulae (2.3.20) and (2.3.21) imply respectively

$$\mathbb{C}^m \langle 0; r; \log|\tilde{g}|\rangle \geq T_g(r) + O(1)$$

and

$$\mathbb{C}^m \left\langle 0; r; \log|\tilde{f}| \right\rangle = T_f(r) + O(1).$$

Obviously,

$$T_g(r) \leq NT_f(r) + O(1),$$

and so (3.8.41) yield

$$I \leq (n+1)dT_f(r) + \frac{M(M-1)}{2K} \log \left\{ \left(\frac{\rho}{r}\right)^{2m-1} \frac{T_f(R)}{\rho - r} \right\} + O(1), \qquad (3.8.42)$$

where $O(1)$ is a constant independent of r, ρ and R, but depends on N and f. Finally, by (3.8.38) and (3.8.42), we obtain the following result (cf. [330]):

Theorem 3.80. *Let* $f : \mathbb{C}^m \longrightarrow \mathbb{P}(V)$ *be an algebraically non-degenerate meromorphic mapping. Fix a positive integer* d. *Let* $\mathscr{A} = \{a_j\}_{j=0}^q$ *be a finite admissible family of points* $a_j \in \mathbb{P}\left(\amalg_d V^*\right)$. *Then for any* $r_0 < r < \rho < R$, *we have*

$$\sum_{j=0}^q m_{f^{\amalg_d}}(r, a_j) \leq (n+1)dT_f(r) + \frac{M(M-1)}{2K} \log \left\{ \left(\frac{\rho}{r}\right)^{2m-1} \frac{T_f(R)}{\rho - r} \right\} + O(1).$$

$$\tag{3.8.43}$$

Related to the above theorem, Conjecture 2.97 assumes the following form:

Conjecture 3.81. *Let* $f : \mathbb{C}^m \longrightarrow \mathbb{P}(V)$ *be an algebraically non-degenerate meromorphic mapping. Fix a positive integer* d. *Let* $\mathscr{A} = \{a_j\}_{j=0}^q$ *be a finite family of points* $a_j \in \mathbb{P}\left(\amalg_d V^*\right)$ *such that the divisor* $\ddot{E}^d[a_0] + \cdots + \ddot{E}^d[a_q]$ *has normal crossings. Then*

$$\sum_{j=0}^q m_{f^{\amalg_d}}(r, a_j) \leq (n+1)T_f(r) + O(\log^+ T_f(r)) + O(\varepsilon \log r). \qquad (3.8.44)$$

This conjecture is an analogue of Conjecture 3.69 in value distribution theory. A part of this result was given by Biancofiore [23].

3.9   Vanishing sums in function fields

In this section, we will introduce some inequalities over algebraic function fields, which will deliver some supporting evidence to generalized *abc*-conjectures discussed in Chapter 4. Further, some methods in this section will be used to prove *abc*-theorems for meromorphic functions.

3.9.1   Algebraic function fields

Let κ be any field. An *algebraic function field* over κ is a finitely generated extension field K of κ which is not algebraic. If the transcendence degree is r, then K is a *field of r variables* over κ. In what follows we shall take $r = 1$ and assume that κ is perfect. By Proposition 1.20, there exists an element $x \in K - \kappa$ such that K is separable over $\kappa(x)$. Thus for the rest of this section, *function field* will mean "algebraic function field of one variable over a perfect ground field". As a finite separable extension $K/\kappa(x)$ can be generated by a single element y; the minimal equation for y over $\kappa(x)$ defines an algebraic curve, and in this sense a function field represents the field of all rational functions on an algebraic curve.

Proposition 3.82 (cf. [63]). *Let K/κ be a function field, where κ is perfect of prime characteristic p. Then for any $x \in K$ the following conditions are equivalent:*

(a) *x is a separating element for K/κ;*

(b) *dx is a K-basis for the space of differentials;*

(c) *x is not of the form y^p for some $y \in K$.*

Let K/κ be a function field. We note that if $x \in K - \kappa$, then all valuations of $\kappa(x)/\kappa$ are discrete (cf. Proposition 1.39), and that all valuations of K/κ are obtained by extending those of $\kappa(x)/\kappa$ (cf. Theorem 1.38), and they are again discrete. Although the latter depend on x, the set of all valuations of K/κ does not depend on the choice of x. Every function field has infinitely many places (cf. [63]). Let $\mathfrak{p}$ be any place of K/κ, with associated valuation $v_{\mathfrak{p}}$ and residue class field $\mathbb{F}(K)$. Then $\mathbb{F}(K)$ is a finite algebraic extension of κ; for we have $v_{\mathfrak{p}}(\alpha) = 0$ for all $\alpha \in \kappa_*$, so the residue class mapping $x \mapsto \bar{x}$ of K into $\mathbb{F}(K) \cup \{\infty\}$ is injective when restricted to κ, hence κ is embedded in $\mathbb{F}(K)$. The restriction to $\kappa(x)$ has as residue class field a finite extension of κ, of degree equal to the degree of the irreducible polynomial defining the place $\mathfrak{q}$ on $\kappa(x)$ such that $\mathfrak{p}|\mathfrak{q}$. It follows that $d_{\mathfrak{p}} = [\mathbb{F}(K) : \kappa]$ is finite too, which is called the *absolute residue degree* or simply the *degree* of $\mathfrak{p}$. Its relation to the relative degree is given by the formula

$$d_{\mathfrak{p}} = [\mathbb{F}(K) : \mathbb{F}(F)]d_{\mathfrak{q}} \qquad (3.9.1)$$

for $\kappa \subset F \subset K$, where $\mathfrak{p}|\mathfrak{q}$ for a place $\mathfrak{q}$ of F/κ.

Given any $f \in K$ and any place $\mathfrak{p}$ of K/κ, with normalized valuation $v_{\mathfrak{p}}$, there are three possibilities:

(α) If $v_{\mathfrak{p}}(f) > 0$, we say that f has a *zero* at $\mathfrak{p}$ of order $v_{\mathfrak{p}}(f)$;

(β) If $v_{\mathfrak{p}}(f) < 0$, we say that f has a *pole* at $\mathfrak{p}$ of order $-v_{\mathfrak{p}}(f)$;

(γ) $v_{\mathfrak{p}}(f) = 0$.

Usually we also denote the number $v_{\mathfrak{p}}(f)$ by $\mathrm{ord}_{\mathfrak{p}} f$. In a function field, any non-constant element has finitely many zeros and at least one zero; similarly for poles. With each element $f \in K_*$ we can associate a divisor

$$(f) = \prod_{\mathfrak{p}} \mathfrak{p}^{\mathrm{ord}_{\mathfrak{p}} f},$$

called a *principal divisor*, which is usually expressed additively

$$(f) = \sum_{\mathfrak{p}} \mathrm{ord}_{\mathfrak{p}}(f)\mathfrak{p}.$$

To examine (f) more closely, let us write it as

$$(f) = (f)_0 (f)_\infty^{-1},$$

where

$$(f)_0 = \prod_{\mathfrak{p}} \left\{ \mathfrak{p}^{\operatorname{ord}_{\mathfrak{p}} f} \mid \operatorname{ord}_{\mathfrak{p}} f > 0 \right\}, \quad (f)_\infty = \prod_{\mathfrak{p}} \left\{ \mathfrak{p}^{-\operatorname{ord}_{\mathfrak{p}} f} \mid \operatorname{ord}_{\mathfrak{p}} f < 0 \right\}.$$

The divisor $(f)_0$ is called the *divisor of zeros* of f, $(f)_\infty$ is the *divisor of poles* of f.

We recall that the divisor group D is the free Abelian group on the set of all places as free generating set; its elements are called divisors. The general divisor has the form

$$\mathfrak{a} = \prod \mathfrak{p}^{v_{\mathfrak{p}}(\mathfrak{a})} = \mathfrak{p}_1^{\alpha_1} \cdots \mathfrak{p}_r^{\alpha_r}. \tag{3.9.2}$$

Each divisor has a *degree*, given by

$$\deg(\mathfrak{a}) = \sum d_{\mathfrak{p}} v_{\mathfrak{p}}(\mathfrak{a}),$$

where $d_{\mathfrak{p}}$ is the degree of $\mathfrak{p}$ as defined earlier. Of course when κ is algebraically closed, then $d_{\mathfrak{p}} = 1$ and it can be omitted. In particular, for $f \in K - \kappa$ (cf. [63]),

$$\deg((f)_0) = \deg((f)_\infty) = [K : \kappa(f)].$$

To express the fact that $f \in K$ has a zero of order at least α_i at $\mathfrak{p}_i$ $(i = 1, \ldots, r)$ we shall write

$$f \equiv 0 (\operatorname{mod} \mathfrak{a}), \tag{3.9.3}$$

where $\mathfrak{a}$ is given by (3.9.2). Associated with the divisor (3.9.2) we have a vector space over κ given by the equation

$$\mathcal{L}(\mathfrak{a}) = \{ f \in K \mid f \equiv 0 (\operatorname{mod} \mathfrak{a}) \}. \tag{3.9.4}$$

Theorem 3.83 (Riemann's theorem, cf. [63]). *Let K/κ be a function field. Then as $\mathfrak{a}$ runs over all divisors, $\dim_\kappa \mathcal{L}(\mathfrak{a}) + \deg(\mathfrak{a})$ is bounded below. Thus there exists an integer g such that*

$$\dim_\kappa \mathcal{L}(\mathfrak{a}) + \deg(\mathfrak{a}) \geq 1 - g. \tag{3.9.5}$$

Moreover, given $f \in K - \kappa$, the lower bound is attained by $\mathfrak{a} = (f)_\infty^{-m}$, for all large m.

The constant g in (3.9.5) is an important invariant of K, called the *genus*. It can assume any non-negative integer value, and it has geometrical and function theoretic interpretations, some of which we shall meet later.

Assume that κ is algebraically closed. Take a separating element x of K. At any place $\mathfrak{p}$ of K with the valuation $v_{\mathfrak{p}}$ choose an element t generating the unique maximal ideal of the valuation ring, which is called a *prime element* or a *uniformizer*, and also is a separating element for K, and define a divisor

$$\mathfrak{K} = \left(\frac{dx}{dt}\right),$$

called the *canonical divisor* of K. This definition is easily seen to be independent of the choice of t. If $\mathfrak{a}$ is any divisor, then we have the *Riemann-Roch formula* (cf. [63])

$$\dim_\kappa \mathcal{L}\left(\mathfrak{a}^{-1}\right) = \deg(\mathfrak{a}) + 1 - g + \dim_\kappa \mathcal{L}\left(\mathfrak{a}\mathfrak{K}^{-1}\right), \tag{3.9.6}$$

which implies specially

$$\deg(\mathfrak{K}) = 2g - 2. \tag{3.9.7}$$

3.9.2 Mason's inequality

Let κ be an algebraically closed field of zero characteristic, and let K be a function field of genus g in one variable over κ. We normalize each valuation v on K so that its order group consists of all rational integers. For $n \geq 1$, $\{f_0, f_1, \ldots, f_n\} \subset K$ not all zero, we define the *projective height* as

$$h(f_0, f_1, \ldots, f_n) = -\sum_v \min\{\text{ord}_v f_0, \text{ord}_v f_1, \ldots, \text{ord}_v f_n\}. \tag{3.9.8}$$

The sum formula on K shows that this is really a height on the projective space $\mathbb{P}(K^{n+1})$. R. C. Mason [253], [255] proved essentially the following result:

Theorem 3.84. *Let* $\{f_0, f_1, \ldots, f_n\} \subset K$ $(n \geq 2)$ *be such that*

$$f_1 + \cdots + f_n = f_0 \tag{3.9.9}$$

but $f_1, \ldots, f_n$ *are linearly independent over* κ. *Then*

$$h(f_0, f_1, \ldots, f_n) \leq 4^{n-1}(\#S + 2g - 2), \tag{3.9.10}$$

where S *is the set of places of* K *where some* f_i *is not a unit.*

By the definition, here $\#S$ is the cardinality of S. Actually Mason's Lemma 2 in [255] is stated differently, since he deals with the inhomogeneous equation $f_1 + \cdots + f_n = 1$. This form is given in [447]. The work of Mason generalized his previous result with $n = 2$ (see [254]) which he had used to solve effectively certain classical Diophantine equations over function fields. Write

$$\gamma_{-1} = 0, \quad \gamma_p = \frac{p(p-1)}{2} \ (p \geq 0).$$

W.D. Brownawell and D.W. Masser [41] improved Theorem 3.84 as follows:

Theorem 3.85. *Let $\{f_0, f_1, \ldots, f_n\} \subset K$ $(n \geq 2)$ be such that $(3.9.9)$ holds but $f_1, \ldots, f_n$ are linearly independent over κ. For each valuation v, let $m = m(v) = m(v; f_0, \ldots, f_n)$ be the number of elements among $f_0, \ldots, f_n$ which are units at v. Then*

$$h(f_0, f_1, \ldots, f_n) \leq \gamma_n(2g - 2) + \sum_v (\gamma_n - \gamma_{m-1}). \tag{3.9.11}$$

Proof. Fix a separating element z of K, and we write $\frac{d}{dz}$ for the corresponding derivation on K. Let

$$\mathbf{W}_z = \mathbf{W}_z(f_1, \ldots, f_n) = \det\left(\frac{d^{j-1} f_i}{dz^{j-1}}\right) \quad (1 \leq i, j \leq n)$$

be the Wronskian of $f_1, \ldots, f_n$ (with respect to z) and define

$$\mathbf{S}_z = \mathbf{S}_z(f_1, \ldots, f_n) = (f_1 \cdots f_n)^{-1} \mathbf{W}_z(f_1, \ldots, f_n).$$

Then $\mathbf{W}_z \neq 0$ since $f_1, \ldots, f_n$ are linearly independent. Taking derivatives on both sides of the identity $(3.9.9)$ yields

$$\frac{d^{j-1} f_1}{dz^{j-1}} + \frac{d^{j-1} f_2}{dz^{j-1}} + \cdots + \frac{d^{j-1} f_n}{dz^{j-1}} = \frac{d^{j-1} f_0}{dz^{j-1}} \quad (j = 1, 2, \ldots, n).$$

The above equation and $(3.9.9)$ yield $\mathbf{W}_z = \mathbf{W}_{zj}$ $(j = 1, 2, \ldots, n)$, where

$$\mathbf{W}_{zj} = \mathbf{W}_z(f_1, \ldots, f_{j-1}, f_0, f_{j+1}, \ldots, f_n).$$

Hence

$$\mathbf{S}_z f_j = \mathbf{S}_{zj} f_0, \quad j = 1, \ldots, n, \tag{3.9.12}$$

where $\mathbf{S}_{zj}$ is defined by $\mathbf{W}_{zj} = \mathbf{S}_{zj} f_1 \cdots f_{j-1} f_0 f_{j+1} \cdots f_n$. The equations $(3.9.12)$ yield

$$h(f_0, f_1, \ldots, f_n) = h(\mathbf{S}_z, \mathbf{S}_{z1}, \ldots, \mathbf{S}_{zn}). \tag{3.9.13}$$

For each valuation v on K there is a separating element $\zeta = \zeta_v$ in K with $\mathrm{ord}_v \zeta = 1$ such that every element of K can be expressed as a Laurent series in ζ with only finitely many negative exponents. Let $\frac{d}{d\zeta}$ denote the corresponding derivation on K. For any non-zero f in K, and any valuation v, we have an expansion

$$f = a\zeta^m + \cdots \quad (a \neq 0).$$

Thus if $m \neq 0$, that is, f is not a unit at v, then for any non-negative integer j we obtain

$$f^{-1} \frac{d^j f}{d\zeta^j} = m(m-1) \cdots (m-j+1)\zeta^{-j} + \cdots,$$

and hence

$$\mathrm{ord}_v\left(f^{-1} \frac{d^j f}{d\zeta^j}\right) \geq -j. \tag{3.9.14}$$

If $m = 0$, that is, f is a unit at v, then it is easier to show that

$$\operatorname{ord}_v \left(f^{-1} \frac{d^j f}{d\zeta^j} \right) \geq 0. \tag{3.9.15}$$

By the formula for change of variable in a Wronskian (see [283], p. 662 or [436], p. 69) we have

$$\mathbf{W}_z = \mathbf{W}_\zeta \left(\frac{d\zeta}{dz} \right)^{\gamma_n}. \tag{3.9.16}$$

Suppose that exactly μ among $f_1, \ldots, f_n$ are units at v. To estimate $\operatorname{ord}_v \mathbf{S}_\zeta$ we apply (3.9.14) and (3.9.15) to each of the columns, and get

$$\operatorname{ord}_v \mathbf{S}_\zeta \geq -\{(n-1) + (n-2) + \cdots + \mu\} = -\{\gamma_n - \gamma_\mu\}.$$

Therefore by (3.9.16), we obtain

$$\operatorname{ord}_v \mathbf{S}_z \geq -\gamma_n \operatorname{ord}_v \frac{dz}{d\zeta} - \{\gamma_n - \gamma_\mu\}. \tag{3.9.17}$$

Now we have by definition

$$h(\mathbf{S}_z, \mathbf{S}_{z1}, \ldots, \mathbf{S}_{zn}) = -\sum_v \min\{\operatorname{ord}_v \mathbf{S}_z, \operatorname{ord}_v \mathbf{S}_{z1}, \ldots, \operatorname{ord}_v \mathbf{S}_{zn}\}.$$

Fix j with $0 \leq j \leq n$ and any valuation v. Since m elements of $f_0, \ldots, f_n$ are units at v, it follows that at least $m-1$ elements of $\{f_0, \ldots, f_n\} - \{f_j\}$ are units at v. Thus in (3.9.17) we have $\mu \geq m-1$, so $\gamma_\mu \geq \gamma_{m-1}$ and we conclude

$$\operatorname{ord}_v \mathbf{S}_{zj} \geq -\gamma_n \operatorname{ord}_v \frac{dz}{d\zeta} - \{\gamma_n - \gamma_{m-1}\},$$

where $\mathbf{S}_{z0} = \mathbf{S}_z$. Therefore

$$h(\mathbf{S}_z, \mathbf{S}_{z1}, \ldots, \mathbf{S}_{zn}) \leq \gamma_n \sum_v \operatorname{ord}_v \frac{dz}{d\zeta} + \sum_v \{\gamma_n - \gamma_{m-1}\}.$$

But we know (see for example [255], equation (6); or [63])

$$\sum_v \operatorname{ord}_v \frac{dz}{d\zeta} = 2g - 2,$$

and we end up with

$$h(\mathbf{S}_z, \mathbf{S}_{z1}, \ldots, \mathbf{S}_{zn}) \leq \gamma_n (2g - 2) + \sum_v \{\gamma_n - \gamma_{m-1}\}.$$

Theorem 3.85 follows from this and (3.9.13). $\qquad\square$

Recall that if S is a finite set of valuations, an element f of K is said to be an S-unit if it is a unit at v for all v not in S.

Corollary 3.86 ([41]). *Let $\{f_0, f_1, \ldots, f_n\} \subset K$ $(n \geq 2)$ be such that (3.9.9) holds but $f_1, \ldots, f_n$ are linearly independent over κ. If $f_0, \ldots, f_n$ are S-units for some finite set S, then*

$$h(f_0, f_1, \ldots, f_n) \leq \frac{n(n-1)}{2} \{\#S + 2g - 2\}. \tag{3.9.18}$$

3.9.3   No vanishing subsums

W.D. Brownawell and D.W. Masser [41] further proved that the inequality of Theorem 3.85 remains true, in slightly modified form, if the assumption of linear independence is replaced by a weaker hypothesis of no vanishing subsums. To introduce this result, according to W.D. Brownawell and D.W. Masser [41], a finite function family $\mathcal{F} = \{f_i \mid i \in J\}$ will be called *minimal* if $\mathcal{F}$ is linearly dependent, and for any proper subset I of J the family $\{f_i \mid i \in I\}$ is linearly independent. For simplicity, we also call the indices J *minimal*.

Lemma 3.87 ([41]). *Assume $f_0 + \cdots + f_n = 0$ but no non-empty proper subsum vanishes. Then there exists a partition of indices*

$$\{0, 1, \ldots, n\} = I_0 \cup \cdots \cup I_k$$

satisfying the following properties:

(i) *I_α are non-empty disjoint sets;*

(ii) *There exist subsets I'_α of $\{0, 1, \ldots, n\}$ with*

$$I'_0 = \emptyset, \quad \emptyset \neq I'_\alpha \subseteq I_0 \cup \cdots \cup I_{\alpha-1} \ (\alpha = 1, \ldots, k)$$

such that the set $I_\alpha \cup I'_\alpha$ is minimal for each $\alpha = 0, \ldots, k$.

Proof. Consider a non-empty subspace of $\mathbb{C}^{n+1}$ as follows:

$$V = \{(a_0, \ldots, a_n) \in \mathbb{C}^{n+1} \mid a_0 f_0 + \cdots + a_n f_n = 0\}.$$

Then each element A in V can be written as

$$A = \sum_J c_J A_J \ (c_J \in \mathbb{C}, \ A_J = (a_{J0}, \ldots, a_{Jn}) \in V) \tag{3.9.19}$$

such that $a_{Ji} = 0$ when $i \notin J$, where J is minimal. Next we prove the expression (3.9.19) by induction on the number $l(A)$ of non-zero components in A. The cases $l(A) = 1, 2$ are trivial. Assume that for some $l > 2$ the fact (3.9.19) holds for all elements A of V with $l(A) < l$. Next take $A \in V$ with $l(A) = l$; without loss of generality, we may assume that

$$A = (a_0, \ldots, a_{l-1}, 0, \ldots, 0), \ a_i \neq 0 \ (i = 0, \ldots, l-1).$$

Then we are done if $J = \{0, \ldots, l-1\}$ is minimal. Otherwise, there is an element

$$B = (b_0, \ldots, b_{l-1}, 0, \ldots, 0) \in V$$

with $2 \leq l(B) < l$. Without loss of generality, we may assume that $b_0 \neq 0$. Then $l(b_0 A - a_0 B) < l$, and so the induction hypothesis can be applied to B and $C = b_0 A - a_0 B$. Therefore

$$A = \frac{a_0}{b_0} B + \frac{1}{b_0} C$$

has the desired decomposition.

Next we show that if

$$\sum_{i \in E} f_i \neq 0$$

for some $E \subset Z = \{0, \ldots, n\}$, then there is a minimal set J such that

$$J \cap E \neq \emptyset, \quad J \cap E^c \neq \emptyset, \tag{3.9.20}$$

where E^c is the complement of E in Z. In fact, if this is false, applying (3.9.19) to $A = (1, \ldots, 1) \in V$, then every J in (3.9.19) is contained either in E or in E^c, and so

$$A = \sum_{J \subset E} c_J A_J + \sum_{J \subset E^c} c_J A_J.$$

Especially, we have

$$\sum_{i \in E} f_i = \sum_{J \subset E} c_J (a_{J0} f_0 + \cdots + a_{Jn} f_n) = 0,$$

which leads to a contradiction.

Now we can prove Lemma 3.87. If each proper subset of $\{f_0, \ldots, f_n\}$ is linearly independent, then we can take $k = 0$ in Lemma 3.87. So we may assume that some proper subset of $\{f_0, \ldots, f_n\}$ is linearly dependent. According to the hypothesis, the set Z is not minimal. Thus applying (3.9.20) to, say, $E = \{0\}$, we may choose a proper minimal subset J_0 of Z, and so

$$\sum_{i \in J_0} f_i \neq 0.$$

Applying (3.9.20) to $E = J_0$, there exists a minimal set J_1 such that

$$J_1 \cap E \neq \emptyset, \quad J_1 \cap E^c \neq \emptyset.$$

Set

$$I_0 = E, \quad I_1 = J_1 \cap E^c, \quad I_1' = J_1 \cap E.$$

If $Z = I_0 \cup I_1$, then we are done. Otherwise, applying (3.9.20) to $E = I_0 \cup I_1$, there exists a minimal set J_2 such that

$$J_2 \cap E \neq \emptyset, \quad J_2 \cap E^c \neq \emptyset.$$

Set

$$I_2 = J_2 \cap E^c, \quad I_2' = J_2 \cap E.$$

If $Z = I_0 \cup I_1 \cup I_2$, then we are done. Otherwise, we repeat the above procedures until the union reaches Z. □

Theorem 3.88. *Let $\{f_0, f_1, \ldots, f_n\} \subset K$ $(n \geq 2)$ be such that (3.9.9) holds but no subsum of (3.9.9) vanishes. For each valuation v, let $m = m(v) = m(v; f_0, \ldots, f_n)$ be the number of elements among $f_0, \ldots, f_n$ which are units at v. Then*

$$h(f_0, f_1, \ldots, f_n) \leq \gamma_n \max\{0, 2g - 2\} + \sum_v (\gamma_n - \gamma_{m-1}). \qquad (3.9.21)$$

Proof. By Theorem 3.85, we can suppose that $f_1, \ldots, f_n$ are linearly dependent over κ. By Lemma 3.87, there exists a partition of indices

$$\{0, 1, \ldots, n\} = I_0 \cup \cdots \cup I_k$$

satisfying the properties (i) and (ii) in Lemma 3.87 with $k \geq 1$. Set

$$n_0 + 1 = \#I_0 \geq 2; \quad n_\alpha = \#I_\alpha \ (\alpha = 1, \ldots, k)$$

and write

$$s_\alpha = 1 + \sum_{\beta=0}^{\alpha} n_\beta, \quad \alpha = 0, 1, \ldots, k.$$

Then

$$n_0 + n_1 + \cdots + n_k = n.$$

Without loss of generality, we may assume that

$$I_0 = \{0, \ldots, n_0\}, \quad I_\alpha = \{s_{\alpha-1}, \ldots, s_\alpha - 1\} \ (\alpha = 1, \ldots, k).$$

Since I_0 is minimal, then $f_0, f_1, \ldots, f_{n_0}$ are linearly dependent, and so there is a linear relation

$$a_{0,0} f_0 + \cdots + a_{0,n_0} f_{n_0} = 0,$$

with $a_{0,i} \neq 0$ for $i \in I_0$. Set $a_{0,i} = 0$ for all $i > n_0$. Then

$$a_{0,0} f_0 + \cdots + a_{0,n} f_n = 0.$$

Similarly, for $\alpha = 1, \ldots, k$, $\{f_i \mid i \in I_\alpha \cup I_\alpha'\}$ are linearly dependent, and so there is a linear relation

$$\sum_{i \in I_\alpha \cup I_\alpha'} a_{\alpha,i} f_i = 0,$$

with $a_{\alpha,i} \neq 0$. Set $a_{\alpha,i} = 0$ for all $i \notin I_\alpha \cup I_\alpha'$. Then

$$a_{\alpha,0} f_0 + \cdots + a_{\alpha,n} f_n = 0.$$

Fix any z as in the proof of Theorem 3.85. We consider the following $n \times (n+1)$ matrix

$$
\mathbf{D}_z =
\begin{pmatrix}
a_{0,0} f_0 & \cdots & c_{0,n} f_n \\
\vdots & \ddots & \vdots \\
a_{0,0} f_0^{(n_0-1)} & \cdots & a_{0,n} f_n^{(n_0-1)} \\
a_{1,0} f_0 & \cdots & a_{1,n} f_n \\
\vdots & \ddots & \vdots \\
a_{1,0} f_0^{(n_1-1)} & \cdots & a_{1,n} f_n^{(n_1-1)} \\
\vdots & \ddots & \vdots \\
a_{k,0} f_0^{(n_k-1)} & \cdots & a_{k,n} f_n^{(n_k-1)}
\end{pmatrix}
$$

Let $\mathbf{D}_{zj}$ be the determinant of the matrix obtained by deleting the jth column of $\mathbf{D}_z$. Note that the sum of each row of $\mathbf{D}_z$ is zero. We have

$$
\mathbf{D}_{zj} = (-1)^j \mathbf{D}_{z0}, \quad j = 1, \ldots, n.
$$

It is easy to show that

$$
\mathbf{D}_{z0} = \lambda \mathbf{W}_{z0} \cdots \mathbf{W}_{zk}, \quad \lambda = a_{0,0}^{-1} \prod_{\alpha=0}^{k} \prod_{i \in I_\alpha} a_{\alpha,i} \neq 0,
$$

where

$$
\mathbf{W}_{z0} = \mathbf{W}_z(f_1, \ldots, f_{n_0}) \neq 0, \quad \mathbf{W}_{z\alpha} = \mathbf{W}_z(f_{s_\alpha-1}, \ldots, f_{s_\alpha-1}) \neq 0 \ (\alpha = 1, \ldots, k)
$$

because the property (ii) in Lemma 3.87. Define

$$
\mathbf{S}_{zj} = \mathbf{S}_{zj}(f_0, \ldots, f_{j-1}, f_{j+1}, \ldots, f_n) = \frac{(-1)^j \mathbf{D}_{zj}}{f_0 \cdots f_{j-1} f_{j+1} \cdots f_n}.
$$

Then

$$
\mathbf{S}_{zj} f_0 = \mathbf{S}_{z0} f_j, \tag{3.9.22}
$$

which means

$$
h(f_0, \ldots, f_n) = h(\mathbf{S}_{z0}, \ldots, \mathbf{S}_{zn}).
$$

With respect to the corresponding local variable ζ, then we also have

$$
\mathbf{S}_{\zeta j} f_0 = \mathbf{S}_{\zeta 0} f_j, \tag{3.9.23}
$$

as in (3.9.22), and the formula (3.9.16) applied to $\mathbf{D}_{z0}$ gives

$$
\mathbf{S}_{zj} = \mathbf{S}_{\zeta j} \left(\frac{d\zeta}{dz} \right)^\gamma, \tag{3.9.24}
$$

where

$$\gamma = \sum_{\alpha=0}^{k} \gamma_{n_\alpha}. \tag{3.9.25}$$

According to the proof of Theorem 3.85, we conclude that

$$h(\mathbf{S}_{z0}, \mathbf{S}_{z1}, \ldots, \mathbf{S}_{zn}) \leq \gamma(2g - 2) + \sum_v \{\gamma_n - \gamma_{m-1}\},$$

and hence

$$h(f_0, f_1, \ldots, f_n) \leq \gamma \max\{0, 2g - 2\} + \sum_v \{\gamma_n - \gamma_{m-1}\}.$$

Finally, it is easily checked that

$$\gamma_p + \gamma_q \leq \gamma_{p+q}$$

for all $p \geq 0$, $q \geq 0$, and repeated application to (3.9.25) gives $\gamma \leq \gamma_n$. This yields the theorem. $\square$

J.F. Voloch [423], independently of W.D. Brownawell and D.W. Masser [41], considered similar questions and, by methods different from Mason's, obtained results which easily implied Corollary 3.86. U. Zannier [447] further proved the following

Theorem 3.89. *Let* $f_1, \ldots, f_n \in K$ *be* S-*units such that* $\sum_{i \in I} f_i \neq 0$ *for every non-empty* $I \subset \{1, \ldots, n\}$. *Put* $f_0 = f_1 + \cdots + f_n$. *Then*

$$\sum_{v \in S} \left\{ \mathrm{ord}_v f_0 - \min_{1 \leq i \leq n} \mathrm{ord}_v f_i \right\} \leq \gamma_d \{ \#S + 2g - 2 \}, \tag{3.9.26}$$

where $d = \dim \sum \kappa f_i$.

Proof. We follow [447] and the proof of Theorem 3.85 to treat the case $d = n$ first. Since $f_1, \ldots, f_n$ are linearly independent over the constant field κ of the derivation $\frac{d}{dz}$, the Wronskian $\mathbf{W}_z$ of $f_1, \ldots, f_n$ does not vanish. Let v be any place of K and choose a local parameter $\zeta = \zeta_v$ at v. Also let $l = l_v$ be an index such that

$$\mathrm{ord}_v f_l = \min\{\mathrm{ord}_v f_1, \ldots, \mathrm{ord}_v f_n\}.$$

By using the equality $\mathbf{W}_z = \mathbf{W}_{zl}$ and applying the inequality (3.9.17) to $\mathbf{S}_{zl}$ with $\mu = 0$, the following inequality holds:

$$\mathrm{ord}_v \mathbf{W}_z + \gamma_n \mathrm{ord}_v \frac{dz}{d\zeta} + \gamma_n \geq \mathrm{ord}_v f_0 + \sum_{i \neq l, 0} \mathrm{ord}_v f_i$$

$$= \mathrm{ord}_v f_0 - \min_{1 \leq i \leq n} \mathrm{ord}_v f_i + \sum_{i=1}^{n} \mathrm{ord}_v f_i.$$

Now sum over $v \in S$ and note that

$$\sum_{v \in S} \mathrm{ord}_v f_i = 0, \quad i = 1, \ldots, n$$

since the f_i are S-units. We have

$$\sum_{v \in S} \left\{ \mathrm{ord}_v f_0 - \min_{1 \le i \le n} \mathrm{ord}_v f_i \right\} \le \gamma_n \#S + \sum_{v \in S} \left\{ \mathrm{ord}_v \mathbf{W}_z + \gamma_n \mathrm{ord}_v \frac{dz}{d\zeta} \right\}. \quad (3.9.27)$$

On the other hand, applying (3.9.17) with $\mu = n$ we get

$$\mathrm{ord}_v \mathbf{W}_z + \gamma_n \mathrm{ord}_v \frac{dz}{d\zeta} \ge 0, \quad v \notin S,$$

whence

$$\sum_{v \in S} \left\{ \mathrm{ord}_v \mathbf{W}_z + \gamma_n \mathrm{ord}_v \frac{dz}{d\zeta} \right\} \le \sum_{v} \left\{ \mathrm{ord}_v \mathbf{W}_z + \gamma_n \mathrm{ord}_v \frac{dz}{d\zeta} \right\}. \quad (3.9.28)$$

Now it suffices to use (3.9.27) and to recall that

$$\sum_{v} \mathrm{ord}_v \mathbf{W}_z = 0, \quad \sum_{v} \mathrm{ord}_v \frac{dz}{d\zeta} = 2g - 2. \quad (3.9.29)$$

To deal with the general case we argue by induction on n, the case $n = 1$ being trivial. Let $f_1, \ldots, f_d$ be a basis for $\kappa f_1 + \cdots + \kappa f_n$ and set, renumbering indices if necessary,

$$f_0 = f_1 + \cdots + f_n = \sum_{i=1}^{\nu} a_i f_i, \quad (3.9.30)$$

where $a_1 \cdots a_\nu \ne 0$, and $1 \le \nu \le d$. Since each f_i is a linear combination of $f_1, \ldots, f_d$ with coefficients in κ, we have

$$\min_{1 \le i \le d} \mathrm{ord}_v f_i = \min_{1 \le i \le n} \mathrm{ord}_v f_i. \quad (3.9.31)$$

If $d = n$ Theorem 3.89 follows at once from the particular case treated above. If $\nu = d$, we could apply the previous result with $a_1 f_1, \ldots, a_d f_d$ in place of $f_1, \ldots, f_n$.

Next we may assume $1 \le \nu < d < n$. By the inductive assumption applied to (3.9.30), we get

$$\sum_{v \in S} \left\{ \mathrm{ord}_v f_0 - \min_{1 \le i \le \nu} \mathrm{ord}_v f_i \right\} \le \gamma_\nu \{ \#S + 2g - 2 \}. \quad (3.9.32)$$

We now construct recursively a finite sequence $\{d_l\}$ of integers such that

(I) $d_0 = \nu$, $d_l > d_{l-1}$ for $l \geq 1$,

(II) $\max\{d_l\} = d$,

(III) there is a renumbering of the indices $\nu + 1, \ldots, d$ such that

$$\sum_{v \in S} \left\{ \mathrm{ord}_v f_0 - \min_{1 \leq i \leq d_l} \mathrm{ord}_v f_i \right\} \leq \gamma_{d_l} \{ \#S + 2g - 2 \}.$$

Clearly this construction, in view of (II) and of (3.9.31), will complete the proof. The first step, namely the construction of d_0, is just (3.9.32). Assume $d_0, \ldots, d_l$ have been constructed. For any index j we have

$$f_j = \sum_{i=1}^{d} b_{ij} f_i = \sum_{i=1}^{d_l} b_{ij} f_i + \sum_{i=d_l+1}^{d} b_{ij} f_i = F_j + G_j$$

say, the b_{ij} being suitable elements of κ. If $d_l = d$, as already observed, we are done, so assume $d_l < d$. We contend that, for some j, both F_j and G_j are non-zero. In fact, assume the contrary. Then either $G_j = 0$ or $f_j = G_j$. Since $d_l \geq \nu$, the equation (3.9.30) clearly implies

$$\sum_{j=1}^{n} G_j = \sum_{i=1}^{\nu} a_i f_i - \sum_{j=1}^{n} F_j = 0,$$

and so

$$\sum_{G_j \neq 0} f_j = 0. \tag{3.9.33}$$

However, the set $I = \{ j \mid G_j \neq 0 \}$ is non-empty. In fact, $d_l < d$ and thus $d \in I$. The equation (3.9.33) would contradict our assumptions.

Pick j_0 such that both F_{j_0} and G_{j_0} are non-zero. Certainly $j_0 > d$. Renumber the indices $d_l + 1, \ldots, d$ to write

$$G_{j_0} = \sum_{i=d_l+1}^{d_{l+1}} b_{ij_0} f_i, \quad b_{ij_0} \neq 0 \ (d_l + 1 \leq i \leq d_{l+1}), \tag{3.9.34}$$

which defines d_{l+1} with $d_l < d_{l+1} \leq d$. Apply the induction assumption to F_{j_0} in place of f_0 and $f_{j_0}, -b_{ij_0} f_i$ $(d_l + 1 \leq i \leq d_{l+1})$ in place of $f_1, \ldots, f_n$. The assumptions are in fact satisfied, for

$$F_{j_0} = f_{j_0} - \sum_{i=d_l+1}^{d_{l+1}} b_{ij_0} f_i,$$

and moreover no non-empty subsum of the right-hand side vanishes since $F_{j_0} \neq 0$, the $f_1, \ldots, f_d$ are linearly independent and (3.9.34) holds. Setting

$$J = \{j_0\} \cup \{d_l + 1, \ldots, d_{l+1}\},$$

we obtain

$$\sum_{v \in S} \left\{ \mathrm{ord}_v F_{j_0} - \min_{i \in J} \mathrm{ord}_v f_i \right\} \leq \gamma_{d_{l+1} - d_l + 1} \{\#S + 2g - 2\}.$$

Adding this inequality to that in (III) above and putting $K = \{1, \ldots, d_l\}$ yields

$$\sum_{v \in S} \left\{ \mathrm{ord}_v f_0 + \mathrm{ord}_v F_{j_0} - \min_{i \in J} \mathrm{ord}_v f_i - \min_{i \in K} \mathrm{ord}_v f_i \right\}$$

$$\leq \{\gamma_{d_l} + \gamma_{d_{l+1} - d_l + 1}\} \{\#S + 2g - 2\}$$

$$\leq \gamma_{d_{l+1}} \{\#S + 2g - 2\}. \tag{3.9.35}$$

Next we will deal with the left-hand side. Since

$$F_{j_0} = \sum_{i=1}^{d_l} b_{ij_0} f_i = f_{j_0} - \sum_{i=d_l+1}^{d_{l+1}} b_{ij_0} f_i,$$

for any v we have

$$\mathrm{ord}_v F_{j_0} \geq \max \left\{ \min_{i \in J} \mathrm{ord}_v f_i, \min_{i \in K} \mathrm{ord}_v f_i \right\}.$$

Hence each term in the sum on the left of (3.9.35) is bounded below by

$$\mathrm{ord}_v f_0 - \min_{i \in J \cup K} \mathrm{ord}_v f_i \geq \mathrm{ord}_v f_0 - \min_{1 \leq i \leq d_{l+1}} \mathrm{ord}_v f_i.$$

Thus we complete the verification of (I), (III) for $l+1$ in place of l (in case $d_l < d$), and so finish the proof of Theorem 3.89. □

Corollary 3.90 ([447]). *If $f_0, f_1, \ldots, f_n \in K$ are S-units such that $f_1 + \cdots + f_n = f_0$ but no subsum of the f_i vanishes, then*

$$h(f_0, f_1, \ldots, f_n) \leq \gamma_d \{\#S + 2g - 2\}, \tag{3.9.36}$$

where $d = \dim \sum \kappa f_i$.

Proof. Now the inequality (3.9.26) holds. On the other hand, we have

$$\mathrm{ord}_v f_i = 0 \ (v \notin S, \ i = 0, 1, \ldots, n).$$

Hence the range of summation in the left-hand side of (3.9.26) may be extended to all v. To get Corollary 3.90, it now suffices to use the equations

$$\sum_v \mathrm{ord}_v f_0 = 0, \quad \min_{1 \leq i \leq n} \mathrm{ord}_v f_i = \min_{0 \leq i \leq n} \mathrm{ord}_v f_i,$$

the last one following from the basic assumption $f_0 = f_1 + \cdots + f_n$. □

Chapter 4

Function Solutions of Diophantine Equations

In this chapter, we will give analogues of the abc-conjecture, Hall's conjecture, Fermat's conjecture and Waring's problem in Nevanlinna theory, and formulate these problems into more general forms accordingly. In Section 3.1, we saw that elliptic curves are modular, that is, the distribution of rational points on elliptic curves have normal properties, we will show correspondingly that mappings of non-constant holomorphic curves into elliptic curves all are surjective. In Section 3.3, we introduced Lang's conjecture that a projective variety is hyperbolic if and only if it is Mordellic. To comprehend the question well, we need to know more examples and properties of hyperbolic spaces. Some Kobayashi hyperbolic spaces of lower dimensions will be exhibited accordingly. Referring to factorization of integers, we will introduce basic notation and questions of factorization of meromorphic functions in Section 4.10. To understand Nevanlinna theory of meromorphic functions over non-Archimedean fields in Chapter 5 well, we simply discuss Wiman-Valiron theory in Section 4.11, which contains a few interesting problems.

4.1 Nevanlinna's third main theorem

In Section 3.2, we introduce the analogy between the abc-conjecture and the Stothers-Mason's Theorem 2.65. To seek generalized abc-conjectures, we first extend Theorem 2.64 to more general cases. A generalization of Theorem 2.64 due to Nevanlinna [292] and its variations will be exhibited. To do this, we will use the following assumption:

(M) Let $f_0, f_1, \ldots, f_n$ $(n \geq 2)$ be non-zero meromorphic functions in $\mathbb{C}^m$ satisfying

$$f_0 = f_1 + f_2 + \cdots + f_n. \tag{4.1.1}$$

Assume that no proper subsum of (4.1.1) is equal to 0, and that the f_j are not all constant.

Under the assumption (M), replacing the projective height (3.9.8) we will use the following two *characteristic functions* to study the equation (4.1.1)

$$m(r) = \mathbb{C}^m \left\langle 0; r; \log \left(|f_0|^2 + \cdots + |f_n|^2\right)^{1/2} \right\rangle, \qquad (4.1.2)$$

$$T(r) = \max_{1 \le k \le n} T\left(r, \frac{f_k}{f_0}\right). \qquad (4.1.3)$$

First of all, we prove the main theorem in this section.

Theorem 4.1. *If the condition* (M) *holds such that* $f_1, f_2, \ldots, f_n$ *are linearly independent, then for* $R > \rho > r > r_0$, *we have*

$$m(r) \le \sum_{k=0}^{n} \left\{ N\left(r, \frac{1}{f_k}\right) - N(r, f_k) \right\} + N(r, \mathbf{W})$$

$$- N\left(r, \frac{1}{\mathbf{W}}\right) + l \log \left\{ \left(\frac{\rho}{r}\right)^{2m-1} \frac{T(R)}{\rho - r} \right\} + O(1), \qquad (4.1.4)$$

where $\mathbf{W} = \mathbf{W}_{\nu_1 \cdots \nu_{n-1}}(f_1, f_2, \ldots, f_n) \not\equiv 0$ *is a Wronskian determinant, and*

$$n - 1 \le l = |\nu_1| + \cdots + |\nu_{n-1}| \le \frac{n(n-1)}{2}. \qquad (4.1.5)$$

Proof. Taking partial derivatives on both sides of the identity (4.1.1) yields

$$\partial^{\nu_k} f_1 + \partial^{\nu_k} f_2 + \cdots + \partial^{\nu_k} f_n = \partial^{\nu_k} f_0 \ (k = 1, 2, \ldots, n-1).$$

Since $f_1(z), f_2(z), \ldots, f_n(z)$ are linearly independent, Lemma 2.51 implies that there exist multi-indices $\nu_i \in \mathbb{Z}_+^m$ in Theorem 4.1. The above equation and (4.1.1) yield

$$\mathbf{W} = \mathbf{W}_j \ (j = 1, 2, \ldots, n),$$

where

$$\mathbf{W}_j = \mathbf{W}_{\nu_1 \cdots \nu_{n-1}}(f_1, \ldots, f_{j-1}, f_0, f_{j+1}, \ldots, f_n).$$

Hence

$$\mathbf{S}_0 f_j = \mathbf{S}_j f_0, \qquad (4.1.6)$$

where

$$\mathbf{S}_0 = \mathbf{S}(f_1, \ldots, f_n) = \frac{\mathbf{W}}{f_1 \cdots f_n} = \begin{vmatrix} 1 & 1 & \cdots & 1 \\ \dfrac{\partial^{\nu_1} f_1}{f_1} & \dfrac{\partial^{\nu_1} f_2}{f_2} & \cdots & \dfrac{\partial^{\nu_1} f_n}{f_n} \\ \vdots & \vdots & \ddots & \vdots \\ \dfrac{\partial^{\nu_{n-1}} f_1}{f_1} & \dfrac{\partial^{\nu_{n-1}} f_2}{f_2} & \cdots & \dfrac{\partial^{\nu_{n-1}} f_n}{f_n} \end{vmatrix},$$

and $\mathbf{S}_j$ is defined by $\mathbf{W}_j = \mathbf{S}_j f_1 \cdots f_{j-1} f_0 f_{j+1} \cdots f_n$. Set

$$g_i = \frac{f_i}{f_0}, \quad i = 1, \ldots, n.$$

Then we also have

$$\mathbf{S}_0 = \mathbf{S}(g_1, \ldots, g_n), \quad \mathbf{S}_j = \mathbf{S}(g_1, \ldots, g_{j-1}, 1, g_{j+1}, \ldots, g_n).$$

By (4.1.6), a simple computation shows

$$m(r) = \mathbb{C}^m \langle 0; r; \log|f_0| \rangle - \mathbb{C}^m \langle 0; r; \log|\mathbf{S}_0| \rangle$$
$$+ \mathbb{C}^m \left\langle 0; r; \log(|\mathbf{S}_0|^2 + \cdots + |\mathbf{S}_n|^2)^{1/2} \right\rangle. \tag{4.1.7}$$

Applying the Jensen formula (2.3.11), we get

$$\mathbb{C}^m \langle 0; r; \log|f_0| \rangle = N\left(r, \frac{1}{f_0}\right) - N(r, f_0) + O(1), \tag{4.1.8}$$

and

$$-\mathbb{C}^m \langle 0; r; \log|\mathbf{S}_0| \rangle = \sum_{k=1}^{n} \mathbb{C}^m \langle 0; r; \log|f_k| \rangle - \mathbb{C}^m \langle 0; r; \log|\mathbf{W}| \rangle$$
$$= \sum_{k=1}^{n} \left\{ N\left(r, \frac{1}{f_k}\right) - N(r, f_k) \right\}$$
$$+ N(r, \mathbf{W}) - N\left(r, \frac{1}{\mathbf{W}}\right) + O(1). \tag{4.1.9}$$

By the concavity of the logarithmic function and Lemma 2.54, for any $r_0 < r < \rho < R$ we have

$$\mathbb{C}^m \left\langle 0; r; \log\sqrt{\sum|\mathbf{S}_j|^2} \right\rangle = \frac{1}{\alpha} \mathbb{C}^m \left\langle 0; r; \log\left(\sum|\mathbf{S}_j|^2\right)^{\frac{\alpha}{2}} \right\rangle$$
$$\leq \frac{1}{\alpha} \log\left\{ \sum \mathbb{C}^m \langle 0; r; |\mathbf{S}_j|^\alpha \rangle \right\}$$
$$\leq l \log\left\{ \left(\frac{\rho}{r}\right)^{2m-1} \frac{T(R)}{\rho - r} \right\} + O(1), \tag{4.1.10}$$

where α is a real number with $0 < \alpha < 1$. Thus (4.1.4) follows from (4.1.7) to (4.1.10). $\square$

For the case $m = 1$, $n = 2$, by using Theorem 4.1, a simple computation shows that

Theorem 4.2. *Assume that f_0, f_1, f_2 are meromorphic functions in $\mathbb{C}$ without common zeros, without common poles and not all constants such that $f_1 + f_2 = f_0$. Then for $r_0 < r < \rho < R$, we have*

$$m(r) \leq \overline{N}\left(r, \frac{1}{f_0 f_1 f_2}\right) + \log\left\{\frac{\rho T(R)}{r(\rho - r)}\right\} + O(1).$$

Theorem 4.2 is a direct generalization of Theorem 2.64 to the case of meromorphic functions. It shows that Theorem 4.1 may be regarded as a *generalized abc-theorem* for meromorphic functions. By using the Jensen formula (2.3.11), we may rewrite the inequality (4.1.4) into the following form:

$$m(r) \leq N\left(r, \frac{1}{\mathbf{H}}\right) - N(r, \mathbf{H}) + l\log\left\{\left(\frac{\rho}{r}\right)^{2m-1} \frac{T(R)}{\rho - r}\right\} + O(1), \qquad (4.1.11)$$

where

$$\mathbf{H} = \frac{f_0 f_1 \cdots f_n}{\mathbf{W}}. \qquad (4.1.12)$$

Observing the formula (2.6.17), the inequality (4.1.11) assumes a similar form with the main inequality in Second Main Theorem 2.56.

Next we continue to study the general equation (4.1.1). By (2.3.40) and (2.3.39), we obtain

$$T\left(r, \frac{f_j}{f_0}\right) = N\left(r, \frac{f_j}{f_0}\right) + \mathbb{C}^m\left\langle 0; r; \log\sqrt{1 + |f_j/f_0|^2}\right\rangle + O(1). \qquad (4.1.13)$$

The Jensen formula (2.3.11) implies

$$\mathbb{C}^m\left\langle 0; r; \log\sqrt{|f_0|^2 + |f_j|^2}\right\rangle = \mathbb{C}^m\left\langle 0; r; \log|f_0|\right\rangle + \mathbb{C}^m\left\langle 0; r; \log\sqrt{1 + |f_j/f_0|^2}\right\rangle$$

$$= N\left(r, \frac{1}{f_0}\right) - N(r, f_0)$$

$$+ \mathbb{C}^m\left\langle 0; r; \log\sqrt{1 + |f_j/f_0|^2}\right\rangle + O(1). \quad (4.1.14)$$

Note that

$$\mathbb{C}^m\left\langle 0; r; \log\sqrt{|f_0|^2 + |f_j|^2}\right\rangle \leq m(r).$$

Applying Theorem 4.1, we obtain the following type of Nevanlinna's theorem [292], sometimes called *Nevanlinna's third main theorem* (cf. Baesch and Steinmetz [10]). For the version of several variables, see Hu-Yang [172] or [168].

Theorem 4.3. *If the condition* (M) *holds such that* $f_1, f_2, \ldots, f_n$ *are linearly independent, then for* $1 \le j \le n$, $R > \rho > r > r_0$,

$$
T\left(r, \frac{f_j}{f_0}\right) \le N\left(r, \frac{f_j}{f_0}\right) + \sum_{k=1}^{n}\left\{N\left(r, \frac{1}{f_k}\right) - N(r, f_k)\right\} + N(r, \mathbf{W})
$$

$$
- N\left(r, \frac{1}{\mathbf{W}}\right) + l \log\left\{\left(\frac{\rho}{r}\right)^{2m-1} \frac{T(R)}{\rho - r}\right\} + O(1), \qquad (4.1.15)
$$

where $\mathbf{W} = \mathbf{W}_{\nu_1 \cdots \nu_{n-1}}(f_1, f_2, \ldots, f_n) \not\equiv 0$ *is a Wronskian determinant, and* l, $T(r)$ *are defined by* (4.1.5) *and* (4.1.3), *respectively.*

Therefore, the *abc*-theorems for meromorphic functions are only variations of Nevanlinna's third main theorem. Note that $f_0 = 1$ in the original theorem of Nevanlinna [292]. In Section 3.5, we showed that the *abc*-conjecture implies Roth's theorem. Here we derive Nevanlinna's second main theorem from Nevanlinna's third main theorem. The former is an analogue of Roth's theorem, the latter is a counterpart of the *abc*-conjecture. For simplicity, we consider only the case of one variable. Let f be a non-constant meromorphic function on $\mathbb{C}$. Then f and $1 - f$ are linearly independent such that the equation $f + (1 - f) = 1$ holds. Note that

$$
\mathbf{W}(f, 1 - f) = -f'.
$$

By Theorem 4.3, for any $R > \rho > r > r_0$, we obtain

$$
T(r, f) \le N(r, f) + N\left(r, \frac{1}{f}\right) + N\left(r, \frac{1}{f - 1}\right)
$$

$$
- N_{\mathrm{Ram}}(r, f) + \log\left\{\frac{\rho T(R, f)}{r(\rho - r)}\right\} + O(1). \qquad (4.1.16)
$$

This is the original *second main theorem* due to R. Nevanlinna [291] with a sharp error term. The general form of the second main theorem was given by Collingwood [64] and Littlewood.

Recall that if $f_1, f_2, \ldots, f_n$ are linearly independent, we may take multi-indices $\nu_i \in \mathbb{Z}_+^m$ such that

$$
0 < |\nu_i| \le i \; (i = 1, \ldots, n - 1), \; |\nu_1| \le |\nu_2| \le \cdots \le |\nu_{n-1}|,
$$

and $\mathbf{W}_{\nu_1 \cdots \nu_{n-1}}(f_1, f_2, \ldots, f_n) \not\equiv 0$. Define two integers

$$
w = |\nu_{n-1}|, \; l = |\nu_1| + \cdots + |\nu_{n-1}|
$$

and set

$$
A_n = \max_{2 \le s \le n}\left\{\frac{1}{s}\sum_{i=1}^{s-1}|\nu_{n-i}|\right\}. \qquad (4.1.17)
$$

By using Theorem 4.3 and a more precise estimation on the zeros and poles of the Wronskian determinant in Theorem 4.3, we can obtain the truncated form of *Nevanlinna's third main theorem:*

Theorem 4.4. *Under the condition* (M), *we further assume that* $f_0 = 1$, *and that* $f_1, f_2, \ldots, f_n$ *are linearly independent. Then the inequalities*

$$T(r, f_j) < \sum_{i=1}^{n} N_w \left(r, \frac{1}{f_i} \right) + \overset{*}{N}_j(r) + l \log \left\{ \left(\frac{\rho}{r} \right)^{2m-1} \frac{T(R)}{\rho - r} \right\} + O(1) \quad (4.1.18)$$

hold for $r_0 < r < \rho < R$ *and for all* $j = 1, 2, \ldots, n$, *where*

$$\overset{*}{N}_j(r) = \overset{*}{N}_j(r, A_n, w) = \min \left\{ A_n \sum_{i=1}^{n} \overline{N}(r, f_i), w \sum_{i \neq j} \overline{N}(r, f_i) \right\}.$$

It is easy to show the estimates

$$1 \leq w \leq n - 1, \quad 1 - \frac{1}{n} \leq A_n \leq \vartheta_n, \quad\quad (4.1.19)$$

where

$$\vartheta_n = \begin{cases} \frac{1}{2}, & n = 2, \\ \frac{2n-3}{3}, & n = 3, 4, 5, \\ \frac{2n+1-2\sqrt{2n}}{2} & n \geq 6. \end{cases} \quad\quad (4.1.20)$$

Specially, if $w = 1$, then

$$A_n = 1 - \frac{1}{n}.$$

We do not know the best upper bound of A_n or the best form of terms $\overset{*}{N}_j(r)$ in the inequality (4.1.18). For the one-variable case, a proof of Theorem 4.4 is given in [245]. In [172], Hu-Yang prove the case of entire functions. This version of Theorem 4.4 is given in Hu-Yang [185] (or see [168]). For the basic methods in the proof of Theorem 4.4, the reader is refereed to the proof of Theorem 4.47.

The assumption of linear independence in Theorem 4.1 is not crucial; it can be replaced by the weaker hypothesis (M) of no vanishing subsums according to the idea of W.D. Brownawell and D.W. Masser [41] introduced in Section 3.9. Here we restate it again. Assume that (M) holds and write the equation (4.1.1) in the form

$$-f_0 + f_1 + f_2 + \cdots + f_n = 0. \quad\quad (4.1.21)$$

By Lemma 3.87, there exists a partition of indices

$$\{0, 1, \ldots, n\} = I_0 \cup \cdots \cup I_k$$

satisfying the properties (i) and (ii) in Lemma 3.87. Set

$$n_0 + 1 = \#I_0; \quad n_\alpha = \#I_\alpha \ (\alpha = 1, \ldots, k)$$

and write

$$s_\alpha = 1 + \sum_{\beta=0}^{\alpha} n_\beta, \quad \alpha = 0, 1, \ldots, k.$$

Then

$$n_0 + n_1 + \cdots + n_k = n.$$

Without loss of generality, we may assume that

$$I_0 = \{0, \ldots, n_0\}, \quad I_\alpha = \{s_{\alpha-1}, \ldots, s_\alpha - 1\} \quad (\alpha = 1, \ldots, k).$$

Since I_0 is minimal, then $f_1, \ldots, f_{n_0}$ are linearly independent. Lemma 2.51 implies that there exist multi-indices $\nu_{0i} \in \mathbb{Z}_+^m$ such that

$$\mathbf{W}_0 = \mathbf{W}_{\nu_{01}\cdots\nu_{0,n_0-1}}(f_1, \ldots, f_{n_0}) \not\equiv 0.$$

Similarly, the functions $f_{s_{\alpha-1}}, \ldots, f_{s_\alpha-1}$ are linearly independent, and so Lemma 2.51 implies that there exist multi-indices $\nu_{\alpha i} \in \mathbb{Z}_+^m$ such that

$$\mathbf{W}_\alpha = \mathbf{W}_{\nu_{\alpha 1}\cdots\nu_{\alpha,n_\alpha-1}}\left(f_{s_{\alpha-1}}, \ldots, f_{s_\alpha-1}\right) \not\equiv 0, \quad \alpha = 1, \ldots, k.$$

Write

$$\mathbf{W} = \mathbf{W}_0 \cdots \mathbf{W}_k, \quad l = \sum_{\alpha=0}^{k} \sum_{i=1}^{n_\alpha-1} |\nu_{\alpha i}|, \quad w = \max_{0 \leq \alpha \leq k} |\nu_{\alpha,n_\alpha-1}| \quad (4.1.22)$$

and similarly define $\mathbf{H}$ by (4.1.12). For convenience, we also call l and w the *index* and the *Wronskian degree* of the family $\{f_1, \ldots, f_n\}$, respectively. Then the following estimates are trivial:

$$w \leq d - 1, \quad w \leq l \leq \sum_{\alpha=0}^{k} \frac{n_\alpha(n_\alpha - 1)}{2} \leq \frac{n(n-1)}{2}, \quad (4.1.23)$$

where d is the dimension of the vector space spanned by the f_j over $\mathbb{C}$.

Theorem 4.5. *If the condition* (M) *holds, then for $R > \rho > r > r_0$, the inequality* (4.1.4) *is true, where $\mathbf{W}$ and l are defined by* (4.1.22).

Proof. If $f_1, f_2, \ldots, f_n$ are linearly independent, then this is a consequence of Theorem 4.1. Next we suppose that $f_1, f_2, \ldots, f_n$ are linearly dependent. Since I_0 is minimal, then $f_0, f_1, \ldots, f_{n_0}$ are linearly dependent, and so there is a linear relation

$$a_{0,0}f_0 + \cdots + a_{0,n_0}f_{n_0} = 0,$$

with $a_{0,i} \neq 0$ for $i \in I_0$. Set $a_{0,i} = 0$ for all $i > n_0$. Then

$$a_{0,0}f_0 + \cdots + a_{0,n}f_n = 0.$$

Similarly, for $\alpha = 1, \ldots, k$, $\{f_i \mid i \in I_\alpha \cup I'_\alpha\}$ are linearly dependent, and so there is a linear relation

$$\sum_{i \in I_\alpha \cup I'_\alpha} a_{\alpha,i} f_i = 0,$$

with $a_{\alpha,i} \neq 0$. Set $a_{\alpha,i} = 0$ for all $i \notin I_\alpha \cup I'_\alpha$. Then

$$a_{\alpha,0} f_0 + \cdots + a_{\alpha,n} f_n = 0.$$

Further, we consider the following $n \times (n+1)$ matrix:

$$\mathbf{D} = \begin{pmatrix} a_{0,0} f_0 & \cdots & c_{0,n} f_n \\ \vdots & \ddots & \vdots \\ a_{0,0} \partial^{\nu_0,n_0-1} f_0 & \cdots & a_{0,n} \partial^{\nu_0,n_0-1} f_n \\ a_{1,0} f_0 & \cdots & a_{1,n} f_n \\ \vdots & \ddots & \vdots \\ a_{1,0} \partial^{\nu_1,n_1-1} f_0 & \cdots & a_{1,n} \partial^{\nu_1,n_1-1} f_n \\ \vdots & \ddots & \vdots \\ a_{k,0} \partial^{\nu_k,n_k-1} f_0 & \cdots & a_{k,n} \partial^{\nu_k,n_k-1} f_n \end{pmatrix}$$

Let $\mathbf{D}_j$ be the determinant of the matrix obtained by deleting the jth column of $\mathbf{D}$. Note that the sum of each row of $\mathbf{D}$ is zero. We have

$$\mathbf{D}_0 = (-1)^j \mathbf{D}_j, \quad j = 1, \ldots, n.$$

It is easy to show that

$$\mathbf{D}_0 = \lambda \mathbf{W}_0 \cdots \mathbf{W}_k, \quad \lambda = a_{0,0}^{-1} \prod_{\alpha=0}^{k} \prod_{i \in I_\alpha} a_{\alpha,i} \neq 0.$$

Define

$$\mathbf{S}_j = \mathbf{S}_j(f_0, \ldots, f_{j-1}, f_{j+1}, \ldots, f_n) = \frac{(-1)^j \mathbf{D}_j}{f_0 \cdots f_{j-1} f_{j+1} \cdots f_n}.$$

Then (4.1.6) still holds with

$$\mathbf{S}_j = \mathbf{S}_j(g_0, \ldots g_{j-1}, g_{j+1}, \ldots, g_n),$$

where

$$g_j = \frac{f_j}{f_0}, \quad j = 0, 1, \ldots, n.$$

Hence (4.1.7) to (4.1.10) hold, and so Theorem 4.5 follows. $\square$

According to the proof of Theorem 4.3, Theorem 4.5 yields immediately

Theorem 4.6. *Under the condition* (M), *the inequality* (4.1.15) *holds for* $1 \leq j \leq n$, $R > \rho > r > r_0$, *where* $\mathbf{W}, l$ *are defined by* (4.1.22).

Multiplying (4.1.1) by a universal denominator, we may change the condition (M) into the following form:

(E) Let $f_0, f_1, \ldots, f_n$ $(n \geq 2)$ be non-zero entire functions in $\mathbb{C}^m$ satisfying (4.1.1). Assume that no proper subsum of (4.1.1) is equal to 0, the f_j are not all constant, and that $\dim I \leq m - 2$, where

$$I = \{z \in \mathbb{C}^m \mid f_0(z) = f_1(z) = \cdots = f_n(z) = 0\}.$$

Note that if $m = 1$, the condition $\dim I \leq m - 2$ in (E) means that $f_1, f_2, \ldots, f_n$ have no common zeros.

Under the condition (E), a meromorphic mapping $f : \mathbb{C}^m \longrightarrow \mathbb{P}^n(\mathbb{C})$ with a reduced representative $\tilde{f} = (f_0, f_1, \ldots, f_n)$ is well defined. Thus the formula (2.3.21) implies

$$m(r) = T_f(r) + O(1). \tag{4.1.24}$$

By (4.1.24) and Lemma 2.37, we obtain the inequality

$$T(r) - O(1) \leq m(r) \leq nT(r) + O(1). \tag{4.1.25}$$

The following *generalized abc-theorem* for entire functions is due to Hu and Yang [180] (or see [168]):

Theorem 4.7. *Under the condition* (E), *for* $r_0 < r < \rho < R$, *we have*

$$m(r) < \sum_{i=0}^{n} N_w \left(r, \frac{1}{f_i}\right) - N(r, \mathbf{H}) + l \log \left\{ \left(\frac{\rho}{r}\right)^{2m-1} \frac{m(R)}{\rho - r} \right\} + O(1), \tag{4.1.26}$$

$$m(r) < N_l \left(r, \frac{1}{f_0 \cdots f_n}\right) - N(r, \mathbf{H}) + l \log \left\{ \left(\frac{\rho}{r}\right)^{2m-1} \frac{m(R)}{\rho - r} \right\} + O(1), \tag{4.1.27}$$

where l, w *are respectively the index and the Wronskian degree of the family* $\{f_1, \ldots, f_n\}$.

Proof. Without loss of generality, we may assume that $f_1, f_2, \ldots, f_n$ are linearly independent. Applying Theorem 4.1 and the inequality (4.1.25), we obtain

$$m(r) \leq \sum_{k=0}^{n} N \left(r, \frac{1}{f_k}\right) - N \left(r, \frac{1}{\mathbf{W}}\right) + l \log \left\{ \left(\frac{\rho}{r}\right)^{2m-1} \frac{m(R)}{\rho - r} \right\} + O(1). \tag{4.1.28}$$

The Jensen formula (2.3.11) implies

$$\sum_{k=0}^{n} N \left(r, \frac{1}{f_k}\right) - N \left(r, \frac{1}{\mathbf{W}}\right) = N \left(r, \frac{1}{\mathbf{H}}\right) - N(r, \mathbf{H}). \tag{4.1.29}$$

According to the proof of (2.6.18), we can obtain

$$N\left(r, \frac{1}{\mathbf{H}}\right) \le \sum_{k=0}^{n} N_w\left(r, \frac{1}{f_k}\right). \tag{4.1.30}$$

Hence (4.1.26) follows from (4.1.28), (4.1.29) and (4.1.30). Similarly, according to the proof of (2.6.19) we can prove the inequality (4.1.27).   □

Under the condition (E), a meromorphic mapping $F : \mathbb{C}^m \longrightarrow \mathbb{P}(V)$ with a reduced representative $\tilde{F} = (f_1, \ldots, f_n)$ is well defined, where $V = \mathbb{C}^n$. By using the formula (2.3.21), it is easy to show that

$$T_F(r) = \mathbb{C}^m \langle 0; r; \log A \rangle + O(1), \tag{4.1.31}$$

where

$$A(z) = \max_{1 \le j \le n} |f_j(z)|.$$

Thus the formulae (2.3.21), (4.1.31) and the equation (4.1.1) imply

$$m(r) = T_F(r) + O(1). \tag{4.1.32}$$

For $j = 1, \ldots, n$, set

$$e_j = (0, \ldots, 0, 1, 0, \ldots, 0) \in V$$

in which 1 is the jth component of e_j. Then e_1, ..., e_n constitute the standard basis of V. Let V^* be the dual space of V and let ϵ_1, ..., ϵ_n be the dual basis of e_1, ..., e_n. Write

$$\epsilon_0 = \epsilon_1 + \cdots + \epsilon_n.$$

Then the family $\mathscr{A} = \{\mathbb{P}(\epsilon_j)\}_{j=0}^n$ is in general position in $\mathbb{P}(V^*)$. Further, if f_1, ..., f_n are linearly independent, then Theorem 4.7 yields immediately a truncated form of the second main theorem of F for the family $\mathscr{A}$ (see Corollary 2.59).

4.2   Generalized Mason's theorem

Following Section 4.1, we particularly study the abc-problem of polynomials in $\mathbb{C}^m$. For convenience, we divide the condition (E) on polynomials into the following two parts:

(P1) Let $f_0, f_1, \ldots, f_n$ $(n \ge 2)$ be non-zero polynomials in $\mathbb{C}[z_1, \ldots, z_m]$ satisfying

$$f_0 = f_1 + f_2 + \cdots + f_n. \tag{4.2.1}$$

Assume that no proper subsum of (4.2.1) is equal to 0, and that the f_j are not all constant.

(P2) Let $f_1, f_2, \ldots, f_n$ $(n \geq 2)$ be polynomials in $\mathbb{C}[z_1, \ldots, z_m]$ such that $\dim I \leq m - 2$, where

$$I = \{z \in \mathbb{C}^m \mid f_1(z) = f_2(z) = \cdots = f_n(z) = 0\}, \qquad (4.2.2)$$

and that the f_j are not all constant.

Theorem 4.7, Lemma 2.44 and (4.1.24) yield immediately the following *generalized abc-theorem* for polynomials (cf. Hu and Yang [180]):

Theorem 4.8. *Under the assumptions* (P1) *and* (P2), *the inequalities*

$$\max_{0 \leq j \leq n} \{\deg(f_j)\} \leq \sum_{k=0}^{n} r_w(f_k) - n(\infty, \mathbf{H}) - l, \qquad (4.2.3)$$

$$\max_{0 \leq j \leq n} \{\deg(f_j)\} \leq r_l(f_0 \cdots f_n) - n(\infty, \mathbf{H}) - l, \qquad (4.2.4)$$

hold, where l, w is the index and the Wronskian degree of $f_1, f_2, \ldots, f_n$, respectively.

Remark 1. For $0 \leq i < j \leq n$, set

$$\xi_{ij} = (\xi_{ij,1}, \ldots, \xi_{ij,n-1}) = (f_0, \ldots, f_{i-1}, f_{i+1}, \ldots, f_{j-1}, f_{j+1}, \ldots, f_n),$$

$$\xi_{ij}' = \begin{pmatrix} \partial_{z_1} \xi_{ij,1} & \partial_{z_1} \xi_{ij,2} & \cdots & \partial_{z_1} \xi_{ij,n-1} \\ \partial_{z_2} \xi_{ij,1} & \partial_{z_2} \xi_{ij,2} & \cdots & \partial_{z_2} \xi_{ij,n-1} \\ \cdots\cdots\cdots\cdots\cdots\cdots\cdots\cdots\cdots\cdots \\ \partial_{z_m} \xi_{ij,1} & \partial_{z_m} \xi_{ij,2} & \cdots & \partial_{z_m} \xi_{ij,n-1} \end{pmatrix},$$

$$\gamma = \max_{z \in \mathbb{C}^m} \max_{0 \leq i < j \leq n} \operatorname{rank}\left(\xi_{ij}'(z)\right). \qquad (4.2.5)$$

If $\gamma = n - 1$, then we can take $w = 1$, $l = n - 1$ in Theorem 4.8. If $f_1, f_2, \ldots, f_n$ are linearly independent, we have

$$1 \leq w \leq n - \gamma, \quad n - 1 \leq l \leq \gamma + \frac{(n - \gamma - 1)(n - \gamma + 2)}{2}. \qquad (4.2.6)$$

Remark 2. Assume $m = 1$. If $n = 2$, Theorem 4.8 yields the Mason Theorem 2.65. If $n \geq 2$, the example

$$f_0(z) = (z + 1)^{n-1}, \quad f_{i+1}(z) = \binom{n-1}{i} z^i \ (i = 0, \ldots, n - 1), \qquad (4.2.7)$$

which obviously satisfies the conditions in Theorem 4.8, shows that the inequalities (4.2.3) and (4.2.4) in fact are equalities for this example, that is, the two inequalities in Theorem 4.8 are sharp.

For any positive integer k, $a \in \mathbb{P}^1$ and any meromorphic function f on $\mathbb{C}^m$, note that

$$n_k \left(\infty, \frac{1}{f - a} \right) \le k\bar{n} \left(\infty, \frac{1}{f - a} \right).$$

Theorem 4.8 yields immediately the following facts:

Corollary 4.9. *Under the assumptions* (P1) *and* (P2), *the inequalities*

$$\max_{0 \le j \le n} \{\deg(f_j)\} \le w \sum_{k=0}^{n} r\left(f_k\right) - l, \tag{4.2.8}$$

$$\max_{0 \le j \le n} \{\deg(f_j)\} \le lr\left(f_0 \cdots f_n\right) - l, \tag{4.2.9}$$

hold, where l, w *denote the index and the Wronskian degree of* $f_1, f_2, \ldots, f_n$, *respectively.*

For the case $m = 1$, it follows that

$$l = \frac{1}{2}n(n - 1).$$

The inequality (4.2.9) was obtained independently by J.F. Voloch [423], W.D. Brownawell and D. Masser [41]. A previous result of R.C. Mason [255] yields this estimate with l replaced by 4^{n-1}.

Corollary 4.10. *Under the assumptions* (P1) *and* (P2), *we have the inequalities*

$$\max_{0 \le j \le n} \{\deg(f_j)\} \le (d - 1) \left(\sum_{k=0}^{n} r(f_k) - 1 \right), \tag{4.2.10}$$

$$\max_{0 \le j \le n} \{\deg(f_j)\} \le \frac{n(n - 1)}{2} \left(r\left(f_0 \cdots f_n\right) - 1 \right), \tag{4.2.11}$$

where d *is the dimension of the vector space spanned by the* f_j *over* $\mathbb{C}$.

Let $\mathrm{rad}(f_i)$ be the *radical* of f_i, which is the product of the distinct irreducible factors of f_i, i. e., $\mathrm{rad}(f_i)$ is the squarefree part of f_i, and define

$$\bar{r}(f_i) = \deg(\mathrm{rad}(f_i)).$$

Proposition 2.43 implies

$$r(f_i) = \bar{n} \left(\infty, \frac{1}{f_i} \right) \le n \left(\infty, \frac{1}{\mathrm{rad}(f_i)} \right) = \bar{r}(f_i).$$

Thus a theorem of Shapiro and Sparer [343] follows from Corollary 4.10:

Theorem 4.11. *If the condition* (P1) *holds such that the* f_j *are relatively prime by pairs, then*

$$\max_{0 \le j \le n} \{\deg(f_j)\} \le (n - 1) \{\bar{r}\left(f_0 \cdots f_n\right) - 1\}. \tag{4.2.12}$$

For polynomials on a field of characteristic zero, J. Browkin and J. Brzeziński [40] proposed a conjecture as follows:

Conjecture 4.12. *Let $f_j(j = 0, \ldots, n)$ be non-zero polynomials on a field K of characteristic zero with $n \geq 2$ such that $f_0, \ldots, f_n$ have no non-constant common divisors, at least one of the f_j is not a constant, (4.2.1) holds and no proper subsum of (4.2.1) is equal to 0. Then*

$$\max_{0 \leq j \leq n} \{\deg(f_j)\} \leq (2n - 3)\left(\bar{r}(f_0 \cdots f_n) - 1\right). \tag{4.2.13}$$

Corollary 4.10 shows that Conjecture 4.12 is true if the number $2n - 3$ in (4.2.13) is replaced by $\frac{n(n-1)}{2}$. For the case $n = 2$, the conditions in Conjecture 4.12 mean that f_1 and f_2 are linearly independent. Hence it follows from (4.2.9). It also is easy to show from (4.2.9) that Conjecture 4.12 is true for the case $n = 3$. Note that these are all cases such that $2n - 3 = \frac{1}{2}n(n - 1)$ holds for a positive integer n.

J. Browkin and J. Brzeziński [40] studied the following example: For every $k \geq 0$, define a polynomial of positive integral coefficients by

$$f_k(z) = \prod_{j=1}^{k}(z + 2 - 2\cos\alpha_j) = \sum_{j=0}^{k} s_j z^j, \quad \alpha_j = \frac{2\pi j}{2k + 1}, \tag{4.2.14}$$

which satisfies (cf. [40])

$$\frac{x^{2k+1} - 1}{x - 1} = x^k f_k\left(\frac{(x - 1)^2}{x}\right). \tag{4.2.15}$$

If in (4.2.15) we put $k = n - 2$ and $x = -b/a$, then, in view of (4.2.14), one gets

$$a^{2n-3} + b^{2n-3} - \sum_{j=0}^{n-2} s_j(a + b)^{2j+1}(-ab)^{n-2-j} = 0. \tag{4.2.16}$$

J. Browkin and J. Brzeziński use the example (4.2.16) by putting $a = r^k + 1\ (k > 0)$ and $b = -1$, that is,

$$(r^k + 1)^{2n-3} - 1 - r^k \sum_{j=0}^{n-2} s_j r^{2kj}(r^k + 1)^{n-2-j} = 0 \tag{4.2.17}$$

to show that the number $2n - 3$ in Conjecture 4.12 is a sharp lower bound (also see [74]).

Based on Corollary 3.90 and Corollary 4.10, we suggest the following estimate:

Conjecture 4.13. *Under the conditions of Conjecture 4.12, the inequality*

$$\max_{0 \leq j \leq n} \{\deg(f_j)\} \leq \frac{d(d-1)}{2}(\bar{r}(f_0 \cdots f_n) - 1) \tag{4.2.18}$$

holds, where d is the dimension of the vector space spanned by the f_i over K.

Now for any positive integer k, it is natural to ask what is the minimal numbers $\mathfrak{x}_{m,n}(k)$ and $\mathfrak{y}_{m,n}(k)$ such that under the assumptions (P1) and (P2), we have the inequalities

$$\max_{0 \leq j \leq n} \{\deg(f_j)\} \leq \sum_{j=0}^{n} \mathfrak{x}_{m,n}(k) r_k(f_j) - l, \tag{4.2.19}$$

$$\max_{0 \leq j \leq n} \{\deg(f_j)\} \leq \mathfrak{y}_{m,n}(k) r_k(f_0 \cdots f_n) - l. \tag{4.2.20}$$

Abbreviate

$$\mathfrak{x}_n(k) = \mathfrak{x}_{1,n}(k), \quad \mathfrak{y}_n(k) = \mathfrak{y}_{1,n}(k).$$

Theorem 4.8, the example (4.2.7) and Corollary 4.10 show that

$$\mathfrak{x}_n(k) \begin{cases} = 1, & \text{if } k \geq n-1; \\ \leq n-1, & \text{if } 1 \leq k < n-1 \end{cases} \tag{4.2.21}$$

and

$$\mathfrak{y}_n(k) \begin{cases} = 1, & \text{if } k \geq \frac{n(n-1)}{2}; \\ \leq \frac{n(n-1)}{2}, & \text{if } 1 \leq k < \frac{n(n-1)}{2}. \end{cases} \tag{4.2.22}$$

We can prove easily

$$\mathfrak{x}_n(k) \leq \frac{n-1}{k}, \quad 1 \leq k \leq n-1,$$

and

$$\mathfrak{y}_n(k) \leq \frac{n(n-1)}{2k}, \quad 1 \leq k \leq \frac{n(n-1)}{2}.$$

Theorem 4.14. *Assume that the condition (P2) holds such that f_1, f_2, $\ldots$, f_n are linearly independent. Take a positive integer $q > n$ and let $[a_{j1}, \ldots, a_{jn}]$ ($j = 1, \ldots, q$) be a family of points of $\mathbb{P}(\mathbb{C}^n)$ in general position. Then the inequalities*

$$(q-n) \max_{1 \leq j \leq n} \{\deg(f_j)\} \leq \sum_{j=1}^{q} r_w(a_{j1}f_1 + \cdots + a_{jn}f_n) - n(\infty, \mathbf{H}) - l, \tag{4.2.23}$$

$$(q-n) \max_{1 \leq j \leq n} \{\deg(f_j)\} \leq r_l \left(\prod_{j=1}^{q} (a_{j1}f_1 + \cdots + a_{jn}f_n) \right) - n(\infty, \mathbf{H}) - l \tag{4.2.24}$$

hold, where l, w is the index and the Wronskian degree of $f_1, f_2, \ldots, f_n$, respectively.

Proof. Let $f : \mathbb{C}^m \longrightarrow \mathbb{P}(\mathbb{C}^n)$ be the meromorphic mapping with a reduced representation $\tilde{f} = (f_1, \ldots, f_n)$. Set $a_j = [a_{j1}, \ldots, a_{jn}]$. Then

$$r_w(a_{j1}f_1 + \cdots + a_{jn}f_n) = \lim_{r \to \infty} n_{f,w}(r, a_j) = \lim_{r \to \infty} \frac{N_{f,w}(r, a_j)}{\log r}.$$

Since the mapping f is linearly non-degenerate under the assumptions of Theorem 4.14, then Corollary 2.59 and (2.4.7) imply the inequality (4.2.23). Similarly, we can prove (4.2.24) by using Theorem 2.56, (2.6.17) and (2.6.19). $\qquad\square$

Theorem 4.15. *Assume that* (P2) *holds. Let d be the dimension of the vector space spanned by the f_j over $\mathbb{C}$. Take a positive integer $q > 2n - d$ and let $[a_{j1}, \ldots, a_{jn}]$ $(j = 1, \ldots, q)$ be a family of points of $\mathbb{P}(\mathbb{C}^n)$ in general position such that*

$$a_{j1}f_1 + \cdots + a_{jn}f_n \not\equiv 0, \quad j = 1, \ldots, q.$$

Then we have the inequality

$$(q - 2n + d) \max_{1 \le j \le n} \{\deg(f_j)\} \le \sum_{j=1}^{q} r_w(a_{j1}f_1 + \cdots + a_{jn}f_n) - l\theta, \qquad (4.2.25)$$

where l, w is the index and the Wronskian degree of $f = [f_1, f_2, \ldots, f_n] : \mathbb{C}^m \longrightarrow \mathbb{P}(\mathbb{C}^n)$, respectively, such that

$$1 \le w \le d - 1 \le l \le \frac{d(d-1)}{2},$$

and where θ is a Nochka constant with $n \le \theta d \le 2n - d$.

Proof. Note that the holomorphic mapping

$$f = [f_1, f_2, \ldots, f_n] : \mathbb{C}^m \longrightarrow \mathbb{P}(\mathbb{C}^n)$$

is $(d - 1)$-flat with $d \ge 2$. By using Theorem 2.70, the proof of Theorem 4.15 can be completed according to that of Theorem 4.14. $\qquad\square$

4.3 Generalized *abc*-conjecture

In this section, we will study the following analogues of conditions (P1) and (P2) in Section 4.2 for integers:

(N1) Let $a_0, a_1, \ldots, a_n$ $(n \ge 2)$ be non-zero integers satisfying

$$a_0 = a_1 + a_2 + \cdots + a_n. \qquad (4.3.1)$$

Assume that no proper subsum of (4.3.1) is equal to 0.

(N2) Let $a_1, a_2, \ldots, a_n$ ($n \geq 2$) be integers satisfying $\gcd(a_1, \ldots, a_n) = 1$, where the symbol $\gcd(a_1, \ldots, a_n)$ denotes the *greatest common divisor* of $a_1, \ldots, a_n$.

For a non-zero integer a, write

$$a = \pm p_1^{i_1} \cdots p_s^{i_s} \tag{4.3.2}$$

for distinct primes $p_1, \ldots, p_s$ and $(i_1, \ldots, i_s) \in (\mathbb{Z}^+)^s$, and define

$$r_k(a) = \prod_{\nu=1}^{s} p_\nu^{\min\{i_\nu, k\}}. \tag{4.3.3}$$

Based on the classical analogy between polynomials and integers, we think that the number $r_k(f)$ of a polynomial f corresponds to $\log r_k(a)$ of an integer a. Thus Theorem 4.8 can be translated into the following *generalized abc-conjecture* for integers (see Hu and Yang [182]).

Conjecture 4.16. *If* (N1) *and* (N2) *are true, then for $\varepsilon > 0$, $k \in \mathbb{Z}^+$, there exists a number $C = C(n, k, \varepsilon)$ satisfying*

$$\max_{0 \leq j \leq n} \{|a_j|\} \leq C \left(\prod_{i=0}^{n} r_k(a_i) \right)^{\mathfrak{x}_n(k) + \varepsilon}, \tag{4.3.4}$$

$$\max_{0 \leq j \leq n} \{|a_j|\} \leq C r_k (a_0 a_1 \cdots a_n)^{\mathfrak{y}_n(k) + \varepsilon}. \tag{4.3.5}$$

In [176], [180] (or see [181]), we proposed the conjecture for the case

$$\mathfrak{x}_n(n-1) = 1, \quad \mathfrak{y}_n \left(\frac{n(n-1)}{2} \right) = 1. \tag{4.3.6}$$

If $n = 2$, Conjecture 4.16 corresponds to the well-known *abc*-conjecture.

We discuss the example (4.2.16) studied by J. Browkin and J. Brzeziński [40]. If we choose $a = 2^i$, where $i > n - 2$, and $b = -1$, then we have

$$a_1 + \cdots + a_n = a_0,$$

where

$$a_{j+1} = s_j (2^i - 1)^{2j+1} 2^{i(n-2-j)} \ (0 \leq j \leq n-2), \ a_n = 1, \ a_0 = 2^{i(2n-3)}.$$

Obviously, it has no proper subsum equal to zero. Since $a_n = 1$, hence the greatest common divisor of all a_j is 1. Therefore the conditions in Conjecture 4.16 are satisfied. Now we have

$$M_n = \max_{0 \leq j \leq n} \{|a_j|\} = a_0 = 2^{i(2n-3)}.$$

A positive integer $\chi_n \geq 2n - 3$ exists such that

$$L_n := \prod_{i=0}^{n} r_{n-1}(a_i) = 2^{n-2} \prod_{j=0}^{n-2} r_{n-1}\left(s_j(2^i - 1)^{2j+1} 2^{i(n-2-j)}\right)$$

$$\geq 2^{(n-2)(n-2)} \prod_{j=0}^{n-2} r_{n-1}\left((2^i - 1)^{2j+1}\right) = 2^{(n-2)(n-2)}(2^i - 1)^{\chi_n}.$$

Since there are infinitely many $i > n-2$ such that the numbers $2^i - 1$ are relatively prime (e.g., all prime $i > n - 2$), there exists a positive constant $C(n)$ which is independent of i such that

$$\frac{2^{i(2n-3)}}{2^{(n-2)(n-2)}(2^i - 1)^{\chi_n}} \leq C(n),$$

that is, $M_n \leq C(n)L_n$. We can also show that for some constant $C(n)$,

$$M_n \leq C(n)r_{\frac{n(n-1)}{2}}(a_0 a_1 \cdots a_n).$$

Thus for the case (4.3.6), Conjecture 4.16 holds for such a_j.

Next we exhibit a few conjectures related to Conjecture 4.16. If $a_j (j = 0, \ldots, n)$ are non-zero integers such that a_i, a_j are coprime for $i \neq j$, then

$$\prod_{i=0}^{n} r_k(a_i) = r_k(a_0 a_1 \cdots a_n) \leq r(a_0 a_1 \cdots a_n)^k.$$

Hence Conjecture 4.16 implies immediately the following conjecture due to W.M. Schmidt [338]:

Conjecture 4.17. *If* (N1) *holds such that a_i and a_j are coprime for $i \neq j$, then for $\varepsilon > 0$, there exists a number $C = C(n, \varepsilon)$ such that*

$$\max_{0 \leq j \leq n} \{|a_j|\} \leq Cr(a_0 a_1 \cdots a_n)^{n-1+\varepsilon}. \tag{4.3.7}$$

It was indicated by Vojta in [419] that Conjecture 3.56 could derive the following conjecture:

Conjecture 4.18. *If $a_0, \ldots, a_n$ are non-zero integers satisfying* (4.3.1) *and* (N2), *then for $\varepsilon > 0$, there exists a number $C = C(n, \varepsilon)$ such that*

$$\max_{0 \leq j \leq n} \{|a_j|\} \leq Cr(a_0 a_1 \cdots a_n)^{1+\epsilon} \tag{4.3.8}$$

hold for all $a_0, \ldots, a_n$ as above outside a proper Zariski-closed subset of the hyperplane $x_1 + \cdots + x_n = x_0$ in $\mathbb{P}^n$.

J. Browkin and J. Brzeziński [40] conjectured as follows:

Conjecture 4.19. *If* (N1) *and* (N2) *are true, then for* $\varepsilon > 0$*, there exists a number* $C = C(n, \varepsilon)$ *such that*

$$\max_{0 \leq j \leq n} \{|a_j|\} \leq Cr \left(a_0 a_1 \cdots a_n\right)^{2n-3+\varepsilon}. \tag{4.3.9}$$

J. Browkin and J. Brzeziński use the above example to show that the number $2n - 3$ is a sharp lower bound. Thus the number $\mathfrak{y}_n(1)$ should satisfy

$$2n - 3 \leq \mathfrak{y}_n(1) \leq \frac{n(n-1)}{2}. \tag{4.3.10}$$

Generally, we think that Theorem 4.14 corresponds to the following problem:

Conjecture 4.20. *Assume that* (N2) *holds and further assume that there exist integers* M *and* N *with* $M < N$ *such that*

$$B_1 a_1 + \cdots + B_n a_n \neq 0, \quad (B_1, \ldots, B_n) \in \mathbb{Z}[M, N]^n - \{0\}.$$

Take an integer q *with* $q > n$ *and let a family* $\{[A_{j1}, \ldots, A_{jn}] \mid j = 1, \ldots, q\}$ *of* $\mathbb{P}(\mathbb{C}^n)$ *be in general position with* $A_{ji} \in \mathbb{Z}[M, N]$*. Then for* $\varepsilon > 0$*,* $k \in \mathbb{Z}^+$*, there exists a number* $C = C(n, k, q, M, N, \varepsilon)$ *satisfying*

$$\max_{1 \leq j \leq n} \{|a_j|^{q-n}\} \leq C \left(\prod_{j=1}^{q} r_k \left(A_{j1} a_1 + \cdots + A_{jn} a_n \right) \right)^{\mathfrak{r}_n(k)+\varepsilon}, \tag{4.3.11}$$

$$\max_{1 \leq j \leq n} \{|a_j|^{q-n}\} \leq Cr_k \left(\prod_{j=1}^{q} \left(A_{j1} a_1 + \cdots + A_{jn} a_n \right) \right)^{\mathfrak{y}_n(k)+\varepsilon}. \tag{4.3.12}$$

The inequalities (4.3.4) and (4.3.5) in Conjecture 4.16 follow from Conjecture 4.20 by taking $M = 0$, $N = 1$ and $q = n + 1$.

4.4   Generalized Hall's conjecture

We continue to study the case (E) in Section 4.1. To compare with the conditions in Section 4.2 and 4.3 accordingly, we divide it into two parts:

(E1) Let $f_0, f_1, \ldots, f_n$ $(n \geq 2)$ be non-zero entire functions in $\mathbb{C}^m$ satisfying

$$f_0 = f_1 + f_2 + \cdots + f_n. \tag{4.4.1}$$

Assume that no proper subsum of (4.4.1) is equal to 0, and that the f_j are not all constant.

(E2) Let $f_1, f_2, \ldots, f_n$ $(n \geq 2)$ be entire functions in $\mathbb{C}^m$ such that $\dim I \leq m - 2$, where

$$I = \{z \in \mathbb{C}^m \mid f_1(z) = f_2(z) = \cdots = f_n(z) = 0\}, \qquad (4.4.2)$$

and that the f_j are not all constant.

In this section, we will study entire functions with higher multiplicity of zeros. Thus we make the following additional assumption:

(E3) Let $f_1, f_2, \ldots, f_n$ $(n \geq 2)$ be entire functions in $\mathbb{C}^m$. Assume that there exist positive integers d_j such that the multiplicity of each root of f_j in $\mathbb{C}^m - I$ is not less than d_j for $j = 1, \ldots, n$.

Theorem 4.21. *Under the conditions* (E1), (E2) *and* (E3), *for* $r_0 < r < \rho < R$ *the meromorphic mapping* $f : \mathbb{C}^m \longrightarrow \mathbb{P}^n$ *with a reduced representative* $\tilde{f} = (f_0, f_1, \ldots, f_n) : \mathbb{C}^m \longrightarrow \mathbb{C}^{n+1}$ *satisfies the following inequalities:*

$$\left\{ 1 - \sum_{j=1}^n \frac{w}{d_j} \right\} T_f(r) < N_w\left(r, \frac{1}{f_0}\right) + l \log \left\{ \left(\frac{\rho}{r}\right)^{2m-1} \frac{T_f(R)}{\rho - r} \right\} + O(1), \quad (4.4.3)$$

where l, w *are respectively the index and the Wronskian degree of the family* $\{f_1, \ldots, f_n\}$.

Proof. Note that for $j = 1, \ldots, n$,

$$N_w\left(r, \frac{1}{f_j}\right) \leq w \overline{N}\left(r, \frac{1}{f_j}\right) \leq \frac{w}{d_j} N\left(r, \frac{1}{f_j}\right) \leq \frac{w}{d_j} T_f(r) + O(1).$$

Hence Theorem 4.21 follows from Theorem 4.7. $\qquad \square$

Theorem 4.21 implies directly the following fact:

Theorem 4.22. *Let* $f_0, f_1, \ldots, f_n$ *be polynomials in* $\mathbb{C}^m$ *satisfying the conditions* (E1), (E2) *and* (E3). *Then the inequality*

$$\left\{ 1 - \sum_{j=1}^n \frac{w}{d_j} \right\} \max_{0 \leq j \leq n} \deg(f_j) \leq r_w(f_0) - l \qquad (4.4.4)$$

holds, where l, w *is the index and the Wronskian degree of* $\{f_1, f_2, \ldots, f_n\}$, *respectively.*

Obviously, Theorem 4.22 also follows directly from Theorem 4.8. As a consequence, we obtain the following fact (see [186]):

Theorem 4.23. *Let* $f_0, f_1, \ldots, f_n$ *be polynomials in* $\mathbb{C}^m$ *satisfying* (E1) *and* (E2). *Assume that there exist positive integers* d_j *and polynomials* P_j *in* $\mathbb{C}^m$ *such that*

$$f_j = P_j^{d_j}, \quad j = 1, \ldots, n.$$

Then the inequality

$$\left\{ 1 - \sum_{j=1}^{n} \frac{w}{d_j} \right\} \max_{0 \le j \le n} d_j \deg\left(P_j\right) \le r_w\left(f_0\right) - l \tag{4.4.5}$$

holds, where l, w is the index and the Wronskian degree of $\{P_1^{d_1}, \ldots, P_n^{d_n}\}$, *respectively.*

Since $w \le n - 1 \le l$, (4.4.5) implies

$$\left\{ 1 - \sum_{j=1}^{n} \frac{n-1}{d_j} \right\} \max_{0 \le j \le n} d_j \deg\left(P_j\right) \le (n-1)\left\{ r\left(\sum_{j=1}^{n} P_j^{d_j}\right) - 1 \right\}. \tag{4.4.6}$$

In particular, if f and g are non-zero polynomials in $\mathbb{C}$ with $f^2 - g^3 \ne 0$, and are not all constant, then (4.4.6) yields

$$\frac{1}{6} \max\{2 \deg(f), 3 \deg(g)\} \le r\left(f^2 - g^3\right) - 1 \tag{4.4.7}$$

when f and g have no common zeros, which provides the inequality in the following *Davenport theorem* (or see [25], [387]):

Theorem 4.24 ([72]). *If f and g are non-constant polynomials in $\mathbb{C}$ with $f^2 - g^3 \ne 0$, then*

$$\frac{1}{2} \deg\left(g\right) \le \deg\left(f^2 - g^3\right) - 1. \tag{4.4.8}$$

The analogue of Theorem 4.24 in number theory is just Hall's conjecture. The inequality (4.4.7) is an analogue of (3.2.6). For the case

$$n = 2, \quad d_1 = k, \quad d_2 = n, \tag{4.4.9}$$

and

$$f_1 = f, \quad f_2 = \lambda g, \tag{4.4.10}$$

where λ is a constant such that $\lambda^n = -1$, the inequality (4.4.6) yields

$$\left\{ 1 - \frac{1}{k} - \frac{1}{n} \right\} \max\{\deg\left(f^k\right), \deg\left(g^n\right)\} \le r\left(f^k - g^n\right) - 1 \tag{4.4.11}$$

when f and g have no common zeros, which implies

$$\left\{ 1 - \frac{1}{k} - \frac{1}{n} \right\} \max\{\deg\left(f^k\right), \deg\left(g^n\right)\} \le \deg\left(f^k - g^n\right) - 1. \tag{4.4.12}$$

This inequality is an analogue of the Hall-Lang-Waldschmidt conjecture (3.2.10) for polynomials.

For integers, the conditions (E1), (E2) and (E3) are respectively replaced by:

(N1) Let $A_1, \ldots, A_n$ $(n \geq 2)$ be fixed non-zero integers and let $x_j (j = 0, 1, \ldots, n)$ be non-zero integers satisfying

$$A_1 x_1 + \cdots + A_n x_n = x_0. \qquad (4.4.13)$$

Assume that no proper subsum of (4.4.13) is equal to 0.

(N2) Let $x_1, x_2, \ldots, x_n$ $(n \geq 2)$ be integers satisfying $\gcd(x_1, \ldots, x_n) = 1$.

(N3) Suppose that there are positive integers d_j such that for each $j = 1, \ldots, n$, $p \mid x_j$ for some prime p implies $p^{d_j} \mid x_j$.

We conjectured that the analogue of Theorem 4.22 in number theory would be the following problem:

Conjecture 4.25. *If* (N1), (N2) *and* (N3) *hold such that for some* $k \in \mathbb{Z}^+$,

$$\alpha = 1 - \sum_{j=1}^{n} \frac{k \mathfrak{r}_n(k)}{d_j} > 0, \qquad (4.4.14)$$

then for $\varepsilon > 0$, *there exists a number* $C = C(n, k, \varepsilon, A_1, \ldots, A_n)$ *such that*

$$\max_{0 \leq j \leq n} \{|x_j|^{\alpha}\} \leq C r_k (x_0)^{\mathfrak{r}_n(k)+\varepsilon}. \qquad (4.4.15)$$

Conjecture 4.25 is a generalization of Conjecture 3.26. Note that when x_0 is fixed, the equation (4.4.13) has integer solutions $x_1, \ldots, x_n$ if and only if the fixed non-zero integers $A_1, \ldots, A_n$ satisfy

$$\gcd(A_1, \ldots, A_n) \mid x_0.$$

According to the discussion in Section 3.2, it is easy to show that Conjecture 4.25 follows from Conjecture 4.16.

Conjecture 4.26. *Assume that* (N1) *and* (N2) *hold and suppose that there are positive integers* d_j *and integers* a_j *such that*

$$x_j = a_j^{d_j}, \quad j = 1, \ldots, n.$$

If (4.4.14) *holds, then for* $\varepsilon > 0$, *there is a number* $C = C(n, k, \varepsilon, A_1, \ldots, A_n)$ *such that*

$$\max_{0 \leq j \leq n} \{|a_j|^{\alpha d_j}\} \leq C r_k (x_0)^{\mathfrak{r}_n(k)+\varepsilon}, \qquad (4.4.16)$$

where $a_0 = x_0$, $d_0 = 1$.

Conjecture 4.26 is a generalization of Conjecture 3.28, and follows easily from Conjecture 4.25. Some special cases of Conjecture 4.26 were suggested in [186] and [181].

4.5   Borel's theorem and its analogues

In this section, we first establish a general Borel theorem. Based on the theorem, we suggest a generalized Fermat conjecture. To do this, we need some notation. Take $f \in \mathcal{M}(\mathbb{C}^m)$ and $a \in \mathbb{P}^1$. For a positive integer k, denote the *truncated multiplicity functions* on $\mathbb{C}^m$ by

$$\overline{\mu}_{f)k}^a(z) = \begin{cases} 1 & \text{if } 0 < \mu_f^a(z) \le k, \\ 0 & \text{otherwise}, \end{cases} \tag{4.5.1}$$

$$\overline{\mu}_{f(k}^a(z) = \begin{cases} 1 & \text{if } \mu_f^a(z) \ge k, \\ 0 & \text{if } \mu_f^a(z) < k, \end{cases} \tag{4.5.2}$$

and further define the *truncated valence functions* by

$$N_\nu(t) = \begin{cases} \overline{N}_{k)}\left(t, \frac{1}{f-a}\right) & \text{if } \nu = \overline{\mu}_{f)k}^a, \\ \overline{N}_{(k}\left(t, \frac{1}{f-a}\right) & \text{if } \nu = \overline{\mu}_{f(k}^a. \end{cases} \tag{4.5.3}$$

Let $\lambda(r)$ be a continuous, increasing non-negative unbounded function of $r \in \mathbb{R}^+$. Let $\mathcal{M}_\lambda(\mathbb{C}^m)$ be the field of meromorphic functions a on $\mathbb{C}^m$ such that

$$\| \quad T(r, a) = o(\lambda(r)).$$

If $T(r)$ is a non-negative function on $\mathbb{R}^+$, we will use the notation

$$\| \quad T(r) \ne o(\lambda(r))$$

to denote that

$$\limsup_{E \not\ni r \to \infty} \frac{T(r)}{\lambda(r)} > 0$$

for any subset $E \subset \mathbb{R}^+$ with $\int_E d\log r < \infty$.

4.5.1   Borel's theorem

In this part, we will study meromorphic functions satisfying the following conditions:

(M1)   Let $f_0, f_1, \ldots, f_n$ $(n \ge 2)$ be non-zero meromorphic functions on $\mathbb{C}^m$. Assume that there exists $a_j \in \mathcal{M}_\lambda(\mathbb{C}^m)$ satisfying the equation

$$a_0 f_0 + a_1 f_1 + \cdots + a_n f_n = 0. \tag{4.5.4}$$

(M2) Take positive integers n and d_j $(j = 0, 1, \ldots, n)$ with $n \geq 2$ and set $d = \max\{d_j\}$. Let $f_0, f_1, \ldots, f_n$ be non-zero meromorphic functions on $\mathbb{C}^m$ satisfying the following condition:

$$\| \sum_{j=0}^{n} \left\{ \overline{N}_{d-1)}(r, f_j) + \overline{N}_{d_j-1)} \left(r, \frac{1}{f_j} \right) \right\} = o(\lambda(r)). \qquad (4.5.5)$$

If the functions a_j in (M1) are constants, we usually replace the condition (M2) by the stronger condition:

(M2') Take positive integers n and d_j $(j = 0, 1, \ldots, n)$ with $n \geq 2$ and set $d = \max\{d_j\}$. Let $f_0, f_1, \ldots, f_n$ be non-zero meromorphic functions on $\mathbb{C}^m$ satisfying the following condition:

$$\overline{\mu}^{\infty}_{f_j)d-1} = \overline{\mu}^{0}_{f_j)d_j-1} = 0, \quad j = 0, 1, \ldots, n. \qquad (4.5.6)$$

Here we give a generalization of a result due to Borel [29] and Nevanlinna [292] as follows:

Theorem 4.27. *If* (M1) *and* (M2) *hold such that*

$$\beta = 1 - \sum_{j=0}^{n} \frac{n-1}{d_j} > 0, \qquad (4.5.7)$$

then we have

1) *the functions* $a_1 f_1, \ldots, a_n f_n$ *are linearly dependent if*

$$\| \max_{1 \leq j \leq n} T \left(r, \frac{f_j}{f_0} \right) \neq o(\lambda(r));$$

2) $a_j = 0$ *for* $j = 0, 1, 2, \ldots, n$ *if*

$$\| T \left(r, \frac{f_j}{f_k} \right) \neq o(\lambda(r)), \quad j \neq k.$$

Proof. We prove Theorem 4.27, 2) by induction on n. First of all, we consider the case $n = 1$. Since $f_j \neq 0$ $(j = 0, 1)$, then $a_0 \neq 0$ and $a_1 \neq 0$ if one of a_0 and a_1 is not zero. Hence

$$\| T \left(r, \frac{f_0}{f_1} \right) = T \left(r, \frac{a_1}{a_0} \right) \leq T(r, a_1) + T(r, a_0) + O(1) = o(\lambda(r))$$

which is a contradiction. Hence $a_0 = a_1 = 0$.

Assume that Theorem 4.27 holds up to $n - 1$. It is sufficient to show that one of $a_0, \ldots, a_n$ is zero. Assume, to the contrary, that $a_j \neq 0$ for $j = 0, 1, \ldots, n$.

Then $a_0 f_0, \ldots, a_{n-1} f_{n-1}$ are linearly independent over $\mathbb{C}$. In fact, if there exists $c_j \in \mathbb{C}$ for $j = 0, \ldots, n-1$ such that

$$c_0 a_0 f_0 + \cdots + c_{n-1} a_{n-1} f_{n-1} = 0,$$

by induction, then $c_j a_j = 0$ $(j = 0, \ldots, n-1)$. Thus $c_j = 0$ $(j = 0, \ldots, n-1)$. Let V be a complex vector space of dimension n. Take a base $e_0, \ldots, e_{n-1}$ of V and let $\epsilon_0, \ldots, \epsilon_{n-1}$ be the dual base in V^*. Let Δ $(\not\equiv 0)$ be a *universal denominator* of $\{a_0 f_0, \ldots, a_{n-1} f_{n-1}\}$, that is, $\Delta a_i f_i$ is holomorphic for each $i = 0, \ldots, n-1$ with

$$\dim\{z \in \mathbb{C}^m \mid (\Delta a_0 f_0)(z) = \cdots = (\Delta a_{n-1} f_{n-1})(z) = 0\} \le m - 2.$$

Then a meromorphic mapping $F : \mathbb{C}^m \longrightarrow \mathbb{P}(V)$ is defined with a reduced representation

$$\tilde{F} = \Delta a_0 f_0 e_0 + \cdots + \Delta a_{n-1} f_{n-1} e_{n-1} : \mathbb{C}^m \longrightarrow V.$$

Obviously, F is linearly non-degenerate. Set

$$b_i = \mathbb{P}(\epsilon_i) \ (0 \le i \le n-1), \quad b_n = \mathbb{P}\left(\sum_{i=0}^{n-1} \epsilon_i\right).$$

Then the family $\{b_0, \ldots, b_n\}$ is in general position. Then Corollary 2.59 implies

$$\| \quad T_F(r) \le \sum_{i=0}^{n} N_{F,n-1}(r, b_i) + O(\log T_F(r)), \tag{4.5.8}$$

where, by definition,

$$N_{F,n-1}(r, b_i) = N_{n-1}\left(r, \frac{1}{\Delta a_i f_i}\right) \le (n-1)\overline{N}\left(r, \frac{1}{\Delta a_i f_i}\right).$$

According to the definition of Δ, we can show easily that the inequality

$$\| \quad \overline{N}_{d_i-1)}\left(r, \frac{1}{\Delta a_i f_i}\right) \le \overline{N}_{d_i-1)}\left(r, \frac{1}{f_i}\right) + \sum_{j=0}^{n-1} \overline{N}_{d-1)}(r, f_j) + o(\lambda(r)) = o(\lambda(r))$$

holds for each $i \in \{0, 1, \ldots, n\}$. Note that

$$\| \quad \overline{N}\left(r, \frac{1}{\Delta a_i f_i}\right) = \overline{N}_{d_i-1)}\left(r, \frac{1}{\Delta a_i f_i}\right) + \overline{N}_{(d_i}\left(r, \frac{1}{\Delta a_i f_i}\right)$$

$$= \overline{N}_{(d_i}\left(r, \frac{1}{\Delta a_i f_i}\right) + o(\lambda(r)),$$

and

$$d_i \overline{N}_{(d_i}\left(r, \frac{1}{\Delta a_i f_i}\right) \le N\left(r, \frac{1}{\Delta a_i f_i}\right) = N_F(r, b_i) \le T_F(r) + O(1)$$

hold for $i = 0, \ldots, n$. Therefore the inequality (4.5.8) yields

$$\| \quad \{\beta - o(1)\} T_F(r) \le o(\lambda(r)). \tag{4.5.9}$$

On another hand, we have

$$\| \quad T\left(r, \frac{f_j}{f_0}\right) \le \begin{cases} T_F(r) + o(\lambda(r)), & 1 \le j \le n-1; \\ (n-1)T_F(r) + o(\lambda(r)), & j = n. \end{cases} \tag{4.5.10}$$

In fact, for each $j = 1, \ldots, n-1$, Lemma 2.37 implies

$$\| \quad T\left(r, \frac{f_j}{f_0}\right) \le T\left(r, \frac{a_j f_j}{a_0 f_0}\right) + o(\lambda(r)) \le T_F(r) + o(\lambda(r)),$$

and so

$$\begin{aligned}
\| \quad T\left(r, \frac{f_n}{f_0}\right) & \le T\left(r, \frac{a_n f_n}{a_0 f_0}\right) + o(\lambda(r)) \\
& \le \sum_{j=0}^{n-1} T\left(r, \frac{a_j f_j}{a_0 f_0}\right) + o(\lambda(r)) \\
& \le (n-1)T_F(r) + o(\lambda(r)).
\end{aligned}$$

According to (4.5.9), (4.5.10) and the assumption

$$\| \quad T\left(r, \frac{f_j}{f_0}\right) \ne o(\lambda(r)),$$

we may choose an unbounded subset $S_j \subset \mathbb{R}^+$ and a positive constant c_j such that (4.5.9) and (4.5.10) hold on S_j, and such that

$$\lambda(r) \le c_j T\left(r, \frac{f_j}{f_0}\right), \quad r \in S_j.$$

Thus we obtain

$$\{\beta - o(1)\} T_F(r) \le o(T_F(r)), \quad r \in S_j, \tag{4.5.11}$$

which implies $\beta \le 0$. This contradicts the assumption.

The claim 1) is trivial if either a_0 is equal to 0 or $a_1 f_1, \ldots, a_n f_n$ are linearly dependent, otherwise, $a_0 f_0, \ldots, a_{n-1} f_{n-1}$ are also linearly independent, and so a contradiction follows from the argument of case 1). $\qquad \square$

Some special cases of Theorem 4.27 were given by Berenstein, Chang and Li [20], Fujimoto [107], Narasimhan [287], and Hu and Yang (cf. [172], [174], [168]).

Corollary 4.28. *If* (M1), (M2) *and* (4.5.7) *hold, then there exists a partition of indices*

$$\{0, 1, \ldots, n\} = I_0 \cup I_1 \cup \cdots \cup I_k$$

such that $I_\alpha \neq \emptyset$ $(\alpha = 0, 1, \ldots, k)$, $I_\alpha \cap I_\beta = \emptyset$ $(\alpha \neq \beta)$,

$$\sum_{i \in I_\alpha} a_i f_i = 0, \quad \alpha = 0, 1, \ldots, k,$$

and $f_i/f_j \in \mathcal{M}_\lambda(\mathbb{C}^m)$ for any $i, j \in I_\alpha$. In particular, if $a_i \neq 0$ for $i = 0, 1, \ldots, n$, each I_α contains at least two indices.

Proof. Consider the partition $\{0, 1, \ldots, n\} = I_0 \cup I_1 \cup \cdots \cup I_k$ such that two indices i and j are in the same class if and only if $f_i/f_j \in \mathcal{M}_\lambda(\mathbb{C}^m)$. Then we have

$$\sum_{i=0}^{n} a_i f_i = \sum_{\alpha=0}^{k} \sum_{i \in I_\alpha} a_i f_i = \sum_{\alpha=0}^{k} c'_\alpha f_{i_\alpha} = 0$$

for any fixed $i_\alpha \in I_\alpha$ and some $c'_\alpha \in \mathcal{M}_\lambda(\mathbb{C}^m)$. Corollary 4.28 is trivial for the case $k = 0$. Suppose $k \geq 1$. Then

$$1 - \sum_{\alpha=0}^{k} \frac{k-1}{d_{i_\alpha}} \geq 1 - \sum_{j=0}^{n} \frac{n-1}{d_j} > 0.$$

Note that $f_{i_\alpha}/f_{i_\beta} \notin \mathcal{M}_\lambda(\mathbb{C}^m)$ $(\alpha \neq \beta)$, that is,

$$\| \ T\left(r, \frac{f_{i_\alpha}}{f_{i_\beta}}\right) \neq o(\lambda(r)), \ \alpha \neq \beta.$$

By Theorem 4.27, we obtain $c'_\alpha = 0$ for $\alpha = 0, 1, \ldots, k$, which yields Corollary 4.28. $\qquad\square$

Corollary 4.28 improves the related result of H. Fujimoto [107].

Corollary 4.29. *Assume that* (M1), (M2') *and* (4.5.7) *hold. If the functions* a_j *in* (M1) *are constants, then there exists a partition of indices*

$$\{0, 1, \ldots, n\} = I_0 \cup I_1 \cup \cdots \cup I_k$$

such that $I_\alpha \neq \emptyset$ $(\alpha = 0, 1, \ldots, k)$, $I_\alpha \cap I_\beta = \emptyset$ $(\alpha \neq \beta)$,

$$\sum_{i \in I_\alpha} a_i f_i = 0, \quad \alpha = 0, 1, \ldots, k,$$

and f_i/f_j *is constant for any* $i, j \in I_\alpha$. *In particular, if* $a_i \neq 0$ *for* $i = 0, 1, \ldots, n$, *each* I_α *contains at least two indices.*

If $d_0 = \cdots = d_n$, Corollary 4.29 was proved by Y. Aihara [3].

Corollary 4.30 (Green [121]). *Take positive integers* n *and* d *with* $d \geq n^2 \geq 4$. *We have the following three equivalent statements:*

(i) *If non-zero meromorphic functions $f_1, f_2, \ldots, f_n$ on $\mathbb{C}^m$ satisfy*

$$f_1^d + f_2^d + \cdots + f_n^d = 1,$$

then $f_1^d, f_2^d, \ldots, f_n^d$ are linearly dependent;

(ii) *Under the same assumption as in* (i), *then at least one of the f_i's is constant;*

(iii) *Assume that non-zero meromorphic functions $f_0, f_1, \ldots, f_n$ on $\mathbb{C}^m$ satisfy*

$$f_0^d + f_1^d + \cdots + f_n^d = 0.$$

Then there exists a partition of indices

$$\{0, 1, \ldots, n\} = I_0 \cup I_1 \cup \cdots \cup I_k$$

such that f_i/f_j is constant for any $i, j \in I_\alpha$, and

$$\sum_{i \in I_\alpha} f_i^d = 0, \quad \alpha = 0, 1, \ldots, k.$$

Proof. The claim (iii) follows directly from Corollary 4.29. We derive (ii) from (i) as follows: Since the functions $f_1^d, \ldots, f_n^d$ are linearly dependent, without loss of generality, we may assume the following linear relation:

$$c_1 f_1^d + \cdots + c_{n-1} f_{n-1}^d + f_n^d = 0.$$

By subtracting this identity from $f_1^d + \cdots + f_n^d = 1$, we have

$$(1 - c_1) f_1^d + \cdots + (1 - c_{n-1}) f_{n-1}^d = 1.$$

We could use the relation to get a shorter linear combination of the f_i^d to equal 1, and hence (i) can be used again. Finally, it follows that a constant c exists such that $c f_i^d = 1$ for some i.

In order to derive (iii) from (ii), we have

$$\sum_{i=0}^n f_i^d = \sum_{\alpha=0}^k \sum_{i \in I_\alpha} f_i^d = \sum_{\alpha=0}^k a_\alpha f_{i_\alpha}^d = 0$$

for some $a_\alpha \in \mathbb{C}$ and any fixed $i_\alpha \in I_\alpha$. Then (iii) follows if $a_\alpha = 0$ for $\alpha = 0, \ldots, k$. Assume that (iii) does not hold. Without loss of generality, we may assume that

$$a_\alpha \neq 0 \ (0 \leq \alpha \leq s, \ s \geq 1), \quad a_\alpha = 0 \ (s+1 \leq \alpha \leq k).$$

We may choose $g_\alpha \in \mathcal{M}(\mathbb{C}^m)$ satisfying

$$g_\alpha^d = -\frac{a_\alpha f_{i_\alpha}^d}{a_0 f_{i_0}^d}, \quad \alpha = 1, \ldots, s.$$

so that $g_1^d + \cdots + g_s^d = 1$. Then one of the g_α's, say g_1, is constant by (ii). This means that f_{i_1}/f_{i_0} is constant, contradicting the definition of I_α.

Finally, we derive (i) from (iii). We choose a constant f_0 satisfying $f_0^d = -1$ so that $f_0^d + f_1^d + \cdots + f_n^d = 0$, and apply (iii) to this identity. Let I_0 be the index set that contains 0. If $I = I_0$, then the functions $f_1^d, \ldots, f_n^d$ are all constant and hence linearly dependent. If $I \neq I_0$, then

$$\sum_{i \in I_\alpha} f_i^d = 0, \quad \alpha \neq 0,$$

thus yielding a non-trivial linear relation. □

Corollary 4.31. *One has the following three equivalent statements:*

(A) *Assume that entire functions $f_1, f_2, \ldots, f_n$ vanish nowhere on $\mathbb{C}^m$ such that*

$$f_1 + f_2 + \cdots + f_n = 1.$$

Then $f_1, f_2, \ldots, f_n$ are linearly dependent.

(B) *Under the same assumption as in (A), then at least one of the f_i's is constant.*

(C) *Assume that entire functions $f_0, f_1, \ldots, f_n$ vanish nowhere on $\mathbb{C}^m$ such that*

$$f_0 + f_1 + \cdots + f_n = 0.$$

Partition the index set $I = \{0, 1, \ldots, n\}$ into subsets I_α, $I = \cup_{\alpha=0}^k I_\alpha$, putting two indices i and j in the same subset I_α if and only if f_i/f_j is constant. Then we have

$$\sum_{i \in I_\alpha} f_i = 0, \quad \alpha = 0, 1, \ldots, k.$$

Corollary 4.31 is the classic Borel theorem. E. Borel [29] originally observed that the Picard theorem of entire functions on $\mathbb{C}$ may be restated in the following form: If two entire functions f and g on $\mathbb{C}$ vanish nowhere and satisfy the identity $f + g = 1$, then they are constant. S. Kobayashi [208] shows equivalence of three statements in Corollary 4.31 (or see [168]).

We may weaken the condition (4.5.7) for the family

$$\mathcal{M}_1(\mathbb{C}^m) = \{f \in \mathcal{M}(\mathbb{C}^m) \mid \mathrm{Ord}(f) < 1\}.$$

We also define

$$\mathcal{A}_1(\mathbb{C}^m) = \{f \in \mathcal{A}(\mathbb{C}^m) \mid \mathrm{Ord}(f) < 1\}.$$

Theorem 4.32. *Assume that (M2′) holds for $f_j \in \mathcal{M}_1(\mathbb{C}^m)$ $(j = 0, 1, \ldots, n)$ and suppose*

$$\beta = 1 - \sum_{j=0}^n \frac{n-1}{d_j} \geq 0. \tag{4.5.12}$$

If the function

$$T(r) = \min_{j \neq k} \left\{ T\left(r, \frac{f_j}{f_k}\right) \right\}$$

is unbounded, then $f_0, \ldots, f_n$ *are linearly independent.*

Proof. Assume that there are constants $a_i \in \mathbb{C}$ satisfying the equation $a_0 f_0 + \cdots + a_n f_n = 0$. We can prove Theorem 4.32 similar to Theorem 4.27, where by using Corollary 2.59, the inequality (4.5.8) is replaced by

$$\| \quad T_F(r) \leq \sum_{i=0}^{n} N_{F,n-1}(r, b_i) + l(\lambda + \varepsilon - 1) \log r + O(1), \qquad (4.5.13)$$

in which $\lambda \, (< 1)$ is the order of F, and l is the index of F. We can take ε sufficiently small such that $\lambda + \varepsilon < 1$. □

Corollary 4.33. *Assume that* (M1), (M2$'$) *and* (4.5.12) *hold for* $f_j \in \mathcal{M}_1(\mathbb{C}^m)$ *and* $a_j \in \mathbb{C}$ $(j = 0, \ldots, n)$. *Then there exists a partition of indices*

$$\{0, 1, \ldots, n\} = I_0 \cup I_1 \cup \cdots \cup I_k$$

such that $I_\alpha \neq \emptyset$ $(\alpha = 0, 1, \ldots, k)$, $I_\alpha \cap I_\beta = \emptyset$ $(\alpha \neq \beta)$,

$$\sum_{i \in I_\alpha} a_i f_i = 0, \quad \alpha = 0, 1, \ldots, k,$$

and f_i/f_j *is constant for any* $i, j \in I_\alpha$. *In particular, if* $a_i \neq 0$ *for* $i = 0, 1, \ldots, n$, *each* I_α *contains at least two indices.*

4.5.2 Siu-Yeung's theorem

Siu and Yeung [367] further extend Corollary 4.30 as follows:

Theorem 4.34. *Let* $P_j(x_0, \ldots, x_n)$ *be a homogeneous polynomial of degree* δ_j *for* $0 \leq j \leq n$. *Let* $f_0, \ldots, f_n$ *be holomorphic functions on* $\mathbb{C}$ *satisfying the following equation:*

$$\sum_{j=0}^{n} f_j^{d-\delta_j} P_j(f_0, \ldots, f_n) = 0.$$

If $d \geq n^2 + \sum_{j=0}^{n} \delta_j$, *there is a non-trivial linear relation among* $f_1^{d-\delta_1} P_1(f_0, \ldots, f_n)$, $\ldots$, $f_n^{d-\delta_n} P_n(f_0, \ldots, f_n)$.

Proof. For convenience, we write

$$F_j = f_j^{d-\delta_j} P_j(f_0, \ldots, f_n), \ j = 0, \ldots, n.$$

Theorem 4.34 is trivial if one of $F_0, \ldots, F_n$ is equal to 0. Assume that

$$F_j \neq 0, \; j = 0, \ldots, n,$$

and without loss of generality, assume that $F_0, \ldots, F_n$ have no common zeros. Thus we obtain a holomorphic curve $f : \mathbb{C} \longrightarrow \mathbb{P}^n$ with a reduced representation

$$\tilde{f} = (f_0, \ldots, f_n) : \mathbb{C} \longrightarrow \mathbb{C}^{n+1}.$$

Assume, to the contrary, that $F_1, \ldots, F_n$ are linearly independent, which means that $F_0, \ldots, F_{n-1}$ also are linearly independent. Using the notations in the proof of Theorem 4.27, we obtain a holomorphic curve $F : \mathbb{C} \longrightarrow \mathbb{P}(V)$ with a reduced representation

$$\tilde{F} = F_0 e_0 + \cdots + F_{n-1} e_{n-1} : \mathbb{C} \longrightarrow V,$$

which satisfies (4.5.8). We claim that

$$T_F(r) = dT_f(r) + O(1). \tag{4.5.14}$$

Define

$$A(z) = \max_{0 \leq j \leq n} |f_j(z)|.$$

Then (2.3.21) implies

$$T_f(r) = \int_{\mathbb{C}\langle 0; r \rangle} \log A \sigma + O(1). \tag{4.5.15}$$

Obviously, there exists a constant $C > 0$ such that $|\tilde{F}| \leq C A^d$, and so (2.3.21) and (4.5.15) yield

$$T_F(r) \leq dT_f(r) + O(1). \tag{4.5.16}$$

On the other hand, by using the formula (4.5.15) and the equation $F_0 + \cdots + F_n = 0$, it is not difficult to show that

$$T_F(r) = \frac{1}{2} \int_{\mathbb{C}\langle 0; r \rangle} \log \left(\sum_{j=0}^{n} |F_j|^2 \right) \sigma + O(1). \tag{4.5.17}$$

Take a point $z \in \mathbb{C}\langle 0; r \rangle$. Then there exists $j_z \in \{0, \ldots, n\}$ such that $A(z) = |f_{j_z}(z)|$. Note that

$$F_{j_z}(z) = f_{j_z}(z)^d Q_{j_z}(z), \; Q_{j_z}(z) = P_{j_z}\left(\frac{f_0(z)}{f_{j_z}(z)}, \ldots, \frac{f_n(z)}{f_{j_z}(z)} \right)$$

with $|Q_{j_z}(z)| \leq C'$ for a constant $C' > 0$. Thus we have

$$\sum_{i=0}^{n} |F_i(z)|^2 \geq A(z)^{2d} |Q_{j_z}(z)|^2,$$

which means

$$T_F(r) \geq dT_f(r) + O(1). \tag{4.5.18}$$

Therefore (4.5.14) follows from (4.5.16) and (4.5.18).

Now we can estimate the terms $N_{F,n-1}(r, b_i)$ in (4.5.8). Note that

$$N_{F,n-1}(r, b_i) = N_{n-1}\left(r, \frac{1}{F_i}\right) \leq N\left(r, \frac{1}{P_i}\right) + (n-1)\overline{N}\left(r, \frac{1}{f_i}\right).$$

The first main theorem (2.3.33) and (2.7.5) imply respectively

$$\overline{N}\left(r, \frac{1}{f_i}\right) \leq T_f(r) + O(1)$$

and

$$N\left(r, \frac{1}{P_i}\right) \leq \delta_i T_f(r) + O(1).$$

Thus by using (4.5.14), we obtain

$$N_{F,n-1}(r, b_i) \leq \frac{\delta_i + n - 1}{d} T_F(r) + O(1). \tag{4.5.19}$$

Therefore (4.5.8) and (4.5.19) yield

$$\| \quad \left(1 - \sum_{i=0}^{n} \frac{\delta_i + n - 1}{d}\right) T_F(r) \leq O(\log T_F(r)),$$

which means $d \leq n^2 - 1 + \sum_{i=0}^{n} \delta_i$. This is a contradiction. $\qquad \square$

4.5.3 Analogue of Borel's theorem

For integers, the conditions (M1) and (M2) are respectively replaced by

(N1) Let $x_0, x_1, \ldots, x_n$ $(n \geq 2)$ be non-zero integers satisfying the equation

$$x_0 + x_1 + \cdots + x_n = 0. \tag{4.5.20}$$

(N2) Take positive integers n and d_j $(j = 0, 1, \ldots, n)$ with $n \geq 2$. Let $x_0, x_1, \ldots,$ x_n be non-zero integers such that for each $i \in \{0, 1, \ldots, n\}$, there is no prime p satisfying

$$0 < v_p(x_i) < d_i. \tag{4.5.21}$$

Note that $\mathbb{C}[z] \subset \mathcal{M}_1(\mathbb{C})$. According to the classic analogy between polynomials and integers, we think that the analogue of Corollary 4.33 in number theory should be the following problem:

Conjecture 4.35. *If* (N1), (N2) *and* (4.5.12) *hold, then either there are a finite number of coprime integers* $x_0, \ldots, x_n$ *satisfying these properties, or there exists a partition of indices*

$$\{0, 1, \ldots, n\} = I_0 \cup I_1 \cup \cdots \cup I_k$$

such that $I_\alpha \neq \emptyset$ $(\alpha = 0, 1, \ldots, k)$, $I_\alpha \cap I_\beta = \emptyset$ $(\alpha \neq \beta)$,

$$\sum_{i \in I_\alpha} x_i = 0, \quad \alpha = 0, 1, \ldots, k,$$

$x_i/x_j \in \{-1, 1\}$ *for any* $i, j \in I_\alpha$, *and each* I_α *contains at least two indices.*

If $d_i = d$ for $i = 0, \ldots, n$, we think that this conjecture can be strengthened as follows:

Conjecture 4.36. *If* (N1) *and* (N2) *hold for* $d_i = d \geq n^2 - 1$ $(i = 0, \ldots, n)$, *then there exists a partition of indices*

$$\{0, 1, \ldots, n\} = I_0 \cup I_1 \cup \cdots \cup I_k$$

such that $I_\alpha \neq \emptyset$ $(\alpha = 0, 1, \ldots, k)$, $I_\alpha \cap I_\beta = \emptyset$ $(\alpha \neq \beta)$,

$$\sum_{i \in I_\alpha} x_i = 0, \quad \alpha = 0, 1, \ldots, k,$$

and $x_i/x_j \in \{-1, 1\}$ *for any* $i, j \in I_\alpha$.

This conjecture yields the following special case:

Conjecture 4.37. *Assume that* (N1) *holds. If there are integers* $d \geq n^2 - 1$, $a_i = \pm 1$ *and* y_i *such that*

$$x_i = a_i y_i^d, \quad i = 0, 1, \ldots, n,$$

then there exists a partition of indices

$$\{0, 1, \ldots, n\} = I_0 \cup I_1 \cup \cdots \cup I_k$$

such that $I_\alpha \neq \emptyset$ $(\alpha = 0, 1, \ldots, k)$, $I_\alpha \cap I_\beta = \emptyset$ $(\alpha \neq \beta)$,

$$\sum_{i \in I_\alpha} a_i y_i^d = 0, \quad \alpha = 0, 1, \ldots, k,$$

and $y_i/y_j \in \{-1, 1\}$ *for any* $i, j \in I_\alpha$.

Obviously, the Fermat-Wiles theorem is a special case of Conjecture 4.37. Assume that $y_0, y_1, \ldots, y_n$ $(n \geq 2)$ are non-zero integers satisfying

$$a_0 y_0^d + a_1 y_1^d + \cdots + a_n y_n^d = 0, \tag{4.5.22}$$

where $d \geq n^2 - 1$ and $a_i = \pm 1$ for each i.

Further we may assume $\gcd(y_0, y_1, \ldots, y_n) = 1$. If $(y_0, y_1, \ldots, y_n)$ is a non-trivial solution of (4.5.22), that is, no proper subsum of (4.5.22) is equal to 0, then Conjecture 4.16 implies that for $\varepsilon > 0$, there exists a number $C = C(n, \varepsilon)$ satisfying

$$\max_{0 \leq j \leq n} \{|y_j|^d\} \leq C \left(\prod_{i=0}^{n} r_{n-1} \left(y_i^d \right) \right)^{1+\varepsilon} \leq C |y_0 y_1 \cdots y_n|^{(n-1)(1+\varepsilon)}$$

which implies

$$\max_{0 \leq j \leq n} \{|y_j|\}^{d - n^2 + 1 - (n^2-1)\varepsilon} \leq C(n, \varepsilon).$$

In particular, taking $\varepsilon = \frac{1}{2(n^2-1)}$ and $d \geq n^2$, so

$$d - n^2 + 1 - (n^2 - 1)\varepsilon \geq \frac{d}{2n^2},$$

we deduce from Conjecture 4.16 that

$$\max_{0 \leq j \leq n} \{|y_j|^d\} \leq C \left(n, \frac{1}{2(n^2 - 1)} \right)^{2n^2}.$$

We have thus proved that in any non-trivial solution of (4.5.22) with $d \geq n^2$, the numbers $|y_j|^d$ are all less than some absolute bound depending only on n, and so there are no more than finitely many such solutions. If we had an explicit version of Conjecture 4.16 (that is, with the values of $C(n, \varepsilon)$ given), then we could give an explicit bound on all non-trivial solutions to the equation (4.5.22) and compute up to that bound to finally determine whether there are any non-trivial solutions.

Euler had a false intuition when he guessed that the Fermat hypersurface

$$y_1^d + \cdots + y_n^d = y_0^d$$

would have no non-trivial rational solutions for $d = n+1$. Lander and Parkin [220] found the solution in degree 5:

$$27^5 + 84^5 + 110^5 + 133^5 = 144^5.$$

Then Elkies [91] found infinitely many solutions in degree 4, including

$$2682440^4 + 15365639^4 + 18796760^4 = 20615673^4.$$

4.6 Meromorphic solutions of Fermat equations

We continue with the situation of Section 3.4. Meromorphic solutions of Fermat-Catalan equations and their general forms will be discussed first in this section. Concretely, it is interesting to find the smallest integer $G_n(\mathcal{F})$ such that when $d \geq G_n(\mathcal{F})$, there do not exist non-constant functions $f_1, \ldots, f_n$ $(n \geq 2)$ in a certain function class $\mathcal{F}$ satisfying

$$f_1^d + f_2^d + \cdots + f_n^d = 1. \tag{4.6.1}$$

First of all, we prove the following fact:

Theorem 4.38 (cf. [168]). *Take positive integers k and l with $k \geq l \geq 2$ and set*

$$\alpha = 1 - \frac{1}{k} - \frac{1}{l}.$$

Then the Fermat-Catalan equation

$$f^k + g^l = 1 \tag{4.6.2}$$

has no solutions $f, g \in \mathcal{F} - \mathbb{C}$ satisfying one of the following cases

(1) $\alpha > \frac{1}{k}$, $\mathcal{F} = \mathcal{M}(\mathbb{C}^m)$;
(2) $\alpha > 0$, $\mathcal{F} = \mathcal{A}(\mathbb{C}^m)$;
(3) $\alpha \geq \frac{1}{k}$, $\mathcal{F} = \mathcal{M}_1(\mathbb{C}^m)$;
(4) $\alpha \geq 0$, $\mathcal{F} = \mathcal{A}_1(\mathbb{C}^m)$.

Proof. Assume, to the contrary, that there exist two non-constant meromorphic functions f and g on $\mathbb{C}^m$ satisfying (4.6.2). Let $a_1, \ldots, a_k$ be the zeros of $z^k - 1$ in $\mathbb{C}$. Then for each $j = 1, \ldots, k$, each zero of $f - a_j$ has order $\geq l$, and hence

$$\overline{N}\left(r, \frac{1}{f - a_j}\right) \leq \frac{1}{l} N\left(r, \frac{1}{f - a_j}\right) \leq \frac{1}{l} T(r, f) + O(1).$$

By using the second main theorem (cf. Corollary 2.59), for any $\delta > 0$ one has

$$\| \quad (k-1)T(r, f) \leq \overline{N}(r, f) + \frac{k}{l} T(r, f) + (1+\delta) \log^+ T(r, f) - \log r + O(1). \tag{4.6.3}$$

Define a characteristic number ϵ for f and g by

$$\epsilon = \begin{cases} 0 & : f, g \in \mathcal{A}(\mathbb{C}^m), \\ 1 & : \text{otherwise.} \end{cases} \tag{4.6.4}$$

Then we have

$$\overline{N}(r, f) \leq \epsilon T(r, f).$$

Thus (4.6.3) yields

$$\| \quad \left(k - 1 - \frac{k}{l} - \epsilon\right) T(r, f) + \log r \leq (1+\delta) \log^+ T(r, f) + O(1), \tag{4.6.5}$$

that is

$$\| \quad (k\alpha - \epsilon)T(r, f) + \log r \le (1 + \delta)\log^+ T(r, f) + O(1).$$

This inequality contradicts any one of the hypothesis (1)-(4). The proof of the theorem is completed.   □

If we replace the second main theorem in the proof of Theorem 4.38 by using abc-theorem on meromorphic functions, say, Theorem 4.4, we obtain

$$\| \quad T(r, f^k) \;\le\; \overline{N}\left(r, \frac{1}{f^k}\right) + \overline{N}\left(r, \frac{1}{g^l}\right) + \frac{1}{2}\overline{N}(r, f^k) + \frac{1}{2}\overline{N}(r, g^l)$$
$$+(1 + \delta)\log^+ T(r, f) - \log r + O(1) \qquad (4.6.6)$$

for small $\delta > 0$. Theorem 2.45 and the first main theorem (2.3.15) further yields the inequality

$$\| \quad kT(r, f) \le \left(1 + \frac{\epsilon}{2}\right)T(r) + (1 + \delta)\log^+ T(r) - \log r + O(1), \qquad (4.6.7)$$

where

$$T(r) = T(r, f) + T(r, g).$$

Similarly, we have

$$\| \quad lT(r, g) \le \left(1 + \frac{\epsilon}{2}\right)T(r) + (1 + \delta)\log^+ T(r) - \log r + O(1). \qquad (4.6.8)$$

Summing (4.6.7) and (4.6.8), we obtain

$$\| \quad \left\{\left(1 + \frac{\epsilon}{2}\right)\alpha - \frac{\epsilon}{2}\right\}T(r) + (1 - \alpha)\log r \le (1 + \delta)(1 - \alpha)\log^+ T(r) + O(1). \quad (4.6.9)$$

Thus we still obtain the cases (2) and (4) in Theorem 4.38. According to (4.6.9), now (1) and (3) in Theorem 4.38 assume the following forms:

(1′) $\alpha > \frac{1}{3}$, $\mathcal{F} = \mathcal{M}(\mathbb{C}^m)$;
(3′) $\alpha \ge \frac{1}{3}$, $\mathcal{F} = \mathcal{M}_1(\mathbb{C}^m)$.

However, we still obtain the same conclusion as Theorem 4.38. We know that the abc-conjecture would imply the asymptotic Fermat conjecture. Here we show that the abc-theorem on meromorphic functions implies that the Fermat equation has no meromorphic solutions.

In particular, if $d \ge 4$ there are no non-constant solutions f and g in the class $\mathcal{M}(\mathbb{C})$ satisfying the Fermat equation

$$f^d + g^d = 1. \qquad (4.6.10)$$

Baker [12] has characterized all f, g in the class $\mathcal{M}(\mathbb{C})$ such that (4.6.10) holds for $d = 3$. Thus we have

$$G_2(\mathcal{M}(\mathbb{C})) = 4.$$

By Theorem 4.38, if $d \geq 3$ there are no non-constant solutions f and g satisfying (4.6.10) in the class $\mathbb{C}[z]$, $\mathbb{C}(z)$, and $\mathcal{A}(\mathbb{C})$, respectively, which are due to Montel [275], Jategaonkar [197], Yang [438], and Gross [130], [131] (or see [33], [35], [132], [176]). Since the rational functions (cf. [130])

$$f(z) = \frac{2z}{z^2 + 1}, \quad g(z) = \frac{z^2 - 1}{z^2 + 1}$$

satisfy the equation

$$f^2 + g^2 = 1, \tag{4.6.11}$$

we have

$$G_2(\mathbb{C}(z)) = 3.$$

On the other hand, if f and g are non-constant entire solutions of the equation (4.6.11), then we have

$$f = \sin w, \quad g = \cos w$$

for some entire function w (see [196]). It is easy to see

$$G_2\left(\mathcal{A}(\mathbb{C})\right) = 3, \quad G_2\left(\mathbb{C}[z]\right) = 2.$$

Related to the equation (3.4.4), Corollary 4.29 yields directly the following result:

Theorem 4.39. *Take positive integers k, l, n satisfying (3.4.5). If f, g and h are non-zero entire functions on $\mathbb{C}^m$ such that the equation*

$$f^k + g^l = h^n \tag{4.6.12}$$

holds, then f^k/h^n and g^l/h^n are constants.

Theorem 4.40 (cf. [168]). *Let k, l and n be positive integers and set*

$$\alpha = 1 - \frac{2}{k} - \frac{2}{l} - \frac{2}{n}.$$

The functional equation

$$f^k + g^l + h^n = 1 \tag{4.6.13}$$

has no solutions $f, g, h \in \mathcal{F} - \mathbb{C}$ satisfying one of the following cases:

(A) $\alpha > \frac{1}{3}$, $\mathcal{F} = \mathcal{M}(\mathbb{C}^m)$;

(B) $\alpha > 0$, $\mathcal{F} = \mathcal{A}(\mathbb{C}^m)$;

(C) $\alpha \geq \frac{1}{3}$, $\mathcal{F} = \mathcal{M}_1(\mathbb{C}^m)$;

(D) $\alpha \geq 0$, $\mathcal{F} = \mathcal{A}_1(\mathbb{C}^m)$.

Proof. Assume, to the contrary, that there exist three non-constant meromorphic functions f, g and h on $\mathbb{C}^m$ satisfying (4.6.13). Then f^k, g^l and h^n are linearly independent. Otherwise, there is $(a, b, c) \in \mathbb{C}^3 - \{0\}$ such that

$$af^k + bg^l + ch^n = 0.$$

Without loss of generality, we may assume $c \neq 0$. Thus

$$\left(1 - \frac{a}{c}\right) f^k + \left(1 - \frac{b}{c}\right) g^l = 1.$$

Since f and g are non-constant, then $a \neq c$ and $b \neq c$. Since we can take $\gamma, \beta \in \mathbb{C}_*$ such that

$$\gamma^k = 1 - \frac{a}{c}, \quad \beta^l = 1 - \frac{b}{c},$$

then

$$(\gamma f)^k + (\beta g)^l = 1,$$

which is impossible by Theorem 4.38.

By Theorem 4.4, for any $\delta > 0$ we have

$$
\begin{aligned}
\| \quad kT(r, f) &= T(r, f^k) \leq N_2\left(r, \frac{1}{f^k}\right) + N_2\left(r, \frac{1}{g^l}\right) + N_2\left(r, \frac{1}{h^n}\right) \\
&\quad + \overline{N}(r, f^k) + \overline{N}(r, g^l) + \overline{N}(r, h^n) \\
&\quad + (1 + \delta) \log^+ T(r) - \nu \log r + O(1) \\
&\leq (2 + \epsilon)T(r) + (1 + \delta) \log^+ T(r) - \nu \log r + O(1),
\end{aligned}
$$

where ν is the index of f^k, g^l and h^n, ϵ is defined by (4.6.4) and

$$T(r) = T(r, f) + T(r, g) + T(r, h).$$

Similarly,

$$\| \quad lT(r, g) \leq (2 + \epsilon)T(r) + (1 + \delta) \log^+ T(r) - \nu \log r + O(1),$$

and

$$\| \quad nT(r, h) \leq (2 + \epsilon)T(r) + (1 + \delta) \log^+ T(r)) - \nu \log r + O(1).$$

Therefore,

$$\| \quad \left(1 - \frac{(2 + \epsilon)(1 - \alpha)}{2}\right) T(r) + \frac{\nu(1 - \alpha)}{2} \log r \leq \frac{(1 + \delta)(1 - \alpha)}{2} \log^+ T(r) + O(1),$$

or, equivalently,

$$\| \quad \frac{(2 + \epsilon)\alpha - \epsilon}{1 - \alpha} T(r) + \nu \log r \leq (1 + \delta) \log^+ T(r) + O(1),$$

which is impossible under one of the conditions (A)-(D). $\quad\square$

Hayman [153] proved that for $d \geq 9$ (resp., $d \geq 6$), there do not exist three non-constant meromorphic (resp., entire) functions f, g and h on $\mathbb{C}$ satisfying the equation

$$f^d + g^d + h^d = 1. \tag{4.6.14}$$

When $d = 3$ ([243]) or $d = 4$ ([131]), the equation (4.6.14) has solutions in $\mathcal{A}(\mathbb{C}) - \mathbb{C}$. Thus we have

$$5 \leq G_3(\mathcal{A}(\mathbb{C})) \leq 6.$$

For the cases $d = 5$ and $d = 6$, Gundersen [133], [134] (or cf. [153]) gave examples of non-constant meromorphic solutions of (4.6.14). Therefore

$$7 \leq G_3(\mathcal{M}(\mathbb{C})) \leq 9.$$

Ishizaki [194] proved that if there exist transcendental meromorphic functions f, g, h over $\mathbb{C}$ satisfying (4.6.14) with $d = 8$, then they also satisfy the differential equation

$$\mathbf{W}(f^8, g^8, h^8) = a(fgh)^6,$$

where a is a small function with respect to f, g and h, and $\mathbf{W}(f^8, g^8, h^8)$ is the Wronskian determinant of f^8, g^8 and h^8. For more examples, also see Tohge [399].

Generally, by induction and Theorem 4.4, we can prove the following version of results due to Toda [398], Hayman [153] and Yu-Yang [444], respectively:

Theorem 4.41 (cf. [168]). *Take positive integers $n(\geq 3)$, $d_1, \ldots, d_n$ and set*

$$\alpha = 1 - \sum_{j=1}^{n} \frac{n-1}{d_j}.$$

If A_n is the number defined in Theorem 4.4, then the functional equation

$$f_1^{d_1} + f_2^{d_2} + \cdots + f_n^{d_n} = 1 \tag{4.6.15}$$

has no solutions $\{f_1, \ldots, f_n\} \subset \mathcal{F} - \mathbb{C}$ satisfying one of the following cases:

(i) $\alpha > \frac{A_n}{n-1+A_n}$, $\mathcal{F} = \mathcal{M}(\mathbb{C}^m)$;
(ii) $\alpha > 0$, $\mathcal{F} = \mathcal{A}(\mathbb{C}^m)$;
(iii) $\alpha \geq \frac{A_n}{n-1+A_n}$, $\mathcal{F} = \mathcal{M}_1(\mathbb{C}^m)$;
(iv) $\alpha \geq 0$, $\mathcal{F} = \mathcal{A}_1(\mathbb{C}^m)$.

For the case (i) of Theorem 4.41 with $m = 1$, Yu and Yang [444] obtained $\frac{3}{4}$ replacing the number $\frac{A_n}{n-1+A_n}$. Obviously,

$$\frac{A_n}{n-1+A_n} \leq \frac{\vartheta_n}{n-1+\vartheta_n} < \frac{3}{4}, \quad n \geq 3.$$

Based on Theorem 4.38 and Theorem 4.40, we conjecture that the restrictions of α in Theorem 4.41, (i) and (iii) would be replaced by $\alpha \geq \frac{1}{n}$. By Corollary 4.29 or Corollary 4.30, we can obtain a partial answer to the problem as follows:

Theorem 4.42. *For $n \geq 2$ and $d \geq n^2$, there do not exist non-constant meromorphic functions $f_1, \ldots, f_n$ on $\mathbb{C}^m$ satisfying (4.6.1).*

Corollary 4.33 yields immediately the following fact:

Theorem 4.43. *For $n \geq 2$ and $d \geq n^2 - 1$, there do not exist non-constant meromorphic functions $f_1, \ldots, f_n$ of order < 1 on $\mathbb{C}^m$ satisfying the equation (4.6.1).*

Theorem 4.42 is a several variable version of a result in [153], which implies

$$G_n(\mathcal{M}(\mathbb{C}^m)) \leq n^2.$$

Theorem 4.43 yields

$$G_n(\mathcal{M}_1(\mathbb{C}^m)) \leq n^2 - 1.$$

By Theorem 4.41, we obtain

$$G_n(\mathcal{A}(\mathbb{C}^m)) \leq n(n-1) + 1, \quad G_n(\mathcal{A}_1(\mathbb{C}^m)) \leq n(n-1).$$

Assume $d \geq n^2$ and assume that there exist non-zero meromorphic functions $f_1, \ldots, f_n$ on $\mathbb{C}^m$ satisfying (4.6.1). M. L. Green [121] observed that the image of the meromorphic mapping

$$[1, f_1, \ldots, f_n] : \mathbb{C}^m \longrightarrow \mathbb{P}^n$$

is contained in a linear subspace of dimension $\left[\frac{n-1}{2}\right]$, where the bracket $[x]$ means the integer with $[x] \leq x < [x] + 1$.

Theorem 4.44. *Take positive integers $n, d_1, \ldots, d_n$ and meromorphic functions $f_1, \ldots, f_n$ in $\mathbb{C}$ satisfying (4.6.15). Then $f_1^{d_1}, \ldots, f_n^{d_n}$ are linearly dependent if*

$$\sum_{j=1}^{n} d_j \geq n^2 + (n-1) \max_{1 \leq j \leq n} \{d_j\}.$$

Proof. Set

$$d = \max_{1 \leq j \leq n} \{d_j\}$$

and choose holomorphic functions $g_0, g_1, \ldots, g_n$ in $\mathbb{C}$ such that $f_j = g_j/g_0$. Then the equation (4.6.15) implies

$$\sum_{j=1}^{n} g_j^{d_j} g_0^{d-d_j} = g_0^d.$$

Theorem 4.34 shows that $g_1^{d_1} g_0^{d-d_1}, \ldots, g_n^{d_n} g_0^{d-d_n}$ are linearly dependent, and so $f_1^{d_1}, \ldots, f_n^{d_n}$ are linearly dependent. $\square$

Given a subset $\mathcal{F}$ of $\mathcal{M}(\mathbb{C}^m)$ such that $\mathbb{C}$ is a proper subfield of $\mathcal{F}$. Related to Waring's problem, it is interesting that for any fixed positive integer d, find the smallest integer $n = G_{\mathcal{F}}(d)$ such that there exist non-constant functions $f_1, \ldots, f_n$ in $\mathcal{F}$ satisfying (4.6.1). Theorem 4.41 implies

$$G_{\mathcal{A}_1(\mathbb{C}^m)}(d) > \frac{1}{2} + \sqrt{d + \frac{1}{4}}, \quad d \geq 2, \tag{4.6.16}$$

$$G_{\mathcal{A}(\mathbb{C}^m)}(d) \geq \frac{1}{2} + \sqrt{d + \frac{1}{4}}, \quad d \geq 2. \tag{4.6.17}$$

Theorem 4.42 and Theorem 4.43 show, respectively

$$G_{\mathcal{M}(\mathbb{C}^m)}(d) \geq \sqrt{d + 1}, \qquad d \geq 2, \tag{4.6.18}$$

$$G_{\mathcal{M}_1(\mathbb{C}^m)}(d) > \sqrt{d + 1}, \qquad d \geq 2. \tag{4.6.19}$$

These estimates are due to Toda [398], Green [121], Hayman [153], and Yu-Yang [444] (or see [168]).

We now give examples for equation (4.6.1). First note that [153]

$$\left(\frac{1+z}{\sqrt{2}}\right)^2 + \left(\frac{1-z}{\sqrt{2}}\right)^2 + (iz)^2 = 1, \tag{4.6.20}$$

which shows that $G_{\mathbb{C}[z]}(2) \leq 3$. By the remark after Theorem 4.38, it follows that

$$G_{\mathbb{C}[z]}(2) = 3, \quad G_{\mathcal{A}(\mathbb{C})}(2) = G_{\mathcal{M}(\mathbb{C})}(2) = G_{\mathbb{C}(z)}(2) = 2.$$

In [274], Molluzzo considered

$$\frac{1}{p} \sum_{j=1}^{p} (1 + \omega_j z^d)^d = 1 + a_1 z^{pd} + a_2 z^{2pd} + \cdots + a_{[d/p]} z^{[d/p]pd}, \tag{4.6.21}$$

where

$$\omega_j = \exp\left(\frac{2\pi j \sqrt{-1}}{p}\right), \quad j = 1, \ldots, p.$$

This gives 1 as the sum of $p + [d/p]$ dth powers. The minimum of $p + [d/p]$ for all p is $[(4d + 1)^{1/2}]$ (see [296]). Therefore

$$G_{\mathbb{C}[z]}(d) \leq \sqrt{4d + 1}, \quad d \geq 2.$$

When $p = 2$ and $d = 3$ in (4.6.21), we obtain the following identity:

$$\frac{1}{2}(1 + z^3)^3 + \frac{1}{2}(1 - z^3)^3 - 3(z^2)^3 = 1. \tag{4.6.22}$$

On the other hand [12], f and g are non-constant meromorphic solutions of the equation

$$f^3 + g^3 = 1 \qquad (4.6.23)$$

if and only if f and g are certain non-constant elliptic functions composed with an entire function. Combining this result with (4.6.22) gives

$$G_{\mathcal{M}(\mathbb{C})}(3) = 2, \quad G_{\mathcal{A}(\mathbb{C})}(3) = G_{\mathbb{C}[z]}(3) = G_{\mathbb{C}(z)}(3) = 3.$$

Newman and Slater [296] studied the following identity

$$\sum_{j=1}^{p} \frac{\omega_j(1 + \omega_j z^d)^d}{z^{(p-1)d}} = b_0 + b_1 z^{pd} + b_2 z^{2pd} + \cdots + b_{[(d+1)/p]-1} z^{([(d+1)/p]-1)pd}, \qquad (4.6.24)$$

where each b_j is a positive integer, that is, there exist non-constant rational functions $f_1, f_2, \ldots, f_n$ satisfying the equation (4.6.1), where

$$n = n(p) = p + [(d+1)/p] - 1.$$

Since the minimum of $n(p)$ for all p is $[(4d+5)^{1/2}] - 1$ (see [153]), we obtain

$$G_{\mathcal{F}}(d) \leq \sqrt{4d+5} - 1, \quad d \geq 2, \quad \mathcal{F} \in \{\mathcal{A}(\mathbb{C}), \mathbb{C}(z), \mathcal{M}(\mathbb{C})\}.$$

When $\mathcal{F} = \mathcal{A}(\mathbb{C})$, we can see this, by replacing z with e^z in (4.6.24). When $p = 3$ and $d = 4$ in (4.6.24), we have the following identity:

$$\frac{1}{18}\left(\frac{1+z^4}{z^2}\right)^4 + \frac{\omega_1}{18}\left(\frac{1+\omega_1 z^4}{z^2}\right)^4 + \frac{\omega_2}{18}\left(\frac{1+\omega_2 z^4}{z^2}\right)^4 = 1. \qquad (4.6.25)$$

By combining (4.6.25) with the estimates (4.6.16) to (4.6.19), we obtain

$$G_{\mathcal{M}(\mathbb{C})}(4) = G_{\mathcal{A}(\mathbb{C})}(4) = G_{\mathbb{C}(z)}(4) = 3.$$

Examples in [133], [134], [136], [320], together with the estimates (4.6.16) to (4.6.19), show that

$$G_{\mathcal{M}(\mathbb{C})}(5) = G_{\mathcal{A}(\mathbb{C})}(5) = G_{\mathbb{C}(z)}(5) = 3,$$

and

$$G_{\mathcal{M}(\mathbb{C})}(6) = 3, \quad G_{\mathcal{A}(\mathbb{C})}(14) = 5.$$

4.7   Waring's problem for meromorphic functions

We continue with the situation of Section 3.4. Meromorphic solutions of Waring's problem will be discussed in this section. Concretely, we may ask whether, for a function f in a certain function class $\mathcal{F}$ of $\mathcal{M}(\mathbb{C})$, there exist non-constant functions $f_1, \ldots, f_n$ in the class $\mathcal{F}$ such that

$$f_1^d + \cdots + f_n^d = f \tag{4.7.1}$$

holds. The problem of representing any function in $\mathcal{F}$ as a sum of dth powers of functions in $\mathcal{F}$ is known as the *Waring problem* for the family $\mathcal{F}$. Here it is enough to suppose that $f(z) = z$, that is, we need to consider only the equation

$$f_1(z)^d + \cdots + f_n(z)^d = z. \tag{4.7.2}$$

To see this, suppose first that $\mathcal{F} = \mathbb{C}[z]$, $\mathbb{C}(z)$ or $\mathcal{A}(\mathbb{C})$ and that the equation (4.7.2) is satisfied by n functions $f_1, \ldots, f_n$ in $\mathcal{F}$. Then for any $f \in \mathcal{F}$, we have

$$f_1(f(z))^d + \cdots + f_n(f(z))^d = f(z).$$

For the class $\mathcal{M}(\mathbb{C})$ this argument must be modified slightly as follows (see [153], [135]): Suppose that (4.7.2) holds for n functions $f_1, \ldots, f_n$ in $\mathcal{M}(\mathbb{C})$. Take $f \in \mathcal{M}(\mathbb{C})$. Then there exist $g, h \in \mathcal{A}(\mathbb{C})$ such that $f = g/h^d$, and so (4.7.2) yields

$$\left(\frac{f_1(g(z))}{h(z)} \right)^d + \cdots + \left(\frac{f_n(g(z))}{h(z)} \right)^d = f(z).$$

Let $n = g_{\mathcal{F}}(d)$ be the smallest number such that there exist non-constant functions $f_1, \ldots, f_n$ in the class $\mathcal{F}$ such that (4.7.2) holds. We have $g_{\mathbb{C}[z]}(2) = 2$ since

$$\left(\frac{z+1}{2} \right)^2 + \left(\frac{z-1}{2i} \right)^2 = z.$$

The following theorem is due to Hayman [153].

Theorem 4.45. *The number $g_{\mathcal{F}}(d)$ satisfies the following estimates:*

$$g_{\mathbb{C}[z]}(d) > \frac{1}{2} + \sqrt{d + \frac{1}{4}}, \quad d \geq 3, \tag{4.7.3}$$

$$g_{\mathcal{A}(\mathbb{C})}(d) \geq \frac{1}{2} + \sqrt{d + \frac{1}{4}}, \quad d \geq 2, \tag{4.7.4}$$

$$g_{\mathcal{M}(\mathbb{C})}(d) \geq \sqrt{d + 1}, \quad d \geq 2, \tag{4.7.5}$$

$$g_{\mathbb{C}(z)}(d) > \sqrt{d + 1}, \quad d \geq 2. \tag{4.7.6}$$

Proof. Here we follow Gundersen and Hayman [135] to prove this theorem. The numbers $g_{\mathcal{F}}(d)$ and $G_{\mathcal{F}}(d)$ are related by the following inequality [153]:

$$G_{\mathcal{F}}(d) \leq g_{\mathcal{F}}(d), \ d \geq 2, \ \mathcal{F} \in \{\mathcal{A}(\mathbb{C}), \ \mathcal{M}(\mathbb{C}), \ \mathbb{C}(z)\}. \qquad (4.7.7)$$

In fact, if there exist n functions $f_1, \ldots, f_n$ in $\mathcal{F}$ satisfying (4.7.2), then by replacing z in (4.7.2) with either z^d or e^{dz}, we deduce that there exist n non-constant functions $h_1, \ldots, h_n$ in $\mathcal{F}$ satisfying $h_1^d + \cdots + h_n^d = 1$. Thus the estimates (4.7.4) to (4.7.6) follow from (4.6.17) to (4.6.19).

To show (4.7.3), take $d \in \mathbb{Z}$ with $d \geq 3$ and suppose that $f_1, \ldots, f_n$ are polynomials satisfying (4.7.2). Then

$$k = \max_{1 \leq j \leq n} \deg(f_j^d) > 0.$$

Since $d \geq 3$, we obtain that $n \geq 3$ from (4.7.6). By Theorem 4.8,

$$k \leq 1 + \sum_{j=1}^{n} r_{n-1}\left(f_j^d\right) - \frac{n(n-1)}{2} \leq 1 + \frac{nk(n-1)}{d} - \frac{n(n-1)}{2} < \frac{nk(n-1)}{d},$$

which yields $d < n^2 - n$. This proves (4.7.3). $\qquad \square$

For $d \in \mathbb{Z}^+$, set

$$\omega_j = \exp\left(\frac{2\pi j \sqrt{-1}}{d}\right), \ j = 1, \ldots, d.$$

Then

$$\sum_{j=1}^{d} \omega_j^m = 0 \ (0 < |m| < d, \ m \in \mathbb{Z}).$$

Thus we obtain the representation given by Heilbronn (cf. [152])

$$\frac{1}{d^2} \sum_{j=1}^{d} \frac{(1 + \omega_j z)^d}{\omega_j} = z,$$

and so

$$g_{\mathcal{A}(\mathbb{C})}(d) \leq g_{\mathbb{C}[z]}(d) \leq d.$$

Further, by using Theorem 4.45, we have

$$g_{\mathbb{C}[z]}(3) = g_{\mathcal{A}(\mathbb{C})}(3) = g_{\mathbb{C}(z)}(3) = 3.$$

On the other hand [132], there exist $f, g \in \mathcal{M}(\mathbb{C})$ satisfying $f(z)^3 + g(z)^3 = z$, which shows that $g_{\mathcal{M}(\mathbb{C})}(3) = 2$.

It is natural to ask how many d_jth powers of positive integers x_j are necessary to represent an arbitrary positive integer N so that

$$x_1^{d_1} + \cdots + x_n^{d_n} = N. \tag{4.7.8}$$

Here we do not study the problem in number theory, but the problem of representing any function in $\mathcal{F}$ as a sum of d_jth powers of functions in $\mathcal{F}$.

Theorem 4.46. *Take positive integers $d_j \geq 2$ $(j = 1, \ldots, n)$ and set*

$$\alpha = 1 - \sum_{j=1}^{n} \frac{n-1}{d_j}.$$

If $n \geq 2$ and if there exist non-constant entire functions $f_1, \ldots, f_n$ on $\mathbb{C}$ satisfying

$$f_1(z)^{d_1} + \cdots + f_n(z)^{d_n} = z, \tag{4.7.9}$$

then $\alpha \leq 0$.

Proof. Assume that there exist non-constant entire functions $f_1, \ldots, f_n$ satisfying (4.7.9). First of all, we consider the case that $f_1^{d_1}, \ldots, f_n^{d_n}$ are linearly independent.

By the equation (4.7.9), it is easy to see that $f_1, \ldots, f_n$ have no common zeros. Hence a non-degenerate holomorphic curve

$$F : \mathbb{C} \longrightarrow \mathbb{P}(\mathbb{C}^n)$$

is well defined with a reduced representation

$$\tilde{F} = (\tilde{F}_1, \ldots, \tilde{F}_n),$$

where

$$\tilde{F}_j = f_j^{d_j}, \quad j = 1, \ldots, n.$$

Set $\tilde{F}_0(z) = z$. By Corollary 2.59, for $R > \rho > r > r_0$ we obtain

$$T_F(r) \leq \sum_{j=0}^{n} N_{n-1}\left(r, \frac{1}{\tilde{F}_j}\right) + \frac{n(n-1)}{2} \log\left\{\frac{\rho T_F(R)}{r(\rho - r)}\right\} + O(1)$$

$$\leq \log r + \sum_{j=1}^{n} \frac{n-1}{d_j} N\left(r, \frac{1}{\tilde{F}_j}\right) + \frac{n(n-1)}{2} \log\left\{\frac{\rho T_F(R)}{r(\rho - r)}\right\} + O(1).$$

The first main theorem implies

$$N\left(r, \frac{1}{\tilde{F}_j}\right) \leq T_F(r) + O(1), \quad j = 1, \ldots, n.$$

Therefore we have

$$\alpha T_F(r) \leq \log r + \frac{n(n-1)}{2} \log \left\{ \frac{\rho T_F(R)}{r(\rho - r)} \right\} + O(1). \qquad (4.7.10)$$

Specially, if $f_1, \ldots, f_n$ are polynomials, then (4.7.10) yields

$$\alpha \max_{1 \leq j \leq n} \{d_j \deg(f_j)\} \leq 1 - \frac{n(n-1)}{2} \leq 0. \qquad (4.7.11)$$

If one of functions f_j is transcendental, then $T_F(r)/\log r \to \infty$ as $r \to \infty$, and so the inequality (4.7.10) yields $\alpha \leq 0$ and Theorem 4.46 follows again.

Finally, we study the case that $f_1^{d_1}, \ldots, f_n^{d_n}$ are linearly dependent. Without loss of generality, we may assume that $f_1^{d_1}, \ldots, f_l^{d_l}$ are linearly independent ($1 \leq l < n$), but $f_1^{d_1}, \ldots, f_l^{d_l}, f_j^{d_j}$ are linearly dependent for each $j = l+1, \ldots, n$. Thus there is a $(a_1, \ldots, a_l) \in \mathbb{C}^l - \{0\}$ such that (4.7.9) becomes the following form:

$$a_1 f_1(z)^{d_1} + \cdots + a_l f_l(z)^{d_l} = z. \qquad (4.7.12)$$

We may assume $a_i \neq 0$ for each $i = 1, \ldots, l$, otherwise, deleting the terms with null coefficients in (4.7.12). Since

$$\alpha \leq 1 - \sum_{i=1}^{l} \frac{l-1}{d_i},$$

the proof of Theorem 4.46 can be completed by applying the above arguments to the equation (4.7.12). $\qquad \square$

To study meromorphic solutions of (4.7.9), we will need the following result:

Theorem 4.47. Let $f_1, f_2, \ldots, f_n$ ($n \geq 2$) be linearly independent meromorphic functions in $\mathbb{C}$ such that

$$f_1(z) + f_2(z) + \cdots + f_n(z) \equiv z.$$

Then for $j = 1, 2, \ldots, n$, the inequalities

$$T(r, f_j) < \sum_{i=1}^{n} N_{n-1} \left(r, \frac{1}{f_i} \right) + \overset{*}{N}_j(r) + \log^+ r$$
$$+ \frac{n(n-1)}{2} \log \left\{ \frac{\rho T(R)}{r(\rho - r)} \right\} + O(1) \qquad (4.7.13)$$

hold for $r_0 < r < \rho < R$, where $T(r) = \max_{1 \leq i \leq n} \{T(r, f_i)\}$, and

$$\overset{*}{N}_j(r) = \min \left\{ \vartheta_n \sum_{i=1}^{n} \overline{N}(r, f_i), (n-1) \sum_{i \neq j} \overline{N}(r, f_i) \right\}.$$

Proof. Set $f_0(z) = z$. Note that

$$T(r, f_j) \leq N(r, f_j) + m\left(r, \frac{f_j}{f_0}\right) + \log^+ r.$$

According to the proof of Theorem 4.3, we can obtain

$$T(r, f_j) \leq N(r, f_j) + \sum_{i=1}^{n}\left\{N\left(r, \frac{1}{f_i}\right) - N(r, f_i)\right\} + N(r, \mathbf{W})$$

$$- N\left(r, \frac{1}{\mathbf{W}}\right) + \frac{n(n-1)}{2}\log\left\{\frac{\rho T(R)}{r(\rho - r)}\right\} + \log^+ r + O(1) \quad (4.7.14)$$

for $r_0 < r < \rho < R$, $j = 1, \ldots, n$, where $\mathbf{W} = \mathbf{W}(f_1, \ldots, f_n)$ is a Wronskian determinant.

Without loss of generality, we consider only the case $j = 1$. Clearly, Theorem 4.47 follows from (4.7.14) and the inequalities

$$\mu = \sum_{i=1}^{n}\mu_{f_i}^0 - \sum_{i=2}^{n}\mu_{f_i}^\infty + \mu_{\mathbf{W}}^\infty - \mu_{\mathbf{W}}^0$$

$$\leq \sum_{i=1}^{n}\mu_{f_i,n-1}^0 + \vartheta_n\sum_{i=1}^{n}\mu_{f_i,1}^\infty = \mu_1, \quad (4.7.15)$$

and

$$\mu \leq \sum_{i=1}^{n}\mu_{f_i,n-1}^0 + (n-1)\sum_{i=2}^{n}\mu_{f_i,1}^\infty = \mu_2. \quad (4.7.16)$$

Take $z_0 \in \mathbb{C}$. We distinguish several cases to show the inequality

$$\mu(z_0) \leq \min\{\mu_1(z_0), \mu_2(z_0)\}.$$

If z_0 is not a pole of f_i for any $i = 1, \ldots, n$, obviously we have

$$\mu_{f_i^{(j)}}^0(z_0) \geq \mu_{f_i}^0(z_0) - \mu_{f_i,j}^0(z_0) \geq \mu_{f_i}^0(z_0) - \mu_{f_i,n-1}^0(z_0), \quad 1 \leq i \leq n, \ 1 \leq j \leq n-1,$$

and hence

$$\mu_{\mathbf{W}}^0(z_0) \geq \sum_{i=1}^{n}\{\mu_{f_i}^0(z_0) - \mu_{f_i,n-1}^0(z_0)\},$$

that is

$$\mu(z_0) = \sum_{i=1}^{n}\mu_{f_i}^0(z_0) - \mu_{\mathbf{W}}^0(z_0) \leq \sum_{i=1}^{n}\mu_{f_i,n-1}^0(z_0) = \mu_1(z_0) = \mu_2(z_0).$$

Next, suppose that z_0 is a pole of f_i for each $i = 1, \ldots, n$. Note that we can find holomorphic functions g_i and h_i such that

$$f_i = \frac{g_i}{h_i}, \quad g_i^{-1}(0) \cap h_i^{-1}(0) = \emptyset.$$

Hence $z_0 \in h_i^{-1}(0)$ $(1 \leq i \leq n)$. Fix one i. We can write

$$f_i(z) = \frac{\hat{f}_i(z)}{(z - z_0)^{l_i}}, \quad l_i \in \mathbb{Z}^+,$$

where $\hat{f}_i$ is a holomorphic function near z_0 which does not vanish at z_0. Thus we have

$$\mu_{f_i^{(j)}}^\infty(z_0) = \mu_{f_i}^\infty(z_0) + j = \mu_{f_i}^\infty(z_0) + j\mu_{f_i,1}^\infty(z_0), \quad j \in \mathbb{Z}^+. \tag{4.7.17}$$

Let D_{ij} be the algebraic minor of $f_i^{(j)}$ in $\mathbf{W}$. Then

$$\mathbf{W} = f_0 D_{i0} + D_{i1}.$$

Therefore

$$\mu_{\mathbf{W}}^\infty \leq \max\{\mu_{D_{i0}}^\infty, \mu_{D_{i1}}^\infty\},$$

and so, for the case $i = 1$,

$$\begin{aligned} \mu_{\mathbf{W}}^\infty(z_0) &\leq \sum_{i=2}^n \mu_{f_i}^\infty(z_0) + \frac{n(n-1)}{2} \\ &= \sum_{i=2}^n \mu_{f_i}^\infty(z_0) + \frac{n}{2}\sum_{i=2}^n \mu_{f_i,1}^\infty(z_0) \\ &= \sum_{i=2}^n \mu_{f_i}^\infty(z_0) + \frac{n-1}{2}\sum_{i=1}^n \mu_{f_i}^\infty(z_0), \end{aligned}$$

which implies

$$\begin{aligned} \mu(z_0) &\leq \mu_{\mathbf{W}}^\infty(z_0) - \sum_{i=2}^n \mu_{f_i}^\infty(z_0) \leq \min\left\{\frac{n-1}{2}\sum_{i=1}^n \mu_{f_i,1}^\infty(z_0), \frac{n}{2}\sum_{i=2}^n \mu_{f_i,1}^\infty(z_0)\right\} \\ &\leq \min\{\mu_1(z_0), \mu_2(z_0)\}. \end{aligned}$$

If $n = 2$, these are all possible cases, and therefore Theorem 4.47 is proved.

Assume $n \geq 3$. Except for the above two cases, the following case may occur:

Case 3: $\mu_{f_i}^\infty(z_0) > 0$, but $\mu_{f_j}^\infty(z_0) = 0$ for some $i, j \in \mathbb{Z}^+$.

To study Case 3, without loss of generality, assume that $\mu_{f_i}^\infty(z_0) > 0$ for $i = 1, \ldots, s(< n)$, and $\mu_{f_i}^\infty(z_0) = 0$ for $i > s$. Obviously, we have $s \geq 2$. Note that

$$\mu_{\mathbf{W}}^\infty \leq \max\{\mu_{D_{10}}^\infty, \mu_{D_{11}}^\infty\} \leq \max_{i \in \mathcal{J}_{n-1}} \mu_{f_2^{(i(1))} \cdots f_n^{(i(n-1))}}^\infty$$

holds, where $\mathcal{J}_{n-1}$ is the permutation group on $\mathbb{Z}[1, n-1]$. Since the poles of

$$f_2^{(i(1))} \cdots f_s^{(i(s-1))}$$

and the zeros of

$$f_{s+1}^{(i(s))} \cdots f_n^{(i(n-1))}$$

may cancel each other, when $\mu_{\mathbf{W}}^{\infty}(z_0) > 0$ we have

$$\mu_{\mathbf{W}}^{\infty}(z_0) \leq \sum_{i=2}^{s} \mu_{f_i}^{\infty}(z_0) + \sum_{i=1}^{s-1}(n-i) - \sum_{i=s+1}^{n} \{\mu_{f_i}^0(z_0) - \mu_{f_i,n-1}^0(z_0)\},$$

which means

$$
\begin{aligned}
\mu(z_0) \quad &\leq \quad \sum_{i=s+1}^{n} \mu_{f_i,n-1}^0(z_0) + \frac{(2n-s)(s-1)}{2} \\
&= \quad \sum_{i=s+1}^{n} \mu_{f_i,n-1}^0(z_0) + \left(n - \frac{s}{2}\right) \sum_{i=2}^{s} \mu_{f_i,1}^{\infty}(z_0) \\
&= \quad \sum_{i=s+1}^{n} \mu_{f_i,w}^0(z_0) + \frac{(2n-s)(s-1)}{2s} \sum_{i=1}^{s} \mu_{f_i,1}^{\infty}(z_0) \\
&\leq \quad \min\{\mu_1(z_0), \mu_2(z_0)\}.
\end{aligned}
$$

Otherwise, when $\mu_{\mathbf{W}}^{\infty}(z_0) = 0$ we have

$$\mu_{\mathbf{W}}^0(z_0) \geq \sum_{i=s+1}^{n} \{\mu_{f_i}^0(z_0) - \mu_{f_i,n-1}^0(z_0)\} - \sum_{i=2}^{s} \mu_{f_i}^{\infty}(z_0) - \sum_{i=1}^{s-1}(n-i),$$

and, similarly, we have the inequality $\mu(z_0) \leq \min\{\mu_1(z_0), \mu_2(z_0)\}$.   □

Theorem 4.48. *Take positive integers $d_j \geq 2$ $(j = 1, \ldots, n)$ and set*

$$\alpha = 1 - \sum_{j=1}^{n} \frac{n-1}{d_j}.$$

If $n \geq 2$ and if there exist non-constant meromorphic functions $f_1, \ldots, f_n$ on $\mathbb{C}$ satisfying the equation(4.7.9), then

$$\alpha \leq \frac{\vartheta_n}{n-1+\vartheta_n}.$$

Proof. Assume that there exist non-constant meromorphic functions $f_1, \ldots, f_n$ satisfying (4.7.9). First of all, we consider the case that $f_1^{d_1}, \ldots, f_n^{d_n}$ are linearly independent. By Theorem 4.47, we have

$$
\begin{aligned}
T\left(r, f_j^{d_j}\right) &< \sum_{i=1}^{n} N_{n-1}\left(r, \frac{1}{f_i^{d_i}}\right) + \vartheta_n \sum_{i=1}^{n} \overline{N}(r, f_i) \\
&+ \frac{n(n-1)}{2} \log\left\{\frac{\rho T(R)}{r(\rho - r)}\right\} + \log^+ r + O(1)
\end{aligned}
$$

for $r_0 < r < \rho < R$, $j = 1, \ldots, n$, where

$$T(r) = \max_{1 \le i \le n} \left\{ T\left(r, f_i^{d_i}\right) \right\}.$$

Since

$$N_{n-1}\left(r, \frac{1}{f_i^{d_i}}\right) \le \frac{n-1}{d_i} N\left(r, \frac{1}{f_i^{d_i}}\right) \le \frac{n-1}{d_i} T(r) + O(1),$$

and

$$\overline{N}(r, f_i) \le \frac{1}{d_i} N\left(r, f_i^{d_i}\right) \le \frac{1}{d_i} T(r),$$

then we obtain the estimate

$$T(r) \le \sum_{i=1}^{n} \frac{n-1+\vartheta_n}{d_i} T(r) + \frac{n(n-1)}{2} \log\left\{\frac{\rho T(R)}{r(\rho-r)}\right\} + \log^+ r + O(1),$$

or equivalently,

$$\left(\alpha - \sum_{i=1}^{n} \frac{\vartheta_n}{d_i}\right) T(r) \le \frac{n(n-1)}{2} \log\left\{\frac{\rho T(R)}{r(\rho-r)}\right\} + \log^+ r + O(1). \qquad (4.7.18)$$

Specially, if $f_1, \ldots, f_n$ are rational, then (4.7.18) yields

$$\left(\alpha - \sum_{i=1}^{n} \frac{\vartheta_n}{d_i}\right) \max_{1 \le j \le n} \left\{d_j \deg(f_j)\right\} \le 1 - \frac{n(n-1)}{2} \le 0, \qquad (4.7.19)$$

and so Theorem 4.48 follows.

If one of functions f_j is transcendental, then $T(r)/\log r \to \infty$ as $r \to \infty$, and so the inequality (4.7.18) yields

$$\alpha - \sum_{i=1}^{n} \frac{\vartheta_n}{d_i} \le 0$$

and Theorem 4.48 follows again.

Finally, if $f_1^{d_1}, \ldots, f_n^{d_n}$ are linearly dependent, we can complete the proof of Theorem 4.48 by applying the arguments in the proof of Theorem 4.46. □

Let a be a non-zero integer. A set of n positive integers $\{a_1, a_2, \ldots, a_n\}$ is said to have the property $D_d(a)$ if $a_i a_j + a$ is a dth power of an integer for all $1 \le i < j \le n$. Such a set is called a Diophantine n-tuple (with the property $D_d(a)$). Diophantus found the quadruple $\{1, 33, 68, 105\}$ with the property $D_2(256)$. The first Diophantine quadruple with the property $D_2(1)$, the set $\{1, 3, 8, 120\}$, was found by Fermat (see [79]). A famous conjecture is that there does not exist a Diophantine quintuple with the property $D_2(1)$. Set

$$\lambda_d(a) = \sup\{\#S \mid S \text{ has the property } D_d(a) \},$$

where $\#S$ denotes the number of elements in the set S. A. Dujella [85] proved that $\lambda_2(a)$ is finite for all $a \in \mathbb{Z} - \{0\}$.

Let $\{a_1, a_2, \ldots, a_n\}$ be a Diophantine n-tuple (with the property $D_d(a)$). Then there exist integers b_{ij} such that

$$a_i a_j + a = b_{ij}^d, \quad 1 \le i < j \le n.$$

Set

$$N = \sum_{1 \le i < j \le n} (a_i a_j + a), \quad s = \binom{n}{2}.$$

Then we have

$$\sum_{1 \le i < j \le n} b_{ij}^d = N.$$

When $d \ge 3$, $N \ge N_0(d)$, it is well known that the integer s satisfies the inequalities

$$s \ge g(d) \ge d.$$

Analogously for a function f in a certain function class $\mathcal{F}$ of $\mathcal{M}(\mathbb{C})$, a set of n functions $\{f_1, f_2, \ldots, f_n\}$ in $\mathcal{F}$ is said to have the property $D_d(f)$ if $f_i f_j + f$ is a dth power of a non-constant function in $\mathcal{F}$ for all $1 \le i < j \le n$. Such a set is called a Diophantine n-tuple (with the property $D_d(f)$) over $\mathcal{F}$. Here it is enough to suppose that $f(z)$ is 1 or the identity function z.

Example 4.49 ([85]). *If $f \in \mathcal{F}$ is non-constant, then $\{f, f+2, 4f+4, 16f^3+48f^2+44f+12\}$ has the property $D_2(1)$.*

Assume that $\{f_1, f_2, \ldots, f_n\} \subset \mathcal{F}$ have the property $D_d(f)$, that is, there exist non-constant functions g_{ij} in $\mathcal{F}$ such that

$$f_i f_j + f = g_{ij}^d, \quad 1 \le i < j \le n.$$

Hence we have

$$\sum_{1 \le i < j \le n} g_{ij}^d = F,$$

where

$$F = \sum_{1 \le i < j \le n} (f_i f_j + f).$$

Thus according to $F = 0$, non-zero constant or non-constant, we may apply Corollary 4.29, (4.6.16) to (4.6.19) or Theorem 4.45 to study Diophantine n-tuples (with the property $D_d(f)$) over $\mathbb{C}[z]$ (resp., $\mathcal{A}(\mathbb{C})$, $\mathbb{C}(z)$, $\mathcal{M}(\mathbb{C})$).

4.8   Holomorphic curves into a complex torus

In this section, we will show that non-constant holomorphic curves into a complex torus have normal properties.

4.8.1 Elliptic curves

Theorem 4.50. *Each non-constant holomorphic curve* $\varphi : \mathbb{C} \longrightarrow E(\mathbb{C})$ *into a smooth elliptic curve* $E(\mathbb{C})$ *is surjective.*

Proof. Assume, to the contrary, that φ is not surjective, say $\varphi(\mathbb{C}) \subset E(\mathbb{C}) - \{O\}$. Let (3.1.3) be the Weierstrass equation of E, that is,

$$y^2 = x^3 + ax + b.$$

Then

$$E(\mathbb{C}) - \{O\} = \{[x, y, 1] \in \mathbb{P}^2 \mid y^2 = x^3 + ax + b\},$$

and so φ can be given by

$$\varphi = [f, g, 1]$$

with two entire functions f and g on $\mathbb{C}$ such that

$$g^2 = f^3 + af + b.$$

Without loss of generality, we may assume that f is not constant. Let e_1, e_2, e_3 be three distinct roots of the equation $x^3 + ax + b = 0$. Then we have

$$g^2 = (f - e_1)(f - e_2)(f - e_3),$$

which means that for each i, $f - e_i$ has no simple zeros. Hence

$$\delta_{f,1}(e_i) \geq \frac{1}{2}, \quad i = 1, 2, 3.$$

Note that $\delta_{f,1}(\infty) = 1$. Thus the defect relation (cf. Corollary 2.60) yields

$$\frac{3}{2} \leq \sum_{i=1}^{3} \delta_{f,1}(e_i) \leq 1,$$

which is impossible. Hence Theorem 4.50 is proved. □

Corollary 4.51. *Each smooth elliptic curve omitting one point is Brody hyperbolic.*

According to the remark in Section 3.1.4, there exists a unique lattice $\Lambda \subset \mathbb{C}$ such that

$$g_2 = g_2(\Lambda) = -4a, \quad g_3 = g_3(\Lambda) = -4b.$$

Hence the equation (3.1.3) has non-constant meromorphic solutions $x = \wp$, $y = \omega\wp'$, where ω is a constant with $\omega^2 = \frac{1}{4}$. Thus we obtain non-constant holomorphic curves

$$[\wp, \omega\wp', 1] : \mathbb{C} \longrightarrow E(\mathbb{C}).$$

A natural question is to characterize all non-constant holomorphic curves into $E(\mathbb{C})$.

S.S. Chern [57] and H. Wu [434] studied systematically Nevanlinna theory of holomorphic mappings of Riemann surfaces, and obtained stronger results than Theorem 4.50. Here we recall some of their results.

Theorem 4.52. *Let Φ be a C^∞ real 2-form on a compact Riemann surface M such that*

$$\int_M \Phi = c > 0.$$

Let $a \in M$ be a fixed point of M. Then there exists a real-valued non-negative function u_a with the following properties:

(1) *u_a is C^∞ in $M - \{a\}$;*

(2) *$dd^c u_a = \frac{1}{c}\Phi$ in $M - \{a\}$;*

(3) *If z is any a-centered holomorphic coordinate function in a neighborhood U of a (i.e., $z(a) = 0$), then $u_a(z) + \log|z|^2$ is C^∞ on U.*

The compact Riemann surface M can be given a Hermitian metric ds_M^2 of constant Gaussian curvature such that its volume element Ω satisfies

$$\int_M \Omega = 1.$$

Fix $a \in M$. By Theorem 4.52, there exists a function u_a such that on $M - \{a\}$,

$$dd^c u_a = \Omega.$$

Let $f : \mathbb{C} \longrightarrow M$ be a non-constant holomorphic mapping. Let $n_f(r, a)$ denote the number of pre-images of a (counting multiplicity) in the closed disc $\mathbb{C}[0; r]$ of radius r and define the *valence function*

$$N_f(r, a) = \int_{r_0}^r n_f(t, a)\frac{dt}{t} \tag{4.8.1}$$

for $r \geq r_0 > 0$. Write

$$A_f(r) = \int_{\mathbb{C}[0;r]} f^*\Omega \tag{4.8.2}$$

and define the *characteristic function* of f by

$$T_f(r) = \int_{r_0}^r A_f(t)\frac{dt}{t}. \tag{4.8.3}$$

Then the *first main theorem*

$$T_f(r) = N_f(r, a) + m_f(r, a) - m_f(r_0, a) \tag{4.8.4}$$

holds, where

$$m_f(r, a) = \frac{1}{2}\int_{\mathbb{C}\langle 0;r\rangle} f^*u_a\sigma \tag{4.8.5}$$

serves as the *proximity function* of f for a. If M is the Riemann sphere equipped with the constant curvature metric, then

$$u_a(x) = -\log\chi_\infty(x, a)^2,$$

where χ_∞ is just the chordal distance defined by (1.5.6) for the ordinary absolute value $v = \infty$. This retrieves Nevanlinna's original result cast in the form of Ahlfors-Shimizu.

Further, if $a_1, \ldots, a_q$ are distinct points of M, Chern [57] proved the *defect relation*

$$\sum_{j=1}^{q} \delta_f(a_j) \leq \chi(M), \tag{4.8.6}$$

where $\chi(M)$ is the Euler characteristic of M, and where $\delta_f(a_j)$ is the *defect* of f for a_j defined by

$$\delta_f(a_j) = 1 - \limsup_{r \to \infty} \frac{N_f(r, a_j)}{T_f(r)}. \tag{4.8.7}$$

S.S. Chern uses the following singular volume form approach towards the second main theorem:

$$\Psi = c_1 \left(\prod_{j=1}^{q} \exp(u_{a_j}) \right)^\lambda \Omega, \quad 1 < \lambda < 1,$$

where c_1 is so chosen that $\int_M \Psi = 1$. S. Lang [230] (or see [235]) pointed out that Chern correctly obtains an inequality of second main theorem type valid outside an exceptional set with what is an inefficient error term in his formula (48). When Chern writes, "Letting $\lambda \to 1$ and using (48) we get (49)," he overlooks the fact that the exceptional set depends on λ, and so he cannot take the limit as $\lambda \to 1$ to get rid of that error term and get (49). Chern's previous inequality is enough, of course, to give the defect relation, which was his main purpose at the time. Chern's idea is however essentially correct; what one really has to do is to allow λ to be a function in the singular volume form (see [433]).

When $\chi(M) < 0$, the defect relation (4.8.6) means that there are no non-constant holomorphic mappings from $\mathbb{C}$ into M, which is just the classic Picard theorem. In particular, when $M = E(\mathbb{C})$, we have $\chi(E(\mathbb{C})) = 0$ and so (4.8.6) implies

$$\limsup_{r \to \infty} \frac{N_f(r, a)}{T_f(r)} = 1$$

for each $a \in E(\mathbb{C})$.

4.8.2 Complex torus

Generally, the growth of holomorphic mappings into a complex torus is described by the following fact (cf. [302]):

Proposition 4.53. *Let $\mathbb{C}^n/\Lambda$ be a complex torus with the Kähler form ψ defined by (2.10.9). If $f : \mathbb{C}^m \longrightarrow \mathbb{C}^n/\Lambda$ is a non-constant holomorphic mapping, then there* *exists a positive constant c such that*

$$T(r, r_0; f^*(\psi)) > cr^2$$

holds for larger r.

Proof. There exists a lifting of f,

$$\tilde{f} = (f_1, \ldots, f_n) : \mathbb{C}^m \longrightarrow \mathbb{C}^n,$$

satisfying $p \circ \tilde{f} = f$, where $p : \mathbb{C}^n \longrightarrow \mathbb{C}^n/\Lambda$ is the projection. Then

$$T(r, r_0; f^*(\psi)) = T(r, r_0; dd^c |\tilde{f}|^2).$$

It follows from (2.9.20) that

$$T(r, r_0; f^*(\psi)) = \frac{1}{2} \mathbb{C}^m \left\langle 0; r; |\tilde{f}|^2 \right\rangle - \frac{1}{2} \mathbb{C}^m \left\langle 0; r_0; |\tilde{f}|^2 \right\rangle.$$

By Lemma 2.40, we have

$$\mathbb{C}^m \left\langle 0; r; |\tilde{f}|^2 \right\rangle = \int_{\mathbb{C}^m \langle 0;1 \rangle} \left(\frac{1}{2\pi} \int_0^{2\pi} |\tilde{f}\left(re^{i\varphi}\xi\right)|^2 d\varphi \right) \sigma(\xi).$$

Since f is not constant, there is some f_j which is not constant. Let

$$f_j(z) = \sum_{k \geq 0} P_k(z_1, \ldots, z_m)$$

be the Taylor expansion with homogeneous polynomials P_k of degree k. Then there is some $P_\mu \neq 0$ with $\mu \geq 1$ and

$$\mathbb{C}^m \left\langle 0; r; |\tilde{f}|^2 \right\rangle \geq \int_{\mathbb{C}^m \langle 0;1 \rangle} \left(\frac{1}{2\pi} \int_0^{2\pi} |f_j\left(re^{i\varphi}\xi\right)|^2 d\varphi \right) \sigma(\xi).$$

However,

$$\frac{1}{2\pi} \int_0^{2\pi} |f_j\left(re^{i\varphi}\xi\right)|^2 d\varphi = \sum_{k,l \geq 0} \frac{r^{k+l}}{2\pi} \int_0^{2\pi} e^{(k-l)\varphi} P_k(\xi) \overline{P_l(\xi)} d\varphi$$

$$= \sum_{k \geq 0} r^{2k} |P_k(\xi)|^2 \geq r^{2\mu} |P_\mu(\xi)|^2.$$

Therefore

$$\mathbb{C}^m \left\langle 0; r; |\tilde{f}|^2 \right\rangle \geq r^{2\mu} \int_{\mathbb{C}^m \langle 0;1 \rangle} |P_\mu|^2 \sigma \geq r^2 \int_{\mathbb{C}^m \langle 0;1 \rangle} |P_\mu|^2 \sigma,$$

and so Proposition 4.53 is proved.   □

By using Lemma 2.88 and Theorem 2.105, we have

Theorem 4.54. *Let D be a divisor of simple normal crossings on a complex torus $\mathbb{C}^n/\Lambda$ such that $L = [D]$ is positive. Suppose that $f : \mathbb{C} \longrightarrow \mathbb{C}^n/\Lambda$ is a non-constant holomorphic mapping such that $f(\mathbb{C}) \not\subseteq \mathrm{supp}(D)$. Then we have*

$$\| \quad T_f(r, L) \quad \leq \quad N_f(r, D) - N_{\mathrm{Ram}}(r, f) + \frac{n-1}{2} R_f(r)$$
$$+ O(\log^+ T_f(r, L)) + O(\log r). \qquad (4.8.8)$$

Siu and Yeung [367], [368] (or see [303]) show a stronger inequality as follows:

Theorem 4.55. *Let A be an Abelian variety of complex dimension n and D be an ample divisor in A. Inductively let $k_0 = 0$, $k_1 = 1$ and*

$$k_{j+1} = k_j + 3^{n-j-1}\{4(k_j + 1)\}^j D^n$$

for $1 \leq j < n$, where D^n is the Chern number of D. Then for any holomorphic mapping $f : \mathbb{C} \longrightarrow A$ whose image is not contained in any translate of D,

$$\| \quad T_f(r, [D]) \leq N_{f,k_n}(r, D) + O(\log^+ T_f(r, [D])) + O(\log r).$$

Thus Theorem 4.55 yields immediately the Lang conjecture, which was proved by Siu and Yeung in [366]:

Theorem 4.56. *Let A be an Abelian variety. Let D be an ample divisor in A. If $f : \mathbb{C} \longrightarrow A$ is a non-constant holomorphic mapping, the image of f must intersect D.*

The analogues of Lang's conjecture in number theory is the Theorem 3.39.

4.9   Hyperbolic spaces of lower dimensions

To understand Lang's Conjecture 3.33 well, one way is to find concrete examples of hyperbolic varieties, and further prove that they are Mordellic. In this section, we will give some hyperbolic curves and surfaces by applying Nevanlinna theory and a generalized Borel theorem, say, Siu-Yeung's theorem. In particular, the important role of Nevanlinna theory and the Borel theorem for finding hyperbolic varieties is presented clearly.

4.9.1   Picard's theorem

The Mordell-Faltings theorem states that on a curve of genus greater than 1 there are only finitely many rational points. According to analogy between Diophantine approximation and Nevanlinna theory, this curve could not contain non-constant holomorphic curves, which is just answered by the classic *Picard theorem*:

Theorem 4.57. *A holomorphic mapping $f : \mathbb{C} \longrightarrow M \subset \mathbb{P}^2$ to an irreducible algebraic curve of genus greater than 1 must be constant.*

Proof. See Green [121]. Here we give a sketchy proof according to the idea from B. Shiffman [345]. The universal covering $\tilde{M}$ of M is a well-defined simply connected Riemann surface with a canonical projection mapping $\pi : \tilde{M} \longrightarrow M$. A lifting mapping $\tilde{f} : \mathbb{C} \longrightarrow \tilde{M}$ of f is defined with $\pi \circ \tilde{f} = f$. However, the *uniformization theorem* states that any simply connected Riemann surface is biholomorphically isomorphic either to the plane $\mathbb{C}$, or to the open unit disk $\mathbb{D}$, or to the Riemann sphere $\mathbb{P}^1 \cong \mathbb{C} \cup \{\infty\}$. It follows that $\tilde{M} \cong \mathbb{D}$ since the genus of M is greater than 1. Thus Liouville's theorem implies that $\tilde{f}$ is constant, and so f does. $\qquad\square$

In fact, the converse of Theorem 4.57 is also true.

Theorem 4.58. *If each holomorphic mapping $f : \mathbb{C} \longrightarrow M \subset \mathbb{P}^2$ to an irreducible algebraic curve is constant, then the genus of M must be greater than 1.*

Proof. By the normalization (or resolution of singularity), there exist a compact Riemann surface $\tilde{M}$ and a holomorphic mapping $\eta : \tilde{M} \longrightarrow \mathbb{P}^2$ such that $\eta(\tilde{M}) = M$, and η is one-to-one on smooth points of M. Note that M has only finitely many singular points. Each holomorphic mapping $h : \mathbb{C} \longrightarrow \tilde{M}$ induces a holomorphic mapping $\tilde{h} = \eta \circ h : \mathbb{C} \longrightarrow M$ which must be constant by the assumption. Therefore h is constant. Thus $\tilde{M}$ is Kobayashi hyperbolic by Brody's theorem [38]. Hence $\tilde{M}$ possesses a unique Poincaré metric (cf. [174]), that is, $\tilde{M}$ is conformally isomorphic to the open unit disc in $\mathbb{C}$. Hence the genus of $\tilde{M}$ is greater than 1. $\qquad\square$

Let $f(x, y)$ be an irreducible polynomial of two variables x and y over $\mathbb{C}$. We consider the algebraic curve M in $\mathbb{P}^2$ defined by

$$F(x, y, z) = z^d f\left(\frac{x}{z}, \frac{y}{z}\right) = 0,$$

where d is the degree of f. If there are meromorphic functions g and h on $\mathbb{C}$ satisfying the equation

$$f(x, y) = 0, \qquad\qquad\qquad (4.9.1)$$

that is, $f(g, h) = 0$, we can find entire functions g_0, g_1 and g_2 on $\mathbb{C}$ such that $g = g_1/g_0$, $h = g_2/g_0$ and g_0, g_1, g_2 have no common zeros. Then

$$F(g_1, g_2, g_0) = g_0^d f(g, h) = 0.$$

By Theorem 4.57, if the genus of M is greater than 1, then g and h must be constant. Hence by counting the genus of a curve, for example, by using Theorem 1.82, we may use Theorem 4.57 to distinguish algebraic curves without meromorphic solutions. Usually it is not easy to find the genus of a general curve. Conversely, we may use Theorem 4.58 to find hyperbolic algebraic curves, which will be done in the sequent by using Nevanlinna theory and a generalized Borel theorem.

Let P and Q be polynomials of respectively degree p and q with $q \geq p$. We will consider the following special affine curve

$$f(x, y) = P(x) - Q(y) = 0, \qquad (4.9.2)$$

and hence an algebraic curve

$$F(x, y, z) = z^q P\left(\frac{x}{z}\right) - z^q Q\left(\frac{y}{z}\right) \qquad (4.9.3)$$

in $\mathbb{P}^2(\mathbb{C})$ is well defined.

Here we make a remark on entire solutions of (4.9.2). Assume that p and q are coprime. If there exist two non-constant entire functions g and h on $\mathbb{C}$ such that $f(g, h) = 0$, that is, $P(g) = Q(h)$, we know (cf. [61], Lemma 3.6) that there are a non-constant entire function β and two polynomials R and S of degree q and p, respectively, such that

$$g = R(\beta), \quad h = S(\beta).$$

Since $\beta(\mathbb{C})$ contains an open subset of $\mathbb{C}$, the identity theorem shows that

$$P(R) = Q(S). \qquad (4.9.4)$$

Further, Ritt (cf. [61], Lemma 3.7) proved that the polynomials P and Q are one of the following cases:

(A) there exist $L_i \in \mathrm{Aut}(\mathbb{C})$ $(i = 1, \ldots, 4)$ such that

$$L_1 \circ P \circ L_2 = T_p, \quad L_1 \circ Q \circ L_3 = T_q,$$

$$L_2^{-1} \circ R \circ L_4 = T_q, \quad L_3^{-1} \circ S \circ L_4 = T_p,$$

where $T_p(z) = \cos(p \arccos z)$ is the Tschebyscheff polynomial.

(B) when $p > q$, there exist $L_i \in \mathrm{Aut}(\mathbb{C})$ $(i = 1, \ldots, 4)$ and $T \in \mathbb{C}[z]$ such that

$$L_1 \circ P \circ L_2(z) = z^r T(z)^q, \quad L_1 \circ Q \circ L_3(z) = z^q,$$

$$L_2^{-1} \circ R \circ L_4(z) = z^q, \quad L_3^{-1} \circ S \circ L_4(z) = z^r T(z^q).$$

4.9.2 Hyperbolic curves

In this part, we study meromorphic solutions of the equation (4.9.2). Hà and Yang [141] obtained the following result:

Theorem 4.59. *Take $P, Q \in \mathbb{C}[z]$ with $q \geq p$, where $p = \deg(P)$ and $q = \deg(Q)$. Assume that the affine curve (4.9.2) is irreducible, and has no singular points. If $q \geq 4$ and if $p = q$ or $p = q - 1$, then there are no non-constant meromorphic functions g and h on $\mathbb{C}$ such that $P(g) = Q(h)$.*

Proof. Let $M \subset \mathbb{P}^2$ be the algebraic curve defined by (4.9.3). If M has a singular point $[x_0, y_0, z_0]$, then $z_0 = 0$ by our assumption. Set

$$P(z) = \sum_{i=0}^{p} a_i z^i, \quad Q(z) = \sum_{j=0}^{q} b_j z^j$$

with $a_p \neq 0$, $b_q \neq 0$. If $q = p$, we have

$$\frac{\partial F}{\partial x}(x_0, y_0, 0) = q a_q x_0^{q-1} = 0, \quad \frac{\partial F}{\partial y}(x_0, y_0, 0) = q b_q y_0^{q-1} = 0,$$

which means $x_0 = 0$ and $y_0 = 0$. This is impossible. If $p = q - 1$, similarly we have $y_0 = 0$. So we may assume $x_0 = 1$. Note that

$$F(1, y, z) = a_p z + \cdots + a_0 z^q + \sum_{j=0}^{q} b_j y^j z^{q-j}.$$

Hence $[1, 0, 0]$ is not a singular point. Therefore there are no singular points on M.

Assume that there are meromorphic functions g and h on $\mathbb{C}$ such that $P(g) = Q(h)$. Then we can find entire functions g_0, g_1 and g_2 on $\mathbb{C}$ such that $g = g_1/g_0$, $h = g_2/g_0$ and g_0, g_1, g_2 have no common zeros. Note that

$$g_0^q P\left(\frac{g_1}{g_0}\right) - g_0^q Q\left(\frac{g_2}{g_0}\right) = 0.$$

Thus we obtain a holomorphic mapping

$$G = [g_0, g_1, g_2] : \mathbb{C} \longrightarrow M.$$

Since the genus of M is $(q - 1)(q - 2)/2 > 1$, the mapping G is constant by Theorem 4.57. Therefore g and h are constant. □

Theorem 4.59 gives conditions in which the curve (4.9.2) is hyperbolic. Note that if $q - p \geq 2$, then the curve (4.9.2) has a singular point at infinity $[1, 0, 0]$, which is a non-ordinary singular point of multiple $q - p$.

A point $(c, d) \in \mathbb{C}^2$ is a singular point of the affine algebraic curve (4.9.2) if and only if P and Q satisfy

$$P(c) = Q(d), \quad P'(c) = Q'(d) = 0,$$

that is, the affine algebraic curve (4.9.2) has no singular points if and only if

$$\{P(c) \mid P'(c) = 0\} \cap \{Q(d) \mid Q'(d) = 0\} = \emptyset.$$

Yang and Li [441] gave more refined results by weakening the condition on singular points. To introduce their results, we first strengthen the lemmas in [441] as follows:

Theorem 4.60. *Suppose that $P(z)$ and $Q(z)$ are respectively two polynomials of degrees p and q with $q \geq p \geq 2$. Let S be a non-empty subset of zeros of P' and let T be the set of zeros of Q' such that $P(S) \cap Q(T) = \emptyset$. If there exist two non-constant meromorphic functions g and h satisfying*

$$P(g) = Q(h), \qquad (4.9.5)$$

then the inequality

$$\frac{lq - p}{q} T(r, g) \leq \overline{N}(r, g) + \log \frac{\rho T(R, g)}{r(\rho - r)} + O(1) \qquad (4.9.6)$$

holds for any $r_0 < r < \rho < R$, where $l = \sum_{c \in S} \mu_{P'}^0(c)$. In particular, we have

(I) *if $q > p$, then $q - p = (p, q)$, S has only one element which is a simple zero of P';*

(II) *if $q = p$ and $\mu_{P'}^0(c) \geq 2$ for some $c \in S$, then $\mu_{P'}^0(c) = 2$, $S = \{c\}$;*

(III) *if $q = p$ and $\#S \geq 2$, then $\#S = 2$, and the elements in S are simple zeros of P'.*

Proof. From (4.9.5), we obtain

$$pT(r, g) = qT(r, h) + O(1). \qquad (4.9.7)$$

Set $n = (p, q)$. Then there exist two coprime integers p_1 and q_1 such that

$$p = p_1 n, \quad q = q_1 n. \qquad (4.9.8)$$

Suppose that z_0 is a pole of g. Then from (4.9.5), z_0 is also a pole of h, and

$$p\mu_g^\infty(z_0) = q\mu_h^\infty(z_0),$$

which means

$$p_1 \mu_g^\infty(z_0) = q_1 \mu_h^\infty(z_0), \quad q_1 \mid \mu_g^\infty(z_0)$$

since p_1 and q_1 are relatively prime to each other. Hence the multiplicity of any pole of g is at least q_1. This implies

$$q_1 \overline{N}(r, g) \leq N(r, g). \qquad (4.9.9)$$

There exists a partition

$$S = S_1 \cup \cdots \cup S_k$$

such that

$$S_\alpha \neq \emptyset \ (\alpha = 1, \ldots, k), \quad S_\alpha \cap S_\beta = \emptyset \ (\alpha \neq \beta),$$

and such that

$$P(c) = P(c') \; (c, c' \in S_\alpha), \; P(c) \neq P(c'') \; (c \in S_\alpha, \; c'' \in S_\beta, \; \alpha \neq \beta).$$

Take $c_\alpha \in S_\alpha$. Since c_α is a root of $P'(z) = 0$, there exists a polynomial $R_\alpha(z)$ such that

$$P(z) - P(c_\alpha) = R_\alpha(z) \prod_{c \in S_\alpha} (z - c)^{s_{\alpha,c}}, \quad R_\alpha(c) \neq 0 \; (c \in S_\alpha). \tag{4.9.10}$$

Consequently,

$$R_\alpha(g) \prod_{c \in S_\alpha} (g - c)^{s_{\alpha,c}} = Q(h) - P(c_\alpha). \tag{4.9.11}$$

We put

$$l = \sum_{\alpha=1}^{k} \sum_{c \in S_\alpha} (s_{\alpha,c} - 1) = \sum_{c \in S} \mu_{P'}^0(c)$$

with $l \geq k$ since by definition

$$s_{\alpha,c} \geq 2, \; c \in S_\alpha, \; \alpha = 1, \ldots, k.$$

Since, by hypothesis, $Q - P(c_\alpha)$ does not vanish at any zero of Q', this implies that it has no multiple root and hence it has a factorization

$$Q(z) - P(c_\alpha) = \lambda_\alpha \prod_{j=1}^{q} (z - b_{\alpha,j}),$$

where $b_{\alpha,j}$ are all distinct for each fixed α.

On the other hand, we claim that

$$b_{\alpha,i} \neq b_{\beta,j}, \; (\alpha, i) \neq (\beta, j).$$

In fact, if $b_{\alpha,i} = b_{\beta,j}$ for some $(\alpha, i) \neq (\beta, j)$, then $\alpha \neq \beta$, and so $P(c_\alpha) \neq P(c_\beta)$, therefore

$$\tilde{Q} := (Q - P(c_\alpha)) - (Q - P(c_\beta)) = \text{const} \neq 0,$$

but since $b_{\alpha,i} = b_{\beta,j}$, we have $\tilde{Q}(b_{\alpha,i}) = 0$, a contradiction. Take

$$a \in \mathbb{C} - Q^{-1}(P(S)).$$

By Nevanlinna's second main theorem (cf. Corollary 2.59), for any $r_0 < r < \rho < R$, we have

$$kqT(r, h) \leq \sum_{\alpha=1}^{k} \sum_{j=1}^{q} \overline{N}\left(r, \frac{1}{h - b_{\alpha,j}}\right) + \overline{N}\left(r, \frac{1}{h - a}\right)$$

$$+ \overline{N}(r, h) + \log \frac{\rho T(R, h)}{r(\rho - r)} + O(1),$$

or, equivalently,

$$
\begin{aligned}
kqT(r,h) \;\le\; & \sum_{\alpha=1}^{k} \overline{N}\left(r, \frac{1}{Q(h) - P(c_\alpha)}\right) + \overline{N}\left(r, \frac{1}{h - a}\right) \\
& + \overline{N}(r,h) + \log \frac{\rho T(R,h)}{r(\rho - r)} + O(1). \quad (4.9.12)
\end{aligned}
$$

By (4.9.11), we have

$$
\overline{N}\left(r, \frac{1}{Q(h) - P(c_\alpha)}\right) \le \sum_{c \in S_\alpha} \overline{N}\left(r, \frac{1}{g - c}\right) + \overline{N}\left(r, \frac{1}{R_\alpha(g)}\right)
$$

$$
\le \left\{ p - \sum_{c \in S_\alpha} (s_{\alpha,c} - 1) \right\} T(r,g) + O(1).
$$

Noting that $\overline{N}(r,h) = \overline{N}(r,g)$, the above inequalities and (4.9.7) lead to

$$
T(r,h) - \overline{N}\left(r, \frac{1}{h - a}\right) + \frac{lq - p}{q} T(r,g) \le \overline{N}(r,g) + \log \frac{\rho T(R,g)}{r(\rho - r)} + O(1). \quad (4.9.13)
$$

Hence the inequality (4.9.6) follows from (4.9.13).

By (4.9.9), we have

$$
\overline{N}(r,g) \le \frac{1}{q_1} N(r,g) \le \frac{1}{q_1} T(r,g). \quad (4.9.14)
$$

Then (4.9.13) yields

$$
\frac{lq - p}{q} T(r,g) \le \frac{1}{q_1} T(r,g) + \log \frac{\rho T(R,g)}{r(\rho - r)} + O(1), \quad (4.9.15)
$$

which implies that

$$
\frac{lq - p}{q} \le \frac{1}{q_1}.
$$

Since $p/q = p_1/q_1$, the above inequality is equivalent to

$$
lq_1 \le p_1 + 1. \quad (4.9.16)
$$

Now we can prove the case (I). When $p < q$, we also have $p_1 + 1 \le q_1$. Therefore from (4.9.16) we can deduce

$$
1 \le k \le l \le 1, \quad q_1 = p_1 + 1.
$$

Furthermore,

$$
q - p = (q_1 - p_1)n = n = (p,q).
$$

Since $l = 1$, from (4.9.10) we see that S has only one element c which just is a simple zero of $P'(z)$. The case (I) in Theorem 4.60 is proved.

Next we prove the case (II). Since $q = p$ and $c \in S$ is a multiple zero of $P'(z)$, then (4.9.16) implies $l = 2$, and so

$$S = S_1 = \{c\}, \quad s_{1,c} = 3, \quad \mu_{P'}^0(c) = 2.$$

Finally, we show the case (III). When $q = p$ and $\#S \geq 2$, then (4.9.16) implies $l = 2$, and so $\#S = 2$. Hence the elements in S are simple zeros of P'. $\square$

Note that the set S in Theorem 4.60 may be a proper subset of zeros of P'. In particular, if S is the set of zeros of P', then the condition $P(S) \cap Q(T) = \emptyset$ in Theorem 4.60 implies that the affine curve (4.9.2) has no singular points. When $q > p \geq 2$, the case (I) in Theorem 4.60 claims that the curve (4.9.2) is elliptic or hyperelliptic (see Section 3.1). In fact, case (I) means that $p = 2$ and $q = 3$ or 4. If $q = p \geq 3$, the case (II) and (III) show that the curve (4.9.2) is of degree 3 (see Section 4.6). If $q = p = 2$, Theorem 4.60 is nothing.

Corollary 4.61 ([441]). *Suppose that $P(z)$ and $Q(z)$ are respectively two polynomials of degrees p and q with $q \geq p \geq 2$. Let S be a non-empty subset of zeros of P' and let T be the set of zeros of Q' such that $P(S) \cap Q(T) = \emptyset$. Then there do not exist non-constant meromorphic functions g and h satisfying (4.9.5) if $P(z)$ and $Q(z)$ satisfy one of the following conditions:*

(a) $p < q$, $q - p \neq (p, q)$.

(b) $p < q$ and $\mu_{P'}^0(c) \geq 2$ *for some* $c \in S$.

(c) $p < q$ and $\#S \geq 2$.

(d) $p \leq q$, $\#S \geq 2$ and $\mu_{P'}^0(c) \geq 2$ *for some* $c \in S$.

(e) $p \leq q$ and $\#S \geq 3$.

(f) $p \leq q$ and $\mu_{P'}^0(c) \geq 3$ *for some* $c \in S$.

Thus when the affine curve (4.9.2) is irreducible, Theorem 4.58 implies that the curve (4.9.3) has genus > 1 if P and Q further satisfy the conditions in Corollary 4.61, that is, the curve (4.9.3) is hyperbolic. A simple example is the *Fermat curve*

$$x^d + y^d = 1.$$

Setting

$$P(x) = x^d, \quad Q(y) = -y^d + 1,$$

when $d \geq 4$ the polynomials P and Q satisfy the conditions in Corollary 4.61. Also note that the Fermat curve has the genus $(d-1)(d-2)/2 > 1$ when $d \geq 4$. Thus there are no non-constant meromorphic functions satisfying the Fermat equation, which re-proves part of Theorem 4.38.

Theorem 4.62. *Let p and q be positive integers with $p > 1$, $q > 1$ and $(p, q) \neq (2, 2)$. Given non-zero complex numbers A, B and C, the equation $Ax^p - By^q = C$ has only finitely many rational solutions except at most the cases*

$$(p, q) = (2, 3), \ (3, 2), \ (3, 3), \ (2, 4), \ (4, 2). \tag{4.9.17}$$

Proof. Without loss of generality, we may assume $p \leq q$. Set

$$P(x) = Ax^p, \ Q(y) = By^q + C.$$

When $(p, q) \neq (2, 3)$, $(3, 3)$, $(2, 4)$, Corollary 4.61 implies that there exist no non-constant meromorphic functions g and h satisfying (4.9.5). Hence the curve (4.9.3) has genus > 1, and so Theorem 4.62 follows from Theorem 3.32. $\square$

Take $b \in \mathbb{C}$ and set $d = 6l + 3$ for a positive integer l. A.M. Nadel [285] proved that if $l \geq 3$, the plane curve $C \subset \mathbb{P}^2(\mathbb{C})$ defined by

$$X^d + Y^d + Z^3(X^{d-3} + Y^{d-3}) = bZ^d$$

is hyperbolic, and that if $l \geq 2$, the plane curve $C \subset \mathbb{P}^2(\mathbb{C})$ defined by

$$X^d + Y^d + Z^3 X^{d-3} = bZ^d$$

is hyperbolic. In fact, taking $\{a, A, b\} \subset \mathbb{C}$ and setting

$$P(x) = x^d - ax^{d-k}, \quad Q(y) = -y^d + Ay^{d-k} + b,$$

if we choose a, b and A such that the sets S and T of zeros of P' and Q' respectively satisfy $P(S) \cap Q(T) = \emptyset$, then Yang-Li's theorem implies that the plane curve $C \subset \mathbb{P}^2(\mathbb{C})$ defined by

$$X^d + Y^d - Z^k(aX^{d-k} + AY^{d-k}) = bZ^d \tag{4.9.18}$$

is hyperbolic when $d > k \geq 3$. As an illustration of another approach to this type of question, we give the following result under the assumption that $d > k \geq 4$.

Theorem 4.63 ([188]). *Take integers d and k with $d > k \geq 4$ and take $\{a, A\} \subset \mathbb{C}$ such that $z^d + 1 = 0$ and $az^{d-k} + A = 0$ have no common zeros. Then for any $b \in \mathbb{C}$, the plane curve $C \subset \mathbb{P}^2(\mathbb{C})$ defined by (4.9.18) is hyperbolic.*

Proof. From the Picard theorem, it is sufficient to show that if there is a holomorphic curve

$$[F_0, F_1, F_2] : \mathbb{C} \longrightarrow \mathbb{P}^2(\mathbb{C})$$

such that

$$F_1^d + F_2^d - F_0^k(aF_1^{d-k} + AF_2^{d-k}) = bF_0^d,$$

then it must be constant. We may assume that F_0, F_1 and F_2 have no common zeros.

If $F_2 = 0$, we have

$$F_1^d - F_0^k(aF_1^{d-k} + bF_0^{d-k}) = 0.$$

Obviously, $F_1 = 0$ if $F_0 = 0$. This is impossible. If $F_0 \neq 0$, we have

$$\left(\frac{F_1}{F_0}\right)^d - a\left(\frac{F_1}{F_0}\right)^{d-k} = b,$$

which implies that F_1/F_0 must be a constant, and so the holomorphic curve $[F_0, F_1, F_2] = [1, F_1/F_0, 0]$ is constant. Assume that $F_2 \neq 0$ and write

$$g = \frac{F_1}{F_2}, \quad h = \frac{F_0}{F_2}.$$

Then the meromorphic functions g, h in $\mathbb{C}$ satisfy

$$g^d - h^k(ag^{d-k} + bh^{d-k} + A) = -1. \tag{4.9.19}$$

It is sufficient to show that g and h are constant.

First of all, we show that one of g and h is constant. Assume, to the contrary, that g and h are non-constant. When $k \geq 4$ Theorem 4.34 means that the functions g^d and $h^k(ag^{d-k} + bh^{d-k} + A)$ are linearly dependent. Then there is an element $c \in \mathbb{C} - \{0\}$ such that

$$h^k(ag^{d-k} + bh^{d-k} + A) = cg^d,$$

and hence $(1 - c)g^d = -1$. This is a contradiction since g is non-constant.

Hence one of g and h is constant. Obviously, (4.9.19) means that g is constant if h is constant. If $g = c$ is a constant, then

$$bh^d + (ac^{d-k} + A)h^k = c^d + 1,$$

which implies that h also is constant since at least one of $c^d + 1$ and $ac^{d-k} + A$ is not zero.   □

4.9.3   Hyperbolic surfaces

In Section 4.9.2, we introduce two methods for finding hyperbolic curves based on the classic Picard theorem. In higher-dimensional spaces, we may use the Brody theorem (cf. Theorem 2.26) to seek hyperbolic varieties. Here we introduce some examples in this direction.

Take positive integers d, k with $d - k \geq 2$ and take $a \in \mathbb{C} - \{0\}$ such that the equation

$$z^d - az^{d-k} + 1 = 0$$

has no multiple roots, that is,

$$a^d \neq c_{d,k} = \frac{d^d}{k^k(d-k)^{d-k}}.$$

Nadel [285] proved the following result:

Theorem 4.64. *Let $e \geq 3$ be an integer and take $k = 3, d = 6e + 3$. Further assume that the non-zero complex number a satisfies*

$$a^{\frac{d}{3}} \notin \left\{ \sqrt[3]{c_{d,3}}, \frac{1}{2} \sqrt[3]{c_{d,3}} \right\}.$$

The surface $M \subset \mathbb{P}^3(\mathbb{C})$ defined by

$$X^d + Z^d + W^d + Y^3(Y^{d-3} - aX^{d-3} - aW^{d-3}) = 0$$

is hyperbolic.

Siu and S.K. Yeung [367] proved that when $d \geq 11$, the surface in $\mathbb{P}^3(\mathbb{C})$ defined by

$$X^d + Z^d + W^d + Y^{d-2}(Y^2 - a_0X^2 - a_1Z^2 - a_2W^2) = 0$$

is hyperbolic if

$$a_i^d \neq (-1)^{d+1}a_j^d \ (i \neq j), \quad a_j^d \neq c_{d,2} \ (0 \leq j \leq 2).$$

R. Brody and M. Green [39] proved that the surface of $\mathbb{P}^3(\mathbb{C})$ defined by

$$X^d + Z^d + W^d + Y^k(Y^{d-k} - aX^{d-k} - bZ^{d-k}) = 0 \qquad (4.9.20)$$

is hyperbolic for even degree $d = 2k \geq 50$ and for generic $a, b \in \mathbb{C} - \{0\}$. Based on this example, K. Azukawa and M. Suzuki [8] constructed a smooth hyperbolic curve of $\mathbb{P}^2(\mathbb{C})$ in which complement is hyperbolic. Related to this kind of surfaces, we prove the following results:

Theorem 4.65 ([188]). *Take $b \in \mathbb{C} - \{0\}$ and take $(a_0, a_1, a_2) \in \mathbb{C}^3 - \{0\}$ satisfying the following conditions:*

(i) *$z^d - b = 0$ and $a_1 z^{d-k} + a_0 = 0$ have no common zeros;*
(ii) *$z^d - b = 0$ and $a_2 z^{d-k} + a_0 = 0$ have no common zeros;*
(iii) *$z^d + 1 = 0$ and $a_2 z^{d-k} + a_1 = 0$ have no common zeros.*

If $d > k \geq 9$, then the surface in $\mathbb{P}^3(\mathbb{C})$ defined by

$$X^d + Z^d + Y^k(Y^{d-k} - a_0W^{d-k} - a_1X^{d-k} - a_2Z^{d-k}) = bW^d$$

is hyperbolic.

Proof. From a theorem due to Brody (cf. Theorem 2.26), it is sufficient to show that if there is a holomorphic curve

$$[F_0, F_1, F_2, F_3] : \mathbb{C} \longrightarrow \mathbb{P}^3(\mathbb{C})$$

such that

$$F_1^d + F_3^d + F_2^k(F_2^{d-k} - a_0 F_0^{d-k} - a_1 F_1^{d-k} - a_2 F_3^{d-k}) = b F_0^d,$$

then it must be constant. We may assume that F_0, F_1, F_2 and F_3 have no common zeros.

If $F_0 = 0$, we have

$$F_1^d + F_2^d + F_3^d - F_2^k(a_1 F_1^{d-k} + a_2 F_3^{d-k}) = 0.$$

If $F_2 = 0$, then $F_1 \neq 0$ and F_3/F_1 is constant. Hence the holomorphic curve $[F_0, F_1, F_2, F_3]$ is constant, and we are done. Assume that $F_2 \neq 0$ and set

$$G = \frac{F_1}{F_2}, \quad H = \frac{F_3}{F_2}.$$

Then we obtain

$$G^d + H^d - (a_1 G^{d-k} + a_2 H^{d-k}) = -1.$$

If G is non-constant, then H also is non-constant. When $k \geq 9$, Theorem 4.34 implies that G^d, H^d and 1 are linearly dependent, that is, there is $(b_0, b_1, b_2) \in \mathbb{C}^3 - \{0\}$ such that

$$b_1 G^d + b_2 H^d = b_0.$$

Theorem 4.38 implies that $b_0 = 0$, and so $H = cG$ for some $c \in \mathbb{C} - \{0\}$. Thus we have

$$(c^d + 1)G^d - (a_2 c^{d-k} + a_1)G^{d-k} = -1,$$

which implies that G is constant since $(c^d + 1, a_2 c^{d-k} + a_1) \in \mathbb{C}^2 - \{0\}$ according to our assumption. This is a contradiction. Hence G is constant, and so H also is constant. We are done.

Assume that $F_0 \neq 0$ and write

$$g = \frac{F_1}{F_0}, \quad h = \frac{F_2}{F_0}, \quad f = \frac{F_3}{F_0}.$$

Then meromorphic functions g, h, f in $\mathbb{C}$ satisfy

$$g^d + f^d + h^k(h^{d-k} - a_1 g^{d-k} - a_2 f^{d-k} - a_0) = b. \tag{4.9.21}$$

It is sufficient to show that g, h and f are constant.

First of all, we show that at least one of g, h and f is constant. Assume, to the contrary, that g, h and f all are non-constant. When $k \geq 9$, Theorem 4.34 implies that g^d, f^d and b are linearly dependent, that is, there is $(c_0, c_1, c_2) \in \mathbb{C}^3 - \{0\}$ such that

$$c_1 g^d + c_2 f^d = c_0.$$

Theorem 4.38 implies that $c_0 = 0$, and so $f = cg$ for some $c \in \mathbb{C} - \{0\}$. Then (4.9.21) has the form

$$(c^d + 1)g^d + h^k(h^{d-k} - (a_1 + a_2 c^{d-k})g^{d-k} - a_0) = b.$$

If $c^d + 1 \neq 0$, then Theorem 4.34 implies that $(c^d+1)g^d$ and b are linearly dependent, that is, g is constant. This is a contradiction. Hence $c^d + 1 = 0$, and so $a_1 + a_2 c^{d-k} \neq 0$. Since

$$h^d - h^k((a_1 + a_2 c^{d-k})g^{d-k} + a_0) = b,$$

Theorem 4.34 again implies that h^d and b are linearly dependent, that is, h is constant. This is a contradiction. Hence at least one of g, h and f is constant.

Assume that $g = c$ is a constant. If h also is a constant, then (4.9.21) means that f is constant, and we are done. Assume, to the contrary, that h is non-constant. Then f also is non-constant. The equation (4.9.21) has the form

$$f^d + h^k(h^{d-k} - a_2 f^{d-k} - a_1 c^{d-k} - a_0) = b - c^d. \tag{4.9.22}$$

It follows that $c^d - b = 0$ from Theorem 4.34. Thus we have

$$f^d + h^d = h^k(a_2 f^{d-k} + a_1 c^{d-k} + a_0).$$

Since $a_1 c^{d-k} + a_0 \neq 0$, Theorem 4.34 implies that $f = \hat{c}h$ for some $\hat{c} \in \mathbb{C} - \{0\}$, and hence

$$(\hat{c}^d - a_2 \hat{c}^{d-k} + 1)h^d = h^k(a_1 c^{d-k} + a_0),$$

which means that h is constant. This is a contradiction.

Assume, to the contrary, that g is not a constant. If h is not a constant, then f must be a constant c. Now the equation (4.9.21) has the form

$$g^d + h^k(h^{d-k} - a_1 g^{d-k} - a_2 c^{d-k} - a_0) = b - c^d.$$

A contradiction follows according to the proof related to the equation (4.9.22). Hence h must be a constant c.

Now f is not a constant, otherwise, it follows that g is a constant. Thus (4.9.21) becomes

$$g^d + f^d - c^k(a_1 g^{d-k} + a_2 f^{d-k}) = b - c^d + a_0 c^k. \tag{4.9.23}$$

Theorem 4.38 implies that $c \neq 0$. Further, by Theorem 4.38 and Theorem 4.34, we can prove $f = \tilde{c}g$ for some $\tilde{c} \in \mathbb{C} - \{0\}$. Thus we obtain

$$(\tilde{c}^d + 1)g^d - c^k(a_1 + a_2 \tilde{c}^{d-k})g^{d-k} = b - c^d + a_0 c^k$$

which implies that g is constant since

$$(\tilde{c}^d + 1, a_1 + a_2\tilde{c}^{d-k}) \in \mathbb{C}^2 - \{0\}.$$

This is a contradiction. Therefore g, h and f all are constant.   □

In [188], we also constructed other hyperbolic surfaces of lower degrees and hyperbolic hypersurfaces in $\mathbb{P}^4(\mathbb{C})$ by using the above methods, say,

Theorem 4.66. *Take positive integers d and k with $d > k$ and $2k \geq d + 16$. Take complex numbers b, a_i and b_i for $i = 0, 1, 2, 3$ with $b \neq 0$ and $a_2 b_2 \neq 1$ satisfying the following conditions:*

(1) $z^d + 1 = 0$ *and* $a_1 z^{d-k} + a_0 = 0$ *have no common zeros*

(2) $z^d + 1 = 0$ *and* $b_1 z^{d-k} + b_0 = 0$ *have no common zeros;*

(3) $z^d - b_0 z^k + 1 = 0$ *and* $a_2 z^{d-k} + a_0 = 0$ *have no common zeros;*

(4) $z^d - a_0 z^k + 1 = 0$ *and* $b_2 z^{d-k} + b_0 = 0$ *have no common zeros;*

(5) $z^d - b = 0$ *and* $a_1 z^{d-k} + a_3 = 0$ *have no common zeros;*

(6) $z^d - b = 0$ *and* $(a_1 + b_1 a_2) z^{d-k} + a_2 b_3 + a_3 = 0$ *have no common zeros;*

(7) $z^d - b = 0$ *and* $b_1 z^{d-k} + b_3 = 0$ *have no common zeros;*

(8) $z^d - b = 0$ *and* $(b_1 + a_1 b_2) z^{d-k} + b_2 a_3 + b_3 = 0$ *have no common zeros;*

(9) $x^d + y^d = b$, $a_0 x^{d-k} + a_1 y^{d-k} + a_3 = 0$ *and* $b_0 x^{d-k} + b_1 y^{d-k} + b_3 = 0$ *have no common zeros;*

(10) $z^d + 1 = 0$ *and* $(a_1 + b_1 a_2) z^{d-k} + b_0 a_2 + a_0 = 0$ *have no common zeros;*

(11) $z^d - a_3 z^k - b = 0$ *and* $b_2 z^{d-k} + b_3 = 0$ *have no common zeros;*

(12) $z^d + 1 = 0$ *and* $(b_1 + a_1 b_2) z^{d-k} + a_0 b_2 + b_0 = 0$ *have no common zeros;*

(13) $z^d - b_3 z^k - b = 0$ *and* $a_2 z^{d-k} + a_3 = 0$ *have no common zeros.*

Then the hypersurface in $\mathbb{P}^4(\mathbb{C})$ defined by

$$\begin{aligned} bV^d \;=\; & X^d + Y^k (Y^{d-k} - a_0 W^{d-k} - a_1 X^{d-k} - a_2 Z^{d-k} - a_3 V^{d-k}) \\ & + Z^k (Z^{d-k} - b_0 W^{d-k} - b_1 X^{d-k} - b_2 Y^{d-k} - b_3 V^{d-k}) + W^d \end{aligned}$$

is hyperbolic.

For example, take non-zero complex numbers b, a_0, b_0, a_3 and b_3 such that $a_3/a_0 \neq b_3/b_0$ and set

$$a_1 = a_2 = b_1 = b_2 = 0.$$

When $d > k$ and $2k \geq d + 16$, the conditions in Theorem 4.66 are satisfied, and so the hypersurface in $\mathbb{P}^4(\mathbb{C})$ defined by

$$\begin{aligned} bV^d \;=\; & X^d + Y^k (Y^{d-k} - a_0 W^{d-k} - a_3 V^{d-k}) \\ & + Z^k (Z^{d-k} - b_0 W^{d-k} - b_3 V^{d-k}) + W^d \end{aligned}$$

is hyperbolic.

4.9.4 Uniqueness polynomials

We make a remark on the equation (4.9.2). Related to finding non-constant meromorphic solutions of (4.9.2), we hope to characterize the polynomials P and Q such that if (4.9.2) has non-constant meromorphic solutions $(x, y) = (g, h)$, then it must be the case that $g = h$. If P and Q are such two polynomials, then $P(g) = Q(g)$ means

$$pT(r, g) = qT(r, g) + O(1),$$

and so $p = q$ since $T(r, g) \to \infty$ as $r \to \infty$. Further, by comparing the multiplicity of zeros of $P(g)$ and $Q(g)$, it is not difficult to show that $Q = \lambda P$ for some $\lambda \in \mathbb{C}_*$. Hence the question turns to characterizing a polynomial P such that if $P(g) = \lambda P(h)$ holds for non-constant meromorphic functions g, h and some non-zero constant λ, then $g = h$. Such a polynomial P is called a *strong uniqueness polynomial* of meromorphic functions or a *uniqueness polynomial* for the case $\lambda = 1$. Similarly, we may define *(strong) uniqueness polynomials* of n variables with respect to a family of non-constant meromorphic mappings $[1, f_1, \ldots, f_n]$ into a complex projective space $\mathbb{P}^n(\mathbb{C})$.

Theorem 4.67 ([4]). *Let P be a polynomial of degree n over $\mathbb{C}$ and*

$$P'(z) = c(z - c_1)^{m_1} \cdots (z - c_l)^{m_l},$$

where c is a non-zero constant and

$$c_i \neq c_j, \ P(c_i) \neq P(c_j), \ 1 \leq i \neq j \leq l.$$

Then P is a uniqueness polynomial for rational functions if and only if $l \geq 3$, or $l = 2$ and $\min\{m_1, m_2\} \geq 2$. Further, P is a strong uniqueness polynomial for rational functions if and only if no non-trivial affine transformation of $\mathbb{C}$ preserves the set of zeros of P and one of the following conditions is satisfied.

(α) $l \geq 3$, except when $n = 4$, $m_1 = m_2 = m_3 = 1$ and

$$\frac{P(c_1)}{P(c_2)} = \frac{P(c_2)}{P(c_3)} = \frac{P(c_3)}{P(c_1)} = \omega, \ \omega^2 + \omega + 1 = 0;$$

(β) $l = 2$ and $\min\{m_1, m_2\} \geq 2$.

Theorem 4.68 ([4]). *Let P be a polynomial of degree n over $\mathbb{C}$ and*

$$P'(z) = c(z - c_1)^{m_1} \cdots (z - c_l)^{m_l},$$

where c is a non-zero constant and

$$c_i \neq c_j, \ P(c_i) \neq P(c_j), \ 1 \leq i \neq j \leq l.$$

Then P is a uniqueness polynomial for meromorphic functions if and only if one of the following conditions is satisfied:

(γ) $l \geq 3$, except when $n = 4$, $m_1 = m_2 = m_3 = 1$;

(δ) $l = 2$ and $\min\{m_1, m_2\} \geq 2$ except when $n = 5$, $m_1 = m_2 = 2$.

Further, P is a strong uniqueness polynomial for meromorphic functions if and only if no non-trivial affine transformation of $\mathbb{C}$ preserves the set of zeros of P and one of the conditions (γ) and (δ) is satisfied.

For more general equations, we mention a result due to Osgood. Let $n \geq 1$ and suppose that $F(y, y_1, \ldots, y_n)$ is a not identically zero irreducible polynomial in the variables indicated, where the coefficients of F are meromorphic functions. Let $g, g_1, \ldots, g_n$ be $n + 1$ meromorphic functions. Suppose that each coefficient, say a, of F satisfies

$$T(r, a) = o\left(\max\{T(r, g), T(r, g_i)\}\right).$$

Let $n(r, g_1, \ldots, g_n)$ be the number of distinct poles of the functions $g_1, \ldots, g_n$, occurring in the disc of radius r centered at the origin, where each pole is counted with the highest multiplicity occurring among the functions in $\{g_1, \ldots, g_n\}$. Write

$$G(z) = \max_{0 \leq i \leq n} |g_i(z)|,$$

where $g_0 = 1$, and define

$$T(r, g_1, \ldots, g_n) = \int_{r_0}^{r} \frac{n(t, g_1, \ldots, g_n)}{t} dt + \frac{1}{2\pi} \int_{0}^{2\pi} \log^+ G\left(re^{2\pi i\theta}\right) d\theta.$$

C. F. Osgood [312] claimed the following result:

Theorem 4.69. *Let F and $g, g_1, \ldots, g_n$ be as above. Let M denote the total degree of F in $y_1, \ldots, y_n$. Suppose that on a sequence of r values tending to infinity one has the limit*

$$\lim_{r} \frac{T(r, g_1, \ldots, g_n)}{T(r, g)} = \mu < \infty.$$

If $\deg_y F > n\mu M$, then $F(g, g_1, \ldots, g_n) \not\equiv 0$.

4.10   Factorization of functions

After being influenced by the unique factorization of integers into prime factors, P.C. Rosenbloom, F. Gross and C.C. Yang defined prime (or pseudo-prime) meromorphic functions, and tried to comprehend the uniqueness of factorization of meromorphic (or entire) functions in the sense of composition. However, the case of functions is very complicated. Some basic questions are not clear. Here we exhibit this topic roughly.

A meromorphic function F on $\mathbb{C}$ is said to have a (*meromorphic*) *factorization* if there exist two meromorphic functions f and g on $\mathbb{C}$ with $F = f \circ g$. Here f and

g are called respectively a *left factor* and a *right factor* of F. To make $f \circ g$ well defined, we have to assume that the left factor f is rational if the right factor g has poles, and that g is entire if f is transcendental. Following P.C. Rosenbloom, the function F is called *prime* (resp., *pseudo-prime*) if every factorization $f \circ g$ of F implies either $f \in \mathrm{Aut}(\mathbb{P}^1)$ (resp., $f \in \mathbb{C}(z)$) or $g \in \mathrm{Aut}(\mathbb{P}^1)$ (resp., $g \in \mathbb{C}(z)$). Thus when F is prime, we obtain only trivial factorizations. The function F is *left* (resp., *right*) *prime* if left (resp. right) factors are in $\mathrm{Aut}(\mathbb{P}^1)$ when right (resp., left) factors are transcendental. Similarly, we can define left (resp., right) pseudo-primeness.

Usually, for entire functions we also use the following definition. An entire function F is called *E-prime* (resp., *E-pseudo-prime*) if every *entire factorization* $f \circ g$ of F, that is, f and g are entire, implies either $f \in \mathrm{Aut}(\mathbb{C})$ (resp., $f \in \mathbb{C}[z]$) or $g \in \mathrm{Aut}(\mathbb{C})$ (resp., $g \in \mathbb{C}[z]$). We also can define left (resp., right) E-prime or E-pseudo-prime.

Lemma 4.70 (Clunie). *Let f be a transcendental meromorphic function on $\mathbb{C}$ and let g be a non-constant entire function on $\mathbb{C}$. Then*

$$\lim_{r \to \infty} \frac{T(r, f \circ g)}{T(r, g)} = \infty.$$

Theorem 4.71 (Gross). *A transcendental non-periodic entire function is prime if and only if it is E-prime.*

Proof. The necessity is trivial. Let F be a transcendental non-periodic entire function on $\mathbb{C}$. Next we assume that F is E-prime and assume, to the contrary, that F is not prime. Thus F has a factorization $f \circ g$ such that $f, g \notin \mathrm{Aut}(\mathbb{P}^1)$, and f is a meromorphic function on $\mathbb{C}$ with poles.

First of all, we suppose that f is transcendental, and so g must be entire. Then f and g must be of the following forms:

$$f(\xi) = (\xi - a)^{-n} f_0(\xi), \quad g(z) = a + e^{\alpha(z)},$$

where n is a positive integer, f_0 is a transcendental entire function, and α is a non-constant entire function. Hence we obtain

$$F(z) = f \circ g(z) = e^{-n\alpha(z)} f_0 \left(a + e^{\alpha(z)} \right) = f_1 \circ \alpha(z),$$

where

$$f_1(\xi) = e^{-n\xi} f_0 \left(a + e^{\xi} \right)$$

is a non-constant entire function. We claim $f_1 \notin \mathrm{Aut}(\mathbb{C})$, otherwise there exists $(b, c) \in \mathbb{C}^2 - \{0\}$ such that $f_1(\xi) = b\xi + c$. Set $g_0(\xi) = a + e^{\xi}$. Then

$$T(r, f_0 \circ g_0) = T\left(r, e^{n\xi}(b\xi + c)\right) \leq nT\left(r, e^{\xi}\right) + O(\log r),$$

and hence

$$\limsup_{r \to \infty} \frac{T(r, f_0 \circ g_0)}{T(r, g_0)} \leq n.$$

This is impossible by Lemma 4.70. Therefore $\alpha \in \mathrm{Aut}(\mathbb{C})$ since F is E-prime, which means that g is a periodic function, and so F is as well. This is a contradiction.

Next assume that f is rational. If g is entire, we can derive a contraction as above. Assume that g has poles. We distinguish two cases as follows:

(I) f has a pole at a. Then f and g assume the following forms:

$$f(\xi) = (\xi - a)^{-n} P_m(\xi), \quad g = a + \frac{1}{\beta},$$

where n is a positive integer, P_m is a polynomial of degree m, and β is a transcendental entire function with zeros. Then

$$F = f \circ g = \beta^n P_m \left(a + \frac{1}{\beta} \right) = P^* \circ \beta,$$

which implies $m \leq n$ since F is entire, where

$$P^*(\xi) = \xi^n P_m \left(a + \frac{1}{\xi} \right) \in \mathbb{C}[\xi].$$

Note that $P^* \notin \mathrm{Aut}(\mathbb{C})$ since $f \notin \mathrm{Aut}(\mathbb{P}^1)$. Then $\beta \in \mathrm{Aut}(\mathbb{C})$ since F is E-prime. This is impossible because g is transcendental.

(II) f has two distinct poles a and b. Then f and g assume the following forms:

$$f(\xi) = (\xi - a)^{-n}(\xi - b)^{-m} Q_k(\xi), \quad g = a + \frac{1}{\beta}, \quad g = b + \frac{1}{\gamma},$$

where n, m are positive integers, Q_k is a polynomial of degree k, and β, γ are two transcendental entire functions with zeros. Then

$$F = f \circ g = \frac{\beta^{n+m}}{\{1 + (a - b)\beta\}^m} Q_k \left(a + \frac{1}{\beta} \right) = Q^* \circ \beta,$$

where

$$Q^*(\xi) = \frac{\xi^{n+m}}{\{1 + (a - b)\xi\}^m} Q_k \left(a + \frac{1}{\xi} \right) \in \mathbb{C}(\xi).$$

We can show $k \leq n + m$, and also obtain a contradiction according to the proof of the first part. □

It is well known that if a ($\neq 0$) and b are constant, and if P is a non-constant polynomial, then $e^{az+b} + P(z)$ is prime. The function $\cos z$ is pseudo-prime. The non-trivial factorizations $f \circ g$ of $\cos z$ have only the following three forms:

(i) $f(\xi) = \cos\sqrt{\xi}$, $g(z) = z^2$;

(ii) $f(\xi) = T_n(\xi) = \cos(n\arccos\xi)$, $g(z) = \cos\frac{z}{n}$, where T_n is the Tschebyscheff polynomial of degree n (≥ 2);

(iii) $f(\xi) = \frac{1}{2}(\xi^n + \xi^{-n})$, $g(z) = e^{\frac{\sqrt{-1}z}{n}}$, $n \in \mathbb{Z} - \{0\}$.

Further, N. Steinmetz [371] obtained the following result:

Theorem 4.72. *Take a positive integer n and let $w = w(z)$ be a transcendental meromorphic function on $\mathbb{C}$ satisfying the differential equation*

$$w^{(n)} + A_1 w^{(n-1)} + \cdots + A_{n-1}w' + A_n w + A_{n+1} = 0,$$

where A_j ($j = 1, \ldots, n+1$) are rational functions. Then w is pseudo-prime.

Let $\pi(x)$ be the number of primes $\leq x$. Then $\pi(x) \to +\infty$ as $x \to +\infty$ (Euclid). C.F. Gauss conjectured

$$\lim_{x \to +\infty} \frac{\pi(x)\log x}{x} = 1. \qquad (4.10.1)$$

At almost the same time (1896), J. Hadamard and Ch. de la Vallée-Poussin proved (4.10.1) which now is called the *prime number theorem*.

The prime number theorem gives a class of density of prime numbers distributed in real numbers. A natural question is how to describe the qualitative or quantitative distribution of prime meromorphic (resp. entire) functions in $\mathcal{M}(\mathbb{C})$ (resp., $\mathcal{A}(\mathbb{C})$)? For example, there is the compact-open topology on $\mathcal{M}(\mathbb{C})$. Are prime meromorphic functions dense in $\mathcal{M}(\mathbb{C})$ under the topology?

We can define the C^0-topology of $\mathcal{M}(\mathbb{C})$ as follows. Let χ be the distance function on $\mathbb{P}^1 \cong \mathbb{C} \cup \{\infty\}$ defined by (1.5.6). For $f, g \in \mathcal{M}(\mathbb{C})$, set

$$\chi(f, g) = \sup_{z \in \mathbb{C}} \chi(f(z), g(z)).$$

It is easy to see that χ is a metric on $\mathcal{M}(\mathbb{C})$. The topology generated by the metric χ is called the C^0-*topology* of $\mathcal{M}(\mathbb{C})$. Given $f \in \mathcal{M}(\mathbb{C})$, let $B_f(x)$ be the closed ball of center f and radius x in $\mathcal{M}(\mathbb{C})$. We hope to know the density of prime meromorphic functions in $B_f(x)$ under the C^0-topology. Related to the above questions, we introduce a result due to Noda. The following two lemmas will be needed:

Lemma 4.73 (Ozawa). *Let F be a transcendental entire function on $\mathbb{C}$ such that the inequality*

$$N\left(r, \frac{1}{F'}\right) > \lambda T(r, F')$$

holds in a subset of $\mathbb{R}^+$ *with infinite measure, where* λ *is a positive constant. Assume that for any constant* c, *the system of equations*

$$\begin{cases} F(z) & = c, \\ F'(z) & = 0 \end{cases} \tag{4.10.2}$$

has only finitely many solutions. Then F *is left E-prime.*

Lemma 4.74. *If* f *is a transcendental entire function on* $\mathbb{C}$, *then there exists a countable subset* S *in* $\mathbb{C}$ *such that for any* $c \in \mathbb{C}$, $a \in \mathbb{C} - S$, *the system of equations*

$$\begin{cases} f(z) - az & = c, \\ f'(z) - a & = 0 \end{cases} \tag{4.10.3}$$

has at most one solution.

Theorem 4.75 (Noda). *If* f *is a transcendental entire function on* $\mathbb{C}$, *then the set*

$$E_f = \{ a \in \mathbb{C} \mid f(z) - az \text{ is not prime} \}$$

is countable.

Proof. Take $\lambda \in \mathbb{R}$ with $0 < \lambda < \frac{1}{2}$. By Lemma 4.74 and the second main theorem, there exists a countable subset S in $\mathbb{C}$ such that Lemma 4.74 holds, and for $a \notin S$, the inequality

$$N\left(r, \frac{1}{f' - a}\right) > \lambda T(r, f') \tag{4.10.4}$$

holds in a subset of $\mathbb{R}^+$ with infinite measure. Thus Lemma 4.73 implies that $f(z) - az$ $(a \notin S)$ is left E-prime.

Next we consider an entire factorization

$$f(z) - az = g(P(z)),$$

where P is a polynomial of degree $d \geq 2$, g is a transcendental entire function on $\mathbb{C}$. Since

$$f'(z) - a = g'(P(z))P'(z),$$

so the inequality (4.10.4) implies that g' has infinitely many zeros, say, $\{w_n\}_{n=1}^{\infty}$. We may choose n sufficiently large such that $P(z) = w_n$ has d distinct roots which also satisfy the equation system

$$\begin{cases} f(z) - az & = g(w_n), \\ f'(z) - a & = 0. \end{cases}$$

This is in contradiction with Lemma 4.74. Hence $f(z) - az$ $(a \notin S)$ is E-prime.

Finally, we prove that $f(z) - az$ $(a \notin S)$ is prime. First of all, we claim that $f(z) - az$ is non-periodic except for at most one exceptional value a in $\mathbb{C}$. Assume, to the contrary, that there are two distinct values $a, b \in \mathbb{C}$ such that $f(z) - az$ and $f(z) - bz$ are periodic with periods μ and ν, respectively. Then μ and ν also are periods of f'. Hence μ/ν is real, and so $f(z) - az$ and $f(z) - bz$ are bounded on the line $\{t\mu \mid t \in \mathbb{R}\}$. That is impossible. Hence Theorem 4.75 follows from Theorem 4.71. $\qquad \square$

In 1742, C. Goldbach proposed the following conjecture in a letter to L. Euler: Each even number (≥ 6) can be expressed as a sum of two odd primes. If f is a transcendental entire function on $\mathbb{C}$, by Theorem 4.75 we can choose $a \in \mathbb{C} - \{0\}$ such that

$$p_1(z) = \frac{1}{2}\{f(z) - az\}, \quad p_2(z) = \frac{1}{2}\{f(z) + az\}$$

are prime. Thus we have $f = p_1 + p_2$, which is a version of Goldbach's conjecture for entire functions.

Finally we consider the question on uniqueness of factorization of entire functions. If a is a positive integer, we know that there are distinct primes $p_1, \ldots, p_n$ and positive integers $l_1, \ldots, l_n$ such that

$$a = p_1^{l_1} \cdots p_n^{l_n}.$$

The factorization is unique up to the ordering of the primes p_i. However, for entire functions, the case is much more complicated.

Let $p_1, p_2, p_3, \ldots$ denote the sequence of primes $2, 3, 5, \ldots$ and set

$$a_n = p_1 \cdots p_n, \quad P_n(z) = z^{p_n}.$$

However, for $n \geq 2$, the polynomial

$$F(z) = z^{a_n} \tag{4.10.5}$$

has many factorizations

$$F = P_{i_1} \circ P_{i_2} \circ \cdots \circ P_{i_n}$$

independent of the ordering of factors, where $(i_1, \ldots, i_n)$ is a permutation of $(1, \ldots, n)$.

For a prime p, the function

$$F(z) = z^p e^{z^p} \tag{4.10.6}$$

has two factorizations

$$F = P \circ f_1 = f_2 \circ P$$

into prime entire functions

$$P(z) = z^p, \quad f_1(z) = z e^{\frac{z^p}{p}}, \quad f_2(z) = z e^z.$$

The example shows that we have to consider the ordering of factors in factorizations of functions. Thus we define the notation of unique factorization of entire functions as follows:

Definition 4.76. *Two entire functions f and g are called equivalent if there are $L_1, L_2 \in \mathrm{Aut}(\mathbb{C})$ such that $g = L_1 \circ f \circ L_2$. A transcendental entire function F is called uniquely factorizable if any two factorizations*

$$F(z) = f_1 \circ f_2 \circ \cdots \circ f_m$$

and

$$F(z) = g_1 \circ g_2 \circ \cdots \circ g_n$$

into non-linear prime entire factors mean that $m = n$, and that f_j is equivalent to g_j for each $j = 1, \ldots, n$. The unique number n is called the length of factorization of F.

According to the definition, the functions F in (4.10.6) and (4.10.5) are not uniquely factorizable. Generally, if any two factorizations of an entire function F into non-linear prime entire factors have the same length, and if the corresponding factors are equivalent up to an interchangeable factorization, F may be called *universally uniquely factorizable*, for example, the function in (4.10.5) is. However, the problem of studying universally uniquely factorizable entire functions becomes more difficult. Urabe [403], Urabe and Yang [404] and Ozawa [315] gave many examples of uniquely factorizable entire functions of length 2 or 3.

To further understand the complication of factorization of entire functions, we check more examples. For any $n \geq 1$, we have factorizations

$$e^z = P_1 \circ \cdots \circ P_n \circ e^{\frac{z}{a_n}}, \tag{4.10.7}$$

that is, e^z can not be factored into a composition of finitely many prime entire functions.

The following example is even stranger. The function

$$F(z) = (\sin z)^2 e^{2\cos z} \tag{4.10.8}$$

has a finite factorization

$$F = P_1 \circ f$$

into two prime entire functions

$$P_1(z) = z^2, \quad f(z) = \sin z e^{\cos z},$$

but it also has factorizations

$$F(z) = g \circ Q^{\circ n} \circ \cos \frac{z}{3^n}, \quad n = 1, 2, \ldots,$$

where $Q^{\circ n}$ means the nth composition of Q, and where

$$g(z) = (1 - z^2)e^{2z}, \quad Q(z) = 4z^3 - 3z.$$

Ng ([297], [298]) confirms that there are convergent infinite compositions of prime entire factors by proving the following result: There exists a sequence of positive real numbers $c_1, c_2, \ldots$ such that the sequence of functions

$$g_n \circ \cdots \circ g_1, \; n = 1, 2, \ldots \tag{4.10.9}$$

converges uniformly on compact subsets to an entire function F, where

$$g_n(z) = c_n e^z + z.$$

Furthermore, for each $n \in \mathbb{Z}^+$, there exists some entire function F_n such that

$$F = F_n \circ g_n \circ \cdots \circ g_1.$$

Based on the above examples, to make the uniquely factorizable problem of entire functions more challenging, we have to rule out polynomial factors and restrict to factorizations of finitely many transcendental prime entire functions. Thus we make the following definition. A factorization of an entire function is called *canonical* if it is a composition of finitely many transcendental prime entire factors. Related to the definitions, the following problems are interesting:

Problem 4.77. *Let f and g be two transcendental prime entire functions such that $f \circ g = g \circ f$. Is it true that f and g must be equivalent to each other?*

Problem 4.78. *Do any two canonical factorizations of a transcendental entire function have the same length?*

Problem 4.79. *If a transcendental entire function F has an canonical factorization, is F uniquely factorizable?*

Entire functions admit other kinds of factorizations. For example, the entire function

$$F(z) = e^{e^z + e^{-z}} \tag{4.10.10}$$

has a factorization

$$F = f \circ g$$

into two functions

$$f(z) = e^{z + \frac{1}{z}}, \quad g(z) = e^z$$

such that f is defined only on the image $g(\mathbb{C}) = \mathbb{C} - \{0\}$ of g, that is, is defined in the complement of Picard values of g. Such factorization will be referred as a *right entire factorization*. Generally, if an entire function F has a right entire factorization $f \circ g$, that is, g is entire and f is a non-constant holomorphic function in the region consisting of the complex plane minus any possible value omitted by g, then we call g a *right factor* of F in the sense of Eremenko and Rubel (cf. [93]). Obviously, if g has no finite Picard value, then f also is entire, and so a right entire factorization becomes an entire factorization. A natural question is to determine a class of entire functions such that any right factor of each function in this family has no finite Picard value.

For entire functions g and F on $\mathbb{C}$, define $g \leq F$ if $g(z) = g(w)$ implies $F(z) = F(w)$, $z, w \in \mathbb{C}$. Eremenko and Rubel (cf. [93]) show that $g \leq F$ if and only if g is an Eremenko-Rubel right factor of F. Thus the non-constant entire functions on $\mathbb{C}$ have been made into a lattice. It is natural to ask about greatest lower bounds within this lattice. For a family $\{F_\alpha\}$ of entire functions, we say that g is a *common right factor* of $\{F_\alpha\}$ in the sense of Eremenko and Rubel if F_α has a right entire factorization of the form $f_\alpha \circ g$ for each α. A non-constant entire function g on $\mathbb{C}$ is said to be a *(strong) greatest common right factor* of $\{F_\alpha\}$ if $g \leq F_\alpha$ for all α, and if $h \leq F_\alpha$ for all α implies $h \leq g$.

Theorem 4.80 ([93]). *Any family of non-constant entire functions of one complex variable has a (unique) strong greatest common right factor.*

4.11   Wiman-Valiron theory

In this part, we introduce the main methods in classic Wiman-Valiron theory. This will deliver the historical background of methods we used in Chapter 5. Some problems will be presented.

Take $f \in \mathcal{A}(\mathbb{C})$ and write

$$f(z) = \sum_{n=0}^{\infty} a_n z^n. \tag{4.11.1}$$

For $r > 0$, we can define the *maximum term*:

$$\mu(r, f) = \max_{n \geq 0} |a_n| r^n,$$

and the *central index*:

$$\nu(r, f) = \max_{n \geq 0} \{n \mid \mu(r, f) = |a_n| r^n\}.$$

We also define

$$\mu(0, f) = \lim_{r \to 0+} \mu(r, f), \quad \nu(0, f) = \lim_{r \to 0+} \nu(r, f).$$

Proposition 4.81. *If f is a non-constant entire function on $\mathbb{C}$, then $\mu(r, f)$ is a continuous increasing function such that $\mu(r, f) \to \infty$ as $r \to \infty$.*

Corresponding to the Jensen formula (2.3.11), one has the following *Valiron formula*:

Theorem 4.82 ([405]). *For the entire function f defined by (4.11.1), one has*

$$\log \mu(r, f) = \log |a_{\nu(0,f)}| + \int_0^r \frac{\nu(t, f) - \nu(0, f)}{t} dt + \nu(0, f) \log r. \qquad (4.11.2)$$

Proof. According to the definition of central index, we may assume $\nu(t, f) = \nu_{k-1}$ for $t \in [r_{k-1}, r_k)$ $(k = 1, 2, \dots)$, where

$$0 \le \nu_0 < \nu_1 < \cdots, \qquad 0 = r_0 < r_1 < \cdots .$$

Thus we have

$$\mu(t, f) = |a_{\nu_{k-1}}| t^{\nu_{k-1}}, \ t \in [r_{k-1}, r_k).$$

Set

$$\mu(t) = \frac{\mu(t, f)}{t^{\nu_0}}.$$

Then we have

$$\mu'(t) = (\nu_{k-1} - \nu_0)|a_{\nu_{k-1}}| t^{\nu_{k-1} - \nu_0 - 1} = \frac{\nu(t, f) - \nu(0, f)}{t} \mu(t), \ t \in [r_{k-1}, r_k).$$

Since $\mu(t, f)$ is continuous, and so is $\mu(t)$, we obtain

$$\log \mu(r) - \log \mu(0) = \int_0^r \frac{\mu'(t)}{\mu(t)} dt = \int_0^r \frac{\nu(t, f) - \nu(0, f)}{t} dt,$$

and so the formula (4.11.2) follows. $\qquad \square$

Fix $r_0 > 0$ and set

$$N(r, f = 0) = \int_{r_0}^r \frac{\nu(t, f)}{t} dt.$$

Then the formula (4.11.2) assumes the form

$$N(r, f = 0) = \log \mu(r, f) - \log \mu(r_0, f). \qquad (4.11.3)$$

Next we assume $f \in \mathcal{M}_1(\mathbb{C})$. Then according to Nevanlinna's factorization theorem (cf. [151]), there exist $g, h \in \mathcal{A}_1(\mathbb{C})$ such that g and h have no common zeros, and $f = g/h$. Thus for $r > 0$, the quotient

$$\mu(r, f) = \frac{\mu(r, g)}{\mu(r, h)} \qquad (4.11.4)$$

is well defined. Take $a \in \mathbb{C} \cup \{\infty\}$. Write

$$\nu(r, f = a) = \begin{cases} \nu(r, g - ah), & \text{if } a \in \mathbb{C}; \\ \nu(r, h), & \text{if } a = \infty, \end{cases} \qquad (4.11.5)$$

and set

$$N(r, f = a) = \int_{r_0}^{r} \frac{\nu(t, f = a)}{t} dt. \tag{4.11.6}$$

We also define

$$m(r, f = a) = \begin{cases} \log^{+} \mu \left(r, \frac{1}{f-a}\right), & \text{if } a \in \mathbb{C}; \\ \log^{+} \mu(r, f), & \text{if } a = \infty. \end{cases} \tag{4.11.7}$$

A *characteristic function* of f can be defined as follows:

$$\overset{*}{T}(r, f) = m(r, f = \infty) + N(r, f = \infty). \tag{4.11.8}$$

Applying the formula (4.11.3) to g and h respectively and then minus each other, we obtain

$$N(r, f = 0) - N(r, f = \infty) = \log \mu(r, f) - \log \mu(r_0, f), \tag{4.11.9}$$

which implies

$$\overset{*}{T}\left(r, \frac{1}{f}\right) = \overset{*}{T}(r, f) - \log \mu(r_0, f). \tag{4.11.10}$$

When $a \in \mathbb{C}$, applying (4.11.9) to $f - a$ and noting that

$$|\mu(r, f - a) - \mu(r, f)| \leq |a|, \quad \mu \left(r, \frac{1}{f-a}\right) = \frac{1}{\mu(r, f - a)},$$

we obtain

$$m(r, f = a) + N(r, f = a) = \overset{*}{T}(r, f) + O(1). \tag{4.11.11}$$

For two distinct $a, b \in \mathbb{C}$, it is not difficult to show that

$$\overset{*}{T}\left(r, \frac{f - b}{f - a}\right) = \overset{*}{T}(r, f) + O(1). \tag{4.11.12}$$

On the other hand, we have

$$\begin{aligned} \max\{\log \mu(r, g), \log \mu(r, h)\} &= \max\{\log \mu(r, f), 0\} + \log \mu(r, h) \\ &= m(r, f = \infty) + \log \mu(r, h) \\ &= m(r, f = \infty) + N(r, f = \infty) + \log \mu(r_0, h), \end{aligned}$$

that is,

$$\overset{*}{T}(r, f) = \max\{\log \mu(r, g), \log \mu(r, h)\} - \log \mu(r_0, h).$$

Further, by the formula (4.11.3), we obtain

$$\overset{*}{T}(r, f) = \max\{N(r, f = 0), N(r, f = \infty)\} + O(1). \tag{4.11.13}$$

By using (4.11.12) and (4.11.13), we can prove

$$\overset{*}{T}(r, f) = \max\{N(r, f = a), N(r, f = b)\} + O(1) \tag{4.11.14}$$

for any distinct $a, b \in \mathbb{C} \cup \{\infty\}$.

In particular, we consider a non-constant polynomial

$$P(z) = a_0 + a_1 z + \cdots + a_n z^n$$

with $a_n \neq 0$. Set

$$r_P = \max\left\{ \frac{|a_0|}{|a_n|}, \frac{|a_1|}{|a_n|}, \ldots, \frac{|a_n|}{|a_n|}, \lambda \right\},$$

where λ is the maximal absolute value of roots of P. Then when $r \geq r_P$, we have

$$\nu(r, P) = n\left(r, \frac{1}{P}\right) = n.$$

Problem 4.83. *Is there a real number r_f such that when $r \geq r_f$, the relation $\nu(r, f) = n\left(r, \frac{1}{f}\right)$ holds for $f \in \mathcal{A}_1(\mathbb{C})$?*

If Problem 4.83 is sure, naturally we can confirm the following question:

Problem 4.84. *Given a non-constant meromorphic function $f \in \mathcal{M}_1(\mathbb{C})$ and take $a \in \mathbb{C} \cup \{\infty\}$. Find sharp relations between $N\left(r, \frac{1}{f-a}\right)$ and $N(r, f = a)$, say,*

$$N(r, f = a) = N\left(r, \frac{1}{f - a}\right) + O(1).$$

We make a few remarks for a transcendental $f \in \mathcal{A}_1(\mathbb{C})$. It is well known that (cf. [154])

$$\limsup_{r \to \infty} \frac{\log \log \mu(r, f)}{\log r} = \limsup_{r \to \infty} \frac{\log \nu(r, f)}{\log r} = \rho \tag{4.11.15}$$

where $\rho = \mathrm{Ord}(f)$, and (cf. [151])

$$\limsup_{r \to \infty} \frac{\log n\left(r, \frac{1}{f}\right)}{\log r} = \rho. \tag{4.11.16}$$

The limits (4.11.15) and (4.11.16) give part of the evidence for the solution of Problem 4.83.

In 1908 Littlewood and Lindelöf conjectured independently that when $0 \leq \rho < 1$,

$$\limsup_{r \to \infty} \frac{\log m(r)}{\log M(r)} \geq \cos \pi \rho, \tag{4.11.17}$$

where

$$M(r) = \max_{0 \le \theta < 2\pi} \left| f\left(re^{i\theta}\right) \right|, \quad m(r) = \min_{0 \le \theta < 2\pi} \left| f\left(re^{i\theta}\right) \right|,$$

and this was later proved independently by Wiman and Valiron; closely allied results have occupied the attention of many authors since (cf. [52], [396]).

When $\rho < \frac{1}{2}$, it follows that there exists a sequence $\{r_n\}$ tending to $+\infty$ such that

$$\lim_{n \to \infty} \frac{\log m(r_n)}{\log M(r_n)} \ge \cos \pi \rho > 0,$$

which means that $\left| f\left(r_n e^{i\theta}\right) \right| \to +\infty$ holds uniformly for $0 \le \theta < 2\pi$ as $n \to +\infty$. Hence for any $a \in \mathbb{C}$, we have

$$m\left(r_n, \frac{1}{f-a}\right) = \frac{1}{2\pi} \int_0^{2\pi} \log^+ \frac{1}{\left| f\left(r_n e^{i\theta}\right) - a\right|} d\theta = 0$$

if n is sufficiently large. Then the first main theorem implies

$$N\left(r_n, \frac{1}{f-a}\right) = T(r_n, f) + O(1). \tag{4.11.18}$$

Note that for $0 < r < R$ (cf. [168]),

$$T(r, f) = m(r, f) \le \log^+ M(r) \le \frac{R+r}{R-r} T(R, f), \tag{4.11.19}$$

and (cf. [154])

$$\lim_{r \to \infty} \frac{\log M(r)}{\log \mu(r, f)} = 1. \tag{4.11.20}$$

Thus by (4.11.18) and (4.11.3), we obtain

$$\limsup_{n \to \infty} \frac{N\left(r_n, \frac{1}{f}\right)}{N(r_n, f = 0)} = \limsup_{n \to \infty} \frac{T(r_n, f)}{\log \mu(r_n, f)} \le 1. \tag{4.11.21}$$

On the other hand, we choose $\varepsilon > 0$ with $\cos \pi \rho - \varepsilon > 0$, and so

$$(\cos \pi \rho - \varepsilon) \log M(r_n) < \log m(r_n) \le T(r_n, f)$$

if n is sufficiently large. Therefore (4.11.18) and (4.11.3) yield

$$\liminf_{n \to \infty} \frac{N\left(r_n, \frac{1}{f}\right)}{N(r_n, f = 0)} \ge \cos \pi \rho. \tag{4.11.22}$$

Pólya [318] (or cf. [157]) obtained the classical inequality

$$\limsup_{r \to \infty} \frac{n\left(r, \frac{1}{f}\right)}{\log M(r)} \ge \frac{\sin \pi \rho}{\pi}. \tag{4.11.23}$$

Based on (4.11.20) and Problem 4.83, we suggest the estimate

$$\limsup_{r\to\infty} \frac{\nu(r,f)}{\log \mu(r,f)} \geq \frac{\sin \pi\rho}{\pi}. \tag{4.11.24}$$

Valiron [407] (or cf. [88]) observed

$$\limsup_{r\to\infty} \frac{N\left(r,\frac{1}{f}\right)}{\log M(r)} \geq \frac{\sin \pi\rho}{\pi\rho} \tag{4.11.25}$$

if $f \in \mathcal{A}_1(\mathbb{C})$ with order $\rho > 0$. Similarly, we conjecture the estimate:

$$\limsup_{r\to\infty} \frac{N(r,f=0)}{\log \mu(r,f)} \geq \frac{\sin \pi\rho}{\pi\rho}. \tag{4.11.26}$$

Problem 4.85. *Given a non-constant meromorphic function $f \in \mathcal{M}_1(\mathbb{C})$, find sharp relations between $T(r,f)$ and $\overset{*}{T}(r,f)$.*

When $\frac{1}{2} < \rho < 1$, the entire function (cf. [442])

$$f(z) = \prod_{n=1}^{\infty} \left(1 + \frac{z}{n^{\frac{1}{\rho}}}\right) \tag{4.11.27}$$

is of order ρ such that

$$\limsup_{r\to\infty} \frac{N\left(r,\frac{1}{f}\right)}{T(r,f)} = \sin \rho\pi.$$

However, the formula (4.11.13) implies

$$\lim_{r\to\infty} \frac{N(r,f=0)}{\overset{*}{T}(r,f)} = 1.$$

Thus if either Problem 4.83 or Problem 4.84 is sure, we should obtain

$$\limsup_{r\to\infty} \frac{\overset{*}{T}(r,f)}{T(r,f)} = \sin \rho\pi.$$

If $f \in \mathcal{M}_1(\mathbb{C})$ with order ρ, Edrei and Fuchs [88] (or cf. [151]) improved sharply a result due to R. Nevanlinna [292] as follows:

$$K(\rho) = \limsup_{r\to\infty} \frac{N\left(r,\frac{1}{f}\right) + N(r,f)}{T(r,f)} \geq k(\rho), \tag{4.11.28}$$

where

$$k(\rho) = \begin{cases} 1, & \text{if } 0 \leq \rho < \frac{1}{2}; \\ \sin \pi\rho, & \text{if } \frac{1}{2} \leq \rho < 1. \end{cases}$$

However, the formula (4.11.13) yields

$$\overset{*}{T}(r, f) \leq N\,(r, f = 0) + N(r, f = \infty) + O(1). \tag{4.11.29}$$

Thus based on Problem 4.83, we suggest the estimate

$$K(\rho) \geq \limsup_{r \to \infty} \frac{\overset{*}{T}(r, f)}{T(r, f)} \geq \frac{1}{2} k(\rho). \tag{4.11.30}$$

Particularly, if f is entire, we conjecture

$$1 \geq \limsup_{r \to \infty} \frac{\overset{*}{T}(r, f)}{T(r, f)} \geq k(\rho). \tag{4.11.31}$$

Chapter 5

Functions over Non-Archimedean Fields

In this chapter, we introduce the value distribution theory of meromorphic functions defined on a non-Archimedean algebraically closed field, and give non-Archimedean analogues of some results and problems in number theory. Wiman-Valiron theory and Nevanlinna theory will be unified. In particular, one basic formula (5.1.5) and three equidistribution formulae (5.1.15), (5.3.7), (5.5.2) illustrating the differences with the ordinary Nevanlinna theory will be exhibited. Some universal properties appeared in height theory and Nevanlinna theory will be further presented.

5.1 Equidistribution formula

Let κ be an algebraically closed field of characteristic $p \geq 0$, complete for a nontrivial non-Archimedean absolute value $|\cdot|$. We will show that the Valiron formula over κ plays the role of Jensen's formula based on the formula (5.1.5), and further derive the important formula (5.1.15) showing equidistribution of values.

Take $a \in \kappa$. For a positive real number r, define

$$\kappa[a; r] = \{z \in \kappa \mid |z - a| \leq r\}.$$

Let ν be a function from κ into $\mathbb{Z}$ such that $\mathrm{supp}\nu \cap \kappa[0; r]$ is finite for any $r > 0$. The *counting function* n_ν for ν is defined by

$$n_\nu(r) = \sum_{z \in \kappa[0;r]} \nu(z). \tag{5.1.1}$$

If ν is non-negative, then n_ν increases. Fix $r_0 > 0$. The *valence function* of ν is defined by

$$N_\nu(r) = N_\nu(r, r_0) = \int_{r_0}^r n_\nu(t) \frac{dt}{t} \quad (r \geq r_0). \tag{5.1.2}$$

Let $\mathcal{A}(\kappa)$ be the set of entire functions on κ. The field of fractions of $\mathcal{A}(\kappa)$ will be denoted by $\mathcal{M}(\kappa)$. An element f in the set $\mathcal{M}(\kappa)$ will be called a *meromorphic function* on κ. Take $f \in \mathcal{M}(\kappa)$. For $a \in \kappa \cup \{\infty\}$, let $\mu_f^a(z_0)$ denote the *a-valued multiplicity* of f at z_0, that is, $\mu_f^a(z_0) = m$ if and only if

$$f(z) = \begin{cases} a + (z - z_0)^m h(z) & : \quad a \neq \infty, \\ (z - z_0)^{-m} h(z) & : \quad a = \infty \end{cases}$$

with $h(z_0) \neq 0, \infty$. By the definition, we obtain a function

$$\mu_f^a : \kappa \longrightarrow \mathbb{Z}_+$$

with $\mu_f^a(z_0) > 0$ for some $z_0 \in \kappa$ if and only if $f(z_0) = a$. Define the *counting function* and the *valence function* of f for poles respectively by

$$n(r, f) = n_{\mu_f^\infty}(r), \quad N(r, f) = N_{\mu_f^\infty}(r).$$

For $a \in \kappa$, the *counting function*

$$n\left(r, \frac{1}{f - a}\right) = n_{\mu_f^a}(r)$$

and the *valence function*

$$N\left(r, \frac{1}{f - a}\right) = N_{\mu_f^a}(r)$$

of f for a are also well defined.

Each $f \in \mathcal{A}(\kappa)$ can be given by a power series

$$f(z) = \sum_{n=0}^\infty a_n z^n, \quad (a_n \in \kappa), \tag{5.1.3}$$

such that

$$\limsup_{n \to \infty} |a_n|^{\frac{1}{n}} = 0,$$

that is, the series (5.1.3) converges for any $z \in \kappa$. For $r > 0$, we can define the *maximum term*:

$$\mu(r, f) = \max_{n \geq 0} |a_n| r^n,$$

and the *central index*:

$$\nu(r, f) = \max_{n \geq 0}\{n \mid |a_n|r^n = \mu(r, f)\}.$$

We also define

$$\mu(0, f) = \lim_{r \to 0+} \mu(r, f), \quad \nu(0, f) = \lim_{r \to 0+} \nu(r, f).$$

The central index $\nu(r, f)$ increases as $r \to \infty$ and satisfies the *Valiron formula*:

$$\log \mu(r, f) = \log |a_{\nu(0,f)}| + \int_0^r \frac{\nu(t, f) - \nu(0, f)}{t} dt + \nu(0, f) \log r. \qquad (5.1.4)$$

See the proof of (4.11.2), or [176], [405]. Basic properties of the maximum term are summarized as follows:

Theorem 5.1 ([176]). *For $r > 0$, the function $\mu(r, \cdot) : A(\kappa) \longrightarrow \mathbb{R}_+$ satisfies properties:*

1) $\mu(r, f) = 0$ *if and only if $f \equiv 0$;*
2) $\mu(r, f + g) \leq \max\{\mu(r, f), \mu(r, g)\}$;
3) $\mu(r, fg) = \mu(r, f)\mu(r, g)$.

By the Weierstrass preparation theorem, we know that the counting function of zeros of f is just the central index (see [176]), that is,

$$n\left(r, \frac{1}{f}\right) = \nu(r, f). \qquad (5.1.5)$$

The formula (5.1.5) is an important connection between Wiman-Valiron theory and Nevanlinna theory. Based on (5.1.5), the Valiron formula (5.1.4) becomes the following *Valiron-Jensen formula* (see [176]):

$$N\left(r, \frac{1}{f}\right) = \log \mu(r, f) - \log \mu(r_0, f). \qquad (5.1.6)$$

The *proximity function* (or *compensation function*)

$$m(r, f) = \log^+ \mu(r, f) = \log \max\{1, \mu(r, f)\}$$

serves as the *characteristic function* $T(r, f)$ of the entire function f.

Lemma 5.2 ([95]). *If $f \in A(\kappa)$ has m zeros in $\kappa[0; r]$ with $m \geq 1$ (taking multiplicities into account), then for $b \in f(\kappa[0; r])$, the function $f - b$ also admits m zeros in $\kappa[0; r]$ (counting multiplicity).*

Proof. We expand f into the series (5.1.3) and set

$$s = \sup_{n \geq 1} |a_n| r^n.$$

We claim $f(\kappa[0; r]) = \kappa[a_0; s]$. On the one hand, it is trivial that $f(\kappa[0; r]) \subseteq \kappa[a_0; s]$.

On the other hand, we prove $f(\kappa[0; r]) \supseteq \kappa[a_0; s]$. If $s = 0$, it is obvious that

$$f(\kappa[0; r]) = \kappa[a_0; s] = \{a_0\}.$$

If $s > 0$, we take $b \in \kappa[a_0; s]$ and consider the function

$$g(z) = f(z) - b = a_0 - b + \sum_{n=1}^{\infty} a_n z^n.$$

It follows that $\nu(r, g) \geq 1$ since

$$|a_0 - b| \leq s = \sup_{n \geq 1} |a_n| r^n.$$

Therefore

$$n\left(r, \frac{1}{g}\right) = \nu(r, g) \geq 1,$$

which means that g admits at least one zero in $\kappa[0; r]$ and hence $b \in f(\kappa[0; r])$. Thus the claim is proved.

Finally, we prove Lemma 5.2. By the assumption, we have

$$m = n\left(r, \frac{1}{f}\right) = \nu(r, f) \geq 1,$$

and hence

$$|a_m| r^m \begin{cases} \geq |a_n| r^n, & \text{if } n < m; \\ > |a_n| r^n, & \text{if } n > m. \end{cases}$$

Take $b \in f(\kappa[0; r])$. By the claim, we obtain

$$|a_0 - b| \leq \sup_{n \geq 1} |a_n| r^n,$$

and hence

$$n\left(r, \frac{1}{f - b}\right) = n\left(r, \frac{1}{g}\right) = \nu(r, g) = m.$$

Therefore $f - b$ also admits m zeros in $\kappa[0; r]$. □

Theorem 5.3. *Assume that f is a non-constant entire function. Then for any $b \in \kappa$, we have*

$$N\left(r, \frac{1}{f - b}\right) = N\left(r, \frac{1}{f}\right) + O(1).$$

Proof. Note that f and $f - b$ each have at least one zero since $f - b$ also is a non-constant entire function. Thus there is an $r' \in \mathbb{R}^+$ such that f has at least one zero in $\kappa[0; r']$ and such that $b \in f(\kappa[0; r'])$. By Lemma 5.2, one obtains

$$n\left(r, \frac{1}{f - b}\right) = n\left(r, \frac{1}{f}\right) \quad (r \geq r').$$

Therefore when $r \geq r'$, we have

$$
\begin{aligned}
N\left(r, \frac{1}{f - b}\right) &= \int_{r'}^{r} n\left(t, \frac{1}{f - b}\right) \frac{dt}{t} + N\left(r', \frac{1}{f - b}\right) \\
&= \int_{r'}^{r} n\left(t, \frac{1}{f}\right) \frac{dt}{t} + N\left(r', \frac{1}{f - b}\right) \\
&= N\left(r, \frac{1}{f}\right) + N\left(r', \frac{1}{f - b}\right) - N\left(r', \frac{1}{f}\right),
\end{aligned}
$$

and Theorem 5.3 follows. $\qquad\square$

Let f be a non-constant entire function in κ. Then

$$N\left(r, \frac{1}{f}\right) = \log \mu(r, f) - \log \mu(r_0, f) \to +\infty$$

as $r \to \infty$, and hence $\mu(r, f) > 1$ when r is sufficiently large. Therefore

$$N\left(r, \frac{1}{f}\right) = T(r, f) + O(1), \tag{5.1.7}$$

and hence Theorem 5.3 implies

$$N\left(r, \frac{1}{f - a}\right) = T(r, f) + O(1) \tag{5.1.8}$$

for all $a \in \kappa$. In particular, if P is a polynomial, the formula (5.1.7) implies

$$T(r, P) = \deg(P) \log r + O(1). \tag{5.1.9}$$

Let f be a non-constant meromorphic function in κ. Since greatest common divisors of any two elements in $\mathcal{A}(\kappa)$ exist, there are $g, h \in \mathcal{A}(\kappa)$ with $f = \frac{g}{h}$ such that g and h do not have common factors in the ring $\mathcal{A}(\kappa)$. We can uniquely extend the *maximum term* μ for entire functions to the meromorphic function $f = \frac{g}{h}$ by defining

$$\mu(r, f) = \frac{\mu(r, g)}{\mu(r, h)}, \quad 0 < r < \infty.$$

Then the *Valiron-Jensen formula*

$$N\left(r, \frac{1}{f}\right) - N(r, f) = \log \mu(r, f) - \log \mu(r_0, f) \tag{5.1.10}$$

follows from (5.1.6), where

$$N(r, f) = N\left(r, \frac{1}{h}\right)$$

is the valence function of f for poles. Note that

$$\mu(r, f_1 f_2) = \mu(r, f_1)\mu(r, f_2), \quad f_1, f_2 \in \mathcal{M}(\kappa).$$

Thus the Valiron-Jensen formula implies

$$N\left(r, \frac{1}{f_1 f_2}\right) - N(r, f_1 f_2) = N\left(r, \frac{1}{f_1}\right) + N\left(r, \frac{1}{f_2}\right)$$
$$- N(r, f_1) - N(r, f_2). \tag{5.1.11}$$

Define the *compensation function* by

$$m(r, f) = \log^+ \mu(r, f) = \log\max\{1, \mu(r, f)\}.$$

As usual, we define the *characteristic function*:

$$T(r, f) = m(r, f) + N(r, f) \ (r_0 < r < \infty).$$

Then the Valiron-Jensen formula (5.1.10) can be rewritten as

$$T\left(r, \frac{1}{f}\right) = T(r, f) - \log\mu(r_0, f). \tag{5.1.12}$$

Usually, the mapping

$$\tilde{f} = (h, g) : \kappa \longrightarrow \kappa^2 - \{0\}$$

is called a *reduced representation* of f. Define

$$\mu(r, \tilde{f}) = \max\{\mu(r, h), \mu(r, g)\}.$$

Noting that

$$
\begin{aligned}
\log\mu(r, \tilde{f}) &= \max\left\{0, \log\frac{\mu(r, g)}{\mu(r, h)}\right\} + \log\mu(r, h) \\
&= \max\{0, \log\mu(r, f)\} + \log\mu(r, h) \\
&= m(r, f) + \log\mu(r, h),
\end{aligned}
$$

and by the Valiron-Jensen formula

$$N(r, f) = N\left(r, \frac{1}{h}\right) = \log\mu(r, h) - \log\mu(r_0, h),$$

we obtain

$$T(r, f) = \log \mu(r, \tilde{f}) - \log \mu(r_0, h),$$

or equivalently

$$T(r, f) = \log \mu(r, \tilde{f}) - \log \mu(r_0, \tilde{f}) + m(r_0, f). \tag{5.1.13}$$

By (5.1.13) and the Valiron-Jensen formula, the formula

$$T(r, f) = \max\left\{ N(r, f), N\left(r, \frac{1}{f}\right) \right\} + O(1) \tag{5.1.14}$$

holds for a non-constant meromorphic function f in κ. Thus it is easy to prove that the *equidistribution formula*

$$T(r, f) = \max\left\{ N\left(r, \frac{1}{f - a}\right), N\left(r, \frac{1}{f - b}\right) \right\} + O(1) \tag{5.1.15}$$

holds for any two distinct elements $a, b \in \kappa \cup \{\infty\}$. Particularly, if P and Q ($\neq 0$) are coprime polynomials, the formula (5.1.14) yields

$$T\left(r, \frac{P}{Q}\right) = \max\{\deg(P), \deg(Q)\} \log r + O(1). \tag{5.1.16}$$

The following result is called the *first main theorem* (cf. [32], [65], [137], [140], [176]).

Theorem 5.4. *Let f be a non-constant meromorphic function in κ. Then for every $a \in \kappa$ we have*

$$m\left(r, \frac{1}{f - a}\right) + N\left(r, \frac{1}{f - a}\right) = T(r, f) + O(1) \quad (r \to +\infty).$$

Proof. By (5.1.12), we have

$$m\left(r, \frac{1}{f - a}\right) + N\left(r, \frac{1}{f - a}\right) = T\left(r, \frac{1}{f - a}\right) = T(r, f - a) - \log \mu(r_0, f - a).$$

Then the theorem follows from the simple properties

$$T(r, f - a) \le T(r, f) + \log^+ |a|,$$
$$T(r, f) \le T(r, f - a) + \log^+ |a|. \qquad \square$$

Theorem 5.5 ([176],[440]). *Take $\{a_0, \ldots, a_k, b_0, \ldots, b_q\} \subset \mathcal{M}(\kappa)$ with $a_k \not\equiv 0$ and $b_q \not\equiv 0$ such that*

$$A(z, w) = \sum_{j=0}^{k} a_j(z) w^j \tag{5.1.17}$$

and

$$B(z, w) = \sum_{j=0}^{q} b_j(z) w^j \tag{5.1.18}$$

are coprime polynomials in w. *Define*

$$R(z, w) = \frac{A(z, w)}{B(z, w)}. \tag{5.1.19}$$

If $f \in \mathcal{M}(\kappa)$ *is non-constant, then*

$$T(r, R \circ f) = \max\{k, q\} T(r, f) + O\left(\sum_{j=0}^{k} T(r, a_j) + \sum_{j=0}^{q} T(r, b_j) \right), \tag{5.1.20}$$

where $R \circ f$ *is defined by* $R \circ f(z) = R(z, f(z))$.

Finally, we make a remark on the first main theorem. Take $f \in \mathcal{M}(\kappa)$. For $a \in \kappa$, set

$$m_f(r, a) = -\log \chi(f(z), a) \ (r = |z|), \tag{5.1.21}$$

where χ is defined by (1.5.5) for $v = |\cdot|$. Fix an element $z \in \kappa$ with $|z| = r > r_0$. Assume $f(z) \neq 0, a, \infty$. Then

$$\begin{aligned} m_f(r, a) &= \log^+ |f(z)| + \log^+ |a| - \log|f(z) - a| \\ &= \log^+ \mu(r, f) + \log^+ |a| - \log \mu(r, f - a). \end{aligned}$$

Applying the Valiron-Jensen formula (5.1.10) to $f - a$, we obtain

$$m_f(r, a) + N\left(r, \frac{1}{f-a}\right) = T(r, f) + \log^+ |a| - \log \mu(r_0, f - a), \tag{5.1.22}$$

which shows that the definition (5.1.21) is independent of the choice of z on the circle $\{|z| = r\}$. Since the set $|\kappa| - \{|z| \mid f(z) = 0, a, \infty\}$ is dense in $\mathbb{R}^+$, we may extend the definition of $m_f(r, a)$ by applying (5.1.22) to all $r > r_0$. We also write

$$m_f(r, \infty) = m(r, f).$$

R. L. Benedetto [18] presents analogs for non-Archimedean (or p-adic) analytic functions of well-known mapping theorems in complex function theory proved by Koebe, Bloch, Schottky, Landau, and Ahlfors. Of particular interest is the author's analogue of the Ahlfors island theorem. In accord once again with the style of this book, we do not discuss these results.

5.2 Second main theorem of meromorphic functions

Let κ be an algebraically closed field of characteristic $p \geq 0$, complete for a non-trivial non-Archimedean absolute value $|\cdot|$. We will show that the second main theorem of meromorphic functions over κ can be simply derived from the equidistribution formula (5.1.15) and the Valiron-Jensen formula.

First of all, we make a simple remark for the case $p > 0$. Then we have the following rules of arithmetic:

$$(\alpha + \beta)^p = \sum_{i=0}^{p} \binom{p}{i} \alpha^{p-i} \beta^i.$$

Note that

$$p \mid \binom{p}{i}, \quad 1 \leq i < p.$$

Therefore, we have the simple rule

$$(\alpha \pm \beta)^p = \alpha^p \pm \beta^p. \tag{5.2.1}$$

As usual, we have

$$(\alpha\beta)^p = \alpha^p \beta^p, \quad (\beta^{-1})^p = (\beta^p)^{-1}. \tag{5.2.2}$$

Hence the mapping ρ defined as

$$\rho(\alpha) = \alpha^p \tag{5.2.3}$$

is a field injection from κ to κ, which is called the *Frobenius mapping of the field*. Note that the Frobenius mapping is also surjective since κ is algebraically closed. Hence the $\mathbb{F}_p$-automorphism Θ can be defined as

$$\Theta(z) = \rho^{-1}(z) := \sqrt[p]{z}. \tag{5.2.4}$$

More generally this mapping has a continuation to a κ-algebra automorphism of $\kappa[X]$ as

$$\Theta\left(a \prod_{j=1}^{n} (X - a_j)\right) = \Theta(a) \prod_{j=1}^{n} (X - \Theta(a_j)).$$

Take $f \in \mathcal{M}(\kappa)$. Set

$$\mu_f = \mu_f^0 - \mu_f^\infty.$$

It is easy to show (cf. [34])

$$\mu_{f'}(z) \begin{cases} = \mu_f(z) - 1, & \text{if } p \nmid \mu_f(z); \\ \geq \mu_f(z), & \text{if } p \mid \mu_f(z). \end{cases}$$

Note that if $p = 0$, then a meromorphic function f on κ is constant if and only if $f' = 0$. For the case $p > 0$, Boutabaa and Escassut [34] give the following fact:

Theorem 5.6. *Assume $p \neq 0$ and let $f \in \mathcal{M}(\kappa)$ be non-constant. Then there exists $g \in \mathcal{M}(\kappa)$ such that $f = g^p$ if and only if $f' = 0$. Moreover, there exist $h \in \mathcal{M}(\kappa)$ and a unique $s \in \mathbb{Z}^+$ such that $f = h^{p^s}$ and $h' \neq 0$.*

Proof. If f is of the form g^p with $g \in \mathcal{M}(\kappa)$, then of course we have $f' = pg^{p-1}g' = 0$. Next assume that $f' = 0$. If $f \in \mathcal{A}(\kappa)$, then it is obvious that all non-zero coefficients of

$$f(z) = \sum_{n=0}^{\infty} a_n z^n$$

have an index that is a multiple of p, hence $f = g^p$ with $g \in \mathcal{A}(\kappa)$.

We now consider the general case when $f \in \mathcal{M}(\kappa)$. Then there are $f_0, f_1 \in \mathcal{A}(\kappa)$ such that $f = \frac{f_1}{f_0}$. Clearly $f_0^p f = f_0^{p-1} f_1 \in \mathcal{A}(\kappa)$ and satisfies

$$(f_0^p f)' = pf_0^{p-1} f_0' f + f_0^p f' = 0.$$

Consequently, $f_0^p f$ is of the form f_2^p with $f_2 \in \mathcal{A}(\kappa)$, and therefore

$$f = \left(\frac{f_2}{f_0}\right)^p.$$

On the other hand, it is immediate that if there exists $h \in \mathcal{M}(\kappa)$ such that $h^n = f$ holds for $n \in \mathbb{Z}^+$, then

$$n \mid \gcd\{\mu_f(z) \mid \mu_f(z) \neq 0\}.$$

Hence the set of integers s such that $f = h^{p^s}$ with $h \in \mathcal{M}(\kappa)$ is bounded and therefore admits a biggest element, which satisfies the property in Theorem 5.6. □

When κ has characteristic $p > 0$, given $f \in \mathcal{M}(\kappa) - \kappa$, we will mean by the *ramification index* of f the integer s such that $\sqrt[p^s]{f} \in \mathcal{M}(\kappa)$ and $(\sqrt[p^s]{f})' \neq 0$, and denote it by $e(f)$. Moreover, in order to simplify notation, if $p = 0$, for any $f \in \mathcal{M}(\kappa) - \kappa$, we put $e(f) = 0$. Thus $e(f) = 0$ if and only if $f' \neq 0$. Assume $p > 0$. In fact,

$$p^{-e(f)} = |d|_p,$$

where

$$d = \gcd(\mu_f(z) \mid \mu_f(z) \neq 0).$$

If $f, g \in \mathcal{M}(\kappa) - \kappa; P \in \kappa(X)$ satisfy an equation

$$g(z) = P(f(z)),$$

then $e(g) \geq e(f)$, and

$$\sqrt[p^s]{g} = Q \circ \sqrt[p^s]{f}, \quad Q = \Theta^s(P), \quad s = e(f).$$

The following fact is usually referred to as the *lemma of the logarithmic derivative* (see [1], [32], [60], [140], [176]):

Lemma 5.7. *Let f be a meromorphic function in κ. Then for any integer $k > 0$,*

$$\mu\left(r, \frac{f^{(k)}}{f}\right) \leq \frac{1}{r^k} \quad (r > 0).$$

Proof. This is trivial if $f^{(k)} = 0$. Next we assume $f^{(k)} \neq 0$. If $f \in \mathcal{A}(\kappa)$, set

$$f(z) = \sum_{n=0}^{\infty} a_n z^n.$$

Then

$$f'(z) = \sum_{n=1}^{\infty} n a_n z^{n-1},$$

and hence for $r > 0$,

$$\mu(r, f') = \max_n |n a_n| r^{n-1} \leq \frac{1}{r} \max_n |a_n| r^n = \frac{1}{r} \mu(r, f)$$

which implies

$$\mu\left(r, \frac{f'}{f}\right) = \frac{\mu(r, f')}{\mu(r, f)} \leq \frac{1}{r}.$$

Therefore

$$\mu\left(r, \frac{f^{(k)}}{f}\right) = \mu\left(r, \prod_{i=1}^{k} \frac{f^{(i)}}{f^{(i-1)}}\right) = \prod_{i=1}^{k} \mu\left(r, \frac{f^{(i)}}{f^{(i-1)}}\right) \leq \frac{1}{r^k},$$

where $f^{(0)} = f$. Now let $f = g/h \in \mathcal{M}(\kappa)$. Then

$$\begin{aligned}
\mu\left(r, \frac{f'}{f}\right) &= \mu\left(r, \frac{hg' - gh'}{h^2} \cdot \frac{h}{g}\right) = \mu\left(r, \frac{g'}{g} - \frac{h'}{h}\right) \\
&\leq \max\left\{\mu\left(r, \frac{g'}{g}\right), \mu\left(r, \frac{h'}{h}\right)\right\} \leq \frac{1}{r},
\end{aligned}$$

and similarly, we can obtain

$$\mu\left(r, \frac{f^{(k)}}{f}\right) \leq \frac{1}{r^k}. \qquad \square$$

Define the *ramification term* of f by

$$N_{\mathrm{Ram}}(r, f) = 2N(r, f) - N(r, f') + N\left(r, \frac{1}{f'}\right).$$

We have the *second main theorem* (cf. [32], [34], [60], [65], [66], [140], [176]):

Theorem 5.8. *Let f be a meromorphic function in κ with $f' \neq 0$ and let $a_1, \ldots, a_q$ be distinct numbers of κ. Then for $r > r_0$, we have*

$$(q-1)T(r,f) \leq N(r,f) + \sum_{j=1}^{q} N\left(r, \frac{1}{f-a_j}\right) - N_{\mathrm{Ram}}(r,f) - \log r + O(1). \quad (5.2.5)$$

Proof. For each $r > r_0$ there exists an index $i(r) \in \{1, \ldots, q\}$ such that

$$N\left(r, \frac{1}{f-a_{i(r)}}\right) = \min_{1 \leq j \leq q} N\left(r, \frac{1}{f-a_j}\right).$$

Then the equidistribution formula (5.1.15) implies

$$(q-1)T(r,f) = \sum_{j=1}^{q} N\left(r, \frac{1}{f-a_j}\right) - N\left(r, \frac{1}{f-a_{i(r)}}\right) + O(1). \quad (5.2.6)$$

By the Valiron-Jensen formula (5.1.10), we have

$$N\left(r, \frac{1}{f-a_{i(r)}}\right) - N(r,f) = \log \mu(r, f-a_{i(r)}) - \log \mu(r_0, f-a_{i(r)}),$$

and

$$N\left(r, \frac{1}{f'}\right) - N(r,f') = \log \mu(r, f') - \log \mu(r_0, f').$$

One formula minus the other yields

$$-N\left(r, \frac{1}{f-a_{i(r)}}\right) = N(r,f') - N(r,f) - N\left(r, \frac{1}{f'}\right)$$

$$+ \log \mu\left(r, \frac{f'}{f-a_{i(r)}}\right) + O(1),$$

and hence Lemma 5.7 implies

$$-N\left(r, \frac{1}{f-a_{i(r)}}\right) \leq N(r,f) - N_{\mathrm{Ram}}(r,f) - \log r + O(1). \quad (5.2.7)$$

Hence (5.2.5) follows from (5.2.6) and (5.2.7).   □

Take $f \in \mathcal{M}(\kappa)$ and take $a \in \kappa \cup \{\infty\}$. For a positive integer k, define a function $\mu_{f,k}^a$ by

$$\mu_{f,k}^a(z) = \begin{cases} \min\{\mu_f^a(z), k\}, & \text{if } p \nmid \mu_f^a(z); \\ 0, & \text{if } p \mid \mu_f^a(z). \end{cases}$$

The *truncated counting function* of f for a by k is defined by

$$n_{\mu_{f,k}^a}(r) = \sum_{|z| \le r} \mu_{f,k}^a(z) = \begin{cases} n_k\left(r, \frac{1}{f-a}\right) & \text{if } a \ne \infty, \\ n_k(r, f) & \text{if } a = \infty, \end{cases} \tag{5.2.8}$$

and denote the *truncated valence function* of f for a by

$$N_{\mu_{f,k}^a}(r) = \begin{cases} N_k\left(r, \frac{1}{f-a}\right) & \text{if } a \ne \infty, \\ N_k(r, f) & \text{if } a = \infty. \end{cases}$$

As usual, we also write

$$n_1(r, f) = \overline{n}(r, f), \quad N_1(r, f) = \overline{N}(r, f),$$

and so on.

Assume that $f' \ne 0$ and let $\mathscr{A} = \{a_1, \ldots, a_q\}$ be distinct numbers of κ. A simple calculation shows

$$\sum_{j=1}^{q} N\left(r, \frac{1}{f - a_j}\right) = \sum_{j=1}^{q} \overline{N}\left(r, \frac{1}{f - a_j}\right) + N\left(r, \frac{1}{f'}\right) - N_{\mu_{f,\mathscr{A}}}(r), \tag{5.2.9}$$

where

$$0 \le \mu_{f,\mathscr{A}}(z) = \begin{cases} \mu_{f'}^0(z), & \text{if } \mu_f^{a_j}(z) = 0 \text{ for all } j; \\ \mu_{f'}^0(z) - \mu_f^{a_j}(z), & \text{if } \mu_f^{a_j}(z) > 0 \text{ and } p \mid \mu_f^{a_j}(z) \text{ for some } j; \\ 0, & \text{otherwise.} \end{cases}$$

We also have the estimate

$$N(r, f') - N(r, f) \le \overline{N}(r, f) - N_{\mu_{f,\infty}}(r),$$

where

$$0 \le \mu_{f,\infty}(z) = \begin{cases} \mu_f^\infty(z) - \mu_{f'}^\infty(z), & \text{if } \mu_f^\infty(z) > 0 \text{ and } p \mid \mu_f^\infty(z); \\ 0, & \text{otherwise.} \end{cases}$$

Particularly, if $f \in A(\kappa)$, by (5.1.8) and (5.2.9), we obtain

$$qT(r, f) = \sum_{j=1}^{q} \overline{N}\left(r, \frac{1}{f - a_j}\right) + N\left(r, \frac{1}{f'}\right) - N_{\mu_{f,\mathscr{A}}}(r) + O(1). \tag{5.2.10}$$

Thus by Theorem 5.8, we obtain the *truncated form of the second main theorem*:

Theorem 5.9. *Let f be a meromorphic function in κ with $f' \neq 0$ and let $a_1, \ldots, a_q$ be distinct numbers of κ. Then for $r > r_0$, we have*

$$(q-1)T(r,f) \leq \overline{N}(r,f) + \sum_{j=1}^{q} \overline{N}\left(r, \frac{1}{f-a_j}\right)$$
$$- N_{\mu_{f,\infty}}(r) - N_{\mu_{f,\mathscr{A}}}(r) - \log r + O(1).$$

Finally, we replace the condition $f' \neq 0$ in Theorem 5.9 by using the ramification index of f. Denote the *characteristic exponent* of κ by

$$\bar{p} = \begin{cases} p, & \text{if } p > 0; \\ 1, & \text{if } p = 0. \end{cases}$$

Take $f \in \mathcal{M}(\kappa) - \kappa$ and set

$$h = \sqrt[\bar{p}^{e(f)}]{f}, \quad b_j = \sqrt[\bar{p}^{e(f)}]{a_j} \quad (j = 1, \ldots, q).$$

Applying Theorem 5.9 to h, and noting that

$$T(r,h) = \frac{1}{\bar{p}^{e(f)}}T(r,f), \quad \overline{N}\left(r, \frac{1}{h-b_j}\right) = \overline{N}\left(r, \frac{1}{f-a_j}\right) \quad (j = 1, \ldots, q),$$

we have the following form of Theorem 5.9 (cf. [34]):

Theorem 5.10. *Let f be a non-constant meromorphic function in κ and let a_1, $\ldots$, a_q be distinct numbers of κ. Then for $r > r_0$, we have*

$$\frac{q-1}{\bar{p}^{e(f)}}T(r,f) \leq \overline{N}(r,f) + \sum_{j=1}^{q} \overline{N}\left(r, \frac{1}{f-a_j}\right)$$
$$- N_{\mu_{h,\infty}}(r) - N_{\mu_{h,\mathscr{B}}}(r) - \log r + O(1), \qquad (5.2.11)$$

where $\mathscr{B} = \{b_1, \ldots, b_q\}$.

5.3   Equidistribution formula for hyperplanes

Let κ be an algebraically closed field, complete for a non-trivial non-Archimedean absolute value $|\cdot|$. We will extend (5.1.15) to the equidistribution formula (5.3.7) of holomorphic curves into projective spaces over κ, and use it to show the second main theorem for hyperplanes.

Lemma 5.11. *Let $f_1, \ldots, f_n$ be non-zero meromorphic functions in κ and let $\mathbf{W} = \mathbf{W}(f_1, \ldots, f_n)$ be the Wronskian determinant of $f_1, \ldots, f_n$. Then for $r > 0$, we have*

$$\mu(r, \mathbf{S}) \leq r^{-\frac{n(n-1)}{2}},$$

where

$$\mathbf{S} = \mathbf{S}(f_1, \ldots, f_n) = \frac{\mathbf{W}}{f_1 \cdots f_n}.$$

Proof. The value of **S** is clearly

$$\sum \pm \frac{f_{j_1}}{f_{j_1}} \cdot \frac{f'_{j_2}}{f_{j_2}} \cdots \frac{f^{(n-1)}_{j_n}}{f_{j_n}}$$

summed for the $n!$ permutations $(j_1, j_2, \ldots, j_n)$ of $(1, 2, \ldots, n)$, the positive sign being taken for a positive permutation, the negative sign for a negative permutation. The relations

$$\mu(r, \mathbf{S}) \leq \max \mu \left(r, \frac{f'_{j_2}}{f_{j_2}} \cdots \frac{f^{(n-1)}_{j_n}}{f_{j_n}} \right) = \max \mu \left(r, \frac{f'_{j_2}}{f_{j_2}} \right) \cdots \mu \left(r, \frac{f^{(n-1)}_{j_n}}{f_{j_n}} \right)$$

follow trivially from the non-Archimedean property. By Lemma 5.7, we obtain

$$\mu(r, \mathbf{S}) \leq \frac{1}{r} \cdot \frac{1}{r^2} \cdots \frac{1}{r^{n-1}} = r^{-\frac{n(n-1)}{2}},$$

and so Lemma 5.11 is proved. $\qquad\square$

Let V be a vector space of dimension $n+1 > 0$ over κ and let $e = (e_0, \ldots, e_n)$ be a basis of V. By a (*non-Archimedean*) *holomorphic curve*

$$f : \kappa \longrightarrow \mathbb{P}(V),$$

we mean an equivalence class of $(n + 1)$-tuples of entire functions

$$(\tilde{f}_0, \ldots, \tilde{f}_n) : \kappa \longrightarrow \kappa^{n+1}$$

such that $\tilde{f}_0, \ldots, \tilde{f}_n$ have no common factors in the ring of entire functions on κ and such that not all of the $\tilde{f}_j$ are identically zero. The mapping

$$\tilde{f} = \tilde{f}_0 e_0 + \cdots + \tilde{f}_n e_n : \kappa \longrightarrow V \tag{5.3.1}$$

will be called a *reduced representation* of f. Define

$$\mu(r, \tilde{f}) = \max_{0 \leq i \leq n} \mu(r, \tilde{f}_i).$$

Then the *characteristic function*

$$T(r, f) = \log \mu \left(r, \tilde{f} \right) \tag{5.3.2}$$

is well defined for all $r > 0$, up to $O(1)$.

By the Valiron-Jensen formula (5.1.6), it is easy to obtain the formula

$$T(r, f) = \max_{0 \leq i \leq n} N \left(r, \frac{1}{\tilde{f}_i} \right) + O(1), \tag{5.3.3}$$

where we think $N\left(r, \frac{1}{\tilde{f}_i}\right) = 0$ if $\tilde{f}_i$ is constant for some i. If f is non-constant, we have $T(r, f) \to \infty$ as $r \to \infty$. Obviously, f is rational if and only if

$$T(r, f) = O(\log r).$$

Let V^* be the dual space of V and let $\epsilon = (\epsilon_0, \dots, \epsilon_n)$ be the dual basis of e. Take $a \in \mathbb{P}(V^*)$ and take $\alpha \in V^*$ with $\mathbb{P}(\alpha) = a$. Write

$$\alpha = a_0 \epsilon_0 + \cdots + a_n \epsilon_n.$$

Then the inner product

$$\langle \tilde{f}, \alpha \rangle = a_0 \tilde{f}_0 + \cdots + a_n \tilde{f}_n$$

defines an entire function on κ. Assume $\langle \tilde{f}, \alpha \rangle \not\equiv 0$. Let $\mu_f^a = \mu_{\langle \tilde{f}, \alpha \rangle}^0$ denote the 0-valued multiplicity of $\langle \tilde{f}, \alpha \rangle$. Then the *valence function* of f for a is well defined by

$$N_f(r, a) = N\left(r, \frac{1}{\langle \tilde{f}, \alpha \rangle}\right).$$

For a positive integer u, we also write

$$N_{f,u}(r, a) = N_u\left(r, \frac{1}{\langle \tilde{f}, \alpha \rangle}\right).$$

For convenience, we define $N_f(r, a) = 0$ if $f(\kappa) \subset \ddot{E}[a]$, that is, $\langle \tilde{f}, \alpha \rangle = 0$.

Take an element $z \in \kappa$ with $|z| = r > r_0$ such that $\langle \tilde{f}(z), \alpha \rangle \neq 0$. Define the *compensation function* of f for a by

$$m_f(r, a) = -\log |f(z), a|. \tag{5.3.4}$$

Then

$$m_f(r, a) = \log |\tilde{f}(z)| + \log |\alpha| - \log |\langle \tilde{f}(z), \alpha \rangle|$$
$$= \log \mu(r, \tilde{f}) + \log |\alpha| - \log \mu(r, \langle \tilde{f}(z), \alpha \rangle).$$

Applying the Valiron-Jensen formula (5.1.10) to $\langle \tilde{f}, \alpha \rangle$, we obtain

$$m_f(r, a) + N_f(r, a) = T(r, f) + \log |\alpha| - \log \mu(r_0, \langle \tilde{f}, \alpha \rangle), \tag{5.3.5}$$

which shows that the definition (5.3.4) is independent of the choice of z on the circle $\{|z| = r\}$. Since the set

$$|\kappa| - \{|z| \mid \langle \tilde{f}(z), \alpha \rangle = 0\}$$

is dense in $\mathbb{R}^+$, we may extend the definition of $m_f(r, a)$ by using (5.3.5) to all $r > r_0$. The equality (5.3.5) is referred to as the *first main theorem*, which yields

$$N_f(r, a) \leq T(r, f) + O(1). \tag{5.3.6}$$

Theorem 5.12 ([176]). *Let $\mathscr{A} = \{a_0, a_1, \ldots, a_q\}$ be a family of points $a_j \in \mathbb{P}(V^*)$ in general position with $q \geq n$. Let $f : \kappa \longrightarrow \mathbb{P}(V)$ be a non-constant holomorphic curve. Then for any $\lambda \in J_n^q$, we have the equidistribution formula*

$$T(r, f) = \max_{0 \leq i \leq n} N_f \left(r, a_{\lambda(i)} \right) + O(1). \tag{5.3.7}$$

Proof. Take $\alpha_j \in V^* - \{0\}$ with $\mathbb{P}(\alpha_j) = a_j$. Write

$$\alpha_j = a_{j0}\epsilon_0 + \cdots + a_{jn}\epsilon_n, \quad j = 0, \ldots, q,$$

where $\epsilon = (\epsilon_0, \ldots, \epsilon_n)$ is the dual of e. For $j = 0, 1, \ldots, q$, set

$$F_j = \left\langle \tilde{f}, \alpha_j \right\rangle = a_{j0}\tilde{f}_0 + a_{j1}\tilde{f}_1 + \cdots + a_{jn}\tilde{f}_n.$$

Then

$$\mu(r, F_j) = O\left(\mu\left(r, \tilde{f}\right) \right), \quad j = 0, \ldots, q.$$

On the other hand, since $\mathscr{A}$ is in general position, for each $\lambda \in J_n^q$ there exist $b_{ij}^\lambda \in \kappa$ such that

$$\tilde{f}_j = \sum_{i=0}^{n} b_{ij}^\lambda F_{\lambda(i)}, \quad j = 0, \ldots, n, \tag{5.3.8}$$

and so we also obtain

$$\mu\left(r, \tilde{f}\right) = O\left(\max_{0 \leq i \leq n} \mu\left(r, F_{\lambda(i)}\right) \right).$$

Thus, by the definition of characteristic functions, we have

$$T(r, f) = \log \mu\left(r, \tilde{f}\right) = \max_{0 \leq i \leq n} \log^+ \mu\left(r, F_{\lambda(i)}\right) + O(1).$$

Finally, by the Valiron-Jensen formula (5.1.6), we obtain

$$T(r, f) = \max_{0 \leq i \leq n} N_f\left(r, a_{\lambda(i)}\right) + O(1),$$

and hence Theorem 5.12 follows. $\qquad\square$

Given a reduced representation (5.3.1) of f, let $\mathbf{W} = \mathbf{W}\left(\tilde{f}_0, \ldots, \tilde{f}_n \right)$ be the Wronskian determinant of $\tilde{f}_0, \ldots, \tilde{f}_n$. When κ has characteristic zero, we know that $\mathbf{W} \not\equiv 0$ if and only if f is linearly non-degenerate (cf. [176]), but this is not so if κ has characteristic $p > 0$. Here the mapping f will be called *analytically non-degenerate* if $\mathbf{W} \not\equiv 0$, so we may define the *ramification term* by

$$N_{\text{Ram}}(r, f) = N\left(r, \frac{1}{\mathbf{W}}\right). \tag{5.3.9}$$

Using the symbols in the proof of Theorem 5.12, for the family $\mathscr{A} = \{a_0, a_1, \ldots, a_q\}$ we can define a meromorphic function by

$$\mathbf{H} = \frac{F_0 F_1 \cdots F_q}{\mathbf{W}(\tilde{f}_0, \ldots, \tilde{f}_n)}. \tag{5.3.10}$$

Now we use the equidistribution formula (5.3.7) to derive the *second main theorem*:

Theorem 5.13 (cf.[176]). *Let* $f : \kappa \longrightarrow \mathbb{P}(V)$ *be an analytically non-degenerate holomorphic curve and let* $\mathscr{A} = \{a_0, a_1, \ldots, a_q\}$ *be a family of points* $a_j \in \mathbb{P}(V^*)$ *in general position. Then*

$$(q - n)T(r, f) \leq \sum_{j=0}^{q} N_{f,n}(r, a_j) - N(r, \mathbf{H}) - \frac{n(n+1)}{2} \log r + O(1).$$

Proof. We will use the symbols in the proof of Theorem 5.12. For $\lambda \in J_n^q$, we abbreviate the Wronskian

$$\mathbf{W}_\lambda = \mathbf{W}\left(F_{\lambda(0)}, F_{\lambda(1)}, \ldots, F_{\lambda(n)}\right).$$

Then

$$\mathbf{W}_\lambda = c_\lambda \mathbf{W}, \quad c_\lambda = \det\left(a_{\lambda(i)j}\right) \neq 0.$$

For each $r > r_0$, we can choose $\lambda_r \in J_n^q$ such that

$$N_f(r, a_j) \geq \max_{0 \leq i \leq n} N_f\left(r, a_{\lambda_r(i)}\right), \quad j \in J_r,$$

where

$$J_r = \{0, 1, \ldots, q\} - \{\lambda_r(0), \ldots, \lambda_r(n)\}.$$

Theorem 5.12 and the Valiron-Jensen formula (5.1.6) imply

$$(q - n)T(r, f) = \sum_{j=0}^{q} N_f(r, a_j) - \sum_{i=0}^{n} N_f\left(r, a_{\lambda_r(i)}\right) + O(1)$$

$$= \sum_{j=0}^{q} N_f(r, a_j) - \sum_{i=0}^{n} \log \mu\left(r, F_{\lambda_r(i)}\right) + O(1).$$

However,

$$-\sum_{i=0}^{n} \log \mu\left(r, F_{\lambda_r(i)}\right) = \log \mu\left(r, \frac{\mathbf{W}_{\lambda_r}}{F_{\lambda_r(0)} \cdots F_{\lambda_r(n)}}\right) - \log \mu(r, \mathbf{W}) + O(1).$$

From Lemma 5.11, we have

$$\log \mu\left(r, \frac{\mathbf{W}_{\lambda_r}}{F_{\lambda_r(0)} \cdots F_{\lambda_r(n)}}\right) \leq -\frac{n(n+1)}{2} \log r.$$

Thus by applying the Valiron-Jensen formula (5.1.6) to $\mathbf{W}$, we obtain

$$(q-n)T(r,f) \leq \sum_{j=0}^{q} N_f(r,a_j) - N_{\mathrm{Ram}}(r,f) - \frac{n(n+1)}{2}\log r + O(1). \quad (5.3.11)$$

By the Valiron-Jensen formula (5.1.10), we also have

$$\sum_{j=0}^{q} N_f(r,a_j) - N_{\mathrm{Ram}}(r,f) = N\left(r,\frac{1}{\mathbf{H}}\right) - N(r,\mathbf{H}). \quad (5.3.12)$$

The inequality in Theorem 5.13 can be derived easily from (5.3.11) and (5.3.12). $\quad\square$

Theorem 5.13 means directly the following *defect relation*:

$$\sum_{j=0}^{q} \delta_{f,n}(a_j) \leq n+1, \quad (5.3.13)$$

where $\delta_{f,n}(a_j)$ is the *defect* of f for a_j defined by

$$\delta_{f,n}(a_j) = 1 - \limsup_{r\to\infty} \frac{N_{f,n}(r,a_j)}{T(r,f)}$$

with $0 \leq \delta_{f,n}(a_j) \leq 1$. However, from the equidistribution formula (5.3.7), we obtain

$$\delta_f(a_j) = 1 - \limsup_{r\to\infty} \frac{N_f(r,a_j)}{T(r,f)} = 0 \quad (5.3.14)$$

except for n elements of $\mathscr{A}$ at most.

5.4 Non-Archimedean Cartan-Nochka theorem

Let κ be an algebraically closed field, complete for a non-trivial non-Archimedean absolute value $|\cdot|$. Let V be a vector space of dimension $n+1 > 0$ over κ. We continue to study a holomorphic curve $f : \kappa \longrightarrow \mathbb{P}(V)$. First of all, we prove the *second main theorem* for a family in subgeneral position.

Theorem 5.14. *Let* $\mathscr{A} = \{a_j\}_{j=0}^{q}$ *be a finite family of points* $a_j \in \mathbb{P}(V^*)$ *in u-subgeneral position with* $u \leq 2u - n < q$. *Let* $f : \kappa \longrightarrow \mathbb{P}(V)$ *be a holomorphic curve which is analytically non-degenerate. Then*

$$(q-2u+n)T(r,f) \leq \sum_{j=0}^{q} \theta\omega(a_j)N_f(r,a_j) - \theta N_{\mathrm{Ram}}(r,f) - \frac{n(n+1)}{2}\theta\log r + O(1),$$

where $\theta \geq 1$ *is a Nochka constant, and* $\omega : \mathscr{A} \longrightarrow \mathbb{R}(0,1]$ *is a Nochka weight.*

Proof. We will adopt the notations that were used in the proof of Theorem 5.13, and without loss of generality, assume $|\alpha_j| = 1$ for $j = 0, \ldots, q$. Lemma 1.61 implies

$$\prod_{j=0}^{q} \left(\frac{1}{|f, a_j|} \right)^{\omega(a_j)} \leq \left(\frac{1}{\Gamma(\mathscr{A})} \right)^{q-u} \max_{\lambda \in J_n(\mathscr{A})} \prod_{j=0}^{n} \frac{1}{|f, a_{\lambda(j)}|},$$

which yields

$$\prod_{j=0}^{q} \left(\frac{1}{|f, a_j|} \right)^{\omega(a_j)} \leq \left(\frac{1}{\Gamma(\mathscr{A})} \right)^{q-u} \max_{\lambda \in J_n(\mathscr{A})} \prod_{j=0}^{n} \frac{|\tilde{f}|}{|F_{\lambda(j)}|}$$

$$= \left(\frac{1}{\Gamma(\mathscr{A})} \right)^{q-u} \frac{|\tilde{f}|^{n+1}}{|\mathbf{W}|} \max_{\lambda \in J_n(\mathscr{A})} \frac{1}{|c_\lambda|} \frac{|\mathbf{W}_\lambda|}{|F_{\lambda(0)} \cdots F_{\lambda(n)}|}.$$

We obtain

$$\sum_{j=0}^{q} \omega(a_j) m_f(r, a_j) \leq (n+1)T(r, f) - N\left(r, \frac{1}{\mathbf{W}}\right) - \frac{n(n+1)}{2} \log r + O(1). \quad (5.4.1)$$

According to the proof of Theorem 2.66 and by the properties of the Nochka weights, we can obtain the inequality in Theorem 5.14.   □

According to the proof of Lemma 2.67, we can also obtain

$$\sum_{j=0}^{q} \omega(a_j) \mu_{F_j}^0 - \mu_{\mathbf{W}}^0 \leq \sum_{j=0}^{q} \omega(a_j) \min\left\{ \mu_{F_j}^0, n \right\}. \quad (5.4.2)$$

Thus Theorem 5.14 yields immediately the truncated form of the *second main theorem* and the *defect relations*.

Corollary 5.15. *Assumptions as in Theorem 5.14. Then*

$$(q - 2u + n)T(r, f) \leq \sum_{j=0}^{q} \theta \omega(a_j) N_{f,n}(r, a_j) - \frac{n(n+1)}{2} \theta \log r + O(1). \quad (5.4.3)$$

Corollary 5.16. *Assumptions as in Theorem 5.14. Then*

$$\sum_{j=0}^{q} \omega(a_j) \delta_f(a_j) \leq \sum_{j=0}^{q} \omega(a_j) \delta_{f,n}(a_j) \leq n + 1, \quad (5.4.4)$$

$$\sum_{j=0}^{q} \delta_f(a_j) \leq \sum_{j=0}^{q} \delta_{f,n}(a_j) \leq 2u - n + 1. \quad (5.4.5)$$

Let $f : \kappa \longrightarrow \mathbb{P}(V)$ be a non-constant holomorphic curve. For a finite family $\mathscr{A} = \{a_j\}_{j=0}^q$ of points $a_j \in \mathbb{P}(V^*)$ in u-subgeneral position with $n \leq u \leq q$, Theorem 5.12 yields immediately

$$T(r, f) = \max_{0 \leq i \leq u} N_f\left(r, a_{\lambda(i)}\right) + O(1) \tag{5.4.6}$$

for any $\lambda \in J_u^q$. According to the proof of Theorem 5.13, by using the formula (5.4.6) we can prove the following *second main theorem* without Nochka weights.

Theorem 5.17. *Let $\mathscr{A} = \{a_j\}_{j=0}^q$ be a finite family of points $a_j \in \mathbb{P}(V^*)$ in u-subgeneral position with $n \leq u \leq q$. Let $f : \kappa \longrightarrow \mathbb{P}(V)$ be a holomorphic curve which is analytically non-degenerate. Then*

$$(q - u)T(r, f) \leq \sum_{j=0}^q N_f(r, a_j) - N_{\mathrm{Ram}}(r, f) - \frac{n(n+1)}{2} \log r + O(1). \tag{5.4.7}$$

Next we assume that the field κ has characteristic zero and eliminate the restriction of non-degeneracy on f. Take a reduced representation $\tilde{f} : \kappa \longrightarrow V_*$ of a non-constant holomorphic curve $f : \kappa \longrightarrow \mathbb{P}(V)$ and define a linear subspace of V^*,

$$E[f] = \{\alpha \in V^* \mid \langle \tilde{f}, \alpha \rangle \equiv 0\},$$

and write

$$\ell_f = \dim E[f], \quad k = n - \ell_f.$$

The number k is non-negative, i.e., $0 \leq \ell_f \leq n$. If $k < 0$, that is, $\ell_f = n+1$, there is $\{\alpha_0, \ldots, \alpha_n\} \subset E[f]$ such that

$$\alpha_0 \wedge \cdots \wedge \alpha_n \neq 0; \quad \langle \tilde{f}, \alpha_j \rangle \equiv 0 \ (0 \leq j \leq n).$$

By Cramer's rule, $\tilde{f} \equiv 0$, which is impossible. Then V^* is decomposed into a direct sum

$$V^* = W^* \oplus E[f],$$

where W^* is a $(k+1)$-dimensional subspace of V^*. The mapping f will be said to be k-flat. In order to simplify our notation, we define $\ell_f = 0$ if f is linearly non-degenerate, that is, $E[f] = \emptyset$, and say that f is n-flat.

Assume that $\mathscr{A} = \{a_j\}_{j=0}^q$ is in general position and assume that f is non-constant and k-flat with $0 \leq k \leq n < q$ such that each pair (f, a_j) is free for $j = 0, \ldots, q$. We take a basis $\epsilon = (\epsilon_0, \ldots, \epsilon_n)$ of V^* such that $\epsilon_0, \ldots, \epsilon_k$ and $\epsilon_{k+1}, \ldots, \epsilon_n$ is a basis of W^* and $E[f]$, respectively. Let $e = (e_0, \ldots, e_n)$ be the dual basis of ϵ. Let W be the vector space spanned by $e_0, \ldots, e_k$ over κ. Thus the reduced representation $\tilde{f} : \kappa \longrightarrow V_*$ is given by

$$\tilde{f} = \sum_{j=0}^k \tilde{f}_j e_j = \sum_{j=0}^k \langle \tilde{f}, \epsilon_j \rangle e_j$$

such that $\langle \tilde{f}, \epsilon_0 \rangle, \ldots, \langle \tilde{f}, \epsilon_k \rangle$ are holomorphic and linearly independent over κ. Hence a linearly non-degenerate holomorphic curve $\hat{f} : \kappa \longrightarrow \mathbb{P}(W)$ is defined with a reduced representation

$$\tilde{\hat{f}} = \tilde{f} = \sum_{j=0}^{k} \langle \tilde{f}, \epsilon_j \rangle e_j : \kappa \longrightarrow W_*.$$

Therefore, we obtain

$$T(r, \hat{f}) = \log \mu(r, \tilde{f}) + O(1) = T(r, f) + O(1). \tag{5.4.8}$$

If $k = 0$, then $T(r, \hat{f})$ is constant. But then the inequality (5.4.8) is violated. Thus, we must have $k \geq 1$. Define

$$N_{\mathrm{Ram}}(r, f) = N_{\mathrm{Ram}}(r, \hat{f}).$$

Now we prove the *second main theorem* of f, which is an analogue of Cartan-Nochka's Theorem 2.70 over κ.

Theorem 5.18. *Let $\mathscr{A} = \{a_j\}_{j=0}^{q}$ be a finite family of points $a_j \in \mathbb{P}(V^*)$ in general position. Take an integer k with $1 \leq k \leq n \leq 2n - k < q$. Let $f : \kappa \longrightarrow \mathbb{P}(V)$ be a non-constant holomorphic curve that is k-flat such that each pair (f, a_j) is free for $j = 0, \ldots, q$. Then*

$$(q - 2n + k)T(r, f) \leq \sum_{j=0}^{q} N_{f,k}(r, a_j) - \frac{k(k+1)}{2}\theta \log r + O(1),$$

where θ is a Nochka constant with

$$\frac{n+1}{k+1} \leq \theta \leq \frac{2n - k + 1}{k+1}.$$

Proof. Take $\tilde{a}_j \in V^* - \{0\}$ with $\mathbb{P}(\tilde{a}_j) = a_j$ and write

$$\tilde{a}_j = \sum_{i=0}^{n} \langle e_i, \tilde{a}_j \rangle \epsilon_i, \quad j = 0, \ldots, q.$$

Define a new family $\hat{\mathscr{A}} = \{\hat{a}_j\}$ as follows:

$$\tilde{\hat{a}}_j = \sum_{i=0}^{k} \langle e_i, \tilde{a}_j \rangle \epsilon_i \in W^* - \{0\}, \quad \hat{a}_j = \mathbb{P}\left(\tilde{\hat{a}}_j\right) \in \mathbb{P}(W^*), \quad j = 0, \ldots, q.$$

Take $\sigma \in J_n^q$. Then $\tilde{a}_\sigma \neq 0$ since $\mathscr{A}$ is in general position, and hence

$$\det(\langle e_i, \tilde{a}_{\sigma(j)} \rangle) \neq 0 \quad (0 \leq i, j \leq n).$$

Therefore, there is a $\lambda \in J_k^q$ with $\text{Im}\lambda \subseteq \text{Im}\sigma$ such that

$$\det(\langle e_s, \tilde{a}_{\lambda(t)}\rangle) \neq 0 \quad (0 \leq s, t \leq k).$$

We have

$$\tilde{\hat{a}}_\lambda = \det(\langle e_s, \tilde{a}_{\lambda(t)}\rangle)\epsilon_0 \wedge \cdots \wedge \epsilon_k \neq 0.$$

Hence $\lambda \in J_k(\hat{\mathscr{A}})$. Thus $\hat{\mathscr{A}}$ is in n-subgeneral position.

Note that

$$\langle \tilde{f}, \tilde{a}_j \rangle = \sum_{i=0}^{n} \langle \tilde{f}, \epsilon_i \rangle \langle e_i, \tilde{a}_j \rangle = \sum_{i=0}^{k} \langle \tilde{f}, \epsilon_i \rangle \langle e_i, \tilde{a}_j \rangle = \langle \tilde{\hat{f}}, \tilde{\hat{a}}_j \rangle.$$

We obtain

$$\mu_f^{a_j} = \mu_{\hat{f}}^{\hat{a}_j}, \ j = 0, \ldots, q.$$

By applying Theorem 5.14 to $\hat{f}$, then

$$(q - 2n + k)T(r, \hat{f}) \leq \sum_{j=0}^{q} \theta\omega(\hat{a}_j)N_{\hat{f}}(r, \hat{a}_j) - \theta N_{\text{Ram}}(r, \hat{f}) - \frac{k(k+1)}{2}\theta \log r + O(1).$$

In particular, Corollary 5.15 and the facts above imply

$$(q - 2n + k)T(r, f) \leq \sum_{j=0}^{q} \theta\omega(\hat{a}_j)N_{f,k}(r, a_j) - \frac{k(k+1)}{2}\theta \log r + O(1),$$

and so Theorem 5.18 follows from 1) in Lemma 1.59. □

Corollary 5.19. *With the assumptions as in Theorem 5.18, we have the defect relation*

$$\sum_{j=0}^{q} \delta_f(a_j) \leq \sum_{j=0}^{q} \delta_{f,k}(a_j) \leq 2n - k + 1. \tag{5.4.9}$$

By Theorem 5.12, it is easy to show that if there exists a family $\mathscr{B} = \{b_0, b_1, \ldots, b_s\}$ of points $b_j \in \mathbb{P}(V^*)$ in general position such that

$$\delta_f(b_j) > 0, \ j = 0, 1, \ldots, s$$

hold for a non-constant holomorphic curve $f : \kappa \longrightarrow \mathbb{P}(V)$, then $s < n$. This property was observed by Cherry and Ye [60]. Thus the left part of (5.4.9) assumes the bound

$$\sum_{j=0}^{q} \delta_f(a_j) \leq n. \tag{5.4.10}$$

If we apply Theorem 5.17 to the mapping $\hat{f}$ and the family $\hat{\mathscr{A}}$ directly, we can obtain the following form of the *second main theorem* without Nochka weights and the condition that the pairs (f, a_j) are free.

Theorem 5.20. *Let $\mathscr{A} = \{a_j\}_{j=0}^q$ be a finite family of points $a_j \in \mathbb{P}(V^*)$ in general position. Take an integer k with $1 \leq k \leq n \leq q$. Let $f : \kappa \longrightarrow \mathbb{P}(V)$ be a non-constant holomorphic curve that is k-flat. Then*

$$(q - n)T(r, f) \leq \sum_{j=0}^{q} N_f(r, a_j) - N_{\mathrm{Ram}}(r, f) - \frac{k(k+1)}{2} \log r + O(1). \quad (5.4.11)$$

5.5 Holomorphic curves into projective varieties

In this section, we further extend (5.1.15) to the equidistribution formula (5.5.2) of holomorphic curves for targets of hypersurfaces, and formulate Griffiths' and Lang's conjectures into a question of holomorphic curves over non-Archimedean fields.

5.5.1 Equidistribution formula for hypersurfaces

Let κ be an algebraically closed field of characteristic p, complete for a non-trivial non-Archimedean absolute value $|\cdot|$. Assume that V is a normed vector space of dimension $n + 1 > 0$ over κ. Let $f : \kappa \longrightarrow \mathbb{P}(V)$ be a non-constant holomorphic curve. For a positive integer d, let $\varphi_d : \mathbb{P}(V) \longrightarrow (\mathrm{II}_d V)$ be the Veronese mapping. Then f induces a holomorphic curve

$$f^{\mathrm{II}d} = \varphi_d \circ f : \kappa \longrightarrow \mathbb{P}(\mathrm{II}_d V)$$

such that the characteristic function of $f^{\mathrm{II}d}$ satisfies

$$T\left(r, f^{\mathrm{II}d}\right) = dT(r, f).$$

Take $a \in \mathbb{P}(\mathrm{II}_d V^*)$ such that the pair $(f^{\mathrm{II}d}, a)$ is free for the interior product $\angle$. Applying (5.3.5) to $f^{\mathrm{II}d}$ and a, we obtain the *first main theorem* for a hypersurface $\ddot{E}^d[a]$,

$$dT(r, f) = N_{f^{\mathrm{II}d}}(r, a) + m_{f^{\mathrm{II}d}}(r, a) + O(1). \quad (5.5.1)$$

Further, we have the following *equidistribution formula*:

Theorem 5.21. *Let $\mathscr{A} = \{a_j\}_{j=0}^q$ be a finite admissible family of points $a_j \in \mathbb{P}(\mathrm{II}_d V^*)$ with $q \geq n$ and $d > 0$. Let $f : \kappa \longrightarrow \mathbb{P}(V)$ be a non-constant holomorphic curve. Then*

$$dT(r, f) = \max_{0 \leq i \leq n} N_{f^{\mathrm{II}d}}\left(r, a_{\lambda(i)}\right) + O(1), \quad \lambda \in J_n^q. \quad (5.5.2)$$

Proof. Take $\alpha_j \in \mathrm{II}_d V^* - \{0\}$ with $\mathbb{P}(\alpha_j) = a_j$, and define

$$\tilde{\alpha}_j(\xi) = \left\langle \xi^{\mathrm{II}d}, \alpha_j \right\rangle, \quad \xi \in V, \; j = 0, 1, \ldots, q.$$

Without loss of generality, assume $|\alpha_j| = 1$ for $j = 0, \dots, q$. Write

$$F_j = \tilde{\alpha}_j \circ \tilde{f} = \left\langle \tilde{f}^{\text{IId}}, \alpha_j \right\rangle, \quad j = 0, 1, \dots, q,$$

where $\tilde{f} : \kappa \longrightarrow V_*$ is a reduced representation of f. Since f is non-constant, then $\mu(r, \tilde{f}) \to \infty$ as $r \to \infty$. So we may assume $\mu\left(r, \tilde{f}\right) \geq 1$ for $r \geq r_0$. Obviously, we have

$$\mu(r, F_j) \leq O\left(\mu\left(r, \tilde{f}\right)^d\right), \quad j = 0, \dots, q. \tag{5.5.3}$$

On the other hand, by using the identity (3.8.8), we obtain

$$\tilde{f}_j^s = \sum_{i=0}^{n} b_{ij}^\lambda(\tilde{f}) F_{\lambda(i)} \quad (\lambda \in J_n^q, \ j = 0, \dots, n) \tag{5.5.4}$$

for some integer $s \geq d$, where $b_{ij}^\lambda \in \kappa[\xi_0, \dots, \xi_n]$ are homogeneous polynomials of degree $s - d$. Note that $(F_{\lambda(0)}, \dots, F_{\lambda(n)}) \neq 0$ since the family $\mathscr{A}$ is admissible. Hence

$$\mu\left(r, \tilde{f}\right)^s \leq O\left(\mu\left(r, \tilde{f}\right)^{s-d}\right) \max_{0 \leq i \leq n} \mu\left(r, F_{\lambda(i)}\right),$$

which implies immediately

$$\mu\left(r, \tilde{f}\right)^d \leq O\left(\max_{0 \leq i \leq n} \mu\left(r, F_{\lambda(i)}\right)\right). \tag{5.5.5}$$

Therefore the inequalities (5.5.3) and (5.5.5) yield

$$dT(r, f) = \max_{0 \leq i \leq n} \log^+ \mu\left(r, F_{\lambda(i)}\right) + O(1).$$

Finally, by using the Valiron-Jensen formula (5.1.6), we derive the formula (5.5.2). $\qquad \square$

For fields of characteristic zero, the equidistribution formula (5.5.2) is due to Hu and Yang [176]. Here we give a simple application of Theorem 5.21. For any fixed $r > 0$, without loss of generality, we may assume that

$$N_{f^{\text{IId}}}(r, a_0) \leq N_{f^{\text{IId}}}(r, a_1) \leq \dots \leq N_{f^{\text{IId}}}(r, a_q).$$

By using (5.5.2), we have

$$dT(r, f) = N_{f^{\text{IId}}}(r, a_j) + O(1), \quad j \geq n,$$

which means (cf. [176], [181])

$$d(q + 1 - n)T(r, f) = \sum_{j=n}^{q} N_{f^{\text{IId}}}(r, a_j) + O(1) \leq \sum_{j=0}^{q} N_{f^{\text{IId}}}(r, a_j) + O(1). \tag{5.5.6}$$

This inequality was also observed by Min Ru [329] by using the first main theorem.

A holomorphic curve $f : \kappa \longrightarrow \mathbb{P}(V)$ is said to be *algebraically non-degenerate of order d* if $f(\kappa) \not\subseteq \ddot{E}^d[a]$ holds for all $a \in \mathbb{P}(\mathrm{II}_d V^*)$, and is called *algebraically non-degenerate* if f is algebraically non-degenerate of order d for each $d > 0$. Now we can prove the *second main theorem* which exhibits a stronger inequality than (5.5.6).

Theorem 5.22 ([176]). *Assume $p = \mathrm{char}(\kappa) = 0$. Let $\mathscr{A} = \{a_j\}_{j=0}^q$ be a finite admissible family of points $a_j \in \mathbb{P}(\mathrm{II}_d V^*)$ with $q \geq n$ and $d > 0$. Let $f : \kappa \longrightarrow \mathbb{P}(V)$ be a holomorphic curve which is algebraically non-degenerate of order d. Then for $l \in \{n - 1, n\}$, we have*

$$d(q - l)T(r, f) \leq \sum_{j=0}^q N_{f^{\mathrm{II}d}}(r, a_j) - N_{\mathrm{Ram}}(r, f) - \frac{l(l+1)}{2} \log r + O(1), \quad (5.5.7)$$

where $N_{\mathrm{Ram}}(r, f)$ is defined in the proof.

Proof. Since f is algebraically non-degenerate of order d, for each $\lambda \in J_l^q$, $F_{\lambda(0)}$, $\ldots$, $F_{\lambda(l)}$ are linearly independent, and hence

$$\mathbf{W}_\lambda = \mathbf{W}(F_{\lambda(0)}, \ldots, F_{\lambda(l)}) \not\equiv 0.$$

Without loss of generality, we may assume $\mu(r, \tilde{f}) \geq 1$ for $r \geq r_0$. Lemma 3.73 implies

$$\prod_{j=0}^q \frac{1}{|f^{\mathrm{II}d}, a_j|} \leq \left(\frac{1}{\Gamma(\mathscr{A})} \right)^{q-l} \max_{\lambda \in J_l^q} \prod_{i=0}^l \frac{1}{|f^{\mathrm{II}d}, a_{\lambda(i)}|},$$

which yields

$$\Gamma(\mathscr{A})^{q-l} \prod_{j=0}^q \frac{1}{|f^{\mathrm{II}d}, a_j|} \leq \max_{\lambda \in J_l^q} \prod_{i=0}^l \frac{|\tilde{f}^{\mathrm{II}d}|}{|F_{\lambda(i)}|}$$

$$\leq |\tilde{f}|^{(l+1)d} \max_{\lambda \in J_l^q} \left| \frac{\mathbf{W}_\lambda}{F_{\lambda(0)} \cdots F_{\lambda(l)}} \right| \frac{1}{|\mathbf{W}_\lambda|}.$$

It follows that

$$\sum_{j=0}^q m_{f^{\mathrm{II}d}}(r, a_j) \leq d(l+1)T(r, f) - N_{\mathrm{Ram}}(r, f)$$

$$- \frac{l(l+1)}{2} \log r + O(1), \quad (5.5.8)$$

where

$$N_{\mathrm{Ram}}(r, f) = \min_{\lambda \in J_l^q} N\left(r, \frac{1}{\mathbf{W}_\lambda} \right). \quad (5.5.9)$$

Then theorem 5.22 follows from (5.5.8) and the first main theorem.   □

According to the proof of Theorem 5.13, by using the equidistribution formula (5.5.2) we can give another proof of Theorem 5.22.

5.5.2 Characteristic functions for divisors

We assume that E is a very ample divisor over a projective variety $M = M(\bar{\kappa}) \subset \mathbb{P}(\bar{\kappa}^N)$. Let $f : \kappa \longrightarrow M$ be a holomorphic curve. Let $\varphi : M \longrightarrow \mathbb{P}(V^*)$ be the dual classification mapping, where $V = \mathcal{L}(E)$. The *characteristic function* of f for E

$$T_E(r) = T_{f,E}(r) = T(r, \varphi \circ f)$$

is well defined, up to $O(1)$. If f is not constant, then $\varphi \circ f$ is also non-constant since $\varphi : M \longrightarrow \mathbb{P}(V^*)$ is an embedding. Hence $T_{f,E}(r) \to \infty$ as $r \to \infty$.

For any $s \in V - \{0\}$, set $a = \mathbb{P}(s)$ and write $D = (s)$. If f satisfies $f(\kappa) \not\subset D$, the functions

$$N_f(r, D) = N_{\varphi \circ f}(r, a), \quad m_f(r, D) = m_{\varphi \circ f}(r, a)$$

are well defined. These functions are related by the first main theorem

$$T_{f,E}(r) = N_f(r, D) + m_f(r, D) + O(1). \tag{5.5.10}$$

Proposition 5.23. *If E and E' are very ample divisors, then*

$$T_{E+E'}(r) = T_E(r) + T_{E'}(r). \tag{5.5.11}$$

Proof. Let $\varphi_E : M \longrightarrow \mathbb{P}(V^*)$ and $\varphi_{E'} : M \longrightarrow \mathbb{P}(V'^*)$ be the dual classification mappings of E and E', respectively, given in projective coordinates by

$$\tilde{\varphi}_E = \sum_{i=0}^{n} \tilde{\varphi}_i e_i, \quad \tilde{\varphi}_{E'} = \sum_{i=0}^{n'} \tilde{\varphi}'_i e'_i,$$

where $e_0, \ldots, e_n$ and $e'_0, \ldots, e'_{n'}$ are bases of V^* and V'^*, respectively. Then the dual classification mapping of $E + E'$, denoted by $\varphi_{E+E'} : M \longrightarrow \mathbb{P}(V^* \otimes V'^*)$, is given in projective coordinates by

$$\tilde{\varphi}_{E+E'} = \sum_{i,j} \tilde{\varphi}_i \tilde{\varphi}'_j e_i \otimes e'_j.$$

Hence

$$
\begin{aligned}
T_{E+E'}(r) &= T(r, (\varphi_{E+E'}) \circ f) = \log \mu(r, \tilde{\varphi}_E \circ f \otimes \tilde{\varphi}_{E'} \circ f) \\
&= \log \left(\max_{i,j} \mu(r, \tilde{\varphi}_i \circ f \tilde{\varphi}'_j \circ f) \right) \\
&= \log \left(\max_i \mu(r, \tilde{\varphi}_i \circ f) \max_j \mu(r, \tilde{\varphi}'_j \circ f) \right) \\
&= \log \left(\mu(r, \tilde{\varphi}_E \circ f) \mu(r, \tilde{\varphi}_{E'} \circ f) \right) \\
&= T_E(r) + T_{E'}(r),
\end{aligned}
$$

and so Proposition 5.23 follows. □

Given any divisor D on M, we can write $D = E - E'$, where E and E' are very ample, and define

$$N_f(r, D) = N_f(r, E) - N_f(r, E'),$$

$$m_f(r, D) = m_f(r, E) - m_f(r, E'),$$

and finally set

$$T_D(r) = T_E(r) - T_{E'}(r).$$

It follows that (5.5.11) holds for arbitrary divisors D and D'. If two divisors D and D' on M are linearly equivalent, that is, $D = D' + (h)$ for some rational function h on M, then

$$T_{D'}(r) = T_D(r) + O(1).$$

Lemma 5.24. *If D is an effective divisor, then $T_D(r) \geq 0$.*

Proof. There exist very ample divisors E and E' such that $D = E - E'$. Let $\{f_0, \ldots, f_n\}$ be a basis for $\mathcal{L}(E')$. Then we can extend this choice of functions to a basis $\{f_0, \ldots, f_{n+l}\}$ for $\mathcal{L}(E)$ because D is effective. Let $\varphi_{E'} = [f_0, \ldots, f_n]$ and $\varphi_E = [f_0, \ldots, f_{n+l}]$ be the dual classification mapping of E' and E, respectively. It then follows from the definition of $T_D(r)$ that

$$T_D(r) = T_E(r) - T_{E'}(r) \geq 0,$$

and so the lemma is proved.   □

We suggested the following problem (cf. [176]):

Conjecture 5.25. *Let M be a non-singular projective variety over κ. Let K be the canonical divisor of M, and let D be a simple normal crossings divisor on M. Let E be a pseudo ample divisor. Let $f : \kappa \longrightarrow M$ be a non-constant holomorphic curve. Then there exists a proper algebraic subset Z_D having the following property: when $f(\kappa) \not\subset Z_D$,*

$$m_f(r, D) + T_K(r) \leq o(T_E(r)).$$

In value distribution theory of complex variables, this corresponds to the conjecture due to P. Griffiths [125] and S. Lang [229].

Finally, we consider the case $M = \mathbb{P}(V)$, where V is a normed vector space of dimension $n + 1 > 0$ over κ.

Example 5.26. *Each hyperplane E of $\mathbb{P}(V)$ is very ample with*

$$\dim \mathcal{L}(E) = n + 1,$$

and hence the dual classification mapping φ is the identity. Thus we have $T_E(r) = T(r, f)$, and hence

$$T_K(r) = -(n + 1)T(r, f)$$

since $K = -(n+1)E$. *Let* $D = \sum_i \ddot{E}[a_i]$ *be a sum of hyperplanes in general position. Then the conjecture reduces to*

$$\sum_i m_f(r, a_i) < (n+1)T(r, f) + o(T(r, f))$$

which follows from Theorem 5.12.

For hypersurfaces, Conjecture 5.25 corresponds to the following form:

Conjecture 5.27. *Take finite many of* $a_i \in \mathbb{P}(\Pi_d V^*)$ *such that* $\sum_i \ddot{E}^d[a_i]$ *have normal crossings. Let* $f : \kappa \longrightarrow \mathbb{P}(V)$ *be an algebraically non-degenerate holomorphic curve. Then*

$$\sum_i m_{f^{\text{IId}}}(r, a_i) \leq (n+1)T(r, f) + o(T(r, f)).$$

5.6   The *abc*-theorem for meromorphic functions

Let κ be an algebraically closed field of characteristic zero, complete for a non-trivial non-Archimedean absolute value $|\cdot|$. For an integer $n \geq 2$, we will study the functional equation

$$f_1 + \cdots + f_n = f_0, \tag{5.6.1}$$

and prove the *generalized abc-theorem* of meromorphic functions over κ as follows:

Theorem 5.28. *Let* $f_j (j = 1, \ldots, n)$ *be linearly independent meromorphic functions on* κ *and define* f_0 *by* (5.6.1). *Then*

$$\max_{0 \leq j \leq n} \log \mu(r, f_j) \leq \sum_{i=0}^{n} \left\{ N\left(r, \frac{1}{f_i}\right) - N(r, f_i) \right\} + N(r, \mathbf{W})$$
$$- N\left(r, \frac{1}{\mathbf{W}}\right) - \frac{n(n-1)}{2} \log r + O(1), \tag{5.6.2}$$

where $\mathbf{W} = \mathbf{W}(f_1, \ldots, f_n)$ *is the Wronskian of* $f_1, \ldots, f_n$.

Proof. Since $f_1, \ldots, f_n$ are linearly independent, the Wronskian $\mathbf{W} \not\equiv 0$. By (5.6.1) and

$$f_1^{(i)} + \cdots + f_n^{(i)} = f_0^{(i)}, \quad i = 1, \ldots, n-1,$$

we have

$$\mathbf{W} = \mathbf{W}_j \ (j = 1, 2, \ldots, n),$$

where

$$\mathbf{W}_j = \mathbf{W}(f_1, \ldots, f_{j-1}, f_0, f_{j+1}, \ldots, f_n).$$

Set

$$\mathbf{S}_0 = \mathbf{S}(f_1, \ldots, f_n) = \frac{\mathbf{W}}{f_1 \cdots f_n} = \det\left(\frac{f_j^{(i)}}{f_j}\right).$$

Then

$$\mathbf{S}_0 f_j = \mathbf{S}_j f_0, \quad j = 1, \ldots, n, \tag{5.6.3}$$

where

$$\mathbf{S}_j = \mathbf{S}(f_1, \ldots, f_{j-1}, f_0, f_{j+1}, \ldots, f_n).$$

By (5.6.3), we have

$$\max_{0 \le j \le n} \log \mu(r, f_j) = \max_{0 \le j \le n} \log \mu\left(r, \frac{f_0}{\mathbf{S}_0}\mathbf{S}_j\right)$$

$$= \log \mu(r, f_0) - \log \mu(r, \mathbf{S}_0) + \max_{0 \le j \le n} \log \mu(r, \mathbf{S}_j). \tag{5.6.4}$$

The Valiron-Jensen formula (5.1.10) gives

$$\log \mu(r, f_0) = N\left(r, \frac{1}{f_0}\right) - N(r, f_0) + O(1), \tag{5.6.5}$$

and

$$\begin{aligned}
- \log \mu(r, \mathbf{S}_0) &= -\log \mu(r, \mathbf{W}) + \sum_{i=1}^{n} \log \mu(r, f_i) \\
&= N(r, \mathbf{W}) - N\left(r, \frac{1}{\mathbf{W}}\right) \\
&\quad + \sum_{i=1}^{n} \left\{ N\left(r, \frac{1}{f_i}\right) - N(r, f_i) \right\} + O(1).
\end{aligned} \tag{5.6.6}$$

By Lemma 5.11, we obtain

$$\max_{0 \le j \le n} \log \mu(r, \mathbf{S}_j) \le -\frac{n(n-1)}{2} \log r. \tag{5.6.7}$$

Hence (5.6.2) follows from (5.6.4) to (5.6.7).   □

By using the Valiron-Jensen formula (5.1.10), we may rewrite the inequality (5.6.2) into the simple form

$$\max_{0 \le j \le n} \log \mu(r, f_j) \le N\left(r, \frac{1}{\mathbf{H}}\right) - N(r, \mathbf{H}) - \frac{n(n-1)}{2} \log r + O(1), \tag{5.6.8}$$

where

$$\mathbf{H} = \frac{f_0 f_1 \cdots f_n}{\mathbf{W}}. \tag{5.6.9}$$

By the definitions of the characteristic functions, we obtain

$$T\left(r, \frac{f_j}{f_0}\right) = N\left(r, \frac{f_j}{f_0}\right) + m\left(r, \frac{f_j}{f_0}\right),$$

and however,

$$
m\left(r, \frac{f_j}{f_0}\right) = \max\left\{0, \log \mu\left(r, \frac{f_j}{f_0}\right)\right\}
$$

$$
= \max\left\{\log \mu(r, f_0), \log \mu(r, f_j)\right\} - \log \mu(r, f_0). \quad (5.6.10)
$$

Applying Theorem 5.28 and using (5.6.10) and (5.6.5), we obtain the *non-Archimedean Nevanlinna third main theorem* as follows (cf. [176]):

Theorem 5.29. *Let $f_j (j = 1, \ldots, n)$ be linearly independent meromorphic functions on κ and define f_0 by (5.6.1). Then for $1 \leq j \leq n$,*

$$
T\left(r, \frac{f_j}{f_0}\right) \leq N\left(r, \frac{f_j}{f_0}\right) + \sum_{i=1}^{n}\left\{N\left(r, \frac{1}{f_i}\right) - N(r, f_i)\right\}
$$

$$
+ N(r, \mathbf{W}) - N\left(r, \frac{1}{\mathbf{W}}\right) - \frac{n(n-1)}{2}\log r + O(1), (5.6.11)
$$

where $\mathbf{W} = \mathbf{W}(f_1, \ldots, f_n)$ is the Wronskian of $f_1, \ldots, f_n$.

Similar to Theorem 4.4, we also have the truncated form of *Nevanlinna's third main theorem* (see [176], Corollary 2.27).

Theorem 5.30. *Suppose that $f_1, \ldots, f_n$ are linearly independent meromorphic functions in κ satisfying the equation (5.6.12) with $f_0 = 1$. Then for $j = 1, \ldots, n$, we have*

$$
T(r, f_j) < \sum_{i=1}^{n} N_{n-1}\left(r, \frac{1}{f_i}\right) + \overset{*}{N}_j(r) - \frac{n(n-1)}{2}\log r + O(1),
$$

where

$$
\overset{*}{N}_j(r) = \min\left\{\vartheta_n \sum_{i=1}^{n} \overline{N}(r, f_i), (n-1)\sum_{i \neq j} \overline{N}(r, f_i)\right\}.
$$

Next we replace the assumption of linear independence in Theorem 5.28 by a weaker hypothesis of no vanishing subsums. Write the equation (5.6.1) in the form

$$
-f_0 + f_1 + f_2 + \cdots + f_n = 0. \quad (5.6.12)
$$

Assume that no proper subsum of (5.6.12) is equal to 0. By Lemma 3.87, there exists a partition of indices

$$
\{0, 1, \ldots, n\} = I_0 \cup \cdots \cup I_k
$$

satisfying the properties (i) and (ii) in Lemma 3.87. Set

$$
n_0 + 1 = \#I_0; \quad n_\alpha = \#I_\alpha \ (\alpha = 1, \ldots, k)
$$

and write

$$s_\alpha = 1 + \sum_{\beta=0}^{\alpha} n_\beta, \quad \alpha = 0, 1, \ldots, k.$$

Then

$$n_0 + n_1 + \cdots + n_k = n.$$

Without loss of generality, we may assume that

$$I_0 = \{0, \ldots, n_0\}, \quad I_\alpha = \{s_{\alpha-1}, \ldots, s_\alpha - 1\} \ (\alpha = 1, \ldots, k).$$

Since I_0 is minimal, $f_1, \ldots, f_{n_0}$ are linearly independent. Hence

$$\mathbf{W}_0 = \mathbf{W}(f_1, \ldots, f_{n_0}) \not\equiv 0.$$

Similarly, the functions $f_{s_{\alpha-1}}, \ldots, f_{s_\alpha-1}$ are linearly independent, and so

$$\mathbf{W}_\alpha = \mathbf{W}\left(f_{s_{\alpha-1}}, \ldots, f_{s_\alpha-1}\right) \not\equiv 0, \quad \alpha = 1, \ldots, k.$$

Write

$$\mathbf{W} = \mathbf{W}_0 \cdots \mathbf{W}_k, \quad l = \sum_{\alpha=0}^{k} \frac{n_\alpha(n_\alpha - 1)}{2}, \quad w = \max_{0 \le \alpha \le k} \{n_\alpha - 1\} \qquad (5.6.13)$$

and similarly define $\mathbf{H}$ by using (5.6.9). Obviously, we have

$$w \le d - 1, \quad w \le l \le \frac{n(n-1)}{2}, \qquad (5.6.14)$$

where d is the dimension of the vector space spanned by the f_j over κ. According to the proof of Theorem 4.5 and Theorem 5.28, we can obtain the following results (cf. [182]):

Theorem 5.31. *Suppose that $f_0, f_1, \ldots, f_n$ are meromorphic functions in κ satisfying the equation (5.6.12). Assume that no proper subsum of (5.6.12) is equal to 0. Then we have*

$$\max_{0 \le j \le n} \log \mu(r, f_j) \le N\left(r, \frac{1}{\mathbf{H}}\right) - N(r, \mathbf{H}) - l \log r + O(1). \qquad (5.6.15)$$

Theorem 5.32. *Suppose that $f_0, f_1, \ldots, f_n$ are meromorphic functions in κ satisfying the equation (5.6.12). Assume that no proper subsum of (5.6.12) is equal to 0. Then for $1 \le j \le n$,*

$$T\left(r, \frac{f_j}{f_0}\right) \le N\left(r, \frac{f_j}{f_0}\right) + \sum_{i=1}^{n} \left\{ N\left(r, \frac{1}{f_i}\right) - N(r, f_i) \right\}$$

$$+ N(r, \mathbf{W}) - N\left(r, \frac{1}{\mathbf{W}}\right) - l \log r + O(1), \qquad (5.6.16)$$

where $\mathbf{W}, l$ are defined by (5.6.13).

5.7 The abc-theorem for entire functions

Let κ be an algebraically closed field of characteristic zero, complete for a nontrivial non-Archimedean absolute value $|\cdot|$. We give the analogue of Theorem 4.7 over κ which further yields counterparts of the main theorems in Section 4.2 and 4.4. The following result extends Stothers-Mason's Theorem 2.65, which can be regarded as an analogue of the abc-conjecture over $\mathcal{A}(\kappa)$.

Theorem 5.33 ([177]). *Let $a(z)$, $b(z)$ and $c(z)$ be entire functions in κ without common zeros and not all constants such that $a + b = c$. Then*

$$\max\{T(r,a), T(r,b), T(r,c)\} \leq \overline{N}\left(r, \frac{1}{abc}\right) - \log r + O(1). \tag{5.7.1}$$

Proof. Write

$$f = \frac{a}{c}, \quad g = \frac{b}{c}.$$

Then f and g are both not constant by our assumptions, and satisfy $f + g = 1$. By the second main theorem, and noting that

$$\overline{N}\left(r, \frac{1}{f-1}\right) = \overline{N}\left(r, \frac{1}{g}\right) = \overline{N}\left(r, \frac{1}{b}\right),$$

we obtain

$$
\begin{aligned}
T(r,f) &\leq \overline{N}(r,f) + \overline{N}\left(r, \frac{1}{f}\right) + \overline{N}\left(r, \frac{1}{f-1}\right) - \log r + O(1) \\
&= \overline{N}\left(r, \frac{1}{c}\right) + \overline{N}\left(r, \frac{1}{a}\right) + \overline{N}\left(r, \frac{1}{b}\right) - \log r + O(1) \\
&= \overline{N}\left(r, \frac{1}{abc}\right) - \log r + O(1).
\end{aligned}
$$

Similarly, we have

$$T(r,g) \leq \overline{N}\left(r, \frac{1}{abc}\right) - \log r + O(1).$$

By (5.1.15) and (5.1.8),

$$
\begin{aligned}
T(r,f) &= \max\left\{ N(r,f), N\left(r, \frac{1}{f}\right) \right\} + O(1) \\
&= \max\left\{ N\left(r, \frac{1}{c}\right), N\left(r, \frac{1}{a}\right) \right\} + O(1) \\
&= \max\{T(r,c), T(r,a)\} + O(1).
\end{aligned}
$$

Similarly,

$$T(r,g) = \max\{T(r,c), T(r,b)\} + O(1),$$

and, hence, the theorem follows from the above estimates. $\square$

Theorem 5.33 implies immediately that Theorem 2.65 holds over the field κ. Boutabaa and Escassut [34] claimed that the term $-N\left(r, \frac{f}{f'}\right)$ occurs in the right side of the inequality (5.7.1) by showing the inequality

$$N_{\mu_{f,\infty}}(r) + N_{\mu_{f,\mathscr{A}}}(r) \geq N\left(r, \frac{f}{f'}\right),$$

where $\mathscr{A} = \{0, 1\}$. However, this is wrong by a simple example,

$$f(z) = 1 - z^n, \quad g(z) = z^n.$$

We can prove that the term $-N\left(r, \frac{abc}{ab'-a'b}\right)$ occurs in the right side of the inequality (5.7.1) (see Theorem 5.34). If the field k has characteristic $p > 0$, noting that

$$e\left(\frac{a}{c}\right) = \min\{e(a), e(c)\}, \quad e(f) = e(g),$$

by applying Theorem 5.10 to the proof of Theorem 5.33, we can obtain

$$p^{-s} \max\{T(r, a), T(r, b), T(r, c)\} \leq \overline{N}\left(r, \frac{1}{abc}\right) - \log r + O(1), \qquad (5.7.2)$$

where $s = \min\{e(a), e(b), e(c)\}$ (cf. [34]). Thus Vaserstein's result [413] about polynomials follows from (5.7.2).

Next we will study the more general functional equation

$$f_1 + \cdots + f_n = f_0, \qquad (5.7.3)$$

and prove the *generalized abc-theorem* for entire functions over κ, which is an analogue of Theorem 4.7 over the non-Archimedean field.

Theorem 5.34 ([178]). *Let $f_j (j = 0, \ldots, n)$ be entire functions on κ such that $f_0, \ldots, f_n$ have no common zeros, $f_j (j = 1, \ldots, n)$ are linearly independent on κ and the equation (5.7.3) holds. Then we have*

$$\max_{0 \leq j \leq n} T(r, f_j) \leq \sum_{i=0}^{n} N_{n-1}\left(r, \frac{1}{f_i}\right) - N(r, \mathbf{H}) - \frac{n(n-1)}{2} \log r + O(1),$$

$$\max_{0 \leq j \leq n} T(r, f_j) \leq N_{\frac{n(n-1)}{2}}\left(r, \frac{1}{f_0 \cdots f_n}\right) - N(r, \mathbf{H}) - \frac{n(n-1)}{2} \log r + O(1).$$

Proof. By Theorem 5.28, we have

$$\max_{0 \leq j \leq n} \log \mu(r, f_j) \leq \sum_{i=0}^{n} N\left(r, \frac{1}{f_i}\right) - N\left(r, \frac{1}{\mathbf{W}}\right) - \frac{n(n-1)}{2} \log r + O(1). \quad (5.7.4)$$

The Valiron-Jensen formula (5.1.6) and (5.1.8) imply

$$\log \mu(r, f_j) = T(r, f_j) + O(1), \quad j = 0, 1, \ldots, n. \tag{5.7.5}$$

By the Valiron-Jensen formula (5.1.6), we also obtain

$$\sum_{i=0}^{n} N\left(r, \frac{1}{f_i}\right) - N\left(r, \frac{1}{\mathbf{W}}\right) = N\left(r, \frac{1}{\mathbf{H}}\right) - N(r, \mathbf{H}). \tag{5.7.6}$$

Thus Theorem 5.34 follows from (5.7.4), (5.7.5), (5.7.6) and the estimates

$$\sum_{i=0}^{n} \mu_{f_i}^0 - \mu_{\mathbf{W}}^0 \leq \sum_{i=0}^{n} \mu_{f_i, n-1}^0, \tag{5.7.7}$$

$$\sum_{i=0}^{n} \mu_{f_i}^0 - \mu_{\mathbf{W}}^0 \leq \mu_{f_0 \cdots f_n, \frac{n(n-1)}{2}}^0. \tag{5.7.8}$$

Take $z_0 \in \kappa$. Then $\mu_{f_s}^0(z_0) = 0$ for some $s \in \{0, \ldots, n\}$ since $f_0, \ldots, f_n$ have no common zeros. Note that, by the identity (5.7.3),

$$\mathbf{W} = \mathbf{W}(f_1, \ldots, f_{s-1}, f_0, f_{s+1}, \ldots, f_n).$$

Obviously we have

$$\mu_{f_i^{(j)}}^0(z_0) \geq \mu_{f_i}^0(z_0) - \mu_{f_i, j}^0(z_0) \geq \mu_{f_i}^0(z_0) - \mu_{f_i, n-1}^0(z_0), \quad i \neq s, \ 1 \leq j \leq n - 1,$$

and, hence,

$$\mu_{\mathbf{W}}^0(z_0) \geq \sum_{i \neq s} \{\mu_{f_i}^0(z_0) - \mu_{f_i, n-1}^0(z_0)\},$$

that is,

$$\sum_{i=0}^{n} \mu_{f_i}^0(z_0) - \mu_{\mathbf{W}}^0(z_0) = \sum_{i \neq s} \mu_{f_i}^0(z_0) - \mu_{\mathbf{W}}^0(z_0)$$

$$\leq \sum_{i \neq s} \mu_{f_i, n-1}^0(z_0) = \sum_{i=0}^{n} \mu_{f_i, n-1}^0(z_0).$$

The inequality (5.7.8) can be obtained similarly by comparing the multiplicities of zeros of $f_0 \cdots f_n$ and $\mathbf{W}$. Then Theorem 5.34 follows from (5.7.4), (5.7.5), (5.7.7) and (5.7.8). $\qquad\square$

Similarly, we can prove the following result:

Theorem 5.35 ([182]). *Let $f_j(j = 0, \ldots, n)$ be entire functions on κ satisfying the equation (5.7.3) in which no proper subsum is equal to 0 such that $f_0, \ldots, f_n$ have no common zeros. Then we have*

$$\max_{0 \le j \le n} T(r, f_j) \le \sum_{i=0}^{n} N_w\left(r, \frac{1}{f_i}\right) - N(r, \mathbf{H}) - l \log r + O(1),$$

$$\max_{0 \le j \le n} T(r, f_j) \le N_l\left(r, \frac{1}{f_0 \cdots f_n}\right) - N(r, \mathbf{H}) - l \log r + O(1),$$

where w, l are defined by (5.6.13).

Hence Theorem 4.8 and Theorem 4.22 also hold over the field κ. These further deliver supporting evidence for the generalized abc-conjectures in Section 4.3 and the generalized Hall's conjecture in Section 4.4.

5.8   Non-Archimedean Borel theorem

Let κ be an algebraically closed field of characteristic zero, complete for a non-trivial non-Archimedean absolute value $|\cdot|$. We prove a counterpart of Theorem 4.27 over κ, and further show that the characterization in Borel's theorem (cf. Corollary 4.33) is "universal", which supports the corresponding conjecture in number theory.

Theorem 5.36. *Take positive integers n, d_j $(j = 0, 1, \ldots, n)$ with*

$$\alpha = 1 - \sum_{j=0}^{n} \frac{n-1}{d_j} \ge 0.$$

Let f_j $(j = 0, 1, \ldots, n)$ be non-zero meromorphic functions on κ satisfying the following condition:

$$\overline{\mu}_{f_j)d-1}^{\infty} = \overline{\mu}_{f_j)d_j-1}^{0} = 0 \ (j = 0, 1, \ldots, n),$$

where $d = \max\{d_j\}$. If the function

$$T(r) = \min_{j \ne k}\left\{T\left(r, \frac{f_j}{f_k}\right)\right\}$$

is unbounded, then $f_0, \ldots, f_n$ are linearly independent.

Proof. We prove Theorem 5.36 by induction on n. First of all, we consider the case $n = 1$. Assume that there are $\{a_0, a_1\} \subset \kappa$ such that $a_0 f_0 + a_1 f_1 = 0$. Since $f_j \ne 0$ $(j = 0, 1)$, then $a_0 \ne 0$ and $a_1 \ne 0$ if one of a_0 and a_1 is not zero. Hence

$$T(r) \le T\left(r, \frac{f_0}{f_1}\right) = T\left(r, \frac{a_1}{a_0}\right) = O(1)$$

which is a contradiction. Hence $a_0 = a_1 = 0$, that is, f_0, f_1 are linearly independent.

Assume that Theorem 5.36 holds up to $n-1$. We will show that Theorem 5.36 holds too for the case $n \geq 2$. Assume that there are $\{a_0, \dots, a_n\} \subset \kappa$ such that $a_0 f_0 + \dots + a_n f_n = 0$. It is sufficient to show that one of $a_0, \dots, a_n$ is zero. Assume, to the contrary, that $a_j \neq 0$ for $j = 0, 1, \dots, n$. Then $f_0, \dots, f_{n-1}$ are linearly independent over κ. Let V be a vector space of dimension n over κ. Take a base $e_0, \dots, e_{n-1}$ of V and let $\epsilon_0, \dots, \epsilon_{n-1}$ be the dual base in V^*. Let $\Delta (\neq 0)$ be a *universal denominator* of $\{f_0, \dots, f_{n-1}\}$, that is, Δf_i is holomorphic for each $i = 0, \dots, n - 1$ such that $\Delta f_0, \dots, \Delta f_{n-1}$ have no common zeros. Then a meromorphic mapping $F : \kappa \longrightarrow \mathbb{P}(V)$ is defined with a reduced representation

$$\tilde{F} = \Delta a_0 f_0 e_0 + \dots + \Delta a_{n-1} f_{n-1} e_{n-1} : \kappa \longrightarrow V.$$

Obviously, F is linearly non-degenerate. Set

$$b_i = \mathbb{P}(\epsilon_i) \ (0 \leq i \leq n - 1), \quad b_n = \mathbb{P}\left(\sum_{i=0}^{n-1} \epsilon_i\right).$$

Then the family $\{b_0, \dots, b_n\}$ is in general position. Then Theorem 5.13 implies

$$T(r, F) \leq \sum_{j=0}^{n} N_{F,n-1}(r, b_j) - \frac{n(n-1)}{2} \log r + O(1), \tag{5.8.1}$$

where, by definition,

$$N_{F,n-1}(r, b_i) = N_{n-1}\left(r, \frac{1}{\Delta f_i}\right).$$

By the formulae (5.1.15) and (5.3.6), for each $i = 1, \dots, n - 1$, we have

$$T(r) \leq T\left(r, \frac{f_i}{f_0}\right) \leq \max\left\{N\left(r, \frac{1}{\Delta f_0}\right), N\left(r, \frac{1}{\Delta f_i}\right)\right\} \leq T(r, F) + O(1).$$

According to the definition of Δ, it is easy to show that the inequality

$$\overline{N}_{d_i-1)}\left(r, \frac{1}{\Delta f_i}\right) \leq \overline{N}_{d_i-1)}\left(r, \frac{1}{f_i}\right) + \sum_{j=0}^{n-1} \overline{N}_{d-1)}(r, f_j) = 0$$

holds for each $i \in \{0, 1, \dots, n\}$. Note that

$$\overline{N}\left(r, \frac{1}{\Delta f_i}\right) = \overline{N}_{d_i-1)}\left(r, \frac{1}{\Delta f_i}\right) + \overline{N}_{(d_i}\left(r, \frac{1}{\Delta f_i}\right)$$

$$= \overline{N}_{(d_i}\left(r, \frac{1}{\Delta f_i}\right),$$

and

$$d_i \overline{N}_{(d_i} \left(r, \frac{1}{\Delta f_i} \right) \le N \left(r, \frac{1}{\Delta f_i} \right) = N_F(r, b_i) \le T(r, F) + O(1)$$

hold for $i = 0, \ldots, n$. Hence for $i = 0, \ldots, n$, we obtain

$$N_{F,n-1}(r, b_i) \le (n-1) \overline{N} \left(r, \frac{1}{\Delta f_i} \right)$$

$$\le (n-1) \overline{N}_{(d_i} \left(r, \frac{1}{\Delta f_i} \right) \le \frac{n-1}{d_i} T(r, F) + O(1).$$

Therefore the inequality (5.8.1) yields

$$\alpha T(r, F) + \frac{n(n-1)}{2} \log r \le O(1).$$

This is a contradiction since $\alpha \ge 0$, $n \ge 2$.   □

Corollary 5.37. *Let $f_0, \ldots, f_n$ $(n \ge 2)$ be non-zero meromorphic functions on κ such that there are constants $a_i \in \kappa$ satisfying the equation $a_0 f_0 + \cdots + a_n f_n = 0$. Assume that there exist positive integers d_i $(i = 0, \ldots, n)$ such that*

$$1 - \sum_{i=0}^{n} \frac{n-1}{d_i} \ge 0, \quad \overline{\mu}_{f_i)d-1}^{\infty} = \overline{\mu}_{f_i)d_i-1}^{0} = 0 \ (i = 0, \ldots, n),$$

where $d = \max d_i$. Then there exists a partition of indices

$$\{0, 1, \ldots, n\} = I_0 \cup I_1 \cup \cdots \cup I_k$$

such that $I_\alpha \ne \emptyset$ $(\alpha = 0, 1, \ldots, k)$, $I_\alpha \cap I_\beta = \emptyset$ $(\alpha \ne \beta)$,

$$\sum_{i \in I_\alpha} a_i f_i = 0, \quad \alpha = 0, 1, \ldots, k,$$

and f_i/f_j is constant for any $i, j \in I_\alpha$. In particular, if $a_i \ne 0$ for $i = 0, 1, \ldots, n$, each I_α contains at least two indices.

Proof. Consider the partition $\{0, 1, \ldots, n\} = I_0 \cup I_1 \cup \cdots \cup I_k$ such that two indices i and j are in the same class if and only if f_i/f_j is constant. Then we have

$$\sum_{i=0}^{n} a_i f_i = \sum_{\alpha=0}^{k} \sum_{i \in I_\alpha} a_i f_i = \sum_{\alpha=0}^{k} c_\alpha' f_{i_\alpha} = 0$$

for any fixed $i_\alpha \in I_\alpha$ and some $c_\alpha' \in \kappa$. Corollary 5.37 is trivial for the case $k = 0$. Suppose $k \ge 1$. Then

$$1 - \sum_{\alpha=0}^{k} \frac{k-1}{d_{i_\alpha}} \ge 1 - \sum_{j=0}^{n} \frac{n-1}{d_j} \ge 0.$$

By Theorem 5.36, we obtain $c_\alpha' = 0$ for $\alpha = 0, 1, \ldots, k$, which yields Corollary 5.37.
     □

Theorem 5.38 ([169]). *Take positive integers n and d with $n \geq 2$, $d \geq n^2 - 1$. The following three statements are equivalent:*

(i) *If meromorphic functions $f_1, f_2, \ldots, f_n$ on κ satisfy*

$$f_1^d + f_2^d + \cdots + f_n^d = 1,$$

then $f_1^d, f_2^d, \ldots, f_n^d$ are linearly dependent;

(ii) *Under the same assumption as in (i), then at least one of the f_i's is constant;*

(iii) *Assume that entire functions $f_0, f_1, \ldots, f_n$ on κ satisfy*

$$f_0^d + f_1^d + \cdots + f_n^d = 0.$$

Partition the index set $I = \{0, 1, \ldots, n\}$ into subsets I_α, $I = \cup_{\alpha=0}^k I_\alpha$, putting two indices i and j in the same subset I_α if and only if f_i/f_j is constant. Then we have

$$\sum_{i \in I_\alpha} f_i^d = 0, \quad \alpha = 0, \ldots, k.$$

Proof. The claim (iii) follows directly from Corollary 5.37. Next we show (ii) $\Rightarrow$ (iii) $\Rightarrow$ (i) $\Rightarrow$ (ii). In order to derive (iii) from (ii), we have

$$\sum_{i=0}^n f_i^d = \sum_{\alpha=0}^k \sum_{i \in I_\alpha} f_i^d = \sum_{\alpha=0}^k a_\alpha f_{i_\alpha}^d = 0$$

for some $a_\alpha \in \mathbb{C}$ and any fixed $i_\alpha \in I_\alpha$. Then (iii) follows if $a_\alpha = 0$ for $\alpha = 0, \ldots, k$. Assume that (iii) does not hold. Without loss of generality, we may assume that

$$a_\alpha \neq 0 \ (0 \leq \alpha \leq s, \ s \geq 1), \quad a_\alpha = 0 \ (s + 1 \leq \alpha \leq k).$$

Choose $\xi_\alpha \in \kappa$ such that

$$\xi_\alpha^d = -\frac{a_\alpha}{a_0}, \quad \alpha = 1, \ldots, s$$

and set

$$g_\alpha = \xi_\alpha \frac{f_{i_\alpha}}{f_{i_0}}, \quad \alpha = 1, \ldots, s$$

so that $g_1^d + \cdots + g_s^d = 1$. Then one of the g_α's, say g_1, is constant by (ii). This means that f_{i_1}/f_{i_0} is constant, contradicting the definition of I_α.

We derive (i) from (iii). We can choose $\xi \in \kappa$ and entire functions $F_0, \ldots, F_n$ in κ such that

$$\xi^d = -1, \quad f_i = \frac{F_i}{\xi F_0}, \quad i = 1, \ldots, m$$

so that $F_0^d + F_1^d + \cdots + F_n^d = 0$, and apply (iii) to this identity. Let I_0 be the index set that contains 0. If $I = I_0$, then the functions $f_1, \ldots, f_n$ are all constant and hence linearly dependent. If $I \neq I_0$, then

$$\sum_{i \in I_\alpha} F_i^d = 0, \quad \alpha \neq 0,$$

thus yielding a non-trivial linear relation of $\{f_i^d \mid i \in I_\alpha\}$.

Finally, we derive (ii) from (i) as follows: Since the functions $f_1^d, \ldots, f_n^d$ are linearly dependent, without loss of generality, we may assume the following linear relation:

$$c_1 f_1^d + \cdots + c_{n-1} f_{n-1}^d + f_n^d = 0.$$

By subtracting this identity from $f_1^d + \cdots + f_n^d = 1$, we have

$$(1 - c_1) f_1^d + \cdots + (1 - c_{n-1}) f_{n-1}^d = 1.$$

We could use the relation to get a shorter linear combination of the f_i^d to equal 1, and hence (i) can be used again. Finally, it follows that a constant c exists such that $c f_i^d = 1$ for some i. $\qquad\square$

Theorem 5.38, (iii) was proved by Hà, Huy Khóai and Mai, Van Tu [139], which is a non-Archimedean version of Green's theorem [121]. Note that, in the complex case, the hypothesis is $d \geq n^2$, but $d \geq n^2 - 1$ for the non-Archimedean case. Similarly, according to the proof of Theorem 4.34, it is not difficult to give the analogue of Theorem 4.34 over κ:

Theorem 5.39. *Let $P_j(x_0, \ldots, x_n)$ be a homogeneous polynomial of degree δ_j over κ for $0 \leq j \leq n$. Let $f_0, \ldots, f_n$ be holomorphic functions on κ satisfying the following equation:*

$$\sum_{j=0}^{n} f_j^{d-\delta_j} P_j(f_0, \ldots, f_n) = 0.$$

There is a non-trivial linear relation among $f_1^{d-\delta_1} P_1(f_0, \ldots, f_n), \ldots, f_n^{d-\delta_n} P_n(f_0, \ldots, f_n)$ if

$$d \geq n^2 - 1 + \sum_{j=0}^{n} \delta_j.$$

5.9   Waring's problem over function fields

Let κ be an algebraically closed field of characteristic zero, complete for a non-trivial non-Archimedean absolute value $|\cdot|$. For Waring's problem in a family $\mathcal{F}$ of $\mathcal{M}(\kappa)$, it is interesting that, for any fixed positive integer d, one can find the smallest integer $n = g_{\mathcal{F}}(d)$ such that there exist non-constant functions $f_1, \ldots, f_n$ in $\mathcal{F}$ satisfying (4.7.2), that is,

$$f_1(z)^d + \cdots + f_n(z)^d = z.$$

Theorem 5.40 ([169]). *The number* $g_{\mathcal{F}}(d)$ *satisfies the following inequalities:*

$$g_{\mathcal{A}(\kappa)}(d) > \frac{1}{2} + \sqrt{d + \frac{1}{4}}, \quad d \geq 3, \tag{5.9.1}$$

$$g_{\mathcal{M}(\kappa)}(d) > \sqrt{d + 1}, \quad d \geq 3. \tag{5.9.2}$$

Proof. Assume that there exist non-constant meromorphic functions $f_1, \ldots, f_n$ satisfying (4.7.2). Assume, to the contrary, that (5.9.1) and (5.9.2) are not true. First of all, we consider the case that $f_1^d, \ldots, f_n^d$ are linearly independent.

If $f_1, \ldots, f_n$ are entire functions, then by the assumption we have

$$n \leq \frac{1}{2} + \sqrt{d + \frac{1}{4}},$$

that is, $d \geq n(n-1)$. Obviously, $n \geq 2$ since $d \geq 3$. By the equation (4.7.2), it is easy to see that $f_1, \ldots, f_n$ have no common zeros. Hence a non-degenerate holomorphic curve

$$F : \kappa \longrightarrow \mathbb{P}(\kappa^n)$$

is well defined with a reduced representation $\tilde{F} = (\tilde{F}_1, \ldots, \tilde{F}_n)$, where

$$\tilde{F}_j = f_j^d, \ j = 1, \ldots, n.$$

By Theorem 5.13, we obtain

$$T(r, F) \leq \sum_{j=0}^{n} N_{n-1}\left(r, \frac{1}{\tilde{F}_j}\right) - \frac{n(n-1)}{2}\log r + O(1)$$

$$\leq \log r + \sum_{j=1}^{n} \frac{n-1}{d} N\left(r, \frac{1}{\tilde{F}_j}\right) - \frac{n(n-1)}{2}\log r + O(1)$$

$$\leq \left(1 - \frac{n(n-1)}{2}\right)\log r + \frac{n(n-1)}{d} T(r, F) + O(1),$$

where $\tilde{F}_0(z) = z$, and hence

$$\left(1 - \frac{n(n-1)}{d}\right) T(r, F) \leq \left(1 - \frac{n(n-1)}{2}\right)\log r + O(1). \tag{5.9.3}$$

Since $T(r, F) \to \infty$ as $r \to \infty$, the inequality (5.9.3) is not true since $d \geq n(n-1)$. Thus (5.9.1) is proved for the special case.

When $f_1, \ldots, f_n$ are meromorphic functions, by the assumption we have $d \geq n^2 - 1 \geq 3$. There exist entire functions $h, g_1, \ldots, g_n$ such that

$$f_j = \frac{g_j}{h}, \ j = 1, \ldots, n.$$

Note that h can be chosen such that $g_1, \ldots, g_n$ have no common zeros. Now (4.7.2) becomes

$$g_1(z)^d + \cdots + g_n(z)^d = zh(z)^d. \tag{5.9.4}$$

Also a non-degenerate holomorphic curve $G : \kappa \longrightarrow \mathbb{P}(\kappa^n)$ is well defined with a reduced representation $\tilde{G} = (\tilde{G}_1, \ldots, \tilde{G}_n)$, where

$$\tilde{G}_j = g_j^d, \quad j = 1, \ldots, n.$$

Set $\tilde{G}_0(z) = zh(z)^d$. Then Theorem 5.13 implies

$$T(r, G) \leq \sum_{j=0}^{n} N_{n-1}\left(r, \frac{1}{\tilde{G}_j}\right) - \frac{n(n-1)}{2} \log r + O(1)$$

$$\leq \frac{n-1}{d} N\left(r, \frac{1}{h^d}\right) + \sum_{j=1}^{n} \frac{n-1}{d} N\left(r, \frac{1}{\tilde{G}_j}\right)$$

$$+ \left(1 - \frac{n(n-1)}{2}\right) \log r + O(1)$$

$$\leq \left(1 - \frac{n-1}{d} - \frac{n(n-1)}{2}\right) \log r$$

$$+ \frac{(n+1)(n-1)}{d} T(r, G) + O(1),$$

and hence

$$\left(1 - \frac{(n+1)(n-1)}{d}\right) T(r, G) \leq \left(1 - \frac{n-1}{d} - \frac{n(n-1)}{2}\right) \log r + O(1). \tag{5.9.5}$$

Since $T(r, G) \to \infty$ as $r \to \infty$, the inequality (5.9.5) is not true sine $d \geq (n+1)(n-1)$. Thus (5.9.2) is proved for the special case.

Finally, we study the case that $f_1^d, \ldots, f_n^d$ are linearly dependent. Without loss of generality, we may assume that $f_1^d, \ldots, f_l^d$ are linearly independent ($1 \leq 1 < n$), but $f_1^d, \ldots, f_l^d, f_j^d$ are linearly dependent for each $j = l+1, \ldots, n$. Thus there is a $(a_1, \ldots, a_l) \in \kappa^l - \{0\}$ such that (4.7.2) becomes the following form:

$$a_1 f_1(z)^d + \cdots + a_l f_l(z)^d = z. \tag{5.9.6}$$

We may assume $a_i \neq 0$ for each $i = 1, \ldots, l$, otherwise, deleting the terms with null coefficients in (5.9.6). Thus the proof of Theorem 5.40 can be completed by applying the above arguments to the equation (5.9.6). $\qquad\square$

By considering $(d-1)$th differences of z^d it can be shown [147] that z is representable as the sum of d dth powers. The construction is as follows:

$$\Delta^{d-1} z^d = \sum_{i=0}^{d-1} \binom{d-1}{i} (-1)^{d-1-i} (z+i)^d = (d-1)! z + a.$$

Setting $z = (w - a)/(d - 1)!$, we have the desired representation. Thus

$$d \geq g_{\kappa[z]}(d) > \frac{1}{2} + \sqrt{d + \frac{1}{4}}, \quad d \geq 3.$$

We may ask for the smallest number $n = G_{\mathcal{F}}(d)$ such that there exist non-constant functions $f_1, \ldots, f_n$ in the class $\mathcal{F}$ such that

$$f_1^d + \cdots + f_n^d = 1. \tag{5.9.7}$$

Following the proof of Theorem 5.40, we can obtain the result:

Theorem 5.41 ([169]). *Over the field κ, the number $G_{\mathcal{F}}(d)$ satisfies the following inequalities:*

$$G_{\mathcal{A}(\kappa)}(d) > \frac{1}{2} + \sqrt{d + \frac{1}{4}}, \quad d \geq 3, \tag{5.9.8}$$

$$G_{\mathcal{M}(\kappa)}(d) > \sqrt{d + 1}, \quad d \geq 3. \tag{5.9.9}$$

Some special cases of Theorem 5.41 can be found in [33], [35], and [176]. For the case $d = 2$, we can prove easily that there do not exist two non-constant entire functions f_1 and f_2 on κ satisfying

$$f_1^2 + f_2^2 = 1.$$

Hence (5.9.8) and (5.9.9) are true for $d = 2$ (cf. [176]).

5.10 Picard-Berkovich's theorem

Let κ be an algebraically closed field of characteristic $p \geq 0$, complete for a non-trivial non-Archimedean absolute value $|\cdot|$. Boutabaa and Escassut [34], [35] obtained the following two results:

Theorem 5.42. *Let A and B be two relatively prime polynomials of degrees k and q over κ, respectively, and assume that B has no factor whose power is multiple of p. Let t be the number of distinct zeros of B, and let $g \in \mathcal{M}(\kappa)$ be such that all poles of g have order either multiple of p, or $\geq m \geq 1$, except maybe a finite number l of them. Suppose that there exists a function $f \in \mathcal{M}(\kappa)$ with $f' \neq 0$ satisfying*

$$g(z)B(f(z)) = A(f(z)), \quad z \in \kappa.$$

Then

(a) *$mt \leq q + m$ if $f \in \mathcal{A}(\kappa) - \kappa[z]$;*

(b) *$mt \leq q + 2m$ if $f \in \mathcal{M}(\kappa) - \kappa(z)$. Moreover, if $k > q$, then $mt \leq \min\{q + 2m, k + m\}$;*

(c) $mt < q + 2m$ if $f \in \mathcal{M}(\kappa)$ and $l = 0$ or 1. Moreover, if $k > q$, then $mt < \min\{q + 2m, k + m\}$;

(d) $mt < q + m$ if $f \in \mathcal{A}(\kappa)$ and $l = 0$ or 1.

Proof. The inequalities are trivial if $t < 2$. So we may suppose $t \geq 2$. Write

$$B(z) = (z - b_1)^{q_1} \cdots (z - b_t)^{q_t}.$$

Since A and B have no common zeros, each zero z_0 of $B(f(z))$ is a pole of $g(z)$. Hence z_0 is a zero of $B(f(z))$ of order either multiple of p, or $\geq m$ except maybe a finite number l of them. Note that $f(z_0) - b_j = 0$ for some $j \in \{1, \dots, t\}$. Then when $\mu_{B(f)}^0(z_0) \geq m$, we also have

$$\frac{q_j}{m} \mu_f^{b_j}(z_0) = \frac{1}{m} \mu_{B(f)}^0(z_0) \geq 1.$$

Let l_j be the number of zeros of $f - b_j$ whose order in $B(f(z))$ is neither a multiple of p nor superior or equal to m. At such a point z_0, we have $\mu_f^{b_j}(z_0) > 0$, and so

$$1 \leq \frac{q_j}{m} \mu_f^{b_j}(z_0) + 1 - \frac{q_j}{m} \leq \frac{q_j}{m} \mu_f^{b_j}(z_0) + 1 - \frac{1}{m}.$$

Therefore

$$\overline{N}\left(r, \frac{1}{f - b_j}\right) \leq \frac{q_j}{m} N\left(r, \frac{1}{f - b_j}\right) + l_j \left(1 - \frac{1}{m}\right) \log r + O(1)$$

$$\leq \frac{q_j}{m} T(r, f) + l_j \left(1 - \frac{1}{m}\right) \log r + O(1),$$

where the last inequality is obtained by the first main theorem, and hence

$$\sum_{j=1}^{t} \overline{N}\left(r, \frac{1}{f - b_j}\right) \leq \frac{q}{m} T(r, f) + \frac{l(m-1)}{m} \log r + O(1). \qquad (5.10.1)$$

On the other hand, by using the second main theorem, one has

$$(t - 1) T(r, f) \leq \overline{N}(r, f) + \sum_{j=1}^{t} \overline{N}\left(r, \frac{1}{f - b_j}\right) - \log r + O(1).$$

The estimate (5.10.1) implies

$$(t - 1) T(r, f) \leq \overline{N}(r, f) + \frac{q}{m} T(r, f) + \left(\frac{l(m-1)}{m} - 1\right) \log r + O(1), \quad (5.10.2)$$

which means

$$\frac{mt - m - q}{m} T(r, f) \leq \overline{N}(r, f) + \left(\frac{l(m-1)}{m} - 1\right) \log r + O(1). \qquad (5.10.3)$$

In particular, this yields

$$\frac{mt - 2m - q}{m}T(r, f) \leq \left(\frac{l(m-1)}{m} - 1\right)\log r + O(1). \tag{5.10.4}$$

If $f \in \mathcal{A}(\kappa) - \kappa[z]$ (resp., $f \in \mathcal{M}(\kappa) - \kappa(z)$), it follows from (5.10.3) (resp, (5.10.4)) that $mt \leq m + q$ (resp., $mt \leq 2m + q$) since

$$\lim_{r \to \infty} \frac{T(r, f)}{\log r} = +\infty.$$

Therefore we have proven (a) and the first part of (b).

Next suppose that $k > q$. Then each pole of f also is a pole of g with $\mu_g^\infty = (k - q)\mu_f^\infty$. Hence

$$\overline{N}(r, f) \leq \frac{k - q}{m}N(r, f) \leq \frac{k - q}{m}T(r, f).$$

Thus the inequality (5.10.3) yields

$$\frac{mt - m - k}{m}T(r, f) \leq \left(\frac{l(m-1)}{m} - 1\right)\log r + O(1). \tag{5.10.5}$$

If $f \in \mathcal{M}(\kappa) - \kappa(z)$, it follows from (5.10.5) that $mt \leq m + k$, and so the second part of (b) is proved.

Suppose $f \in \mathcal{M}(\kappa)$ and that $0 \leq l \leq 1$. Since the term about $\log r$ is negative as r is sufficiently large, then $mt < 2m + q$ follows from (5.10.4). In the same way, if $k > q$, we obtain $mt < m + k$ by (5.10.5), which proves (c).

Suppose $f \in \mathcal{A}(\kappa)$ and that $0 \leq l \leq 1$. Since the term about $\log r$ is negative as r is sufficiently large, it follows from (5.10.3) that $mt < m + q$, which proves (d). $\qquad\square$

Theorem 5.43. *Given two relatively prime polynomials*

$$A(z) = a\prod_{i=1}^{s}(z - a_i)^{k_i}, \quad B(z) = b\prod_{j=1}^{t}(z - b_j)^{q_j}$$

of respective degrees k and q over κ, where all a_i and b_j are distinct. Let $m \in \mathbb{Z}^+$ be relatively prime with p. Suppose that there exist two functions $f, g \in \mathcal{M}(\kappa)$ satisfying

$$(g(z))^m B(f(z)) = A(f(z)), \quad z \in \kappa.$$

Then both f and g are constant if

1) $f \in \mathcal{A}(\kappa)$ *and if*

$$s + t > 1 + \frac{1}{m}\left(\sum_{i=1}^{s}(m, k_i) + \sum_{j=1}^{t}(m, q_j)\right),$$

where (m, n) is the largest common factor of m and n, or

2) $f, g \in \mathcal{M}(\kappa)$ *and if*

$$s + t \geq 1 + \frac{1}{m}\left((m, |k - q|_\infty) + \sum_{i=1}^{s}(m, k_i) + \sum_{j=1}^{t}(m, q_j)\right).$$

Proof. It is clear that if f is constant so is g. Suppose, to the contrary, that f is not constant. Then

$$k = \deg(A) = \sum_{i=1}^{s} k_i, \quad q = \deg(B) = \sum_{j=1}^{t} q_j.$$

If f' is identically zero, then so is g'. Let

$$f_1 = \sqrt[p]{f}, \ g_1 = \sqrt[p]{g}, \ A_1 = \Theta(A), \ B_1 = \Theta(B),$$

where Θ is defined by (5.2.4). Then we can see that

$$(g_1(z))^m B_1(f_1(z)) = A_1(f_1(z)),$$

and that A_1, B_1, f_1, g_1 respectively respect the hypotheses of A, B, f, g. So we are led to the same problem, and therefore we can go on $e(f)$ times, until we get a similar equation where the function playing the role of f has a non-identically zero derivative. Thus we can assume that f' is not identically zero.

Obviously, we have

$$k_i \mu_f^{a_i} = m \mu_g^0, \quad i = 1, \ldots, s,$$

and

$$q_j \mu_f^{b_j} = m \mu_g^\infty, \quad j = 1, \ldots, t.$$

It follows that

$$\mu_f^{a_i} \geq \frac{m}{(m, k_i)}, \quad i = 1, \ldots, s,$$

and

$$\mu_f^{b_j} \geq \frac{m}{(m, q_j)}, \quad j = 1, \ldots, t.$$

Hence we obtain

$$\overline{N}\left(r, \frac{1}{f - a_i}\right) \leq \frac{(m, k_i)}{m} N\left(r, \frac{1}{f - a_i}\right) \leq \frac{(m, k_i)}{m} T(r, f) + O(1), \quad i = 1, \ldots, s,$$

and

$$\overline{N}\left(r, \frac{1}{f - b_j}\right) \leq \frac{(m, q_j)}{m} N\left(r, \frac{1}{f - b_j}\right) \leq \frac{(m, q_j)}{m} T(r, f) + O(1), \quad j = 1, \ldots, t.$$

By using the second main theorem, one has

$$
\begin{aligned}
(s+t-1)T(r,f) \;\leq\; & \sum_{i=1}^{s} \overline{N}\left(r, \frac{1}{f-a_i}\right) + \sum_{j=1}^{t} \overline{N}\left(r, \frac{1}{f-b_j}\right) \\
& + \overline{N}(r,f) - \log r + O(1) \\
\leq\; & \frac{1}{m}\left(\sum_{i=1}^{s}(m,k_i) + \sum_{j=1}^{t}(m,q_j)\right) T(r,f) \\
& + \overline{N}(r,f) - \log r + O(1). \qquad (5.10.6)
\end{aligned}
$$

Note that if $\mu_f^\infty(z_0) > 0$,

$$
(k-q)\mu_f^\infty(z_0) = m\left(\mu_g^\infty(z_0) - \mu_g^0(z_0)\right).
$$

We have

$$
\overline{N}(r,f) \leq \frac{(m,|k-q|_\infty)}{m} N(r,f) \leq \frac{(m,|k-q|_\infty)}{m} T(r,f),
$$

which with (5.10.6) yield

$$
\begin{aligned}
(s+t-1)T(r,f) \;\leq\; & \frac{1}{m}\left((m,|k-q|_\infty) + \sum_{i=1}^{s}(m,k_i) + \sum_{j=1}^{t}(m,q_j)\right) T(r,f) \\
& - \log r + O(1). \qquad (5.10.7)
\end{aligned}
$$

Then (5.10.7) implies

$$
s+t-1 < \frac{1}{m}\left((m,|k-q|_\infty) + \sum_{i=1}^{s}(m,k_i) + \sum_{j=1}^{t}(m,q_j)\right) \qquad (5.10.8)
$$

since $\log r \to +\infty$, which contradicts with the hypothesis in 2). Hence f (and so g) is constant.

In particular, if f lies in $\mathcal{A}(\kappa)$, then (5.10.6) implies

$$
s+t-1 \leq \frac{1}{m}\left(\sum_{i=1}^{s}(m,k_i) + \sum_{j=1}^{t}(m,q_j)\right).
$$

This is a contradiction with the hypothesis in 1). Hence f is constant. $\qquad \square$

Particularly, an elliptic curve defined by Weierstrass equation (3.1.3) under the condition (3.1.4) has no non-constant meromorphic solutions $x, y \in \mathcal{M}(\kappa)$. Theorem 5.43 yields easily the following fact:

Corollary 5.44. *Let M be an algebraic curve of genus 1 or 2 on κ and let $f, g \in \mathcal{M}(\kappa)$ be such that $(f(z), g(z)) \in M$ when $z \in \kappa$. Then f and g are constant.*

In fact, according to Picard [316] (or Theorem 3.7, Proposition 3.23), every algebraic curve of genus 1 (resp. 2) is birationally equivalent to a smooth elliptic (resp. hyperelliptic) curve. Hence one can apply Theorem 5.43 with $m = 2, \deg(B) = 0, \deg(A) = s = 3$ and $m = 2, \deg(B) = 0, \deg(A) = s \geq 4$ in Corollary 5.44, respectively (see [35]). According to the proof of Theorem 4.60, we can obtain the following result:

Theorem 5.45. *Suppose that $P(z), Q(z) \in \kappa[z]$ are respectively two polynomials of degrees k and q with $q \geq k \geq 2$ and $P'Q'$ not identically 0. Let S be a non-empty subset of zeros of P' and let T be the set of zeros of Q' such that $P(S) \cap Q(T) = \emptyset$. If there exist two meromorphic functions f and g in κ with $f' \neq 0$ satisfying*

$$P(f) = Q(g), \tag{5.10.9}$$

then

$$\frac{lq - k}{q} T(r, f) \leq \overline{N}(r, f) - \log r + O(1), \tag{5.10.10}$$

where $l = \sum_{c \in S} \mu_{P'}^0(c)$. Further, we have $q = k$, $l = 1$, and S has only one element which is a simple zero of P'.

Theorem 5.45 improves a result in [96]. Note that when $p > 0$, if there exist two non-constant meromorphic functions f and g in κ satisfying (5.10.9), then f and g have the same ramification index s and

$$\sqrt[p^s]{P}\left(\sqrt[p^s]{f}\right) = \sqrt[p^s]{Q}\left(\sqrt[p^s]{g}\right).$$

Thus when $p \neq 2$, 3, we obtain Corollary 5.44 again by using Theorem 5.45.

Picard-Berkovich's theorem claims that the conclusion in Corollary 5.44 holds if the genus of M is not less than 1:

Theorem 5.46 (Berkovich [21]). *Let M be an irreducible curve of genus ≥ 1 of equation $F(x, y) = 0$, where $F(x, y)$ is an irreducible polynomial in two variables with coefficients in κ. There do not exist non-constant meromorphic functions f and g in κ such that $F(f(z), g(z))$ is identically zero.*

Cherry [58] shows that this also holds if M is an Abelian variety, and obtains a non-Archimedean analogue of Bloch's conjecture. Further, in [59] Cherry proves that each Abelian variety over κ carries a Kobayashi distance. Here we suggest an analogue of Green-Griffiths' conjecture (cf. Chapter 6) as follows:

Conjecture 5.47. *If $f : \kappa \longrightarrow M$ is a holomorphic curve into a non-singular pseudo canonical projective variety, then the image of f is contained in a proper algebraic subvariety.*

Finally, we list two analogues of Theorem 4.67 and Theorem 4.68 in non-Archimedean fields:

Theorem 5.48 ([426]). *Let $P \in \kappa[z]$ be of degree n, have no multiple zeros, and*

$$P'(z) = c(z - c_1)^{m_1} \cdots (z - c_l)^{m_l},$$

where c is a non-zero constant, such that

$$c_i \neq c_j, \ P(c_i) \neq P(c_j), \ 1 \leq i \neq j \leq l.$$

Moreover, if $p > 0$, we assume that the multiplicity of the factor $z - c_i$ in $P(z) - P(c_i)$ is $m_i + 1$ for $1 \leq i \leq l$, and if $p \mid n$, we also assume that the coefficient of z^{n-1} in $P(z)$ is not zero. Then the following are equivalent:

(i) *P is a uniqueness polynomial for $\mathcal{M}(\kappa)$;*

(ii) *P is a uniqueness polynomial for $\kappa(z)$;*

(iii) *$(n - 2)(n - 3) > \sum_{i=1}^{l} m_i(m_i - 1)$;*

(iv) *$l \geq 2$ if $p \mid n$; and $l \geq 3$ or $l = 2$ and $\min\{m_1, m_2\} \geq 2$ if $p = 0$ or $p \nmid n$.*

Theorem 5.49 ([426]). *Let $P \in \kappa[z]$ be of degree n, have no multiple zeros, and*

$$P'(z) = c(z - c_1)^{m_1} \cdots (z - c_l)^{m_l},$$

where c is a non-zero constant, such that

$$c_i \neq c_j, \ P(c_i) \neq P(c_j), \ 1 \leq i \neq j \leq l.$$

Moreover, if $p > 0$, we assume that the multiplicity of the factor $z - c_i$ in $P(z) - P(c_i)$ is $m_i + 1$ for $1 \leq i \leq l$, and if $p \mid n$, we also assume that the coefficient of z^{n-1} in $P(z)$ is not zero. Then the following are equivalent:

(I) *P is a strong uniqueness polynomial for $\mathcal{M}(\kappa)$;*

(II) *P is a strong uniqueness polynomial for $\kappa(z)$;*

(III) *no non-trivial affine transformation of $\mathbb{C}$ preserves the set of zeros of P, and $l \geq 2$ if $p \mid n$; and $l \geq 3$ if $p = 0$ or $p \nmid n$, except P satisfies*

$$n = 4, \ m_1 = m_2 = m_3 = 1, \ \frac{P(c_1)}{P(c_2)} = \frac{P(c_2)}{P(c_3)} = \frac{P(c_3)}{P(c_1)} = \omega, \ \omega^2 + \omega + 1 = 0;$$

or $l = 2$ and $\min\{m_1, m_2\} \geq 2$ if $p = 0$ or $p \nmid n$, except P satisfies

$$n = 5, \ m_1 = m_2 = 2, \ P(c_1) = -P(c_2).$$

Chapter 6

Holomorphic Curves in Canonical Varieties

In this chapter, we will prove the degeneracy of holomorphic curves into pseudo canonical projective varieties by using meromorphic connections introduced by Siu, which is originally a conjecture due to Green and Griffiths. It can be regarded as the analogue of Bombieri-Lang's Conjecture 3.37 for holomorphic curves. A counterpart of Vojta's $(1,1)$-form conjecture in value distribution theory will be exhibited (see Theorem 6.2). We also discuss conjectures of Griffiths and Lang by applying meromorphic connections.

6.1 Variations of the first main theorem

To show the Green-Griffiths' conjecture, in this section we first prove an inequality of characteristic functions of holomorphic mappings for line bundles by using the first main theorem. A special integral term related to Jacobian sections appears in this inequality. We will use Siu's inequality to estimate the integral term.

6.1.1 Green's formula

We begin with the Green formula which will be used to prove Siu's inequality. First of all, we introduce some notation. Let $z = x + iy$ be the standard coordinate of $\mathbb{C}$, where $x, y \in \mathbb{R}$, and i is the imaginary unit. By a *regular arc* in $\mathbb{C}$ we mean a complex function

$$\lambda(t) = x(t) + iy(t), \quad t \in I$$

of t in an interval I with the property that $\lambda(t)$ is of class C^1 in I, and $\lambda'(t) \neq 0$ for all t in I. A finite sequence of regular arcs $C_j, j = 1, \ldots, n$, of class C^k placed end to end is called a *Jordan arc of class C^k*. If the end points of a Jordan arc are equal, the Jordan arc is said to be *closed*. A *simple closed* Jordan arc or *curvilinear polygon* is a closed Jordan arc which has no multiple points. A regular component of a curvilinear polygon is called an *edge* of the polygon and the point between two edges is called a *vertex* of the polygon.

Let D be a bounded connected open set in $\mathbb{C}$ such that the boundary ∂D of D consists of finitely many simple closed Jordan arcs of class C^1. Let $\frac{\partial}{\partial n}$ be the operator of directional derivative along the unit normal vector to ∂D pointed to the interior of D. Assume that ∂D has a positive orientation induced from D, i.e., the rotation direction from the unit tangent direction to ∂D to the unit normal vector to ∂D pointed to the interior of D is same as that from the x-axis to the y-axis. For $P, Q \in C^1(\overline{D})$, Green's formula

$$\int_D \left(\frac{\partial Q}{\partial x} - \frac{\partial P}{\partial y} \right) dx dy = \int_{\partial D} P dx + Q dy$$

holds, where $\overline{D}$ is the closure $D \cup \partial D$ of D. Taking $u, w \in C^2(\overline{D})$ and setting

$$P = -u \frac{\partial w}{\partial y}, \quad Q = u \frac{\partial w}{\partial x}$$

in Green's formula, we have

$$\int_D \left(u \Delta w + \frac{\partial u}{\partial x} \frac{\partial w}{\partial x} + \frac{\partial u}{\partial y} \frac{\partial w}{\partial y} \right) dx dy = \int_{\partial D} u \frac{\partial w}{\partial x} dy - u \frac{\partial w}{\partial y} dx$$

$$= - \int_{\partial D} u \frac{\partial w}{\partial n} ds$$

where ds is the arc element, and

$$\Delta = \frac{\partial^2}{\partial x^2} + \frac{\partial^2}{\partial y^2}.$$

By symmetry, we can obtain a similar formula by changing the order of u and w. Thus we easily obtain

$$\int_D (u \Delta w - w \Delta u) dx dy = - \int_{\partial D} \left(u \frac{\partial w}{\partial n} - w \frac{\partial u}{\partial n} \right) ds. \qquad (6.1.1)$$

Further, we let $D = \mathbb{C}(0; r)$ be the disc of radius $r > 0$ in $\mathbb{C}$. Set

$$\partial_z = \frac{\partial}{\partial z}, \quad \bar{\partial}_z = \frac{\partial}{\partial \bar{z}}, \quad \tau(z) = |z|^2, \quad v = dd^c \tau.$$

Then we have

$$\partial_z \bar{\partial}_z u = \frac{1}{4} \Delta u.$$

Note that along the circle $\partial D = \mathbb{C}\langle 0; r \rangle$,

$$\frac{\partial}{\partial n} = -\frac{\partial}{\partial r}.$$

We apply Green's formula to $w = -1$ as follows:

$$\mathbb{C}[0; r; \Delta uv] = \frac{1}{\pi} \int_{\mathbb{C}[0;r]} \Delta u dx dy = \frac{1}{\pi} \int_{\mathbb{C}\langle 0;r \rangle} \frac{\partial u}{\partial r} ds$$

$$= \frac{r}{\pi} \frac{\partial}{\partial r} \int_0^{2\pi} u(re^{i\theta}) d\theta,$$

and hence

$$\mathcal{T}(r, r_0; \Delta uv) = \frac{1}{\pi} \int_0^{2\pi} u(re^{i\theta}) d\theta - \frac{1}{\pi} \int_0^{2\pi} u(r_0 e^{i\theta}) d\theta. \qquad (6.1.2)$$

Note that

$$\Delta uv = 4\partial_z \bar{\partial}_z uv = 4 dd^c u, \qquad \sigma = \frac{d\theta}{2\pi}.$$

We can change the formula (6.1.2) into the following form:

$$\mathcal{T}(r, r_0; dd^c u) = \frac{1}{2} \mathbb{C}\langle 0; r; u \rangle - \frac{1}{2} \mathbb{C}\langle 0; r_0; u \rangle. \qquad (6.1.3)$$

The formula (6.1.3) further explains the formula (2.9.20).

6.1.2   Analogue of Vojta's conjecture

We will assume that M is a parabolic complex manifold of dimension m, N is a compact complex manifold of dimension n, and there exists a positive holomorphic line bundle $L > 0$ on N. Take a metric κ of L such that the Chern form

$$\psi = c_1(L, \kappa) > 0.$$

Let $B_k(L)$ be the base locus of the linear system $\Gamma(N, K_N^k \otimes L^*)$ and set

$$B(L) = \bigcap_k B_k(L), \qquad (6.1.4)$$

where the intersection goes over all positive integers k with

$$P_k(L) = \dim \Gamma(N, K_N^k \otimes L^*) > 0.$$

If $P_k(L) = 0$ for all $k > 0$, we define $B(L) = N$.

We give some examples with $B(L) = N$ based on the following *Kodaira-Nakano vanishing theorem* (cf. [127]):

Theorem 6.1. *If $L \to N$ is a positive line bundle, then*

$$H^q(N, \Omega^p(L)) = 0, \ p + q > n.$$

By *Kodaira-Serre duality*

$$H^q(N, \Omega^p(L)) \cong H^{n-q}(N, \Omega^{n-p}(L^*)),$$

we have

$$\Gamma(N, L^*) = H^0(N, \mathcal{O}(L^*)) = 0.$$

In particular, when $N = \mathbb{P}^n$ and H is the hyperplane section line bundle of the complex projective space $\mathbb{P}^n$, we have $B(H) = \mathbb{P}^n$. If $N = \mathbb{C}^n/\Lambda$ is a complex torus, then $B(L) = N$ since K_N is trivial.

If N is a pseudo canonical non-singular projective variety, without loss of generality, we may assume that L is very ample. Then Lemma 2.30 implies that $P_k(L) \gg k^n$ when $k \to_{div} \infty$, and hence $B(L) \neq N$. Conversely, if $B(L) \neq N$, then $P_k(L) > 0$ for some $k > 0$, which means that there exists an effective divisor Z on N such that $K_N^k \otimes L^* = [Z]$. By Theorem 1.77 or the remark after Lemma 2.30, N is pseudo canonical.

Assume $B(L) \neq N$ and let $f : M \longrightarrow N$ be a holomorphic mapping with $f(M) \not\subset B(L)$. Then there exists a positive integer k such that

$$P_k(L) \neq 0, \quad f(M) \not\subset B_k(L).$$

We can take a non-trivial section $\alpha \in \Gamma(N, K_N^k \otimes L^*)$ with $f(M) \not\subset \alpha^{-1}(0)$, and a metric ρ of $K_N^k \otimes L^*$ such that $|\alpha(x)|_\rho \leq 1$ for all $x \in N$. Then there exists a volume form Ω on N such that

$$c_1(K_N^k \otimes L^*, \rho) = k\mathrm{Ric}(\Omega) - \psi.$$

Denote the zero divisor (α) of α by D_α. Then the first main theorem (2.9.19) for the divisor D_α implies

$$kT_f(r, K_N) - T_f(r, L) = N_f(r, D_\alpha) + m_f(r, D_\alpha) - m_f(r_0, D_\alpha). \tag{6.1.5}$$

For a Jacobian section F of f, define a non-negative function g by

$$F[\psi^n] = g^2 f^*(\psi^b) \wedge v^{m-b}, \quad b = \min\{m, n\}. \tag{6.1.6}$$

Theorem 6.2. *If F is effective and if $f(M) \not\subset B(L) \neq N$, then there exists a positive constant c such that*

$$\| \quad cT_f(r, L) < \mathrm{Ric}_\tau(r, r_0) - N_{\mathrm{Ram}}(r, f) + M\langle O; r; \log g\rangle + O(\varepsilon \log r)$$

holds for any $\varepsilon > 0$.

Proof. By Theorem 2.90 and (6.1.5), we have

$$\frac{1}{k}T_f(r,L) \leq \mathrm{Ric}_\tau(r,r_0) - N_{\mathrm{Ram}}(r,f) + M\langle O; r; \log h\rangle + O(1)$$

for $r, r_0 \in \mathcal{R}_\tau^0$ with $r > r_0 > 0$, where h is defined by

$$F[\psi^n] = h^2 v^m. \tag{6.1.7}$$

Set

$$M^+ = \{x \in M \mid v(x) > 0\}.$$

For an integer i with $1 \leq i \leq m$, define a function ρ_i on M^+ by

$$f^*(\psi^i) \wedge v^{m-i} = \rho_i v^m. \tag{6.1.8}$$

Then a pointwise relation among the ρ_i's is provided by the inequality (2.1.82) or (2.12.18)

$$\rho_j^{\frac{1}{j}} \leq c_{ij}\rho_i^{\frac{1}{i}} \quad (j \geq i). \tag{6.1.9}$$

Thus we obtain

$$\left(\frac{h^2}{g^2}\right)^{\frac{1}{b}} = \rho_b^{\frac{1}{b}} \leq c_{1b}\rho_1,$$

which implies

$$\left(\frac{h^2}{g^2}\right)^{\frac{1}{b}} v^m \leq c_{1b}f^*(\psi) \wedge v^{m-1}.$$

By Lemma 2.92, for any $\varepsilon > 0$ we obtain

$$\| \quad \int_{M\langle O;r\rangle} \log\left(\frac{h^2}{g^2}\right)^{\frac{1}{b}} \sigma \leq \varsigma(1 + 2\varepsilon)\log T_f(r,L) + O(\varepsilon \log r).$$

Hence Theorem 6.2 follows from these estimates. $\qquad\square$

When $m \geq n = \mathrm{rank}(f)$, Stoll [380] proved a version of Theorem 6.2. Further, Hu [167] formulated Stoll's theorem into the form of Theorem 6.2.

According to the proof of Theorem 6.2, we see that the condition $B(L) \neq N$ can be replaced by $B_k(L) \neq N$ for a positive integer k. It is sufficient to assume that Chern classes of L and K_N in de Rham cohomology $H_{\mathrm{DR}}^2(N, \mathbb{R})$ satisfy $kc_1(K_N) > c_1(L)$, that is,

$$c_1(K_N) > \frac{1}{k}c_1(L), \tag{6.1.10}$$

the constant c in Theorem 6.2 assumes the form $\frac{1}{k} - \delta$ for sufficiently small $\delta > 0$. Hence Theorem 6.2 may be viewed as an analogue of Conjecture 3.61 in value distribution theory. Speaking correctly, the counterpart of Theorem 6.2 in number theory should be the following form:

Conjecture 6.3. *Let X be a complete non-singular variety over a number field κ. Let D be a divisor on X satisfying $|kK - D| \neq \emptyset$ for a positive integer k. Let E be a pseudo ample divisor on X. Let $\varepsilon > 0$. Then there exists a proper Zariski closed subset Z such that for all points $P \in X(\kappa) - Z$ we have*

$$h_D(P) \leq kd(P) + \varepsilon h_E(P) + O(1). \tag{6.1.11}$$

Recall that $|kK - D|$ is the complete linear system of $kK - D$ in which K is a canonical divisor of X. Conjecture 6.3 could be derived by Conjecture 3.56 simply. In fact, since $|kK - D|$ contains at least one effective divisor, say, D', which is linearly equivalent to $kK - D$, we have

$$kh_K - h_D = h_{D'} + O(1) \geq -O(1).$$

Thus (6.1.11) follows from (3.7.5) by taking $D = 0$.

Here we exhibit some examples in which the term $M\langle O; r; \log g \rangle$ in Theorem 6.2 is bounded from above. Note that $\mathrm{Ric}_\tau(r, r_0) = 0$ for the special case $M = \mathbb{C}^m$.

Corollary 6.4. *If $f : \mathbb{C}^m \longrightarrow N$ is a holomorphic mapping of rank n, then we have $B(L) = N$.*

Proof. By Lemma 2.83, there exists a holomorphic form φ of degree $m - n$ on $\mathbb{C}^m$ such that the induced Jacobian section F_φ is effective for f, and

$$i_{m-n}\varphi \wedge \bar{\varphi} \leq v^{m-n}.$$

Thus the function g defined by $F = F_\varphi$ satisfies the estimate (2.12.20). If $B(L) \neq N$, then $f(\mathbb{C}^m) \not\subseteq B(L)$ since $B(L)$ is a proper subvariety of N and $f(\mathbb{C}^m)$ contains an open set of N. Theorem 6.2 implies

$$\| \quad cT_f(r, L) < O(\varepsilon \log r) + O(1).$$

Thus we obtain

$$A_f(\infty, L) = \lim_{r \to \infty} \frac{T_f(r, L)}{\log r} = 0,$$

that is, f is constant (see Proposition 2.78). This is a contradiction. Hence it follows that $B(L) = N$. $\qquad\square$

Corollary 6.5. *Let N be a smooth pseudo canonical projective algebraic variety. Then any holomorphic mapping $f : \mathbb{C}^m \longrightarrow N$ has everywhere rank less than $n = \dim_\mathbb{C} N$.*

Corollary 6.5 is due to Kodaira [213], which was further improved by Griffiths [124]. These results all are special cases of the following Griffiths-King's theorem (cf. [128], Proposition 8.1 or [380], Corollary 21.3):

Corollary 6.6. *Let M be an algebraic variety. Let N be a smooth pseudo canonical projective algebraic variety. Then any holomorphic mapping $f : M \longrightarrow N$ whose image contains an open set is necessarily rational.*

Proof. Obviously it will suffice to assume that M is smooth and affine. Then (2.11.5) implies that there exists a parabolic exhaustion τ of M satisfying

$$\mathrm{Ric}_\tau(r, r_0) = O(\log r).$$

By Lemma 2.83, there exists a holomorphic form φ of degree $m - n$ on M such that the induced Jacobian section F_φ is effective for f, and

$$i_{m-n}\varphi \wedge \bar{\varphi} \leq v^{m-n}.$$

Thus the function g defined by $F = F_\varphi$ satisfies the estimate (2.12.20). Theorem 6.2 implies

$$\|\quad cT_f(r, L) < O(\log r) + O(1).$$

By Proposition 2.82, f is rational. $\qquad\square$

Based on Problem 2.32, there is a result related closely to the above corollary. That is *Kwack's theorem* which shows that in case N is negatively curved and projective, any holomorphic mapping $f : M \longrightarrow N$ from an algebraic variety M into N is rational. Griffiths and King (cf. [128], Proposition 9.20) further extends Kwack's theorem to a quasi-projective, negatively curved complex manifold N having a *bounded ample line bundle* $L \to N$, say, $N = \mathcal{D}/\Gamma$ for a bounded symmetric domain $\mathcal{D}$ of $\mathbb{C}^n$ (cf. 132; [128], Lemma 9.19). Here L is an ample line bundle on N satisfying the condition that there exist a metric ρ in L and sections $s_0, \ldots, s_k \in \Gamma(N, L)$ such that (i) $0 < c_1(L, \rho) < A\omega$, where A is constant, and ω is the (1,1)-form associated to the Hermitian metric ds_N^2 with negative holomorphic sectional curvatures; (ii) the sections $s_0, \ldots, s_k$ have bounded length and $[s_0, \ldots, s_k] : N \longrightarrow \mathbb{P}^k$ induces an algebraic embedding of N. This theorem supports Vojta's conjecture using (1,1)-forms (cf. [415]): if there are a complete variety X and a divisor D on X such that $N = X - D$ as above, then any set of (S, D)-integralizable points on X is finite. If $N = \mathcal{D}/\Gamma$, this is just the proved Shafarevich conjecture.

Finally we give a property of the function h in the proof of Theorem 6.2.

Theorem 6.7 (Hu [167]). *Take $M = \mathbb{C}^m$. If F is effective and if $f(\mathbb{C}^m) \not\subseteq B(L) \neq N$, for any $p > 0$ then there exists a positive constant c such that*

$$\mathbb{C}^m \left[0; r; h^{1/p}\right] > c$$

holds for larger r.

Proof. By using (6.1.5) and Theorem 2.90, we obtain

$$\mathbb{C}^m \langle 0; r; \log h \rangle \geq \mathbb{C}^m \langle 0; r_0; \log h \rangle + \frac{1}{k} m_f(r_0, D_\alpha) = c_1,$$

and so

$$c_1 \leq p \log \mathbb{C}^m \left\langle 0; r; h^{1/p} \right\rangle$$

for any positive number p. Note that

$$r^{2m} \mathbb{C}^m \left[0; r; h^{1/p} \right] = m \int_{\mathbb{C}^m[0;r]} h^{1/p} \tau^{m-1} d\tau \wedge \sigma$$

$$= 2m \int_0^r t^{2m-1} \mathbb{C}^m \langle 0; r; h^{1/p} \rangle dt.$$

Therefore we have

$$r^{2m} \mathbb{C}^m \left[0; r; h^{1/p} \right] \geq c_2 + 2m \exp(c_1/p) \int_{r_0}^r t^{2m-1} dt$$

$$= c_3 + \exp(c_1/p) r^{2m},$$

and hence Theorem 6.7 follows.   □

Theorem 6.7 generalizes a theorem due to Kodaira [213].

Let $f : \mathbb{C}^m \longrightarrow N$ be a holomorphic mapping of rank $\min\{m, n\}$. There exist effective Jacobian sections for f. If we can find one effective Jacobian section F for f such that $g^2 \leq c$ for some constant $c > 0$, we have

$$h^2 v^m = F[\psi^n] \leq c f^*(\psi^b) \wedge v^{m-b} = c \rho_b v^m$$

which means

$$h^{2/b} v^m \leq c^{1/b} \rho_b^{1/b} v^m \leq c' \rho_1 v^m = c' f^*(\psi) \wedge v^{m-1}$$

for some constant $c' > 0$. Hence we have

$$\mathbb{C}^m \left[0; r; h^{2/b} \right] \leq c' r^{-2} \mathbb{C}^m [0; r; f^*(\psi)] = c' r^{-2} A_f(r, L).$$

Thus if $\mathbb{C}^m \left[0; r; h^{2/b} \right]$ is bounded low by a positive number, there exists a constant $c > 0$ such that

$$A_f(r, L) > c r^2.$$

6.2   Meromorphic connections

Let N be a compact complex manifold of dimension n. Suppose L is a holomorphic line bundle over N and L carries a Hermitian metric κ with Chern form

$$\psi = c_1(L, \kappa).$$

From the Hermitian metric κ of L, we have a Hermitian connection ∇ for the holomorphic line bundle L which is compatible with the metric κ and agrees with $\bar{\partial}$ in the $(0,1)$-direction. Let $\varphi_U : U \times \mathbb{C} \longrightarrow L_U$ be a trivialization of L over an open set U of N. Set

$$\xi_U(x) = \varphi_U(x, 1).$$

Then a section $s \in \Gamma(U, L)$ can be represented by $s = s_U \xi_U$, where $s_U : U \longrightarrow \mathbb{C}$ is a holomorphic function such that

$$|s|_\kappa^2 = |s_U|^2 |\xi_U|_\kappa^2 = \kappa_U |s_U|^2$$

holds for some positive smooth function κ_U. Then we have

$$\nabla s = (ds_U)\xi_U + (\partial \log \kappa_U) s_U \xi_U.$$

In general, ∇s is not holomorphic. As a matter of fact,

$$\bar{\partial}\nabla s = (\bar{\partial}\partial \log \kappa_U)s = -2\pi i \psi s,$$

where $i = \sqrt{-1}$ is the imaginary unit.

Let L_0 be a holomorphic line bundle over N which is generated by its global holomorphic sections such that

$$H^1(N, \Omega^1(L_0)) = 0.$$

If L_0 is a negative line bundle, when $n > 2$ this is true by the Kodaira-Nakano vanishing theorem. Bott's formula (see [30], or [364]) implies

$$\dim H^q(\mathbb{P}^n, \Omega^p(H^d)) = \begin{cases} \binom{d+n-p}{d}\binom{d-1}{p}, & \text{if } q = 0,\ 0 \le p \le n,\ d > p; \\ 1, & \text{if } d = 0,\ 0 \le p = q \le n; \\ \binom{-d+p}{-d}\binom{-d-1}{n-p}, & \text{if } q = n,\ 0 \le p \le n,\ d < p - n; \\ 0, & \text{otherwise,} \end{cases}$$

where H is the hyperplane section line bundle of the complex projective space $\mathbb{P}^n$. In particular, one has

$$H^1(\mathbb{P}^n, \Omega^1(H^d)) = 0,\ d > 0,\ n \ge 2.$$

Generally, we can take $L_0 = L^d$ for a positive line bundle L and a large integer d based on the following theorem:

Theorem 6.8 (cf. [127]). *Let N be a compact complex manifold and let $L \to N$ a positive line bundle. Then for any holomorphic vector bundle E, there exists d_0 such that*

$$H^q(N, \mathcal{O}(L^d \otimes E)) = 0, \quad q > 0, \quad d \geq d_0.$$

Take $t_0 \in \Gamma(N, L_0)$ with $t_0 \not\equiv 0$. From the vanishing of $H^1(N, \Omega^1(L_0))$ we have a smooth section η of $\Omega^1(L_0)$ over N such that

$$\bar{\partial}\eta = -2\pi i \psi t_0.$$

For $s \in \Gamma(U, L)$, define

$$t_0 \mathscr{D}s = t_0 \nabla s - \eta s.$$

Then $t_0 \mathscr{D}s$ is a holomorphic section of $\Omega^1(L_0 \otimes L)$ over U for $s \in \Gamma(U, L)$. We obtain a connection for L defining $\mathscr{D}$, but it is not a smooth connection. However, $t_0 \mathscr{D}$ is a smooth operator. According to Siu ([363], [364], [365]), the connection $\mathscr{D}$ is called a *meromorphic connection* for L with pole order L_0.

We give N a Kähler metric ds_N^2. For a local coordinate system $(U; z_1, \ldots, z_n)$ on N, it can be expressed as

$$ds_N^2 = 2 \sum_{\alpha, \beta} h_{\alpha\beta} dz_\alpha d\bar{z}_\beta.$$

Then the metric determines uniquely the Hermitian connection ∇ such that the connection coefficients $\Gamma_{\alpha\beta}^\gamma$ are given by

$$\Gamma_{\alpha\beta}^\gamma = \sum_\nu h^{\nu\gamma} \partial_\beta h_{\alpha\nu}, \quad \partial_\beta = \frac{\partial}{\partial z_\beta}.$$

However, we will use also another connection $\mathbf{D}$ for the holomorphic tangent bundle $\mathbf{T}(N)$ of N with the coefficients $\Gamma_{\alpha\beta}^\gamma$ of the connection $\mathbf{D}$ over U. It is not necessarily symmetric and may not even be smooth. It is assumed to satisfy the following property:

(i) There exists a holomorphic line bundle E over N with a global holomorphic section $t \not\equiv 0$ such that $t\mathbf{D}$ is holomorphic, i.e., for each Christoffel symbol $\Gamma_{\alpha\beta}^\gamma$ of $\mathbf{D}$, $t\Gamma_{\alpha\beta}^\gamma$ is holomorphic on U.

According to Siu ([363], [364], [365]), a connection $\mathbf{D}$ satisfying the condition (i) is called a *meromorphic connection* for $\mathbf{T}(N)$ with pole order E (or pole order d in the case when $N = \mathbb{P}^n$ and E is the dth power H^d of the hyperplane section line bundle H of $\mathbb{P}^n$).

Siu [363], [364] and Nadel [285] estimated the pole order of a meromorphic connection on projective space in terms of its Christoffel symbols relative to one given inhomogeneous coordinate system. Let $\mathbf{D}$ be a meromorphic connection on

$\mathbb{P}^n$ and let $\mathbf{\Gamma}^\gamma_{\alpha\beta}$ be the Christoffel symbols of $\mathbf{D}$ relative to the inhomogeneous coordinates $z_1, \ldots, z_n$. Let $X_0, \ldots, X_n$ be homogeneous coordinates on $\mathbb{P}^n$ so that

$$z_i = \frac{X_i}{X_0}, \quad i = 1, \ldots, n.$$

Assume that $A^\gamma_{\alpha\beta}(X_0, \ldots, X_n)$ and $B(X_0, \ldots, X_n)$ are homogeneous polynomials of degree g such that

$$\mathbf{\Gamma}^\gamma_{\alpha\beta} = \frac{A^\gamma_{\alpha\beta}}{B}$$

as rational functions on $\mathbb{P}^n$ for each fixed α, β, γ.

Theorem 6.9 ([285]). *Let t be the global holomorphic section of the line bundle H^{g+3} defined by the homogeneous polynomial*

$$t = X_0^3 B(X_0, \ldots, X_n).$$

Then $t\mathbf{D}$ is holomorphic. In particular, the pole order of $\mathbf{D}$ is $\leq g + 3$.

Proof. We must check that $t\mathbf{D}$ is holomorphic on each inhomogeneous coordinate chart. Set

$$U_i = \{X_i \neq 0\} \subset \mathbb{P}^n.$$

First we note that $t\mathbf{D}$ is holomorphic on the coordinate chart U_0 on which $z_1, \ldots, z_n$ are defined because $B(1, z_1, \ldots, z_n)\mathbf{\Gamma}^\gamma_{\alpha\beta}$ is a holomorphic function on U_0 for each fixed α, β, γ.

Next we check $t\mathbf{D}$ on some other coordinate chart, say U_n. On U_n we have inhomogeneous coordinates

$$w_i = \frac{X_{i-1}}{X_n}, \quad i = 1, \ldots, n.$$

The relation between the two inhomogeneous coordinate systems is given by

$$w_i = \frac{z_{i-1}}{z_n}, \quad 1 \leq i \leq n,$$

where $z_0 = 1$, and

$$z_i = \frac{w_{i+1}}{w_1}, \quad 1 \leq i \leq n,$$

where $w_{n+1} = 1$. Let $\mathbf{\Xi}^\lambda_{\mu\nu}$ be the Christoffel symbols of $\mathbf{D}$ relative to the inhomogeneous coordinates $w_1, \ldots, w_n$. We must show that

$$w_1^3 B(w_1, \ldots, w_n, 1)\mathbf{\Xi}^\lambda_{\mu\nu}$$

is a holomorphic function on U_n for each fixed μ, ν, λ.

Recalling the following transformation law for Christoffel symbols:

$$\Xi_{\mu\nu}^{\lambda} = \sum_{\alpha,\beta,\gamma} \Gamma_{\alpha\beta}^{\gamma} \frac{\partial z_\alpha}{\partial w_\mu} \frac{\partial z_\beta}{\partial w_\nu} \frac{\partial w_\lambda}{\partial z_\gamma} + \sum_{\alpha} \frac{\partial^2 z_\alpha}{\partial w_\mu \partial w_\nu} \frac{\partial w_\lambda}{\partial z_\alpha},$$

by assumption we know that

$$B(w_1,\ldots,w_n,1)\Gamma_{\alpha\beta}^{\gamma} = A_{\alpha\beta}^{\gamma}(w_1,\ldots,w_n,1)$$

is holomorphic on U_n. Note that

$$\frac{\partial w_\lambda}{\partial z_\gamma} = \begin{cases} 0 \\ \frac{1}{z_n} \\ -\frac{z_{\lambda-1}}{z_n^2} \end{cases} = \begin{cases} 0 & \text{if } \gamma \neq n, \lambda - 1, \\ w_1 & \text{if } \gamma = \lambda - 1, \\ -w_1 w_\lambda & \text{if } \gamma = n, \end{cases}$$

and

$$\frac{\partial z_\alpha}{\partial w_\mu} = \begin{cases} 0 & \text{if } \mu \neq 1, \alpha+1, \\ \frac{1}{w_1} & \text{if } \mu = \alpha + 1, \\ -\frac{w_{\alpha+1}}{w_1^2} & \text{if } \mu = 1. \end{cases}$$

For fixed $2 \le \mu \le n$ all entries of the matrix $(\partial^2 z_\alpha / \partial w_\mu \partial w_\nu)$ with $1 \le \alpha \le n$ as the row index and $1 \le \nu \le n$ as the column index are zero except the $(\mu-1)$th entry in the first column which is $-1/w_1^2$. For the case $\mu = 1$, we have

$$\frac{\partial^2 z_\alpha}{\partial w_1 \partial w_\nu} = \begin{cases} 0 & \text{if } \nu \neq 1, \alpha+1, \\ -\frac{1}{w_1^2} & \text{if } \nu = \alpha + 1, \\ -\frac{2w_{\alpha+1}}{w_1^3} & \text{if } \nu = 1. \end{cases}$$

Therefore, $\frac{\partial w_\lambda}{\partial z_\gamma}$ is smooth on U_n and vanishes to at least first order along $w_1 = 0$. However, $\frac{\partial z_\alpha}{\partial w_\mu}$ (resp. $\frac{\partial^2 z_\alpha}{\partial w_\mu \partial w_\nu}$) has poles of order at most 2 (resp. 3) along $w_1 = 0$ and is otherwise smooth on U_n. The proof now follows. □

An analytic subset M of N is said to be *totally geodesic* (relative to **D**) if the following two conditions hold:

(ii) No component of M is contained identically in the support of pole divisor (t) of **D**;

(iii) For any two holomorphic vector fields X and Y defined locally on N and tangent to M, the meromorphic vector field $\mathbf{D}_X Y$ is also tangent to M.

If M is a totally geodesic submanifold of N, then it is clear that **D** restricts to a meromorphic connection $\mathbf{D}|_M$ on M, and moreover that $t|_M \mathbf{D}|_M$ is holomorphic on M.

Let $\mathscr{D}$ be a meromorphic connection for the holomorphic line bundle L over N. Take $s \in \Gamma(N, L) - \{0\}$. Following Siu [363] and Nadel [285], we will use the following assumption:

(iv) There exist meromorphic tensors A_α, B_β, $C_{\alpha\beta}$ on N such that

$$\mathscr{D}_\alpha\mathscr{D}_\beta s - \sum_\gamma \Gamma^\gamma_{\beta\alpha}\mathscr{D}_\gamma s = A_\alpha\mathscr{D}_\beta s + B_\beta\mathscr{D}_\alpha s + C_{\alpha\beta}s \qquad (6.2.1)$$

holds, where $\mathscr{D}_\alpha s$ is the covariant derivative of s in the direction of $\frac{\partial}{\partial z_\alpha}$.
The left-hand side of (6.2.1) is called the *Hessian* of s and sometimes written

$$\mathrm{Hess}(s)_{\alpha\beta} = \mathrm{Hess}(s)\left(\frac{\partial}{\partial z_\alpha}, \frac{\partial}{\partial z_\beta}\right).$$

Locally, the condition (6.2.1) is clearly equivalent to the condition

$$\partial_\alpha\partial_\beta s - \sum_\gamma \Gamma^\gamma_{\beta\alpha}\partial_\gamma s = A_\alpha\partial_\beta s + B_\beta\partial_\alpha s + C_{\alpha\beta}s, \qquad (6.2.2)$$

or in invariant notation,

$$\mathscr{D}_X\mathscr{D}_Y s - \mathscr{D}_{\mathbf{D}_X Y}s = A(X)\mathscr{D}_Y s + B(Y)\mathscr{D}_X s + C(X,Y)s \qquad (6.2.3)$$

for any local holomorphic vector fields X, Y.

Let M be the support of the zero divisor (s) of s. If no component of M is contained identically in the pole sets of $\mathscr{D}$, A_α, B_β and $C_{\alpha\beta}$, then M is totally geodesic. In fact, if there are two holomorphic vector fields X and Y defined locally on N and tangent to M, then all terms in (6.2.3) other than $\mathscr{D}_{\mathbf{D}_X Y}s$ clearly vanish along M. Therefore $\mathscr{D}_{\mathbf{D}_X Y}s$ also vanishes along M. Hence $\mathbf{D}_X Y$ is tangent to M. This explains the geometric meaning of (6.2.1). Siu [363] actually assumes that tA_α, tB_β and $tC_{\alpha\beta}$ are smooth.

Next we introduce some local results due to Nadel [285], and so we work with holomorphic connections rather than meromorphic ones. Let N be a complex manifold of dimension $n \geq 2$, $\mathbf{D}$ a symmetric holomorphic connection on N with Christoffel symbols $\Gamma^\gamma_{\alpha\beta}$, L a holomorphic line bundle over N, and $\mathscr{D}$ a holomorphic connection for L. Let $f : \Delta \longrightarrow N$ be a holomorphic mapping from the unit disc Δ whose *Wronskian* (relative to $\mathbf{D}$) vanishes identically:

$$f' \wedge f'' \wedge \cdots \wedge f^{(n)} \equiv 0. \qquad (6.2.4)$$

Here f' is the derivative of f with respect to the coordinate z of Δ which is a mapping from Δ to the holomorphic tangent bundle $\mathbf{T}(N)$ of N and

$$f^{(k+1)} = \mathbf{D}_{f'} f^{(k)}, \ k \geq 1.$$

By (6.2.4), we can write

$$f^{(k)} = h_1 f' + \cdots + h_{k-1}f^{(k-1)} \qquad (6.2.5)$$

for some k $(2 \leq k \leq n)$, where $h_1, \ldots, h_{k-1}$ are meromorphic functions. By re-placing Δ by a subdisc (not necessarily centered at the origin) if necessary, we can assume without loss of generality that $h_1, \ldots, h_{k-1}$ are in fact holomorphic. Let z denote the Euclidean coordinate on Δ.

Lemma 6.10 ([285]). *Fix* $s \in \Gamma(N, L) - \{0\}$ *and assume that (6.2.1) holds for holomorphic tensor fields* A_α, B_β *and* $C_{\alpha\beta}$ *on* N. *Then for* $i = 2, 3, \ldots$, $\mathscr{D}_{f^{(i)}}s - (\mathscr{D}_{f'})^i s$ *is a linear combination of* s, $\mathscr{D}_{f'}s$, $(\mathscr{D}_{f'})^2 s$, $\ldots$, $(\mathscr{D}_{f'})^{i-1}s$ *with coefficients which are holomorphic functions of* z.

Proof. Note that
$$\mathrm{Hess}(s)(X, Y) = \mathscr{D}_X \mathscr{D}_Y s - \mathscr{D}_{\mathbf{D}_X Y} s$$
for any vector fields X, Y. Take $X = f'$, $Y = f^{(i-1)}$ and combine this formula with (6.2.3) to obtain
$$\mathscr{D}_{f'} \mathscr{D}_{f^{(i-1)}} s - \mathscr{D}_{f^{(i)}} s = A(f') \mathscr{D}_{f^{(i-1)}} s + B(f^{(i-1)}) \mathscr{D}_{f'} s + C(f', f^{(i-1)}) s.$$
This formula gives us the desired result for $i = 2$, and gives us also the inductive step to get from $i - 1$ to i. The lemma follows. $\qquad\square$

Theorem 6.11 ([285]). *Fix* $s \in \Gamma(N, L) - \{0\}$ *and assume that (6.2.1) holds for holomorphic tensor fields* A_α, B_β *and* $C_{\alpha\beta}$ *on* N. *Let* $f : \Delta \longrightarrow N$ *be a holomorphic mapping which satisfies (6.2.4) and (6.2.5). Assume that*

(1) $s(f(0)) = 0$, *and*
(2) $(\mathscr{D}s)(f^{(i)}(0)) = 0$ *for* $i = 1, 2, \ldots, n - 1$.

Then $s(f(z)) = 0$ *for all* $z \in \Delta$.

Proof. We must show that $S = f^*s$ is the zero section of $f^*(L)$. By (6.2.5), we can write
$$\mathscr{D}_{f^{(k)}} S = \sum_{i=1}^{k-1} h_i \mathscr{D}_{f^{(i)}} S.$$
By Lemma 6.10, we see that
$$(\mathscr{D}_{f'})^k S = \sum_{i=0}^{k-1} H_i (\mathscr{D}_{f'})^i S, \tag{6.2.6}$$
where $H_0, H_1, \ldots, H_{k-1}$ are holomorphic functions. Therefore S is a holomorphic solution of a kth order linear homogeneous holomorphic ordinary differential equation. Furthermore, we have by assumption the zero initial conditions
$$s(f(0)) = 0, \quad (\mathscr{D}s)(f^{(i)}(0)) = 0 \ (1 \leq i \leq k - 1).$$
By Lemma 6.10, we can convert these initial conditions into the more familiar
$$(\mathscr{D}_{f'})^i S|_{z=0} = 0, \quad i = 0, 1, \ldots, k - 1. \tag{6.2.7}$$
Hence by uniqueness of solutions to (6.2.6) and (6.2.7), we obtain $S \equiv 0$. $\qquad\square$

6.3 Siu theory

Siu [363], [364] [365] proved a defect relation of holomorphic curves into projective varieties. By using meromorphic connections, Siu constructed a class of associated curves of holomorphic curves, and estimated its growth. In this section, we will introduce a main method and result of Siu which will be used in the proof of Green-Griffiths' conjecture.

6.3.1 Siu's inequality

Let N be a compact complex manifold of dimension n. Suppose L is a holomorphic line bundle over N and L carries a Hermitian metric κ whose *Chern form*

$$\psi = c_1(L, \kappa) > 0.$$

For a local coordinate system $(W; w_1, \ldots, w_m)$ of N, the Chern form ψ can be expressed as

$$\psi = \frac{\sqrt{-1}}{2\pi} \sum_{\alpha, \beta} h_{\alpha\beta} dw_\alpha \wedge d\bar{w}_\beta,$$

and hence define a *Hermitian metric*

$$ds_N^2 = 2 \sum_{\alpha, \beta} h_{\alpha\beta} dw_\alpha d\bar{w}_\beta$$

such that ψ is the *Kähler form*. Let ∇ be the Hermitian connection of N. We will write

$$\partial_\alpha = \frac{\partial}{\partial w_\alpha}, \quad \nabla_\alpha = \nabla_{\partial_\alpha}.$$

Let $f : \mathbb{C} \longrightarrow N$ be a non-constant holomorphic mapping. Let z be the coordinate of $\mathbb{C}$ and f' be the derivative of f with respect to z which is a mapping from $\mathbb{C}$ to the holomorphic tangent bundle $\mathbf{T}(N)$ of N. Under the local coordinates $w = (w_1, \ldots, w_n)$ on N near $f(z)$, writing $f_\alpha = w_\alpha \circ f$, that is,

$$f(z) = (f_1(z), \ldots, f_n(z)),$$

then f' can be expressed as

$$f' = \sum_{\alpha=1}^{n} f'_\alpha \partial_\alpha,$$

where

$$f'_\alpha = \partial_z f_\alpha, \quad \partial_z = \frac{\partial}{\partial z}.$$

Let $\mathbf{D}$ be a meromorphic connection for $\mathbf{T}(N)$ with pole order E such that $t\mathbf{D}$ is holomorphic for some $t \in \Gamma(N, E) - \{0\}$. Fix a Hermitian metric l in E. For a positive integer k, define

$$f^{(k+1)} = (t\mathbf{D}_z)^k f' : \mathbb{C} \longrightarrow E^k \otimes \mathbf{T}(N),$$

where $\mathbf{D}_z$ is the covariant derivative with respect to the connection $\mathbf{D}$ and the coordinate z of $\mathbb{C}$ (i.e., in the direction of $f'(z)$). In the term of local coordinates, we have

$$\mathbf{D}_z f' = \sum_{\alpha=1}^{n} \mathbf{D}_z f'_\alpha \partial_\alpha,$$

where

$$\mathbf{D}_z f'_\alpha = \sum_\beta f'_\beta \mathbf{D}_\beta f'_\alpha = \partial_z f'_\alpha + \sum_{\beta,\gamma} \Gamma^\alpha_{\beta\gamma} f'_\beta f'_\gamma.$$

Then a holomorphic mapping

$$f' \wedge f^{(2)} \wedge \cdots \wedge f^{(n)} : \mathbb{C} \longrightarrow E^{\frac{n(n-1)}{2}} \otimes K_N^*$$

is well defined, where the exterior product is taken in the holomorphic tangent bundle $\mathbf{T}(N)$ of N. In particular, we obtain a holomorphic mapping

$$f^{(2)} \wedge \cdots \wedge f^{(n)} : \mathbb{C} \longrightarrow E^{\frac{n(n-1)}{2}} \otimes \bigwedge_{n-1} \mathbf{T}(N). \tag{6.3.1}$$

The image of f is called *autoparallel with respect to the connection* $\mathbf{D}$ if

$$f' \wedge f^{(2)} \wedge \cdots \wedge f^{(n)} \equiv 0. \tag{6.3.2}$$

Next we prove *Siu's inequality* (cf. [363]):

Theorem 6.12. *Let* $f : \mathbb{C} \longrightarrow N$ *be a non-constant holomorphic mapping. Take* $\mu \in \mathbb{R}$ *with* $0 < \mu < \frac{1}{4^k-1}$. *Then*

$$\| \quad T\left(r, r_0; |f^{(k)}|^{2\mu} v\right) \le c \left\{ T_f(r, L)^{4^k - \frac{1}{2}\mu} + r^4 \right\},$$

where c is some positive constant.

Proof. Since f is non-constant, then $T_f(r, L) \to +\infty$ as $r \to +\infty$. Without loss of generality, we may assume $T_f(r, L) \ge 1$. First of all, we consider the case $k = 1$. Now when $0 < \mu < 1$ the Hölder inequality implies

$$T\left(r, r_0; |f'|^{2\mu} v\right) \le T\left(r, r_0; |f'|^2 v\right)^\mu T\left(r, r_0; v\right)^{1-\mu}.$$

Note that

$$f^*(\psi) = \frac{\sqrt{-1}}{2\pi} \sum_{\alpha,\beta} h_{\alpha\beta} f'_\alpha \overline{f'_\beta} dz \wedge d\bar{z} = |f'|^2 v.$$

We have

$$\mathcal{T}\left(r, r_0; |f'|^2 v\right) = T_f(r, L), \quad \mathcal{T}(r, r_0; v) = \frac{1}{2}(r^2 - r_0^2), \qquad (6.3.3)$$

and hence when $r \geq 1$,

$$\begin{aligned}
\mathcal{T}\left(r, r_0; |f'|^{2\mu} v\right) &\leq T_f(r, L)^{2\mu} + \mathcal{T}\left(r, r_0; v\right)^{2-2\mu} \\
&\leq T_f(r, L)^{2\mu} + r^4.
\end{aligned}$$

Therefore Theorem 6.12 holds for the case $k = 1$.

Following Siu [363], we then study the case $k = 2$. Since $t\Gamma^\gamma_{\alpha\beta}$ is smooth, it is clear that

$$|f^{(2)}|^2 = |t\mathbf{D}_z f'|^2 \leq c\{|\nabla_z f'|^2 + |f'|^4\}, \qquad (6.3.4)$$

where

$$\nabla_z f'_\alpha = \sum_\beta f'_\beta \nabla_\beta f'_\alpha$$

is the covariant derivative of f'_α in the direction of f' with respect to the Kähler metric $h_{\alpha\beta}$ of N. Note that

$$\begin{aligned}
&\partial_z \bar{\partial}_z \log\left(1 + \sum_{\alpha,\beta} h_{\alpha\beta} f'_\alpha \overline{f'_\beta}\right) \\
&= -\frac{\sum_{\alpha,\beta,\gamma,\delta} K_{\alpha\beta\gamma\delta} f'_\alpha \overline{f'_\beta} f'_\gamma \overline{f'_\delta}}{1 + \sum_{\alpha,\beta} h_{\alpha\beta} f'_\alpha \overline{f'_\beta}} + \frac{\sum_{\alpha,\beta} h_{\alpha\beta} \nabla_z f'_\alpha \overline{\nabla_z f'_\beta}}{\left(1 + \sum_{\alpha,\beta} h_{\alpha\beta} f'_\alpha \overline{f'_\beta}\right)^2} \\
&\quad + \frac{\left(\sum_{\alpha,\beta} h_{\alpha\beta} f'_\alpha \overline{f'_\beta}\right)\left(\sum_{\alpha,\beta} h_{\alpha\beta} \nabla_z f'_\alpha \overline{\nabla_z f'_\beta}\right) - \left|\sum_{\alpha,\beta} h_{\alpha\beta} f'_\alpha \overline{\nabla_z f'_\beta}\right|^2}{\left(1 + \sum_{\alpha,\beta} h_{\alpha\beta} f'_\alpha \overline{f'_\beta}\right)^2}.
\end{aligned}$$

The last term on the right-hand side is non-negative because of the Schwarz inequality. Thus

$$\begin{aligned}
\partial_z \bar{\partial}_z \log\left(1 + \sum_{\alpha,\beta} h_{\alpha\beta} f'_\alpha \overline{f'_\beta}\right) &\geq -\frac{\sum_{\alpha,\beta,\gamma,\delta} K_{\alpha\beta\gamma\delta} f'_\alpha \overline{f'_\beta} f'_\gamma \overline{f'_\delta}}{1 + \sum_{\alpha,\beta} h_{\alpha\beta} f'_\alpha \overline{f'_\beta}} \\
&\quad + \frac{\sum_{\alpha,\beta} h_{\alpha\beta} \nabla_z f'_\alpha \overline{\nabla_z f'_\beta}}{\left(1 + \sum_{\alpha,\beta} h_{\alpha\beta} f'_\alpha \overline{f'_\beta}\right)^2}.
\end{aligned}$$

From Green's formula (6.1.3) it follows that

$$-T\left(r, r_0; \frac{\sum_{\alpha,\beta,\gamma,\delta} K_{\alpha\beta\gamma\delta} f'_\alpha \overline{f'_\beta} f'_\gamma \overline{f'_\delta}}{1 + \sum_{\alpha,\beta} h_{\alpha\beta} f'_\alpha \overline{f'_\beta}} v\right) + T\left(r, r_0; \frac{\sum_{\alpha,\beta} h_{\alpha\beta} \nabla_z f'_\alpha \overline{\nabla_z f'_\beta}}{\left(1 + \sum_{\alpha,\beta} h_{\alpha\beta} f'_\alpha \overline{f'_\beta}\right)^2} v\right)$$

$$\leq \frac{1}{2}\mathbb{C}\left\langle 0; r; \log\left(1 + \sum_{\alpha,\beta} h_{\alpha\beta} f'_\alpha \overline{f'_\beta}\right)\right\rangle$$

$$- \frac{1}{2}\mathbb{C}\left\langle 0; r_0; \log\left(1 + \sum_{\alpha,\beta} h_{\alpha\beta} f'_\alpha \overline{f'_\beta}\right)\right\rangle.$$

By Lemma 2.92, we obtain

$$\| \quad \mathbb{C}\left\langle 0; r; \log\left(1 + \sum_{\alpha,\beta} h_{\alpha\beta} f'_\alpha \overline{f'_\beta}\right)\right\rangle \leq (1 + 2\varepsilon) \log T\left(r, r_0; v + \sum_{\alpha,\beta} h_{\alpha\beta} f'_\alpha \overline{f'_\beta} v\right)$$

$$+ 4\varepsilon \log r.$$

Hence

$$\| \quad T\left(r, r_0; \frac{\sum_{\alpha,\beta} h_{\alpha\beta} \nabla_z f'_\alpha \overline{\nabla_z f'_\beta}}{\left(1 + \sum_{\alpha,\beta} h_{\alpha\beta} f'_\alpha \overline{f'_\beta}\right)^2} v\right) \leq CT\left(r, r_0; \frac{\left(\sum_{\alpha,\beta} h_{\alpha\beta} f'_\alpha \overline{f'_\beta}\right)^2}{1 + \sum_{\alpha,\beta} h_{\alpha\beta} f'_\alpha \overline{f'_\beta}} v\right)$$

$$+ \left(\frac{1}{2} + \varepsilon\right) \log T\left(r, r_0; v + \sum_{\alpha,\beta} h_{\alpha\beta} f'_\alpha \overline{f'_\beta} v\right) + 2\varepsilon \log r, \tag{6.3.5}$$

where

$$C = \sup_{N, \xi} \frac{\sum_{\alpha,\beta,\gamma,\delta} K_{\alpha\beta\gamma\delta} \xi_\alpha \overline{\xi_\beta} \xi_\gamma \overline{\xi_\delta}}{\left|\sum_{\alpha,\beta} h_{\alpha\beta} \xi_\alpha \overline{\xi_\beta}\right|^2} < +\infty. \tag{6.3.6}$$

It follows that

$$\| \quad cT\left(r, r_0; \frac{\sum_{\alpha,\beta} h_{\alpha\beta} \nabla_z f'_\alpha \overline{\nabla_z f'_\beta}}{\left(1 + \sum_{\alpha,\beta} h_{\alpha\beta} f'_\alpha \overline{f'_\beta}\right)^2} v\right) \leq T_f(r, L) + r^2,$$

where we have used (6.3.3).

Take $\mu \in \mathbb{R}$ with $0 < \mu < \frac{1}{4}$. By using elementary inequality

$$2|\nabla_z f'|^{2\mu} \leq \left(\frac{|\nabla_z f'|^2}{(1 + |f'|^2)^2}\right)^{2\mu} + \left((1 + |f'|^2)^2\right)^{2\mu}$$

and the Hölder inequality, we have

$$2\mathcal{T}\left(r, r_0; |\nabla_z f'|^{2\mu} v\right) \le \mathcal{T}\left(r, r_0; \left(\frac{|\nabla_z f'|^2}{(1+|f'|^2)^2}\right)^{2\mu} v\right)$$

$$+ \mathcal{T}\left(r, r_0; ((1+|f'|^2)^2)^{2\mu} v\right)$$

$$\le \mathcal{T}\left(r, r_0; \frac{|\nabla_z f'|^2}{(1+|f'|^2)^2} v\right)^{2\mu} \mathcal{T}(r, r_0; v)^{1-2\mu}$$

$$+ \mathcal{T}\left(r, r_0; (1+|f'|^2) v\right)^{4\mu} \mathcal{T}(r, r_0; v)^{1-4\mu},$$

and hence

$$\| \quad c\mathcal{T}\left(r, r_0; |\nabla_z f'|^{2\mu} v\right) \le (T_f(r, L) + r^2)^{2\mu} r^{2(1-2\mu)}$$

$$+ (r^2 + T_f(r, L))^{4\mu} r^{2(1-4\mu)}.$$

Therefore

$$\| \quad c\mathcal{T}\left(r, r_0; |\nabla_z f'|^{2\mu} v\right) \le T_f(r, L)^{8\mu} + r^4. \tag{6.3.7}$$

So our estimate for $\mathcal{T}\left(r, r_0; |f^{(2)}|^{2\mu} v\right)$ is given by

$$\| \quad \mathcal{T}\left(r, r_0; |f^{(2)}|^{2\mu} v\right) \le c\left\{\mathcal{T}\left(r, r_0; |\nabla_z f'|^{2\mu} v\right) + \mathcal{T}\left(r, r_0; |f'|^{4\mu} v\right)\right\}$$

$$\le c'\left\{T_f(r, L)^{8\mu} + r^4 + T_f(r, L)^{2\mu} \mathcal{T}(r, r_0; v)^{1-2\mu}\right\}$$

$$\le c''\left\{T_f(r, L)^{8\mu} + r^4\right\}. \tag{6.3.8}$$

Now we use induction to prove Theorem 6.12. Assume that Theorem 6.12 holds up to $k \ (\ge 1)$. Next we consider the case $k + 1$. By simple calculation, we can prove the following inequality:

$$\left|f^{(k+1)}\right| \le c\left\{\left|\nabla_z f^{(k)}\right| + |f'|^2 + \left|f^{(k)}\right|^2\right\}. \tag{6.3.9}$$

Locally, we can write

$$f^{(k)} = \xi^{k-1} \otimes \sum_{\alpha=1}^{n} f_{\alpha k} \frac{\partial}{\partial w_\alpha},$$

where ξ is a local holomorphic frame for E, and so

$$|f^{(k)}|^2 = a \sum_{\alpha, \beta} h_{\alpha\beta} f_{\alpha k} \overline{f_{\beta k}}$$

with $a = |\xi|_l^{2(k-1)}$. Without loss of generality, we may assume $a = 1$ since $|f^{(k)}|^2$

is locally equivalent to $\sum h_{\alpha\beta} f_{\alpha k}\overline{f_{\beta k}}$. Since

$$\partial_z \overline{\partial}_z \log\left(1 + \sum_{\alpha,\beta} h_{\alpha\beta} f_{\alpha k}\overline{f_{\beta k}}\right)$$

$$= -\frac{\sum_{\alpha,\beta,\gamma,\delta} K_{\alpha\beta\gamma\delta} f_{\alpha k}\overline{f_{\beta k}} f'_\gamma \overline{f'_\delta}}{1 + \sum_{\alpha,\beta} h_{\alpha\beta} f_{\alpha k}\overline{f_{\beta k}}} + \frac{\sum_{\alpha,\beta} h_{\alpha\beta} \nabla_z f_{\alpha k}\overline{\nabla_z f_{\beta k}}}{\left(1 + \sum_{\alpha,\beta} h_{\alpha\beta} f_{\alpha k}\overline{f_{\beta k}}\right)^2}$$

$$+ \frac{\left(\sum_{\alpha,\beta} h_{\alpha\beta} f_{\alpha k}\overline{f_{\beta k}}\right)\left(\sum_{\alpha,\beta} h_{\alpha\beta} \nabla_z f_{\alpha k}\overline{\nabla_z f_{\beta k}}\right) - \left|\sum_{\alpha,\beta} h_{\alpha\beta} f_{\alpha k}\overline{\nabla_z f_{\beta k}}\right|^2}{\left(1 + \sum_{\alpha,\beta} h_{\alpha\beta} f_{\alpha k}\overline{f_{\beta k}}\right)^2},$$

following the argument of the inequality (6.3.5), we can also obtain

$$\|\quad T\left(r, r_0; \frac{\sum_{\alpha,\beta} h_{\alpha\beta} \nabla_z f_{\alpha k}\overline{\nabla_z f_{\beta k}}}{\left(1 + \sum_{\alpha,\beta} h_{\alpha\beta} f_{\alpha k}\overline{f_{\beta k}}\right)^2} v\right)$$

$$\le 2\varepsilon \log r + K^+ T\left(r, r_0; |f'|^2 v\right) + \frac{1}{\mu}\left(\frac{1}{2} + \varepsilon\right)\log T\left(r, r_0; v + |f^{(k)}|^{2\mu} v\right),$$

where $K^+ = \max\{K, 0\}$, and

$$K = \sup_{N,\xi,\eta} \frac{\sum_{\alpha,\beta,\gamma,\delta} K_{\alpha\beta\gamma\delta}\xi_\alpha\overline{\xi_\beta}\eta_\gamma\overline{\eta_\delta}}{\left(\sum_{\alpha,\beta} h_{\alpha\beta}\xi_\alpha\overline{\xi_\beta}\right)\left(\sum_{\gamma,\delta} h_{\gamma\delta}\eta_\gamma\overline{\eta_\delta}\right)} < +\infty, \qquad (6.3.10)$$

and where Lemma 2.92 yields

$$\|\quad \mathbb{C}\left\langle 0; r; \log\left(1 + |f^{(k)}|^2\right)\right\rangle \le \frac{1}{\mu}\mathbb{C}\left\langle 0; r; \log\left(1 + |f^{(k)}|^{2\mu}\right)\right\rangle$$

$$\le \frac{1 + 2\varepsilon}{\mu}\log T\left(r, r_0; v + |f^{(k)}|^{2\mu} v\right)$$

$$+ \frac{4\varepsilon}{\mu}\log r.$$

By (6.3.3) and the induction assumption, it follows that

$$\|\quad cT\left(r, r_0; \frac{\sum_{\alpha,\beta} h_{\alpha\beta} \nabla_z f_{\alpha k}\overline{\nabla_z f_{\beta k}}}{\left(1 + \sum_{\alpha,\beta} h_{\alpha\beta} f_{\alpha k}\overline{f_{\beta k}}\right)^2} v\right) \le T_f(r, L) + r^2. \qquad (6.3.11)$$

Take $\mu \in \mathbb{R}$ with $0 < \mu < \frac{1}{4k}$. Using the elementary inequality

$$2\left|\nabla_z f^{(k)}\right|^{2\mu} \le \left(\frac{|\nabla_z f^{(k)}|^2}{(1 + |f^{(k)}|^2)^2}\right)^{2\mu} + \left((1 + |f^{(k)}|^2)^2\right)^{2\mu}$$

and the Hölder inequality, we have

$$2\mathcal{T}\left(r, r_0; |\nabla_z f^{(k)}|^{2\mu} v\right) \leq \mathcal{T}\left(r, r_0; \left(\frac{|\nabla_z f^{(k)}|^2}{(1+|f^{(k)}|^2)^2}\right)^{2\mu} v\right)$$

$$+ \mathcal{T}\left(r, r_0; \left(1+|f^{(k)}|^2\right)^{4\mu} v\right)$$

$$\leq \mathcal{T}\left(r, r_0; \frac{|\nabla_z f^{(k)}|^2}{(1+|f^{(k)}|^2)^2} v\right)^{2\mu} \mathcal{T}\left(r, r_0; v\right)^{1-2\mu}$$

$$+ \mathcal{T}\left(r, r_0; v + |f^{(k)}|^{8\mu} v\right),$$

and hence

$$\| \quad c\mathcal{T}\left(r, r_0; |\nabla_z f^{(k)}|^{2\mu} v\right) \leq (T_f(r, L) + r^2)^{2\mu} r^{2(1-2\mu)}$$

$$+ r^4 + T_f(r, L)^{4^{k+\frac{1}{2}}\mu}.$$

Therefore

$$\| \quad c\mathcal{T}\left(r, r_0; |\nabla_z f^{(k)}|^{2\mu} v\right) \leq T_f(r, L)^{4^{k+\frac{1}{2}}\mu} + r^4. \qquad (6.3.12)$$

The induction assumption implies

$$\| \quad \mathcal{T}\left(r, r_0; |f^{(k)}|^{4\mu} v\right) \leq c\left\{T_f(r, L)^{4^k \mu} + r^4\right\}.$$

So our estimate for $\mathcal{T}\left(r, r_0; |f^{(k+1)}|^{2\mu} v\right)$ is given by

$$\| \quad \mathcal{T}\left(r, r_0; |f^{(k+1)}|^{2\mu} v\right) \leq c\left\{\mathcal{T}\left(r, r_0; |\nabla_z f^{(k)}|^{2\mu} v\right) + \mathcal{T}\left(r, r_0; |f'|^{4\mu} v\right)\right.$$

$$\left. + \mathcal{T}\left(r, r_0; |f^{(k)}|^{4\mu} v\right)\right\}$$

$$\leq c'\left\{T_f(r, L)^{4^{k+\frac{1}{2}}\mu} + T_f(r, L)^{2\mu}\mathcal{T}\left(r, r_0; v\right)^{1-2\mu} + r^4\right\}$$

$$\leq c''\left\{T_f(r, L)^{4^{k+\frac{1}{2}}\mu} + r^4\right\},$$

and hence Theorem 6.12 is proved. $\square$

6.3.2 Generalization of Siu's theorem

Siu [363] uses the inequality in Theorem 6.12 to prove a defect relation. A special case of *Siu's Theorem* is presented by Nadel [285] as follows:

Theorem 6.13. *Let $f : \mathbb{C} \longrightarrow N$ be a transcendental holomorphic mapping into a smooth projective algebraic variety N of dimension n. Let $K_N \otimes E^{-n(n-1)/2}$ be ample. Then either $f(\mathbb{C})$ is contained in the pole divisor (t) of the meromorphic connection $\mathbf{D}$ or the image of f is autoparallel with respect to $\mathbf{D}$.*

Further, Nadel notes that in Theorem 6.13, the hypothesis on transcendence of f is not needed because one may always replace $f(z)$ by $f(e^z)$. We can omit the ampleness condition of the line bundle in Theorem 6.13 and obtain the following result:

Theorem 6.14 ([167]). *Let* $f : \mathbb{C} \longrightarrow N$ *be a transcendental holomorphic mapping into a smooth projective algebraic variety* N *of dimension* n. *Then either* $f(\mathbb{C}) \subseteq B(L)$ *for any positive line bundle* L *on* N, *or the image of* f *is autoparallel with respect to meromorphic connections on* N.

Proof. Assume, to the contrary, that $f(\mathbb{C}) \not\subseteq B(L)$ for a positive line bundle L on N and

$$f' \wedge f^{(2)} \wedge \cdots \wedge f^{(n)} \not\equiv 0$$

for a meromorphic connection $\mathbf{D}$ on N. The condition $f(\mathbb{C}) \not\subseteq B(L)$ implies $B(L) \neq N$. Since f is not rational, by Proposition 2.82, we have

$$A_f(\infty, L) = \lim_{r \to \infty} A_f(r, L) = \infty. \tag{6.3.13}$$

Since any line bundle on $\mathbb{C}$ is trivial (cf. Theorem 2.16), the pullback $f_f^{(k)}$ of $f^{(k)}$ under f can be identified with a section of $f^*(\mathbf{T}(N))$ under the identification

$$f^*(E^{k-1} \otimes \mathbf{T}(N)) \cong f^*(\mathbf{T}(N))$$

such that

$$f_f' \wedge f_f^{(2)} \wedge \cdots \wedge f_f^{(n)} \not\equiv 0.$$

By Lemma 2.86 and (6.3.1), the holomorphic field

$$\varphi = f_f^{(2)} \wedge \cdots \wedge f_f^{(n)}$$

on f over $\mathbb{C}$ of degree $n - 1$ is effective, that is, φ induces an effective Jacobian section F_φ of f. Define a non-negative function g by

$$F_\varphi[\psi^n] = g^2 f^*(\psi), \tag{6.3.14}$$

where $\psi = c_1(L, \kappa) > 0$. By Lemma 2.87, we have $g \leq |\varphi|$, where the metric of $f^*(\mathbf{T}(N))$ is induced by the metric on N defined by ψ.

By Lemma 1.55, there exists a constant $c > 0$ such that

$$\left| f_f^{(2)} \wedge \cdots \wedge f_f^{(n)} \right| \leq c \left| f_f^{(2)} \right| \cdots \left| f_f^{(n)} \right| = c \left| f^{(2)} \right| \cdots \left| f^{(n)} \right|,$$

and so an elementary inequality implies

$$g^{\frac{2}{n-1}} \leq \left| f_f^{(2)} \wedge \cdots \wedge f_f^{(n)} \right|^{\frac{2}{n-1}} \leq \frac{c^{\frac{2}{n-1}}}{n-1} \sum_{k=2}^{n} \left| f^{(k)} \right|^2. \tag{6.3.15}$$

Take $\mu \in \mathbb{R}$ with $0 < \mu < \frac{1}{4^n - 1}$. By Lemma 2.92, we obtain

$$
\| \quad \mathbb{C}\langle 0; r; \log g \rangle = \frac{n-1}{2\mu} \mathbb{C}\left\langle 0; r; \log g^{\frac{2\mu}{n-1}} \right\rangle
$$

$$
\leq \frac{n-1}{2\mu} \mathbb{C}\left\langle 0; r; \log \sum_{k=2}^{n} \left| f^{(k)} \right|^{2\mu} \right\rangle + O(1)
$$

$$
\leq \frac{n-1}{2\mu} \left\{ (1 + 2\varepsilon) \log T \left(r, r_0; \sum_{k=2}^{n} |f^{(k)}|^{2\mu} v \right) + 4\varepsilon \log r \right\}.
$$

$$(6.3.16)$$

Further, by Theorem 6.12, we have

$$
\| \quad \mathbb{C}\langle 0; r; \log g \rangle = O(\log T_f(r, L)) + O(\log r). \qquad (6.3.17)
$$

By Theorem 6.2, we obtain

$$
\| \quad T_f(r, L) = O(\log r). \qquad (6.3.18)
$$

Therefore we have

$$
A_f(\infty, L) = \lim_{r \to \infty} \frac{T_f(r, L)}{\log r} < \infty
$$

which contradicts (6.3.13). Hence Theorem 6.14 is proved. $\qquad \square$

Theorem 6.15 ([167]). *Let $f : \mathbb{C} \longrightarrow N$ be a holomorphic mapping into a pseudo-canonical smooth projective algebraic variety N. Then either f is algebraically degenerate, or the image of f is autoparallel with respect to meromorphic connections on N.*

Proof. If $\dim N = 1$, Theorem 6.15 follows from Theorem 4.57. Thus we may assume $\dim N > 1$. If f is rational, then it is degenerate, so we may assume that f is transcendental. It is well known that there exist very ample line bundles over the projective algebraic variety N, which are also positive. Since N is pseudo canonical (or general type), according to Lemma 2.30 (or see Kodaira [213], Kobayashi and Ochiai [210], Lang [228]) any very ample line bundle L satisfies $P_k(L) > 0$ for some sufficiently large k. Hence we have $B(L) \neq N$, and consequently Theorem 6.15 follows from Theorem 6.14. $\qquad \square$

If we use Lemma 2.85 to construct effective Jacobian sections by replacing Lemma 2.86 in the proof of Theorem 6.14, we can prove the following fact: If $f : \mathbb{C} \longrightarrow N$ is a non-constant holomorphic mapping into a pseudo canonical smooth projective algebraic variety N of dimension n with holomorphic vector fields $Z_1, \ldots, Z_n$ on N satisfying

$$
Z = Z_1 \wedge \cdots \wedge Z_n \not\equiv 0,
$$

then we have either $f(\mathbb{C}) \subset \mathrm{supp}((Z))$ or $f(\mathbb{C}) \subset B(L)$ for any positive line bundle L on N, where (Z) is the zero divisor of Z.

In fact, assume, to the contrary, that $f(\mathbb{C}) \not\subset \mathrm{supp}((Z))$ and $f(\mathbb{C}) \not\subset B(L)$ for a positive line bundle L on N. By using Lemma 2.85, then there exists $\lambda \in J_{1,n-1}^n$ such that a holomorphic field

$$\varphi = Z_{\lambda f} = (Z_{\lambda(1)} \wedge \cdots \wedge Z_{\lambda(n-1)})_f$$

on f over $\mathbb{C}$ of degree $n-1$ is effective for f. Hence φ induces an effective Jacobian section F_φ of f which, further, defines a non-negative function g by (6.3.14). By Lemma 2.87, we have $g \le |\varphi| \le c$ for a constant c. Then a contradiction follows from Theorem 6.2.

Generally, there is no such $Z \in \Gamma(N, K_N^*)$. For example, when N is canonical, that is, K_N is ample, then K_N is positive, and so $\Gamma(N, K_N^*) = 0$ by the remark after Theorem 6.1.

6.4 Bloch-Green's conjecture

Let H be the hyperplane bundle on $\mathbb{P}^n$ and let $X_0, \ldots, X_n$ be homogeneous coordinates on $\mathbb{P}^n$. Fix a positive integer d. A holomorphic section of H^d over $\mathbb{P}^n$ will be identified with a homogeneous polynomial of degree d in homogeneous coordinates $X_0, \ldots, X_n$.

Take a family

$$\{s_0, s_1, \ldots, s_n\} \subset \Gamma(\mathbb{P}^n, H^d)$$

such that

$$B_0 = \det\left(\frac{\partial s_i}{\partial X_j}\right) \neq 0. \tag{6.4.1}$$

Next we follow Siu [363] and Nadel [285] to construct a meromorphic connection $\mathbf{D}$ on $\mathbb{P}^n$. Fix α, β with $1 \le \alpha, \beta \le n$ and consider the system of equations

$$\partial_\alpha \partial_\beta \begin{pmatrix} s_0 \\ \vdots \\ s_n \end{pmatrix} = \begin{pmatrix} s_0 & \partial_1 s_0 & \cdots & \partial_n s_0 \\ \vdots & \vdots & & \vdots \\ s_n & \partial_1 s_n & \cdots & \partial_n s_n \end{pmatrix} \begin{pmatrix} \Gamma_{\alpha\beta}^0 \\ \vdots \\ \Gamma_{\alpha\beta}^n \end{pmatrix}, \tag{6.4.2}$$

where

$$\partial_i = \frac{\partial}{\partial z_i}, \quad z_i = \frac{X_i}{X_0}, \quad i = 1, \ldots, n.$$

It is clear that the determinant of the square matrix in (6.4.2) is equal to

$$B := \frac{1}{d} X_0^{-(n+1)(d-1)} B_0 \neq 0.$$

Thus we can solve $\Gamma^\gamma_{\alpha\beta}$ for each choice of α, β to obtain a symmetric meromorphic connection $\mathbf{D}$ on the coordinate patch $U_0 = \{X_0 \neq 0\} \subset \mathbb{P}^n$. Note that we do not use $\Gamma^0_{\alpha\beta}$ to define $\mathbf{D}$. By using the transformation law for a connection we can get easily all the components of the connection for other coordinate charts $U_i = \{X_i \neq 0\} \subset \mathbb{P}^n$, $i = 1, \ldots, n$.

We use Cramer's rule to write

$$\Gamma^\gamma_{\alpha\beta} = \frac{A^\gamma_{\alpha\beta}}{B},$$

where $A^\gamma_{\alpha\beta}$, B are polynomials of degree $\leq (d-1)(n+1)+1$ in the inhomogeneous coordinates $z_1, \ldots, z_n$ on $\mathbb{P}^n$. By Theorem 6.9, there exists a global holomorphic section

$$t \in \Gamma(\mathbb{P}^n, H^{g+3}) - \{0\}, \quad g \leq (d-1)(n+1)+1 \tag{6.4.3}$$

such that $t\mathbf{D}$ is holomorphic. If s is any linear combination of $s_0, \ldots, s_n$, then (6.4.2) gives

$$\partial_\alpha \partial_\beta s = \sum_\gamma \Gamma^\gamma_{\alpha\beta} \partial_\gamma s + \Gamma^0_{\alpha\beta} s, \tag{6.4.4}$$

which is the same as (6.2.2) with $C_{\alpha\beta} = \Gamma^0_{\alpha\beta}$ and $A_\alpha = 0 = B_\alpha$.

Theorem 6.16 ([167]). *Let W be a hypersurface of degree d in $\mathbb{P}^n$ with $d \geq n+2$ and let $f : \mathbb{C} \longrightarrow W$ be a holomorphic mapping. Then f is algebraically degenerate.*

Proof. Without loss of generality, we may assume that W is irreducible. The line bundle $[W]$ is of the form H^d. The adjunction formula immediately implies that

$$K_W = (K_{\mathbb{P}^n} \otimes [W])|_W = L^{d-n-1},$$

where $L = H|_W$. Then $d \geq n+2$ is precisely the condition that makes the canonical bundle K_W ample. The line bundle

$$K_W^k \otimes L^* = L^{k(d-n-1)-1}$$

also is ample for $d \geq n+2$, $k \geq 2$. Hence

$$B(L) \subseteq B_2(L) \neq W.$$

The hypersurface W is the zero locus of a section s' of H^d. However, all sections of H^d are defined by homogeneous polynomials of degree d in homogeneous coordinates $X_0, \ldots, X_n$ on $\mathbb{P}^n$ and so

$$W = (s') = (G(X_0, \ldots, X_n) = 0)$$

for some homogeneous irreducible polynomial G of degree d. Observing that

$$\dim \Gamma(\mathbb{P}^n, H^d) = \binom{n+d}{n} > n+1,$$

we can choose $s_0, s_1, \ldots, s_n$ in $\Gamma(\mathbb{P}^n, H^d)$ such that

$$s' = c_0 s_0 + c_1 s_1 + \cdots + c_n s_n, \tag{6.4.5}$$

where $c_0, \ldots, c_n$ are complex numbers, not all zero, and such that

$$B_0 = \det\left(\frac{\partial s_i}{\partial X_j}\right) \not\equiv 0.$$

For example, without loss of generality we may assume that

$$\frac{\partial s'}{\partial X_0} \not\equiv 0,$$

and so we can choose

$$s_0 = s', \quad s_i = X_i^d \ (1 \le i \le n).$$

Thus we have

$$B_0 = d^n (X_1 \cdots X_n)^{d-1} \frac{\partial s'}{\partial X_0} \not\equiv 0.$$

If $f(\mathbb{C}) \subseteq \left\{\frac{\partial s'}{\partial X_0} = 0\right\}$ or $f(\mathbb{C}) \subseteq \{X_i = 0\}$ for some $i \in \{0, 1, \ldots, n\}$, then we are done. So we may assume that

$$f(\mathbb{C}) \not\subseteq \left\{\frac{\partial s'}{\partial X_0} = 0\right\}; \quad f(\mathbb{C}) \not\subseteq \{X_i = 0\} \ (i = 0, 1, \ldots, n).$$

By using (6.4.2), we can define a meromorphic connection $\mathbf{D}$ such that $t\mathbf{D}$ is holomorphic, where t is given in (6.4.3). In particular, s' satisfies the equation (6.4.4). Hence W is totally geodesic. Then $\mathbf{D}$ restricts to a meromorphic connection $\mathbf{D}|_W$ on W, and moreover that $(t|_W)\mathbf{D}|_W$ is holomorphic on W.

If $f(\mathbb{C}) \subseteq B(L)$ or $f(\mathbb{C}) \subseteq (t) \cap W$, then we are done. Assume now that

$$f(\mathbb{C}) \not\subseteq B(L), \quad f(\mathbb{C}) \not\subseteq (t) \cap W.$$

Theorem 6.14 implies

$$f' \wedge f'' \wedge \cdots \wedge f^{(n-1)} \equiv 0.$$

Pick $z_0 \in \mathbb{C}$ such that $f(z_0) \in W - (t)$. Let V be the vector space of all $s \in \Gamma(\mathbb{P}^n, H^d)$ such that the following conditions hold:

(i) s is a complex linear combination of $s_0, \ldots, s_n$;
(ii) $s(f(z_0)) = 0$;
(iii) $(\mathscr{D}s)(f^{(i)}(z_0)) = 0$ for $i = 1, 2, \ldots, n - 2$.

Here $\mathscr{D}$ is any meromorphic connection on H^d which is holomorphic in a neighborhood of $f(z_0)$. Since there are $n+1$ linearly independent sections $s_0, \ldots, s_n$ from which to form s (see the condition (i) above) but only $n-1$ constraints (given by the conditions (ii) and (iii)), we see that $\dim V \geq 2$. Pick one particular $s'' \in V$ such that s' and s'' are linearly independent. By Theorem 6.11, we have

$$s''(f(z)) = 0, \ z \in \mathbb{C}.$$

Thus $f(\mathbb{C}) \subseteq W \cap (s'')$. The proof is completed.   □

Classically, Theorem 6.16 was known for $n = 1$ and $n = 2$ respectively. For the case $n = 3$, this is a question due to A. Bloch [27]. Generally, Theorem 6.16 was stated as a conjecture by Mark Lee Green [121], who proved it for Fermat hypersurfaces of degree at least n^2. A.M. Nadel [285] also obtained it for certain hypersurfaces of high degree.

6.5 Green-Griffiths' conjecture

In this section, we prove a more general result than Theorem 2.99.

Theorem 6.17 ([183]). *Let $f : \mathbb{C} \longrightarrow N$ be a holomorphic mapping into a pseudo-canonical projective algebraic variety N. Then f is algebraically degenerate.*

Proof. In general, the variety N may have singularities. However, by resolution of singularities, we may assume that N is smooth. Also it is sufficient to work on a component of N. Without loss of generality, we may assume that N is connected, and so is irreducible.

Let N be of the dimension n. We know that N can be embedded into a complex projective space, say, $N \subset \mathbb{P}^{n+m}$ for some positive integer $m \geq 1$. Let H be the hyperplane bundle on $\mathbb{P}^{n+m}$ and let $X_0, \ldots, X_{n+m}$ be homogeneous coordinates on $\mathbb{P}^{n+m}$. By Theorem 1.65 or Theorem 1.64, we may assume that the variety N is expressible as the zero locus of m linearly independent sections $s'_1, \ldots, s'_m$ of $H^{d_1}, \ldots, H^{d_m}$. Here s'_i is given by a homogeneous polynomial in $X_0, \ldots, X_{n+m}$ of degree d_i for each $i = 1, \ldots, m$ such that on N,

$$\mathrm{rank}\left(\frac{\partial s'_i}{\partial X_j}\right) = m$$

since N is smooth. Further we may assume that each s'_i is irreducible since N is irreducible.

There exists an integer i_0 with $0 \leq i_0 \leq n + m$ such that $s'_1, \ldots, s'_m, X_{i_0}$ are linearly independent. Without loss of generality, we may assume $i_0 = 0$. If $f(\mathbb{C}) \subseteq N \cap \{X_0 = 0\}$, we are done. So we may assume that $f(\mathbb{C}) \not\subseteq N \cap \{X_0 = 0\}$. Further we can assume that

$$d_1 = d_2 = \cdots = d_m := d,$$

otherwise it is sufficient to consider

$$s_i'' = X_0^{d-d_i} s_i', \quad i = 1, \ldots, m, \quad d = \max_i \{d_i\}.$$

Observing that

$$\dim \Gamma(\mathbb{P}^{n+m}, H^d) = \binom{n+m+d}{n+m} > n+m+1,$$

we can choose $s_0, s_1, \ldots, s_{n+m}$ in $\Gamma(\mathbb{P}^{n+m}, H^d)$ such that

$$s_i' = c_{i0} s_0 + c_{i1} s_1 + \cdots + c_{i,n+m} s_{n+m}, \tag{6.5.1}$$

where $c_{i0}, \ldots, c_{i,n+m}$ are complex numbers, not all zero, and such that on N,

$$B_0 = \det\left(\frac{\partial s_i}{\partial X_j}\right) \not\equiv 0.$$

Thus according to the construction in Section 6.4, we can define a meromorphic connection $\mathbf{D}$ on $\mathbb{P}^{n+m}$ such that $t\mathbf{D}$ is holomorphic, where

$$t \in \Gamma(\mathbb{P}^{n+m}, H^{g+3}) - \{0\}, \quad g \leq (d-1)(n+m+1) + 1. \tag{6.5.2}$$

In particular, by the construction of $\mathbf{D}$ we have

$$\partial_\alpha \partial_\beta s_i' = \sum_\gamma \mathbf{\Gamma}_{\alpha\beta}^\gamma \partial_\gamma s_i' + \mathbf{\Gamma}_{\alpha\beta}^0 s_i', \quad i = 1, \ldots, m. \tag{6.5.3}$$

By using (6.5.3), similar to the argument in Section 6.2, we may show that N is totally geodesic. Thus $\mathbf{D}$ restricts to a meromorphic connection $\mathbf{D}|_N$ on N, and moreover that $(t|_N)\mathbf{D}|_N$ is holomorphic on N.

Set $L = H|_N$. If $f(\mathbb{C}) \subseteq B(L)$ or $f(\mathbb{C}) \subseteq (t) \cap N$, then we are done. Assume now that

$$f(\mathbb{C}) \not\subseteq B(L), \quad f(\mathbb{C}) \not\subseteq (t) \cap N.$$

Theorem 6.14 implies

$$f' \wedge f'' \wedge \cdots \wedge f^{(n)} \equiv 0.$$

Pick $z_0 \in \mathbb{C}$ such that $f(z_0) \in N - (t)$. Let V be the vector space of all $s \in \Gamma(\mathbb{P}^{n+m}, H^d)$ such that the following conditions hold:

(i) s is a complex linear combination of $s_0, \ldots, s_{n+m}$;

(ii) $s(f(z_0)) = 0$;

(iii) $(\mathcal{D}s)(f^{(i)}(z_0)) = 0$ for $i = 1, 2, \ldots, n-1$.

Here $\mathscr{D}$ is any meromorphic connection on H^d which is holomorphic in a neighborhood of $f(z_0)$. Since there are $n+m+1$ linearly independent sections $s_0, \ldots, s_{n+m}$ from which to form s (see the condition (i) above) but only n constraints (given by the conditions (ii) and (iii)), we see that $\dim V \geq m + 1$. Pick one particular $s'_{m+1} \in V$ such that $s'_1, \ldots, s'_{m+1}$ are linearly independent. By Theorem 6.11, we have

$$s'_{m+1}(f(z)) = 0, \ z \in \mathbb{C}.$$

Thus $f(\mathbb{C}) \subseteq N \cap (s'_{m+1})$. The proof is completed. $\square$

Originally, Theorem 6.17 was a conjecture due to M. Green and P. Griffiths [123]. In [228], S. Lang formulated the Green-Griffiths conjecture into the form in Theorem 6.17. McQuillan [265], Lu and Yau [249] proved Theorem 6.17 for surfaces with $c_1^2 > c_2$. In fact, according to the proof of Theorem 6.17 we can obtain the following result:

Theorem 6.18. *If $f : \mathbb{C} \longrightarrow N$ is a holomorphic curve into a projective algebraic variety such that the image of f is autoparallel with respect to a meromorphic connection of N, then f is algebraically degenerate.*

Theorem 6.17 implies directly the following result:

Theorem 6.19. *If N is a pseudo-canonical projective algebraic variety such that all subvarieties are pseudo-canonical, then N is hyperbolic.*

Proof. By Theorem 6.17 and the assumptions of N, each holomorphic curve $f : \mathbb{C} \longrightarrow N$ must be constant. By a theorem of Brody [38], then N is hyperbolic. $\square$

Theorem 6.19 is Lang's conjecture [228] in which he observed that Theorem 6.19 can be deduced simply by Theorem 6.17.

Let N be a non-singular complex projective algebraic variety and $q(N)$ the dimension of the vector space of holomorphic 1-forms on N, that is,

$$q(N) = \dim \Gamma(N, \mathbf{T}^*(N)),$$

and $q(N)$ is called the *irregularity* of N. Related to Theorem 6.17, there is the following result:

Theorem 6.20. *Let N be a non-singular complex projective algebraic variety with $q(N) > \dim N$. Then any holomorphic curve $f : \mathbb{C} \longrightarrow N$ is algebraically degenerate.*

Theorem 6.20 was called Bloch's conjecture. Bloch [27] stated the theorem with an incomplete sketchy proof. Ochiai [307] filled the gaps substantially and proved it in several special cases. The final step for the general case is due to M. Green. Lately, M. Green and P. Griffiths [123] gave another proof by using certain metrics. Kawamata [200] proved the same result by a different method. A new proof was given by McQuillan [264].

In fact, according to Ochiai [307], if N is a connected projective algebraic manifold of dimension n with irregularity $> n$, then we can construct a regular rational mapping (Albanese map)

$$\alpha : N \longrightarrow A$$

into a certain Abelian variety A such that $X = \alpha(N)$ is in *good position* in A, that is, $X \neq A$ and X satisfies the following conditions:

(I) if ω is a non-zero regular rational 1-form on A, then the restriction $\omega|_{X_{\mathrm{reg}}}$ is non-zero, where X_{reg} is the set of regular points of X;

(II) if B is a connected algebraic subgroup of A such that B leaves X invariant, then B is either $\{0\}$ or A.

By Ueno's theorem (see Theorems 1.78, 1.79), the condition (II) is equivalent to saying that the proper subvariety X of A is pseudo-canonical, and so Theorem 6.17 can be applied.

On the other hand, by Kawamata [200] or Smyth [369] for non-singular X, there exists a system $\{\omega_1, \ldots, \omega_{l+1}\}$ $(l = \dim X)$ of regular 1-forms on A such that the restriction of the system of l-forms

$$\{\omega_1 \wedge \cdots \wedge \omega_{j-1} \wedge \omega_{j+1} \wedge \cdots \wedge \omega_{l+1} \mid j = 1, \ldots, l+1\}$$

onto X_{reg} is linearly independent. Let $\iota : \tilde{X} \longrightarrow X$ be the resolution of the singularity of X. If $\alpha \circ f(\mathbb{C})$ is in the singular locus of X, f is clearly algebraically degenerate. If

$$\alpha \circ f(\mathbb{C}) \cap X_{\mathrm{reg}} \neq \emptyset,$$

there exists a holomorphic curve $\tilde{f} : \mathbb{C} \longrightarrow \tilde{X}$ such that $\iota \circ \tilde{f} = \alpha \circ f$. Hence, by Ochiai [307], the image of the holomorphic curve $\tilde{f} : \mathbb{C} \longrightarrow \tilde{X}$ is contained in a canonical divisor of the form

$$\sum_{j=1}^{l+1} a_j \iota^* \omega_1 \wedge \cdots \wedge \iota^* \omega_{j-1} \wedge \iota^* \omega_{j+1} \wedge \cdots \wedge \iota^* \omega_{l+1} = 0.$$

Therefore f is algebraically degenerate.

6.6   Notes on Griffiths' and Lang's conjectures

We continue with the situation $f : M \longrightarrow N$ of Section 2.12, where we will take $M = \mathbb{C}$. In terms of a local coordinate system $w_1, \ldots, w_n$ of N we may write

$$\psi = \mathrm{Ric}(\Psi) = \frac{\sqrt{-1}}{2\pi} \sum_{\alpha, \beta} h_{\alpha\beta} dw_\alpha \wedge d\bar{w}_\beta.$$

The singular positive form ψ induces a metric on $N - D$ defined by

$$h_\Psi = 2 \sum_{\alpha,\beta} h_{\alpha\beta} dw_\alpha d\bar{w}_\beta$$

which has a "singularity" on D. Further, we may define the curvature tensor of the singular metric h_Ψ on N as

$$K_{\alpha\beta\gamma\delta} = -\frac{\partial^2 h_{\alpha\beta}}{\partial w_\gamma \partial \bar{w}_\delta} + \sum_{\mu,\nu} h^{\mu\nu} \frac{\partial h_{\alpha\mu}}{\partial w_\gamma} \frac{\partial h_{\nu\beta}}{\partial \bar{w}_\delta},$$

where $(h^{\mu\nu})$ is the inverse matrix of $(h_{\alpha\beta})$.

Lemma 6.21 ([184]). *The holomorphic bisectional curvatures of h_Ψ is bounded above.*

Proof. By using (2.12.4), we have

$$\mathrm{Ric}(\Psi) = \chi + 2 \sum_{j=1}^{q} \frac{c_1(L,\kappa)}{\log |s_j|_\kappa^2} + 2 \sum_{j=1}^{q} \frac{d \log |s_j|_\kappa^2 \wedge d^c \log |s_j|_\kappa^2}{(\log |s_j|_\kappa^2)^2}, \tag{6.6.1}$$

where

$$\chi = \lambda q c_1(L,\kappa) + \mathrm{Ric}(\Omega).$$

Around any point $x \in D$, one can choose coordinates $(w_1, \ldots, w_n)$ in a neighborhood U of x such that $x = (0, \ldots, 0)$ and for instance $D_j = (w_j)$ in U, this being because D has normal crossings. Hence we can write $s_j = w_j \chi_j$, where χ_j is a non-zero holomorphic section of L over U. Thus

$$\log |s_j|_\kappa^2 = \log |w_j|^2 + \log b_j,$$

where $b_j = |\chi_j|_\kappa^2 > 0$ is a C^∞ function. We can choose w_j and χ_j properly such that the connection matrix

$$\theta = \partial \log b_j = \sum_{i=1}^{n} \partial_{w_i} \log b_j dw_i$$

vanishes at x, which also implies

$$\bar\theta = \bar\partial \log b_j = \sum_{i=1}^{n} \partial_{\bar{w}_i} \log b_j d\bar{w}_i$$

vanishes at x.

Note that

$$d \log |s_j|_\kappa^2 \wedge d^c \log |s_j|_\kappa^2 = \frac{\sqrt{-1}}{2\pi} \frac{dw_j \wedge d\bar{w}_j}{|w_j|^2} + \varrho_j. \tag{6.6.2}$$

The form

$$\varrho_j = \frac{\sqrt{-1}}{2\pi} \left(\frac{\partial b_j \wedge \bar{\partial} b_j}{b_j^2} + \frac{\partial b_j \wedge d\bar{w}_j}{b_j \bar{w}_j} + \frac{dw_j \wedge \bar{\partial} b_j}{w_j b_j} \right)$$

has the property that $|w_j|^2 \varrho_j$ is a smooth form whose coefficients vanish on D_j. Thus the matrix $(h_{\alpha\beta})$ has the following form:

$$\begin{pmatrix} * & \cdots & \frac{2\partial_{w_1} \log b_j}{\bar{w}_j u_j^2} + * & \cdots & * \\ \cdots\cdots\cdots\cdots\cdots\cdots\cdots\cdots\cdots\cdots\cdots\cdots\cdots\cdots\cdots\cdots \\ \frac{2\partial_{\bar{w}_1} \log b_j}{w_j u_j^2} + * & \cdots & \frac{2}{|w_j|^2 u_j^2} + * & \cdots & \frac{2\partial_{\bar{w}_n} \log b_j}{w_j u_j^2} + * \\ \cdots\cdots\cdots\cdots\cdots\cdots\cdots\cdots\cdots\cdots\cdots\cdots\cdots\cdots\cdots\cdots \\ * & \cdots & \frac{2\partial_{w_n} \log b_j}{\bar{w}_j u_j^2} + * & \cdots & * \end{pmatrix}, \qquad (6.6.3)$$

where

$$u_j = \log |s_j|_\kappa^2 = \log |w_j|^2 + \log b_j.$$

Simple calculations show that

$$\frac{\partial}{\partial \bar{w}_j} \left(\frac{1}{|w_j|^2 u_j^2} \right) = -\frac{1}{w_j \bar{w}_j^2 u_j^2} - \frac{2}{w_j \bar{w}_j^2 u_j^3} - \frac{2}{|w_j|^2 u_j^3 b_j} \frac{\partial b_j}{\partial \bar{w}_j},$$

$$\frac{\partial}{\partial w_j} \left(\frac{1}{|w_j|^2 u_j^2} \right) = -\frac{1}{w_j^2 \bar{w}_j u_j^2} - \frac{2}{w_j^2 \bar{w}_j u_j^3} - \frac{2}{|w_j|^2 u_j^3 b_j} \frac{\partial b_j}{\partial w_j},$$

and

$$\frac{\partial^2}{\partial w_j \partial \bar{w}_j} \left(\frac{1}{|w_j|^2 u_j^2} \right) = \frac{1}{|w_j|^4 u_j^2} + \frac{4}{|w_j|^4 u_j^3} + \frac{6}{|w_j|^4 u_j^4} + \eta_j,$$

where

$$\eta_j = \frac{2}{w_j \bar{w}_j^2 u_j^3 b_j} \frac{\partial b_j}{\partial w_j} + \frac{6}{w_j \bar{w}_j^2 u_j^4 b_j} \frac{\partial b_j}{\partial w_j} + \frac{2}{w_j^2 \bar{w}_j u_j^3 b_j} \frac{\partial b_j}{\partial \bar{w}_j} + \frac{6}{w_j^2 \bar{w}_j u_j^4 b_j} \frac{\partial b_j}{\partial \bar{w}_j}$$
$$- \frac{2}{|w_j|^2 u_j^3} \frac{\partial^2 \log b_j}{\partial w_j \partial \bar{w}_j} + \frac{6}{|w_j|^2 u_j^4} \frac{\partial \log b_j}{\partial w_j} \frac{\partial \log b_j}{\partial \bar{w}_j}.$$

Since $\theta(x) = 0 = \bar{\theta}(x)$, the matrix (6.6.3) yields

$$\det(h_{\alpha\beta}) = \frac{2H_{jj}}{|w_j|^2 u_j^2} \left\{ 1 + o\left(\frac{1}{u_j^2} \right) \right\}$$

near x, where H_{jj} is the algebraic minor of h_{jj}. Thus

$$h^{jj} = \frac{H_{jj}}{\det(h_{\alpha\beta})} = \frac{1}{2} |w_j|^2 u_j^2 \left\{ 1 + o\left(\frac{1}{u_j^2} \right) \right\}.$$

Then

$$\frac{\partial h_{jj}}{\partial w_j}\frac{\partial h_{jj}}{\partial \bar{w}_j} = \frac{4}{|w_j|^6 u_j^4} + \frac{16}{|w_j|^6 u_j^5} + \frac{16}{|w_j|^6 u_j^6} + O\left(\frac{1}{|w_j|^4 \bar{w}_j u_j^5}\right),$$

and so

$$K_{j\bar{j}j\bar{j}} = \frac{1}{|w_j|^4 u_j^4}\left\{-4 + o(1)\right\}.$$

We consider cases $(\alpha, \beta) \neq (j, j)$, $(\gamma, \delta) \neq (j, j)$. Since

$$H_{\nu j} = o\left(\frac{1}{|w_j| u_j^2}\right), \quad \nu \neq j,$$

then

$$h^{j\nu} = \frac{H_{\nu j}}{\det(h_{\alpha\beta})} = o(|w_j|) \ (\nu \neq j),$$

and hence

$$K_{\alpha\beta\gamma\delta} = O(1)$$

near $w_j = 0$.

Next assume $(\gamma, \delta) \neq (j, j)$. It is easy to show

$$K_{j\beta\gamma\delta} = O\left(\frac{1}{|w_j| u_j^2}\right) \ (\beta \neq j)$$

but

$$K_{jj\gamma\delta} = O(1)$$

near $w_j = 0$.

Take two non-zero continuous tangent vector fields X and Y of $\mathbf{T}(M)$ over U, and write

$$X = \sum_k \xi_k \frac{\partial}{\partial w_k}, \quad Y = \sum_k \eta_k \frac{\partial}{\partial w_k}.$$

Then when $p \in U - D$, the holomorphic bisectional curvature determined by X_p and Y_p,

$$K(X_p, Y_p) = \left.\frac{\sum_{\alpha,\beta,\gamma,\delta} K_{\alpha\beta\gamma\delta} \xi_\alpha \overline{\xi_\beta} \eta_\gamma \overline{\eta_\delta}}{\left(\sum_{\alpha,\beta} h_{\alpha\beta} \xi_\alpha \overline{\xi_\beta}\right)\left(\sum_{\gamma,\delta} h_{\gamma\delta} \eta_\gamma \overline{\eta_\delta}\right)}\right|_p,$$

is well defined. The above argument shows

$$K(X_x, Y_x) = \limsup_{U - D \ni p \to x} K(X_p, Y_p) < +\infty.$$

When $x \in N - D$, $s_j = \chi_j$ vanishes nowhere near x, but we have to check the case $b_j(x) = |\chi_j(x)|_\kappa^2 \to 0$. Now we have

$$d \log |s_j|_\kappa^2 \wedge d^c \log |s_j|_\kappa^2 = \frac{\sqrt{-1}}{2\pi} \partial \log b_j \wedge \bar{\partial} \log b_j.$$

Note that

$$\frac{\partial}{\partial \bar{w}_\delta} \left(\frac{\partial_{w_\nu} b_j \partial_{\bar{w}_\beta} b_j}{b_j^2 u_j^2} \right) = -\frac{2 \partial_{w_\nu} b_j \partial_{\bar{w}_\beta} b_j \partial_{\bar{w}_\delta} b_j}{b_j^3 u_j^2} \left(1 + \frac{1}{u_j} \right) + \frac{\partial_{\bar{w}_\delta}(\partial_{w_\nu} b_j \partial_{\bar{w}_\beta} b_j)}{b_j^2 u_j^2},$$

$$\frac{\partial}{\partial w_\gamma} \left(\frac{\partial_{w_\alpha} b_j \partial_{\bar{w}_\mu} b_j}{b_j^2 u_j^2} \right) = -\frac{2 \partial_{w_\alpha} b_j \partial_{\bar{w}_\mu} b_j \partial_{w_\gamma} b_j}{b_j^3 u_j^2} \left(1 + \frac{1}{u_j} \right) + \frac{\partial_{w_\gamma}(\partial_{w_\alpha} b_j \partial_{\bar{w}_\mu} b_j)}{b_j^2 u_j^2},$$

vanish at x, and

$$\frac{\partial^2}{\partial w_\gamma \partial \bar{w}_\delta} \left(\frac{\partial_{w_\alpha} b_j \partial_{\bar{w}_\beta} b_j}{b_j^2 u_j^2} \right) = \frac{1}{b_j^2 u_j^2} (\partial_{w_\gamma} \partial_{\bar{w}_\beta} b_j \partial_{\bar{w}_\delta} \partial_{w_\alpha} b_j + \partial_{w_\gamma} \partial_{w_\alpha} b_j \partial_{\bar{w}_\delta} \partial_{\bar{w}_\beta} b_j) + \zeta_j,$$

where ζ_j vanishes at x. Denote the quantity in brackets by $\tilde{K}_{\alpha\beta\gamma\delta}$. Then

$$\sum_{\alpha,\beta,\gamma,\delta} \tilde{K}_{\alpha\beta\gamma\delta} \xi_\alpha \overline{\xi_\beta} \eta_\gamma \overline{\eta_\delta} \geq 0,$$

and so

$$\limsup_{b_j(x) \to 0} K(X_x, Y_x) < +\infty.$$

Finally, we have

$$K = \sup_{N,\xi,\eta} \frac{\sum_{\alpha,\beta,\gamma,\delta} K_{\alpha\beta\gamma\delta} \xi_\alpha \overline{\xi_\beta} \eta_\gamma \overline{\eta_\delta}}{\left(\sum_{\alpha,\beta} h_{\alpha\beta} \xi_\alpha \overline{\xi_\beta} \right) \left(\sum_{\gamma,\delta} h_{\gamma\delta} \eta_\gamma \overline{\eta_\delta} \right)} < \infty,$$

and so Lemma 6.21 follows.   □

Let $\mathbf{D}$ be a meromorphic connection for $\mathbf{T}(N)$ with pole order E such that $t\mathbf{D}$ is holomorphic for some $t \in \Gamma(N, E) - \{0\}$. We will use the notions in Section 6.3.

Lemma 6.22 ([184]). *For $k \geq 1$, we have*

$$T\left(r, r_0; dd^c \log(1 + |f^{(k)}|^2)\right) \leq N_f(r, D) + \frac{1}{2} \mathbb{C}\langle 0; r; \log(1 + |f^{(k)}|^2)\rangle$$

$$- \frac{1}{2} \mathbb{C}\langle 0; r_0; \log(1 + |f^{(k)}|^2)\rangle.$$

Proof. Here we show only the case $k = 1$. Other cases can be proved similarly. Note that $f^*(D)$ consists of discrete points in $\mathbb{C}$. Without loss of generality, we may assume that $f^*(D) \cap \mathbb{C}\langle 0; r \rangle = \emptyset$. Set

$$\operatorname{supp}(f^*(D)) \cap \mathbb{C}[0; r] = \{z_1, \ldots, z_l\}.$$

Let $\mathbb{C}(z_i; \varepsilon)$ denote the open disc centered at z_i with radius ε. Choose a positive ε sufficient small such that

$$\mathbb{C}(z_i; \varepsilon) \subset \mathbb{C}[0; r], \quad i = 1, \ldots, l.$$

By using the symbols in the proof of Lemma 6.21, and set

$$f_\alpha = w_\alpha \circ f, \quad \alpha = 1, \ldots, n.$$

There exist integers l_i and m_i with

$$m_i > 0, \quad 0 \le l_i \le 2\mu_{f^*(D)}(z_i), \quad i = 1, \ldots, l,$$

such that

$$|f'(z)|^2 = \frac{G_i(z)}{|z - z_i|^{l_i}(\log|z - z_i|^{2m_i})^2}, \quad i = 1, \ldots, l$$

hold on $\mathbb{C}(z_i; \varepsilon)$, by choosing ε sufficiently small if necessary, where G_i is a continuous positive function. Applying (6.1.1) to the domain

$$A = \mathbb{C}(0; r) - \bigcup_{i=1}^{l} \mathbb{C}[z_i; \varepsilon],$$

we obtain

$$\int_A \Delta \log(1 + |f'|^2) dx dy = \int_{\mathbb{C}\langle 0; r \rangle} \frac{\partial \log(1 + |f'|^2)}{\partial r} ds$$
$$- \sum_{i=1}^{l} \int_{\partial \mathbb{C}[z_i; \varepsilon]} \frac{\partial \log(1 + |f'|^2)}{\partial \varepsilon} ds. \qquad (6.6.4)$$

Simple calculation shows that

$$\lim_{\varepsilon \to 0} \int_{\partial \mathbb{C}[z_i; \varepsilon]} \frac{\partial \log(1 + |f'|^2)}{\partial \varepsilon} ds = -2\pi l_i, \quad i = 1, \ldots, l.$$

Thus from (6.6.4), we obtain

$$\int_{\mathbb{C}(0; r)} \Delta \log(1 + |f'|^2) dx dy \le \int_{\mathbb{C}\langle 0; r \rangle} \frac{\partial \log(1 + |f'|^2)}{\partial r} ds + 4\pi n_f(r, D),$$

and hence our claim follows according to the proof of (6.1.3). □

We choose a finite open covering $\mathcal{B} = \{W, U, \dots\}$ of N with coordinates $w = (w_1, \dots, w_n)$, $u = (u_1, \dots, u_n), \dots$, and the coefficients $\boldsymbol{\Gamma}_{\gamma\delta}^{\alpha}$, $\boldsymbol{\Xi}_{\gamma\delta}^{\alpha}, \dots$, of the meromorphic connection $\mathbf{D}$, respectively. Let χ_W be a holomorphic local frame of E over W and write $t = h_W \chi_W$. We can define holomorphic vector fields of $\mathbf{T}(N)$ over W by

$$X_{\gamma\delta}^{W} = \sum_{\alpha} h_W \boldsymbol{\Gamma}_{\gamma\delta}^{\alpha} \frac{\partial}{\partial w_{\alpha}},$$

and then extend $X_{\gamma\delta}^{W}$ continuously to vector fields of $\mathbf{T}(N)$ over N.

Theorem 6.23 ([184]). *Let $f : \mathbb{C} \longrightarrow N$ be a non-constant holomorphic mapping. Then there exists a positive constant c such that*

$$\| \ \mathbb{C}\left\langle 0; r; \log^+ |f^{(k)}| \right\rangle \leq \max\left\{ \frac{5}{3}(4^{k-2} - 1) + 1, 0 \right\} R_f(r) + O\left\{ \log(rT_f(r, L)) \right\}$$

$$+ c \max\{k - 2, 0\} \log^+ R_f(r). \tag{6.6.5}$$

Proof. Since f is non-constant, then $T_f(r, L) \to +\infty$ as $r \to +\infty$. Without loss of generality, we may assume that $T_f(r, L) \geq 1$, and that

$$\mathfrak{R} = \operatorname{tr}(\psi) = \operatorname{tr}(\operatorname{Ric}(\Psi)) \geq 1.$$

First of all, we consider the case $k = 1$. Lemma 2.94 yields

$$\mathcal{T}\left(r, r_0; |f'|^2 v\right) \leq c T_f(r, L).$$

By Lemma 2.92, we obtain

$$\| \ \mathbb{C}\left\langle 0; r; \log^+ |f'| \right\rangle \ \leq \ \frac{1}{2} \mathbb{C}\left\langle 0; r; \log(1 + |f'|^2) \right\rangle$$

$$\leq \ \left(\frac{1}{2} + \varepsilon \right) \log \mathcal{T}\left(r, r_0; (1 + |f'|^2)v \right) + 2\varepsilon \log r$$

$$\leq \ \left(\frac{1}{2} + \varepsilon \right) \log T_f(r, L) + O(\log r),$$

and hence the inequality (6.6.5) follows with $k = 1$.

Next we consider the case $k = 2$. The inequality (6.3.4) has the following form:

$$|f^{(2)}|^2 = |t\mathbf{D}_z f'|^2 \leq c \left\{ |\nabla_z f'|^2 + |f'|_{\kappa}^8 + \sum_{W \in \mathcal{B}} \left(\sum_{\gamma, \delta} |X_{\gamma\delta f}^W| \right)^4 \right\}, \tag{6.6.6}$$

where $|f'|_{\kappa}$ is the norm of f' with respect to the Hermitian metric on N induced by $c_1(L, \kappa)$. By Lemma 6.21 and Lemma 6.22, the estimate (6.3.6) holds for the singular metric, and so we can obtain (6.3.7) again. Thus Lemma 2.92 implies

$$\| \ \mathbb{C}\left\langle 0; r; \log^+ |\nabla_z f'| \right\rangle \leq O\{\log T_f(r, L)\} + O(\log r).$$

Note that when $0 < 4\mu < 1$, by the Hölder inequality we have

$$
\begin{aligned}
\mathcal{T}\left(r, r_0; |f'|_\kappa^{8\mu} v\right) &\leq \mathcal{T}\left(r, r_0; |f'|_\kappa^2 v\right)^{4\mu} \mathcal{T}\left(r, r_0; v\right)^{1-4\mu} \\
&\leq T_f(r, L)^{8\mu} + r^4.
\end{aligned}
\tag{6.6.7}
$$

Then Lemma 2.92 yields

$$
\| \quad \mathbb{C}\left\langle 0; r; \log^+ |f'|_\kappa\right\rangle \leq O\{\log T_f(r, L)\} + O(\log r).
$$

Hence by using (6.6.6) and Lemma 2.104, we can prove (6.6.5) for the case $k = 2$.

When using induction to consider the case $k + 1$, the inequality (6.3.9) becomes the following form:

$$
\begin{aligned}
\left|f^{(k+1)}\right| &\leq c\left\{\left|\nabla_z f^{(k)}\right| + \left|f^{(k)}\right|^2 + \sum_{W \in \mathcal{B}}\left(\sum_{\gamma, \delta} |X_{\gamma\delta f}^W|\right)^4\right\} \\
&\quad + c\left\{\left|f^{(k)}\right|_\kappa^2 + |f'|_\kappa^2 + |f'|_\kappa^4\right\}.
\end{aligned}
\tag{6.6.8}
$$

According to the remarks for the case $k = 2$ and the proof in Theorem 6.12, the inequality (6.3.11) has the form

$$
\| \quad c\mathcal{T}\left(r, r_0; \frac{|\nabla_z f^{(k)}|^2}{\left(1 + |f^{(k)}|^2\right)^2} v\right) \leq T_f(r, L) + R_f(r) + r^2
\tag{6.6.9}
$$

if we do not use Lemma 2.92, but directly estimate $\mathbb{C}\left\langle 0; r; \log\left(1 + |f^{(k)}|^2\right)\right\rangle$ by using the induction assumption. Since

$$
\begin{aligned}
\| \quad \mathbb{C}\left\langle 0; r; \log^+ |\nabla_z f^{(k)}|\right\rangle &\leq \frac{1}{2}\mathbb{C}\left\langle 0; r; \log^+ \frac{|\nabla_z f^{(k)}|^2}{(1 + |f^{(k)}|^2)^2}\right\rangle \\
&\quad + \mathbb{C}\left\langle 0; r; \log\left(1 + |f^{(k)}|^2\right)\right\rangle,
\end{aligned}
$$

then (6.6.9), Lemma 2.92 and the induction assumption imply

$$
\begin{aligned}
\| \quad \mathbb{C}\left\langle 0; r; \log^+ |\nabla_z f^{(k)}|\right\rangle &\leq 2\left(\frac{5}{3}(4^{k-2} - 1) + 1\right) R_f(r) \\
&\quad + O\{\log(rT_f(r, L))\} + O(\log^+ R_f(r)).
\end{aligned}
\tag{6.6.10}
$$

Hence the inequality (6.6.5) of case $k+1$ follows from (6.6.8), (6.6.10), Lemma 2.92, Theorem 6.12, Lemma 2.104 and the induction assumption. $\qquad\square$

Now we can estimate the term $M\langle O; r; \log g\rangle$ in Theorem 2.95 for the case $M = \mathbb{C}$.

Theorem 6.24 ([184]). *Let N be a compact complex manifold of dimension n. Let L be a positive holomorphic line bundle on N. Take $0 \neq s_j \in \Gamma(N, L), j = 1, \ldots, q$, such that the divisor $D = D_1 + \cdots + D_q$ has simple normal crossings in N, where $D_j = (s_j)$. Let $f : \mathbb{C} \longrightarrow N$ be a holomorphic curve such that $f(\mathbb{C}) \not\subseteq s_j^{-1}(0)$ for all j. Assume that the image of f is not autoparallel with respect to a meromorphic connection of N. Then*

$$\| \quad qT_f(r, L) + T_f(r, K_N) \leq \sum_{j=1}^{q} N_f(r, D_j) - N_{\mathrm{Ram}}(r, f) + c_n R_f(r)$$

$$+ O(\log^+ R_f(r)) + O\{\log(rT_f(r, L))\}, \qquad (6.6.11)$$

where

$$c_n = \frac{1}{9}(5 \times 4^{n-1} - 6n + 1).$$

Proof. Since the image of f is not autoparallel with respect to a meromorphic connection $\mathbf{D}$ on N, then

$$f' \wedge f^{(2)} \wedge \cdots \wedge f^{(n)} \not\equiv 0 \qquad (6.6.12)$$

where the exterior product $f' \wedge f^{(2)} \wedge \cdots \wedge f^{(n)}$ is taken in the holomorphic tangent bundle $\mathbf{T}(N)$ of N so that it defines a holomorphic mapping from $\mathbb{C}$ to the line bundle $E^{\frac{n(n-1)}{2}} \otimes K_N^*$ of N.

Since any line bundle on $\mathbb{C}$ is trivial, the pullback $f_f^{(k)}$ of $f^{(k)}$ under f can be identified with a section of $f^*(\mathbf{T}(N))$ under the identification

$$f^*(E^{k-1} \otimes \mathbf{T}(N)) \cong f^*(\mathbf{T}(N))$$

such that

$$f_f' \wedge f_f^{(2)} \wedge \cdots \wedge f_f^{(n)} \not\equiv 0.$$

By Lemma 2.86, the holomorphic field

$$\varphi = f_f^{(2)} \wedge \cdots \wedge f_f^{(n)}$$

on f over $\mathbb{C}$ of degree $n - 1$ is effective, that is, φ induces an effective Jacobian section F_φ of f. Define a non-negative function g by

$$F_\varphi[\psi^n] = g^2 f^*(\psi), \qquad (6.6.13)$$

where $\psi = \mathrm{Ric}(\Psi) > 0$. Lemma 2.87 also is true for the singular metric h_Ψ on N. Thus we have $g \leq |\varphi|$, where the metric of $f^*(\mathbf{T}(N))$ is induced by the singular metric on N defined by the form ψ.

By Lemma 1.55, there exists a constant $c > 0$ such that

$$g \leq \left| f_f^{(2)} \wedge \cdots \wedge f_f^{(n)} \right| \leq c \prod_{k=2}^{n} \left| f^{(k)} \right|. \qquad (6.6.14)$$

Thus we obtain an inequality

$$\mathbb{C}\langle 0; r; \log g \rangle \leq \sum_{k=2}^{n} \mathbb{C}\left\langle 0; r; \log \left|f^{(k)}\right|\right\rangle + O(1). \tag{6.6.15}$$

The inequality (6.6.5)) implies

$$\| \quad \mathbb{C}\langle 0; r; \log g \rangle \leq c_n R_f(r) + O(\log^+(rT_f(r, L))) + O(\log^+ R_f(r)), \tag{6.6.16}$$

and hence Theorem 6.24 follows from Theorem 2.95. $\qquad \square$

If there exists a meromorphic connection of N, Theorem 6.18 implies that there is a proper algebraic subset $Z_{D,f}$ of N depending on f such that the image of f is not autoparallel with respect to the meromorphic connection of N if $f(\mathbb{C}) \not\subset Z_{D,f}$. Thus Theorem 6.24 implies the following result:

Theorem 6.25. *Let N be an algebraic manifold of dimension n with a meromorphic connection and let D be an ample divisor on N with normal crossings. Let $f : \mathbb{C} \longrightarrow N$ be a holomorphic curve. There exists a proper algebraic subset $Z_{D,f}$ of N such that when $f(\mathbb{C}) \not\subset Z_{D,f}$, we have*

$$\| \quad T_f(r, [D]) + T_f(r, K_N) \leq N_f(r, D) - N_{\mathrm{Ram}}(r, f) + c_n R_f(r)$$
$$+ O\{\log(rT_f(r, [D]))\} + O(\log^+ R_f(r)). \tag{6.6.17}$$

Chapter 7

Riemann's ζ-function

In this chapter, we will give a few necessary and sufficient conditions on Riemann's hypothesis by using several formulae in analytic function theory, say, Nevanlinna formula, Carleman formula, Levin formula, and so on. Since these formulae can be used to define characteristic functions similar to Nevanlinna's fashion based on Jensen's formula, we hope to use value distribution theory derived from these formulae to study Riemann's hypothesis. The main idea will be exhibited in the following sections. It is interesting to find connections between Diophantine approximation and value distribution theory derived from these formulae.

7.1 Riemann's functional equation

This tract is intended for readers who already have some knowledge of the zeta-function and its role in the analytical theory of numbers; but for the sake of completeness we give a brief sketch of its elementary properties (or see [394]).

The function is defined by the Dirichlet series

$$\zeta(s) = \sum_{n=1}^{\infty} n^{-s}, \quad s \in \mathbb{C}. \tag{7.1.1}$$

The series is convergent, and the function analytic, for $\mathrm{Re}(s) > 1$. We have also Euler's infinite product representation

$$\zeta(s) = \prod_{p} (1 - p^{-s})^{-1}, \quad \mathrm{Re}(s) > 1, \tag{7.1.2}$$

where p runs though all prime numbers. From the convergence of the product (7.1.2) one deduces that $\zeta(s)$ has no zeros for $\mathrm{Re}(s) > 1$.

Summing

$$n^{-s}\Gamma(s) = \int_0^\infty x^{s-1} e^{-nx} dx$$

with respect to n and inverting the order of summation and integration, a third representation of the function is obtained as follows:

$$\zeta(s) = \frac{1}{\Gamma(s)} \int_0^\infty \frac{x^{s-1}}{e^x - 1} dx, \quad \mathrm{Re}(s) > 1. \tag{7.1.3}$$

By using (7.1.3), it is easy to show the following representation:

$$\zeta(s) = \frac{1}{\Gamma(s)} \int_0^\infty \left(\frac{1}{e^x - 1} - \frac{1}{x} \right) x^{s-1} dx, \quad 0 < \mathrm{Re}(s) < 1. \tag{7.1.4}$$

Further, (7.1.3) can be used to extend the domain of definition of ζ to all $\mathbb{C}$. We replace the integral by the contour integral

$$\zeta(s) = -\frac{\Gamma(1-s)}{2\pi i} \int_C \frac{(-w)^{s-1}}{e^w - 1} dw, \tag{7.1.5}$$

where the contour C starts at infinity on the positive real axis, encircles the origin once in the positive direction (but excludes all the poles of $1/(e^w - 1)$ other than 0, i.e., the points $\pm 2i\pi, \pm 4i\pi, \dots$) and returns to its starting point. Here we define

$$(-w)^{s-1} = \exp\{(s-1)\log(-w)\},$$

where the logarithm is real on the negative real axis. This formula is deduced from (7.1.3) in the case $\mathrm{Re}(s) > 1$ by shrinking the contour C into the real axis described twice, and taking into account the different value of the logarithm on the two parts. This integral is uniformly convergent in any finite region, and so represents an entire function of s. This enables us to continue $\zeta(s)$ over the whole plane $\mathbb{C}$. Hence $\zeta(s)$ is analytic for all values of s except for a simple pole at $s = 1$, with residue 1. Since

$$\frac{w}{e^w - 1} = 1 - \frac{1}{2} w + \sum_{n=1}^\infty \frac{(-1)^{n-1} B_n}{(2n)!} w^{2n},$$

where the coefficients B_n are rational numbers, called Bernoulli numbers, by the theorem of residues one finds the following values of $\zeta(s)$:

$$\zeta(0) = -\frac{1}{2}, \quad \zeta(-2m) = 0, \quad \zeta(1 - 2m) = (-1)^m \frac{B_m}{2m} \quad (m = 1, 2, \dots). \tag{7.1.6}$$

The points $z = -2, -4, \dots$ are called the *trivial zeros* of $\zeta(z)$. Euler proved

$$\zeta(2m) = \frac{2^{2m-1} \pi^{2m} B_{2m}}{(2m)!} \quad (m = 1, 2, \dots). \tag{7.1.7}$$

Deform the contour C into the contour C_n consisting of the square with centre the origin and sides parallel to the axes, length of side $(4n+2)\pi$, together with the positive real axis from $(2n+1)\pi$ to infinity, in which it includes the poles of the integrand at $\pm 2i\pi, \dots, \pm 2ni\pi$. The sum of the residues at these points is found to be

$$2 \sum_{m=1}^{n} (2m\pi)^{s-1} \sin \frac{\pi s}{2}.$$

If $\operatorname{Re}(s) < 0$ we can make $n \to \infty$; the integral around C_n tends to zero, and one obtains *Riemann's functional equation*

$$\zeta(s) = 2^s \pi^{s-1} \sin \frac{\pi s}{2} \Gamma(1-s)\zeta(1-s),$$

which is equivalent to the functional equation

$$\zeta(1-s) = 2^{1-s} \pi^{-s} \cos \frac{\pi s}{2} \Gamma(s)\zeta(s). \tag{7.1.8}$$

By continuation, this holds for all values of s. It follows from this that $\zeta(s)$ has no zeros in the half-plane $\operatorname{Re}(s) < 0$ except simple zeros at $s = -2, -4, \dots$. Writing

$$\xi(s) = \frac{s}{2}(s-1)\pi^{-\frac{s}{2}} \Gamma\left(\frac{s}{2}\right) \zeta(s), \tag{7.1.9}$$

the functional equation (7.1.8) takes the simple form

$$\xi(1-s) = \xi(s). \tag{7.1.10}$$

Next we introduce the original proof of the functional equation (7.1.10) due to Riemann. Riemann wrote only a single, ten-page paper in number theory [324]. In it he not only initiated the study of $\zeta(s)$ as a function of a complex variable, but also introduced the Riemann Hypothesis and outlined the eventual proof of the prime number theorem. At the core of Riemann's paper is the *Poisson summation formula*

$$\sum_{n \in \mathbb{Z}} f(n) = \sum_{n \in \mathbb{Z}} \hat{f}(n), \tag{7.1.11}$$

which relates the sum over the integers of a function f and its Fourier transform

$$\hat{f}(x) = \int_{\mathbb{R}} f(y) e^{-2\pi i x y} dy. \tag{7.1.12}$$

The Poisson summation formula is valid for functions f with suitable regularity, such as Schwartz functions: smooth functions which, along with all their derivatives, decay faster than any power of $\frac{1}{|x|}$ as $|x| \to \infty$ (cf. [212], II, § 4, Proposition 6).

Riemann's own, rigorous argument proceeds by applying the Poisson summation formula (7.1.11) to the Gaussian

$$f(x) = e^{-\pi x^2 t}, \quad t > 0,$$

whose Fourier transform is

$$\hat{f}(x) = \frac{1}{\sqrt{t}} e^{-\frac{\pi x^2}{t}}.$$

The Gaussian is a Schwartz function and can be legitimately inserted in (7.1.11). Thus one obtains *Jacobi's transformation identity*

$$\theta(ix) = \frac{1}{\sqrt{x}} \theta\left(\frac{i}{x}\right), \tag{7.1.13}$$

where

$$\theta(z) = \frac{1}{2} \sum_{n \in \mathbb{Z}} e^{\pi i n^2 z} = \frac{1}{2} + \sum_{n=1}^{\infty} e^{\pi i n^2 z} \tag{7.1.14}$$

is holomorphic in the upper half-plane $\mathbb{H} = \{z \mid \text{Im}(z) > 0\}$; moreover,

$$\theta\left(\frac{-1}{z}\right) = \left(\frac{z}{i}\right)^{\frac{1}{2}} \theta(z), \quad \theta(z+2) = \theta(z). \tag{7.1.15}$$

By using the expression,

$$\Gamma(s) = \int_0^\infty x^{s-1} e^{-x} dx, \quad \text{Re}(s) > 0,$$

Riemann then obtained an integral representation as follows:

$$
\begin{aligned}
\pi^{-s} \Gamma(s) \zeta(2s) &= \sum_{n=1}^{\infty} \int_0^\infty (\pi n^2)^{-s} x^{s-1} e^{-x} dx \\
&= \int_0^\infty x^{s-1} \left(\theta(ix) - \frac{1}{2}\right) dx, \quad \text{Re}(s) > \frac{1}{2}. \tag{7.1.16}
\end{aligned}
$$

Note that

$$
\begin{aligned}
\int_0^1 x^{s-1} \left(\theta(ix) - \frac{1}{2}\right) dx &= \int_0^1 x^{s-1} \theta(ix) dx - \frac{1}{2s} \\
&= \int_1^\infty x^{-s-1} \theta\left(\frac{i}{x}\right) dx - \frac{1}{2s} \\
&= \int_1^\infty x^{\frac{1}{2}-s-1} \left(\theta(ix) - \frac{1}{2}\right) dx - \frac{1}{2s} - \frac{1}{1-2s}.
\end{aligned}
$$

Indeed, replacing s by $s/2$, the above expressions read

$$\xi(s) = \frac{1}{2} + \frac{s}{2}(s-1) \int_1^\infty \left(x^{\frac{s}{2}-1} + x^{\frac{1-s}{2}-1}\right) \left(\theta(ix) - \frac{1}{2}\right) dx. \tag{7.1.17}$$

Further for $x \geq 1$, we note that

$$\theta(ix) - \frac{1}{2} \leq \sum_{n=1}^{\infty} e^{-\pi n x} = \frac{e^{-\pi x}}{1 - e^{-\pi x}} = O\left(e^{-\pi x}\right) \qquad (7.1.18)$$

and

$$\int_1^{\infty} \left| x^s e^{-\pi x} \right| dx \leq \int_1^{\infty} x^b e^{-\pi x} dx < \infty, \ \text{Re}(s) \leq b.$$

Hence for any value of s, the integral in (7.1.17) converges to an entire function which is bounded for s in vertical strips. Obviously, Riemann's functional equation (7.1.10) follows from (7.1.17).

For further discussion, we will need the *partial summation formula* (cf. [199]), which is also called *Abel's transformation*:

Lemma 7.1. *Suppose that $f \in C^1([a,b])$ and $e(n)$ are arbitrary complex numbers. Then we have the relation*

$$\sum_{a < n \leq b} e(n)f(n) = E(b)f(b) - \int_a^b E(x)f'(x)dx, \qquad (7.1.19)$$

where

$$E(x) = \sum_{a < n \leq x} e(n).$$

Proof. Obviously, we have

$$I = E(b)f(b) - \sum_{a < n \leq b} e(n)f(n) = \sum_{a < n \leq b} e(n)\{f(b) - f(n)\}$$

$$= \sum_{a < n \leq b} e(n) \int_n^b f'(x)dx = \sum_{a < n \leq b} e(n) \int_a^b \chi_n(x)f'(x)dx,$$

where

$$\chi_n(x) = \begin{cases} 0 & : a < x < n, \\ 1 & : n \leq x \leq b, \end{cases}$$

and hence

$$I = \int_a^b f'(x) \left\{ \sum_{a < n \leq b} e(n)\chi_n(x) \right\} dx = \int_a^b f'(x) \left\{ \sum_{a < n \leq x} e(n) \right\} dx. \qquad \square$$

Let $[x]$ denote the maximal integer $\leq x$, and set

$$\{x\} = x - [x].$$

Next we show *Euler's summation formula* (cf. [199]):

Lemma 7.2. *Suppose that $f \in C^1([a,b])$. Then*

$$\sum_{a<n\leq b} f(n) = \{a\}f(a) - \{b\}f(b) + \int_a^b f(x)dx + \int_a^b \{x\}f'(x)dx. \qquad (7.1.20)$$

Proof. Take $e(n) = 1$ in Lemma 7.1. Then we obtain

$$F(b) = \sum_{a<n\leq b} f(n) = E(b)f(b) - \int_a^b E(x)f'(x)dx,$$

where

$$E(x) = \sum_{a<n\leq b} 1 = [x] - [a] = x - \{x\} - a + \{a\}.$$

Therefore

$$F(b) = (b - \{b\} - a + \{a\})f(b)$$
$$- \int_a^b (x - \{x\})f'(x)dx + (a - \{a\}) \int_a^b f'(x)dx$$
$$= \{a\}f(a) - \{b\}f(b) + \int_a^b f(x)dx + \int_a^b \{x\}f'(x)dx.$$

We have thus proved the statement of the lemma. $\square$

Now for $\operatorname{Re}(s) > 0$, we show the formula

$$\zeta(s) = \sum_{n=1}^N \frac{1}{n^s} + \frac{1}{(s-1)N^{s-1}} - s \int_N^\infty \frac{x - [x]}{x^{s+1}} dx. \qquad (7.1.21)$$

Take positive integers M and N with $M > N$ and apply (7.1.20) to the sum

$$R_{NM} = \sum_{N<n\leq M} \frac{1}{n^s}.$$

Then we have

$$R_{NM} = \int_N^M x^{-s}dx - s \int_N^M \{x\}x^{-s-1}dx$$
$$= \frac{1}{1-s}M^{1-s} + \frac{1}{s-1}N^{1-s} - s \int_N^M \{x\}x^{-s-1}dx.$$

Suppose firstly that $\operatorname{Re}(s) > 1$. Then we obtain

$$\zeta(s) = \sum_{n=1}^N \frac{1}{n^s} + \lim_{M\to+\infty} R_{NM}$$
$$= \sum_{n=1}^N \frac{1}{n^s} + \frac{1}{s-1}N^{1-s} - s \int_N^\infty \{x\}x^{-s-1}dx.$$

But the last improper integral defines an analytic function in the half-plane $\mathrm{Re}(s) > 0$. From this, by virtue of the analytic continuation principle, we get the formula (7.1.21).

One shows that the order of the entire function $\xi(s)$ defined by (7.1.9) is 1. By (7.1.10) it is sufficient to consider the half-plane $\mathrm{Re}(s) \geq \frac{1}{2}$. Set $s = \sigma + it$ and take $0 < \delta < 1$. We deduce from (7.1.21) that for $\sigma \geq \delta$, $|t| \geq 1$,

$$\zeta(s) = O\left(N^{1-\delta}\right) + O\left(tN^{-\delta}\right) = O\left(t^{1-\delta}\right) \tag{7.1.22}$$

on taking $N = [t]$. Since

$$\left| \Gamma\left(\frac{s}{2}\right) \right| \leq \left| \Gamma\left(\frac{\sigma}{2}\right) \right| = O\left(e^{A\sigma \log \sigma}\right),$$

it follows from (7.1.9) that $\xi(s)$ is of order 1 at most. By considering positive real values of s it is easily seen that the order is exactly 1.

7.2 Converse theorems

The following *Phragmen-Lindelöf principle* (cf. [52], [396], [233]) can be used to obtain estimates on $\zeta(s)$ in vertical strips from ones on their edges:

Theorem 7.3. *Let $f(s)$ be meromorphic in a strip*

$$\Omega = \{s \in \mathbb{C} \mid a \leq \mathrm{Re}(s) \leq b\}, \ \{a, b\} \subset \mathbb{R}.$$

Suppose that there exists some $\rho > 0$ such that $f(s)$ satisfies the inequality

$$f(s) = O\left(e^{|s|^\rho}\right)$$

on Ω for $|\mathrm{Im}(s)|$ large and obeys the estimate

$$f(\sigma + it) = O\left(|t|^M\right), \ \sigma \in \{a, b\}, \ |t| \to \infty$$

for some positive integer M. Then

$$f(\sigma + it) = O\left(|t|^M\right), \ a \leq \sigma \leq b, \ |t| \to \infty.$$

A *modular form of weight k* and *multiplier condition C* for the group $G(\lambda)$ of substitutions generalized by

$$z \mapsto z + \lambda, \quad z \mapsto -\frac{1}{z}$$

is a holomorphic function $f(z)$ on the upper half-plane $\mathbb{H}$ satisfying

(i) $f(z + \lambda) = f(z)$;

(ii) $f\left(\frac{-1}{z}\right) = C\left(\frac{z}{i}\right)^k f(z)$;

(iii) $f(z)$ has a Taylor expansion in $e^{\frac{2\pi i z}{\lambda}}$:

$$f(z) = \sum_{n=0}^{\infty} a_n e^{\frac{2\pi i n z}{\lambda}}.$$

We denote the space of such f by $\mathcal{A}_{\lambda,k,C}(\mathbb{H})$; f is a *cusp form* if $a_0 = 0$. The conditions (i) and (ii) mean

$$f\left(\frac{az+b}{cz+d}\right) = C_\gamma(cz+d)^k f(z), \quad \gamma = \left(\begin{array}{cc} a & b \\ c & d \end{array}\right) \in G(\lambda) \qquad (7.2.1)$$

with $|C_\gamma| = 1$. Obviously, one has $C_\gamma^4 = 1$ if k is an integer, and $C_\gamma = \pm 1$ if k is even. In particular, when $\lambda = 1$, $G(1)$ is the modular group $SL(2,\mathbb{Z})$.

Example 7.4. *Those two equations in (7.1.15) say that $\theta(z)$ is a modular form of weight $\frac{1}{2}$ for the group $G(2)$. The space $\mathcal{A}_{2,1/2,1}(\mathbb{H})$ is one-dimensional and consists of multiples of the θ-function.*

Given a sequence $\{a_0, a_1, a_2, \dots\}$ of complex numbers and given

$$\lambda > 0, \; k > 0, \; C = \pm 1,$$

set

$$f(z) = \sum_{n=0}^{\infty} a_n e^{\frac{2\pi i n z}{\lambda}}, \qquad (7.2.2)$$

$$L(f,s) = \sum_{n=1}^{\infty} \frac{a_n}{n^s}, \qquad (7.2.3)$$

and

$$\Lambda_f(s) = \left(\frac{2\pi}{\lambda}\right)^{-s} \Gamma(s) L(f,s). \qquad (7.2.4)$$

First of all, we assume that for some $d > 0$,

$$a_n = O(n^d). \qquad (7.2.5)$$

Then $L(f,s)$ converges for $\mathrm{Re}(s) > d+1$ and f is holomorphic in $\mathbb{H}$. The following *Hecke converse theorem* ([155], [156]) gives a deep relation between Λ_f and f:

Theorem 7.5. *Under the above conditions, the function f belongs to $\mathcal{A}_{\lambda,k,C}(\mathbb{H})$ if and only if the function*

$$\Lambda_f(s) + \frac{a_0}{s} + \frac{Ca_0}{k-s} \qquad (7.2.6)$$

is an entire function which is bounded in vertical strips and satisfies

$$\Lambda_f(s) = C\Lambda_f(k-s). \qquad (7.2.7)$$

Proof. Here we follow S.S. Gelbart and S.D. Miller [112] to prove Theorem 7.5. According to the proof of (7.1.16), one also has

$$
\Lambda_f(s) = \sum_{n=1}^{\infty} \int_0^{\infty} a_n \left(\frac{2\pi n}{\lambda} \right)^{-s} x^{s-1} e^{-x} dx
$$

$$
= \int_0^{\infty} x^{s-1} \left(f(ix) - a_0 \right) dx, \quad \mathrm{Re}(s) > d+1. \tag{7.2.8}
$$

Then Mellin inversion implies

$$
f(ix) - a_0 = \frac{1}{2\pi i} \int_{\mathrm{Re}(s)=c} x^{-s} \Lambda_f(s) ds, \tag{7.2.9}
$$

for $x > 0$, $c > d+1$.

Now assume that Λ_f satisfies the conditions in Theorem 7.5. Take $c > k$ and consider the contour $\partial \Omega_r$ consisting of the boundary of the region

$$
\Omega_r = \{ \sigma + it \in \mathbb{C} \mid k-c \le \sigma \le c, \ -r \le t \le r \}.
$$

Then the residue theorem yields

$$
\frac{1}{2\pi i} \int_{\partial \Omega_r} x^{-s} \Lambda_f(s) ds = C a_0 x^{-k} - a_0. \tag{7.2.10}
$$

Next we show that the integrals of $x^{-s} \Lambda_f(s)$ over the horizontal paths tend to zero as $r \to \infty$. When $\mathrm{Re}(s) \ge c > d+1$, $L(f, s)$ converges absolutely, and so

$$
|L(f, s)| \le \sum_{n=1}^{\infty} \frac{|a_n|}{n^c} = \sum_{n=1}^{\infty} O \left(\frac{1}{n^{c-d}} \right) = O(1).
$$

Stirling's formula gives that, for $\mathrm{Re}(s) \ge 1/2$,

$$
\Gamma(s) \sim \sqrt{2\pi} e^{-s} s^{s-\frac{1}{2}}, \quad |s| \to \infty, \tag{7.2.11}
$$

which implies that $\Lambda_f(s)$ satisfies the estimate

$$
|\Lambda_f(s)| = O \left(e^{|s|^{1+\varepsilon}} \right), \quad \mathrm{Re}(s) \ge c, \quad \varepsilon > 0.
$$

By the functional equation, $\Lambda_f(s)$ does as well in the reflected region $\mathrm{Re}(s) \le k-c$, and the bounded assumption in Theorem 7.5 handles the missing strip. Therefore the function in (7.2.6) is of order 1 on $\mathbb{C}$. Since $\frac{1}{s\Gamma(s)}$ is entire and of order 1, the function

$$
(s-k) L(f, s) = (s-k) \left(\frac{2\pi}{\lambda} \right)^s \Gamma(s)^{-1} \Lambda_f(s)
$$

is also entire and of order 1. Another form of Stirling's formula states that

$$|\Gamma(\sigma + it)| \sim \sqrt{2\pi}|t|^{\sigma-\frac{1}{2}}e^{-\frac{\pi|t|}{2}}, \quad |t| \to \infty \qquad (7.2.12)$$

holds uniformly for $a \le \sigma \le b$. Thus the functional equation

$$L(f, s) = C\left(\frac{2\pi}{\lambda}\right)^{2s-k}\frac{\Gamma(k-s)}{\Gamma(s)}L(f, k-s)$$

shows that

$$|L(f, \sigma + it)| = O\left(t^{2c-k}\right), \quad \sigma = k - c < 0. \qquad (7.2.13)$$

We conclude from the Phragmen-Lindelöf principle that there exists some $K > 0$ such that

$$|L(f, \sigma + it)| = O\left(|t|^K\right), \quad |t| \to \infty$$

holds uniformly in the strip $k - c \le \sigma \le c$. Thus the claim follows easily.

Letting $r \to \infty$ in (7.2.10) and using (7.2.9), one obtains

$$f(ix) - Ca_0 x^{-k} = \frac{1}{2\pi i}\int_{\mathrm{Re}(s)=k-c} x^{-s}\Lambda_f(s)ds. \qquad (7.2.14)$$

The functional equation (7.2.7) implies

$$
\begin{aligned}
f(ix) - Ca_0 x^{-k} &= \frac{C}{2\pi i}\int_{\mathrm{Re}(s)=k-c} x^{-s}\Lambda_f(k-s)ds \\
&= \frac{C}{2\pi i}\int_{\mathrm{Re}(s)=c} x^{s-k}\Lambda_f(s)ds \\
&= Cx^{-k}\left(f\left(\frac{i}{x}\right) - a_0\right),
\end{aligned}
$$

that is,

$$f(ix) = Cx^{-k}f\left(\frac{i}{x}\right),$$

which is the property (ii) of the definition of $\mathcal{A}_{\lambda,k,C}(\mathbb{H})$. Properties (i) and (iii) are immediate from the definition of $f(z)$.

Next suppose $f \in \mathcal{A}_{\lambda,k,C}(\mathbb{H})$. Note that

$$
\begin{aligned}
\int_0^1 x^{s-1}\left(f(ix) - a_0\right)dx &= \int_0^1 x^{s-1}f(ix)dx - \frac{a_0}{s} \\
&= \int_1^\infty x^{-s-1}f\left(\frac{i}{x}\right)dx - \frac{a_0}{s} \\
&= C\int_1^\infty x^{k-s-1}\left(f(ix) - a_0\right)dx - \frac{a_0}{s} - \frac{Ca_0}{k-s}.
\end{aligned}
$$

The equation (7.2.8) yields

$$\Lambda_f(s) + \frac{a_0}{s} + \frac{Ca_0}{k-s} = \int_1^\infty \left(x^{s-1} + Cx^{k-s-1}\right)\left(f(ix) - a_0\right) dx, \qquad (7.2.15)$$

which implies clearly the conclusions in Theorem 7.5. $\qquad\square$

The following *Hamburger converse theorem* [145] shows that ζ is uniquely determined by the Riemann functional equation subject to a certain regularity condition.

Theorem 7.6. *Let*

$$L(f, s) = \sum_{n=1}^\infty \frac{a_n}{n^s}, \quad L(g, s) = \sum_{n=1}^\infty \frac{b_n}{n^s}$$

be an absolutely convergent Dirichlet series for $\mathrm{Re}(s) > 1$, *and suppose that both* $(s-1)L(f, s)$ *and* $(s-1)L(g, s)$ *are entire functions of finite order. Assume the functional equation*

$$\pi^{-\frac{s}{2}}\Gamma\left(\frac{s}{2}\right)L(f, s) = \pi^{-\frac{1-s}{2}}\Gamma\left(\frac{1-s}{2}\right)L(g, 1-s). \qquad (7.2.16)$$

Then $L(f, s) = L(g, s) = a_1\zeta(s)$.

Proof. Here we also follow [112]. For simplicity, suppose that

$$a_n = b_n, \quad n = 1, 2, \ldots.$$

Now the function

$$\Lambda\left(\frac{s}{2}\right) = \pi^{-\frac{s}{2}}\Gamma\left(\frac{s}{2}\right)L(f, s)$$

is holomorphic in $\mathrm{Re}(s) > 0$, except perhaps for a simple pole at $s = 1$. The functional equation (7.2.16), that is,

$$\Lambda(s) = \Lambda\left(\frac{1}{2} - s\right), \qquad (7.2.17)$$

means that $\Lambda(s)$ has an analytic continuation to $\mathbb{C}$ except for potential simple poles at $s = 0$ and $1/2$. Because of (7.2.17) the residues of $\Lambda(s)$ at those points are negatives of each other, and thus

$$\Lambda(s) + \frac{a_0}{s} + \frac{a_0}{k-s}$$

is entire, where a_0 is the residue of $\Lambda(s)$ at $s = k = \frac{1}{2}$.

On the other hand, the assumption of absolute convergence implies that

$$|L(f, \sigma + it)| \leq \sum_{n=1}^\infty \frac{|a_n|}{n^\sigma} < \infty, \quad \sigma > 1.$$

Then for any $\varepsilon > 0$, $|L(f, s)|$ is uniformly bounded in the region $\mathrm{Re}(s) \geq 1 + \varepsilon$. Using the functional equation, we see that

$$|L(f, \sigma + it)| = O\left(|t|^{\frac{1}{2} - \sigma}\right), \quad |t| \to \infty$$

for $\sigma < -\varepsilon$, and uniformly so in vertical strips. Thus the Phragmen-Lindelöf principle shows that

$$|L(f, \sigma + it)| = O\left(|t|^{\frac{1}{2} + \varepsilon}\right), \quad -\varepsilon < \sigma < 1 + \varepsilon.$$

Stirling's estimate (7.2.12) yields that $\Lambda(s)$ decays rapidly as $|t| \to \infty$ in any vertical strip $a \leq \sigma \leq b$, and hence is bounded there. Theorem 7.5 produces a modular form f in $\mathcal{A}_{2,1/2,1}(\mathbb{H})$, which is a one-dimensional space spanned by θ. So f must in fact be a multiple of the θ-function, and we conclude that $L(f, s)$ is a multiple of $\zeta(s)$.   □

Morduhaĭ-Boltovskoĭ [279], and Ostrowski [313] proved Hilbert's conjecture by showing that the Riemann ζ-function does not satisfy any algebraic differential equation with rational functions as the coefficients. Liao and Yang [247] confirmed that $\zeta(s)$ is a prime function, and there does not exist a non-constant polynomial $P \in \mathbb{C}[x, y, z]$ such that

$$P(\Gamma(z), \zeta(z), z) \equiv 0.$$

Hence Riemann's functional equation is almost the unique relation satisfied by $\zeta(s)$.

In 1967, A. Weil completed Hecke's theory by similarly characterizing modular forms for *congruence subgroups*, such as $\Gamma_0(N) \subset SL(2, \mathbb{Z})$. By the definition, a *modular form of weight $k > 0$* for the group $\Gamma_0(N)$ is a holomorphic function $f(z)$ on the upper half-plane $\mathbb{H}$ satisfying

(I) $f\left(\frac{az+b}{cz+d}\right) = (cz + d)^k f(z)$ for all $\begin{pmatrix} a & b \\ c & d \end{pmatrix} \in \Gamma_0(N)$;

(II) $f(z)$ has a Taylor expansion in $e^{2\pi i z}$: $f(z) = \sum_{n=0}^{\infty} a_n e^{2\pi i n z}$.

We denote the space of such f by $\mathcal{A}_{k,N}(\mathbb{H})$. Weil's breakthrough was to twist the series $L(f, s)$ by Dirichlet characters.

Let r be a positive integer. A *Dirichlet character modulo r* is a non-trivial function $\chi : \mathbb{Z} \longrightarrow \mathbb{C}$ satisfying the following conditions:

(1) χ is a periodic function with period r;
(2) χ is completely multiplicative, i.e., $\chi(nm) = \chi(n)\chi(m)$ for any $n, m \in \mathbb{Z}$;
(3) $\chi(1) = 1$, $\chi(n) = 0$ if $(n, r) > 1$.

A Dirichlet character χ modulo r is called *non-primitive* if there exist a proper factor d ($\neq r$) of r and a Dirichlet character χ^0 modulo d such that

$$\chi(n) = \chi^0(n), \ (n, r) = 1.$$

If there exists no such χ^0, then χ is called a *primitive character*.

Theorem 7.7 ([427]). *Fix positive integers N and k. Then*

$$f(z) = \sum_{n=1}^{\infty} a_n e^{2\pi i n z} \in \mathcal{A}_{k,N}(\mathbb{H})$$

if $L(f, s)$ satisfies the following conditions:

(a) *$L(f, s)$ is absolutely convergent for $\mathrm{Re}(s)$ sufficiently large;*

(b) *for each primitive character χ of modulus r with $(r, N) = 1$,*

$$\Lambda_{f,\chi}(s) = (2\pi)^{-s} \Gamma(s) \sum_{n=1}^{\infty} \frac{a_n \chi(n)}{n^s}$$

continues to an entire function of s, bounded in vertical strips;

(c) *each such $\Lambda_{f,\chi}(s)$ satisfies the functional equation*

$$\Lambda_{f,\chi}(s) = w_\chi r^{-1} (r^2 N)^{\frac{k}{2}-s} \Lambda_{f,\bar{\chi}}(k - s),$$

where

$$w_\chi = i^k \chi(N) g(\chi)^2$$

and the Gauss sum

$$g(\chi) = \sum_{n(\mathrm{mod}\ r)} \chi(n) e^{\frac{2\pi i n}{r}},$$

where n in the sum runs on a complete residue system modulo r, that is, $\mathbb{Z}/r\mathbb{Z}$.

For a proof, see [43], [195] or [309].

7.3 Riemann's hypothesis

Writing

$$\Xi(t) = \xi\left(\frac{1}{2} + it\right), \tag{7.3.1}$$

the functional equation (7.1.8) takes the simple form

$$\Xi(-t) = \Xi(t). \tag{7.3.2}$$

It follows that $\Xi(\sqrt{z})$ is an entire function of order $\frac{1}{2}$, and so has an infinity of zeros. From this, one deduces that $\zeta(s)$ has an infinity of zeros other than the real ones already observed. These zeros must be complex and lie in the strip $0 \leq \operatorname{Re}(s) \leq 1$, called the *critical strip* of $\zeta(s)$. Further G. H. Hardy proved that there are an infinity of zeros on $\operatorname{Re}(s) = \frac{1}{2}$ (see [394]).

Since

$$(1 - 2^{1-s})\zeta(s) = 1 - \frac{1}{2^s} + \frac{1}{3^s} - \frac{1}{4^s} + \cdots > 0$$

for $0 < s < 1$ and $\zeta(0) \neq 0$, it follows that $\zeta(s)$ has no zeros on the real axis between 0 and 1. Therefore all possible zeros of $\zeta(s)$ in the critical strip are complex numbers. If s_0 is a zero of $\zeta(s)$ located in the strip, then the functional equation (7.1.8) implies that $1 - s_0$ also is a zero of $\zeta(s)$. Since $\zeta(\sigma)$ is real when $\sigma \in \mathbb{R}$, Schwarz's reflection principle shows that $\overline{\zeta(\overline{s})} = \zeta(s)$. Hence $\overline{s}_0$ also is a zero of $\zeta(s)$. Therefore zeros of $\zeta(s)$ in the strip are symmetric on the lines $\operatorname{Re}(s) = \frac{1}{2}$ and $\operatorname{Im}(s) = 0$.

Conjecture 7.8 (Riemann hypothesis [324]). *If s_0 is a zero of $\zeta(s)$ in the critical strip, then $\operatorname{Re}(s_0) = \frac{1}{2}$.*

As to the zeros of $\zeta(s)$, it is known that there are none on the line $\sigma = \operatorname{Re}(s) = 1$. We derive from (7.1.2) the formula

$$\log \zeta(s) = \sum_p \sum_{m=1}^{\infty} \frac{1}{mp^{ms}}, \tag{7.3.3}$$

which implies

$$\zeta^3(\sigma)|\zeta(\sigma + it)|^4|\zeta(\sigma + 2it)| = \exp\left\{\sum_p \sum_m \frac{3 + 4\cos\theta + \cos 2\theta}{mp^{m\sigma}}\right\}, \tag{7.3.4}$$

where $\theta = mt \log p$. Since

$$3 + 4\cos\theta + \cos 2\theta = 2(1 + \cos\theta)^2 \geq 0,$$

every term in the exponent on the right of (7.3.4) is positive, and hence the left-hand side is not less than 1. Putting $\sigma = 1 + \epsilon$ $(0 < \epsilon < 1)$, and noting that

$$\zeta(1 + \epsilon) = \sum_{n=1}^{\infty} n^{-1-\epsilon} < 1 + \int_1^{\infty} x^{-1-\epsilon}dx = 1 + \frac{1}{\epsilon} < \frac{2}{\epsilon},$$

we have

$$\frac{|\zeta(1 + \epsilon + it)|}{\epsilon} > \frac{1}{2\epsilon^{\frac{1}{4}}|\zeta(1 + \epsilon + 2it)|^{\frac{1}{4}}}.$$

Since $\zeta(s)$ is analytic, the left-hand side would tend to $|\zeta'(1 + it)|$ as $\epsilon \to 0$, if $1 + it$ were a zero of $\zeta(s)$. But the right-hand side tends to infinity. Hence $\zeta(s)$

cannot have a zero on $\text{Re}(s) = 1$. The smallest zeros of Ξ in the horizontal strip $|\text{Im}(z)| < \frac{1}{2}$, $\text{Re}(z) > 0$ have been calculated with great accuracy. They are

$$t_1 = 14.13\cdots, \quad t_2 = 21.02\cdots, \quad t_3 = 25.01\cdots,$$
$$t_4 = 30.42\cdots, \quad t_5 = 32.93\cdots, \quad t_6 = 37.58\cdots,$$
$$t_7 = 40.91\cdots, \quad t_8 = 43.32\cdots, \quad t_9 = 48.00\cdots,$$

$t_{10} = 49.77\cdots$, and so on. They are all real.

The number $N(T)$ of zeros of $\zeta(s)$ between $\text{Im}(s) = 0$ and $\text{Im}(s) = T$ is given approximately by the formula

$$N(T) = \frac{T}{2\pi}\log T - \frac{1 + \log 2\pi}{2\pi}T + O(\log T). \tag{7.3.5}$$

An immediate consequence of (7.3.5) is

$$N(T + 1) - N(T) = +O(\log T). \tag{7.3.6}$$

Next we will need *Perron's summation formula* (cf. [199]):

Lemma 7.9. *Suppose that the Dirichlet series*

$$f(s) = \sum_{n=1}^{\infty} \frac{a(n)}{n^s} \tag{7.3.7}$$

converges absolutely for $\text{Re}(s) > 1$, $|a(n)| \le A(n)$, *where* $A(n)$ *is a positive monotonically increasing function of* n, *and for* $\alpha > 0$,

$$\sum_{n=1}^{\infty} \frac{|a(n)|}{n^\sigma} = O\left(\frac{1}{(\sigma - 1)^\alpha}\right)$$

as $1 < \sigma \to 1$. *Then*

$$\sum_{n \le x} a(n) = \frac{1}{2\pi i}\int_{b-iT}^{b+iT} f(s)\frac{x^s}{s}\,ds + O\left(\frac{x^b}{T(b-1)^\alpha}\right) + O\left(\frac{xA(2x)\log x}{T}\right),$$

where the constant under the symbol O *depends only on* b_0 *for any* $b_0 \ge b > 1$, $T \ge 1$, $x = N + \frac{1}{2}$ $(N \in \mathbb{Z})$.

For $\text{Re}(s) > 1$, if we take logarithms in (7.1.2) and differentiate it, we have

$$-\frac{\zeta'(s)}{\zeta(s)} = \sum_{n=1}^{\infty} \frac{\Lambda(n)}{n^s}, \tag{7.3.8}$$

where $\Lambda(n)$ is *von Mangoldt's function*:

$$\Lambda(n) = \begin{cases} \log p & : \text{if } n = p^k, \, p \text{ is prime,} \\ 0 & : \text{if } n \neq p^k. \end{cases}$$

Define

$$\psi(x) = \sum_{n \leq x} \Lambda(n),$$

which is called *Chebyshev's function*. Chebyshev proved that if one of the limits

$$\lim_{x \to +\infty} \frac{\psi(x)}{x}$$

or

$$\lim_{x \to +\infty} \frac{\pi(x) \log x}{x} \tag{7.3.9}$$

existed, then so would the other limit and these limits would be equal to 1. The exact behavior is given by the *prime number theorem* (4.10.1). In fact, Ch. de la Vallée Poussin proved a stronger claim, that is, there is a constant $c > 0$ such that

$$\psi(x) = x + O\left(x e^{-c\sqrt{\log x}}\right),$$

$$\pi(x) = \int_2^x \frac{du}{\log u} + O\left(x e^{-\frac{c}{2}\sqrt{\log x}}\right).$$

Theorem 7.10. *The Riemann hypothesis is true if and only if the estimate*

$$\psi(x) = x + O\left(x^{\frac{1}{2}+\varepsilon}\right) \tag{7.3.10}$$

is valid for any positive ε.

Proof. Note that

$$\sum_{n=1}^{\infty} \frac{\Lambda(n)}{n^\sigma} \leq \sum_{n=1}^{\infty} \frac{\log n}{n^\sigma} \leq \int_1^{\infty} \frac{\log x}{x^\sigma} dx = O\left(\frac{1}{(\sigma-1)^2}\right).$$

Take

$$b = 1 + \frac{1}{\log x}, \; 2 \leq T \leq x, \; A(n) = \log n, \; \alpha = 2$$

in Lemma 7.9. We obtain

$$\psi(x) = -\frac{1}{2\pi i} \int_{b-iT}^{b+iT} \frac{\zeta'(s)}{\zeta(s)} \frac{x^s}{s} ds + O\left(\frac{x \log^2 x}{T}\right). \tag{7.3.11}$$

Transferring the line of integration in (7.3.11) to the left, onto the straight line $\text{Re}(s) = -\frac{1}{2}$ by using a rectangular contour, the residue theorem implies

$$\psi(x) = x - \sum_{|\text{Im}(\rho)| \leq T} \frac{x^\rho}{\rho} + O\left(\frac{x \log^2 x}{T}\right) \qquad (7.3.12)$$

for $2 \leq T \leq x$, where the summation in the last sum is carried out over all the complex zeros of $\zeta(s)$. Thus if Riemann's hypothesis is true, we can take $\text{Re}(\rho) = \frac{1}{2}$ in (7.3.12) with $T = \sqrt{x}$, and so

$$\psi(x) = x + O\left(\sqrt{x} \log^2 x\right). \qquad (7.3.13)$$

Conversely, suppose that (7.3.13) or a slightly weaker statement (7.3.10) holds for an arbitrary $\varepsilon > 0$. Let us prove that Riemann's hypothesis is true. Applying Lemma 7.1 to (7.3.8) with

$$e(n) = \Lambda(n), \ \ E(x) = \psi(x), \ \ f(n) = n^{-s},$$

for $\text{Re}(s) > 1$ we get

$$-\frac{\zeta'(s)}{\zeta(s)} = s \int_1^\infty \frac{\psi(x)}{x^{s+1}} dx = \frac{1}{s-1} + 1 + s \int_1^\infty \frac{\psi(x) - x}{x^{s+1}} dx$$

or

$$-\frac{\zeta'(s)}{\zeta(s)} - \frac{1}{s-1} = 1 + s \int_1^\infty \frac{\psi(x) - x}{x^{s+1}} dx.$$

By the assumption, the improper integral on the right-hand side converges absolutely and uniformly in the half-plane $\text{Re}(s) \geq \frac{1}{2} + \delta$ for any $\delta > 0$, and consequently, it is a regular function by Weierstrass's theorem. It follows that all singular points of the left-hand side, including the complex zeros of $\zeta(s)$, lie in the half-plane $\text{Re}(s) \leq \frac{1}{2}$, and so they all lie in $\text{Re}(s) = \frac{1}{2}$ since they are symmetrical with respect to the straight line $\text{Re}(s) = \frac{1}{2}$. We have thus proved Riemann's hypothesis. $\qquad \square$

For $\text{Re}(s) > 1$ we have

$$\frac{1}{\zeta(s)} = \sum_{n=1}^\infty \frac{\mu(n)}{n^s},$$

where $\mu(n)$ is the *Möbius' function*,

$$\mu(n) = \begin{cases} (-1)^k, & \text{if } n = p_1 \cdots p_k, \ p_j \text{ are different prime numbers;} \\ 0, & \text{if } p^2 \mid n \text{ for some prime number } p; \\ 1, & \text{if } n = 1. \end{cases}$$

Similar to the proof above, one has the following fact:

Theorem 7.11. *For Riemann's hypothesis to hold true, it is necessary and sufficient that the estimate*

$$\sum_{n \leq x} \mu(n) = O\left(x^{\frac{1}{2}+\varepsilon}\right)$$

is valid for every positive ε.

Let $\rho(x)$ be the function defined for $x > 0$ equal to the representative of x modulo 1; thus

$$x = [x] + \rho(x)$$

where $[x]$ is the largest integer $\leq x$. The linear space $\mathscr{M}$ consisting of functions of the form

$$f(x) = \sum_{k=1}^{N} a_k \rho\left(\frac{\theta_k}{x}\right),$$

where

$$0 < \theta_k \leq 1 \ (k = 1, \ldots, N), \ \sum_{k=1}^{N} a_k \theta_k = 0,$$

then consists of bounded, measurable functions vanishing for

$$x > \max_{k}[\theta_k]$$

and therefore vanishing for $x > 1$. Nyman [306] and Beurling [22] established the following criterion:

Theorem 7.12. $\mathscr{M}$ *is dense in $L^p(0,1)$, $1 \leq p \leq \infty$, if and only if the Riemann ζ-function has no zeros in the half-plane $\sigma > \frac{1}{p}$.*

A complete proof can be found in [82]. The necessity follows from the relation

$$s \int_0^1 x^{s-1} f(x) dx = -\zeta(s) \sum_{k=1}^{N} a_k \theta_k^s$$

valid for any $f \in \mathscr{M}$.

Thus Riemann's hypothesis holds if and only if $\mathscr{M}$ is dense in $L^2(0,1)$. Nyman-Beurling's criterion has been extended by Báez-Duarte, who showed that one may restrict attention to integral values of $1/\theta_k$. Balazard and Saias [13] have rephrased this by showing that Riemann's hypothesis holds if and only if

$$\inf_A \int_{-\infty}^{\infty} \left| 1 - A\left(\frac{1}{2} + it\right) \zeta\left(\frac{1}{2} + it\right) \right|^2 \frac{dt}{\frac{1}{4} + t^2} = 0,$$

where the infimum is over all Dirichlet polynomials A.

Further, if d_N denotes the infimum over all Dirichlet polynomials

$$A(s) = \sum_{n=1}^{N} \frac{a_n}{n^s}$$

of length N, they conjecture that

$$d_N \sim \frac{1}{\log N} \sum_{\rho} \frac{1}{|\rho|^2}.$$

Burnol has proved that

$$d_N \geq \frac{1}{\log N} \sum_{\mathrm{Re}(\rho)=\frac{1}{2}} \frac{m_\rho}{|\rho|^2},$$

where m_ρ is the multiplicity of the zero ρ.

Here are a few other easy-to-state equivalences of Riemann's hypothesis:

(i) Hardy and Littlewood (1918): Riemann's hypothesis holds if and only if

$$\sum_{n=1}^{\infty} \frac{(-\chi)^n}{n!\zeta(2n+1)} = O\left(\chi^{-\frac{1}{4}}\right).$$

(ii) Redheffer (1977): Riemann's hypothesis holds if and only if for every $\varepsilon > 0$ there is a $C(\varepsilon) > 0$ such that

$$|\det(A(n))| < C(\varepsilon)n^{\frac{1}{2}+\varepsilon},$$

where $A(n)$ is the $n \times n$ matrix of 0's and 1's defined by $A(i,j) = 1$ if $j = 1$ or if i divides j, and $A(i,j) = 0$ otherwise.

(iii) Lagarias ([219], 2002): Riemann's hypothesis holds if and only if

$$\sigma_1(n) \leq H_n(0) + \exp(H_n(0))\log H_n(0), \quad n = 1, 2, \ldots,$$

where $\sigma_1(n)$ denotes the sum of the positive divisors of n.

Related to a Dirichlet character χ modulo r, one has the *Dirichlet L-function*:

$$L(\chi, s) = \sum_{n=1}^{\infty} \frac{\chi(n)}{n^s}, \quad \mathrm{Re}(s) > 1.$$

The analog of Euler's formula

$$L(\chi, s) = \prod_{p} \left(1 - \frac{\chi(p)}{p^s}\right)^{-1}, \quad \mathrm{Re}(s) > 1,$$

is valid. If χ is non-primitive, that is, there exist a proper factor d of r and a Dirichlet character χ^0 modulo d such that $\chi(n) = \chi^0(n)$ when $(n, r) = 1$, then

$$L(\chi, s) = L(\chi^0, s) \prod_{p|r} \left(1 - \frac{\chi^0(p)}{p^s}\right).$$

Assume that χ is a primitive character modulo r. Obviously, we have $L(\chi, s) = \zeta(s)$ if $r = 1$. When $r > 1$, it is well known that $L(\chi, s)$ is entire with $L(\chi, 1) \neq 0$. Further, setting

$$a = \begin{cases} 0, & \text{if } \chi(-1) = 1; \\ 1, & \text{if } \chi(-1) = -1 \end{cases}$$

and

$$\Lambda_\chi(s) = \left(\frac{r}{\pi}\right)^{\frac{s}{2}} \Gamma\left(\frac{s+a}{2}\right) L(\chi, s), \qquad (7.3.14)$$

the Dirichlet L-function of a primitive character χ modulo r satisfies the functional equation

$$\Lambda_\chi(s) = \frac{(-1)^a}{\sqrt{r}} g(\chi) \Lambda_{\bar{\chi}}(1 - s), \qquad (7.3.15)$$

where $g(\chi)$ is the Gauss sum. The proof of (7.3.15) is the same as for zeta and is based on the Poisson summation formula (cf. [73]).

The *generalized Riemann hypothesis* states that if $L(\chi, s) = 0$, then either s is a negative integer (a "trivial zero") or $\text{Re}(s) = \frac{1}{2}$. It had been shown, for a sufficiently small constant $c > 0$, that if $L(\chi, s) = 0$ with

$$\text{Re}(s) > 1 - \frac{c}{\log r},$$

then s is real, χ is a quadratic real character, and there is at most one such value of r between R and R^2 for any sufficiently large R. Such zeros are known as *Siegel zeros*. In 1995, Granville and Stark proved, assuming the *abc*-conjecture, that $L(\chi, s)$ has no Siegel zeros for all $\chi \pmod{r}$ with $r \equiv 3 \pmod 4$.

7.4 Hadamard's factorization

We can now write

$$\zeta(s) = \frac{e^{bs}}{2(s-1)\Gamma\left(\frac{s}{2}+1\right)} \prod_\rho \left(1 - \frac{s}{\rho}\right) e^{\frac{s}{\rho}}, \qquad (7.4.1)$$

where ρ runs through the complex zeros of $\zeta(s)$ in the strip $0 \leq \text{Re}(s) \leq 1$. For by *Hadamard's factorization theorem*

$$\xi(s) = \xi(0) e^{b_0 s} \prod_\rho \left(1 - \frac{s}{\rho}\right) e^{\frac{s}{\rho}}.$$

Here $\xi(0) = -\zeta(0) = \frac{1}{2}$, and

$$b = b_0 + \frac{1}{2}\log\pi = \log 2\pi - 1 - \frac{\gamma}{2}, \qquad (7.4.2)$$

where γ is *Euler's constant*

$$\gamma = \lim_{n\to+\infty}\left(1 + \frac{1}{2} + \cdots + \frac{1}{n} - \log n\right).$$

To prove (7.4.2), we first show *Kronecker's limit formula*

$$\lim_{s\to1}\left\{\zeta(s) - \frac{1}{s-1}\right\} = \gamma. \qquad (7.4.3)$$

For $\mathrm{Re}(s) > 0$, by (7.1.21) we have

$$\zeta(s) - \frac{1}{s-1} = \sum_{n=1}^{N}\frac{1}{n} - \int_1^N \frac{dx}{x} + \sum_{n=1}^{N}\left(\frac{1}{n^s} - \frac{1}{n}\right)$$
$$- \int_1^N\left(\frac{1}{x^s} - \frac{1}{x}\right)dx - s\int_N^\infty \frac{x - [x]}{x^{s+1}}\,dx.$$

We can choose N so large that the last term is as small as we please, and at the same time so that the first term differs from γ by as little as we please, and this independently of $\mathrm{Re}(s)$. Having fixed N, the remaining terms tend to zero with $s \to 1$. This proves (7.4.3). We now calculate $\zeta'(0)$ by differentiating (7.1.8) logarithmically, making $s \to 0$, and using (7.1.6) and (7.4.3) and the fact that

$$\Gamma'(1) = -\gamma.$$

We obtain

$$\zeta'(0) = -\frac{1}{2}\log 2\pi. \qquad (7.4.4)$$

Finally $b_0 = \xi'(0)/\xi(0)$, and (7.4.2) follows from (7.4.4).

We will use the following value (see [164], [165]):

$$\frac{\zeta'\left(\frac{1}{2}\right)}{\zeta\left(\frac{1}{2}\right)} = \frac{\pi}{4} + \frac{\gamma}{2} + \frac{1}{2}\log 8\pi. \qquad (7.4.5)$$

This can be seen as follows: Differentiating logarithmically the *Weierstrass formula*

$$\frac{1}{\Gamma(s+1)} = e^{\gamma s}\prod_{n=1}^{\infty}\left(1 + \frac{s}{n}\right)e^{-\frac{s}{n}}, \qquad (7.4.6)$$

we have

$$\frac{\Gamma'(s+1)}{\Gamma(s+1)} = -\gamma + \sum_{n=1}^{\infty}\left(\frac{1}{n} - \frac{1}{n+s}\right).$$

Set

$$H_n(s) = \sum_{k=1}^{n} \frac{1}{k+s}.$$

By the definition of Euler's constant γ, we know

$$H_n(0) = \log n + \gamma + o(1)$$

as $n \to \infty$. Thus we have

$$H_n(s) = \log n - \frac{\Gamma'(s+1)}{\Gamma(s+1)} + o(1). \tag{7.4.7}$$

Moreover, we obtain directly

$$H_n\left(-\frac{1}{2}\right) = 2 \sum_{k=1}^{n} \frac{1}{2k-1} = 2\left\{H_{2n}(0) - \frac{1}{2}H_n(0)\right\} = \log n + \gamma + \log 4 + o(1).$$

Comparing with (7.4.7) for $s = -\frac{1}{2}$, we have

$$\frac{\Gamma'\left(\frac{1}{2}\right)}{\Gamma\left(\frac{1}{2}\right)} = -\gamma - \log 4. \tag{7.4.8}$$

Differentiating logarithmically the functional equation (7.1.8), we obtain

$$-\log 2 + \frac{\Gamma'(s)}{\Gamma(s)} + \frac{\zeta'(s)}{\zeta(s)} - \frac{\pi}{2}\tan\frac{\pi s}{2} = \log\pi - \frac{\zeta'(1-s)}{\zeta(1-s)},$$

and hence

$$2\frac{\zeta'\left(\frac{1}{2}\right)}{\zeta\left(\frac{1}{2}\right)} = \log 2\pi - \frac{\Gamma'\left(\frac{1}{2}\right)}{\Gamma\left(\frac{1}{2}\right)} + \frac{\pi}{2}\tan\frac{\pi}{4}.$$

Thus (7.4.5) follows.

Differentiating logarithmically *Legendre's formula*

$$\Gamma(s)\Gamma\left(s + \frac{1}{2}\right) = \frac{\sqrt{\pi}}{2^{2s-1}}\Gamma(2s)$$

and the functional equation

$$\Gamma(s)\Gamma(1-s) = \frac{\pi}{\sin\pi s},$$

we have

$$\frac{\Gamma'(s)}{\Gamma(s)} + \frac{\Gamma'\left(s + \frac{1}{2}\right)}{\Gamma\left(s + \frac{1}{2}\right)} = -\log 4 + 2\frac{\Gamma'(2s)}{\Gamma(2s)},$$

and

$$\frac{\Gamma'(s)}{\Gamma(s)} - \frac{\Gamma'(1-s)}{\Gamma(1-s)} = -\pi \cot \pi s.$$

In particular, setting $s = \frac{1}{4}$, we obtain

$$\frac{\Gamma'\left(\frac{1}{4}\right)}{\Gamma\left(\frac{1}{4}\right)} = -\frac{\pi}{2} - \gamma - \log 8. \tag{7.4.9}$$

Denote the zeros of $\zeta(s)$ on the semi-line $\{\mathrm{Re}(s) = \frac{1}{2}, \ \mathrm{Im}(s) > 0\}$ by

$$s_\nu = \frac{1}{2} + it_\nu = |s_\nu|e^{i\alpha_\nu}, \quad \nu = 1, 2, \ldots. \tag{7.4.10}$$

Then we have

$$t_\nu > 0, \quad 0 < \alpha_\nu < \frac{\pi}{2}, \quad \nu = 1, 2, \ldots.$$

Assume that

$$0 < t_\nu \le t_{\nu+1} \le \cdots, \quad \nu = 1, 2, 3, \ldots.$$

Further, denote the zeros of $\zeta(s)$ on the region $\{1 > \mathrm{Re}(s) > \frac{1}{2}, \ \mathrm{Im}(s) > 0\}$ (if they exist) by

$$z_\mu = x_\mu + iy_\mu = |z_\mu|e^{i\beta_\mu}, \quad \mu = 1, 2, \ldots. \tag{7.4.11}$$

Thus all zeros of $\zeta(s)$ in the critical strip are

$$\{s_\nu, \bar{s}_\nu \mid \nu \ge 1\} \cup \{z_\mu, \bar{z}_\mu, 1 - z_\mu, 1 - \bar{z}_\mu \mid \mu \ge 1\}.$$

Theorem 7.13 ([165]). *Riemann's hypothesis is true if and only if*

$$\sum_{\nu=1}^{\infty} \frac{1}{|s_\nu|^2} = 1 + \frac{\gamma}{2} - \frac{1}{2}\log 4\pi.$$

Proof. Differentiating logarithmically the formula (7.4.1), we have

$$\frac{\zeta'(s)}{\zeta(s)} + \frac{\Gamma'\left(\frac{s}{2}\right)}{2\Gamma\left(\frac{s}{2}\right)} + \frac{1}{s} + \frac{1}{s-1} - b = \sum_{\rho}\left(\frac{1}{\rho} - \frac{1}{\rho - s}\right).$$

Setting $s = \frac{1}{2}$, and using (7.4.2), (7.4.5) and (7.4.9), we obtain

$$\sum_{\rho}\left(\frac{1}{\rho} - \frac{1}{\rho - \frac{1}{2}}\right) = 1 + \frac{\gamma}{2} - \frac{1}{2}\log 4\pi.$$

Note that the series in the left-hand side is absolutely convergent so that the order of terms in the sum can be changed arbitrarily. Since

$$\frac{1}{s_\nu - \frac{1}{2}} + \frac{1}{\bar{s}_\nu - \frac{1}{2}} = 0,$$

$$\frac{1}{z_\mu - \frac{1}{2}} + \frac{1}{\bar{z}_\mu - \frac{1}{2}} + \frac{1}{(1 - z_\mu) - \frac{1}{2}} + \frac{1}{(1 - \bar{z}_\mu) - \frac{1}{2}} = 0,$$

we obtain

$$\sum_{\rho}\left(\frac{1}{\rho}-\frac{1}{\rho-\frac{1}{2}}\right) = \sum_{\nu=1}^{\infty}\left(\frac{1}{s_\nu}+\frac{1}{\bar{s}_\nu}\right)+\sum_{\mu}\left(\frac{1}{z_\mu}+\frac{1}{\bar{z}_\mu}+\frac{1}{1-z_\mu}+\frac{1}{1-\bar{z}_\mu}\right)$$

$$=\sum_{\nu=1}^{\infty}\frac{1}{|s_\nu|^2}+2\sum_{\mu}\left(\frac{x_\mu}{|z_\mu|^2}+\frac{1-x_\mu}{|1-z_\mu|^2}\right)$$

$$=1+\frac{\gamma}{2}-\frac{1}{2}\log 4\pi.$$

Thus Riemann's hypothesis is true if and only if

$$\sum_{\mu}\left(\frac{x_\mu}{|z_\mu|^2}+\frac{1-x_\mu}{|1-z_\mu|^2}\right)=0,$$

that is, z_μ do not exist, or equivalently

$$\sum_{\nu=1}^{\infty}\frac{1}{|s_\nu|^2}=1+\frac{\gamma}{2}-\frac{1}{2}\log 4\pi.$$

$\square$

Let θ be defined by (7.1.14) and consider the function

$$\Theta(x)=\theta(ix)-\frac{1}{2}=\sum_{n=1}^{\infty}e^{-n^2\pi x}. \tag{7.4.12}$$

Then Riemann's formula (7.1.17) yields

$$\Xi(t)=\frac{1}{2}-\left(t^2+\frac{1}{4}\right)\int_1^{\infty}\Theta(x)x^{-\frac{3}{4}}\cos\left(\frac{t}{2}\log x\right)dx. \tag{7.4.13}$$

If one integrates by parts and uses the relation

$$4\Theta'(1)+\Theta(1)=-\frac{1}{2},$$

which follows at once from (7.1.13), one obtains

$$\Xi(t)=4\int_1^{\infty}\frac{d}{dx}\{x^{\frac{3}{2}}\Theta'(x)\}x^{-\frac{1}{4}}\cos\left(\frac{t}{2}\log x\right)dx \tag{7.4.14}$$

and can write it in the form (see [394])

$$\Xi(t)=8\int_0^{\infty}\Phi(x)\cos 2txdx, \tag{7.4.15}$$

where

$$\Phi(x)=\sum_{n=1}^{\infty}\left(2\pi^2 n^4 e^{9x}-3\pi n^2 e^{5x}\right)e^{-n^2\pi e^{4x}}. \tag{7.4.16}$$

The Taylor series expansion of Ξ about the origin can be written in the form

$$\Xi(t) = 8 \sum_{k=0}^{\infty} \frac{(-1)^k B_k}{(2k)!} (2t)^{2k}, \qquad (7.4.17)$$

where the constants B_k are defined by

$$B_k = \int_0^{\infty} x^{2k} \Phi(x) dx, \quad k = 0, 1, 2, \ldots . \qquad (7.4.18)$$

Theorem 7.14 ([70]). *Let $\{t_\nu\}_{\nu=1}^{\infty}$, $0 < t_1 \le t_2 \le \cdots$, denote the real zeros of Ξ in the right half-plane. Then the Riemann hypothesis is true if and only if*

$$\sum_{\nu=1}^{\infty} \frac{1}{t_\nu^2} = \frac{2B_1}{B_0}. \qquad (7.4.19)$$

Proof. The Riemann hypothesis is the statement that all zeros of Ξ are real. Note that Ξ is an even entire function of order 1, all of whose zeros lie in the horizontal strip $|\mathrm{Im}(z)| < \frac{1}{2}$. Thus if $z = a + ib$ is a zero of Ξ, then $-a + ib$, $-a - ib$ and $a - ib$ are also zeros of Ξ. By the Hadamard factorization theorem Ξ can be represented in the form

$$\Xi(z) = \Xi(0) \prod_{k=1}^{\infty} \left(1 - \frac{z^2}{w_k^2} \right) \qquad (7.4.20)$$

where the zeros w_k of Ξ lie in the right half-plane, and are numbered according to increasing moduli, i.e., $0 < |w_1| \le |w_2| \le \cdots$. Thus the derivative of the logarithmic derivative of $\Xi(z)$ is

$$\frac{\Xi(z)\Xi''(z) - \{\Xi'(z)\}^2}{\Xi(z)^2} = -\sum_{k=1}^{\infty} \left\{ \frac{1}{(z - w_k)^2} + \frac{1}{(z + w_k)^2} \right\}. \qquad (7.4.21)$$

Obviously, $\Xi'(0) = 0$ since Ξ is an even function, and

$$\Xi(0) = 8B_0 > 0, \quad \Xi''(0) = -32B_1$$

by using (7.4.17). Setting $z = 0$ in (7.4.21), we obtain

$$\frac{2B_1}{B_0} = \sum_{k=1}^{\infty} \frac{1}{w_k^2} = \sum_{\nu=1}^{\infty} \frac{1}{t_\nu^2} + 2 \sum_{\mathrm{Im}(w_k) > 0} \frac{\mathrm{Re}(w_k)^2 - \mathrm{Im}(w_k)^2}{|w_k|^4}. \qquad (7.4.22)$$

But if Ξ has zeros w_k with $\mathrm{Im}(w_k) > 0$, then the second summand on the right-hand side of (7.4.22) is positive since all zeros of Ξ lie in the strip $|\mathrm{Im}(z)| < \frac{1}{2}$ and Ξ has no zeros in the closed unit square centered at the origin. Thus we infer from (7.4.22) that the Riemann hypothesis is true if and only if (7.4.19) holds. $\square$

G. Csordas and C.C. Yang [70] gave the following numerical calculations:

$$1 + \frac{\gamma}{2} - \frac{1}{2}\log 4\pi = 0.023095708\ldots,$$

$$\sum_{\nu=1}^{100000} \frac{1}{|s_\nu|^2} = 0.023073645\ldots,$$

$$\frac{2B_1}{B_0} = 0.023104993\ldots,$$

$$\sum_{\nu=1}^{100000} \frac{1}{t_\nu^2} = 0.023082929\ldots.$$

7.5 Nevanlinna's formula

Let D be a bounded connected open set in $\mathbb{C}$ such that the boundary ∂D of D consists of a finite number of simple closed Jordan arcs of class C^1. We will use *Green's function* $G_z(\zeta)$ of the domain D, which is uniquely defined for $\zeta \in \overline{D}$, $z \in \overline{D}$, $\zeta \neq z$ such that

(i) when $z \in D$ is fixed, $G_z(\zeta)$ can be expressed as

$$G_z(\zeta) = -\log|\zeta - z| + h_z(\zeta),$$

   where $h_z(\zeta)$ is harmonic on D and continuous on $\overline{D}$;

(ii) when $\zeta \in \partial D$, $z \in \overline{D}$ or $\zeta \in \overline{D}$, $z \in \partial D$, one has $G_z(\zeta) = 0$.

It is easy to show $G_z(\zeta) > 0$ when $\zeta \in D$, $z \in D$ by the minimum modulus principle. If D is simply connected, the Riemann mapping theorem shows that for $z \in D$, there is a unique analytic function $\mathcal{G}_z : D \longrightarrow \mathbb{C}$ having the properties:

(a) $\mathcal{G}_z(z) = 0$ and $\mathcal{G}_z'(z) > 0$;

(b) $\mathcal{G}_z$ is one-one;

(c) $\mathcal{G}_z(D) = \{z \in \mathbb{C} \mid |z| < 1\}$.

Then Green's function of D is given by

$$G_z(\zeta) = -\log|\mathcal{G}_z(\zeta)|. \tag{7.5.1}$$

Next we will assume that the domain D is simply connected. Let $\gamma_1, \ldots, \gamma_p$ be the edges of the curvilinear polygon ∂D. Let A_l be the common vertex of γ_l and γ_{l+1}, where $\gamma_{p+1} = \gamma_1$. Let $\alpha_l \pi$ $(0 < \alpha_l \leq 2$, $l = 1, \ldots, p)$ be the angle between γ_l and γ_{l+1}. By the Riemann-Schwarz symmetric principle, the function $\mathcal{G}_z$ can be analytically continued to a domain containing $\overline{D} - \{A_1, \ldots, A_p\}$. We will assume that the function $\mathcal{G}_z$ can be expressed as

$$\mathcal{G}_z(\zeta) = (\zeta - A_l)^{\frac{1}{\alpha_l}} \varphi_l(\zeta) + w_l \qquad (7.5.2)$$

near A_l for $l = 1, \ldots, p$, where φ_l is analytic near A_l with $\varphi_l(A_l) \neq 0$, and $|w_l| = 1$.

Theorem 7.15. *Let D be a bounded simple connected open set in $\mathbb{C}$ such that the boundary ∂D of D is a curvilinear polygon of class C^1. Take distinct points $c_1, \ldots, c_q$ in $\overline{D}$ and a function $u \in C^2(\overline{D} - \{c_1, \ldots, c_q\})$. Assume*

$$u(z) = d_k \log|z - c_k| + u_k(z), \quad k = 1, \ldots, q,$$

near c_k, where d_k are constant, and u_k are of class C^2 near c_k. Then for $z \in D - \{c_1, \ldots, c_q\}$, one has

$$u(z) + \frac{1}{2\pi} \int_D G_z \Delta u \, dx dy = \frac{1}{2\pi} \int_{\partial D} \frac{\partial G_z}{\partial n} u \, ds - \sum_{c_k \in D} d_k G_z(c_k).$$

Proof. For $a \in \mathbb{C}$, $\varepsilon > 0$, denote

$$\mathbb{C}(a; \varepsilon) = \{\zeta \in \mathbb{C} \mid |\zeta - a| < \varepsilon\}, \quad \mathbb{C}\langle a; \varepsilon\rangle = \{\zeta \in \mathbb{C} \mid |\zeta - a| = \varepsilon\}.$$

Set $S = \{z, c_1, \ldots, c_q, A_1, \ldots, A_p\}$ and choose ε sufficient small such that

$$\mathbb{C}(a; \varepsilon) \cap \mathbb{C}(b; \varepsilon) = \emptyset, \quad \{a, b\} \subset S, \ a \neq b,$$

$\mathbb{C}(a; \varepsilon) \subset D$ if $a \in D \cap S$, and such that $\mathcal{G}_z$ is analytic in $\mathbb{C}(a; \varepsilon) - \{a\}$ if $a \in \{A_1, \ldots, A_p\}$. Write

$$\Gamma_\varepsilon = \partial D - \bigcup_{l=1}^{p} \{\partial D \cap \mathbb{C}(A_l; \varepsilon)\},$$

$$D_\varepsilon = D - \bigcup_{a \in S} \mathbb{C}(a; \varepsilon).$$

Applying Green's formula (6.1.1) to $u = u(\zeta)$, $w = G_z(\zeta)$ on D_ε and noting $\Delta w = 0$, one has

$$\int_{D_\varepsilon} G_z \Delta u \, dx dy = \left\{ \int_{\Gamma_\varepsilon} + \sum_{a \in S \cap D} \int_{\mathbb{C}\langle a; \varepsilon\rangle} + \sum_{a \in S - D} \int_{\mathbb{C}\langle a; \varepsilon\rangle \cap \partial D_\varepsilon} \right\} \left(u \frac{\partial G_z}{\partial n} - G_z \frac{\partial u}{\partial n} \right) ds. \qquad (7.5.3)$$

Since $G_z(\zeta) = 0$ when $\zeta \in \partial D$, we immediately obtain

$$\lim_{\varepsilon \to 0} \int_{\Gamma_\varepsilon} \left(u \frac{\partial G_z}{\partial n} - G_z \frac{\partial u}{\partial n} \right) ds = \int_{\partial D} \frac{\partial G_z}{\partial n} u \, ds.$$

When $\zeta \in \mathbb{C}\langle z; \varepsilon \rangle$, one has the following estimates:

$$G_z(\zeta) = -\log \varepsilon + O(1), \qquad \frac{\partial G_z}{\partial n}(\zeta) = -\frac{1}{\varepsilon} + O(1),$$

$$u(\zeta) = u(z) + o(1), \qquad \frac{\partial u}{\partial n} = O(1)$$

as $\varepsilon \to 0$. By the integral mean value theorem, there is $\zeta^* \in \mathbb{C}\langle z; \varepsilon \rangle$ such that

$$\int_{\mathbb{C}\langle z;\varepsilon \rangle} \left(u \frac{\partial G_z}{\partial n} - G_z \frac{\partial u}{\partial n} \right) ds = 2\pi\varepsilon \left(u \frac{\partial G_z}{\partial n} - G_z \frac{\partial u}{\partial n} \right) (\zeta^*),$$

and hence

$$\lim_{\varepsilon \to 0} \int_{\mathbb{C}\langle z;\varepsilon \rangle} \left(u \frac{\partial G_z}{\partial n} - G_z \frac{\partial u}{\partial n} \right) ds = -2\pi u(z).$$

When $\zeta \in \mathbb{C}\langle c_k; \varepsilon \rangle$ for some $c_k \in D$, one has the following estimates:

$$G_z(\zeta) = G_z(c_k) + o(1), \qquad \frac{\partial G_z}{\partial n}(\zeta) = O(1),$$

$$u(\zeta) = d_k \log \varepsilon + O(1), \qquad \frac{\partial u}{\partial n} = \frac{d_k}{\varepsilon} + O(1)$$

as $\varepsilon \to 0$. Similarly, by using the integral mean value theorem one can obtain

$$\lim_{\varepsilon \to 0} \int_{\mathbb{C}\langle c_k;\varepsilon \rangle} \left(u \frac{\partial G_z}{\partial n} - G_z \frac{\partial u}{\partial n} \right) ds = -2\pi d_k G_z(c_k).$$

When $\zeta \in \mathbb{C}\langle a; \varepsilon \rangle$ for some $a \in S - D$, one has the following estimates:

$$G_z(\zeta) = o(1), \qquad \frac{\partial G_z}{\partial n}(\zeta) = O\left(\frac{1}{\sqrt{\varepsilon}} \right),$$

$$u(\zeta) = O(|\log \varepsilon|), \qquad \frac{\partial u}{\partial n} = O\left(\frac{1}{\varepsilon} \right)$$

as $\varepsilon \to 0$, where the relations (7.5.1) and (7.5.2) are used. Thus, by using the integral mean value theorem one can obtain

$$\lim_{\varepsilon \to 0} \int_{\mathbb{C}\langle a;\varepsilon \rangle \cap \partial D_\varepsilon} \left(u \frac{\partial G_z}{\partial n} - G_z \frac{\partial u}{\partial n} \right) ds = 0.$$

Therefore the proof of Theorem 7.15 is completed by letting $\varepsilon \to 0$ in (7.5.3). $\square$

Applying Theorem 7.15 to the function $u = \log |f|$, we obtain *Nevanlinna's formula* (cf. [290], [52]):

Theorem 7.16. *Let D be a bounded simple connected open set in $\mathbb{C}$ such that the boundary ∂D of D is a curvilinear polygon of class C^1. Let f $(\not\equiv 0)$ be meromorphic in $\overline{D}$. Let a_μ $(\mu = 1, \ldots, m)$ and b_ν $(\nu = 1, \ldots, n)$ be respectively zeros and poles of f in D. Then for $z \in D - \{a_1, \ldots, a_m, b_1, \ldots, b_n\}$, one has*

$$\log|f(z)| = \frac{1}{2\pi}\int_{\partial D}\log|f|\frac{\partial G_z}{\partial n}ds - \sum_{\mu=1}^{m}G_z(a_\mu) + \sum_{\nu=1}^{n}G_z(b_\nu).$$

In particular, for the disc $D = \mathbb{C}(0; R)$ $(R > 0)$, we know

$$G_z(\zeta) = \log\left|\frac{R^2 - \overline{z}\zeta}{R(\zeta - z)}\right|,$$

and hence

$$\frac{\partial G_z}{\partial n}(\zeta)ds = \frac{R^2 - |z|^2}{|\zeta - z|^2}d\theta = \mathrm{Re}\left(\frac{\zeta + z}{\zeta - z}\right)d\theta, \quad \zeta = Re^{i\theta}.$$

Therefore the *Poisson-Jensen formula* follows.

Theorem 7.17. *Let f $(\not\equiv 0)$ be meromorphic in $\mathbb{C}[0; R]$. Let a_μ $(\mu = 1, \ldots, m)$ and b_ν $(\nu = 1, \ldots, n)$ be respectively zeros and poles of f in $\mathbb{C}(0; R)$. Then for $z \in \mathbb{C}(0; R) - \{a_1, \ldots, a_m, b_1, \ldots, b_n\}$, one has*

$$\log|f(z)| = \frac{1}{2\pi}\int_0^{2\pi}\log\left|f\left(Re^{i\theta}\right)\right|\mathrm{Re}\left(\frac{Re^{i\theta} + z}{Re^{i\theta} - z}\right)d\theta$$

$$- \sum_{\mu=1}^{m}\log\left|\frac{R^2 - \overline{a}_\mu z}{R(z - a_\mu)}\right| + \sum_{\nu=1}^{n}\log\left|\frac{R^2 - \overline{b}_\nu z}{R(z - b_\nu)}\right|.$$

Theorem 7.17 implies that there is a real number c such that

$$\log f(z) = \frac{1}{2\pi}\int_0^{2\pi}\log\left|f\left(Re^{i\theta}\right)\right|\frac{Re^{i\theta} + z}{Re^{i\theta} - z}d\theta$$

$$- \sum_{\mu=1}^{m}\log\frac{R^2 - \overline{a}_\mu z}{R(z - a_\mu)} + \sum_{\nu=1}^{n}\log\frac{R^2 - \overline{b}_\nu z}{R(z - b_\nu)} + ic. \tag{7.5.4}$$

Differentiating (7.5.4), one obtains

$$\left(\frac{d}{dz}\right)^{q-1}\frac{f'(z)}{f(z)} = \frac{q!}{2\pi}\int_0^{2\pi}\log\left|f\left(Re^{i\theta}\right)\right|\frac{2Re^{i\theta}}{(Re^{i\theta} - z)^{q+1}}d\theta$$

$$+ (q-1)!\sum_{\mu=1}^{m}\left\{\frac{\overline{a}_\mu^q}{(R^2 - \overline{a}_\mu z)^q} - \frac{(-1)^q}{(z - a_\mu)^q}\right\}$$

$$- (q-1)!\sum_{\nu=1}^{n}\left\{\frac{\overline{b}_\nu^q}{(R^2 - \overline{b}_\nu z)^q} - \frac{(-1)^q}{(z - b_\nu)^q}\right\}. \tag{7.5.5}$$

Further if f is meromorphic in $\mathbb{C}$ such that

$$\limsup_{R \to \infty} \frac{T(R, f)}{R^q} = 0,$$

then (see [151])

$$\frac{1}{(q-1)!} \left(\frac{d}{dz}\right)^{q-1} \frac{f'(z)}{f(z)} = \lim_{R \to \infty} \left\{ \sum_{|b_\nu| < R} \frac{1}{(b_\nu - z)^q} - \sum_{|a_\mu| < R} \frac{1}{(a_\mu - z)^q} \right\}. \quad (7.5.6)$$

Theorem 7.18. *Let f be a meromorphic function of finite order ρ such that $f(0) \neq 0, \infty$. Then for any integer $q \geq [\rho] + 1 > \rho$, we have*

$$\sum_\nu \frac{1}{b_\nu^q} - \sum_\mu \frac{1}{a_\mu^q} = \frac{1}{(q-1)!} \left(\frac{d}{dz}\right)^{q-1} \left(\frac{f'(z)}{f(z)}\right)_{z=0},$$

where a_μ and b_ν are respectively zeros and poles of f.

Proof. It is trivial to show

$$\lim_{R \to \infty} \frac{T(R, f)}{R^q} = 0.$$

Let ρ_0 and ρ_∞ be the convergence exponents of the zeros and poles of f, respectively. Then

$$\rho_0 = \limsup_{R \to \infty} \frac{\log n\left(R, \frac{1}{f}\right)}{\log R} \leq \rho < q,$$

$$\rho_\infty = \limsup_{R \to \infty} \frac{\log n(R, f)}{\log R} \leq \rho < q.$$

By the definition of convergence exponent, we know that two series in Theorem 7.18 are convergent, and hence the theorem follows from (7.5.6). $\qquad\square$

Using the notations of § 7.4 to denote the zeros of $\zeta(s)$, we prove the following result:

Theorem 7.19 ([165]). *Riemann's hypothesis is true if and only if there exists an integer $q \geq 2$ such that $\zeta(s)$ has no zeros in the region*

$$\left\{ s \in \mathbb{C} \mid \frac{1}{2} < \mathrm{Re}(s) < 1, \ 0 \leq \arg(s) \leq \frac{(q-1)\pi}{2q} \right\},$$

and

$$2 \sum_{\nu=1}^\infty \frac{\cos q\alpha_\nu}{|s_\nu|^q} = 1 - \left(-\frac{1}{2}\right)^q \zeta(q) - \frac{1}{(q-1)!} \left(\frac{d}{ds}\right)^{q-1} \left(\frac{\zeta'(s)}{\zeta(s)}\right)_{s=0}.$$

Proof. Since $\zeta(s)$ is of order 1, Theorem 7.18 implies that the equation

$$1 - \sum_{m=1}^{\infty} \frac{1}{(-2m)^q} - \sum_{\rho} \frac{1}{\rho^q} = \frac{1}{(q-1)!} \left(\frac{d}{ds} \right)^{q-1} \left(\frac{\zeta'(s)}{\zeta(s)} \right)_{s=0}$$

holds for any integer $q \geq 2$, where ρ runs through the complex zeros of $\zeta(s)$ in the strip $0 < \text{Re}(s) < 1$, which yields

$$2 \sum_{\nu=1}^{\infty} \frac{\cos q\alpha_\nu}{|s_\nu|^q} + 2 \sum_{\mu} \left\{ \frac{\cos q\beta_\mu}{|z_\mu|^q} + \frac{\cos q\varphi_\mu}{|1 - z_\mu|^q} \right\}$$

$$= 1 - \left(-\frac{1}{2} \right)^q \zeta(q) - \frac{1}{(q-1)!} \left(\frac{d}{ds} \right)^{q-1} \left(\frac{\zeta'(s)}{\zeta(s)} \right)_{s=0},$$

where

$$1 - \overline{z}_\mu = |1 - z_\mu| e^{i\varphi_\mu} \quad \left(0 < \varphi_\mu < \frac{\pi}{2} \right).$$

Hence the necessity follows easily.

Next we show the sufficiency. The sufficiency condition implies

$$\sum_{\mu} \left\{ \frac{\cos q\beta_\mu}{|z_\mu|^q} + \frac{\cos q\varphi_\mu}{|1 - z_\mu|^q} \right\} = 0,$$

and

$$\frac{\pi}{2} > \varphi_\mu > \beta_\mu > \frac{(q-1)\pi}{2q}, \quad \mu \geq 1.$$

Set

$$\theta_\mu = q \left(\beta_\mu - \frac{\pi}{2} \right), \quad \psi_\mu = q \left(\varphi_\mu - \frac{\pi}{2} \right)$$

with $0 > \psi_\mu > \theta_\mu > -\frac{\pi}{2}$. Then the above equation becomes

$$\sum_{\mu} \left\{ \frac{\cos \theta_\mu}{|z_\mu|^q} + \frac{\cos \psi_\mu}{|1 - z_\mu|^q} \right\} = 0 \quad (q \text{ even }),$$

$$\sum_{\mu} \left\{ \frac{\sin \theta_\mu}{|z_\mu|^q} + \frac{\sin \psi_\mu}{|1 - z_\mu|^q} \right\} = 0 \quad (q \text{ odd }).$$

This is impossible if some z_μ exist. Hence Theorem 7.19 is proved. $\square$

Similar to the proof of Theorem 7.19, applying Theorem 7.18 to $\zeta\left(s + \frac{1}{2}\right)$ we can prove the following result:

Theorem 7.20 ([165]). *Let* $\{\frac{1}{2} + it_\nu\}_{\nu=1}^{\infty}$, $0 < t_1 \leq t_2 \leq \cdots$, *denote the zeros of* $\zeta(s)$ *on the semi-line* $\text{Re}(s) = \frac{1}{2}$, $\text{Im}(s) > 0$. *Then the Riemann hypothesis is true if and only if there exists an even* q *such that* $\zeta(s)$ *has no zeros in the region*

$$\left\{ s \in \mathbb{C} \mid 0 \leq \arg\left(s - \frac{1}{2} \right) \leq \frac{(q-1)\pi}{2q} \right\},$$

and

$$2(-1)^{\frac{q}{2}} \sum_{\nu=1}^{\infty} \frac{1}{t_\nu^q} = 2^q - 2^q \sum_{m=1}^{\infty} \frac{1}{(4m+1)^q} - \frac{1}{(q-1)!} \left(\frac{d}{ds}\right)^{q-1} \left(\frac{\zeta'(s)}{\zeta(s)}\right)_{s=\frac{1}{2}}.$$

We consider zeros $\rho_j = \sigma_j + it_j$ of ζ with $t_j > 0$ and assume that the zeros ρ_j are counted according to their multiplicities and ordered so that $0 < t_j \leq t_{j+1}$ (and $\sigma_j \leq \sigma_{j+1}$ if $t_j = t_{j+1}$) for $j \geq 1$. By "the first n zeros of ζ" we mean $\rho_1, \ldots, \rho_n$. For brevity we let $H(n)$ denote the statement that the first n zeros of ζ are simple and lie on the critical line. Gram [118], Backlund [9], Hutchinson [189], and Titchmarsh and Comrie [395] established $H(10)$, $H(79)$, $H(138)$ and $H(1,041)$, respectively. For a description of these computations see Edwards [89].

D.H. Lehmer [241], [242] performed the first extensive computation of zeros of ζ on a digital computer and established $H(25,000)$. Using similar methods, Meller [267], Lehman [239], and Rosser, Yohe and Schoenfeld [326] established

$$H(35,337), \quad H(250,000), \quad \text{and } H(3,500,000),$$

respectively. Using essentially the method introduced by Lehmer, R.P. Brent [36] has established $H(75,000,001)$. Moreover, there are precisely $75,000,000$ zeros with $0 < t_j < 32,585,736.4$. Thus the condition without zeros in the regions of Theorem 7.19 and Theorem 7.20 is satisfied for small q.

Similar to the proof of Theorem 7.19, applying Theorem 7.18 to $\Xi(z)$ we can prove the following result:

Theorem 7.21. *Let $\{t_\nu\}_{\nu=1}^{\infty}$, $0 < t_1 \leq t_2 \leq \cdots$, denote the real zeros of Ξ in the right half-plane. Then the Riemann hypothesis is true if and only if there exists an even q such that $\Xi(z)$ has no zero in the region*

$$\left\{z \in \mathbb{C} \mid \frac{\pi}{2q} \leq \arg(z) \leq \frac{\pi}{2}\right\},$$

and

$$2\sum_{\nu=1}^{\infty} \frac{1}{t_\nu^q} = -\frac{1}{(q-1)!} \left(\frac{d}{dz}\right)^{q-1} \left(\frac{\Xi'(z)}{\Xi(z)}\right)_{z=0}.$$

If $q = 2$, this is just Theorem 7.14. X. J. Li [246] proved that Riemann's hypothesis is equivalent to

$$\lambda_q := \frac{1}{(q-1)!} \left(\frac{d}{ds}\right)^q \left(s^{q-1} \log \xi(s)\right)\Big|_{s=1} \geq 0$$

for each $q = 1, 2, \ldots$. The necessity of the non-negativity condition is immediate from the equation

$$\lambda_q = \sum_{\rho} \left\{1 - \left(1 - \frac{1}{\rho}\right)^q\right\},$$

where ρ runs over the complex (non-trivial) zeros of the zeta-function. The sufficiency argument is based on considerations of the Taylor series ([201], [246])

$$\frac{\varphi'(z)}{\varphi(z)} = \sum_{q=1}^{\infty} \lambda_q z^{q-1} \quad \text{and} \quad 2\varphi(z) = 1 + \sum_{j=1}^{\infty} a_j z^j$$

satisfying the recurrence relation

$$\lambda_q = qa_q - \sum_{j=1}^{q-1} a_{q-j}\lambda_j, \quad \text{where} \quad \varphi(z) = \xi\left(\frac{1}{1-z}\right).$$

7.6 Carleman's formula

Take $R, \rho \in \mathbb{R}^+$ with $R > \rho > 0$. We will consider the domain

$$D = \{z \in \mathbb{C} \mid \rho < |z| < R, \ \text{Im}(z) > 0\}.$$

Theorem 7.22. *Take distinct points $c_1, \ldots, c_q$ in $\overline{D} - \mathbb{C}\langle 0; \rho \rangle$ and a function $u \in C^2(\overline{D} - \{c_1, \ldots, c_q\})$. Assume*

$$u(z) = d_k \log|z - c_k| + u_k(z), \ k = 1, \ldots, q,$$

near c_k, where d_k are constant, and u_k are of class C^2 near c_k. Then one has

$$-\frac{1}{2\pi} \int_D \text{Im}\left(\frac{1}{\zeta} + \frac{\zeta}{R^2}\right) \Delta u(\zeta) dx dy = \frac{1}{\pi R} \int_0^{\pi} u\left(Re^{i\theta}\right) \sin\theta d\theta$$

$$+ \frac{1}{2\pi} \int_{\rho}^{R} \left(\frac{1}{t^2} - \frac{1}{R^2}\right)[u(t) + u(-t)] dt$$

$$+ \sum_{c_k \in D} d_k \text{Im}\left(\frac{1}{c_k} + \frac{c_k}{R^2}\right) + Q(R, \rho; u),$$

$$(7.6.1)$$

where $Q(R, \rho; u) = O(1)$ as $R \to \infty$.

Proof. Applying the proof of Theorem 7.15 to the functions

$$u = u(\zeta), \quad w = -\text{Im}\left(\frac{1}{\zeta} + \frac{\zeta}{R^2}\right),$$

and noting that

(a) $\Delta w = 0$;

(b) $w(\zeta) = 0$, $\frac{\partial w}{\partial n}(\zeta) = \frac{2}{R^2}\sin\theta$, $\zeta = Re^{i\theta}$;

(c) $w(t) = 0$, $\frac{\partial w}{\partial n}(t) = \frac{1}{t^2} - \frac{1}{R^2}, t \in \mathbb{R}$, then the formula (7.6.1) follows, where

$$Q(R, \rho; u) = \frac{1}{2\pi} \int_{\overline{D} \cap \mathbb{C}\langle 0; \rho\rangle} \left(u \frac{\partial w}{\partial n} - w \frac{\partial u}{\partial n} \right) ds \qquad (7.6.2)$$

$$= -\frac{\rho}{2\pi} \int_0^\pi \left\{ \left(\frac{1}{\rho^2} + \frac{1}{R^2} \right) u \left(\rho e^{i\theta} \right) \right.$$

$$\left. + \left(\frac{1}{\rho} - \frac{\rho}{R^2} \right) \frac{\partial u}{\partial \rho} \left(\rho e^{i\theta} \right) \right\} \sin \theta d\theta.$$

$\square$

Now we can prove *Carleman's formula* (cf. [46], [52], [396]):

Theorem 7.23. *Let $f(z)$ be meromorphic in $\mathbb{C}[0; R] \cap \{\mathrm{Im}(z) \geq 0\}$ with $f(0) = 1$, and suppose that it has the zeros $r_1 e^{i\theta_1}$, $r_2 e^{i\theta_2}$, $\ldots$, $r_m e^{i\theta_m}$ and the poles $s_1 e^{i\varphi_1}$, $s_2 e^{i\varphi_2}$, $\ldots$, $s_n e^{i\varphi_n}$ inside $\mathbb{C}(0; R) \cap \{\mathrm{Im}(z) > 0\}$. Then*

$$\sum_{\mu=1}^m \left(\frac{1}{r_\mu} - \frac{r_\mu}{R^2} \right) \sin \theta_\mu - \sum_{\nu=1}^n \left(\frac{1}{s_\nu} - \frac{s_\nu}{R^2} \right) \sin \varphi_\nu = \frac{1}{\pi R} \int_0^\pi \log \left| f \left(R e^{i\theta} \right) \right| \sin \theta d\theta$$

$$+ \frac{1}{2\pi} \int_0^R \left(\frac{1}{t^2} - \frac{1}{R^2} \right) \log |f(t) f(-t)| dt + \frac{1}{2} \mathrm{Im}(f'(0)).$$

Proof. We can choose $\rho \in \mathbb{R}^+$ sufficiently small such that $\rho < R$, and f has no zeros or poles in $\mathbb{C}[0; \rho]$. Applying Theorem 7.22 to the function $u = \log |f|$, we obtain

$$\sum_{\mu=1}^m \left(\frac{1}{r_\mu} - \frac{r_\mu}{R^2} \right) \sin \theta_\mu - \sum_{\nu=1}^n \left(\frac{1}{s_\nu} - \frac{s_\nu}{R^2} \right) \sin \varphi_\nu = \frac{1}{\pi R} \int_0^\pi \log \left| f \left(R e^{i\theta} \right) \right| \sin \theta d\theta$$

$$+ \frac{1}{2\pi} \int_\rho^R \left(\frac{1}{t^2} - \frac{1}{R^2} \right) \log |f(t) f(-t)| dt + Q(R, \rho; u).$$

For $\zeta = \rho e^{i\theta}$, we have

$$u(\zeta) = \mathrm{Re}(f'(0)\zeta) + O(\rho^2), \quad \frac{\partial u}{\partial \rho}(\zeta) = \mathrm{Re}\left(f'(0) e^{i\theta} \right) + O(\rho).$$

Hence (7.6.2) implies

$$\lim_{\rho \to 0} Q(R, \rho; u) = \frac{1}{2} \mathrm{Im}(f'(0)),$$

and the theorem follows. $\square$

Theorem 7.24. *Let $f(z)$ be meromorphic in $\mathbb{C}[0; R] \cap \{\mathrm{Re}(z) \geq 0\}$ with $f(0) = 1$, and suppose that it has the zeros $r_1 e^{i\theta_1}$, $r_2 e^{i\theta_2}$, $\ldots$, $r_m e^{i\theta_m}$ and the poles $s_1 e^{i\varphi_1}$, $s_2 e^{i\varphi_2}$, $\ldots$, $s_n e^{i\varphi_n}$ inside $\mathbb{C}(0; R) \cap \{\mathrm{Re}(z) > 0\}$. Then*

$$\sum_{\mu=1}^m \left(\frac{1}{r_\mu} - \frac{r_\mu}{R^2} \right) \cos \theta_\mu - \sum_{\nu=1}^n \left(\frac{1}{s_\nu} - \frac{s_\nu}{R^2} \right) \cos \varphi_\nu = C_f(R) - \frac{1}{2} \mathrm{Re}(f'(0)),$$

where

$$C_f(R) = \frac{1}{\pi R} \int_{-\frac{\pi}{2}}^{\frac{\pi}{2}} \log \left| f\left(Re^{i\theta}\right)\right| \cos\theta d\theta$$

$$+ \frac{1}{2\pi} \int_0^R \left(\frac{1}{y^2} - \frac{1}{R^2}\right) \log|f(iy)f(-iy)|dy. \qquad (7.6.3)$$

Proof. Theorem 7.24 follows from Theorem 7.23 applied to $f(-iz)$.   □

Theorem 7.25 ([164]). *Riemann's hypothesis is true if and only if*

$$\lim_{R\to\infty} C_{\zeta_0}(R) = \frac{\pi}{8} + \frac{\gamma}{4} + \frac{1}{4}\log 8\pi - 2,$$

where γ is Euler's constant, and

$$\zeta_0(s) = \frac{\zeta\left(s + \frac{1}{2}\right)}{\zeta\left(\frac{1}{2}\right)}. \qquad (7.6.4)$$

Proof. Note that

$$z_\mu - \frac{1}{2} = r_\mu e^{i\theta_\mu} \left(r_\mu > 0,\ 0 < \theta_\mu < \frac{\pi}{2}\right),\ \bar{z}_\mu - \frac{1}{2}$$

would be the zeros of ζ_0 in the half-plane $\mathrm{Re}(s) > 0$, and $s = \frac{1}{2}$ is the unique pole of ζ_0 in $\mathrm{Re}(s) > 0$. Hence Theorem 7.24 implies

$$2 \sum_{r_\mu < R} \left(\frac{1}{r_\mu} - \frac{r_\mu}{R^2}\right) \cos\theta_\mu - \left(2 - \frac{1}{2R^2}\right) = C_{\zeta_0}(R) - \frac{1}{2}\mathrm{Re}(\zeta_0'(0)).$$

Since ζ_0 is of order 1, then the convergence exponent of zeros for ζ_0 is at most 1. Hence the series

$$\sum_\mu \frac{1}{r_\mu^{1+\varepsilon}}$$

is convergent for any $\varepsilon > 0$, and so

$$0 \leq \sum_\mu \frac{\cos\theta_\mu}{r_\mu} = \sum_\mu \frac{r_\mu \cos\theta_\mu}{r_\mu^2} \leq \frac{1}{2}\sum_\mu \frac{1}{r_\mu^2} < \infty.$$

By (7.3.5), we find

$$0 \leq \sum_{r_\mu < R} \frac{r_\mu \cos\theta_\mu}{R^2} \leq \frac{N(R)}{2R^2} \to 0\ (R \to \infty).$$

Thus by using (7.4.5) we obtain

$$\lim_{R\to\infty} C_{\zeta_0}(R) = 2 \sum_\mu \frac{\cos\theta_\mu}{r_\mu} + \frac{\pi}{8} + \frac{\gamma}{4} + \frac{1}{4}\log 8\pi - 2.$$

Riemann's hypothesis is true if and only if the zeros z_μ do not exist, that is,

$$\sum_{\mu} \frac{\cos\theta_\mu}{r_\mu} = 0,$$

equivalently,

$$\lim_{R\to\infty} \mathcal{C}_{\zeta_0}(R) = \frac{\pi}{8} + \frac{\gamma}{4} + \frac{1}{4}\log 8\pi - 2. \qquad \square$$

Further, Theorem 7.25 yields

Theorem 7.26 ([164]). *The Riemann hypothesis is true if and only if*

$$\frac{1}{\pi}\int_0^\infty \log|\zeta_0(it)|\frac{dt}{t^2} = \frac{\pi}{8} + \frac{\gamma}{4} + \frac{1}{4}\log 8\pi - 2.$$

Proof. Next we estimate the integral $\mathcal{C}_{\zeta_0}(R)$ in Theorem 7.25. We know (cf. [394])

$$\int_0^R \left|\zeta\left(\frac{1}{2} + it\right)\right|^2 dt = O(R\log R).$$

By the concavity of the logarithmic function, we obtain

$$\frac{1}{R^2}\int_0^R \log|\zeta_0\left(it\right)|^2 dt \le \frac{1}{R}\log\left\{\frac{1}{R}\int_0^R |\zeta_0\left(it\right)|^2 dt\right\} = O\left(\frac{\log\log R}{R}\right). \quad (7.6.5)$$

On the other hand, for

$$-1 \le \sigma = \mathrm{Re}(s) \le 2, \quad T - \frac{1}{2} \le t = \mathrm{Im}(s) \le T + \frac{1}{2},$$

we have (cf. [394])

$$\log|\zeta(s)| = \sum_{|T-\beta|<1} \log|s - \rho| + O(\log T), \qquad (7.6.6)$$

where $\rho = \alpha + i\beta$ runs through zeros of $\zeta(s)$, which implies

$$\log|\zeta(s)| > \sum_{|T-\beta|<1} \log|t - \beta| + O(\log T).$$

It is easily seen from a graph that the integral

$$\int_{T-\frac{1}{2}}^{T+\frac{1}{2}} \log|t - \beta|dt,$$

considered as a function of β, is a minimum when $\beta = T$; and it is then equal to $-\log 2 - 1$. Since there are $O(\log T)$ terms in the sum, it follows that

$$\int_{T-\frac{1}{2}}^{T+\frac{1}{2}} \sum_{|T-\beta|<1} \log|t - \beta| dt > -A \log T,$$

where A is a positive constant. Thus we have

$$\int_{T-\frac{1}{2}}^{T+\frac{1}{2}} \log|\zeta(\sigma + it)| dt > O(\log T).$$

Hence

$$\int_{\frac{1}{2}}^{[R]-\frac{1}{2}} \log|\zeta(\sigma + it)| dt = \sum_{k=1}^{[R]-1} \int_{k-\frac{1}{2}}^{k+\frac{1}{2}} \log|\zeta(\sigma + it)| dt > O(\log([R]-1)!).$$

Similarly we can show

$$\int_{[R]-\frac{1}{2}}^{R} \sum_{|[R]-\beta|<1} \log|t - \beta| dt > -A \log[R],$$

where $R \le [R] + \frac{1}{2}$, and

$$\int_{[R]+\frac{1}{2}}^{R} \sum_{|[R]+1-\beta|<1} \log|t - \beta| dt > -A \log([R] + 1),$$

where $R > [R] + \frac{1}{2}$. Then we also have

$$\int_{[R]-\frac{1}{2}}^{R} \log|\zeta(\sigma + it)| dt > O(\log[R]).$$

Therefore

$$\int_{\frac{1}{2}}^{R} \log|\zeta(\sigma + it)| dt > O(\log[R]!).$$

Stirling's formula yields

$$\log[R]! = \left([R] + \frac{1}{2}\right) \log[R] - [R] + O(1).$$

Hence we have

$$\int_{\frac{1}{2}}^{R} \log|\zeta(\sigma + it)| dt > O(R \log R).$$

Particularly,

$$\frac{1}{R^2} \int_0^R \log |\zeta_0\,(it)|^2 \, dt > O\left(\frac{\log R}{R}\right). \tag{7.6.7}$$

Finally, the inequalities (7.6.5) and (7.6.7) imply

$$\frac{1}{R^2} \int_0^R \log |\zeta_0\,(it)|^2 \, dt = O\left(\frac{\log R}{R}\right). \tag{7.6.8}$$

Set

$$\delta = \arcsin \frac{1}{R}.$$

Then we can take R sufficiently large such that

$$R - R\cos \delta \le \frac{1}{2}.$$

For $\frac{\pi}{2} - \delta \le \theta \le \frac{\pi}{2}$, the estimate (7.6.6) implies

$$\log \left|\zeta \left(\frac{1}{2} + Re^{i\theta}\right)\right| > \sum_{|R-\beta|<1} \log |R\sin \theta - \beta| + O(\log R),$$

where $\rho = \alpha + i\beta$ runs though zeros of $\zeta(s)$. Note that

$$\int_{\frac{\pi}{2}-\delta}^{\frac{\pi}{2}} \sum_{|R-\beta|<1} \log |R\sin \theta - \beta| \cos \theta d\theta = \frac{1}{R} \int_{R\cos \delta}^R \sum_{|R-\beta|<1} \log |t - \beta| dt > -\frac{A\log R}{R}.$$

Then

$$\int_{\frac{\pi}{2}-\delta}^{\frac{\pi}{2}} \log \left|\zeta \left(\frac{1}{2} + Re^{i\theta}\right)\right| \cos \theta d\theta > O(\log R).$$

We also know

$$\zeta(s) = O(\log t)$$

uniformly in any region

$$1 - \frac{A}{\log t} \le \sigma = \text{Re}(s) \le 2, \ t = \text{Im}(s) > t_0,$$

and

$$\zeta(s) = O\left(t^{\frac{1}{2}-\frac{\sigma}{2}} \log t\right)$$

uniformly in $0 \le \sigma \le 1$ (cf. [394]), which yield immediately

$$\int_{\frac{\pi}{2}-\delta}^{\frac{\pi}{2}} \log \left|\zeta \left(\frac{1}{2} + Re^{i\theta}\right)\right| \cos \theta d\theta < O(\log R).$$

Thus we obtain

$$\int_{\frac{\pi}{2}-\delta}^{\frac{\pi}{2}} \log\left|\zeta\left(\frac{1}{2} + Re^{i\theta}\right)\right| \cos\theta d\theta = O(\log R).$$

If $\sigma > 1$, then

$$|\zeta(s)| \le \zeta(\sigma), \quad \frac{1}{|\zeta(s)|} \le \frac{\zeta(\sigma)}{\zeta(2\sigma)}$$

for all values of t (cf. [394]). Therefore

$$\int_{0}^{\frac{\pi}{2}-\delta} \log\left|\zeta\left(\frac{1}{2} + Re^{i\theta}\right)\right| \cos\theta d\theta = O(1).$$

Since $\log\left|\zeta\left(\frac{1}{2} + Re^{i\theta}\right)\right|$ is an even function of θ, these estimates give

$$\int_{-\frac{\pi}{2}}^{\frac{\pi}{2}} \log\left|\zeta\left(\frac{1}{2} + Re^{i\theta}\right)\right| \cos\theta d\theta = O(\log R). \tag{7.6.9}$$

Therefore Theorem 7.26 follow from Theorem 7.25, (7.6.8) and (7.6.9).   □

Take $a \in \mathbb{R}$ with $\frac{1}{2} \le a < 1$. According to the proof of Theorem 7.25, we can show that Riemann's ζ-function has no zeros in $\mathrm{Re}(s) > a$ if and only if

$$\lim_{R\to\infty} C_{\zeta_a}(R) = \frac{\zeta'(a)}{2\zeta(a)} - \frac{1}{1-a},$$

where

$$\zeta_a(s) = \frac{\zeta(s+a)}{\zeta(a)}. \tag{7.6.10}$$

We also have

$$\int_{-\frac{\pi}{2}}^{\frac{\pi}{2}} \log\left|\zeta\left(a + Re^{i\theta}\right)\right| \cos\theta d\theta = O(\log R)$$

and

$$\frac{1}{R^2}\int_{0}^{R} \log\left|\zeta_a(it)\right|^2 dt = O\left(\frac{\log R}{R}\right) \tag{7.6.11}$$

by using the estimate (cf. [394])

$$\lim_{R\to\infty} \frac{1}{R}\int_{0}^{R} |\zeta(\sigma + it)|^2 dt = \zeta(2\sigma) \quad \left(\sigma > \frac{1}{2}\right).$$

Thus we obtain the following result:

Theorem 7.27. *Take $a \in \mathbb{R}$ with $\frac{1}{2} \le a < 1$. Riemann's ζ-function has no zeros in $\mathrm{Re}(s) > a$ if and only if*

$$\frac{1}{\pi}\int_{0}^{\infty} \log\left|\frac{\zeta(a+it)}{\zeta(a)}\right| \frac{dt}{t^2} = \frac{\zeta'(a)}{2\zeta(a)} - \frac{1}{1-a}.$$

Related to Theorems 7.25, 7.26, Fu Traing Wang [424] proved the following formula:

$$\int_1^T \log\left|\zeta\left(\frac{1}{2}+it\right)\right|\frac{dt}{t^2} = \int_0^{\frac{\pi}{2}} \operatorname{Re}\left\{e^{-i\theta}\log\zeta\left(\frac{1}{2}+e^{i\theta}\right)\right\}d\theta$$

$$+2\pi\sum_{\mu=1}^{\infty}\frac{\cos\theta_\mu}{r_\mu}+O\left(\frac{\log T}{T}\right). \qquad (7.6.12)$$

Consequently, a necessary and sufficient condition for the truth of the Riemann hypothesis is that

$$\int_1^\infty \log\left|\zeta\left(\frac{1}{2}+it\right)\right|\frac{dt}{t^2} = \int_0^{\frac{\pi}{2}} \operatorname{Re}\left\{e^{-i\theta}\log\zeta\left(\frac{1}{2}+e^{i\theta}\right)\right\}d\theta. \qquad (7.6.13)$$

V.V. Volchkov [422] proved that Riemann's hypothesis is equivalent to the equality

$$\int_0^\infty \frac{1-12t^2}{(1+4t^2)^3}\int_{\frac{1}{2}}^\infty \ln|\zeta(\sigma+it)|d\sigma dt = \frac{\pi(3-\gamma)}{32}. \qquad (7.6.14)$$

M. Balazard, E. Saias and M. Yor [14] obtained that

$$\frac{1}{2\pi}\int_{\operatorname{Re}(s)=\frac{1}{2}}\frac{\log|\zeta(s)|}{|s|^2}|ds| = \sum_{\operatorname{Re}(\rho)>\frac{1}{2}}\log\left|\frac{\rho}{1-\rho}\right|, \qquad (7.6.15)$$

where ρ's are zeros of the Riemann zeta function (counted with multiplicity). In particular, the Riemann hypothesis is equivalent to the vanishing of the integral on the left (above).

Andriy A. Kondratyuk [216] proved a Carleman-Nevanlinna theorem for a rectangle, which is close to Littlewood's proof of a counterpart of the Jensen theorem for a rectangle [248]. The theorem is applied to the summation of the logarithm of the Riemann zeta-function on the critical and other vertical lines. In particular, for $\varepsilon > 0$, set

$$I(\varepsilon) = \int_0^\infty e^{-\varepsilon t}\log\left|\zeta\left(\frac{1}{2}+it\right)\right|dt$$

and let $\{\rho_j\}$ be non-trivial zeros of $\zeta(s)$, then

$$\frac{\pi}{2}\sum_j\left|\operatorname{Re}(\rho_j)-\frac{1}{2}\right| = I(+0)+\frac{\pi}{2}, \qquad (7.6.16)$$

where $I(+0) = \lim_{\varepsilon\to 0}I(\varepsilon)$ exists (not necessarily finitely). Thus, the Riemann hypothesis holds if and only if $I(+0) = -\frac{\pi}{2}$.

The size of $|\zeta(s)|$ in the critical strip is quite difficult to pin down. In fact, one of the central unsolved problems in analytic number theory is the following *Lindelöf hypothesis* and its generalizations.

Conjecture 7.28. *For any fixed $\varepsilon > 0$ and $\sigma \geq \frac{1}{2}$,*

$$\zeta(\sigma + it) = O(|t|^{\varepsilon}), \quad |t| \to \infty. \tag{7.6.17}$$

The implied constant in the O-notation here depends implicitly on the value of ε. In particular,

$$\zeta\left(\frac{1}{2} + it\right) = O(|t|^{\varepsilon})$$

for $|t|$ large. This case turns out to be equivalent to (7.6.17) via the Phragmen-Lindelöf principle (cf. Theorem 7.3). The Lindelöf hypothesis is implied by the Riemann hypothesis and conversely implies that very few zeros disobey it (see [397], Section 13). Hardy and Littlewood proved that

$$\zeta\left(\frac{1}{2} + it\right) = O\left(|t|^{\frac{1}{4}+\varepsilon}\right), \quad |t| \to \infty.$$

Weyl improved the bound to $|t|^{\frac{1}{6}+\varepsilon}$ with his new ideas for estimating special trigonometric sums, now called Weyl sums.

7.7 Levin's formula

Take $R, \rho \in \mathbb{R}^{+}$ with $R > \rho > 0$. We will consider the domain

$$D = \left\{ z \in \mathbb{C} \mid \left| z - \frac{R}{2}i \right| < \frac{R}{2}, \ |z| > \rho \right\}.$$

Theorem 7.29. *Take distinct points $c_1, \ldots, c_q$ in $\overline{D} - \mathbb{C}\langle 0; \rho \rangle$ and a function $u \in C^2(\overline{D} - \{c_1, \ldots, c_q\})$. Assume*

$$u(z) = d_k \log|z - c_k| + u_k(z), \quad k = 1, \ldots, q,$$

near c_k, where d_k are constant, and u_k are of class C^2 near c_k. Then one has

$$-\frac{1}{2\pi} \int_D \operatorname{Im}\left(\frac{1}{\zeta} + \frac{i}{R}\right) \Delta u(\zeta) dx dy = \frac{1}{2\pi R} \int_\delta^{\pi-\delta} u\left(R \sin \theta e^{i\theta}\right) \frac{d\theta}{\sin^2 \theta}$$

$$+ \sum_{c_k \in D} d_k \operatorname{Im}\left(\frac{1}{c_k} + \frac{i}{R}\right) + Q(R, \rho; u), \quad (7.7.1)$$

where $\delta = \arcsin \frac{\rho}{R}$, and $Q(R, \rho; u) = O(1)$ as $R \to \infty$.

Proof. Applying the proof of Theorem 7.15 to the following functions

$$u = u(\zeta), \quad w = -\mathrm{Im}\left(\frac{1}{\zeta} + \frac{i}{R}\right),$$

and noting that

(a) $\Delta w = 0$;

(b) $w(\zeta) = 0$, $\frac{\partial w}{\partial n}(\zeta) = \frac{1}{R^2 \sin^2 \theta}$, $\zeta = R \sin \theta e^{i\theta}$ $(\delta \le \theta \le \pi - \delta)$,

then the formula (7.7.1) follows, where

$$Q(R, \rho; u) = \frac{1}{2\pi} \int_{\overline{D} \cap C\langle 0; \rho\rangle} \left(u \frac{\partial w}{\partial n} - w \frac{\partial u}{\partial n}\right) ds \tag{7.7.2}$$

$$= -\frac{1}{2\pi} \int_\delta^{\pi - \delta} \left\{\frac{\sin \theta}{\rho} u\left(\rho e^{i\theta}\right) + \left(\sin \theta - \frac{\rho}{R}\right) \frac{\partial u}{\partial \rho}\left(\rho e^{i\theta}\right)\right\} d\theta. \qquad \square$$

Next we prove *Levin's formula*:

Theorem 7.30. *Let $f(z)$ be meromorphic in $\mathbb{C}[iR/2; R/2]$ with $f(0) = 1$, and suppose that it has the zeros $r_1 e^{i\theta_1}, r_2 e^{i\theta_2}, \ldots, r_m e^{i\theta_m}$ and the poles $s_1 e^{i\varphi_1}, s_2 e^{i\varphi_2},$ $\ldots, s_n e^{i\varphi_n}$ inside $\mathbb{C}(iR/2; R/2)$. Then*

$$\sum_{\mu=1}^m \left(\frac{\sin \theta_\mu}{r_\mu} - \frac{1}{R}\right) - \sum_{\nu=1}^n \left(\frac{\sin \varphi_\nu}{s_\nu} - \frac{1}{R}\right) = \frac{1}{2} \mathrm{Im}(f'(0))$$

$$+ \frac{1}{2\pi R} \lim_{\delta \to 0} \int_\delta^{\pi - \delta} \log \left|f\left(R \sin \theta e^{i\theta}\right)\right| \frac{d\theta}{\sin^2 \theta}.$$

Proof. We can choose $\rho \in \mathbb{R}^+$ sufficiently small such that $\rho < R$, and f has no zeros or poles in $\mathbb{C}[0; \rho]$. Applying Theorem 7.29 to the function $u = \log |f|$, we obtain

$$\sum_{\mu=1}^m \left(\frac{\sin \theta_\mu}{r_\mu} - \frac{1}{R}\right) - \sum_{\nu=1}^n \left(\frac{\sin \varphi_\nu}{s_\nu} - \frac{1}{R}\right) = \frac{1}{2\pi R} \int_\delta^{\pi - \delta} \log \left|f\left(R \sin \theta e^{i\theta}\right)\right| \frac{d\theta}{\sin^2 \theta}$$

$$+ Q(R, \rho; u).$$

For $\zeta = \rho e^{i\theta}$, we have

$$u(\zeta) = \mathrm{Re}(f'(0)\zeta) + O(\rho^2), \quad \frac{\partial u}{\partial \rho}(\zeta) = \mathrm{Re}\left(f'(0) e^{i\theta}\right) + O(\rho).$$

Hence (7.7.2) implies

$$\lim_{\rho \to 0} Q(R, \rho; u) = \frac{1}{2} \mathrm{Im}(f'(0)),$$

and the theorem follows. $\square$

Theorem 7.31. *Let $f(z)$ be meromorphic in $\mathbb{C}[R/2; R/2]$ with $f(0) = 1$, and suppose that it has the zeros $r_1 e^{i\theta_1}, r_2 e^{i\theta_2}, \ldots, r_m e^{i\theta_m}$ and the poles $s_1 e^{i\varphi_1}, s_2 e^{i\varphi_2}, \ldots, s_n e^{i\varphi_n}$ inside $\mathbb{C}(R/2; R/2)$. Then*

$$\sum_{\mu=1}^{m} \left(\frac{\cos \theta_\mu}{r_\mu} - \frac{1}{R} \right) - \sum_{\nu=1}^{n} \left(\frac{\cos \varphi_\nu}{s_\nu} - \frac{1}{R} \right) = \mathcal{L}_f(R) - \frac{1}{2} \mathrm{Re}(f'(0)),$$

where

$$\mathcal{L}_f(R) = \frac{1}{2\pi R} \lim_{\delta \to 0} \int_{-\frac{\pi}{2}+\delta}^{\frac{\pi}{2}-\delta} \log \left| f \left(R \cos \theta e^{i\theta} \right) \right| \frac{d\theta}{\cos^2 \theta}.$$

Proof. Theorem 7.31 follows from Theorem 7.30 applied to $f(-iz)$. $\qquad\square$

Theorem 7.32 ([165]). *Riemann's hypothesis is true if and only if*

$$\lim_{R \to \infty} \mathcal{L}_{\zeta_0}(R) = \frac{\pi}{8} + \frac{\gamma}{4} + \frac{1}{4} \log 8\pi - 2,$$

where

$$\zeta_0(s) = \frac{\zeta \left(s + \frac{1}{2} \right)}{\zeta \left(\frac{1}{2} \right)}.$$

Proof. Note that

$$z_\mu - \frac{1}{2} = r_\mu e^{i\theta_\mu} \left(r_\mu > 0, \ 0 < \theta_\mu < \frac{\pi}{2} \right), \ \bar{z}_\mu - \frac{1}{2}$$

would be the zeros of ζ_0 in the half-plane $\mathrm{Re}(s) > 0$, and $s = \frac{1}{2}$ is the unique pole of ζ_0 in $\mathrm{Re}(s) > 0$. Hence Theorem 7.31 implies

$$2 \sum_{r_\mu < R \cos \theta_\mu} \left(\frac{\cos \theta_\mu}{r_\mu} - \frac{1}{R} \right) - \left(2 - \frac{1}{R} \right) = \mathcal{L}_{\zeta_0}(R) - \frac{1}{2} \mathrm{Re}(\zeta_0'(0)).$$

By (7.3.5), we find

$$\sum_{r_\mu < R \cos \theta_\mu} \frac{1}{R} \leq \frac{N(\sqrt{R})}{R} \to 0 \ (R \to \infty).$$

Thus by using (7.4.5) we obtain

$$\lim_{R \to \infty} \mathcal{L}_{\zeta_0}(R) = 2 \sum_{\mu} \frac{\cos \theta_\mu}{r_\mu} + \frac{\pi}{8} + \frac{\gamma}{4} + \frac{1}{4} \log 8\pi - 2.$$

Riemann's hypothesis is true if and only if the zeros z_μ do not exist, that is,

$$\sum_{\mu} \frac{\cos \theta_\mu}{r_\mu} = 0,$$

equivalently,

$$\lim_{R \to \infty} \mathcal{L}_{\zeta_0}(R) = \frac{\pi}{8} + \frac{\gamma}{4} + \frac{1}{4} \log 8\pi - 2. \qquad\square$$

7.8   Notes on Nevanlinna's conjecture

It is easy to verify that for the sector

$$D = \mathbb{C}(0; R) \cap \{z \in \mathbb{C} \mid \operatorname{Re}(z) > 0\},$$

its Green's function is

$$G_z(\zeta) = \log \left| \frac{R^2 - \bar{z}\zeta}{\zeta - z} \frac{\zeta + \bar{z}}{R^2 + z\zeta} \right|.$$

Let H_z be a conjugate of G_z such that

$$G_z(\zeta) + i H_z(\zeta) = \log \left(\frac{R^2 - \bar{z}\zeta}{\zeta - z} \frac{\zeta + \bar{z}}{R^2 + z\zeta} \right),$$

where $i = \sqrt{-1}$ is the imaginary unit. Putting $\zeta = Re^{i\varphi}$, $z = re^{i\theta}$, on $\mathbb{C}\langle 0; R \rangle$ we have

$$\frac{dH_z}{d\varphi} = \operatorname{Im} \left(\frac{d(G_z + i H_z)}{d\varphi} \right)$$

$$= \frac{R^2 - r^2}{R^2 + 2Rr\cos(\theta + \varphi) + r^2} - \frac{R^2 - r^2}{R^2 - 2Rr\cos(\theta - \varphi) + r^2}$$

$$= \operatorname{Re} \left(\frac{\bar{\zeta} - z}{\bar{\zeta} + z} \right) - \operatorname{Re} \left(\frac{\zeta + z}{\zeta - z} \right).$$

On $\operatorname{Re}(z) = 0$, putting $\zeta = it$ we have

$$\frac{dH_z}{dt} = \operatorname{Im} \left(\frac{d}{dt} \log \frac{it + \bar{z}}{it - z} + \frac{d}{dt} \log \frac{R^2 - it\bar{z}}{R^2 + itz} \right)$$

$$= \frac{2r\cos\theta}{r^2 - 2tr\sin\theta + t^2} - \frac{2R^2 r\cos\theta}{R^4 - 2R^2 tr\sin\theta + t^2 r^2}$$

$$= 2\operatorname{Re} \left(\frac{1}{z - it} \right) - 2\operatorname{Re} \left(\frac{z}{R^2 + itz} \right).$$

Note that along the boundary of D, we have

$$\frac{\partial G_z}{\partial n} = -\frac{\partial H_z}{\partial s}.$$

Thus Theorem 7.16 implies the following fact:

Theorem 7.33. *Let f ($\not\equiv 0$) be meromorphic in $\bar{D} = \mathbb{C}[0; R] \cap \{\operatorname{Re}(z) \geq 0\}$. Let a_μ ($\mu = 1, \ldots, m$) and b_ν ($\nu = 1, \ldots, n$) be respectively zeros and poles of f in D.*

Then for $z \in D - \{a_1, \ldots, a_m, b_1, \ldots, b_n\}$, one has

$$\log f(z) = \frac{1}{2\pi} \int_{-\frac{\pi}{2}}^{\frac{\pi}{2}} \left(\frac{Re^{i\varphi} + z}{Re^{i\varphi} - z} - \frac{Re^{-i\varphi} - z}{Re^{-i\varphi} + z} \right) \log \left| f\left(Re^{i\varphi}\right) \right| d\varphi$$

$$+ \frac{1}{\pi} \int_{-R}^{R} \left(\frac{1}{z - it} - \frac{z}{R^2 + itz} \right) \log |f(it)| dt$$

$$- \sum_{\mu=1}^{m} \log \left(\frac{R^2 - \bar{a}_\mu z}{z - a_\mu} \frac{z + \bar{a}_\mu}{R^2 + a_\mu z} \right) + \sum_{\nu=1}^{n} \log \left(\frac{R^2 - \bar{b}_\nu z}{z - b_\nu} \frac{z + \bar{b}_\nu}{R^2 + b_\nu z} \right) + ic,$$

where c is constant.

Let $f(z)$ be a non-constant meromorphic function in $\{\text{Re}(z) \geq 0\}$ with $f(0) = 1$, and suppose that it has the poles $s_1 e^{i\varphi_1}$, $s_2 e^{i\varphi_2}$, ..., $s_n e^{i\varphi_n}$ inside $\mathbb{C}(0; r) \cap \{\text{Re}(z) > 0\}$ for some $r > 0$. Set

$$A(r, f) = \frac{1}{2\pi} \int_0^r \left(\frac{1}{t^2} - \frac{1}{r^2} \right) \log^+ |f(it)| dt$$

$$+ \frac{1}{2\pi} \int_0^r \left(\frac{1}{t^2} - \frac{1}{r^2} \right) \log^+ |f(-it)| dt,$$

$$B(r, f) = \frac{1}{\pi r} \int_{-\frac{\pi}{2}}^{\frac{\pi}{2}} \log^+ \left| f\left(re^{i\theta}\right) \right| \cos\theta d\theta,$$

$$C(r, f) = \sum_{\nu=1}^{n} \left(\frac{1}{s_\nu} - \frac{s_\nu}{r^2} \right) \cos\varphi_\nu,$$

and define a *characteristic function* of f by

$$S(r, f) = A(r, f) + B(r, f) + C(r, f).$$

Theorem 7.24 yields

$$S\left(r, \frac{1}{f}\right) = S(r, f) - \frac{1}{2}\text{Re}(f'(0)).$$

In 1925, Nevanlinna proposed the following conjecture on the logarithmic derivative:

$$A\left(r, \frac{f'}{f}\right) + B\left(r, \frac{f'}{f}\right) = o(S(r, f))$$

as $r \to \infty$ outside a set of r which has finite linear measure. Generally, we will show that it is not true.

Next we estimate $S(r, \zeta_0)$ for the function ζ_0 defined in Section 7.6. Since $\overline{\zeta_0(-it)} = \zeta_0(it)$, we have

$$A(r, \zeta_0) = \frac{1}{\pi} \int_0^r \left(\frac{1}{t^2} - \frac{1}{r^2} \right) \log^+ |\zeta_0(it)| dt.$$

By using Weyl's estimation in Section 7.6, we obtain

$$\int_1^r \frac{1}{t^2} \log^+ |\zeta_0(it)| dt = O(1).$$

Hence the estimations in Section 7.6 imply

$$A(r, \zeta_0) = O(1), \quad B(r, \zeta_0) = O\left(\frac{\log r}{r}\right).$$

Obviously, when $r > 1/2$ we have

$$C(r, \zeta_0) = 2 - \frac{1}{2r^2}.$$

Therefore we obtain

$$S(r, \zeta_0) = O(1). \tag{7.8.1}$$

This example shows that the characteristic function $S(r, f)$ usually grows slowly. To make sense for Nevanlinna's conjecture, we have to assume growth conditions of f, say, $S(r, f) \to \infty$ as $r \to \infty$. Theorem 7.33 can be applied to study Nevanlinna's conjecture (see Hu [163]).

Bibliography

[1] Adams, W.W. and Straus, E.G., Non-Archimedean analytic functions taking the same values at the same points, Illinois J. Math. 15(1971), 418–424.

[2] Ahlfors, L.V., The theory of meromorphic curves, Acta Soc. Sci. Fenn. Nova Ser. A 3(4)(1941), 1–31.

[3] Aihara, Y., Finiteness theorems for meromorphic mappings, Osaka J. Math. 35 (1998), 603–616.

[4] An, T.T.H., Wang, J.T.Y. and Wong, P.M., Strong uniqueness polynomials: the complex case, Complex Variables 49 (1) (2004), 25–54.

[5] Andreotti, A. and Stoll, W., Analytic and algebraic dependence of meromorphic functions, Lecture Notes in Math. 234, Springer-Verlag, 1971.

[6] Apostol, T., Modular functions and Dirichlet series in number theory, Springer-Verlag, 1976.

[7] Atiyah, M.F. and Macdonald, I.G., Introduction to commutative algebra, Addison-Wesley, 1969.

[8] Azukawa, K. and Suzuki, M., Some examples of algebraic degeneracy and hyperbolic manifolds, Rocky Mountain J. Math. 10 (1980), 655–659.

[9] Backlund, R., Sur les zéros de la fonction $\zeta(s)$ de Riemann, C.R. Acad. Sci. Paris 158 (1914), 1979–1982.

[10] Baesch, A. and Steinmetz, N., Exceptional solutions of nth order periodic linear differential equations, Complex Variables 34 (1997), 7–17.

[11] Baily, W. and Borel, A., Compactification of arithmetic quotients of bounded symmetric domains, Ann. of Math. 84 (1966), 442–528.

[12] Baker, I.N., On a class of meromorphic functions, Proc. Amer. Math. Soc. 17 (1966), 819–822.

[13] Balazard, M., Saias, E., The Nyman-Beurling equivalent form for the Riemann hypothesis, Expo. Math. 18 (2000), 131–138.

[14] Balazard, M., Saias, E. and Yor, M., Notes sur la fonction ζ de Riemann. II, Adv. Math. 143 (1999), no. 2, 284–287.

[15] Ballico, E., Algebraic hyperbolicity of generic high degree hypersurfaces, Arch. Math. 63 (1994), 282-2-83.

[16] Bass, H., Connell, E.H. and Wright, D., The Jacobian conjecture: reduction of degree and formal expansion of the inverse, Bull. Amer. Math. Soc. 7(1982), 287–330.

[17] Beckenbach, E.F. and Bellman, R., Inequalities, Springer, Berlin-Heidelberg-New York, 1971.

[18] Benedetto, R.L., Non-Archimedean holomorphic maps and the Ahlfors islands theorem, Amer. J. Math. 125 (2003), no. 3, 581–622.

[19] Berenstein, C., Chang, D.C. and Li, B.Q., A note on Wronskians and linear dependence of entire functions in $\mathbb{C}^n$, Complex Variables 24(1993), 131–144.

[20] Berenstein, C., Chang, D.C. and Li, B.Q., Exponential sums and shared values for meromorphic functions in several complex variables, Advances of Mathematics 115 (1995), 201–220.

[21] Berkovich, V., Spectral theory and analytic geometry over non-Archimedean fields, Math. surveys and monographs, Coll. Amer. Math. Soc. 33, 1990.

[22] Beurling, A., A closure problem related to the Riemann zeta-function, Proc. Nat. Acad. Sci. U.S.A. 41 (1955), 312–314.

[23] Biancofiore, A., A hypersurface defect relation for a class of meromorphic maps, Trans. Amer. Math. Soc. 270 (1982), 47–60.

[24] Biancofiore, A. and Stoll, W., Another proof of the lemma of the logarithmic derivative in several complex variables, Annals of Math. Studies 100 (1981), 29–45, Princeton Univ. Press.

[25] Birch, B.J., Chowla, S., M. Hall Jnr. and Schinzel, A., On the difference $x^3 - y^2$, Norske Vid. Selsk. Forh. (Trondheim) 38 (1965), 65–69.

[26] Birch, B., and Swinnerton-Dyer, H.P.F., Notes on elliptic curves. II, J. Reine Angew. Math. 218 (1965), 79–108.

[27] Bloch, A., Sur les système de fonctions uniformes satisfaisant à l'équation d'une variété algébrique dont l'irrégularité dépasse la dimension, J. Math. Pures Appl. 5 (1926), 9–66.

[28] Bloch, A., Sur les système de fonctions holomorphes à variétés linéaires lacunaires, Ann. Ecole. Norm. Sup. 43 (1926), 309–362.

[29] Borel, E., Sur les zéros des fonctions entières, Acta Math. 20 (1897), 357–396.

[30] Bott, R., Homogeneous vector bundles, Ann. of Math. 66 (1957), 203–248.

[31] Bott, R. and Chern, S.S., Hermitian vector bundles and the equidistribution of the zeros of their holomorphic sections, Acta Math. 114 (1965), 71–112.

[32] Boutabaa, A., Applications de la théorie de Nevanlinna p-adic, Collect. Math. 42(1991), 75–93.

[33] Boutabaa, A., On some p-adic functional equations, Lecture Notes in Pure and Applied Mathematics N. 192 (1997), 49–59, Marcel Dekker.

[34] Boutabaa, A. and Escassut, A., Nevanlinna theory in characteristic p, and applications, preprint.

[35] Boutabaa, A. and Escassut, A., Applications of the p-adic Nevanlinna theory to functional equations, Ann. Inst. Fourier(Grenoble) 50 (3) (2000), 751–766.

[36] Brent, R.P., On the zeros of the Riemann zeta function in the critical strip, Math. Comp. 33 (1979), 1361–1372.

[37] Breuil, C., Conrad, B., Diamond, F. and Taylor, R., On the modularity of elliptic curves over $\mathbb{Q}$: wild 3-adic exercises, J. Amer. Math. Soc. 14 (2001), 843–939.

[38] Brody, R., Compact manifolds and hyperbolicity, Trans. Amer. Math. Soc. 235 (1978), 213–219.

[39] Brody, R. and Green, M., A family of smooth hyperbolic hypersurfaces in $\mathbb{P}_3$, Duke Math. J. 44 (1977), 873–874.

[40] Browkin, J. and Brzezinski, J., Some remarks on the abc-conjecture, Mathematics of Computation 62 (1994), 931–939.

[41] Brownawell, W.D. and Masser, D., Vanishing sums in function fields, Math. Proc. Cambridge Philos. Soc. 100 (1986), 427–434.

[42] Bryuno, A.D., Continued fraction expansion of algebraic numbers, USSR Comput. Math. and Math. Phys 4(1964), 1–15.

[43] Bump, D., Automorphic forms and representations, Cambridge Studies in Advanced Mathematics, Vol. 55, Cambridge University Press, Cambridge, 1997.

[44] Burns, D., Curvature of the Monge-Ampère foliations and parabolic manifolds, Ann. of Math. 115 (1982), 349–373.

[45] Carayol, H., Sur les représentations ℓ-adiques associées aux formes modulaires de Hilbert, Ann. Sci. Éc. Norm. Sup. 19 (1986), 409–468.

[46] Carleman, T., Über die Approximation analytischer Funktionen durch lineare Aggregate von vorgegebenen Potenzen, Arkiv för Mat. Astr. o. Fys. 17 (1923), No. 9, 1–30.

[47] Carlson, J. and Griffiths, P., Defect relation for equidimensional holomorphic mappings between algebraic varieties, Ann. of Math. 95 (1972), 557–584.

[48] Carlson, J. and Griffiths, P., The order function for entire holomorphic mappings, In: R.O. Kujala and A.L. Vitter (eds.), Value distribution theory, Part A. (Pure Appl. Math. vol. 25, pp. 225–248), New York: Dekker 1974.

[49] Cartan, H., Sur les systèmes de fonctions holomorphes à variétés linéaires lacunaires, Ann. Sci. Ecole Norm. Sup. 45 (1928), 255–346.

[50] Cartan, H., Sur la fonction de croissance attachée à une fonction méromorphe de deux variables et ses applications aux fonctions méromorphes d'une variable, C.R. Acad. Sci. Paris 189 (1929), 521–523.

[51] Cartan, H., Sur les zéros des combinaisons linéaires de p fonctions holomorphes données, Mathematica Cluj 7 (1933), 5–31.

[52] Cartwright, M.L., Integral functions, Cambridge University Press, 1956.

[53] Cassels, J.W.S., On the equation $a^x - b^y = 1$, II, Proc. Cambridge Philos. Soc. 56 (1960), 97–103.

[54] Catalan, E., Note extraite d'une lettre adressée à l'éditeur, J. Reine Angew. Math. 27 (1844), p. 192.

[55] Chein, E.Z., A note on the equation $x^2 = y^q + 1$, Proc. Amer. Math. Soc. 56 (1976), 83–84.

[56] Chen, W., Cartan's conjecture: defect relations for meromorphic maps from parabolic manifold to projective space, University of Notre Dame Thesis, 1987.

[57] Chern, S.S., Complex analytic mappings of Riemann surfaces. I, Amer. J. Math. 82(1960), 323–337.

[58] Cherry, W., Non-Archimedean analytic curves in Abelian varieties, Math. Ann. 300 (1994), 393–404.

[59] Cherry, W., A non-Archimedean analogue of the Kobayashi semi-distance and its non-degeneracy on Abelian varieties, Illinois Journal of Mathematics 40 (1) (1996), 123–140.

[60] Cherry, W. and Ye, Z., Non-Archimedean Nevanlinna theory in several variables and the Non-Archimedean Nevanlinna inverse problem, Trans. Amer. Math. Soc. 349(1997), 5043–5071.

[61] Chuang, C.T. and Yang, C.C., Fix-points and factorization of meromorphic functions, Word Scientific, 1990.

[62] Ciliberto, C. and Zaidenberg, M., 3-fold symmetric products of curves as hyperbolic hypersurfaces in $\mathbb{P}^4$, International J. Math. 14 (4) (2003), 413–436.

[63] Cohn, P.M., Algebraic numbers and algebraic functions, Chapman & Hall, 1991.

[64] Collingwood, E.F., Sur quelques théorèmes de M.R. Nevanlinna, C.R. Acad. Sci. 179 (1924), 955–957.

[65] Corrales-Rodrigáñez, C., Nevanlinna theory in the p-adic plane, Ph.D. Thesis, University of Michigan, 1986.

[66] Corrales-Rodrigáñez, C., Nevanlinna theory on the p-adic plane, Annales Polonici Mathematici LVII (1992), 135–147.

[67] Corvaja, P. and Zannier, U., A subspace theorem approach to integral points on curves, C.R. Acad. Sci. Paris (Ser. I, Math.) 334(4) (2002), 267–271.

[68] Corvaja, P. and Zannier, U., On a general Thue's equation, Amer. J. Math. 126 (5) (2004), 1033–1055.

[69] Corvaja, P. and Zannier, U., On integral points on surfaces, Ann. of Math. 160(2) (2004), 705–726.

[70] Csordas, G. and Yang, C.C., On the zeros of the Riemann ξ-function, Electronic Journal: Southwest Journal of Pure and Applied Mathematics (Internet: http://rattler.cameron.edu/swjpam.html), Issue 1, July 2003, pp. 33–42.

[71] Danilov, L.V., Diophantine equation $x^3 - y^2 = k$ and Hall's conjecture, Mat. Zametki 32 (1982), 273–275; English translation, Math. Notes of the USSR 32 (1982), 617–618.

[72] Davenport, H., On $f^3(t) - g^2(t)$, Norske Vid. Selsk. Forh. (Trondheim) 38 (1965), 86–87.

[73] Davenport, H., Multiplicative number theory, Springer Verlag, G.T.M., 1974.

[74] Davies, D., A note on the limit points associatedd with the generalized abc-conjecture for $\mathbb{Z}[t]$, Colloquium Mathematicum 71 (2) (1996), 329–333.

[75] Deligne, P., Formes modulaires et représentatione ℓ-adiques, Lecture Notes in Math. 179, Springer-Verlag, 1971.

[76] Deligne, P. and Serre, J.P., Forms modulaires de poids 1, Ann. Sci. Ec. Norm. Sup. 7 (1974), 507–530.

[77] Demailly, J.-P., Algebraic criteria for Kobayashi hyperbolic projective varieties and jet differentials, Proc. Sympos. Pure Math. 62, Part 2, Amer. Math. Soc., Providence, RI, 1997, 285–360.

[78] Demailly, J.-P. and El Goul, J., Hyperbolicity of generic surfaces of high degree in projective 3-space, Amer. J. Math. 122 (2000), 515–546.

[79] Dickson, L. E., History of the theory of numbers, Vol. 2 (Chelsea, 1966), pp. 513–520.

[80] Dirichlet, L.G.P., Verallgemeinerung eines Satzes aus der Lehre von den Kettenbrüchen nebst einigen Anwendungen auf die Theorie der Zahlen, S.B. Preuss. Akad. Wiss., 1842, 93–95.

[81] Dobrowolski, E., On a question of Lehmer and the number of irreducible factors of a polynomial, Acta Arith. 34 (1979), 391–401.

[82] Donoghue, W.F., Distributions and Fourier transforms, Academic Press, New York, 1969.

[83] Drouilhet, S.J., A unicity theorem for meromorphic mappings between algebraic varieties, Trans. Amer. Math. Soc. 265 (1981), 349–358.

[84] Dufresnoy, H., Théorie nouvelle des familles complexes normales. Applications à l'étude des fonctions algébroïdes, Ann. Sci. Ecole Norm. Sup. 61 (1944), 1–44.

[85] Dujella, A., On the size of Diophantine m-tuples, Math. Proc. Camb. Phil. Soc. 132 (2002), 23–33.

[86] Duval, J., Une sextique hyperbolique dans $\mathbf{P}^3(\mathbf{C})$, Math. Ann. 330 (2004), no. 3, 473–476.

[87] Dyson, F.J., The approximation to algebraic numbers by rationals, Acta Math. 79 (1947), 225–240.

[88] Edrei, A. and Fuchs, W.H.J., The deficiencies of meromorphic functions of order less than one, Duke Math. J. 27 (1960), 233–249.

[89] Edwards, H.M., Riemann's zeta function, Academic Press, New York, 1974.

[90] El Goul, J., Algebraic families of smooth hyperbolic surfaces of low degree in P_C^3, Manuscripta Math. 90 (1996), 521–532.

[91] Elkies, N., On $A^4 + B^4 + C^4 = D^4$, Math. Comp. 51, No. 184 (1988), 825–835.

[92] Elkies, N., ABC implies Mordell, Int. Math. Res. Not. 7 (1991), 99–109; Duke Math. J. 64 (1991).

[93] Eremenko, A.E. and Rubel, L.A., The arithmetic of entire functions under composition, Advances in Math. 124 (1996), 334–354.

[94] Eremenko, A.E. and Sodin, M.L., The value distribution of meromorphic functions and meromorphic curves from the point of view of potential theory, St. Petersburg Math. J. 3 (1)(1992), 109–136.

[95] Escassut, A., Analytic elements in p-adic analysis, World Scientific Publishing Co. Pte. Ltd., 1995.

[96] Escassut, A. and Yang, C.C., The functional equation $P(f) = Q(g)$ in a p-adic field, J. Number Theory 105 (2004), 344–360 (http://www.sciencedirect.com).

[97] Faltings, G., Arakelov's theorem for abelian varieties, Invent. Math. 73(1983), 337–347.

[98] Faltings, G., Endlichkeitssätze für abelsche Varietäten über Zahlkörpern, Invent. Math. 73(1983), 349–366.

[99] Faltings, G., Diophantine approximation on abelian varieties, Ann. of Math. (2) 133 (1991), no. 3, 549–576.

[100] Forster, O., Lectures on Riemann surfaces, GTM 81, Springer-Verlag, 1981.

[101] Frey, G., Links between stable elliptic curves and certain Diophantine equations, Annales Universitatis Saraviensis, Series Mathematicae, 1 (1986), 1–40.

[102] Frey, G., Links between elliptic curves and solutions of $A - B = C$, J. Indian Math. Soc. 51 (1987), 117–145.

[103] Fubini, G., Sulle metriche definite da una forma Hermitiana, Atti Ist. Veneto 6 (1903), 501–513.

[104] Fujimoto, H., Extensions of the big Picard theorem, Tohoku Math. J. 24 (1972), 415–422.

[105] Fujimoto, H., Families of holomorphic maps into the projective space omitting some hyperplanes, J. Math. Soc. Japan 25 (1973), 235–249.

[106] Fujimoto, H., Non-integrated defect relation for meromorphic maps of complete Kähler manifolds into $\mathbb{P}^{N_1}(\mathbb{C}) \times \cdots \times \mathbb{P}^{N_k}(\mathbb{C})$, Japan. J. Math. 11 (2) (1985), 233–264.

[107] Fujimoto, H., Value distribution theory of the Gauss map of minimal surfaces in $\mathbb{R}^m$, Aspects of Mathematics E21, Vieweg, 1993.

[108] Fujimoto, H., A family of hyperbolic hypersurfaces in the complex projective space, Complex Variables Theory Appl. 43 (2001), 273–283.

[109] Fulton, W., Algebraic curves, Benjamin, 1969.

[110] Fulton, W., Intersection theory, Springer-Verlag, 1984.

[111] Gackstatter, F. and Laine, I., Zur Theorie der gewöhnlichen Differentialgleichungen im Komplexen, Ann. Polon. Math. 38(1980), 259–287.

[112] Gelbart, S.S. and Miller, S.D., Riemann's zeta function and beyond, Bull. Amer. Math. Soc. 41 (1) (2004), 59–112.

[113] Gel'fond, A.O. and Linnik, Yu.V., On Thue's method in the problem of effectiveness in quadratic fields (Russian), Doklady Akad. Nauk SSSR (N.S.) 61 (1948), 773–776.

[114] Gol'dberg, A. and Grinshtein, A., The logarithmic derivative of a meromorphic function, Matematicheskie Zametki 19 (1976), 523–530.

[115] Goldberg, S.I. and Kobayashi, S., On holomorphic bisectional curvature, J. Diff. Geometry 1 (1967), 225–233.

[116] Goldfeld, D., Sur les produits partiels Eulériens attachés aux courbes elliptiques, C. R. Acad. Sci. Paris Sér. I Math. 294 (1982), 471–474.

[117] Gouvêa, F. Q., p-adic numbers, Springer, 1997.

[118] Gram, J., Sur les zéros de la fonction $\zeta(s)$ de Riemann, Acta Math. 27 (1903), 289–304.

[119] Granville, A. and Tucker, T.J., It's as easy as abc, Notices Amer. Math. Soc. 49 (2002), 1224–1231.

[120] Green, M.L., Holomorphic maps into complex projective space omitting hyperplanes, Trans. Amer. Math. Soc. 169 (1972), 89–103.

[121] Green, M.L., Some Picard theorems for holomorphic maps to algebraic varieties, Amer. J. Math. 97 (1) (1975), 43–75.

[122] Green, M.L., Some examples and counter examples in value distribution theory of several complex variables, Composito Math. 39 (1975), 317–322.

[123] Green, M. and Griffiths, P., Two applications of algebraic geometry to entire holomorphic mappings, The Chern Symposium 1979 (Proc. Internat. Sympos., Berkeley, Calif., 1979), Springer-Verlag, New York, 1980, pp. 41–74.

[124] Griffiths, P., Holomorphic mappings into canonical algebraic varieties, Ann. of Math. 93(1971), 439–458.

[125] Griffiths, P., Holomorphic mappings: Survey of some results and discussion of open problems, Bull. Amer. Math. Soc. 78(1972), 374–382.

[126] Griffiths, P., Algebraic curves (Chinese), Beijing Univ. Press, 1981.

[127] Griffiths, P. and Harris, J., Principles of algebraic geometry, John Wiley and Sons,1978.

[128] Griffiths, P. and King, J., Nevanlinna theory and holomorphic mappings between algebraic varieties, Acta Math. 130(1973), 145–220.

[129] Gross, B.H. and Zagier, D.B., Heegner points and derivatives of L-series, Invent. Math. 84 (1986), 225–320.

[130] Gross, F., On the equation $f^n + g^n = 1$, Bull. Amer. Math. Soc. 72 (1966), 86–88. Correction: 72 (1966), p. 576.

[131] Gross, F., On the functional equation $f^n + g^n = h^n$, Amer. Math. Monthly 73 (1966), 1093–1096.

[132] Gross, F. and Osgood, C.F., On the functional equation $f^n + g^n = h^n$, and a new approach to a certain class of more general functional equations, Indian J. Math. 23 (1981), 17–39.

[133] Gundersen, G.G., Meromorphic solutions of $f^6 + g^6 + h^6 = 1$, Analysis 18 (1998), 285–290.

[134] Gundersen, G.G., Meromorphic solutions of $f^5 + g^5 + h^5 \equiv 1$, Complex Variables 43 (2001), 293–298.

[135] Gundersen, G.G. and Hayman, W.K., The strength of Cartan's version of Nevanlinna theory, preprint.

[136] Gundersen, G.G. and Tohge, K., Entire and meromorphic solusions of $f^5 + g^5 + h^5 = 1$, preprint.

[137] Hà, Huy Khóai, On p-adic meromorphic functions, Duke Math. J. 50(1983), 695–711.

[138] Hà, Huy Khóai, Hyperbolic surfaces in $\mathbf{P}^3(\mathbf{C})$, Proc. Amer. Math. Soc. 125 (1997), 3527–3532.

[139] Hà, Huy Khóai and Mai, Van Tu, p-adic Nevanlinna-Cartan theorem, International J. of Math. 6 (1995), 719–731.

[140] Hà, Huy Khóai and My, Vinh Quang, On p-adic Nevanlinna theory, Lecture Notes in Math. 1351(1988), 146–158, Springer-Verlag.

[141] Hà, Huy Khóai and Yang, C.C., On the functional equation $P(f) = Q(g)$, Value Distribution Theory and Related Topics (edited by G. Barsegian, I. Laine and C.C. Yang), Kluwer Academic Publishers, 2004.

[142] Hadamard, J., Résolution d'une question relative aux déterminants, Bull. Sci. Math. 2 (1893), 240–248.

[143] Hadamard, J., The psychology of invention in the mathematical field, Princeton, N.J., Princeton University Press, 1949.

[144] Hall, Jr., Marshall, The diophantine equation $x^3 - y^2 = k$, in Computers in Number Theory, ed. by A.O.L. Atkin and B.J. Birch, Academic Press, London, 1971, pp. 173–198.

[145] Hamburger, H., Über die Funktionalgleichung der ζ-Funktion, Math. Z. 10 (1921), 240–258; 11 (1921), 224–245; 13 (1922), 283–311.

[146] Hardy, G.H., Littlewood, J.E. and Pólya, G., Inequalities, Cambridge, 1934.

[147] Hardy, G.H. and Wright, E.M., An introduction to the theory of numbers, Oxford Univ. Press, London/New York, 1960, pp. 325–328.

[148] Hartshorne, R., Algebraic geometry, Springer-Verlag, 1977.

[149] Hasse, H., Beweis des Analogons der Riemannschen Vermutung für die Artinschen und F.K. Schmidtschen Kongruenzzetafunktionen in gewissen elliptischen Fällen. Vorläufige Mitteilung, Nachr. Ges. Wiss. Göttingen I, Math.-phys. Kl. Fachgr. I Math. Nr. 42 (1933), 253–262 (# 38 in H. Hasse, Mathematische Abhandlungen, Band 2, Walter de Gruyter, Berlin-New York, 1975).

[150] Hasse, H., Abstrakte Begründung der komplexen Multiplikation und Riemannsche Vermutung in Funktionenkörpern, Abh. Math. Sem. Univ. Hamburg. 10 (1934), 325–348 (# 40 in H. Hasse, Mathematische Abhandlungen, Band 2, Walter de Gruyter, Berlin-New York, 1975).

[151] Hayman, W.K., Meromorphic Functions, Oxford, Clarendon Press, 1964.

[152] Hayman, W.K., Research problems in function theory, Athlone Press, University of London, 1967.

[153] Hayman, W.K., Waring's Problem für analytische Funktionen, Bayer. Akad. Wiss. Math.-Natur. Kl. Sitzungsber. 1984 (1985), 1–13.

[154] He, Y.Z. and Xiao, X.Z., Algebroid functions and ordinary differential equations (Chinese), Science Press, Beijing, 1988.

[155] Hecke, E., Lectures on Dirichlet series, modular functions and quadratic forms, Vandenhoeck & Ruprecht, Göttingen, 1983, Edited by Bruno Schoeneberg.

[156] Hecke, E., Mathematische Werke, 3rd ed., Vandenhoeck & Ruprecht, Göttingen, 1983.

[157] Hellerstein, S. and Shea, D.F., Bounds for the deficiencies of meromorphic functions of finite order, Proc. Symposia in Pure Math. 11 (1968), 214–239.

[158] Hindry, M. and Silverman, J.H., Diophantine geometry: An introduction, GTM 201, Springer, 2000.

[159] Hinkkanen, A., A sharp form of Nevanlinna's second fundamental theorem, Invent. Math. 108 (1992), 549–574.

[160] Hironaka, H., Resolution of singularities of an algebraic variety over a field of characteristic zero, I, II, Ann. Math. 79 (1964), 109–326.

[161] Hirzebruch, F., Topological methods in algebraic geometry, Springer-Verlag, 1966.

[162] Hörmander, L., An introduction to complex analysis in several variables, North-Holland, 1990.

[163] Hu, P.C., Meromorphic functions in angular domains (Chinese), J. of Shandong University (3)23(1988), 1–8.

[164] Hu, P.C., An application of the Carleman formula, Hunan Annals of Mathematics (Chinese) 10 (1–2)(1990), 152–156.

[165] Hu, P.C., The properties of zero point distribution for Riemann's zeta-function, Pure and Applied Mathematics (Chinese) 6 (2) (1990), 6–12.

[166] Hu, P.C., Holomorphic mapping into algebraic varieties of general type, Nagoya Math. J. 120 (1990), 155–170.

[167] Hu, P.C., Holomorphic mappings between spaces of different dimensions I, Math. Z. 214 (1993), 567–577; II, Math. Z. 215 (1994), 187–193.

[168] Hu, P.C., Li, P. and Yang, C.C., Unicity of meromorphic mappings, Advances in Complex Analysis and Its Applications, Kluwer Academic Publishers, 2003.

[169] Hu, P.C. and Wang, X.L., Waring's problem over non-Archimedean meromorphic function fields, Mathematica Applicata (Special Issue, Chinese) 16(2003), 18–21.

[170] Hu, P.C. and Yang, C.C., Malmquist type theorem and factorization of meromorphic solutions of partial differential equations, Complex Variables 27(1995), 269–285.

[171] Hu, P.C. and Yang, C.C., Factorization of holomorphic mappings, Complex Variables 27 (1995), 235–244.

[172] Hu, P.C. and Yang, C.C., Uniqueness of meromorphic functions on $\mathbb{C}^m$, Complex Variables 30 (1996), 235–270.

[173] Hu, P.C. and Yang, C.C., Notes on second main theorems, Complex Variables 37 (1–4) (1998), 251–277.

[174] Hu, P.C. and Yang, C.C., Differentiable and complex dynamics of several variables, Mathematics and Its Applications 483, Kluwer Academic Publishers, 1999.

[175] Hu, P.C. and Yang, C.C., A note on Shiffman's conjecture, a talk presented at 2nd ISAAC's Congress, Fukuoka, 1999.

[176] Hu, P.C. and Yang, C.C., Meromorphic functions over non-Archimedean fields, Mathematics and Its Applications 522, Kluwer Academic Publishers, 2000.

[177] Hu, P.C. and Yang, C.C., The "abc" conjecture over function fields, Proc. Japan Acad. 76, Ser. A(2000), 118–120.

[178] Hu, P.C. and Yang, C.C., Notes on a generalized abc-conjecture over function fields, Annales Mathématiques Blaise Pascal 8 (1) (2001), 61–71.

[179] Hu, P.C. and Yang, C.C., A generalized abc-conjecture over function fields, Journal of Number Theory 94 (2002), 286–298.

[180] Hu, P.C. and Yang, C.C., A note on the abc-conjecture, Communications on Pure and Applied Mathematics, Vol. LV (2002), 1089–1103.

[181] Hu, P.C. and Yang, C.C., Some progress in non-Archimedean analysis, Contemporary Mathematics Vol. 303 (2002), 37–50.

[182] Hu, P.C. and Yang, C.C., A note on Browkin-Brzeziński conjecture, Contemporary Mathematics, 384 (2005), 101–109.

[183] Hu, P.C. and Yang, C.C., Degeneracy of holomorphic curves into pseudo-canonical projective varieties, preprint.

[184] Hu, P.C. and Yang, C.C., The second main theorem of holomorphic curves, preprint.

[185] Hu, P.C. and Yang, C.C., Unique range sets of meromorphic functions on $\mathbb{C}^m$, preprint.

[186] Hu, P.C. and Yang, C.C., Generalized Fermat and Hall's conjectures, in: Methods of Complex and Clifford Analysis (proceedings of ICAM Hanoi 2004).

[187] Hu, P.C. and Yang, C.C., Subspace theorems for homogeneous polynomial forms, Israel Journal of Mathematics (to appear).

[188] Hu, P.C. and Yang, C.C., Hyperbolic hypersurfaces of lower degrees, preprint.

[189] Hutchinson, J.I., On the roots of the Riemann zeta-function, Trans. Amer. Math. Soc. 27 (1925), 49–60.

[190] Iitaka, S., On D-dimensions of algebraic varieties, J. Math. Soc. Japan 23 (1971), 356–373.

[191] Iitaka, S., Logarithmic forms of algebraic varieties, J. Fac. Sci. Univ. Tokyo Sect. IA Math 23 (1976), 525–544.

[192] Iitaka, S., On logarithmic Kodaira dimension of algebraic varieties, Complex analysis and algebraic geometry, Iwanami Shoten, Tokyo, 1977, 175–189.

[193] Iitaka, S., Algebraic geometry, Grad. Texts Math. 76, Springer-Verlag, 1982.

[194] Ishizaki, K., A note on the functional equation $f^n + g^n + h^n = 1$ and some complex differential equations, Comput. Methods Funct. Theory 2 (2002), 67–85.

[195] Iwaniec, H., Fourier coefficients of modular forms of half-integral weight, Invent. Math. 87 (1987), 385–401.

[196] Iyer, G., On certain functional equations, J. Indian Math. Soc. 3 (1939), 312–315.

[197] Jategaonkar, A.V., Elementary proof of a theorem of P. Montel on entire functions, J. London Math. Soc. 40 (1965), 166–170.

[198] Jung, H.W.E., Über ganze birationale Transformationen der Ebene, J. Reine Angew. Math. 184(1942), 161–174.

[199] Karatsuba, A.A., Complex analysis in number theory, CRC Press, 1995.

[200] Kawamata, Y., On Bloch's conjecture, Invent. Math. 57 (1980), 97–100.

[201] Keiper, J.B., Power series expansions of Riemann's ξ function, Math. Comput. 58 (1992), 765–773.

[202] Khintchine, A., Zur metrischen Theorie der diophantischen Approximationen, Math. Z. 24 (1926), 706–714.

[203] Kiernan, P.J., Hyperbolic submanifolds of complex projective space, Proc. Amer. Math. Soc. 22 (1969), 603–606.

[204] Kiernan, P. and Kobayashi, S., Holomorphic mappings into projective space with lacunary hyperplanes, Nagoya Math. J. 50 (1973), 199–216.

[205] Kneser, H., Zur Theorie der gebrochenen Funktionen mehrerer Veränderlicher, Jber. Deutsch. Math. Verein. 48 (1938), 1–38.

[206] Ko, C., On the Diophantine equation $x^2 = y^n + 1$, $xy \neq 0$, Sci. Sinica 14 (1964), 457–460.

[207] Kobayashi, S., Hyperbolic manifolds and holomorphic mappings, New York, Marcel Dekker, 1970.

[208] Kobayashi, S., Hyperbolic complex spaces, Springer, 1998.

[209] Kobayashi, S. and Nomizu, K., Foundations of differential geometry, John Wiley & Sons, 1969.

[210] Kobayashi, S. and Ochiai, T., Mappings into compact complex manifolds with negative first Chern class, J. Math. Soc. Japan 23 (1971), 137–143.

[211] Kobayashi, S. and Ochiai, T., Meromorphic mappings into compact complex spaces of general type, Invent. Math. 31 (1975), 7–16.

[212] Koblitz, N., Introduction to elliptic curves and modular forms, GTM 97, Springer-Verlag, 1984.

[213] Kodaira, K., On holomorphic mappings of polydiscs into compact complex manifolds, J. Differ. Geom. 6 (1971), 33–46.

[214] Kolyvagin, V.A., Finiteness of $E(\mathbb{Q})$ and $\amalg(E,\mathbb{Q})$ for a subclass of Weil curves, Izv. Akad. Nauk SSSR Ser. Mat. 52 (1988), 522–540, 670–671 (= Math. USSR - Izvestija 32 (1989), 523–541).

[215] Kolyvagin, V.A., Euler systems, in The Grothendieck Festschrift (Vol. II), P. Cartier et al., eds., Prog. in Math. 89, Birkhäuser, Boston (1990), 435–483.

[216] Kondratyuk, Andriy A., A Carleman-Nevanlinna theorem and summation of the Riemann zeta-function logarithm, Computational Methods and Function Theory 4 (2) (2004), 391–403.

[217] Krull, W., Allgemeine Bewertungstheorie, J. reine u. angew. Math. 167 (1931), 160–196.

[218] Kubert, D., Universal bounds on the torsion of elliptic curves, Proc. London Math. Soc. 33 (1976), 193–237.

[219] Lagarias, J.C., An elementary problem equivalent to the Riemann hypothesis, Amer. Math. Monthly 109 (2002), 534–543.

[220] Lander, L. and Parkin, T., Counterexamples to Euler's conjecture on sums of the powers, Bull. Amer. Math. Soc. 72 (1966), p. 1079.

[221] Lang, S., Integral points on curves, Pub. Math. IHES, 1960.

[222] Lang, S., Some theorems and conjectures in diophantine equations, Bull. Amer. Math. Soc. 66 (1960), 240–249.

[223] Lang, S., Report on Diophantine approximations, Bull. Soc. Math. France 93 (1965), 177–192.

[224] Lang, S., Higher-dimensional Diophantine problems, Bull. Amer. Math. Soc. 80 (1974), 779–787.

[225] Lang, S., Fundamentals of Diophantine geometry, Springer-Verlag, New York, 1983.

[226] Lang, S., Conjectured Diophantine estimates on elliptic curves, in Arithmetic and Geometry, M. Artin and J. Tate, eds., Birkhäuser, Boston, 1983; pp. 155–172.

[227] Lang, S., Algebra, Addison-Wesley Publishing Company, Inc., 1984.

[228] Lang, S., Hyperbolic and Diophantine analysis, Bull. Amer. Math. Soc. 14 (1986), no. 2,159–205.

[229] Lang, S., Introduction to complex hyperbolic spaces, Springer-Verlag, 1987.

[230] Lang, S., The error term in Nevanlinna theory, Duke Math. J. 56(1988), 193–218.

[231] Lang, S., Old and new conjectured Diophantine inequalities, Bull. Amer. Math. Soc. 23 (1990), 37–75.

[232] Lang, S., Number theory III, Encyclop. Math. Sc. vol. 60 (1991), Springer-Verlag.

[233] Lang, S., Complex analysis, 4th ed., Graduate Texts in Mathematics, Vol. 103, Springer-Verlag, New York, 1999.

[234] Lang, S., Algebra, revised third edition, Springer-Verlag, New York, 2002.

[235] Lang, S. and Cherry, W., Topics in Nevanlinna theory, Lecture Notes in Math. 1433 (1990), Springer.

[236] Langevin, Michel, Partie sans facteur carré d'un produit d'entiers voisins. (Square-free divisor of a product of neighbouring integers). Approximations diophantiennes et nombres transcendants, C.-R. Colloq., Luminy/ Fr. 1990, (1992), 203–214.

[237] Langevin, M., Cas d'égalite pour le théorème de Mason et applications de la conjecture (abc). (Extremal cases for Mason's theorem and applications of the (abc) conjecture). C.R. Acad. Sci., Paris, Ser. I 317, No. 5, (1993), 441–444.

[238] Lebesgue, V.A., Sur l'impossibilité en nombres entiers de l'équation $x^m = y^2 + 1$, Nouv. Ann. Math. 9 (1850), 178–181.

[239] Lehman, R.S., Separation of zeros of the Riemann zeta-function, Math. Comp. 20 (1966), 523–541.

[240] Lehmer, D.H., Factorization of certain cyclotomic functions, Ann. Math. 34(2) (1933), 461–479.

[241] Lehmer, D.H., Extended computation of the Riemann zeta-function, Mathematika 3 (1956), 102–108; RMT 108, MTAC 11 (1957), p. 273.

[242] Lehmer, D.H., On the roots of the Riemann zeta-function, Acta Math. 95 (1956), 291–298; RMT 52, MTAC 11 (1957), 107–108.

[243] Lehmer, D.H., On the Diophantine equation $x^3 + y^3 + z^3 = 1$, J. London Math. Soc. 31 (1965), 275–280.

[244] Levin, A., Generalizations of Siegel's and Picard's theorem, preprint.

[245] Li, P. and Yang, C.C., Some further results on the unique range sets of meromorphic functions, Kodai Math. J. 18 (1995), 437–450.

[246] Li, X.J., The positivity of a sequence of numbers and the Riemann hypothesis, J. Number Theory 65 (1997), no. 2, 325–333.

[247] Liao, L.W. and Yang, C.C., Some new properties of the gamma function and the Riemann zeta function, Math. Nachr. 257 (2003), 1–8.

[248] Littlewood, J.E., On the zeros of the Riemann zeta-function, Proc. Cambr. Phil. Soc. 22 (1924), 295–318.

[249] Lu, S. and Yau, S.T., Holomorphic curves in surfaces of general type, Proc. Nat. Acad. Sci. U.S.A. 87 (1990), no. 1, 80–82.

[250] Maclaurin, C., A second letter to Martin Folges, Esq.; concerning the roots of equations with the demonstration of other rules in algebra, Phil. Trans. 36 (1729), 59–96.

[251] Manin, Ju., The p-torsion of elliptic curves is uniformly bounded, Izv. Akad. Nauk SSSR 33 (1969), AMS Transl., 433–438.

[252] Mason, R.C., The hyperelliptic equation over function fields, Math. Proc. Cambridge Philos. Soc. 93(1983), 219–230.

[253] Mason, R.C., Equations over function fields, Lecture Notes in Math. 1068 (1984), 149–157, Springer.

[254] Mason, R.C., Diophantine equations over function fields, London Math. Soc. Lecture Note Series, Vol. 96, Cambridge University Press, United Kingdom, 1984.

[255] Mason, R.C., Norm form equations I, J. Number Theory 22 (1986), 190–207.

[256] Masser, D.W., Open problems, Proc. Symp. Analytic Number Th., W.W.L. Chen (ed), London: Imperial College, 1985.

[257] Masuda, K. and Noguchi, J., A construction of hyperbolic hypersurface of $\mathbb{P}^n(\mathbb{C})$, Math. Ann. 304 (1996), 339–362.

[258] Matiyasevich, Yu.V., Enumerable sets are diophantine, Soviet Math. Dokl. 11 (1970), 354–358.

[259] Matiyasevich, Yu.V., Hilbert's tenth problem, Foundations of Computing Series, MIT Press, Cambridge, Mass., 1993.

[260] Matsumura, H., Commutative algebra, W. A. Benjamin Co., New York, 1970.

[261] Mazur, B., Modular curves and the Eisenstein ideal, Publ. Math. IHES 47 (1977), 33–186.

[262] Mazur, B., Rational isogenies of prime degree, Invent. Math. 44 (1978), 129–162.

[263] Mazur, B., Number theory as gadfly, Amer. Math. Monthly 98 (1991), 593–610.

[264] McQuillan, M., A new proof of the Bloch conjecture, J. Algebraic Geom. 5 (1996), no. 1, 107–117.

[265] McQuillan, M., Diophantine approximations and foliations, Inst. Hautes Études Sci. Publ. Math. No. 87 (1998), 121–174.

[266] McQuillan, M., Holomorphic curves on hyperplane sections of 3-folds, Geom. Funct. Anal. 9 (1999), 370–392.

[267] Meller, N.A., Computations connected with the check of Riemann's hypothesis, Dokl. Akad. Nauk SSSR 123 (1958), 246–248.

[268] Metsänkylä, T., Catalan's conjecture: another old Diophantine problem solved, Bull. Amer. Math. Soc. 41 (1) (2004), 43–57.

[269] Mihăilescu, P., A class number free criterion for Catalan's conjecture, J. Number Theory 99 (2003), 225–231.

[270] Mihăilescu, P., Primary cyclotomic units and a proof of Catalan's conjecture, preprint.

[271] Milovanović, G.V., Mitrinović, D.S. and Rassias, Th.M., Topics in polynomials: extremal problems, inequalities, zeros, World Scientific Publishing Co. Pte. Ltd., 1994.

[272] Moh, T.T., Algebra, Series on University Mathematics Vol. 5, World Scientific Publishing Co. Pte. Ltd., 1992.

[273] Mohon'ko, A.Z., On the Nevanlinna characteristics of some meromorphic functions, in "Theory of functions, functional analysis and their applications", Izd-vo Khar'kovsk. Un-ta 14(1971), 83–87.

[274] Molluzzo, J., Doctoral Thesis, Yeshiva University, 1972.

[275] Montel, P., Leçons sur les familles normales de fonctions analytiques et leurs applications, Gauthier-Villars, Paris 1927.

[276] Mordell, L.J., On Mr. Ramanujan's Empirical Expansions of Modular Functions, Proc. Cambridge Phil. Soc. 19 (1917), 117–124.

[277] Mordell, L.J., Note on the integer solutions of the equation $Ey^2 = Ax^3 + Bx^2 + Cx + D$, Messenger Math. 51 (1922), 169–171.

[278] Mordell, L.J., On the rational solutions of the indeterminate equations of the third and fourth degrees, Proc. Cambridge Philos. Soc. 21(1922), 179–192.

[279] Morduhaĭ-Boltovskoĭ, D.D., On hypertranscendence of the function $\zeta(x,s)$, Izv. Politekh. Inst. Warsaw 2 (1914), 1–16.

[280] Mori, S., Threefolds whose canonical bundles are not numerically effective (2), Ann. of Math. 116 (1) (1982), 133–176.

[281] Mori, S., Classification of higher-dimensional varieties, Proc. of Symp. in Pure Math. AMS 46 (1987), 269–288.

[282] Mori, S. and Mukai, S., The uniruledness of the moduli space of curves of genus 11, Algebraic Geometry Conference, Tokyo-Kyoto, 1982; Lecture Notes in Mathematics, Vol. 1016, 334–353.

[283] Muir, T. and Metzler, W.H., A treatise on the theory of Determinants, Dover, 1960.

[284] Musili, C., Algebraic geometry for beginners, Texts and Readings in Mathematics 20, Hindustan Book Agency (India), 2001.

[285] Nadel, A.M., Hyperbolic surfaces in $\mathbb{P}^3$, Duke Math. J. 58 (1989), 749–771.

[286] Nadel, A.M., The non-existence of certain level structures on abelian varieties over complex function fields, Ann. Math. 129(1989), 161–178.

[287] Narasimhan, R., Un analogue holomorphe du théorème de Lindemann, Ann. Inst. Fourier, Grenoble 21 (1971), 271–278.

[288] Nathanson, M.B., Elementary methods in number theory, Springer-Verlag, New York, 2000.

[289] Nevanlinna, F., Über eine Klasse meromorpher Funktionen, C.R. 7e Congr. Math. Scand. Oslo (1929), 81–83.

[290] Nevanlinna, F. and Nevanlinna, R., Über die Eigenschaften analytischer Funktionen in der Umgebung einer singulären Stelle oder Linie, Acta Soc. Sci. Fenn. 50 (1925), no. 5.

[291] Nevanlinna, R., Zur Theorie der meromorphen Funktionen, Acta Math. 46 (1925), 1–99.

[292] Nevanlinna, R., Le théorme de Picard-Borel et la théorie der fonctions méromorphes, Gauthier-Villars, Paris, 1929.

[293] Nevanlinna, R., Über Riemannsche Flächen mit endlich vielen Windungspunkten, Acta Math. 58 (1932), 295–373.

[294] Nevanlinna, R., Eindeutige analytische Funktionen, Die Grundl. d. Math. Wiss. XLVC Springer-Verlag, Berlin-Göttingen-Heidelberg 2 ed. 1953.

[295] Nevanlinna, R., Analytic functions, Springer-Verlag, New York-Heidelberg-Berlin, 1970.

[296] Newman, D.J. and Slater, M., Waring's problem for the ring of polynomials, J. Number Theory 11 (1979), 477–487.

[297] Ng, T.W., An example concerning infinite factorizations of transcendental entire functions, Expo. Math. 18 (2000), 127–129.

[298] Ng, T.W., Recent progress in unique factorization of entire functions, Proc. of the Second ISAAC Congress (eds. H.G.W. Begehr et al.), Vol.2, 1187–1199, Kluwer Academic Publishers, 2000.

[299] Nochka, E.I., Defect relations for meromorphic curves, Izv. Akad. Nauk. Moldav. SSR Ser. Fiz.-Teklam. Mat. Nauk 1 (1982), 41–47.

[300] Nochka, E.I., On a theorem from linear algebra, Izv. Akad. Nauk. Moldav. SSR Ser. Fiz.-Teklam. Mat. Nauk 3 (1982), 29–33.

[301] Nochka, E.I., On the theory of meromorphic curves, Dokl. Akad. Nauk SSSR 269 (1983), 377–381.

[302] Noguchi, J. and Ochiai, T., Geometric function theory in several complex variables, Translations of Mathematical Monographs, Vol. 80, AMS, 1990.

[303] Noguchi, J., Winkelmann, J. and Yamanoi, K., The second main theorem for holomorphic curves into semi-Abelian varieties, Acta Math. 188 (2002), 129–161.

[304] Northcott, D.G., An inequality in the theory of arithmetic on algebraic varieties, Proc. Cambridge Philos. Soc. 45(1949), 502–509.

[305] Northcott, D.G., A further inequality in the theory of arithmetic on algebraic varieties, Proc. Cambridge Philos. Soc. 45(1949), 510–518.

[306] Nyman, B., On the one-dimensional translation group and semi-group in certain function spaces, Thesis, University of Uppsala, 1950.

[307] Ochiai, T., On holomorphic curves in algebraic varieties with ample irregularity, Invent. Math. 43 (1977), 83–96.

[308] Oesterlé, J., Nouvelles approches du "théorème" de Fermat (New approaches to Fermat's last theorem), Semin. Bourbaki, 40ème Année, Vol. 1987/88, Exp. No. 694, Astérisque 161/162 (1988), 165–186.

[309] Ogg, A., Modular forms and Dirichlet series, W. A. Benjamin, Inc., New York, 1969.

[310] Osgood, C.F., A number theoretic-differential equations approach to generalizing Nevanlinna theory, Indian J. of Math. 23 (1981), 1–15.

[311] Osgood, C.F., Sometimes effective Thue-Siegel Roth-Schmidt-Nevanlinna bounds, or better, J. Number Theory 21 (1985), 347–389.

[312] Osgood, C.F., Meromorphic points on curves, preprint.

[313] Ostrowski, A., Über Dirichletsche Reihen und algebraische Differentialgleichungen, Math. Z. 8 (1920), 241–298.

[314] Ostrowski, A., Untersuchungen zur arithmetischen Theorie der Körper, Math. Zeit. 39 (1935), 269–404.

[315] Ozawa, M., On uniquely factorizable entire functions, Kodai Math. Sem. Rep. 8 (1977), 342–360.

[316] Picard, E., Traité d'analyse II, Gauthier-Villars, Paris, 1925.

[317] Pillai, S.S., On the equation $2^x - 3^y = 2^X + 3^Y$, Bull. Calcutta Math. Soc. 37 (1945), 15–20.

[318] Pólya, G., Bemerkungen über unendliche Folge und ganze Funktionen, Math. Ann. 88 (1923), 169–183.

[319] Remmert, R., Holomorphe und meromorphe Abbildungen komplexer Räume. Math. Annalen 133 (1957), 328–370.

[320] Reznick, B., Patterns of dependence among powers of polynomials, in "Algorithmic and quantitative real algebraic geometry", DIMACS: Series in Discrete Mathematics and Theoretical Computer Science, Vol. 60, pp. 101–121, AMS, 2003.

[321] Ribet, K.A., The ℓ-adic representations attached to an eigenform with Nebentypus: a survey, Lecture Notes in Math. 601, Springer-Verlag, 1977.

[322] Ribet, K.A., On modular representations of $\mathrm{Gal}(\bar{\mathbb{Q}}/\mathbb{Q})$ arising from modular forms, Invent. Math. 100 (1990), 431–476.

[323] Richtmyer, R., Devaney, M. and Metropolis, N., Continued fraction expansions of algebraic numbers, Numer. Math. 4 (1962), 68–84.

[324] Riemann, B., Über die Anzahl der Primzahlen unter einer gegebenen Grösse, Mon. Not. Berlin Akad (Nov. 1859), 671–680.

[325] Robert, A., Elliptic curves, Lecture Notes in Math. 326, Springer-Verlag, 1973.

[326] Rosser, J.B., Yohe, J.M. and Schoenfeld, L., Rigorous computation and the zeros of the Riemann zeta-function, Information Processing 68 (Proc. IFIP Congress, Edinburgh, 1968), vol. 1, North-Holland, Amsterdam, 1969, pp. 70–76. Errata: Math. Comp. 29 (1975), p. 243.

[327] Roth, K.F., Rational approximations to algebraic numbers, Mathematika 2 (1955), 1–20.

[328] Ru, M., On a general form of the second main theorem, Trans. Amer. Math. Soc. 349(1997), 5093–5105.

[329] Ru, M., A note on p-adic Nevanlinna theory, Proc. Amer. Math. Soc. 129 (2001), 1263–1269.

[330] Ru, M., A defect relation for holomorphic curves intersecting hypersurfaces, Amer. J. Math. 126 (2004), 215–226.

[331] Rubin, K. and Silverberg, A., Ranks of elliptic curves, Bull. Amer. Math. Soc. 39 (2002), 455–474.

[332] Rutishauser, H., Über Folgen und Scharen von analytischen und meromorphen Funktionen mehrerer Variabeln, sowie von analytischen Abbildungen, Acta Math. 83 (1950), 349–325.

[333] Schanuel, S., Heights in number fields, Bull. Soc. Math. France 107 (1979), 433–449.

[334] Schlickewei, H.P., Linearformen mit algebraischen Koeffizienten, Manuscripta Math. 18 (1976), 147–185.

[335] Schlickewei, H.P., On products of special linear forms with algebraic coefficients, Acta Arith. 31 (1976), 389–398.

[336] Schlickewei, H.P., The $\wp$-adic Thue-Siegel-Roth-Schmidt theorem, Archiv der Math. 29 (1977), 267–270.

[337] Schmidt, W.M., Diophantine approximation, Lecture Notes in Math. 785(1980), Springer.

[338] Schmidt, W.M., Diophantine approximations and Diophantine equations, Lecture Notes in Math. 1467(1991), Springer.

[339] Selberg, A., Discontinuous groups and harmonic analysis, Proc. Internat. Congr. Mathematicians (Stockholm, 1962), Inst. Mittag-Leffler, Djursholm, 1963, pp. 177–189.

[340] Serre, J.P., Groupes Algébriques et Corps de Classes, Hermann, 1959.

[341] Serre, J.P., A course in arithmetic, Springer-Verlag, 1973.

[342] Shafarevich, I.R., Basic algebraic geometry, Springer-Verlag, 1994.

[343] Shapiro, H.N. and Sparer, G.H., Extension of a theorem of Mason, Communications on Pure and Applied Mathematics Vol. XLVII (1994), 711–718.

[344] Shiffman, B., Nevanlinna defect relations for singular divisors, Invent. Math. 31 (1975), 155–182.

[345] Shiffman, B., Holomorphic curves in algebraic manifolds, Bull. Amer. Math. Soc. 83 (1977), 553–568.

[346] Shiffman, B., On holomorphic curves and meromorphic maps in projective space, Indiana Math. J. 28 (1979), 627–641.

[347] Shiffman, B., Introduction to Carlson-Griffiths equidistribution theory, Lecture Notes in Math. 981 (1983), 44–89, Springer-Verlag.

[348] Shiffman, B. and Zaidenberg, M., Two classes of hyperbolic surfaces in $\mathbb{P}^3$, International J. Math. 11 (1) (2000), 65–101.

[349] Shiffman, B. and Zaidenberg, M., Hyperbolic hypersurfaces in $\mathbb{P}^n$ of Fermat-Waring type, Proc. Amer. Math. Soc. 130 (2002), 2031–2035.

[350] Shimura, G., On elliptic curves with complex multiplication as factors of the Jacobians of modular function fields, Nagoya Math. J. 43(1971), 199–208.

[351] Shimura, G., Introduction to the arithmetic theory of automorphic functions, Princeton Univ. Press, Princeton, 1971.

[352] Shimura, G., Response to 1996 Steele Prize, Notes of the AMS 43 (1996), 1344–1347.

[353] Shirosaki, M., Hyperbolic hypersurfaces in the complex projective spaces of low dimensions, Kodai Math. J. 23 (2000), 224–233.

[354] Shirosaki, M., A hyperbolic hypersurface of degree 10, Kodai Math. J. 23 (2000), 376–379.

[355] Siegel, C.L., Approximation algebraischer Zahlen, Math. Z. 10 (1921), 173–213.

[356] Siegel, C.L., The integer solutions of the equation $y^2 = ax^n + bx^{n-1} + \cdots + k$, J. London Math. Soc. 1 (1926), 66–68.

[357] Siegel, C.L., Über einige Anwendungen diophantischer Approximationen, Abh. Preuss. Akad. d. Wiss., Math. Phys. Kl., Nr. 1 = Ges. Abh. I (1929), 209–266.

[358] Siegel, C.L., Topics in complex function theory, Vol. I, Elliptic functions and uniformization theory, John Wiley & Sons Inc., 1969.

[359] Siegel, C.L., Topics in complex function theory, Vol. II, Automorphic functions and Abelian integrals, John Wiley & Sons Inc., 1971.

[360] Siegel, C.L., Topics in complex function theory, Vol. III, Abelian functions and modular functions of several variables, John Wiley & Sons Inc., 1973.

[361] Silverman, J.H., Lower bounds for height functions, Duke Math. J. 51 (2) (1984), 395–403.

[362] Silverman, J.H., The arithmetic of elliptic curves, Springer-Verlag, 1986.

[363] Siu, Y.T., Defect relations for holomorphic maps between spaces of different dimensions, Duke Math. J. 55 (1) (1987), 213–251.

[364] Siu, Y.T., Non-equidimensional value distribution theory and meromorphic connections, Duke Math. J. 61 (2) (1990), 341–367.

[365] Siu, Y.T., Non-equidimensional value distribution theory, Lecture Notes in Math. 1351, 285–311, Springer-Verlag.

[366] Siu, Y.T. and Yeung, S.K., A generalized Bloch's theorem and the hyperbolicity of the complement of an ample divisor in an abelian variety, Math. Ann. 306 (4) (1996), 743–758.

[367] Siu, Y.T. and Yeung, S.K., Defects for ample divisors of Abelian varieties, Schwarz lemma, and hyperbolic hypersurfaces of low degrees, Amer. J. Math. 119 (1997), 1139–1172.

[368] Siu, Y.T. and Yeung, S.K., Addendum to "Defects for ample divisors of Abelian varieties, Schwarz lemma, and hyperbolic hypersurfaces of low degrees", Amer. J. Math. 125 (2003), 441–448.

[369] Smyth, B., Weakly ample Kaehler manifolds and Euler numbers, Math. Ann. 224 (1976), 269–279.

[370] Song, X.J. and Tucker, T.J., Dirichlet's theorem, Vojta's inequality, and Vojta's conjecture, Compositio Math. 116 (2) (1999), 219–238.

[371] Steinmetz, N., Eigenschaften eindeutiger Lösungen gewöhnlicher Differentialgleichungen im Komplexen, Karlsruhe, Dissertation, 1978.

[372] Stewart, C.L. and Tijdeman, R., On the Oesterlé-Masser conjecture, Monatsh. Math. 102 (1986), 251–257.

[373] Stewart, C.L. and Yu, Kunrui, On the abc conjecture, Math. Ann. 291(2) (1991), 225–230.

[374] Stewart, C.L. and Yu, Kunrui, On the abc conjecture. II, Duke Math. J. 108 (1) (2001), 169–181.

[375] Stoll, W., Mehrfache Integrale auf komplexen Mannigfaltigkeiten, Math. Zeitschr. 57 (1952), 116–152.

[376] Stoll, W., Ganze Funktionen endlicher Ordnung mit gegebenen Nullstellenflächen, Math. Z. 57 (1953), 211–237.

[377] Stoll, W., Einige Bemerkungen zur Fortsetzbarkeit analytischer Mengen, Math. Z. 60 (1954), 287–304.

[378] Stoll, W., Deficit and Bezout estimates, Value-Distribution Theory Part B (Edited by Robert O. Kujola and Albert L. Vitter III), Pure and Appl. Math. 25, Marcel Dekker, New York, 1973.

[379] Stoll, W., Holomorphic functions of finite order in several complex variables, Conference Board of the Mathematical Sciences, Regional Conference Series in Mathematics 21, Amer. Math. Soc., 1974.

[380] Stoll, W., Value distribution on parabolic spaces, Lecture Notes in Math. 600(1977), Springer-Verlag.

[381] Stoll, W., The characterization of strictly parabolic manifolds, Ann. Scuola. Norm. Sup. Pisa 7 (1980), 87–154.

[382] Stoll, W., The characterization of strictly parabolic manifolds, Compositio Math. 44 (1981), 305–373.

[383] Stoll, W., Introduction to value distribution theory of meromorphic maps, Lecture Notes in Math. 950 (1982), 210–359, Springer-Verlag.

[384] Stoll, W., The Ahlfors-Weyl theory of meromorphic maps on parabolic manifolds, Lecture Notes in Math. 981 (1983), 101–219, Springer-Verlag.

[385] Stoll, W., Value distribution theory of meromorphic maps, Aspects of Math. E 7 (1985), pp. 347, Vieweg-Verlag.

[386] Stoll, W., An extension of the theorem of Steinmetz-Nevanlinna to holomorphic curves, Math. Ann. 282 (1988), 185–222.

[387] Stothers, W.W., Polynomial identities and Hauptmoduln, Quart. J. Math. Oxford Ser. (2) 32 (1981), no. 127, 349–370.

[388] Study, E., Kürzeste Wege im komplexen Gebiete, Math. Ann. 60 (1905), 321–377.

[389] Tate, J.T., The arithmetic of elliptic curves, Invent. Math. 23 (1974), 179–206.

[390] Taylor, R. and Wiles, A., Ring-theoretic properties of certain Hecke algebras, Ann. of Math. 141 (1995), 553–572.

[391] Thue, A., Über Annäherungswerte algebraischer Zahlen, J. reine angew. Math. 135 (1909), 284–305.

[392] Tijdeman, R., On the equation of Catalan, Acta Arith. 29 (1976), 197–209.

[393] Tijdeman, R., In Number Theory and Applications, ed. by R.A. Mollin, Kluwer, 1989, p.234.

[394] Titchmarsh, E.C., The zeta-function of Riemann, Cambridge University Press, 1930.

[395] Titchmarsh, E.C., The zeros of the Riemann zeta-function, Proc. Roy. Soc. London 151 (1935), 234–255; also ibid. 157 (1936), 261–263.

[396] Titchmarsh, E.C., The theory of functions, 2nd edition, Oxford, 1939.

[397] Titchmarsh, E.C., The theory of the Riemann zeta-function, 2nd ed., The Clarendon Press Oxford University Press, New York, 1986.

[398] Toda, N., On the functional equation $\sum_{i=0}^{p} a_i f_i^{n_i} = 1$, Tohoku, Math. J. 23 (1971), 289–299.

[399] Tohge, K., On meromorphic functions satisfying Fermat type functional equations, preprint.

[400] Totaro, B., Proof of a conjecture of Lang, preprint, 1986.

[401] Ueno, K., Classification of algebraic varieties I, Comp. Math. 27 (1973), 277–342.

[402] Ueno, K., Classification theory of algebraic varieties and compact complex spaces, Lecture Notes in Math. 439, Springer, 1975.

[403] Urabe, H., Uniqueness of the factorization under composition of certain entire functions, J. Math. Kyoto Univ. 18 (1978), 95–120.

[404] Urabe, H. and Yang, C.C., Uniquely factorizable entire functions under composition, Forum Math. 1 (1989), 309–313.

[405] Valiron, G., Lectures on the general theory of integral functions, Toulouse: Édouard privat, 1923.

[406] Valiron, G., Sur la dérivée des fonctions algébroides, Bull. Soc. Math. France 59(1931), 17–39, Jbuch 57, 371.

[407] Valiron, G., Sur le minimum du module des fonctions entières d'ordre inférieur é un, Mathematica 11 (1935), 264–269.

[408] van der Poorten, A.J., Additive relations in number fields, Seminar on number theory, Paris 1982–83, 259–266, Progr. Math. 51, Birkhäuser Boston, Boston, MA, 1984.

[409] Van der Waerden, B.L., Algebra, Vol. 2, 7th ed., Springer-Verlag, 1991.

[410] van Frankenhuysen, M., Hyperbolic spaces and the abc conjecture, Katholieke Universiteit Nijmegen, Thesis, 1995.

[411] van Frankenhuysen, M., The abc conjecture implies Roth's theorem and Mordell's conjecture, Math. Contemporanea 76 (1999), 45–72.

[412] van Frankenhuysen, M., The abc conjecture implies Vojta's height inequality for curves, J. Number Theory 95 (2002), no. 2, 289–302.

[413] Vaserstein, L.N., Quantum (abc)-theorems, Journal of Number Theory 81 (2000), 351–358.

[414] Vitter, A., The lemma of the logarithmic derivative in several complex variables, Duke Math. J. 44 (1977), 89–104.

[415] Vojta, P., Diophantine approximation and value distribution theory, Lecture Notes in Math. 1239, Springer, 1987.

[416] Vojta, P., Arithmetic discriminants and quadratic points on curves, In Arithmetic algebraic geometry (Texel, 1989), volume 89 of Progr. Math., pages 359–376, Birkhäuser Boston, Boston, MA, 1991.

[417] Vojta, P., A generalization of theorems of Faltings and Thue-Siegel-Roth-Wirsing, J. Amer. Math. Soc. 5(4) (1992), 763–804.

[418] Vojta, P., On Cartan's theorem and Cartan's conjecture, Amer. J. Math. 119 (1997), 1–17.

[419] Vojta, P., A more general abc conjecture, International Mathematics Research Notices 1998 (1998), 1103–1116.

[420] Vojta, P., On the ABC conjecture and Diophantine approximation by rational points, Amer. J. Math. 122 (2000), 843–872.

[421] Vojta, P., Diagonal quadratic forms and Hilbert's Tenth Problem, Contemporary Mathematics 270 (2000), 261–274.

[422] Volchkov, V.V., On an equality equivalent to the Riemann hypothesis, Ukraïn. Mat. Zh. 47 (1995), no. 3, 422–423; translation in Ukrainian Math. J. 47 (1995), no. 3, 491–496.

[423] Voloch, J. F., Diagonal equations over function fields, Bol. Soc. Brasil. Mat. 16 (1985), 29–39.

[424] Wang, F.T., A note on the Riemann zeta-function, Bull. Amer. Math. Soc. 52 (1946), 319–321.

[425] Wang, H.C., Complex parallisable manifolds, Proc. Amer. Math. Soc. 5 (1954), 771–776.

[426] Wang, J.T.Y., Uniqueness polynomials and bi-unique range sets for rational functions and non-Archimedean meromorphic functions, Acta Arithmetica (2002).

[427] Weil, A., Über die Bestimmung Dirichletscher Reihen durch Funktionalgleichungen, Math. Ann. 168 (1967), 149–156.

[428] Weil, A., Scientific works. Collected papers. III (1964–1978), Springer-Verlag, 1979.

[429] Weil, A., Number theory, an approach through history, Birkhäuser, Boston, 1984.

[430] Weyl, H. and Weyl, J., Meromorphic functions and analytic curves, Annals of Math. Studies 12, Princeton Univ. Press, Princeton, N.J., 1943.

[431] Wiles, A., Modular elliptic curves and Fermat's Last Theorem, Ann. of Math. 141 (1995), 443–551.

[432] Wong, P.M., Geometry of the homogeneous complex Monge-Ampère equation, Invent. Math. 67 (1982), 261–274.

[433] Wong, P.M., On the second main theorem in Nevanlinna theory, Amer. J. Math. 111 (1989), 549–583.

[434] Wu, H., Mapping of Riemann surfaces (Nevanlinna theory), Proc. Sympos. Pure Math. Vol. XI, "Entire Functions and Related Parts of Analysis". Amer. Math. Soc. 1968, 480–532.

[435] Wu, H., Remarks on the first main theorem in equidistribution theory I, II, III, IV, J. Diff. Geom. 2 (1968), 197–202, 369–384, ibid 3 (1969), 83–94, 433–446.

[436] Wu, H., The equidistribution theory of holomorphic curves, Princeton University Press, Princeton, New Jersey, 1970.

[437] Yamanoi, K., The second main theorem for small functions and related problems, Acta. Math. 2004.

[438] Yang, C.C., A generalization of a theorem of P. Montel on entire functions, Proc. Amer. Math. Soc. 26 (2) (1970), 332–334.

[439] Yang, C.C. and Hu, P.C., Value distribution theory and its applications to meromorphic mappings (Part 1), Journal of the National Institute for Compilation and Translation (Taiwan) 25 (2) (1996), 321–388; Part 2, 26 (1) (1997), 355–427.

[440] Yang, C.C. and Hu, P.C., A survey on p-adic Nevanlinna theory and its applications to differential equations, Taiwanese J. of Math. 3 (1) (1999), 1–34.

[441] Yang, C.C. and Li, P., Some further results on the functional equation $P(f) = Q(g)$, Value Distribution Theory and Related Topics (edited by G. Barsegian, I. Laine and C.C. Yang), Kluwer Academic Publishers, 2004.

[442] Yang, L., Value distribution theory and its new researches (Chinese), Science Press, Beijing, 1982.

[443] Ye, Z., On Nevanlinna's second main theorem in projective space, Invent. Math. 122 (1995), 475–507.

[444] Yu, K.W. and Yang, C.C., On the meromorphic solutions of $\sum_{j=1}^{p} a_j(z) f_j^{k_j}(z) = 1$, Indian J. Pure Appl. Math. 33 No. 10 (2002), 1495–1502.

[445] Zagier, D., Algebraic numbers close to both 0 and 1, Math. Computation 61 (1993), 485–491.

[446] Zaidenberg, M., Stability of hyperbolic embeddedness and construction of examples, Math. USSR-Sb. 63 (1989), 351–361.

[447] Zannier, U., Some remarks on the S-unit equation in function fields, Acta Arithmetica LXIV.1(1993), 87–98.

[448] Zariski, O. and Samuel, P., Commutative algebra (Vol. I, II), Van Nostrand, Princeton (1958, 1960).

[449] Zhang, S., Positive line bundles on arithmetic surfaces, Ann. of Math. 136(1992), 569–587.

Symbols

Index